Advances in Human Aspects of Healthcare

Advances in Human Aspects of Healthcare

Edited by

Vincent G. Duffy

CRC Press is an imprint of the
Taylor & Francis Group, an **informa** business

CRC Press
Taylor & Francis Group
6000 Broken Sound Parkway NW, Suite 300
Boca Raton, FL 33487-2742

First issued in paperback 2019

ISBN-13: 978-1-4398-7021-1 (hbk)
ISBN-13: 978-0-367-38109-7 (pbk)

Visit the Taylor & Francis Web site at
http://www.taylorandfrancis.com

and the CRC Press Web site at
http://www.crcpress.com

Table of Contents

Section IV. Physical Aspects and Risk Factors for Patients and Caregivers

Section V. Patient Care, Patient Safety and Medical Error

Section VI. Medical User Centered Design

Section VII. Human Modeling and Patient Users of Medical Devices

Section XI. Organizational Aspects in Healthcare

Preface

This book is concerned with human factors and ergonomics in healthcare. The utility of this area of research is to aid the design of systems and devices for effective and safe healthcare delivery. New approaches are demonstrated for improving healthcare devices such as portable ultrasound systems. Research findings for improved work design, effective communications and systems support are also included. Healthcare informatics for the public and usability for patient users are considered separately but build on results from usability studies for medical personnel.

Quality and safety are emphasized and medical error is considered for risk factors and information transfer in error reduction. Physical, cognitive and organizational aspects are considered in a more integrated manner so as to facilitate a systems approach to implementation. New approaches to digital human healthcare modeling, human factors and ergonomics measurement and model validation are included. Recent research on special populations, collaboration and teams, as well as learning and training allow practitioners to gain a great deal of knowledge overall from this book.

Each of the chapters of the book were either reviewed by the members of Scientific Advisory and Editorial Board or germinated by them. Our sincere thanks and appreciation goes to the Board members listed below for their contribution to the high scientific standard maintained in developing this book.

Explicitly, the book is organized into sections that contain the following subject areas:

I: Assessing Ergonomic Characteristics in Biomedical Technologies
II: Communications, Systems Support and Healthcare Informatics
III: How to Improve Quality of Ergonomics in Healthcare
IV: Physical Aspects and Risk Factors for Patients and Caregivers
V: Patient Care, Patient Safety and Medical Error

VI: Medical User Centered Design
VII: Human Modeling and Patient Users of Medical Devices
VIII: Measures and Validation in Healthcare
IX: Medical Devices and Special Populations
X: Collaboration and Learning in Healthcare Systems
XI: Organizational Aspects in Healthcare

This book would be of special value internationally to those researchers and practitioners involved in various aspects of healthcare delivery.

March 2012

Vincent G. Duffy
Fulbright Scholar, The Russian Federation
Faculty of Engineering Business and Management
Bauman Moscow State Technical University
Moscow, Russia
and School of Industrial Engineering
Department of Agricultural & Biological Engineering
Purdue University
West Lafayette, Indiana USA

Editor

Section I

Assessing Ergonomic Characteristics in Biomedical Technologies

CHAPTER 1

A Multifactorial Approach and Method for Assessing Ergonomic Characteristics in Biomedical Technologies

Giuseppe Andreoni, Fiammetta Costa, Marco Mazzola, Marcello Fusca, Maximiliano Romero, Elena Carniglia, Daniela Zambarbieri, and Giorgio C. Santambrogio

Dip. INDACO, Politecnico di Milano
Dip. Bioingegneria, Politecnico di Milano
Dip. Informatica e Sistemistica, Università di Pavia
giuseppe.andreoni@polimi.it

ABSTRACT

The ergonomic assessment of healthcare products is becoming regulatory, but current state of art relies on checklist of end-users interview about general or limited aspects. Ergonomics deals with the human as a whole and as a part of a more complex system; instead the assessment of objects and products is often treated separating the different components of the interaction: physical, sensory, cognitive. This research aims to develop an integrated method and a protocol for qualitative and quantitative evaluation of ergonomic features in healthcare products. The integration of methods for a global and more comprehensive ergonomic assessment can be used in a proactive way in the early stages of development. Basic methodological approaches refer to biomechanics and product usability assessment techniques. The first one is based on the measurement of angular excursions of the joints associated with the implementation of the human motion detectable in dedicated laboratory; the second method rely on direct observations and on site experiments supported by questionnaires/interviews to quantify ease of use and user satisfaction by means of special scales of assessments.

The methodological approach here proposed is aimed to integrate ethnographic analysis, biomechanical analysis, and cognitive usability assessment within a multifactorial approach for the evaluation of ergonomic characteristics.

Keywords: Human behavior analysis, observational methods, quantitative movement analysis, HMI assessment, Ergonomics in Healthcare

INTRODUCTION

Designing for the medical domain is challenging. Health is for everyone a primary, reference, essential and indispensable value. For this reason it is always at the center, at the top of the list of both individual and social goals.

The term HealthCare is proposing a process that provides a clinical service not exclusively addressed to security and provision of treatment to the subject (which is and remains the main point), but which supports the concept of quality of life for the same patient, his/her family and all the health professionals who interact with him/her every day. In a simple word, we have to move from the concept of "cure" to the concept of "Care", term that embeds the philosophy of quality of care or of "taking care" of the whole person. Thus it is to create a new complex and multifactorial process in which technological factors, organizational, and human dimensions must find a balanced mix for a full success. To provide safe and high quality care for patients the healthcare industry requires clinically effective and well-designed medical devices (Martin *et al.*, 2008). Medical devices are a diverse group of products that ranges from simple items to complex devices. According to the international definition, and in particular with the one used by the U.S. Food and Drug Administration (FDA), the federal agency responsible for the oversight of medical devices - as well as national and international consensus standards-making bodies – a medical device is "an instrument, apparatus, implement, machine, contrivance, implant, in vitro reagent, or other similar or related article, including a component part, or accessory which is intended for use in the diagnosis of disease or other conditions, or in the cure, mitigation, treatment, or prevention of disease, in man or other animals, or intended to affect the structure or any function of the body of man or other animals, and which does not achieve any of it's primary intended purposes through chemical action within or on the body of man or other animals and which is not dependent upon being metabolized for the achievement of any of its primary intended purposes" (http://www.fda.gov).

Traditionally, many errors associated with medical devices were blamed on "user error". Poorly designed medical devices can cause clinicians to make errors that lead to adverse patient outcomes. Even simple tasks such as inadvertently pushing the wrong buttons, loading infusion pump cartridges incorrectly, or misinterpreting onscreen information could be fatal (Wiklund and Wilcox, 2005). With modern advances in sophisticated technology, it is troubling that many medical device user interfaces are poorly designed, fail to adequately support the clinical tasks for which they are intended, and frequently contribute to medical error. Thus the design of devices should take account of the environment in which they are required to function and should support the working patterns of professional users and the lifestyles of patients and carers. In the recent vision of collaborative Health, also the patient and/or his/her own familiars are to be considered "users". Thus, similarly, devices intended for use by patients can similarly foster adverse outcomes if not designed properly to making medical devices safe and effective by involving users in the design process (Martin *et al.*, 2008; Wiklund, Kendler and Strochlic, 2011; Weinger, Wiklund and Gardner-Bonneau, 2011). For this reason, recently, international organizations have adopted specific rules for ergonomic design of medical devices. FDA has mandated that medical device manufacturers use HFE design principles and adhere to standard "Good Manufacturing Practices" (Weinger, Wiklund and Gardner-Bonneau, 2011; U.S. Department of Health and Human Services, 2011]. Similarly other national and international committees are acting (EN 62366, 2008). From now on, most design and manufacturing companies that gave only limited attention to the human factors engineering of medical devices, have to undergo a radical change in developing new devices.

HealthCare Product Design, and in particular the development of bio-electronic device for diagnosis and monitoring, is usually very complex because it concerns many different disciplines (e.g. Medicine, Electronics, Computer Science, Product Design, etc.). Moreover Healthcare products are almost always used by many different actors, for example caregivers, physicians, patients and their relatives. They can be therefore defined as multidisciplinary products. This implies great difficulties during the design and development of new products because it is necessary to consider many different points of view, in particular without forgetting the diverse users' needs (Romero *et al.*, 2010).

Ergonomics may be defined as the application of knowledge about human characteristics and abilities (physical, emotional, and intellectual) to the design of tools, devices, systems, environments, and organizations. The development of usable medical devices requires the adherence to ergonomic principles and processes throughout the entire design cycle, beginning with the earliest concepts and continuing after the device is released for commercial use (Wiklund and Wilcox, 2005).

In the "user-centered design" approach the user is the focus of the design process, and user input and formal user testing starts at the earliest conceptual stages and then continues throughout the design process, thereby facilitating iterative design improvement. In the previous comma, we already discussed how a single medical device can be used by a wide variety of users, from highly educated physicians to patients and caregivers but we could also include also those who unpack, transport, maintain, clean, and test the device, as well as other individuals who are directly affected by the device's use. Therefore, when trying to design the best device possible, it is advantageous to define device users broadly as all individuals who may interact with the device. This means that a wide analysis should be carried out. Here start this process by focusing on the clinical application of medical devices and we will discuss a multidisciplinary approach to collect user information, where user is intend to be both the patient and, above all, the clinician using the device for diagnosis.

METHODS

The idea of merging qualitative and quantitative methods has become increasingly popular, in particular in areas of applied research. Human-Machine-Interaction (HMI) and ergonomics are multifaceted issues so it is important to approach the phenomenon under investigation from diverse sides and to combine data resulting from diverse methods. Based on a previous experience on white goods ergonomic analysis, the proposed method integrates different research methods into a research strategy increasing the quality of final results and to provide a more comprehensive understanding of analyzed phenomena (Andreoni *et al.*, 2010).

The methodological approach here proposed follows three steps: a preliminary ethnographic analysis have been performed to design and drive the following phases consisting into two in-depth analyses regarding usability and biomechanics of the interaction with Portable Ultrasound system (Figure 1).

Figure 1: The portable ultrasound system used in the methodological set-up of the study.

Ethnographic Analysis

Ethnographic analysis have been performed to design and drive the following phases consisting into two in-depth analyses regarding usability, cognitive and biomechanical interaction with the sonograph which is the test-bed used for this research. Contextual inquiry (Beyer and Holtzblatt, 1997), based on an ethnographic approach was conducted by observing the interaction of expert and unskilled sonographer in clinical laboratory to detect user habits (Amit, 2000) and better define the objectives and procedures of the next analyses. In our case study we observed 5 subjects-sonographers. The ethnographic analysis was carried out through the non-interference principle: we minimized the observer influence thanks to video recording and to avoiding the entrance in the clinical space during the investigation. This allowed us to have the maximum likelihood of the real situation. These tests lead us to identify and isolate the main tasks to be considered in the next biomechanical and cognitive studies.

The outcome of this step was the identification of the two classes of subjects to be analysed (skilled and inexperienced sonographer with respect to the Portable ultrasound system to be investigated in its ergonomic features) and a structured clinical protocol for the investigation of the HMI of the Portable ultrasound device and the sonographers in two standard clinical conditions regarding two applications (vascular and abdominal). An example is reported in Figure 2.

VASCULAR APPLICATION:
Probe choice, application and user preset
Start exam: Real-time on
IMAGING

- visualization of Common Carotid Artery (CCA)
- parameters adjiustment: Depth, Frequency, General Gain and TGC.
- Eventually XView (Speckle Reduction) and MView (Spatial Compound).
- Distance Measurement of CCA width
- image saving.

COLOR DOPPLER

- visualization of Color Doppler image in CCA;
- adjustment of color box dimensions, of its position and of steering
- optimization of PRF, of Doppler frequency and of Gain to avoid Aliasing

PULSED WAVE DOPPLER

- signal sampling in CCA;
- line-of-sight position adjustment and sample volume adjustment;
- adjustment of correction angle teta, of velocity scale, of baseline and of PW General Gain;
- Maximum Velocity measurement on the PW Doppler track
- Image saving

Close exam and saving in local DB.

Figure 2: Example of the single-operation protocol for the analysis of the vascular application.

Biomechanical Analysis

The biomechanical analysis was conducted at the instrumental laboratory of human motion analysis to quantitatively measure by means of an ergonomic index for the measurement of comfort/discomfort of human movement involving a statistically suitable sample of subjects eventually clustered in different categories (e.g. expert or novice as in this case).

Upper-body kinematics was recorded through a six cameras optoelectronic system while the subject performed the clinical examination with the Portable ultrasound system. The cameras were placed so that a volume of about 3 x 2 x 2 m was covered. A set of 32 passive and reflective markers, placed on the subject's body surface, were used for the kinematic computation (Figure 3) (Schmidt *et al.*, 1999). Calibration procedures were carried out before each experimental session. Ergonomic evaluation of postures and tasks was carried out with the MMGA index of discomfort (Table 1) (Andreoni *et al.*, 2010). Also time duration of each task and spatial excursions of the fingertip as end-effector could be extracted and analyzed (Figure 4).

A Body Discomfort Assessment questionnaire combined with a Visual-Analogue Scale was administered too in order to have also a subjective assessment of the perceived discomfort. Subjective scores were re-analyzed in a graphical representation for a better and immediate understanding (Figure 5).

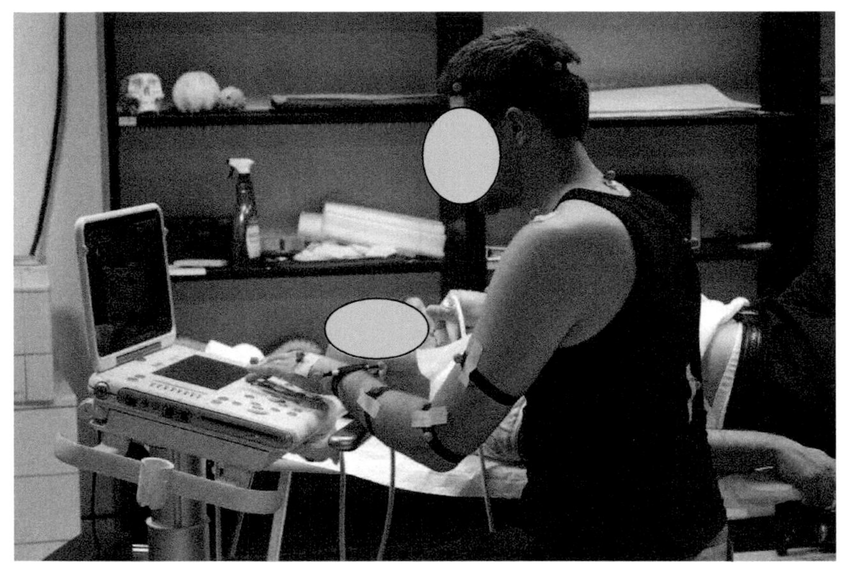

Figure 3: Protocol for the kinematic acquisition of the operator's postures and movements.

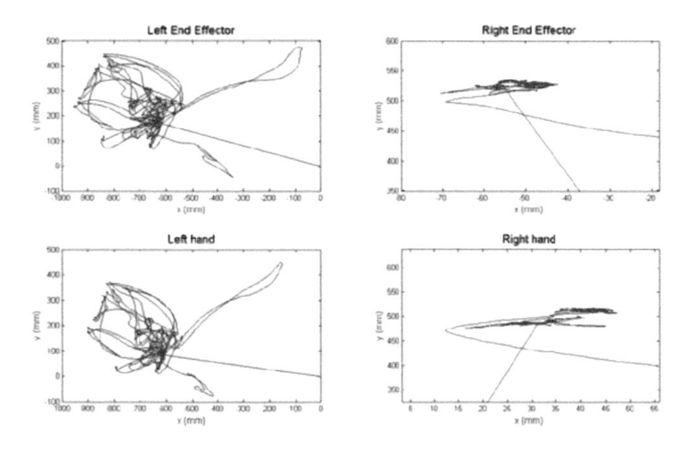

Figure 4: Analysis of end-effector trajectories during the Abdomen Exam.

SUBJ – EXAM	Avg	STD	Min	Max
S1 - ABDOMEN	4,83	0,34	0,93	6,14
S1 - VASCULAR	3,12	0,20	0,93	3,34
S2 – ABDOMEN	3,66	0,56	1,61	4,55
s2 – VASCULAR	2,47	0,13	1,94	3,33

Table 1: MMGA Index values in the analyzed conditions

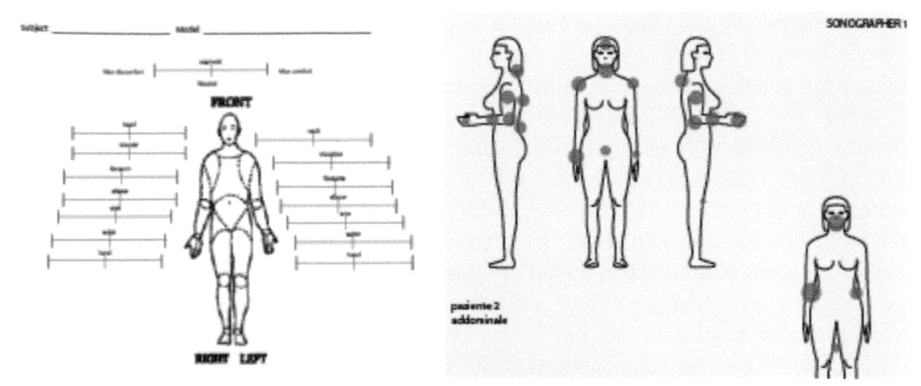

Figure 5: Subjective evaluation of perceived discomfort and the graphical results of the analysis.

Cognitive Usability Assessment

Cognitive usability assessment of communication interfaces (design and flow of operations) is carried out by means of visual, subjective and eye-tracking based techniques according to the desired level of accuracy or needed feedback information to improve HCI design. Eye movements analysis represents a powerful tool in all studies dealing with visual attention and exploration, as in the case of a human-machine interaction (Zambarbieri, 2007). The analysis of subject's eye movements provides quantitative and objective information on where the subject is looking at and for how long. Clear vision of an object is guaranteed only when its image falls within the central part of the retina which is called fovea and remains stable. Therefore, in order to explore a visual scene, the eyes have to move in such a way as to bring the image of an object of interest onto the fovea and to maintain it stable during any relative motion between the scene and the observer.

During visual exploration the eyes make saccades and fixations. Saccades are fast movements of the eyes that rapidly shift the gaze from one position to another. Vision is suppressed, at least partially, during the execution of saccades since the high velocity reached by the eyes (up to 500 deg/sec) would cause blurred vision. Fixation is the time between two successive saccades and it represents the period of time during which visual information can actually be acquired. The two-dimensional movement composed of alternating saccades and fixations that the eyes execute during the exploration of a scene is normally called "scanpath". Since visual information is acquired by the CNS only during fixations, whereas saccades are used to shift the gaze from one point to another, it is reasonable to infer that the analysis of the scanpath is a powerful tool for the study of exploration strategies and the underlying cognitive processes. Thus, when a subject is exploring visual scene eye movements supply information about the focus of subject's attention.

Among the different techniques currently available to detect the rotation of the eyes in the orbit, video-oculography (VOG) is the one more suited for the purpose of studying subjects' behavior in HMI. VOG makes use of infrared emitting sources to produce corneal reflexes and the measure of eye position is made through the processing of the eye image taken by a video camera. In this way eye movements can be detected simultaneously in two dimensions. There are two types of systems that implement video-oculographic technique: remote systems and wearable systems. In remote systems a camera is usually positioned below the computer display and it records gaze position with respect to the image that is presented on the display. Head movements are tolerated within the limits in which the eye remains in the camera field of view. Wearable systems, also called head-mounted, are placed on the subject's head, for instance by using a bicycle helmet or a baseball cap, or they are worn through a kind of eyeglass frame. A head-mounted device can be equipped with two recording cameras: one for recording the movement of the eyes, the other to detect the scene in front of the subjects that is completely free to move. Specialized software processes the information from the two cameras and produces a video in which a mobile cursor identifies, frame by frame, the point of the visual scene the subject is looking at (Figure 6).

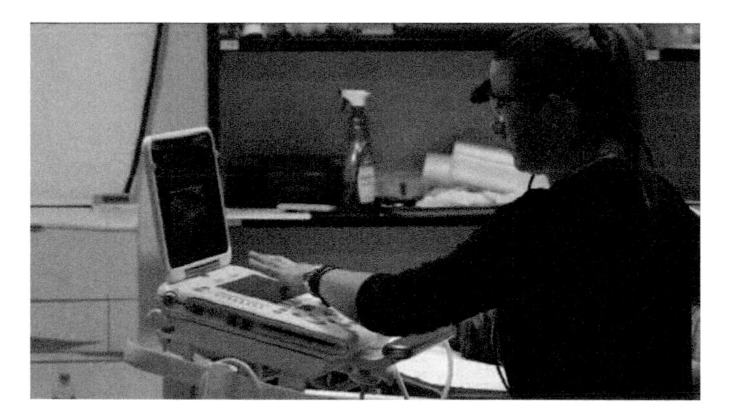

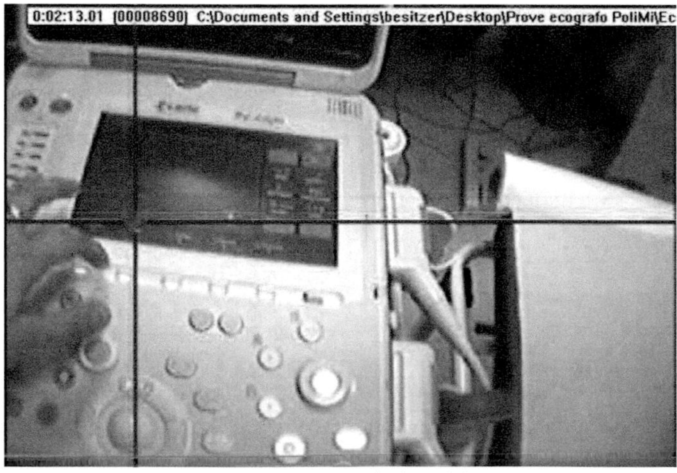

Figure 6: the acquisition setup of eye-tracking system (up) and the corresponding output (below). The center of the red cross indicate fixation position.

The purpose of data analysis is to identify within the recorded video the actions taken by the subject. It is particularly suitable for behavioral studies such as ergonomic assessment in the field of HMI since it allows identifying what the subject was doing at any instant. To achieve that result, the first step of analysis is to define a list of actions and to assign them codes (letters or numbers). Subsequently, by sliding the images of the video, the user has to identify what the subject is doing by assigning the relevant code through the keyboard. Based on these results the information regarding the occurrence of any event can be viewed in terms of duration, and in terms of timeline chart (Figure 7).

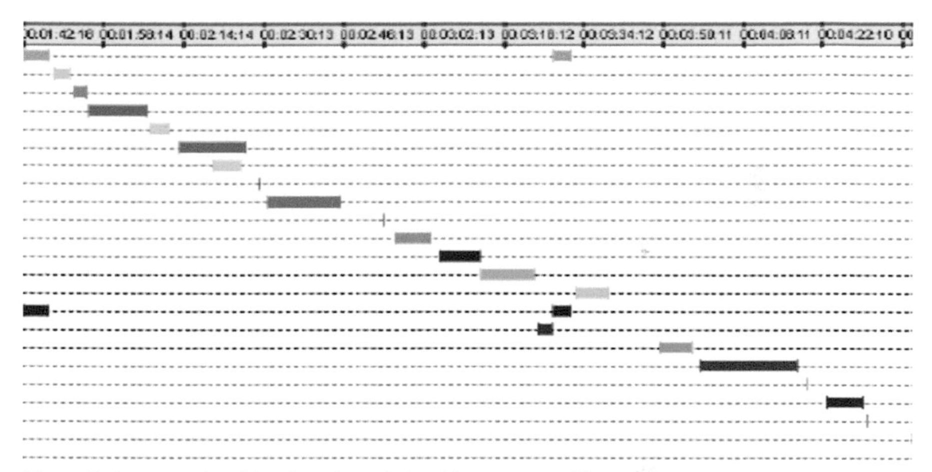

Figure 7: An example of timeline chart derived from eye-tracking analysis.

The first provide a global distribution of actions duration among the different kind of actions, whereas the second represent the evolution during time of subject behavior. In other words results can be described in terms of static or dynamical observation of human behavior.

CONCLUSIONS

Ergonomic and usability testing calls for representative users to perform representative tasks as a means to reveal the interactive strengths and opportunities for improvement of a device. You can think of the activity as pressure testing or debugging the user interface of a device in terms of how it serves the users' needs, a critical need being safe operation. Tests may focus on early design concept models, more advanced prototypes, and even production units (Wiklund, Kendler and Strochlic, 2011). The described protocol represent a feasible, structured approach to collect quantitative user data of physical and cognitive usability of a medical device. The approach has been developed on a ultrasound system but it could be easily applied in any other case, both simpler and more complex. To complete the physical aspects we can add total body kinematics and EMG evaluation to better evidence eventual musculo-skeletal risks. The next step will be a real application with a comparison of the ergonomic features of the current device and the new one that it is still a prototype and will be released next year. This could be though as an example of proactive ergonomics, that is applied early up in the design phase of a new medical device through a real "user-centered-design" approach.

REFERENCES

Amit, V. 2000, *Constructing the Field: Ethnographic Fieldwork in the Contemporary World.* Routledge.

Andreoni, G., L. Anselmi, F. Costa, M. Mazzola, E. Preatoni, M. Romero, and B. Simionato. 2010. Human Behaviour Analysis and Modelling: A Mixed Method Approach. In. *Advances in Applied Digital Human Modeling*, Vincent G. Duffy ed., CRC Press, Taylor and Francis Group, 77 – 83.

Andreoni, G., M. Mazzola, O. Ciani, M. Zambetti, M. Romero, F. Costa, and E. Preatoni. 2010. Method for Movement and Gesture Assessment (MMGA) in ergonomics. *International Journal of Human Factors Modeling and Simulation*, 1(4): 309-405.

Beyer, H., and K. Holtzblatt. 1997. *Contextual Design: Defining Customer-Centered Systems.* Morgan Kaufmann.

European Standard EN 62366 January 2008, Medical devices – Application of usability engineering to medical devices (IEC 62366:2007).

http://www.fda.gov (accessed 26.02.2012).

Martin, J.L., B.J. Norris, E. Murphy, and J.A. Crowe. 2008. Medical Device Development: The Challenge for Ergonomics, *Applied Ergonomics*, 39(3): 271-283.

Romero, M., P. Perego, G. Andreoni, and F. Costa. 2010. New strategies for technology products development in HealthCare. In. *New Trends in Technologies: Control, Management,* ed. Meng Joo Er. Computational Intelligence and Network Systems, Sciyo, 131 – 142.

Schmidt, R., C. Disselhorst-Klug, J. Silny, and G. Rau. 1999. A marker-based measurement procedure for unconstrained wrist and elbow motions. *Journal of Biomechanics*, 32(6): 615-621.

Wiklund, M.E., S.B. Wilcox (eds.). 2005. *Designing usability into medical products.* CRC Press, Taylor and Francis Group.

Wiklund, M.E., J. Kendler, and A.Y. Strochlic. 2011. *Usability testing of medical devices.* CRC Press, Taylor and Francis Group.

Weinger, M.B., M.E. Wiklund, and D.J. Gardner-Bonneau (eds.). 2011. Handbook of Human Factors in Medical Device Design, CRC Press, Taylor and Francis Group.

U.S. Department of Health and Human Services, Food and Drug Administration, Center for Devices and Radiological Health, Office of Device Evaluation – Draft Guidance for Industry and Food and Drug Administration Staff – Applying Human Factors and Usability Engineering to Optimize Medical Device Design, June 22, 2011.

Zambarbieri, D. 2007. Eye movements analysis in the evaluation of human computer interaction. In. *Atti Congresso Nazionale AICA. Cittadinanza e Democrazia Digitale*, 291-301.

CHAPTER 2

Case Study of Integrated Ergonomic Assessment of a Portable Ultrasound System

Leonardo Forzoni [1], Nicola Guraschi [1], Claudio Fertino [1], Marco Delpiano [1], Giorgio Santambrogio [2], Giuseppe Baselli [2], Fiammetta Costa [3], Maximiliano Romero [3], Giuseppe Andreoni [3]

Esaote S.p.A. Firenze, Italy. [1]
Politecnico di Milano, Dip. di Bioingegneria, Milano, Italy [2]
Politecnico di Milano, Dip. INDACO, Milano, Italy [3]
leonardo.forzoni@esaote.com

ABSTRACT

In recent years, the portable ultrasound (PU) systems have worldwide increased their importance both economically (number of systems sold) and clinically. The major task for user interface (UI) designers is represented by the concentration of the necessary controls for advanced technical and clinical features within reduced dimensions. The MyLab30CV (Esaote S.p.A., Firenze, Italy) PU, was evaluated during clinical examination sessions by expert sonographers in different clinical applications. This PU has a typical portable systems graphical user interface (GUI) menu by means of soft keys in the lower part of the monitor with direct controls right under the screen. The outputs of this analysis were the design inputs for the UI of a new PU system, the MyLabAlpha (Esaote S.p.A., Firenze, Italy), characterized by the integration of a high definition touch screen (TS) within the control panel, a reduced number of physical controls, compact dimensions and decreased weight. A detailed description and an evaluation of the new UI and its features are provided.

Keywords: portable ultrasound system, TS, user interface

1 INTRODUCTION

In the last years, the PUs have increased their importance on the worldwide market both regarding number of systems sold pcr year (56.214 in 2011; InMedica – May 2011) and annual market grow rate (23.4%; InMedica – May 2011).

The number of different products, from diverse manufacturers, available today is increased year after year. Nowadays, the portable ultrasound systems (PU) can be divided, as per the console-based systems, in low, middle and high level, regarding the diagnostic capabilities and technological features. The high level segment is the most interesting from the performances and ergonomics point of view: small systems which are highly transportable (weight under the 12kg) and which need to have clinical performances comparable to the ones of premium level console-based systems. A PU can be also used mounted on its proper cart. In this case it can be intended as a small console-based system with also plugs, holders for echographic gel and probes and supports for peripherals. The PU is nowadays not considered as a second best choice, with respect to a console based system as a main unit, but it is many times the first choice, being the only system available in the imaging lab.

Recently, always more functions and tools were added to the PU systems. The concentration of the necessary controls for thc advanced technical and clinical features within the reduced dimensions of the PU control panel has to be properly achieved. At the same time the number/kind of users is increased (also non-sonographers): the same system has to house more "things" and they have to be organized in the easiest and most accessible way possible.

In the design process of any ultrasound (US) system it has to be kept in mind that during the real-time acquisition, the US transducer is a "fixed" interface for the operator, therefore the US system during scanning has to be usable with only one hand, while the other one has to handle the probe.

As console-based system, also the PUs have usually to be shared-service units able to perform examinations in all the main Clinical Applications (i.e. Cardiology - Car, Abdomen - Abd, Vascular - Vas and Obstetrics and Gynecology - Ob-Gyn). PUs are usually characterized by a physical control panel with buttons, toggles, encoders, sliders and a trackball, a qwerty keyboard and a LCD screen. One of the techniques to integrate such a large amount of commands within a compact panel is the use of the direct controls of a GUI menu shown on the lower part of the screen. This menu is re-configurable and usually organized on more levels due to the fact that many controls are needed in any real-time acquisition modality (see Figure 1).

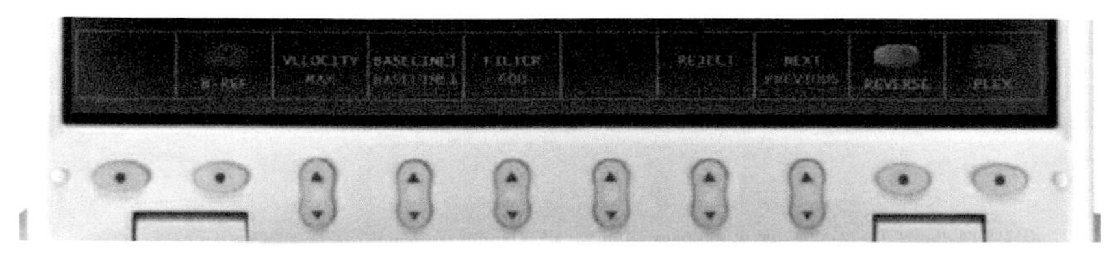

Figure 1: Example of a reconfigurable GUI menu on the lower part of the main screen and buttons and toggles to operate it.

The issue of US systems ergonomics, well known among sonographers (Craig, 1985; Atjak and Gattinella, 1989; Pike *et al*.m 1997; Smith *et al*., 1997; Gregory, 1998; Schoenfeld, 1998; OHS Update, 1999; Magnavita *et al*., 1999; Evans, Roll, and Baker, 2009; Sommerich *et al*., 2011), has been described and treated in many Standards and Guidance Documents from Regulatory Organizations, Healthcare Institutions and Sonographers Associations (Habes and Baron, 1999; Society of Diagnostic Medical Sonography, 2003; DHHS (NIOSH), 2006; The Society of Radiographers, 2007;European Standard EN 62366, 2008). However, today no Industrial Standards are available dedicated to the design of US systems (control panels, GUI and probes). Some recommendations on how the systems need to have certain characteristics (height adjustment, monitor features, size, shape and spacing of controls) are given, but no detailed design and implementation rules and guidelines are available. The lack of detailed UI design rules, missing also for console-based US systems, is even more stringent for the portable units, used by many sonographers even outside the imaging lab, also at the patient bedside (both at the hospital or at home). Hereunder we consider some ergonomics issues analyzed in PU systems and, as an improvement, a new user interface will be presented and analyzed.

2 METHODS

A PU (MyLab30CV, Esaote S.p.A.. Firenze, Italy – see Figure 2), was evaluated during clinical examination sessions by four expert sonographers in the four main clinical applications (Vas, Card, Abd, Ob-Gyn).

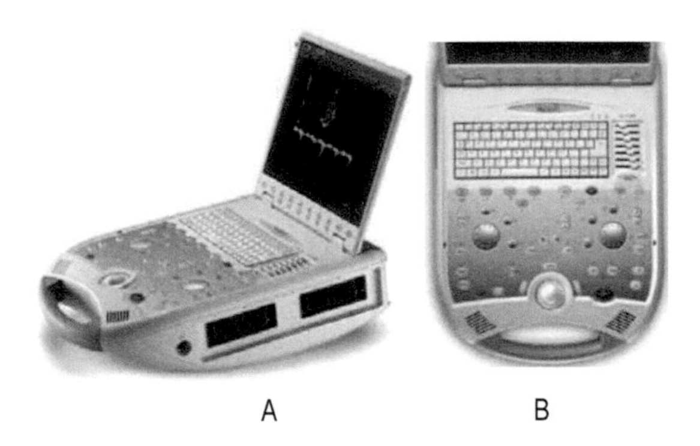

Figure 2: A) MyLab30CV; B) MyLab30CV Control Panel and qwerty keyboard

The MyLab30CV has a typical UI for PU systems: a reconfigurable, multi level, GUI (Soft key) menu is present in the lower part of the monitor with direct controls right under the screen. A physical control panel with encoders for the modality general gain, plus Time Gain Compensation (TGC) controls and modality, measure, annotations, body-marks, saving options and line/update buttons are placed around the trackball. Between the Soft key menu controls and the physical control panel, a qwerty keyboard is present, for Patient ID data entering, free text annotations and Report fulfilling and comments.

The used controls and commands in each of the four clinical applications were counted by an independent observer. Moreover, through an ad hoc interview, the expert sonographer reported the encountered difficulties while performing the examination. For the simulation phases of different possible UIs, a proper portable US system (MyLabOne, Esaote Europe, Maastricht, The Netherlands) with a TS GUI integrated in the same (single) monitor utilized also for the echo image presentation, was used.

3 RESULTS

The MyLab30CV US systems was used by four expert sonographers with different settings according to the specific application (Vas, Abd, Car and OB-Gyn were considered) and to the different body regions to be examined. The different settings called "Presets" were related to the application and the chosen probe. The most used modalities were B-Mode, Color Doppler (CD), Pulsed Wave Doppler (PW) and, only in Cardiology Application, Continuous Wave Doppler (CW) and M-Mode. After the Preset choice, the most used controls and frequent adjustments were:

- In B-Mode: Depth, General Gain, Time Gain Compensation controls;
- In CD modality: General Gain, Region of Interest (ROI) size and position, Pulsed Repetition Frequency (PRF or Scale);
- In PW Doppler modality: PRF, PW line of sight and Sample Volume (SV) position adjustments, Steering (if a linear array probe was used), Doppler Angle Correction (θ) if blood flow velocity measurements were performed.

The subjective analysis with sonographers demonstrated that the highest level of discomfort was reported from the distance between the trackball (which can be considered the center of the US system UI) and the Soft key menu controls. Starting from the user position, the analyzed classical UI has the trackball with the physical control panel, then the qwerty keyboard, then the Soft key controls and finally the main screen with the Soft key menu and the echographic image. The second critical point was the necessity to find the desired command/adjustment within the Soft key menu (organized as a single line), and the presence of controls not intuitively positioned on the panel (many controls for different applications/modalities can be properly positioned for a certain application but not for another: the simplest example is the CW Doppler which is widely used in Cardiology but completely unused in any other Application).

The points reported as utilization discomforts by the sonographer interviews were:

- Soft key menu controls distant from the trackball;
- Qwerty keyboard less frequently used than the Soft key menu controls but positioned closer to the user;

- High number and relative closeness of physical buttons/controls concentrated in a small area (also related to conceptually different functions);
- Necessity to scroll more pages of the Soft key menu (organized in levels) to find the desired control/function.

4 DISCUSSION

The outputs of the conventional PU UI analysis were the input for the design of a new concept of UI for a portable US system, the MyLabAlpha (Esaote S.p.A., Firenze, Italy – see Figure 3).

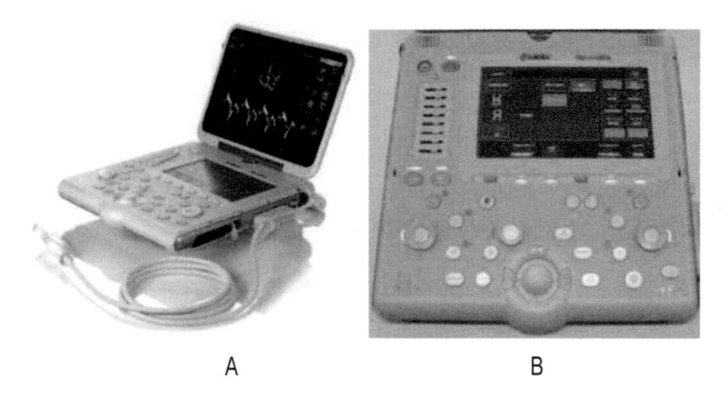

A B

Figure 3: A) MyLabAlpha; B) MyLabAlpha Control Panel.

A high definition TS is integrated within the physical control panel at a shorter distance from the trackball. A reduced number of well-spaced physical controls (buttons, toggles, sliders and encoders) is available on the panel with the intention to facilitate their detection and use by the operator. Buttons and controls which can be activated (or are already active in a certain system status) are lighted. Some of the most frequently used controls/adjustments are physical in order to be activated/changed while the eyes of the user are focused on the main screen where the echo image is shown: Freeze, General Gain, TGC, Automatic Adjustment of imaging and Doppler traces, line/update, saving options (image/clip/print). The same can be said regarding modality buttons like B-Mode, CD, PW, CW, M-Mode. The level of the UI are three on the MyLab30CV (Physical control panel, qwerty keyboard and Soft key menu controls), on the MyLabAlpha are two (physical control panel and TS), having the Soft key menu controls integrated within the TS, where also other ex-physical buttons/controls are grouped. Moreover, the qwerty keyboard is integrated within the TS and can be activated by touching a proper icon on it.

Multiple possible choices of UI were evaluated and tested prior to the developing phase of the new system. Just to name a few it was discussed the possibility to have the whole controls integrated within a single, wider TS separated from the main screen. That is, no physical controls and buttons at all. The simulation of this scenario was performed using a TS based system (MyLabOne, Esaote Europe, Maastricht, The Netherlands) connected to an external monitor to duplicate the echo image. In the simulation the user had to watch the gain controls while changing them and their effect on the main screen image, alternatively and repeatedly depending on the total amount of increase/decrease needed. The same happened regarding other modality-related controls such as frequency, depth, PRF/scale and angle correction.

Another possible choice would have been to integrate the control panel as a TS input device on the same display of echographic image thus solving the possible limit of eye movements between the main screen and the TS. As a drawback, being the task the integration of a multi-application high level system, the total amount of controls and features would have led to obtaining a multi-level GUI with the controls necessary spread over more windows and tabs. This would have been the only possible choice being a constrain the portability of the system, and therefore having proper specifications for the screen size. On the other hand the echographic image had to be the maximum size possible, this maintaining limited the expandability of the GUI portion on the main screen and forcing to have multiple GUI layers. Those considerations led to the decision to maintain a limited physical control panel with the reconfigurable GUI within a TS separated from the main screen on which the echo image was shown.

The concept of TS is widely used in everyday life within lots of technological gadgets and input devices. Anyway, the US system interface is widely different with respect to the one of the other "consumer electronics" systems. The US device is used, at least during the real-time acquisition, with only one hand (the other has to handle the transducer) with the operator usually not positioned frontally to the US system control panel: therefore, the US UI cannot follow all the same design rules of the everyday life technological gadgets. Additionally, the US system UI has to be designed in order to be used also in situations of low level of illumination (as many imaging labs are) in order to see the echographic image in a better way. The TS as input device is anyway common among console-based US systems: in the last ten years almost all the manufacturers have integrated TSs at least in their high level systems, but it has been introduced only recently among the PU units.

On the MyLabAlpha most of the GUI is present within the TS where direct toggles are used for parameter adjustments. Virtual buttons or functions/controls which can be activated are in dark blue, light blue is for controls/commands already activated and grey for not-available ones The GUI is organized in only two layers maximum for each tab. On the physical control panel controls and commands are more grouped and positioned closer to the trackball.

The ergonomic integration was obtained with no reduction of functionalities with respect to the previous system.

Table 1 presents the direct dimensional comparison between the MyLabAlpha PU (new UI) and a conventional PU UI, here represented by the MyLab30CV system:

Table 1 New UI and conventional PU UI direct dimensional comparison

	New UI	Conventional PU UI	
Control panel dimensions	34 x 36.5cm (1241cm²)	35.5 x 47.5cm (1686,3cm²)	27% smaller control panel
Number of controls present on the panel	44 (27 buttons 6 toggles 3 encoders 8 sliders	59 (41 buttons 9 toggles 2 encoders 7 sliders)	25% reduction in the number of controls present on the physical control panel
Distance trackball – modality gains	13cm	14cm	Trackball closer to the modality gains
Distance trackball GUI controls	19cm (MIN distance 13)	34.5cm (MIN distance 33)	45% reduction
Distance trackball – toggles	11cm (MIN distance) – 15cm (MAX distance)	33cm (MIN) – 34.5cm (MAX)	67% reduction
Distance Trackball – TGC	22cm	27cm	19% reduction
Distance trackball – Qwerty	19cm (Virtual Qwerty keyboard)	23cm	Trackball 17% closer to the Qwerty keyboard

In order to have an initial evaluation of the UI quality of MyLabAlpha, Tests were performed with expert sonographers in the four main clinical applications (Vas, Abd, Car, Ob-Gyn). The sonographer used the MyLabAlpha system in a routine examination without any indication regarding the UI or the functionalities of the PU.

After the use of the system the sonographers were interviewed reporting to be sufficiently confident about the utilization of the system since the first approach regarding the localization and use of the most common real time controls and, after the creation of a proper Preset for the chosen application and probe, they generally needed to operate only on the Basic page of the last selected modality. An independent observer was present during the test confirming the comments of the sonographers.

Some interesting outputs regarding possible future improvements of the GUI came out from the interviews and the reports of the independent observer:

- PW SV when the PW acquisition is active, could be positioned in the Basic page of the PW acquisition TS GUI;
- Sub-windows on the TS with a time defined presence (this point less stringent in case of an already properly set system regarding the chosen probe/application): after a fixed period of time, the sub-window, once activated and not "touched" has to disappear, coming back to the original window on the TS;
- Increment of the font size/readability of the measurement on the main screen and on the black and white prints.

Moreover, the transducers of the new system were designed following the Industry Standards for the Prevention of Musculoskeletal Disorders in Sonography. They are light (100-150gr, for linear and convex probes) and with a dual-possibility hand grip (AppleProbe Design) made available (pinch grip and palmar grip) in order to provide a neutral wrist position (see Figure 4).

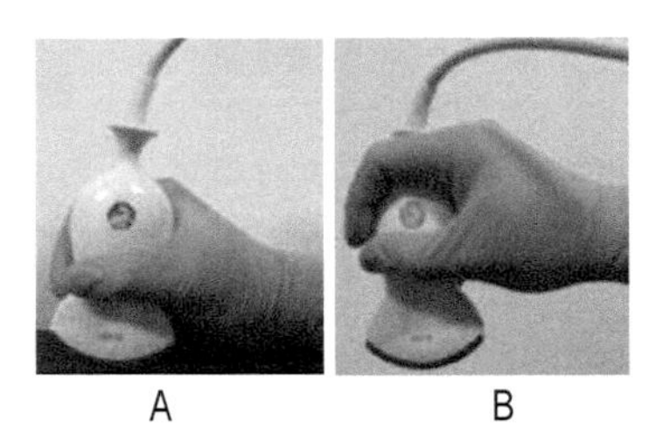

Figure 4: A) palm hold; B) pincer hold

In addition, the novel eTouch concept was implemented on the new PU.

The eTouch represents a tool for the creation of a personalized, user-targeted workflow of the system. It allows the user to record sequences of key mixing functions of the TS and of the control panel. Each recorded sequence (macro) can be named and saved to be available as a customized button in customized TS (eTouch environment). The recording of the macro is performed during the normal use of the system.

Pressing the eTouch button on the physical Control Panel, the user enters in the eTouch environment on the TS (see Figure 5).

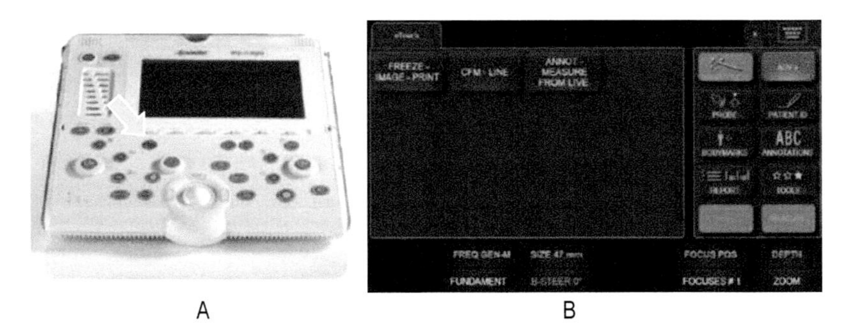

Figure 5: A) eTouch button; B) eTouch environment TS page layout.

Examples of eTouch macros can be the following:

- "Freeze-image-print" macro: from Real Time (live), the activation of this macro, drives the system in Freeze, saves an image (the last image of the cine loop) and prints the image (if a printer is connected and active the image will be printed, otherwise, the macro will end with the saving of the image);
- "CFM and line" macro: from Real Time (live), the CFM modality is activated and the Doppler line placed;

- "Annot-measure from live" macro: from Real Time (live), the system is forced in Freeze status, the annotation "tricuspid" is placed on the image and a generic distance measure is activated (the first caliper of the measurement is shown on screen, linked to trackball movements but not fixed).

For any macro the eTouch environment is characterized by the TS encoders' controls (shown in the lower part of the TS) which are those of the Basic TS page of the last activated mode (B-Mode, CD, PW, etc.).

The eTouch environment is a finite-state machine where the macro has to be recalled while the system is in the same situation when it was when recorded, otherwise the system will perform the exact sequence of tasks, but starting from a different initial condition, which may lead to a different output than the expected one.

5 CONCLUSION

From the analysis of the conventional UI of a PU, performed on the MyLab30CV, the points of not optimized usability were:

- Distance between the trackball (which can be considered the center of the US system UI) and the Soft key menu controls;
- Qwerty keyboard less frequently used than the Soft key menu controls but positioned closer to the user;
- High number and relative closeness of physical buttons/controls concentrated in a compact surface area;
- Necessity to scroll more pages of the Soft key menu levels to find the desired control/function.

The new portable US system MyLabAlpha is intended to offer an easier to be operated UI, a reduced number of controls in order to have a few commands only when needed and only what needed.

A smaller physical control panel allows smaller dimensions to increase portability and it has the aim to lessen hand/arm movements for the user. The reduction in the number of controls present on the physical control panel wants to reduce the learning curve related to the system use. The trackball closer to the modality general gain encoders, TGC sliders and the qwerty keyboard (virtual qwerty keyboard on the MyLabAlpha), the reconfigurable controls (toggles) and reconfigurable GUI (TS) are solutions which go in the direction of less hand/arm movements while performing an examination.

Some interesting outputs regarding a possible improvement of the GUI, which came out from the interviews and the examination of the independent observer, were collected as well. A control panel with reduced dimensions but with well-spaced controls has the aim to increase the readability: the purpose is to reduce the strength and the time-o-flight of the operator hand and arm, reducing the possible erroneous command selection. The use of a high definition TS is intended to concentrate all the main features of the re-configurable GUI in a well defined area close to the trackball. The light transducers designed with a choice between two different grips want to reduce hand, arm and shoulder stress.

ACKNOWLEDGEMENTS

The authors wish to thank G. Altobelli, S. D'Onofrio, M. Moglia, L. Bigi (Esaote S.p.A., Italy), Maria Marcella Laganà (Fondazione Don Carlo Gnocchi ONLUS, IRCCS S. Maria Nascente, Milano, Italy) for the valuable contribution given to this work.

REFERENCES

Craig M. 1985, Sonography: an occupational health hazard, *Journal of Diagnostic Medical Sonography*. 1, 121-125.

Atjak A, Gattinella JA 1989. Ergonomics in ultrasound equipment: productivity and patient throughput, *Radiol Manage*. Fall;11(4): 38-40.

Pike I, et al. 1997. The prevalence of musculoskeletal disorders among diagnostic medical sonographers, *Journal of Diagnostic Medical Sonography*, 13(5):219-227.

Smith AC, Wolf JG, Xie GY, and Smith MD 1997. Musculoskeletal Pain in Cardiac Ultrasonographers: Results of a Random Survey, *J. Am. Soc Echocardiogr*, May; 10(4):357-62.

Gregory V., 1998, Musculoskeletal Injuries: Occupational Health and Safety Issues in Sonography. Sound Effects 30, September 1998.

Schoenfeld A 1998. Ultrasonographer's wrist – an occupational hazard, *Ultrasound Obstet Gynecol*, 11:313–316.

OHS Update, Report on the results of the Australian Sonography Survey on the prevalence of musculoskeletal disorders among Sonographers., Val Gregory, Sound Effects 42. December 1999.

Magnavita N, Bevilacqua L, Mirk P, Fileni A, and Castellino N, 1999. Work-related Musculoskeletal Complaints in Sonologists, *J Occup Environ Med*, Nov; 41(11):981-8.

Evans K, Roll S, Baker J 2009. Work-Related Musculoskeletal Disorders (WRMSD) Among Registered Diagnostic Medical Sonographers and Vascular Technologists. A Representative Sample, *Journal of Diagnostic Medical Sonography*, 25(6): 287-299 .

Sommerich C et al. 2011, Participatory Ergonomics Applied to Sonographers' Work, Proceedings of the Human Factors and Ergonomics Society Annual Meeting, September 2011, 55(1): 1067-1070.

Habes DJ, Baron S, Health Hazard Evaluation Report, HETA 99-0093-2749, NIOSH Ergonomics evaluation of sonographers at St. Peter's University Hospital.

Society of Diagnostic Medical Sonography, Plano, Texas, USA, 2003. Industry Standards for the Prevention of Work-Related Musculoskeletal Disorders in Sonography.

DHHS (NIOSH), 2006. Preventing Work-Related Musculoskeletal Disorders in Sonography, Publication No. 2006–148.

The Society of Radiographers, 2007, Prevention of Work Related Musculoskeletal Disorders in Sonography, UK. March 2007.

European Standard EN 62366 January 2008, Medical devices – Application of usability engineering to medical devices (IEC 62366:2007).

CHAPTER 3

Co-designing Better Work Organization in Healthcare

Julia A. Garde, Mascha C. van der Voort

University of Twente
Enschede, The Netherlands
j.a.garde@utwente.nl

ABSTRACT

Hospital care changes in the wake of social and technical developments. When a hospital is renewing buildings, in-house logistics or ICT as an adaption to the changing environment, it is faced with a demand for care that is more human-centered as well as the necessity to save costs In facing these challenges, including hospital staff in the re-design of the work organization has two advantages. First, the staff has valuable experience with everyday care that will help to create a new situation that is workable. Second, the consultation of the staff and serious efforts to co-design a new work organization will increase staff commitment. In this paper we present the application of tools for co-design in a large health-care project. The project is performed in a hospital that will move to a completely new building. It involves the redesign of work organization concepts for nurses with regard to ICT, ward communication, supply logistics, catering and visiting policies for the general wards. Four co-design workshops on different redesign aspects were set up. The center-piece of the workshops are scenario based design games that promote an overview of complete care processes. In this paper the workshops are described and evaluated concerning their applicability to develop work organization concepts, elicit requirements for related products and systems and to be reproduced in future work organization re-design. The chosen set-up enabled participants to generate new situations and walkthrough imaginative work processes and thereby address work organization related problems.

Keywords: co-design, work organization, participatory design, design game

1 INTRODUCTION

Continuous development of new cure possibilities and competition between hospitals force hospitals to reorganize themselves. When a hospital is renewing buildings, in-house logistics or ICT as an adaption to the changing environment, it is faced with a demand for care that is more human-centered as well as the necessity to save costs. Additional challenges are the ageing population and the progressive replacement of treatments with hospitalization by policlinic treatments. In the Netherlands the number of hospitalizations has increased, whereas the average time of hospitalization is decreased (VTV, 2010). These developments result in a higher turnover of patients that stay only a day or two and an older and sicker patient population that remains in the hospital. Therefore, the composition of tasks and work load for nurses changes. In the struggle of reorganization and redesign the ergonomics of new nursing work organization must not be overlooked, as they form the center of day-to day hospital care practice. In the daily nursing work organization, ICT, supply logistics, medication and catering come together and need to leave enough time and space to provide patients with personal attention. Therefore, a good organization of all nursing tasks should form an important point of departure to formulate requirements for hospital renewal.

We present a project that aims to co-design nursing work organization in order to cope with the developments. In the scope of the project, work organization is defined as the way roles, responsibilities, tasks, tools and material resources are distributed over individuals, time and space in an organization. This definition covers four of the five elements of work systems as described in the Balance Theory of job design (Smith and Carayon-Sainfort, 1989); tasks, tools and technologies, physical environment (space) and organization. Redesigning the organization, as an institution with policies, goes beyond the scope of the presented project. Also, the general distribution of space (as in architectural space) will not be covered, because the architectural drawings for the new hospital were already completed, when the project started. Adaptions on a smaller scale are still possible.

The change of work organization implies changes to work ergonomics. A fundamental redesign offers the opportunity to improve ergonomics and thereby health and safety of the staff. However, this is not easily achieved in a top-down approach. In hospitals, ergonomics has a "socially situated practice" (Hignett, 2001), i.e. ergonomics is embedded in the organization in the way staff handles their daily tasks. Beyond the structures that are often defined top-down (schedules, material, tasks), staff do (and should) shape their own tasks in detail. Changing the way work is done can be difficult, as staff does not always see the necessity to change the way they handled things for years. The participation of staff in the design process can help to grow feelings of self-determination (Carayon and Smith, 2000) and enthusiasm for a new situation.

2 CO-DESIGN

Several approaches to redesigning work methods can be found in literature and the participatory approach is a successful one (Vink et al., 2008). Including hospital staff in the re-design of the work organization has two advantages. First, the staff has valuable experience with everyday care that will help to create a new situation that is workable. Second, the consultation of the staff and serious efforts to co-design new work organization will increase staff commitment to a new work situation (Davies, 2004).

Participation in a co-design process is different from participatory ergonomics. In participatory ergonomics typically a group of staff with differing functions get a training in ergonomic principles and then try to use the training to improve the ergonomics of the work situation (Rivilis et al., 2008). In contrast to participatory ergonomics the participation of staff in co-design of the work organization includes not only implementation of ergonomic principles but active designing by the participants, as new ideas are needed to cope with a changing environment. Furthermore co-design of the work organization exceeds ergonomic questions, as participants have to address logistics, styling and experience aspects as well.

2 CO-DESIGN PROJECT

2.1 Project

In this paper we present the application and assessment of co-design tools in a hospital renewal project. The project is performed at the Medisch Spectrum Twente hospital (MST). The MST will move to a newly-built hospital, comprising 620 beds, in 2015. Currently the hospital has 4000 employees, including 250 medical specialist. The migration will entail five important changes that effect the workflow of the nursing staff:

1. in the new building there will only be single-person rooms, while in the current building there are one-, two-, and four-person rooms;
2. medical specialties will not have a pre-defined number of beds in the new wards but will be assigned beds according to demand; hence, their ward unit size and the amount of staff will fluctuate;
3. the hospital management has the ambition to create a paper-free hospital, including digital patient records;
4. the visitor policy will change. Today the hospital works with visiting hours, but single-person rooms offer the possibility to allow visitors to be around 24/7 and for family to stay overnight;
5. the catering will change from a single central kitchen that serves at predefined times to a kitchen on every floor that can serve food continuously and to order.

The new plans mean that the way nurses and ward assistants work, how they organize their shifts and use materials, must change to adapt to the new situation.

To use this change as a chance and create a positive impulse to adapt and improve the work organization, a co-design project was started in cooperation with researchers from the Industrial Design Department of the University of Twente. The co-design project is called project "SWING", which (in Dutch) stands for "designing work processes for the new building together". It involves the redesign of work organization concepts with regard to ICT, ward communication, supply logistics, catering and visiting policies for the general wards.

2.2 Project organization

Co-design projects need careful management to ensure participants' commitment and motivation and the implementation of resulting concepts. Success factors for the process are a good project inventory, a steering committee and a step-by step approach (Vink et al., 2006). Swing has a steering committee consisting of an experienced product designer/researcher, experienced in co-design (one of the authors), a business process redesign manager, a building project manager, an ICT advisor, a nurse who is also a member of the nursing advise council, two ward team leaders, a nursing process specialist and a business unit manager. The project is managed by the author and the business process redesign manager, together. Furthermore, there are about 40 SWING workshop participants, mostly nurses, but also nurse practitioners, ward assistants, secretaries and a physiotherapist. These participants come from all over the hospital.

In summer 2011 SWING started with the project inventory by a series of interviews with key information holders in the hospital and a series of visioning workshops, with the goal to inventory concerns and visions for the future of the hospital wards. The next step was a series of design workshops with the goal to look into the re-design of work processes and related products as well as responsibilities and rules. This series consisted of four workshops, each with a dedicated topic, i.e. "ICT & communication", "catering concept", "nursing tasks & visitors" and "material logistics". The ideas resulting from the four dedicated workshops will be elaborated during a third series of workshops. Finally, the project results will be evaluated by means of evaluation workshops as well as a second series of interviews. In this paper we present set-up and results from the second series of workshops.

2.3 Work-shop set-up

The four topics of the second series design workshops were "ICT & communication", "catering concept", "nursing tasks & visitors" and "material logistics". For every design workshop topic about 10 participants were invited. During the workshop the participants were split into groups of five. The workshops took about 3 hours each. The general technique used in all four workshops is a scenario based design game that promotes an overview of complete care processes. The game is a combination of a board game and a task card analysis used by Garde and van der Voort (2009) in earlier projects. The two components strengthen and complement each other and serve as mutual verification tools. The task flow analysis is a way to structure the work processes concerning chronology, time management, staff deployment and information flow. The board game supports participants to imagine the procedure in a realistic setting. By enacting the tasks with board game figures, a developed procedure can be assessed. The chosen workshop set-up has several qualities, that were described earlier by Garde and van der Voort (2009):

1. *it enables the users to invent and design a new procedure,*
2. *it includes all different users at the same time, so that it can be discussed immediately what a change in one user's domain of responsibility means for the domains of others,*
3. *it gives a clear overview of a procedure and the consequences that changes to this procedure have,*
4. *it triggers the participants to empathize the new situation,*
5. *it includes all possible appliances that could be involved in the procedure,*
6. *it is time efficient in view of the limited availability of time medical specialists [and nurses] have.*

Materials used to support the chosen technique in the SWING project were:

- task cards for building the task flow
- a large "game board" consisting of the architectural drawings of a ward in the new building,
- playing figures in different colors representing the different functions of the staff,
- playing figures depicting materials such as trolleys,
- problem cards describing problems that had to be solved with the new work organization.

The general set up of de workshops can be described in three steps:
First participants explored the design of the new hospital building: participants were asked to play out their current nursing work procedure on the game board, i.e. the large architectural plan of the new ward, with the playing figures. This technique (partly also used under the name "living blueprint" by Dalsgaard (2010)), is a way to visualize changes in a future workflow by collating them to the current workflow, and by this means prevents unfit solutions (Jalote-Parmar and Badke-Schaub, 2008); playing out the work procedures, participants encountered problems with the combination of today's work practice and the future ward design (see Fig. 1, left).

Second, the problems found in the first step than had to be solved and the work organization improved. Participants redesigned their work processes and decided upon what kind of technology or materials they want to use in their daily work. Participants used task cards (similar to CUTA (Lafrenière, 1996)) to fixate a task flow proposal (see Fig. 1, right). When creating a task flow, participants could also set-up new rules and assign responsibilities. Doing this, participants created a scenario that had to be tested repeatedly by playing out scenes on the game board.

Third, the designed set-up had to be put through a stress-test. For this purpose, the steering committee had created "problem cards" in advance. These cards contained descriptions of possible problems that were related to the four topics and had to be solvable in the context of the self-created scenario. By this means the new scenario was tested for feasibility and flexibility.

Minor variations on the general workshop set-up were applied in order to dedicate the set-up of each workshop to its specific focus. For the ICT & communication workshop we used additional mock-ups of mobile appliances in real size (e.g. tablets and smart phones) to evaluate what kind of appliances would be useful for what kind of task (e.g. reading, ticking off or writing at different places in the ward) (see Fig. 1, left). The catering concept workshop had a strong accent on responsibilities and therefore we had prepared special “hats” for the playing figures that symbolized responsibilities with respect to diet, feeding, and serving. In addition a number of patient cases with special diet requirements were prepared, that the participants had to deal with. In the nursing tasks & visitors workshop the first emphasis was on the scheduling of nursing tasks over the day. Then, tasks during which visitors were not allowed to be around were marked. Visiting rules and regulatory tasks were discussed and written on special cards. For the material logistics workshop we had prepared lists of materials used at the wards, playing figures of trolleys and cards that had to be assigned to storages (e.g. a trolley, the patient room or on-site storages) and filled in with the materials that should be stocked.

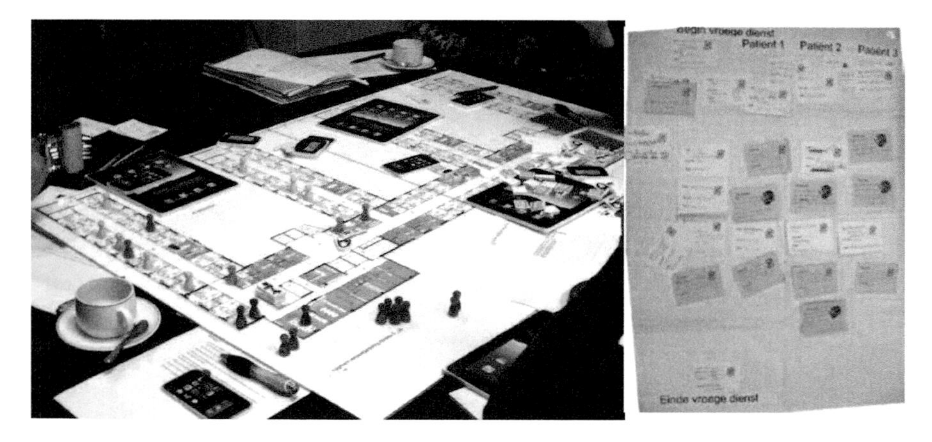

Figure 1 Left: photo from the SWING workshop on ICT & communication, showing the game board, playing figures and appliance mock-ups (October 2011). Right: photo of a task flow scheme, created during SWING workshop on nursing tasks & visitors (October 2011).

3 RESULTS

3.1 Workshop-results

The types of results generated in the four dedicated workshops by eight workshop groups were product requirements, rules, regulatory tasks, assignments of responsibilities, task flows, product requirements, and questions that had to be answered by the building committee. An important overall result was the grow of commitment for the SWING project and engagement with the building process. A few examples for all the types of results will be given.

Product requirements that were generated concerned for the most part ICT solutions. For example, participants wanted all staff members to use a small tablet PC to access and modify information about patients at the patient room. They wanted the application used on the tablets to show a general overview of a patient as first page.

Rules that were generated concerned e.g. the catering concepts stated that there has to be a time-window for ordering food, instead of enabling patients to order food all around the clock. Only then control-moments to check how much everybody has been eating can be created for the ward-assistants.

In every workshop participants created new task flows by placing task cards in chronological order. The existing task flows were altered to adapt to the new situation. For example, single-person rooms require that nursing tasks with one patient are completed before proceeding to a next patient whereas in the former situation, a nurse would first wash all patients in a room and then fill in the measurement scores for all patients. However, a major problem was identified in the task flow for the early shift, i.e. the gross of tasks needs to be accomplished in the first two hours of the shift. The expectation is that the workload will increase beyond acceptable levels due to the longer walking distances and single-person rooms in the new building. A solution, presumably in the form of reorganizing the shift will be explored during the third series of workshops.

The project participants discovered a few problems in the current building plans, e.g. that there was no storage room planned for decubitus matrasses. Questions about such problems were forwarded to the building committee and the SWING participants will be informed on the developments.

A very important result of the workshops was not content related. In the beginning of the project the main tenor of the participants was concerned about the hospital plans and the decisions already made. The project managers were questioning whether the whole project could have a useful outcome at all with this attitude. However, in the second series of workshops people were actively involved in the creation of solutions, which left little room for keeping up an attitude to complain. Since participants felt listened to regarding both their concerns and ideas, the workshops generated commitment for the project, a pro-active attitude of participants, and a more positive view on the developments. This was nicely framed by one participant: *"I think we see a lot of threads at the moment, but I believe that it will all work out in the end".*

3.2 Applicability to develop work organization concepts

The workshop set-up enabled participants to generate new work situations and to walkthrough imaginative work processes. It proved to be suitable for investigating different work organization related problems. It considers both, actions that can be planned (e.g. washing patients) as well as actions that emerge in unplanned situations (e.g. patient calls nurse). Roles and tasks were distributed over individuals, time and space in the created task flows. Responsibilities were assigned and tools and material resources were chosen and tested by use of the playing figures and playing out scenarios. Therefore all aspects of work organization are covered. In addition, product requirements were created. As the chosen technique aims at stimulating creativity, this series of workshops resulted in a rough sketch of the desired work organization. Two sequential workshop series will be organized to detail the work organization and to ensure that all relevant aspects are addressed.

3.3 Reproducibility

The general work-shop set up has proven its power to elicit useful results by each of the eight different groups addressing four different topics. These results confirm the results found in an earlier project that focused on the development of a new treatment procedure by physicians and technicians (Garde and van der Voort, 2009). Therefore, the applied combinatory technique is proven to be successful for the development of a new work organization in combination with relating appliances and ICT solutions. However adaptation of the technique to the specific development situation is a prerequisite (e.g. by creating a game board that reflects the project situation).

Co-design projects regarding work organization often have to deal with political sensitivity. Participants can be afraid to speak out or change the work situation, whereas others have been frustrated with preceding reorganizations or re-design projects. These aspects need to be dealt with carefully. To ensure an open and inspirational environment, transparency and good information management (e.g. keeping participants up-to date about what happens with their input) is of high importance. In the "SWING" project, the first workshop had the function to identify all concerns, discuss them and to jointly formulate visions for the project.

4.3 Follow-up

Sanders et. al. (2010) stated that, in setting up a participatory design method, it is important to keep an eye on the complete experience participants have during a project, and therefore each activity should prepare participants for the next one. The next series of workshops within SWING will build on the outcome of the presented workshops.

The created workflow must be revisited and to achieve this, participants must disconnect even more with their current work-flow and try to think "out of the box". We believe that, with the second series of workshops the participants have achieved a pro-active and open state, so that the third workshop can force them even more out of their comfort-zone.

In the four workshops, the four topics were addressed isolated. In the progression of the project, these topics must be united, and their interplay and mutual influence need to be explored. In order to generate feasible solutions, emergency scenarios must become part of the project to test the created solutions under extreme circumstances.

5 DISCUSSION

When SWING started, the architectural drawings of the new hospital were already finished. They were created around a vision that contained hospitality amongst other focus points. Therefore, the work organization processes had to be adjusted to the building. An ideal situation with respect to the processes would be, that a building would be built around the processes instead of vice-versa.

A general challenge with co-design is, that on the one-hand participants must be enabled to come up with creative solutions, and on the other they must be restricted by boundaries, to ensure that created solutions are feasible (e.g. financially). It is important to find the appropriate degree of freedom, which does not hinder creativity but still leads to useful results. In the second series of SWING we worked with a set of game-rules to ensure that created concepts would fit to the decisions that had already been made in the building process.

Although, participation of employees in the shaping of their own work situation is mostly assumed to be beneficial for job satisfaction Carayon and Smith (2000) stress that there has to be done more research on the potential negative effect of participation, i.e. increased workload In project SWING participants could participate during working hours, potential additional workload was thereby minimized. However, the different wards faced additional effort to plan the shift schedules. Therefore, some participants did miss a workshop due to being scheduled to work at the ward at the same time and not being able to leave due to intense activity at the ward.

6 CONCLUSIONS

We presented a project that aims to co-design nursing work organization for the wards within a hospital building under development. Four workshops with the topics “ICT & communication”, “catering concept”, “nursing tasks & visitors” and “material logistics” were set-up. As work-shop technique a scenario based design game that promotes an overview of complete care processes has been used. The game is a combination of a board game and a task card analysis. The workshop set-up enabled participants to generate new work situations and to walkthrough imaginative work processes. It proved to be suitable for investigating different work organization related problems. It considers both actions that can be planned (e.g. washing patients) as well as actions that emerge in unplanned situations (e.g. patient calls nurse). Roles and tasks were distributed over individuals, time and space in the created task flows. Responsibilities were assigned and tools and material resources were chosen and tested by playing out scenarios on the game board. The types of results generated in the workshops were product requirements, rules, regulatory tasks, assignment of responsibilities, task flows, and questions that had to be answered by the building committee. An important overall result was the grow of commitment for the SWING project and engagement with the building process.

The applied, combinatory technique is proven to be successful for the development of a new work organization in combination with relating appliances and ICT solutions. However adaptation of the technique to the specific development situation is a prerequisite (e.g. by creating a game board that reflects the project situation).

ACKNOWLEDGMENTS

The authors would like to acknowledge the project team and participants of the SWING project of the Medisch Spectrum Twente in Enschede.

REFERENCES

Carayon, P. & Smith, M. J. 2000. Work organization and ergonomics. *Applied Ergonomics*, 31, 649-662.

Dalsgaard, P. 2010. Challenges of participation in large-scale public projects. In: Bodker, K., Bratteteig, T., Loi, D. & Robertson, T. (eds.) Proceedings oft the Participatory Design Conference. Sidney: ACM.

Davies, R. C. 2004. Adapting virtual reality for the participatory design of work environments. *Computer Supported Cooperative Work*, 13, 1-33.

Garde, J. A. & van der Voort, M. C. 2009. The procedure usability game: A participatory game for the development of complex medical procedures & products In: Roy, R. & Shebab, E. (eds.). Proceedings of the CIRP IPS2 Conference. Cranfield: Cranfield University Press.

Hignett 2001. Embedding ergonomics in hospital culture: top-down and bottom-up strategies. *Applied Ergonomics*, 32, 61-69.

Jalote-Parmar, A. & Badke-Schaub, P. 2008. Workflow Integration Matrix: a framework to support the development of surgical information systems. *Design Studies*, 29, 338-368.

Lafrenière, D. 1996. CUTA: A simple, practical, low-cost approach to task analysis. *interactions*, september + october, 35-39.

Rivilis, I., Eerd, D., van, Cullen, K., Cole, D. C., Irvin, E., Tyson, J. & Mahood, Q. 2008. Effectiveness of participatory ergonomic interventions on health outcomes: A systematic review. *Applied Ergonomics*, 39, 342-358.

Sanders, E. B.-N., Brandt, E. & Binder, T. 2010. A Framework for Organizing the Tools and Techniques of Participatory Design. In: Bodker, K., Bratteteig, T., LOI, D. & Robertson, T. (eds.) Proceedings oft he Participatory Design Conference. Sidney: ACM.

Smith, M.J., Carayon-Saintfort , P. 1989. A balance theory of job design for stress reduction. *Industrial Ergonomics* 4, 67 -79.

Vink, P., Imada, A. S. & Zink, K. J. 2008. Defining stakeholder involvement in participatory design processes. *Applied Ergonomics*, 39, 519–526.

Vink, P., Koningsveld, E. A. P. & Mo, J. F. 2006. Positive outcomes of participatory ergonomics in terms of greater comfort and higher productivity. *Applied Ergonomics*, 37, 537-546.

VTV 2010. *Tijd en toekomst,* Deelrapport van de VTV 2010. In: Luijben, A. H. P. & Kommer, G. J. (eds.) Van gezond naar beter. Bithoven, Netherlands: Rijksinstituut voor Volksgezondheid en Milieu, Ministerie van Volksgezondheid, Welzijn en Sport.

CHAPTER 4

A Smart Wearable Prototype for Fetal Monitoring

A. Fanelli[a], *M.G. Signorini*[a], *P. Perego*[b], *G. Magenes*[c], *G. Andreoni*[b]

[a]Department of Bioengineering, Politecnico di Milano, Milano, Italy
[b]Indaco Department, Politecnico di Milano, Milano, Italy
[c]Department of Computer Engineering and Systems Science, Università di Pavia, Pavia, Italy
andrea.fanelli@mail.polimi.it

ABSTRACT

Fetal monitoring during pregnancy is extremely important to identify risky conditions for the fetus. The standard approach to assess fetal well-being in the uterus is represented by cardiotocography (CTG). Unluckily CTG can only be used in a clinical setting, because it requires expert clinicians and cumbersome equipment to be performed. In order to overcome the disadvantages of CTG, which prevent a close and continuous monitoring during the last weeks of the pregnancy, a new prototype for fetal monitoring is proposed. The device we present is a home wearable fetal monitor. Instead of detecting fetal heartbeats using Doppler Ultrasounds, as CTG does, it is based on the recording of abdominal ECG. The system measures the maternal and fetal ECG by 8 leads embedded in a wearable belt. A custom algorithm was developed for recognition of maternal beats, their subtraction from the whole signal and extraction of fetal heart rate. The algorithm is based on an averaging and subtracting process. Afterwards, the beat-to-beat signal of both mother and fetus is computed. The relevant information extracted from the abdominal recordings is then transmitted to a laptop using a Bluetooth connection. The prototype we have developed will change the way in which fetal monitoring is accomplished, allowing comfortable, close and continuous monitoring of fetal well-being.

Keywords: Fetal monitoring, wearable technologies, fetal ECG

1 INTRODUCTION

Fetal monitoring during pregnancy is extremely important to prevent and diagnose life-threatening conditions for the fetus. Approximately 85% of fetuses in the United States are assessed using fetal monitoring (Sameni and Clifford, 2010), in order to identify and treat pathological conditions and ante-partum complications, such as fetal hypoxia and asphyxia, intrauterine growth restrictions, hypertensive disease, placenta abruptio. The early diagnosis of fetal distress is needed to prevent fetal morbidity and mortality.

In particular, IUGR is one of the most severe and common causes of perinatal morbidity and mortality. It represents an inhibition of fetal growth and the failure of the fetus to attain its growth potential. The incidence of IUGR is approximately 5% of all pregnancies (Peleg, Kennedy and Hunter, 1998). In this way, the monitoring of FHR and FHR variability is extremely important to detect IUGR and helps the clinician in the decision process.

The standard approach to monitor fetal well-being in the uterus is the cardiotocography (CTG). Nonetheless this technique requires expert clinicians and/or nurses for the skillful placement of the ultrasound probe on the maternal abdomen, in order to detect the fetal heart rate. Moreover the equipment is cumbersome and a hospital setting is mandatory. The monitoring sessions are rare during the pregnancy and the pregnant woman is required to reach the hospital. The main alternative to cardiotocography is represented by the recording of fetal electrocardiogram (FECG) using electrodes placed on the maternal abdomen. This technique is simple and absolutely non-invasive as it does not require expert clinicians to be performed and can be accomplished using compact smart devices. It also provides additional information about fetal heart, e.g. morphology of P-wave, QRS-complex, QT duration measurement (Karvounis et al., 2004).

Unluckily the recording and preprocessing of FECG is extremely complex because of the intrinsic low signal-to-noise ratio of FECG. The main source of noise is represented by the maternal ECG: in abdominal recordings, the amplitude of the maternal QRS usually ranges from 100 to 150µV, while the fetal QRS ranges from 5 to 60 µV (Hasan et al., 2009). In addition, FECG signal is often hidden by electrical noise from other sources. Common ECG noise sources, such as power line interference, muscle contractions, breathing, skin resistance interference, and instrumental noise, in addition to electromyogram and electrohysterogram due to uterine contractions, can corrupt FECG signals considerably. Mainly power line interference represents a very difficult noise source to deal with.

The device we present here, named *Telefetal Care*, is a home wearable fetal monitor (Fanelli et al., 2010). Instead of detecting fetal heart beats using Doppler Ultrasounds, as CTG does, it is based on the recording of abdominal ECG. We exploited the capabilities offered by wearable technologies to design a patient-oriented fetal monitor that pregnant women can use at home, in the last trimester of the pregnancy, without support by clinicians. The maternal abdominal ECG is recorded using a wearable garment and the FECG is extracted using an optimized algorithm of extraction. Extracted data are then sent to the laptop using a Bluetooth connection and visualized on a Graphical User Interface. Fetal ECG and fetal heart rate variability are used to quantify fetal well-being and to detect risky conditions for the fetus, such as IUGR.

2 TELEFETALCARE

2.1 System Overview

Telefetal Care is characterized by a wearable unit and a compact device for data preprocessing and transmission.

The wearable unit is made of an elastic bodysuit endowed with textile electrodes that are sewed in the garment. The bodysuit is made of cotton and lycra. At the abdomen level, 9 silver electrodes allow the recording of 8 abdominal ECG leads. The elastic properties of the bodysuit allow an efficient contact between the skin and the electrodes, without any need of gel or other medium. The sensors are placed on the bodysuit as shown in Fig.1. Eight electrodes are placed around the navel, with a reference electrode placed in the navel itself. The garment is comfortable and is compatible with FECG recording sessions during everyday activity.

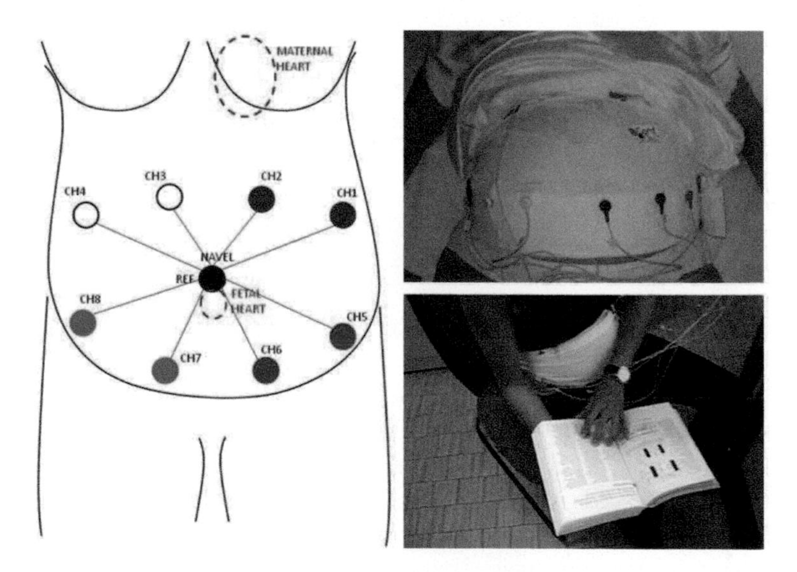

Figure 1 On the left, a scheme represents the position of the electrodes on the maternal abdomen. 8 sensing electrodes are placed around the navel and 1 reference electrode is placed in the navel itself. The device records the differential voltage between the sensing electrodes and the reference one. On the right, two pictures show possible use scenarios, in the hospital during hospitalization or at home during everyday life.

The device for data recording and transmission is characterized by an analog preprocessing stage, which consists of pass-band filtering and amplification. Signals are sampled with a sampling frequency of 256 Hz. The 16 bit ADC has a voltage resolution of 50μV, corresponding to a 30nV resolution before amplification. The power supply is provided by a 4.3V Li-ion battery.

After preprocessing and sampling, signals are transmitted to the laptop using a Bluetooth connection. A GUI allows data real time visualization and memorization, signals preprocessing and extraction of FECG and fetal heart rate (FHR).

2.2 Miniaturization and performances improvement

Many efforts were spent in reducing dimensions and power consumption of the device. The internal circuits were redesigned to improve the performances and reduce noise in recorded signals. Figure 2 shows the different prototypes which were developed, starting from the very first one (a), to the most recent one (c). As the pictures show, the dimension of the case was progressively reduced, and now the device is extremely compact and wearable (96mm x 96mm x 35mm).

Figure 2 Evolution of *TeleFetal Care*. Fig. 2a shows the first prototype. Form factor was changed and dimensions were reduced in the second developed prototype (Fig. 2b). The last prototype (Fig. 2c) was characterized by a further reduction in dimensions and by better performances in terms of power consumption and signal quality.

In the same time, the miniaturization process also involved an increase of the performances and an improvement of signal quality. The power consumption of the preprocessing boards was reduced by 85%, lowering the current absorption from 5.3mA to 0.8mA. Their surface occupation was reduced from 8.3cm^2 to 3.7cm^2. The introduction of a driven-leg electrode also allowed canceling the power line interference.

2.3 Preprocessing and FECG extraction

After preprocessing and sampling, data are transmitted by *TeleFetal Care* to the laptop using a Bluetooth connection. Fig 3 shows the GUI which was developed for data visualization and saving.

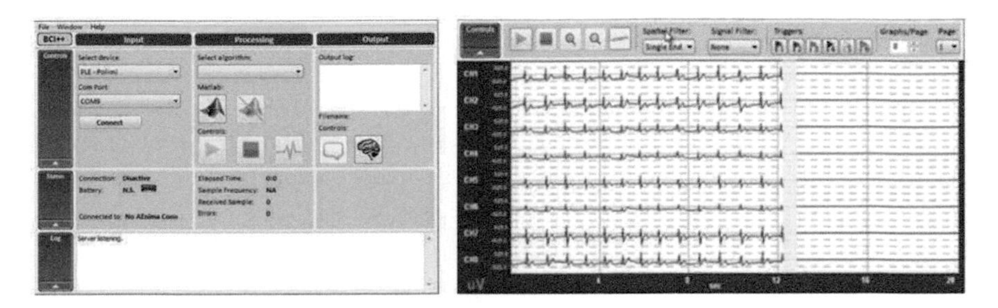

Figure 3 The picture shows the GUI developed for data visualization and memorization.

A set of off-line algorithms were developed for signal preprocessing, FECG extraction and FHR computation. The preprocessing stage consists of a smoothing filter for baseline wonder removal and SNR increase. As Fig. 4 shows, the raw signal (Fig. 4a) quality is improved thanks to baseline wander removal (Fig 4b). After that, maternal QRS are removed thanks to an averaging and subtracting process. The signal shown in Fig. 4c is the extracted FECG signal. After that, fetal and maternal QRS are detected using a template matching approach (Fig. 4d). For a detailed description of the algorithm of extraction, please refer to (Fanelli et al., 2011).

3 DATA ACQUISITION

Telefetal Care was tested at the Università degli Studi Federico II, Napoli, Italy, on 4 pregnant subjects between the 30th and the 34th week of gestation, after informed consent. The subjects were sitting on a chair and were asked to stay still. The system allowed a reliable recording of Fetal ECG and precise detection of fetal and maternal QRS. For maternal QRS detection, we obtained an overall accuracy of 98.52% and sensitivity of 99.5%. For fetal QRS detection, overall accuracy was 91,26% and sensitivity was 92,94%. Fig. 4d shows how maternal and fetal QRS were correctly detected in the recorded signal.

The most relevant problem we are now trying to solve is that the quality of recordings and the accuracy of fetal QRS detection strongly depend on the presence of power line interference. A strong AC interference might corrupt the signal, causing the total loss of the relevant information. Moreover, since the fetal ECG is very low in amplitude, the superimposed noise should not be larger than few μV. A proper isolation of packaging and wires, combined with a smart and efficient design of analog boards, is needed to completely remove this interference from the acquired recordings. We are also trying to reduce the artifacts introduced by movements of the patient during the acquisition.

3 CONCLUSIONS

We believe *TeleFetal Care* will improve the quality of fetal monitoring, allowing a continuous and accurate monitoring of fetal well-being. Moreover, it will be possible for pregnant women to undergo a screening process at home, without moving to the hospital or requiring clinical assistance. The possibility to monitor fetal well-being at home will reduce the number of hospitalization during pregnancy as well as it will have a positive effect on cost reduction in fetal monitoring, thus improving overall quality fetal wellbeing assessment.

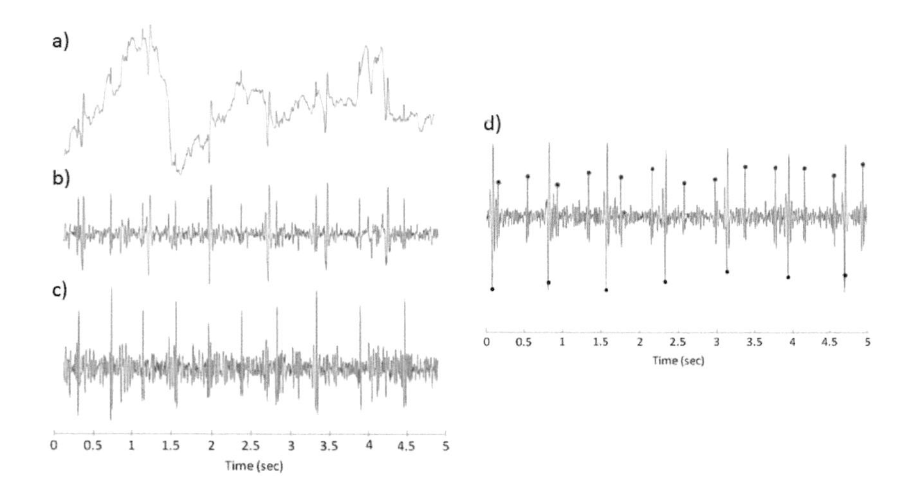

Figure 4 The raw signal acquired by the device in shown in the upper trace (a). The baseline wander removal allows improving signal quality (Fig. 4b). Finally, an algorithm based on averaging and subtraction is used to remove the maternal QRS and obtain the fetal ECG (Fig. 4c). Maternal and Fetal QRS are detected using a template matching approach (Fig. 4d).

REFERENCES

Fanelli A., M. Ferrario, L. Piccini, G. Andreoni, G. Matrone, G. Magenes, M. G. Signorini, 2010. Prototype of a wearable system for remote fetal monitoring during pregnancy. *EMBC 2010.* 2010: 5815-8.

Fanelli A., M.G. Signorini, M. Ferrario, P. Perego, L. Piccini, G. Andreoni and G. Magenes, 2011.Telefetalcare: a first prototype of a wearable fetal electrocardiograph. *EMBC 2011.* 2011:6899-902.

Hasan M.A. et al., 2009. Detection and Processing Techniques of FECG Signal for Fetal Monitoring. *Biological Procedures Online*. 11(1): 263-295.

Karvounis, E.C., C, Papaloukas, D.I. Fotiadis, L.K. Michalis, 2004. Fetal heart rate extraction from composite maternal ECG using complex continuous wavelet transform. *Computers in Cardiology*, 737- 740.

Peleg, D., C.M. Kennedy, and K.S. Hunter, 1998. Intrauterine Growth Restriction: Identification and Management. *Am Fam Physician.* 58(2): 453-460.

Peters, C., R. Vullings, J. Bergmans, G. Oei and P. Wijn, 2006. Heart Rate Detection in Low Amplitude Non-Invasive Fetal ECG Recordings. *EMBS 2006*, 6092-6094.

Sameni, R. and G.D. Clifford, 2010. A review of Fetal ECG Signal Processing: Issues and Promising Directions. *The Open Pacing, Electrophysiology & Therapy Journal*, 3: 4-20.

Section II

Communications, Systems Support and Healthcare Informatics

CHAPTER 5

Human Factors Considerations for a Reusable Medical Equipment Reprocessing Support System

Nancy J. Lightner[a], *R. Darin Ellis*[b], *Serge Yee*[c], *Kai Yang*[b] *and Will Jordan*[a]

[a] Department of Veterans Affairs, Veterans Engineering Resource Center, VA-Center for Applied Systems Engineering, Indianapolis, IN, 46222, USA
Nancy.Lightner@va.gov
Will.Jordan@va.gov

[b] Department of Industrial & Systems Engineering, Wayne State University, Detroit, MI 48202, USA
RDEllis@wayne.edu
Kai.Yang@wayne.edu

[c] Department of Veterans Affairs, Veterans Engineering Resource Center, VA-Center for Applied Systems Engineering, Detroit, MI, 48201, USA
Serge.Yee@va.gov

ABSTRACT

The Interactive Visual Navigator[1] (IVN™) is a computer system consisting of hardware, software, network, data that provides step-by-step reprocessing direction

---

[1] IVN™ is the property of the Department of Veterans Affairs and Wayne State University. Any use of the content presented in this paper without the express written consent of the IVN™ Program Manager is strictly prohibited.

on the specific model of endoscope to technicians. It gathers data on technician certification and on endoscope decontamination dates. IVN™ consists of a central processing unit, a wall-mounted touch screen and a keyboard that reside in a Sterile Processing Service (SPS) area for access by the technicians that reprocess Reusable Medical Equipment (RME). This paper discusses the human factors considerations when making selections of IVN™ hardware and facility layout. It presents findings from the pilot site implementation on those selections and provides lessons learned.

Keywords: endoscope, reusable medical equipment, sterile processing, infection control, touch-screen

1 INTRODUCTION

Flexible fiberoptic endoscopes (FFE) are sophisticated reusable medical equipment (RME) used to identify and evaluate the function of internal organs and cavities, and to locate and biopsy tumors in them. FFEs have lighted cameras that allow visual examination and may have internal channels for the application of suction, delivery of air or water, and for biopsy and other operative procedures. See Figure 1 for a diagram of an example endoscope. Once the tube is inserted into the body, it is exposed to various pathogens and potentially harmful bacteria. Before reuse, endoscopes must undergo a decontamination process to remove these microorganisms so the next patient is not exposed to them. There are over 20 million gastroenterology (GI) endoscopic procedures performed annually in the United States (Everhart, 2008). It is estimated that infections associated with GI endoscopy occur at a rate of 1 in almost 2 million procedures (Schembre, 2000) and that the source of these infections is either failure to follow established reprocessing guidelines, or use of defective equipment (ASGE Quality Assurance in Endoscopy Committee, 2011).

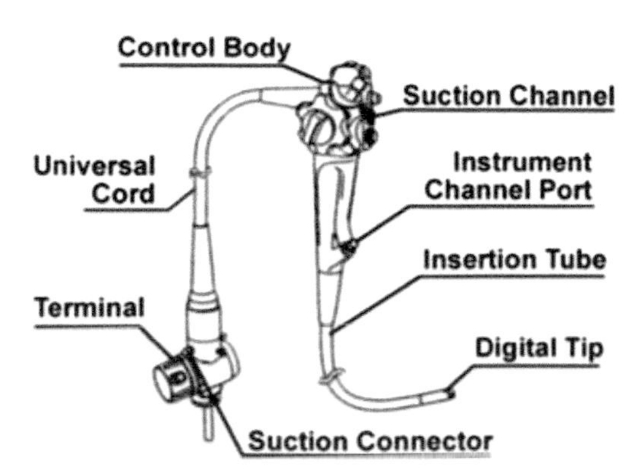

Figure 1: Schematic view of flexible endoscope, showing ports and channels (from Jolly, et. al., 2012) used with permission

Endoscope reprocessing is a complex process typically consisting of precleaning, leak testing, manual cleaning, high-level disinfection/sterilization, and rinsing and drying (with alcohol flush) (Pennsylvania Patient Safety Advisory, 2010). The specific steps within this process depend on make and model of the endoscope, and the type of sterilization used. Because there are a number of endoscope manufacturers producing a variety of models for different procedures, correct reprocessing according to manufacturers specifications presents a challenge to technicians. A study that investigated the usability of reprocessing instructions indicated that novice technicians were unable to correctly disinfect and sterilize FFEs using traditionally-supplied instructions (Jolly, et. al., 2012).

Between January 1 and September 30, 2010, the Veterans Affairs (VA) Office of Inspector General reviewed 45 VA facilities for compliance with VA standards for RME practices and identified six areas for improvement (House Committee on Veterans Affairs, 2011). The Interactive Visual Navigator (IVN™) system was developed by the Veterans Administration in collaboration with Wayne State University to address three of these six areas. They are: standard operating procedures (SOPs) are current, consistent with manufacturers' instructions, and located within the reprocessing areas; employees consistently follow SOPs, supervisors monitor compliance, and annual training and competency assessments are completed and documented; and processes for consistent internal oversight of RME activities are established to ensure senior management involvement (House Committee on Veterans Affairs, 2011).

IVN™ consists of software, networked hardware, and data that display instructions for correct reprocessing on a 17 inch (minimum) hermetically sealed touch screen. The instructions are based on manufacturer's instructions for the specific endoscope model. Each endoscope is tagged with a unique identifier. IVN™ collects data on the time to process steps and provides reports on technician certifications and dates when endoscopes were decontaminated. Based on this date, IVN™ identifies endoscopes that require reprocessing due to the length of time they are in storage. IVN™ is installed in the Sterile Processing Service (SPS) area of facilities where reprocessing technicians can interact with it.

This paper presents the human factors considerations involved in the selection of the hardware, including screen mounting, and controller. It also describes the human factors considerations when an SPS facility is designed for efficiency and effectiveness. It reports on the use of IVN™ at the pilot site and additional considerations as a result of the pilot.

2 BACKGROUND

Insufficient decontamination of RME presents a risk for widespread infection to patient populations. Although rarely resulting in infection, the incidents of improper reprocessing are well publicized once they are known. For example, eight incidents with endoscopes were reported in California in 18 months in 2003 and 2004 (Torassa, 2004). Forbes Regional Hospital contacted 200 patients who received colonoscopies in a four month time frame concerning their risk due to improperly cleaned colonoscopies (Fahy and Spice, 2005). Between 2004 and 2009, the Pennsylvania Patient Safety Authority received 107 reports of potential patient contamination due to inadequate endoscope reprocessing (Pennsylvania Patient Safety Advisory, 2010). Because of the severity of the potential risk of cross contamination from FFEs, the Emergency Care Research Institute (ECRI Institute) designated it as the number one hazard of the top ten medical technology hazards for 2010 (ECRI Institute, 2009).

The risk of improper reprocessing of FFE has prompted organizations around the world to develop guidelines and recommendations in an effort to reduce associated infection (ASGE Quality Assurance in Endoscopy Committee, 2011, O'Brien, 2009, Rey, et al., 2005, Rutala, et. al., 2008, Society of Gastroenterology Nurses and Associates, Inc., 2007, WHO, 2004). Although these guidelines specify characteristics of proper decontamination, they do not indicate how to achieve it. This requires facilities to implement the recommendations in whatever manner they can. The guidelines consistently emphasis 1) staff training, 2) following manufacturers' instructions, and 3) proper storage of clean, sterilized and disinfected equipment.

IVN™ was conceived as a tool to implement a standardized method of following established guidelines and especially those specified by the VA House Committee. It incorporates technology that allows for the collection of information and dissemination of updates in an efficient manner. IVN™ is currently implemented at a pilot site with a planned roll out to all facilities in the network over the next several years.

3 DESCRIPTION OF PROCESSING IN AN SPS AREA

While some departments in a hospital have individual sterilization capability, it is more common that an SPS processes all RME. These areas are typically divided in to decontamination, assembly and packing, sterile storage and distribution areas (University of Rochester Medical Center, 2012). See Figure 2 for a recommended layout design. Used RME is collected and taken to the decontamination area where it is sorted and soaked, if required. Personnel working in the decontamination area should wear protective clothing including a scrub uniform covered by a moisture-resistant barrier, shoe covers, rubber or plastic gloves and a hair covering. During manual cleaning processes, personnel should wear safety goggles and a face mask. Once the RME is cleaned and decontaminated, it is moved to the assembly and packaging area where it is put back together and prepared for either issue (for immediate use), storage, or further processing, such as high-level disinfection. Once the RME is cleaned, decontaminated and sterilized, it is placed in a distribution area for immediate use or storage.

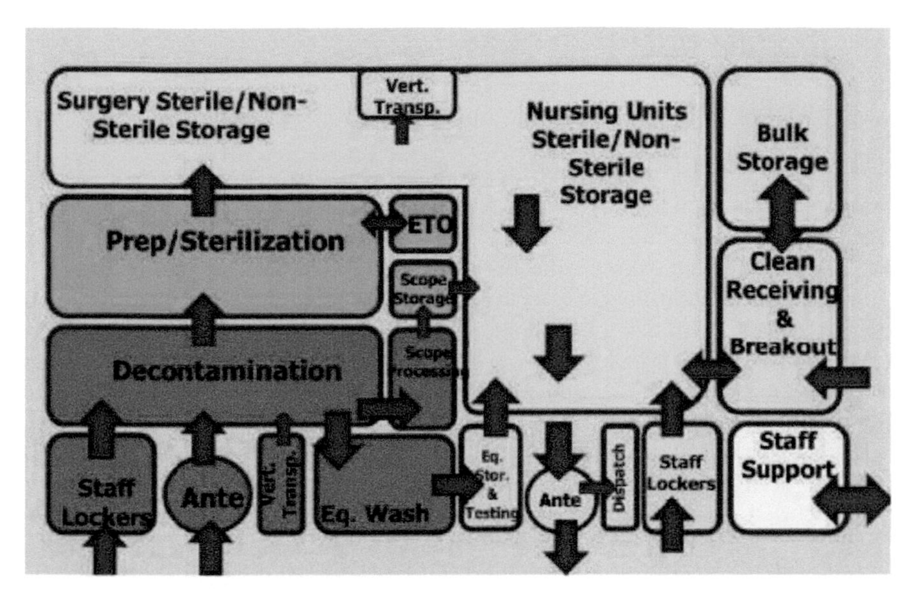

Figure 2: Recommended SPS layout design (Department of Veterans Affairs Office of Construction and Facilities Management, 2010)

Each endoscope make and model is cleaned according to specific manufacturer's instructions. Prior to IVN™, areas would make these instructions accessible in a variety of manners. Figure 3 shows laminated instructions posted above the sink in an actual decontamination area. IVN™ replaces these displays in a more convenient and usable form.

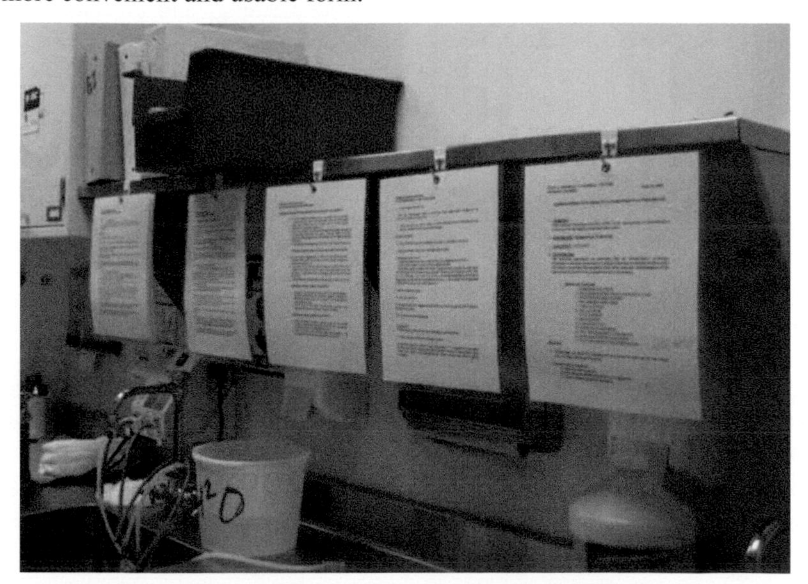

Figure 3: Posted reprocessing instructions in a decontamination area.

4 DESIGN CONSIDERATIONS

4.1 Facility layout

A decontamination suite typically consists of two areas separated by walls, with a door or a pass-through allowing for material and personnel flow (see Figure 2). Additionally, a separate area is reserved for endoscope storage. Reprocessing involves a minimum of two technicians; one on the "dirty' side to perform all steps through detergent flushing, and one on the "clean" side to perform the remaining steps including High Level Disinfection (HLD). There is one set of instructions for each endoscope that encompasses the entire reprocessing workflow. It is expected that each technician follow the appropriate steps for their area. In addition, since reprocessing involves some waiting due to soaking and other steps performed by machines, a technician may perform reprocessing on more than one endoscope at a time.

SPS areas contain reprocessing and storage equipment such as sinks, sprayers, cleaners, washers and disinfectors, drying cabinets, steam guns, and sterilizers. Adequate room for circulation is needed to accommodate the flow of staff, supplies and equipment. Hospitals always have space efficiency in mind, because medical center facility space is sometimes difficult to allocate.

4.2 Technician considerations

Job qualifications for Sterile Processing Technicians include "Ability to perform work that requires frequent standing, bending, reaching and lifting of 25 lbs." (JobTarget.com, 2012). Endoscope reprocessing involves walking to retrieve contaminated RME, standing at a basin for manual cleaning, and movement to retrieve tools and process the equipment. Movement is also required to exchange the RME with other areas and to store it in the distribution area. Technicians are always wearing protective clothing, which constricts movement to an extent. Safety goggles may alter the ability to view small text on instructions as well. Gloved hands which are often wet also affect use of tools and equipment.

4.3 Hardware considerations

IVN™ is an interactive system that displays reprocessing instructions and gathers data for endoscope storage limits. The considerations made while analyzing system requirements were based on the biomechanics of the human body, job design based on the requirements of an SPS, and climate and clothing.

IVN™ contains a screen for the display of instructions and other information. It also requires input such as a technician login and scope model number. Based on site visits, it was determined that in the decontamination area, reprocessing instructions are best viewed when they are close to the sinks. Many SPS areas do not have necessary floor space for a desktop unit. Because technicians vary in height and viewing preference, and are typically standing, a wall-mounted screen was considered the most effective display modality.

Controller alternatives considered were keyboard, mouse, touch screen, game pad and pedal. Preliminary investigation showed that of those choices, the touch screen and mouse were considered the easiest to use. Additional deliberation was given to the location requirements of a keyboard and a mouse, since these devices require a stable base on which to rest. Technician's hands are wet most of the time, so the ability of a device to withstand liquid was evaluated. Technicians wear foot protection at all times, reducing their ability to properly operate a pedal.

Table 1: Results of initial controller considerations.

Controller	Operation Difficulty	Pros	Cons
Keyboard	medium	OK to navigate	Not hand free
Mouse	easy	easy to navigate	Not hand free
Touch screen	easiest	easy to navigate	Not hand free
Gamepad	medium	OK to navigate	Not hand free
Pedal	difficult	Hand free	Difficult to navigate, very few controls

5 RESULTS AND APPLICATION

When the requirements for a system used in an SPS area were considered, it was decided that a hermetically-sealed touch screen secured with an adjustable wall mount bracket best meets the height and location needs of the variety of technicians in a facility. There are several acceptable medical grade personal computers with a 17 inch display available that are sealed for dust-free and waterproof operation. Some casings contain an antimicrobial additive to fight surface bacteria (Tangent, Inc., 2010).

The reason a sealed unit was selected for IVN™ is to reduce contamination if it is ever taken from the SPS area. For example, if it is removed for a repair or other reason, technicians must decontaminate it before it is returned to the SPS area. This is possible with the a hermetically sealed unit by spraying with a decontaminant and wiping it clean. It is recommended that each IVN™ unit also include a medical grade keyboard and mouse.

A wall mounted, adjustable arm is recommended to secure the IVN™ unit to the wall for technician use. One available unit has an arm that extends 19 inches from the wall with a wide range of motion up, down and tilting 70 degrees up and 5 degrees down (Ergotron, 2010). The recommended height for placement for the center of the bracket on the wall is 157.5 centimeters or 5 feet, 2 inches. This is based on the average eye height of United States Army personnel in 1994 (Kroemer et al., 1994). Although it is possible that the average height has changed, the range of adjustment possible with these units accommodates height differences.

The IVN™ system was installed in a pilot location early 2011. It included decontamination instructions for the 13 endoscope models used in the facility. During site visits, the technicians reported that IVN™ greatly improved their ability to reprocess endoscopes correctly. Because of the location in the decontamination area (behind the sinks), movement was required to refer to it between instruction steps. The location in the assembly and packaging area was more accessible; allowing technicians to reprocesses multiple endoscopes with relative ease. The technicians commented about how some adjust the height and location of the screen, but that most typically do not move it around. There is a clear personal preference for keyboard and mouse use, as supplements to the touch screen. All were satisfied with the interactive performance of IVN™ to the point that they were willing to accommodate what they viewed as small inconveniences.

The following is recommended for each IVN™ installation.

- Touch screen capable, 17-inch monitor screen with a dual core 2 processor combined in a single sealed, fan-less casing
- Adequate graphics card to allow a quick response when images are displayed
- Wall-mounted adjustable arm
- Medical-grade wireless mouse
- Medical-grade wireless keyboard

6 CONCLUSION

FFEs are used in over 20 million procedures annually in the United States. Decontamination and sterilization of this equipment is necessary between procedures to reduce the risk of infection to the next patient. Because reprocessing of each model follows potentially different instructions, it is challenging for technicians to properly reprocess FFEs. The IVN™ system, developed by the Department of Veterans Affairs and Wayne State University, is a method of presenting manufacturer's instructions to technicians as they reprocess endoscopes in a sterile environment. Hardware selection considered the SPS environment as well as the requirements of the work. A medical grade unit suspended from the wall near the decontamination and packaging and assembling areas was selected because of its properties of having a touch screen that will operate in an environment that regularly includes liquid. A medical grade mouse and keyboard were also installed in the pilot site. Technicians at the initial installation consider IVN™ successful in directing proper reprocessing of endoscopes. They much preferred having instructions specific to the endoscope model presented to them in a convenient and usable form instead of searching through paperwork for correct guidance.

ACKNOWLEDGMENTS

The authors would like to acknowledge the support of the Veterans Health Administration.

IVN™ is the property of the Department of Veterans Affairs and Wayne State University. Any use of the content presented in this paper without the express written consent of the IVN™ Program Manager is strictly prohibited.

REFERENCES

ASGE Quality Assurance in Endoscopy Committee (2011). Multisociety guideline on reprocessing flexible gastrointestinal endoscopes: 2011, *Gastrointestinal Endoscopy, 73(6)*, 1075-1084.

ECRI Institute. (2009). Top 10 technology hazards: high-priority risks and what to do about them. *Health Devices, 38(11),* 364-73. 2009 Nov;38(11):364-73.

Department of Veterans Affairs Office of Construction & Facilities Management. (2010). Supply Processing & Distribution design guide. Retrieved February 13, 2012 from http://www.cfm.va.gov/til/dGuide/dgSPD-01.pdf

Ergotron. (2010). *Ergotron® LX Desk and Wall Mount LCD Arms.* Retrieved February 16, 2012 from http://www.ergotron.com/Portals/0/literature/productSheets/english/05-056-EA.pdf

Everhart, JE. (2008). The burden of digestive disease in the United States. *NIH publication no. 09-6443.* Washington (DC): U.S. Department of Health and Human Services.

Fahy J, Spice B. (March 31, 2005). Monroeville hospital urges 200 colonoscopy patients to get checked for hepatitis, HIV. *Pittsburgh Post-Gazette.* Retrieved February 9, 2012 from http://www.gpwlaw.com/news/news/personal-injury/2005/03/31/forbes-regional-hospital-alerts-200-patients-who-received-colonoscopies-that-they-may-be-at-risk-for-hepatitis-hiv/

Health Canada. (2007) Infection Control Guidelines: Best Practices Guidelines for the Cleaning, Disinfection and Sterilization of Medical Devices in Health Authorities. Patient Safety Branch Ministry of Health. Retrieved February 9, 2012 from http://www.health.gov.bc.ca/library/publications/year/2007/BPGuidelines_Cleaning_Disinfection_Sterilization_MedicalDevices.pdf

House Committee on Veterans Affairs. (May 3, 2011). Witness Testimony of Hon. Robert A. Petzel, M.D., Under Secretary for Health, Veterans Health Administration, U.S. Department of Veterans Affairs. Retrieved February 9, 2012 from http://veterans.house.gov/prepared-statement/prepared-statement-hon-robert-petzel-md-under-secretary-health-veterans-health-0

JobTarget.com (2012). Job Description: Sterile Processing Tech-PT Evenings Cpd/Central Sterile (Job Number : 00144-5376). Retrieved February 13, 2012 from https://hca-taleo-net.careerliaison.com/careersection/newlargomedctrlargo/jobdetail.ftl?lang=en&job=00144-5376&_CL_SOURCE=559489&jtsrc=http%3A%2F%2Fwww%2Esimplyhired%2Ecom&jtsrcid=12169&jtrfr=http%3A%2F%2Fwww%2Esimplyhired%2Ecom%2Fa%2Fjobs%2Flist%2Fq%2Dsterile%2Bprocessing%2Btechnician

Jolly, J.D., Hildebrand, E.A, Branaghan, R.J., Garland, T.B., Epstein, D., Babcock-Parziale, J., and Brown, V. (2012). Patient Safety and Reprocessing: A Usability Test of the Endoscope Reprocessing Procedure, *Human Factors and Ergonomics in Manufacturing & Service Industries*, 22(1), 39-51.

Kroemer, K. H. E., Kroemer, H. J., and Kroemer-Elbert, K. E. (1994). *Ergonomics. How to Design for Ease and Efficiency*. Englewood Cliffs, NJ: Prentice Hall.

O'Brien, V. (2009). Controlling the process: legislation and guidance regulating the decontamination of medical devices, *Journal of Perioperative Practice, 19(12),* 428-432.

Pennsylvania Patient Safety Advisory. (2010). The Dirt on Flexible Endoscope Reprocessing, *Pa Patient Saf Advis, 7(4),* 135-40. Retrieved February 9, 2012 from http://patientsafetyauthority.org/ADVISORIES/AdvisoryLibrary/2010/dec7(4)/Pages/135.aspx#bm9

Rey, J.F., Bjorkman, D., Duforest-Rey, D. Axon, A., Saenz, R., Fried, M.. et al., (2005). *WGO practice guideline endoscopy disinfection, world gastroenterology organization (WGO).* Retrieved February 8, 2012 from http://www.worldgastroenterology.org/assets/downloads/en/pdf/guidelines/09_endoscope_disinfection_en.pdf

Rutala, W.A., Weber, D.J. and Healthcare Infection Control Practices Advisory Committee. (2008). *Guideline for disinfection and sterilization in health care facilities, 2008.* Atlanta, GA: Centers for Disease Control and Prevention.

Schembre DB. (2000). Infectious complications associated with gastrointestinal endoscopy, *Gastrointest Endosc Clin N Am, 10(2),* 215-32.

Society of Gastroenterology Nurses and Associates, Inc. (2007). Standards of Infection Control in Reprocessing of Flexible Gastrointestinal Endoscopes. Chicago, Illinois. Retrieved February 9, 2011 from http://www.sgna.org/Portals/0/Education/Practice%20Guidelines/InfectionControlStandard.pdf.

Tangent, Inc. (2010). Medix 1700SF Medical Grade Touch Screen Antimicrobial, Retrieved February 14, 2012 from http://www.medicalcomputer.tangent.com/medix1700healthcarepc.htm

Torassa, U. (September 10, 2004). Endoscopy hygiene a growing problem, *San Francisco Chronicle.* Retrieved February 9, 2012 from http://www.sfgate.com/cgi-bin/article.cgi?f=/c/a/2004/09/10/BAG8O8MKOS1.DTL

University of Rochester Medical Center. (2012). Sterile & Materials Processing Department. Retrieved February 10, 2012 from http://www.urmc.rochester.edu/sterile/basics.cfm

World Health Organization (WHO). (2004). Practical guidelines for infection control in health care facilities. Retrieved February 9, 2010 from http://www.searo.who.int/LinkFiles/publications_PracticalguidelinSEAROpub-41.pdf

CHAPTER 6

Contemporary Configuration of the Medicine Package Leaflet: Changes in the Ergonomics of a Technical Scientific Object

Patricia Lopes Fujita,

Carlos José Saldanha Machado,

Márcia de Oliveira Teixeira

Fundação Oswaldo Cruz
Rio de Janeiro, Brazil
patricia.fujita@gmail.com
saldanha@fiocruz.br
marciat@fiocruz.br

ABSTRACT

The aim of this paper was to investigate the construction process of the MPL regulation in Brazil through an informational ergonomics perspective, in order to verify the influence of these processes in a historical period of the existing standards and structuring content of the MPL. With the purpose to verify the influence of these processes in a historical period of the existing standards and structuring content of MPL, a qualitative study was conducted using the technique of document analysis. This in order to assess the medicine MPL regulation in the historical period from 1946 to 2009 (date of first publication of the regulation until the earliest one) on informational ergonomics aspects of form and layout.

Keywords: medicine package leaflet, regulation, technical scientific information

1 INTRODUCTION

The National Agency of Sanitary Surveillance (ANVISA), since September 1999, is the institution of the Brazilian Ministry of Health responsible for the supervision and regulation of health information documents about medicines, especially for medicines package leaflets (MPL), which are produced and launched in the market by the pharmaceutical industry. The content of the MPL has to be submitted for approval of ANVISA, before it´s printed and published. In first instance, this information is of technical scientific nature, and once authorized, they need to be presented in a manner intelligible to users, as a printed document in the form of a medicine leaflet. This makes MPL a legal printed document directed for patients and health professionals, containing technical scientific information and orientation about the use of a pharmaceutical drug, in order to assure patient safety. Thus, visual presentation of information is essential to guarantee safe and proper use of a medicine, by those who provide health care and by patients.

In this context, the aim of this paper was to investigate the construction process of the MPL regulation in Brazil through an informational ergonomics perspective, in order to verify the influence of these processes in a historical period of the existing standards and structuring content of the MPL. With the purpose to verify the influence of these processes in a historical period of the existing standards and structuring content of MPL, a qualitative study was conducted using the technique of document analysis. This in order to assess the medicine MPL regulation in the historical period from 1946 to 2009 (date of first publication of the regulation until the earliest one) on informational ergonomics aspects of form and layout.

2 METHOD

A qualitative study was conducted using the technique of document analysis of secondary sources of digitized publications governing the MPL. The data was collected at the official websites: Diário Oficial da União (http://portal.in.gov.br) and JusBrasil (http://www.jusbrasil.com.br/diarios)

It was found 9 regulation arrangements/documents concerning the MPL. The documents reviewed were published in the historical period of 1946 until 2009 (date of first publication in the Diário Oficial da União), and so were structured in chronological order (publication year) divided into: “Date/Year”; "Type of legal arrangement", "Governments institution" (regulator). Table 1 presents the structure with the data collected in item 4 (Results and discussion) below.

3 RESULTS AND DISCUSSION

The Table 1, presents the legal arrangements/documents concerning the MPL, published in the period of 1946 until 2009:

Table 1: MPL regulation and arrangements published in the period of 1946 until 2006

Date/Year	Type of legal arrangement	Government institution
1946	Decree n. 20.391	Ministry of Health
1959	Ordinance n. 49	National Supervisory Bureau of Medicine and Pharmacy - SNFMF
1984	Ordinance n. 65	National Sanitary Surveillance Secretariat - SNVS
1997	Ordinance n.110	Sanitary Surveillance Secretariat - SVS
2001	Public Inquiry n.95	ANVISA
2002	Public Inquiry n. 2	ANVISA
2003	Resolution n. 140	ANVISA
2009	Public Inquiry n. 1	ANVISA
2009	Resolution n. 47	ANVISA

According to Table 1, in 1946 the Ministry of Health approved by Decree n. 20.391, the Regulation of the Pharmaceutical Industry in Brazil. This Decree, also approaches the regulation of information to compose the MPL, which were addressed in separate sections, depending on type of laboratory and/or pharmaceutical product.

In 1959, the National Supervisory Bureau of Medicine and Pharmacy - SNFMF, published Ordinance n. 49, specifically to regulate the presentation and examination of labels on pharmaceuticals, workshop, dietary supplements, cosmetics, toiletries, toilet. This Ordinance makes reference to the Decree 20.391 (1946), regarding the models of labels and package leaflets for pharmaceuticals, which must contain the legal sayings and attend the requirements of this Decree.

After twenty-three years of publication of Ordinance n.49, in 1984, the National Sanitary Surveillance Secretariat - SNVS, (another agency of the Ministry of Health responsible for the regulation of specific information to be contained in a drug label that year), established by Ordinance n. 65, the first script for the MPL approved by SNVS. The script for the first medicine leaflet was presented in Ordinance No. 65 in 1984, was structured in four parts: (I) Drug Identification, (II) Patient Information (III) Technical Information, (IV) Legal sayings. This general sequence is in force ever since.

Only in 1997, the Sanitary Surveillance Secretariat - SVS publishes another regulation, Ordinance n. 110, keeping the structure and content established in 1984, only modifying the sequence of some items in the (III) Technical Information and adding and changing the language of some observations on (II) Patient Information item, emphasizing two requirements: the information has to be compulsory and uniform, written in language understandable to the general consumer; add a few warnings in the Technical and Patient Information.

In 2001 and 2002, since the existence of ANVISA, there were two Public Inquiries for comments and suggestions were put forward the proposal in effect until that moment (Ordinance n. 110, 1997), in order to review the text of the leaflet free of drugs Prescription ready market due to the heterogeneity of information to consumers (patients) and health professionals. It should be noted in Table 1, which were held two Public Inquiries consecutively, one in 2001 (Public Inquiry n. 95) and another in 2002 (Public Inquiry n.2), but only in 2003 ANVISA prepared and published a new resolution (Resolution n. 140) expanded and more specific regarding MPL´s content (explaining how it should be written on each item) and considering rules established by the Code of Consumer Rights and the World Health Organization - WHO.

In Resolution n.140 (2003), it was mentioned for the first time an item regarding the graphic presentation of the MPL content: the font size has to be 1.5 mm at least. It was also possible to observe that this regulation distinguished three types of content conveyed in the MPL:

- For health professionals: legal document that contains health information and technical-scientific guidelines on medicines for their rational use, which are available to health professionals
- For patients: health legal document that contains technical and scientific information and guidance on medicines, which are available to users on appropriate language, ie, easy to understand, with activity in establishments dispensing medicines, as current law;

It was also possible to observe in Resolution n. 140 (2003) that the required textual content and structure, is very similar to the first script published in 1984 (I. Drug identification; II. Patient Information; III. Technical Information; IV. Legal sayings). On the topic (II) Patient Information it was suggested the presentation and organization of information could be presented in questions and answers, optionally.

The Pictures 3 and 4, illustrate two examples of the most recent MPL in Brazil following the requirements of Resolution n. 47 (2009):

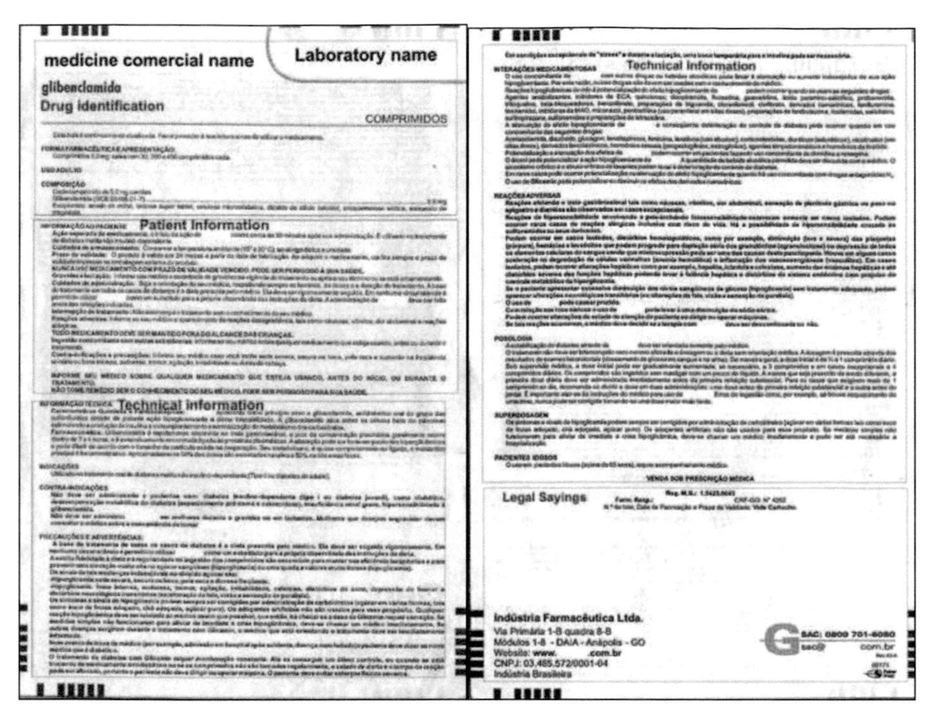

medicine comercial name

Laboratory name

glibenclamida

Drug identification

COMPRIMIDOS

Patient Information

Technical information

Technical Information

VENDA SOB PRESCRIÇÃO MÉDICA

Legal Sayings

Indústria Farmacêutica Ltda.
Via Primária 1-B quadra 8-B
Módulos 1-8 - DAIA - Anápolis - GO
Website: www. .com.br
CNPJ: 03.485.572/0001-04
Indústria Brasileira

SAC: 0800 701-6080
sac@ com.br

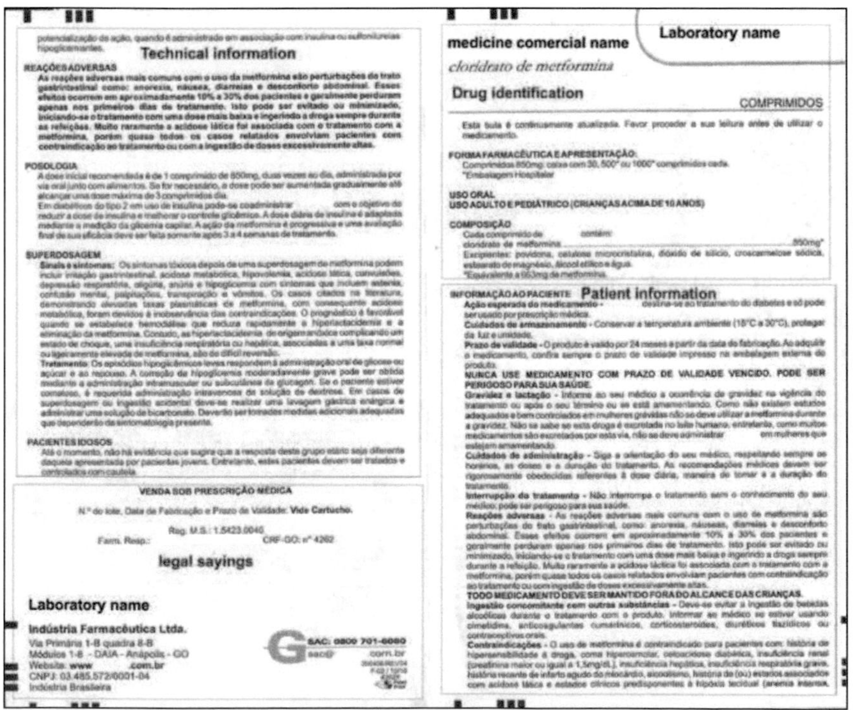

potencialização da ação, quando é administrado em associação com insulina ou sulfoniluréias hipoglicemiantes.

Technical information

REAÇÕES ADVERSAS

As reações adversas mais comuns com o uso da metformina são perturbações do trato gastrintestinal como: anorexia, náusea, diarreias e desconforto abdominal. Esses efeitos ocorrem em aproximadamente 10% a 30% dos pacientes e geralmente perduram apenas nos primeiros dias de tratamento. Isto pode ser evitado ou minimizado, iniciando-se o tratamento com uma dose mais baixa e ingerindo a droga sempre durante as refeições. Muito raramente a acidose lática foi associada com o tratamento com a metformina, porém quase todos os casos relatados envolviam pacientes com contraindicação ao tratamento ou com a ingestão de doses excessivamente altas.

POSOLOGIA

A dose inicial recomendada é de 1 comprimido de 850mg, duas vezes ao dia, administrada por via oral junto com alimentos. Se for necessário, a dose pode ser aumentada gradualmente até alcançar uma dose máxima de 3 comprimidos dia.

Em diabéticos do tipo 2 em uso de insulina pode-se coadministrar com o objetivo de reduzir a dose de insulina e melhorar o controle glicêmico. A dose diária de insulina é adaptada mediante a medição da glicemia capilar. A ação da metformina é progressiva e uma avaliação final de sua eficácia deve ser feita somente após 3 a 4 semanas de tratamento.

SUPERDOSAGEM

Sinais e sintomas: Os sintomas tóxicos depois de uma superdosagem de metformina podem incluir irritação gastrintestinal, acidose metabólica, hipovolemia, acidose lática, convulsões, depressão respiratória, oligúria, anúria e hipoglicemia com sintomas que incluem astenia, confusão mental, palpitações, transpiração e vômitos. Os casos citados na literatura, demonstrando elevadas taxas plasmáticas de metformina, com consequente acidose metabólica, foram devidos à inobservância das contraindicações. O prognóstico é favorável quando se estabelece hemodiálise que reduza rapidamente a hiperlactacidemia e a eliminação da metformina. Contudo, as hiperlactacidemia de origem anóxica complicando um estado de choque, uma insuficiência respiratória ou hepática, associadas a uma taxa normal ou ligeiramente elevada de metformina, são de difícil reversão.

Tratamento: Os episódios hipoglicêmicos leves respondem à administração oral de glicose ou açúcar e ao repouso. A correção da hipoglicemia moderadamente grave pode ser obtida mediante a administração intramuscular ou subcutânea de glucagon. Se o paciente estiver comatoso, é requerida administração intravenosa de solução de dextrose. Em casos de superdosagem ou ingestão acidental deve-se realizar uma lavagem gástrica enérgica e administrar uma solução de bicarbonato. Deverão ser tomadas medidas adicionais adequadas que dependerão da sintomatologia presente.

PACIENTES IDOSOS

Até o momento, não há evidência que sugira que a resposta deste grupo etário seja diferente daquela apresentada por pacientes jovens. Entretanto, estes pacientes devem ser tratados e controlados com cautela.

VENDA SOB PRESCRIÇÃO MÉDICA

N.º do lote, Data de Fabricação e Prazo de Validade: **Vide Cartucho.**

Reg. M.S.: 1.5423.0040

Farm. Resp.: CRF-GO: nº 4262

legal sayings

Laboratory name

Indústria Farmacêutica Ltda.
Via Primária 1-B quadra 8-B
Módulos 1-8 - DAIA - Anápolis - GO
Website: www .com.br
CNPJ: 03.485.572/0001-04
Indústria Brasileira

SAC: 0800 701-6080
sac@ com.br

medicine comercial name

Laboratory name

cloridrato de metformina

Drug identification

COMPRIMIDOS

Esta bula é continuamente atualizada. Favor proceder a sua leitura antes de utilizar o medicamento.

FORMA FARMACÊUTICA E APRESENTAÇÃO:

Comprimidos 850mg: caixa com 30, 500* ou 1000* comprimidos cada.
*Embalagem Hospitalar

USO ORAL
USO ADULTO E PEDIÁTRICO (CRIANÇAS ACIMA DE 10 ANOS)

COMPOSIÇÃO

Cada comprimido de contém:
cloridrato de metformina 850mg*
Excipientes: povidona, celulose microcristalina, dióxido de silício, croscarmelose sódica, estearato de magnésio, álcool etílico e água.
*Equivalente a 663mg de metformina.

INFORMAÇÃO AO PACIENTE Patient information

Ação esperada do medicamento - destina-se ao tratamento do diabetes e só pode ser usado por prescrição médica.

Cuidados de armazenamento - Conservar à temperatura ambiente (15°C a 30°C), proteger da luz e umidade.

Prazo de validade - O produto é válido por 24 meses a partir da data de fabricação. Ao adquirir o medicamento, confira sempre o prazo de validade impresso na embalagem externa do produto.

NUNCA USE MEDICAMENTO COM PRAZO DE VALIDADE VENCIDO. PODE SER PERIGOSO PARA SUA SAÚDE.

Gravidez e lactação - Informe ao seu médico a ocorrência de gravidez na vigência do tratamento ou após o seu término ou se está amamentando. Como não existem estudos adequados e bem controlados em mulheres grávidas não se deve utilizar a metformina durante a gravidez. Não se sabe se esta droga é excretada no leite humano, entretanto, como muitos medicamentos são excretados por esta via, não se deve administrar em mulheres que estejam amamentando.

Cuidados de administração - Siga a orientação do seu médico, respeitando sempre os horários, as doses e a duração do tratamento. As recomendações médicas devem ser rigorosamente obedecidas referentes à dose diária, maneira de tomar e a duração do tratamento.

Interrupção do tratamento - Não interrompa o tratamento sem o conhecimento do seu médico: pode ser perigoso para sua saúde.

Reações adversas - As reações adversas mais comuns com o uso de metformina são perturbações do trato gastrintestinal, como: anorexia, náuseas, diarreias e desconforto abdominal. Esses efeitos ocorrem em aproximadamente 10% a 30% dos pacientes e geralmente perduram apenas nos primeiros dias de tratamento. Isto pode ser evitado ou minimizado, iniciando-se o tratamento com uma dose mais baixa e ingerindo a droga sempre durante a refeição. Muito raramente a acidose láctica foi associada com o tratamento com a metformina, porém quase todos os casos relatados envolviam pacientes com contraindicação ao tratamento ou com ingestão de doses excessivamente altas.

TODO MEDICAMENTO DEVE SER MANTIDO FORA DO ALCANCE DAS CRIANÇAS.

Ingestão concomitante com outras substâncias - Deve-se evitar a ingestão de bebidas alcoólicas durante o tratamento com o produto. Informar ao médico se estiver usando cimetidina, anticoagulantes cumarínicos, corticosteroides, diuréticos tiazídicos ou contraceptivos orais.

Contraindicações - O uso de metformina é contraindicado para pacientes com: história de hipersensibilidade à droga, coma hiperosmolar, cetoacidose diabética, insuficiência renal (creatinina maior ou igual a 1,5mg/dL), insuficiência hepática, insuficiência respiratória grave, história recente de infarto agudo do miocárdio, alcoolismo, história de (ou) estados associados com acidose lática e estados clínicos predisponentes à hipóxia tecidual (anemia intensa,

Figure 1 and 2: Examples of MPL according to requirements made by Resolution n. 140 (2003)

According to Table 1, a further public inquiry was made in the beginning of 2009 (Public Inquiry n. 1, 2009), and afterwards ANVISA released on its official website (www.anvisa.gov.br) a report containing the results of the last public consultation ("Contributions Analysis Report "). In this report it was observed that among the issues and proposals addressed by the public proposals, "Form and Content" of medicine leaflets was the most commented (130 out of 531 issues and suggestions). Those results evidence the importance of graphic presentation of medicine information for its users, contributing to this aspect to be treated with more specificity by the current regulator in Brazil, the ANVISA.

In September of 2009, the last and current regulation was published (Resolution n. 47, 2009) with more specifics about the content and graphic presentation of the MPL. The requirements on the "Form and presentation" are more specific and detailed regarding aspects of graphical presentation (type and font size, line spacing, print , etc..) depending on the type of medication. According to Resolution no. 47 (2009), the main rules on the layout are:

- Times New Roman font in the text with a minimum of 10 points, non-condensing and non-expanded;
- Text with space between letters of at least 10%;
- Text with line spacing of at least 12 points;
- Columns of text with at least 80 mm wide;
- Text aligned left, hyphenated or not;
- Capital letters and bold type to highlight the questions and MPL items;
- Text underline and italics only for scientific names;
- The MPL must be printed in black on white paper that does not allow print preview on the other side, when the bull is over a surface.
- For the printing of leaflets in special format with extended source, you should use the Verdana font with a minimum of 24 points with plain text and no columns.

It was also observed some changes on the textual structure (script) concerning the content on 'Patient Information'. The patient information must be written and structured in nine questions and answers according to a questionnaire previously stipulated, presented in Resolution n. 47 (2009). This requirement in Resolution n.140 (2003) was presented as an option for the pharmaceutical industry. Thus, despite the intention of ANVISA in making the content focused on patients simply as possible through questions and answers, it should be judged whether the content addressed by questions and answers presented above is sufficient and adequate to meet patients information needs.

The Pictures 3 and 4, illustrate two examples of the most recent MPL in Brazil following the requirements of Resolution n. 47 (2009):

Drug identification

atenolol

Medicamento genérico Lei nº 9.787, de 1999

LEIA COM ATENÇÃO ANTES DE USAR O MEDICAMENTO

FORMA FARMACÊUTICA E APRESENTAÇÕES
Comprimido de 25 mg, 50 mg e 100 mg: caixa com 30, 300, 600 e 800 comprimidos.

USO ORAL
USO ADULTO

FÓRMULA
Atenolol 25 mg:
Cada comprimido contém:
atenolol 25mg
excipientes q.s.p. .... 1 comprimido
Excipientes: carbonato de magnésio, amido de milho, laurilsulfato de sódio, gelatina, croscarmelose sódica, estearato de magnésio e água deionizada.

Atenolol 50 mg:
Cada comprimido contém:
atenolol 50mg
excipientes q.s.p. .... 1 comprimido
Excipientes carbonato de magnésio, amido de milho, laurilsulfato de sódio, gelatina, croscarmelose sódica, estearato de magnésio e água deionizada.

Atenolol 100 mg:
Cada comprimido contém:
atenolol 100mg
excipientes q.s.p. .... 1 comprimido
Excipientes carbonato de magnésio, amido de milho, laurilsulfato de sódio, gelatina, croscarmelose sódica, estearato de magnésio e água deionizada.

Patient information

INFORMAÇÕES AO PACIENTE

COMO ESTE MEDICAMENTO FUNCIONA?
O atenolol comprimido reduz a pressão arterial elevada já que age no coração e na circulação. Seu efeito é observado por pelo menos 24 horas após administração do fármaco.

POR QUE ESTE MEDICAMENTO FOI INDICADO?
O atenolol comprimido é indicado para controlar a pressão arterial elevada, arritmias cardíacas e angina pectoris, também é indicado para auxiliar no tratamento (precoce ou tardio) do infarto do miocárdio.

QUANDO NÃO DEVO USAR ESTE MEDICAMENTO?
O atenolol comprimido não deve ser utilizado por crianças ou por pacientes que possuem alergia ao atenolol ou a outros componentes da fórmula. O medicamento também não deve ser administrado a pacientes que apresentam batimentos cardíacos lentos, choque cardiogênico, pressão arterial baixa, acidose metabólica, distúrbios graves da circulação arterial periférica ou insuficiência cardíaca descompensada. Pacientes diabéticos, hipoglicêmicos ou que apresentam problemas cardíacos, circulatórios, renais e pulmonares e ainda pacientes grávidas ou que estejam amamentando devem ter cuidados especiais ao utilizar atenolol.

Patient information

"Informe ao médico ou cirurgião-dentista o aparecimento de reações indesejáveis."
"Informe ao seu médico ou cirurgião-dentista se você está fazendo uso de algum outro medicamento."
"Não use medicamento sem o conhecimento do seu médico. Pode ser perigoso para a saúde."
Dificilmente o uso do atenolol irá prejudicar a habilidade de dirigir ou operar máquinas, porém, deve-se levar em consideração a possibilidade de ocorrer tonturas ou fadiga. Informe seu médico se estiver tomando outros antihipertensivos (inclusive colírios) ou outros medicamentos para tratamento de problemas do coração e circulação, inibidores da MAO (monoaminoxidase), álcool, antiinflamatórios, descongestionantes nasais, medicamentos para gripe, para diabetes, para tratamentos de úlcera e antidepressivos. O resultado do tratamento pode ser alterado se o atenolol for ingerido ao mesmo tempo que estes medicamentos. Enquanto estiver utilizando o atenolol, não tome nenhum outro medicamento sem o conhecimento de seu médico.
"No caso de gravidez durante ou após o tratamento informe seu médico. Informe também se estiver amamentando."

COMO DEVO USAR ESTE MEDICAMENTO?
O atenolol é um comprimido circular e de cor branca.
Posologia
Adultos:
Pressão alta: 1 comprimido de 50 mg a 100 mg, 1 vez ao dia. O efeito pleno é atingido em 1 a 2 semanas de tratamento.
Angina: dose única diária de 100 mg, ou de 50 mg duas vezes ao dia. Dificilmente se obterá benefício adicional pelo aumento da dose.
Arritmias cardíacas: 1 comprimido de 50 mg ou 100 mg, 1 vez ao dia.
Infarto do miocárdio: 1 comprimido de 100 mg, 1 vez ao dia na prevenção em longo prazo do infarto do miocárdio.
O comprimido de 25 mg deve ser engolido inteiro, com o auxílio de um copo de água, já os comprimidos de 50 e 100 mg podem ser partidos caso necessite de ajuste de dose.
"Siga a orientação de seu médico, respeitando sempre os horários, as doses e a duração do tratamento."
"Não interrompa o tratamento sem o conhecimento de seu médico."
"Não use o medicamento com o prazo de validade vencido. Antes de usar observe o aspecto do medicamento."

QUAIS OS MALES QUE ESTE MEDICAMENTO PODE CAUSAR?
O atenolol comprimido é geralmente bem tolerado, porém as seguintes reações adversas podem ocorrer: batimento mais lento do coração, piora da insuficiência cardíaca, mãos e pés frios, confusão, tontura, dor de cabeça, alterações de humor, pesadelos, boca seca, distúrbios gastrintestinais, impotência, distúrbios da visão e cansaço.

O QUE FAZER SE ALGUÉM USAR UMA GRANDE QUANTIDADE DESTE MEDICAMENTO DE UMA SÓ VEZ?
Caso o paciente ingira uma quantidade exagerada de atenolol comprimido poderá apresentar os seguintes sintomas: batimento mais lento do coração, pressão arterial baixa, insuficiência cardíaca aguda e crise de asma. Devendo buscar orientações médicas imediatamente.

ONDE E COMO DEVO GUARDAR ESTE MEDICAMENTO?
Este medicamento deve ser armazenado na sua embalagem original, em temperatura ambiente (entre 15 e 30 °C), em lugar seco, fresco e ao abrigo da luz. Nestas condições a validade do produto é de dois anos a contar da data de fabricação.

TODO MEDICAMENTO DEVE SER MANTIDO FORA DO ALCANCE DAS CRIANÇAS

INFORMAÇÕES TÉCNICAS AOS PROFISSIONAIS DE SAÚDE

Technical information

CARACTERÍSTICAS FARMACOLÓGICAS
O atenolol age preferencialmente sobre os receptores adrenérgicos beta-1 do coração, sendo um bloqueador beta-1 seletivo. Com o aumento da dose a seletividade diminui. O atenolol não possui atividade simpatomimética intrínseca nem atividade estabilizadora de membrana. Como outros beta-bloqueadores, o atenolol é contra-indicado em insuficiência cardíaca descompensada devido aos seus efeitos inotrópicos negativos. O mecanismo de ação do atenolol no tratamento da hipertensão não está completamente elucidado. Em pacientes com angina, é provável que o atenolol reduza a freqüência e a contratilidade cardíaca, o que faz com que ele se mostre eficaz na eliminação ou na redução dos sintomas. Em comparação com a mistura racêmica é improvável que quaisquer propriedades adicionais do S (-) atenolol originem efeitos terapêuticos diferentes.
O atenolol é efetivo e bem tolerado na maioria das populações étnicas, apesar da possibilidade de sua resposta ser menor em pacientes negros.
O atenolol pode ser administrado com outros agentes anti-hipertensivos, diuréticos e anti-anginosos.
Propriedades Farmacocinéticas
Após administração oral de atenolol, sua absorção é consistente, porém, incompleta (aproximadamente 40-50%), ocorrendo picos de concentração plasmática em 2 a 4 horas após a administração da dose. Os níveis sanguíneos do atenolol são consistentes e sujeitos a pequena variabilidade. O metabolismo hepático não é significativo, e mais de 90% da quantidade absorvida alcança a circulação sistêmica inalterada. A meia-vida plasmática do atenolol é de aproximadamente 6 horas, podendo se elevar na presença de insuficiência renal grave, pois os rins constituem a principal via de eliminação do fármaco. Devido a sua baixa solubilidade lipídica, o atenolol penetra pouco nos tecidos, e sua concentração no tecido cerebral é baixa. Sua taxa de ligação às proteínas plasmáticas é baixa (aproximadamente 3%).
O atenolol proporciona uma boa adesão do paciente ao tratamento pela simplicidade de sua posologia, sendo efetivo por pelo menos 24 horas após administração diária da dose oral.

INDICAÇÕES
O atenolol é indicado no controle da hipertensão arterial, angina pectoris e arritmias cardíacas, sendo também utilizado no tratamento do infarto do miocárdio e na intervenção precoce e tardia após infarto do miocárdio.

CONTRA-INDICAÇÕES
A administração do atenolol em crianças é contra-indicada. Este medicamento, assim como os demais beta-bloqueadores, encontra-se contra-indicado para pacientes que apresentam hipersensibilidade ao atenolol ou a qualquer outro componente da fórmula. Também é contra-indicado em casos de bradicardia, choque cardiogênico, hipotensão, acidose metabólica, distúrbios graves da circulação arterial periférica, bloqueio cardíaco de segundo ou terceiro grau, síndrome do nodo sinusal, feocromocitoma não tratado e insuficiência cardíaca descompensada.

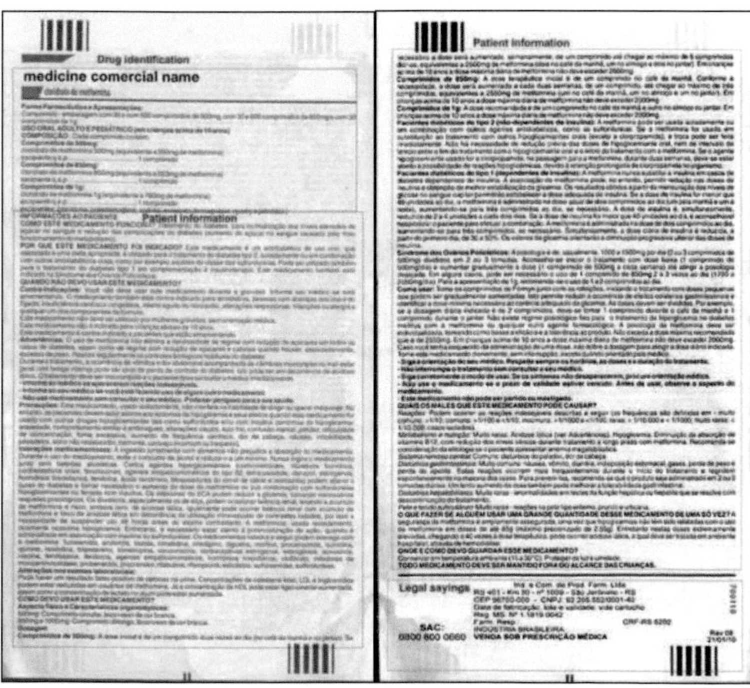

Drug identification

medicine comercial name

Patient information

Patient information

Legal sayings

SAC:
0800 600 0060

INDÚSTRIA BRASILEIRA

VENDA SOB PRESCRIÇÃO MÉDICA

Figure 3 and 4: Examples of MPL according to requirements made by Resolution n. 47 (2009)

4 CONCLUSIONS

Through the analysis of the Brazilian regulatory MPL it was possible to observe and understand how and when changes occurred in the graphic visual aspects in the regulatory process of a structuring element that accompanies the pharmaceutical drug. It is a regulatory process where changes occur between long stretches of historical time.

The results revealed that even though MPL´s regulation has been existant in Brazil for over six decades, informational ergonomics aspects on graphic presentation started to be taking into account and more frequently updated only since 2003. It was also found that since 2001, the development of those rules started to be based on Public Inquiries directed to all Brazilian citizens in order to involve social actors/users in the construction process of the MPL in Brazil, signaling a relationship approach between State and contemporary brazilian Society.

The last and current regulation standard published in September 2009 by ANVISA, has focused on aspects related to language and graphical presentation of the MPL´s content. The new regulation also requires that information should be distinguished on the content and language contained in separate documents each depending on the type of user (health professional and patient). On the other hand, the adequacy of MPL by the pharmaceutical industry with these new standards implies the restructuring of the current pattern of the MPL based on regulation that had not previously considered public opinion. At the same time it is important to consider that the more visual information on medicines are available to users over time, the greater the need for adequacy of technical scientific content.

The document 'medicine package leaflet' has been nowadays treated with more specificity, especially in the matters of graphic presentation aspects, since the "form and content" of the medicine leaflet has received increased attention in both the standards published in the current regulation and by its users.

ACKNOWLEDGMENTS

We would like to thank Fundação Oswaldo Cruz for supporting this research through a full grant for the first author of this paper.

REFERENCES

Brasil. Decreto n. 20.397, de 14 de janeiro de 1946. Aprova o Regulamento da Indústria Farmacêutica no Brasil. Brasília: Diário Oficial da União, 19 jan, 1946.

Brasil. Decreto n. 5.053, de 22 de abril de 2004. **Estabelece** regras para elaboração, de rotulagem e folhetos informativos para produtos veterinários. Brasília: Diário Oficial da União, 23 abr. 2004. p. 1-8.

Brasil. Ministério da Saúde. Agência Nacional de Vigilância Sanitária. Consulta pública n. 96, de 19 de novembro de 2001. Diário Oficial da União, Brasília, DF, 21 nov. 2001. Seção 1, p. 150.

Brasil. Ministério da Saúde. Agência Nacional de Vigilância Sanitária. Consulta pública n. 1, de 8 de janeiro de 2002. Diário Oficial da União, Brasília, DF, 9 jan. 2002. Seção 1, p.189-190.

Brasil. Ministério da Saúde. Agência Nacional de Vigilância Sanitária. Consulta pública n. 1, de 23 de janeiro de 2009. Diário Oficial da União, Brasília, DF, 26 jan. 2009. Seção 1, p. 31-36.

Brasil. Ministério da Saúde. Agência Nacional de Vigilância Sanitária. Resolução n. 140, de 29 de maio de 2003. Diário Oficial da União, Brasília, DF, 24 set. 2003. Seção p, 53-54.

Brasil. Ministério da Saúde. Agência Nacional de Vigilância Sanitária. Relatório de Análise de Contribuições, 2009.

Brasil. Ministério da Saúde. Agência Nacional de Vigilância Sanitária. Resolução n. 47, de setembro de 2009. Diário Oficial da União, Brasília, DF, 9 set. 2009. Seção 1, p. 40.

Brasil. Ministério da Saúde. Conselho Nacional das Secretarias Municipais de Saúde. *O SUS de A a Z : garantindo saúde nos municípios.* 3 ed. Brasília : Editora do Ministério da Saúde, 2009.

Brasil. Ministério da Saúde. Resolução RDC N° 140, de 29 de maio de 2003. Disponível: http://e-legis.anvisa.gov.br/leisref/public/showAct.php?id=6311.

Brasil. Ministério da Saúde. Serviço Nacional de Fiscalização da Medicina e Farmácia. Portaria n°. 49, de 10 de agosto de 1959. Diário Oficial da União, Brasília, DF, 17 ago.1959. Seção 1, p. 1777.

Brasil. Ministério da Saúde. Secretaria Nacional de Vigilância Sanitária. Portaria n°. 65, de 28 de dezembro de 1984. Diário Oficial da União, Brasília, DF, 31 dez.1984. Seção 1, p. 19931-19932.

Brasil. Ministério da Saúde. Secretaria Nacional de Vigilância Sanitária. Portaria n°. 110, de 10 de março de 1997. Diário Oficial da União, Brasília, DF, 8 mar. 1997. Seção 1, p. 5332.

Caldeira, T. R.; Neves, E. R. Z.; Perini, E. (2008). Evolução histórica das bulas de medicamentos no Brasil. *Cadernos de Saúde Pública,* v. 24, n. 4, p.737-743.

Van der Waarde, K. (2006). *Visual information about medicines for patients.* In: FRASCARA, J. (Eds). Designing Effective Communications: Creating contexts for clarity and meaning. New York: Allworth Press, p. 38-50.

World Health Organization. (1998). The role of the pharmacist in self-care and self-medication. *Report of the 4th WHO Consultive Group on the Role of the Pharmacist.* Hague, Netherlands: WHO.

CHAPTER 7

The Construction of the Educational Program of the Problem Solving Thinking Based on the Management System (HACCP)

Manami Nozaki[a], *Yumi Kikuchi*[a], *Saori Yasuoka*[a], *Naomi Imajo*[a], *Shinji Kojima*[b],

[a]Toho University, Tokyo, Japan
[b]Ines Clinic, Hiroshima, Japan
fnozaki@med.toho-u.ac.jp

ABSTRACT

About the nursing student having various backgrounds, it is necessary to make an effort to arrive at the goal for the every educational institution. Therefore we reached the introduction of the management system.

Ordinarily, it is desirable to follow it for Plan-Do-Check-Act (PDCA) cycle to do Total Quality Control (TQC). We decided to introduce Hazard Analysis Critical Control Point (HACCP), which is expected to supplement the cycle. We built educational program while following 7 principles of the HACCP introduction and five procedures. As a result, 98% of attendance was passed. Checking a record thing submitted by an attendance functioned a monitoring and adjustment. For conclusion, I completed the education pass which I could step on for PDCA cycle by introducing HACCP.

Keywords: Nursing Education, Management System, PDCA-Cycle

INTRODUCTION

In Japan changing disease structure, the ability for practice of high quality is expected to the clinical nurses. Under the basic education program, ability for assessment and the specific acquisition of the nursing technique are demanded.

About the nursing student having various backgrounds, it is necessary to make an effort to arrive at the goal for the every educational institution.

However, as for the current Japanese university student, basic scholastic ability deteriorates. They make it bipolar in base layer and the high-ranking layer. Both belong to the present nursing University. The learning degree of progress cannot get the base layer for the base layer and they are at great risk of falling off. While the difference of the high-ranking layer and base layer is small, the subject persons in charge hope to deal and narrow the difference. We thought that the perfection acquisition learning, so to speak, was desirable.

Therefore we reached the introduction of the management system.

In late years the standardization of duties is demanded when we pay more attention to effectiveness and safety of the duties management of every social field. Because examples in response to the certification of the ISO appear in the business establishments performing the duties of various fields. So the management of the quality that the introduction of the clinical pass is performed flourishingly in many medical facilities to secure the effectiveness of duties, the safety to a patient in the medical industry, and is higher TQC (Total Quality Control) is demanded.

Generally the Plan-Do-Check- Act (PDCA) cycle suggested by Deming (Deming 1982) is placed as the heart in the TQC. We intended to introduce a PDCA cycle into the management of duties for the first step of the quality control of the nursing education.

However, the procedure to introduce a PDCA cycle into duties is vague, and the introduction is not easy at all. Therefore Hazard Analysis Critical Control Point which is the existing management system that we are used for hygiene management of the food widely (HACCP), and we paid our attention to it (Hulebak 2002).

Nowdays as for HACCP, the examples in the duties management of the medical field appear(Baird D.R. 2001) (Kojima 2008) and 12 processes to introduce HACCP are stated clearly in, the field of the hygiene management of the food. We can bring a PDCA cycle into the duties naturally and then we build the use of duties following this.

Purpose of this study

In nursing basic education, we introduced system of administration Hazard Analysis Critical Control Point (HACCP), and this study was intended that we built educational program to bring up a solution of the problem thought according to seven principles in five procedures.

METHODOLOGY

2.1. The summary of the pertinence subject

The pertinence subject is nursing process methodology for first graders. It consists of the seven times of the lecture. The aim of the subject is to get a solution of the problem intending to lead a nursing diagnosis.

We think the handling of ability acquisition program of the solution of the problem based on a system of administration.

2.2. The construction of the subject based on HACCP

At first it was necessary to define something as the quality of the instructional activity to make a purpose to build a regime to secure the quality of the making of subject instructional activity based on HACCP clearly. We defined the quality of the nursing education in the subject concerned as "letting the students think by themselves" after the many arguments. And we tried the rebuilding of education duties along HACCP introduction process aiming at the quality control of the instructional activity. 5 procedures, 12 processes consisting of 7 principles are shown to introduce HACCP into duties in the hygiene management of the food. (Table 1)

Table1. Process for implementation of HACCP

Procedures	Principles
1. Assemble HACCP team	1. Hazard analysis
2. Describe product	2. Determine Critical Control Points
3. Identify intended use	3. Establish critical limits for each CCP
4. Construct flow diagram	4. Establish a monitoring system for each CCP
5. On-site confirmation of flow diagram	5. Establish corrective actions
	6. Establish verification procedures
	7. Establish Documentation and Record Keeping

We considered it ethically after having got permission of the attendance students by treating a record thing as data.

RESULTS

3.1. The making of the 'The education pass'

1) The formation of the HACCP team (procedure 1)

The process that you should have performed was formation of the HACCP team at first. The HACCP team constituted it with a teacher, TA (teaching assistant), a doctor. A teacher, TA developed it and carrying it out of the instructional activity and the doctor contributed to use of HACCP. The team held a meeting regularly, and the argument by the team determined all the topics for discussion.

2) Work process listing (procedure, from 2 to 5)

In the meeting of the HACCP team, we compiled all work processes of the instructional activity of subject "nursing process". So we decided the seven times of the contents of the lecture, contents of the group learning, a problem of the self-learning and the last problem. We made a pass of so-called at one time-like education. We extracted the schedule of the learner and made the orientation tool to a learner. This was equivalent to a clinical pass for learners. We made a work process of a teacher and TA in this schedule. We show the work process of the teacher to Table 2.

Table 2. The work progress schedule of the teacher
1. Orientation
(1)The explanation of the subject summary including the learning target
(2)The explanation of how to lead
(3)The explanation of the problem
(4)The explanation of the evaluation method
2.A lecture
(1)The explanation of the summary of the nursing process
(2)The explanation of each step of the nursing process
(3)The explanation of the example
(4)An instruction / promotion of the example development
3. The trial and error of the thought
(1)Instructions of the intra-class personal work
(2)Promotion of the intra-class discussion
(3)Instructions of the homework until the next time
4. Grasp of the reaction
(1)Grasp of the understanding degree based on the quiz at every class initiation
(2)Grasp of the response based on the reaction paper at every class end
5. Explanation for revisions
(1)Additional explanation based on the result of the quiz
(2)The change of the professor method by the reaction paper
(3)Reinforcement and intervention for revisions based on intra-class reaction

3) Harm analysis (principle 1)

We found out harm which is hidden behind in an instructional activity from the work progress schedule of the instructional activity by the argument of the HACCP team. We examined a management method for the prevention classifying it in pedagogics-like harm (a psychogenic harmful phenomenon), businesslike harm (Table 3). We can prevent the trouble concerning the right or wrong of the unit acquisition to be included in businesslike harm by the confirmation of syllavas contents and enough explanation to a learner. However, it is often that the pedagogics-like harm is caused by human error and a relationship, and it is difficult to prevent this with a manual. Therefore we decided to prevent it by managing it for continuation in Critical Control Point (CCP).

Table3. An analysis of the harm to a learner in the instructional activity

Harm of the pedagogics
1. The thing which finishes having difficulty with understanding
The explanation that does not promote understanding
The problem of an inappropriate degree of difficulty
The inappropriate teaching materials
Low basics scholastic ability
The low preparation situation
2. The thing that learning will does not improve
The bad relations between the teacher and student
The development that does not rouse interest / interest
The development that does not get sense of accomplishment
Businesslike harm
1.The nonfulfillment of the problem by the misunderstanding
2.The dissatisfaction to an evaluation
3.Non-acquisition of the unit

4) Decision of CCP and the management standard and the improvement step (principles 2,3 and 5) (Table 4)

The desirable program consists of security of the understanding of contents and learning eagerness. Therefore, as for the arrival to a standard of the understanding degree, as for the maintenance of the motivation, it was lecturing with a management standard after having donated education contents depending on ladyness of the student who attended.

We could find out CCP naturally by performing harm analysis. It was necessary that the sake to let a learner think by himself in learning activity met two conditions equal to or less than and thought that the checkpoint that met them was CCP. Actually it was maintenance / an increase of the security (CCP 1), learning eagerness of the contents understanding degree (CCP 2).

Table4. Two conditions and the CCP to let him think by himself and the management standard (CL),the improvement step

1. Security of the contents understanding degree	
[a concreteness-like action]	reply all items of the quiz
[CCP 1]	check the correct answer rate of five times of quizzes
[CL]	high score, more than 60%
[improvement step]	comment the correct answer next class
2.Maintenance / an increase of the learning eagerness	
[concrete action]	participate in seven times of classes eagerly
[CCP 3]	confirm the description contents of the reaction paper quiz
[CL]	an affirmative description
[improvement step]	return an affirmative feedback next class

5) The decision of the method of the monitoring (principle 4)

During a program period, we established the checkpoint that accorded with five times of CCP. We collected them as a quiz and reaction paper and checked it.

All processes from learner begin to take course to submit the last subject are shown on the clinical pass.

6) Standard duties procedure (Standard Operation Procedure;SOP) which can evade harm

By HACCP introduction, we usually devise procedure (SOP) of the standard duties that can evade an error. We took a work process into consideration and enumerated the procedures of duties. This is equivalent to duties manual.

7) A standard activity procedure

We generalized that I got an activity result of SOP and organized it as SOP.

3.2. The achievement

As a result, as for the dropout, three people, 98% of the registrant passed by the arrival degree of the achievement subject. In other words it was recognized being to when we learned a solution of the problem thought. The mean and SD of results was 81.3 ∓ 4.3. The satisfaction in the class evaluation was also high.

DISCUSSION

4.1. An effect this time when I introduced a management system into an instructional activity

We left a dropout on the way as a few and attendance true 98% passed it. It was judged that the student learned a solution of the problem thought.

Furthermore, it brought the effect for the teacher and the effect for the student to introduce a management system into an instructional activity.

At first we showed intentional intervention by the teacher to another person by describing an instructional activity as a work progress schedule one by one. We could plan the mutual understanding of the teacher team, and anyone was able to do the intervention of the same quality. It can inflect for the guidance of the new face teacher in future.

Secondly, it was with an opportunity to confirm what we set CCP and CL and checked continually whether explanation of teacher oneself promoted the understanding of the student. A teacher reviewed the education technique of own and led to revising it. We became the help of the improvement of the ability for education of the teacher.

Finally the results of the passer were high with 81.3∓4.3 points of mean and SD, and unevenness was small. It is possible for a fine adjustment and thinks that I was able to lift the understanding degree of the class generally by setting a checkpoint little by little while a difference of the understanding degree of the student is small.

On the other hand, for a student, we can confirm the understanding degree of own by a quiz and think that the narrowing of the target of the learning was possible. To find the score of the quiz, the wish, the incentive to learning rose to take high score on the next time. And a classmate marked it by the exchange of the answer sheet. This was an opportunity evaluated by another person, and it was with stimulation for the learning activity of own by knowing the understanding degree of another person. In addition, the praise from another person about the result led to improvement of the learning eagerness.

4.2. Introduction of HACCP to nursing education

We chose HACCP used for food management to fix a PDCA cycle in a nursing instructional activity this time. HACCP is based on the harm analysis and at important management point manages the quality. Heaping up of the result was possible for the seven times lessons that was applied 12 processes to the progress of the class of the series. It may be said that we organized a PDCA cycle for an instructional activity. We showed intentional intervention, a checkpoint, a revision method to another person, besides, by showing CCP and CL. It is useful for the guidance of the younger student.

We targeted the lecture of the solution of the problem thought this time. For the improvement of the nursing practice ability, it is necessary to polish knowledge, technology, a manner. As well as the reinforcement of the thought, it is necessary to plan improvement of the quality of the technical aspect. It means that technical offer is repeating small judgment, and to moving the fingers while handling feeling of strain. In a nursing technology-related subject, we hope to try the introduction of a management system. It will be good to weigh it against "Crew Resource Management" attaching great importance under a high risk, a role of human factor under the high stress bottom more.

CONCLUSION

We completed the education pass which we were able to step on for a PDCA cycle by introducing HACCP.

REFERENCES

Baird, D.R., Henry M., Liddell K.G. et al. 2001. Post-operative endophthalmitis- the application of hazard analysis critical control points (HACCP) to an infection control problem. *Journal of Hospital Infection* 49: 14-22.

Deming, WE. 1982. *Out of the Crisis. 1st ed.* 86-90. Cambridge: MIT Press.

Hulebak, K.L., and Schlosser W.2002. Hazard analysis and critical control point (HACCP) history and conceptual overview. *Risk Analysis* 22: 547-552.

Kojima, S., Kato M., Wang D.H., et al. 2008. Implementation of HACCP in the management of medical waste generated from endoscopy. Journal of Risk Research 11: 925-936.

Robert, M. 2005. Crew resource management and its applications in medicine. Making health care safer- a critical analysis of patient safety practices. 366-373.Tokyo: Igaku-shoin(in Japanese).

CHAPTER 8

Food for the Heart: Lessons Learned while Designing a Visual Decision Support System for Patients with Coronary Heart Disease

Bum chul Kwon, Zhihua Dong, Karen S. Yehle, Kimberly S. Plake, Lane M. Yahiro, Sibylle Kranz, and Ji Soo Yi

Purdue University
West Lafayette, IN, USA
{kwonb, dong17, kyehle, kplake, lyahiro, kranz, yij}@purdue.edu

ABSTRACT

Designing and evaluating a web-based behavioral intervention system is not a trivial task. Such a system has become more and more sophisticated, and understanding the impacts of a web-based intervention often requires multi-phased, longitudinal evaluation studies. In this paper, we would like to share lessons that we have learned while designing and evaluating a web-based dietary intervention system for patients with coronary heart disease, called "Food for the Heart." The three main lessons are as follows: 1) Health intervention requires multi-faceted support; 2) Design should mimic the existing mental model; and 3) Different evaluation methods serve different purposes. In addition, we share some minor technical lessons learned while implementing the website. These lessons may be helpful when designing and evaluating a web-based healthcare intervention system.

Keywords: visualization, decision support system, guideline, coronary heart disease, website design

1 INTRODUCTION

In order to educate and assist the general public in adopting and sustaining healthy behaviors, such as healthy food choices, to prevent chronic disease (e.g., diabetes, obesity, and coronary heart disease), web technologies have been frequently employed. Web technologies have obvious benefits: they can disseminate health information instantly at a relatively low cost to a wide range of populations. Moreover, the recent evolution of web technologies make it possible for web technology to serve other roles beyond simple information dissemination, such as delivering tailored information (De Bourdeaudhuij et al. 2007; Kroeze et al. 2008), supporting monitoring (Beach et al. 2006), and providing social supports among patients (McKay et al. 2001; Tate et al. 2006).

Though web technologies have generated several success stories, designing and evaluating such complex systems is very challenging for the following reasons:

First, increasing complexity becomes a design challenge. Dietary intake choices are a complex matter, thus, more and more features need to be included in a single system to support multiple intervention strategies and numerous design decisions are needed to build a well-integrated system. In addition, since the target population could be people with low computer literacy (e.g., older adults and young children), the design space for a system could be restricted. In other words, if one would like to employ innovative web techniques (e.g., novel interaction and/or visualization technique), an inappropriate level of novelty may become a barrier for some target users, rather than benefiting them.

Second, the complexity of a system also complicates the evaluation process. For example, if a system contains multiple features, it is challenging to understand which features play a more significant role in achieving the goal of web-based intervention. Particularly when the outcomes of the intervention are measured by behavioral changes or health outcomes, it is even more challenging to measure the effects of various features. The measured effects could be the result of the combination of different features, and confounding factors (e.g., learning effects and familiarity) could add noise to the collected data, which complicates data analysis.

Thus, the goal of this paper is to summarize the lessons we learned while we have designed and evaluated the Food for the Heart (FFH) website (https://engineering.purdue.edu/FFH/) in the past two years and to discuss the implications of these lessons. Though the lessons were only obtained from studies on a single system, we believe that they are applicable to many other web-based intervention studies, so other researchers and practitioners can learn from us.

2 FOOD FOR THE HEART

"Food for the Heart (FFH)," is an interactive, visual, web-based dietary intervention system for patients with coronary heart disease (CHD). The primary motivation of this research project is to help patients with CHD overcome the difficulty in planning meals and snacks while satisfying specific nutrition requirements. Patients with CHD have to simultaneously consider multiple aspects of nutrition, such as total energy consumed, specific nutrient intakes (sodium, fiber, saturated fat, and cholesterol), and consciously plan all meals and snacks throughout a day so they collectively result in a balanced daily diet. If CHD patients have additional comorbidities, secondary diseases to CHD (e.g., diabetes and obesity), those further increase the level of difficulty for total diet planning.

The FFH website was developed in the following steps. First, before designing the FFH website, we have conducted a series of focus group studies with 20 CHD patients and 7 informal caregivers (Study 0) with a preliminary prototype. Based on the collected design requirements, we designed and implemented the FFH website (version 0.1) as shown in Figure 1(a) (FFH v0.1). FFH v0.1 was evaluated using an interview study with 6 participants after introducing the system to them (Study 1). Though the interview study was conducted in a short period of time (1 hour per participant), this study helped us see the main design flaws. After the evaluation study, the whole website was redesigned as shown in Figure 1(b) (FFH v0.2) (Zerlina et al. 2011). FFH v0.2 was evaluated through an interview study with 6 participants after they actually used the system for a week within their own daily lives (Study 2). Study 2 provided additional insights, which lead to add additional features to the websites without modifying the overall look and feel (FFH v0.3). Currently, the replication of Study 2 with a larger population is in progress (Study 3). The list of evaluation studies and developed systems were summarized in Table 1.

Table 1 Timeline of the Food for the Heart study.

System	Evaluation	Descriptions
Prototype		An initial prototype used for Study 0
	Study 0	Focus group study (N = 27)
FFH v0.1		Figure 1(a)
	Study 1	Interview study after one-time use (N = 6)
FFH v0.2		Figure 1(b)
	Study 2	Interview study with one-week experiences of the FFH site (N = 6)
FFH v0.3		Implement customization functions
	Study 3	Interview study with one-week experiences of the FFH site (under progress)

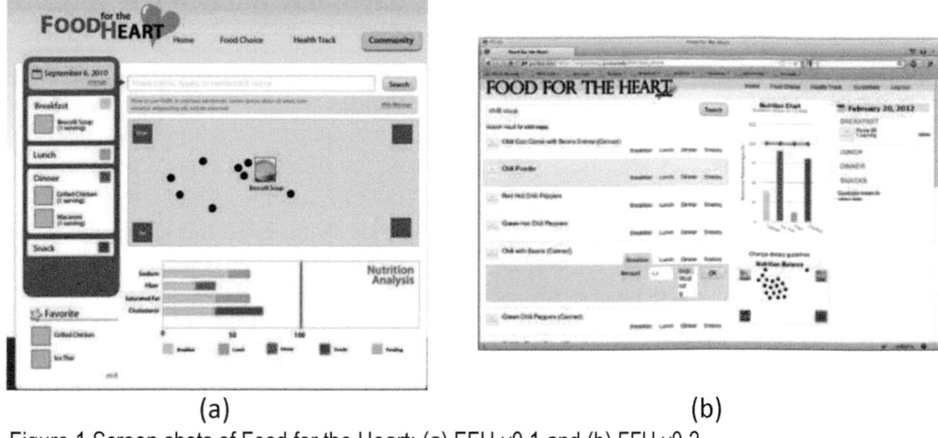

(a) (b)

Figure 1 Screen shots of Food for the Heart: (a) FFH v0.1 and (b) FFH v0.2

3 LESSONS LEARNED

3.1 Health Intervention requires multi-faceted support

The focus group study (Study 0) with 27 participants in Study 0 helped us see the patients' perspectives. Though focus group participants generally perceived the initial prototype positively, it quickly became obvious that a website like this required much more than a simple visualization technique. Additional requirements that emerged from Study 0 included needs to receive recommendations from healthcare professionals, the ability to record and track other health records (e.g., weight, exercise, and blood pressure), the function for healthcare providers to monitor the patient's history, the capability to individualize nutrition components and criteria to each person's chronic disease(s), and the information about interactions between medicine in combination with supplements.

However, we found that implementing all of these features at the same time was beyond the capability of a research team with limited number of developers. Thus, we took a strategy of implementing the most distinctive feature out of many requested features (visualization of multiple nutrition elements). Because some features, such as tracking health records, are already widely adopted in other health-related commercial websites, implementing and evaluating such features does not necessarily increase the merit of the project as academic research, especially from the perspective of human-computer interaction research. However, the evaluation outcomes (e.g., behavioral changes and health outcomes) of this focused approach could be marginal or limited due to a lack of some non-innovative, but required features, which can drastically lower the usability of the system.

In order to avoid this problem, we decided to employ user-centered design approach (Vredenburg et al. 2002 for review), so that we can capture and implement the indispensable requirements while focusing on the visualization feature of the website. We believe that this approach helped us achieve some goals as academic researchers with limited resources.

3.2 Design should mimic the existing mental model

In the process of implementation, it is typical that developers fall into the trap of designing products perfect in their own mind that fail to address users' needs. It is especially the case when the developer's thinking is pre-occupied by technological innovations and ignores the development outcomes from users' perspectives. As developers of FFH, we also learned this lesson in a painful way through our user study in different developing stages. So far, we have conducted three user studies (Study 0, Study 1, and Study 2) corresponding to three developing stages and product versions (Prototype, FFH_v0.1, and FFH_v0.2). It was an iterative process of recognizing and reflecting the users' mental model to newer versions of the tool.

Through focus group studies with our prototype (Study 0), we gained the confidence that there were needs for online intervention tools like FFH. Thus we developed FFH v0.1, offering basic functions of exploring and choosing food for each meal, and recording diet plans. As shown in Figure 1(a), we implemented 'Dust & Magnet' view (Ji Soo Yi et al. 2005) as the major visualization and interaction tool for users to browse and choose food items. To the left of the major view, an uninteractive window was showed recording meals of the day.

From the interview study (Study 1) conducted to assess FFH v0.1, we realized that this design violated the basic usability heuristics and also challenged the users' mental model. First, people liked to read from left to right and top to bottom. Due to such reading conventions of people, the contents in the leftmost column were perceived as the most important information. However, as the diet-recording window was an information display, requiring no interaction or primary attention from users, the leftmost position misled users in attempting to interact with and manipulate the window. Second, the "Dust and Magnet" interactive interface is a metaphoric interface design. It was supposed to provide natural and easy-to-learn interaction through its intuitive connections between interaction cues and their functions in the interface. However, through interviews, we learned that users failed to appreciate this interaction as the major browsing tool to find food options. Though they did comprehend the basic idea of the tool, they found it difficult to use, as it could not directly provide visible food options until extra interactions took place. Furthermore, this interface could only display one food option at a time, which requires users to remember information or execute extra interactions. Using this 'Dust & Magnet' metaphoric interface as major browsing tool violated the two usability heuristics: visibility, and preventing users from recalling important information (Nielsen 1994). Furthermore, one natural limitation of metaphoric interface is that though it can enhance novel users' understanding towards the tool, it inhibits intermediates' efficiency. This also limits the use of metaphoric interface as the only or major interaction tool, as we did in FFH v0.1.

Through Study 1, we gained great insights of how to improve a tool by pushing the presented model towards users' mental model. In FFH v0.2, we addressed these issues by moving the meal recording visualization to the right side of the window, introducing conventional searching-result list as the major tool to explore food options and reducing the size of the 'Dust & Magnet' interaction as a secondary tool. FFH v0.2 thus offers better visibility of important information and requires less unnecessary interactions.

3.3 Different evaluation methods serve different purposes

We learned that conducting a large-scale randomized control study, which is largely preferred in the medical domain, is not always the best option. Results may demonstrate the effectiveness of a certain intervention, but such results cannot reveal the insight of why and how the intervention works. We found that some qualitative studies with a small number of participants (e.g., in-depth interview and field studies) provide more insightful information needed at the developing stage. Through the three studies, we designed, conducted, and analyzed the data with a curious mind, which enabled us to see both expected and unexpected results. As in the focus group study, we confirmed our speculation that systems like FFH is needed among CHD patients, while Study 1 provided us vital lessons to improve the layout of FFH dramatically. It is critical for developers not to hold any predetermined perceptions and beliefs at the product development stages. This mindset is also the most important characteristics of qualitative research. Thus, we found it a good cognitive fit to conduct more open-ended, in-depth qualitative research at this stage.

4 OTHER TECHNICAL LESSONS

4.1 Nutrition Databases

Initially, we failed to identify an appropriate nutrition database that provides nutrition information for food items available in grocery stores and restaurants, which would be essential elements for FFH. The USDA National Nutrient Database for Standard Reference (http://www.ars.usda.gov/Services/docs.htm?docid=8964), which is considered the most comprehensive nutrient database available for the diets of Americans, was the obvious first choice, but we learned that the database did not contain all the restaurant foods that some patients preferred. Thus, we comprehensively reviewed commercial databases, such as the Nutritional Coordinating Center Food and Nutrient Database at the University of Minnesota (http://www.ncc.umn.edu/products/database.html), foodfacts.com, glondon.com, and gladson.com. However, all these databases are relatively expensive and incomprehensive. Though each company has different licensing policies, it generally costs $35,000 - $50,000 to have a copy of database, and the regular charges of additional several thousand dollars for required periodic upgrades. More importantly, no single vendor seems to provide a comprehensive database. A research team led by Dr. Carol Byrd-Bredbenner at Rutgers University constructed a database with UPC codes and nutrition information, but the group is contractually bound with the database vendors and was not at liberty to share the database. We sought advice from researchers at Purdue University and other institutes to identify affordable and comprehensive databases that would serve our purpose, but we failed to find one appropriate database. Therefore, in consultation with our team member, who is a Dietitian and understands the dietary aspects of disease management and who has also expertise in the use of national nutrient databases, we decided to peruse the most comprehensive database that included the foods most commonly eaten by our target population (www.fatsecret.com). The database provides limited free access through a web API. The database owners will charge some license fees for unlimited accesses, but it is still relatively affordable when educational licenses are purchased. In addition, we also constructed a custom databases when the fatsecret.com database did not provide nutrition information for certain restaurants which we wanted to offer in our system. Since nutrition labeling for foods purchased at restaurants is mandatory, we were able to use web crawling technologies to collect the publicly available nutrition information from those restaurant's websites.

4.2 Make the website publicly available as soon as possible

This lesson might be counter-intuitive to those who think that protecting the intellectual property is important. However, what we learned through this research project is that opening up the technologies to a selected group of trusted people as soon as possible is crucial to solicit new ideas and to get feedback in the early phase of the development. Having multiple eyes on a product drastically increases the chance to identify obvious errors. It also motivates the development team to deliver a high quality website. However, there are some moments in the development phases that websites under development are unstable. Thus, we operate two separate websites: one is open to the general public, and the other one is open only to the research team. We used Ruby on Rails as web-application framework, which has a feature to streamline the deployment procedure.

4.3 Browser compatibility

Since this project has been targeted to a large population, we have paid attention to developing a website that is compatible with a majority of popular web browsers (e.g., Microsoft Internet Explorer 7 or higher, Mozilla Firefox, Google Chrome, and Apple Safari & iOS browsers). Though maintaining such high compatibility severely burdened our development team (build one website and test it on different browsers and devices), we believe that we made the right decision. Through this approach, we could select web technologies that have high compatibility in the earlier phase in the development process, which shortened our development time. For example, we did not use Flash-based web visualization technologies (Flex) even though it provides finer controls over visualization and interaction because iOS-based web browsers do not support it. We also did not use various advanced JavaScript-based visualization frameworks (e.g., protovis and d3.js) because older versions of Microsoft Internet Explorer do not support them. Not selecting those advanced technologies obviously limited the capabilities of our system, but it definitely increases the compatibility of FFH. We also felt fortunate to make these decisions earlier in the development phases because changing underlying visualization in the later development phases often results in complete re-development of the system.

5 CONCLUSIONS

Throughout this article, we shared lessons we learned from the experience of building a visual decision support system for patients with CHD. Our limitations include: 1) our users have limitations in their abilities, 2) we should address those limitations fully into the design of the system, and 3) we should acknowledge the inherent limitations in using nutrition databases. This article was not intended to include all the possible issues that can arise from the development of a visual decision support system. Rather, we hope that sharing our experience may be helpful to future designers and practitioners of web-based intervention systems.

6 ACKNOWLEDGEMENTS

This study was partially funded by a seed grant from the Regenstrief Center for Healthcare Engineering at Purdue University. We also appreciate many student developers, Fransisca V. Zerlina, Teik-Ming Lee, Sung-Hee Kim, and Wellars Muhoza, who helped implement and evaluate the Food for the Heart website.

7 REFERENCES

Beach, J. et al., 2006. Health view: a simple and subtle approach to monitoring nutrition. In *Extended Abstracts on Human Factors in Computing Systems*. pp. 1801–1806.

De Bourdeaudhuij, I. et al., 2007. Evaluation of an interactive computer-tailored nutrition intervention in a real-life setting. *Annals of Behavioral Medicine: A Publication of the Society of Behavioral Medicine*, 33(1), pp.39–48.

Kroeze, W. et al., 2008. Comparison of use and appreciation of a print-delivered versus cd-rom-delivered, computer-tailored intervention targeting saturated fat intake: randomized controlled trial. *Journal of Medical Internet Research*, 10(2).

McKay, H.G. et al., 2001. The diabetes network internet-based physical activity intervention: a randomized pilot study. *Diabetes Care*, 24(8), pp.1328–1334.

Nielsen, J., 1994. Enhancing the explanatory power of usability heuristics. In *Proceedings of the SIGCHI conference on Human factors in computing systems: celebrating interdependence*. CHI '94. New York, NY, USA: ACM, pp. 152–158.

Tate, D.F., Jackvony, E.H. & Wing, R.R., 2006. A randomized trial comparing human e-mail counseling, computer-automated tailored counseling, and no counseling in an Internet weight loss program. *Archives of Internal Medicine*, 166(15), pp.1620–1625.

Vredenburg, K. et al., 2002. A survey of user-centered design practice. In *Proceedings of the SIGCHI conference on Human factors in computing systems: Changing our world, changing ourselves*. pp. 471–478.

Yi, J.S. et al., 2005. Dust & Magnet: multivariate information visualization using a magnet metaphor. *Information Visualization*, 4, pp.239–256.

Zerlina, F.V. et al., 2011. Food For The Heart: Visualizing Nutritional Contents for Food Items for Patients with Coronary Heart Disease. In Demonstration at Workshop on Visual Analytics in Healthcare (VAHC 2011) in conjunction with IEEE VisWeek 2011.

CHAPTER 9

Social Media for the eHealth Context. A Requirement Assessment

Anne Kathrin Schaar, André Calero Valdez, Martina Ziefle
Human Interaction Center, RWTH Aachen University
Aachen, Germany
{schaar, calero-valdez, ziefle}@humtec.rwth-aachen.de

ABSTRACT

There were 940 million social network users in 2010 in the world (Belleghem Van 2011). This fact makes social network sites (SNS) an integral part of the Internet. Knowing that health affairs are a central topic in the Internet makes this issue into an addressee for SNS. So-called health social network sites promise to be beneficial by lining up with the success of SNS like Facebook. Due to the fact that many promising technical approaches fail at the market, because of a lack of user acceptance we started an exploratory study to find out criteria for a successful health social network approach, improving health promotion. Being aware that health and health related information reveal special acceptance patterns, especially when provided via Internet, we wanted to define user-centered design criteria and acceptance requirements for a health SNS.

Central findings show that neither age, gender, field of work nor private social media usage impact willingness to disclose personal health related information. Health related SNS are generally perceived positively (77%) but are more likely to be used as a source for professional information. Key issues like data security create a gap between willingness to share and willingness to consume information. Recommending a health SNS could only work if done by personal friends or the private physician. Commercial advertising or insurance company recommendations are rather distrusted.

Keywords: social media, social network sites, ehealth, acceptance, user-centred-design

1 INTRODUCTION

Today online support is no longer a big deal. On the Internet there are platforms, communities, mailing lists etc. for every imaginable topic accessible from every country in the world at any time. Among this vast amount of topics health is one of the top-scorer topic. For every disease and ailment there is a place you can go to in the "virtual world." But not only the range of health related topics are widespread, the same breadth applies for the quality of the provided information and services. Earlier studies about medical technology acceptance revealed that in the context of health and healthcare special demands on technology and technical applications exist that are related to trust, privacy and intimacy (Lahlou et al. 2005; Moturu et al. 2008; Scheermesser et al. 2008; Arning & Ziefle 2009; Ziefle & Schaar 2010; Ziefle & Wilkowska 2010; Ziefle & Schaar 2011).

Healthcare services provided with anonymity, ubiquity and mobility, which are the epitome of the Internet, accentuate contrasting aspects for the users. The sensitivity of health related data and Internet technology highlight the aspect of data safety and data theft. But the aforementioned properties of the Internet also lead to increased accessibility, which becomes especially relevant when providing services for a decreasingly mobile worldwide user base. Additionally the Internet offers people a feeling of protection through its anonymity, especially when the disease is related to feelings of stigmatization (e.g. STDs).

The current paper presents a research approach that is investigating acceptance aspects and design criteria in the context of health SNS. The idea presented in this paper is based on the assumption that applying a user-centred approach from the beginning could avoid misguided and superfluous products in the field of SNS to guarantee a consideration of user-diversity for universal access to future products.

To work out the special character of health the paper is structured as follows: Section 1.1 gives an overview about a range of different health related applications and services on the Internet. Furthermore a definition of health SNS is specified. Afterwards Section 1.2 points out the special character of acceptance in the context of medical technology. Section 2 contains the questions addressed in this research. The study's methodology is presented in Section 3, including a presentation of: variables (Section 3.1); questionnaire instrument (Section 3.2); sample characteristics (Section 3.3). The central results are presented in Section 4. And finally the paper is completed with a discussion of the studies central findings and limitations (Section 5).

1.1 Health affairs in the Web 2.0

Since the development from Web 1.0 to the so called Web 2.0 health topics increased enormously within this domain (Boyd & Ellison 2007). At least the special character of Web 2.0 with its facilitated participatory information-sharing, user centred design and its collaboration on the World Wide Web (O'Reilly 2005) paved the way for a mutual exchange of information by users.

The central problem within health affairs in the Internet is the broad range of products, users and topics as well as a divergence of quality. Hartman et al. are differentiating at least nine different technologies that can be used for health related exchange of information in the Internet (Hartmann et al. 2011). The list of Hartman et al. ranges form *wikis*, that are web pages which can be structured or completed by everyone who has access to the current page (Ward 2006) over *blogs*, *social bookmarking* and *tagging*, *web 2.0 search* and *social search*, *RSS Feeds*, *social communities*, *evaluation and reputations systems* up to *online social gaming* and *virtual worlds*. Each application allows action related to health within the Internet. In the following the focus of our research is set on health SNS, which are defined for our use case in the following section.

Health social networks as a special use case

The focus of interest in this research is set on health SNS and their potential benefit. There is no unique definition of SNS in general and health SNS in particular. As presented in Section 1.2 we are confronted with a number of different technologies and applications that are gathering under keywords like *social media, social software* and/or *Web 2.0*.

Making the subject of this research more precise, we define social media as online social network sites. According to Boyd and Ellison (2007) SNS are defined as

> "web-based services that allow individuals to (1) construct a public or semi-public profile within a bounded system, (2) articulate a list of other users with whom they share a connection, and (3) view and traverse their list of connections and those made by others within the system. The nature and nomenclature of these connections may vary from site to site (Boyd & Ellison 2007)."

In the case of health SNS – health relevant topics are the constitutional topic that connects participants, although the individual motif and the specific subject may differ.

1.2 (Medical) Technology acceptance and acceptance of social networks

(Medical) Technology acceptance

Former studies revealed that the old technology acceptance models are inapplicable to analyze medical technology acceptance. Established Theories like Technology Acceptance Model (TAM) (Davis 1989) and the Unified Theory of Acceptance and Use of Technology (UTAUT) (Venkatesh et al. 2003) were primarily developed for the acceptance of Information and Communication Technologies (ICT) which were basically focused on the job context. The central aspects that generate technology acceptance in the context of the TAM were the perceived ease of use and perceived usefulness of the respective technologies (Davis 1989). The advancement of the TAM the UTAUT additionally included personality traits like age and gender as acceptance influencing aspects

(Venkatesh et al. 2003). But modern medical technology generated new demands: The rapid improvement and development in the branch of medical technology in the last decade transferred medical technology in an increased measure into the home environment. The combination of modern ICT and medical technology supported the so called electronic health sector (ehealth) which combined with Ambient Assisted Living (AAL) components supports the supply of old and frail people in the home environment (Calero Valdez et al. 2009; Calero Valdez et al. 2010). These new forms of healthcare are principally based on the Internet as a provider of data, information and communication. Unlike the parameters in the job context medical technologies and information reveal their own demands: Aspects like trust, privacy and security as well as gender, age and technical expertise are playing a central role (Ziefle & Röcker 2010; Ziefle & Schaar 2010; Ziefle & Wilkowska 2010; Schaar & Ziefle 2011; Wilkowska & Ziefle 2011; Ziefle & Schaar 2011).

2 MAIN FOCUS OF THE STUDY AND QUESTIONS ADDRESSED

The current study basically addresses five central questions:
(a) What information do persons want to disclose on health SNS?
(b) What are the most required functions?
(c) What are the most important characteristics for a health SNS?
(c) Which general associations dominate the attitude towards health SNS?
(d) Do users characteristics influence the attitude towards health SNS?

Generating an answer to these questions should be a first step for the definition of a general health social network approach.

3 METHODOLOGY

The main goal of this study was to find out first general user demands of health SNS. To reach a large number of participants and to take diversity into the group (potential) health SNS user, the questionnaire method was chosen. The questionnaire was delivered electronically. Before distributing the questionnaire it was revised by a sample of differently aged adults. Working on the final version of the questionnaire took about 20 minutes. In the following sections the design and sample of the study is presented: Section 3.1 presents the central variables of the study. Section 3.2 contains a description of the questionnaires' design. Section 3.3 presents the sample of the survey to work out the character and characteristics of the sample. An effort was made to keep the methodology similar to earlier research (Calero Valdez et al. 2012).

3.1 Variables

As *independent variables* we have chosen the participants age, gender and self-reported health status as well as field of work and expertise with social media. *Dependent variables* are an evaluation of information possibly provided via health SNS, evaluation of functions of health SNS as well as an assessment of central characteristics of such networks, as well as a pool of adjectives that could describe the character of a health SNS.

3.2 The Questionaire

The questionnaire used in this survey is divided into four subparts: (a) demographic data and health status, (b) SNS expertise, (c) evaluation of design criteria for health SNS, (d) adjective association about health SNS.

Demographic data and health status

To get information about the participants' characteristics they were asked to answer questions about their gender, age, field of work (technical or non-technical) and health status. The health status had to be evaluated on a five-point Likert scale. Answers had to be selected out of the following statements: *I am and feel completely healthy*; *I have colds regularly*; *My chronic disease influences my life slightly*; *My chronic disease influences my life to a great extent*; *I am gravely ill.*

Social network expertise

This section asks for experience with SNS and its modalities in the form of number of *contacts/friends within the network*, as well as *the place of usage* and the *number of regular interaction to the contacts/friends within and out of the network.*

Evaluations of design criteria of health social networks

To gather more information about design guidelines for health SNS Section 4 contains the evaluation of design criteria. In this context we asked for (a) *the profile information a potential user would disclose,* (b) *desired functions* as well as (c) *importance of special characteristics.*

(a) The profile information are including: *name*, *last name*, *date of birth*, *common interests*, *profile image*, *diseases*, *medical fields of interest*, *health relevant data.* According to the different information the participants had to evaluate the degree of visibility. The rating scale contained: visible for: *nobody*, *my contacts*, *everybody*.

(b) Out of the following functions the participants were asked to pick those they would like to have in a community: *common information about diseases posted by medical staff*, *common information about diseases posted by community members*, *individual forums on diseases and disease related problems*, *self-help forum*, *chat with medical staff*, *chat with other community members*, *messaging function*, *non-disease-related forums*, *interest and thematic search for other community members*, *common search func*tion *within the network*, *diary for health parameter*, *diary for daily support with e.g. calorie calculator*, *diary for personal well being.* In this context multiple answer were allowed.

(c) The evaluation of the networks characteristics were addressed by the following items: *data security, control of articles through medical staff, code of behaviour, no advertisement, special time slots for conversation with medical staff, attractive design, user-friendly design.* Each item had to be evaluated on a six-point Likert scale from 1 (very important) to 6 (very unimportant).

Adjective association

The last task of the questionnaire was a selection of positive and negative attributes that were taken from a former focus group discussion about SNS. These attributes were: *useful, useless, exciting, boring, secure, insecure, suitable for daily use, impractical, connecting, isolating, interesting, superfluous, valuable, simple, complicated, practical, untrustworthy.* The participants were instructed that three replies were possible.

3.3 The Sample

Demografic data

The questionnaire presented in Section 3.2 was answered by a total of 53 participants in an age range from 19 years of age up to 65 (M=34.3, SD=11.9). The sample reveals a gender distribution of 34% (N=18) male and 66% (N=35) female participants.

The educational level of the sample is quite high: Over 70% of the sample reported to have completed their university education (67.9%; N=36) or PhD/habilitation (5.7%; N=3). According to the character of the profession (technical vs. non-technical) we can reveal, that 41.5% (N=22) have a technical background and 57.4% (N=31) a non-technical.

Health status

In terms of the health status we can report that the majority of the sample stated to be completely healthy (73.6%; N=39). Further 13.2% (N=7) reported to suffer form colds frequently. 11.3% (N=6) are affected by a chronic disease, which has no strong influence on daily life. Only one participant is affected by a chronically disease that influences his life enormously.

Social network expertise

76.3% (N=29), of the of the sample confirmed to use a form of SNS. Regarding the question where they use social media we reveal that the majority uses social media at home (83.7%) as well as at work (55.8%) Followed by on the run on Smartphone or notebook (Figure 1). Most participants reported to use Facebook as their main SNS (69.7%).

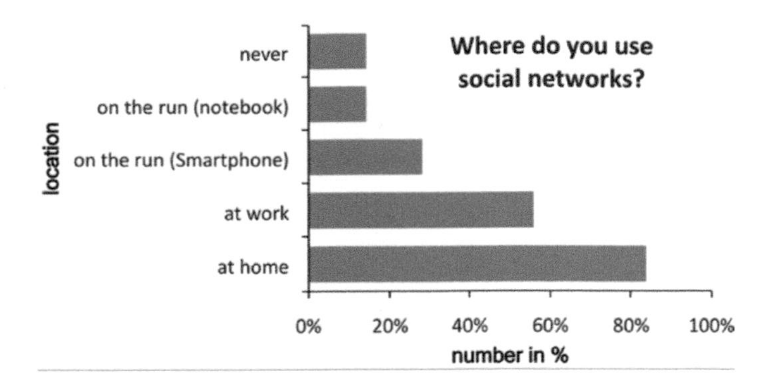

Figure 1 Location of social network usage (multiple answers possible)

In the context of interacting within the pool of SNS friends we can report that the majority of the sample only has regular contact to 10% of their friends list, within the SNS or out of the SNS.

Interactions of independent variables

Bivariate analysis of correlation reveals, that there was a strong effect of age on the usage frequency of social networks (ρ=.452 p=0.02 1-β=.913) as in that older users use Facebook less frequently. Gender in this sample has a strong effect on the field of work. A Chi-square test reveals, that females tend to work in non-technical jobs, where males work in technical jobs (χ^2 = 14.767, df=1, p<0.000 1-β=1).

Summing this section up, we can state that we have a healthy sample, which is distinguished, by a quite high level of technical expertise and SNS usage.

4 RESULTS

Results in this study have been analyzed with bivariate correlation analyses (Pearson's r), chi-square tests, t-Test and ANOVA with a significance level set at 5%. For ordinal-scaled variables Spearman's ρ was used for correlation analysis. Where possible Type-2-Error probability β is reported as power (1-β).

4.1 SNS profile information

The willingness to disclose personal information on a health SNS is relatively low. No single information item was mentioned more than 10 times (N=48) with a visibility to a public audience (see Figure 2). Customizable visibility changes this willingness drastically but only for certain information items. Both parts of the name, profile picture and date of birth are more likely to be shared if the audience is selectable by the user. Nonetheless most users are not willing to disclose medical information on a health SNS.

Almost no trait of user characteristics (age, gender, health status, field of work, social media usage) shows any significant differences in mean for desired disclosure of information (age: ANOVA $p>.313$; gender: t-Test $p>.405$, health status: ANOVA $p>.105$; field of work: $p>.189$, social media usage: ANOVA $p>.304$; all reported significance values are minima for the most significant item). The only item that shows differences in means is "diseases". People that work in a field of technical work are more likely to disclose disease data (t-Test: $T=2.071$ $df=46$ $p=.044$ $1-\beta=.90$). ANOVA analysis also showed that people that use Facebook more regularly are more likely to disclose their first name ($p<.033$, $1-\beta=.952$).

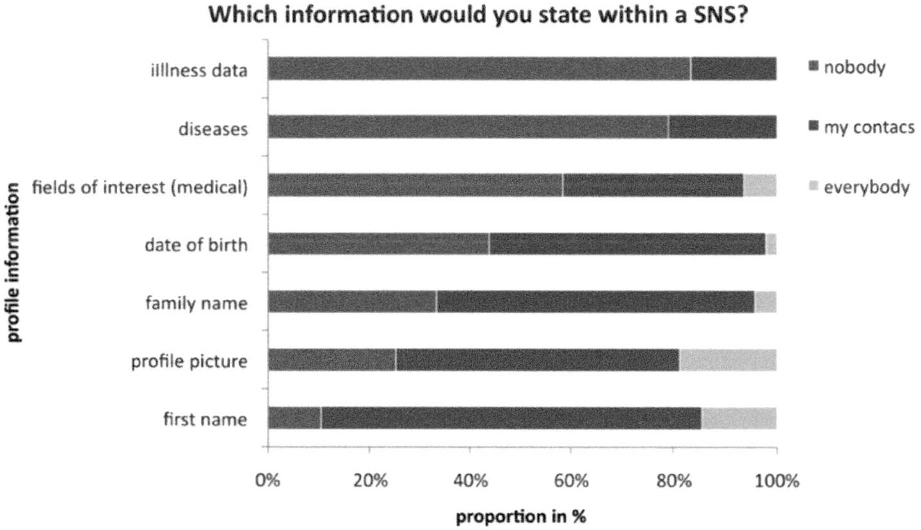

Figure 2 Willingness to disclose personal information to different audiences on a health SNS

4.2 Desired functions

The low willingness to publicly share health related data is also reflected in the desired functions for a potential health SNS. The three most important functions were "*general information about diseases through medical staff*", "*individual disease and special interest forums*" and "*chat with medical staff*" (see Figure 3). The least important functions were "*non-disease oriented forum*" (N=6), "*diary for reporting well being*" (N=8) and "*diary for medical data*" (N=11).

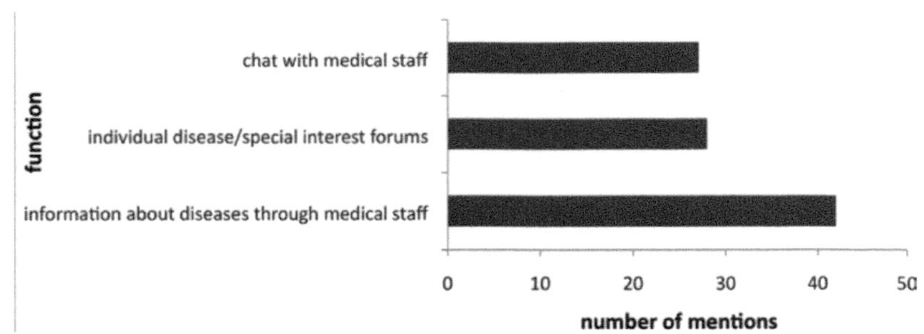

Figure 3 Most desired functions in a potential health SNS

Characteristics that a potential health SNS should incorporate are most importantly data security, medical accuracy and user-friendly design. Members of this sample rated data security as the most important characteristic for a potential health SNS (see Figure 4).

Both desired functions and community characteristics show almost no dependency on user characteristics (age, gender, health status, field of work, social media usage). Only the wish for a medical data diary is higher in the group of people that work in technical a job (t-Test $T=-4.269$ $df=51$ $p<.001$ $1-\beta=.999$) than for those within a non-technical job.

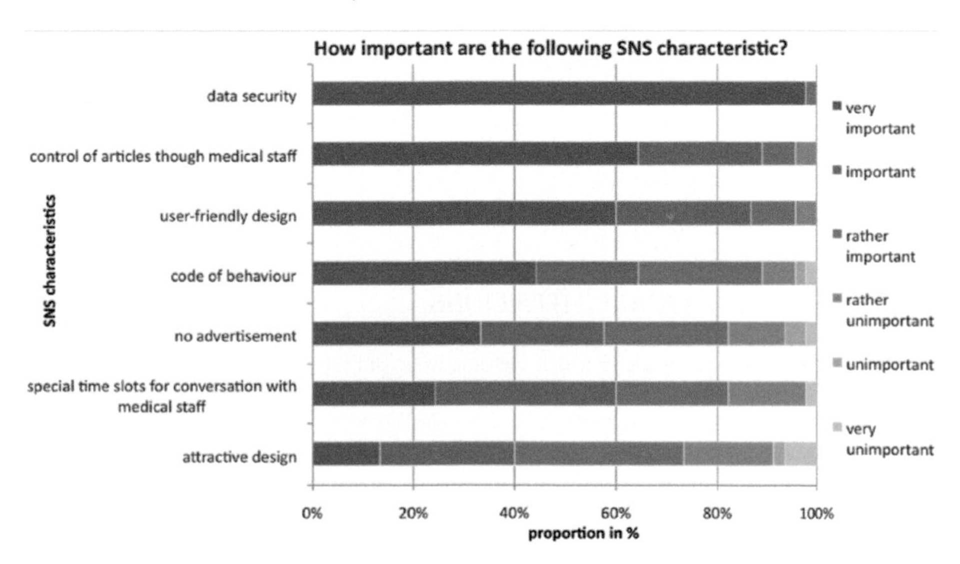

Figure 4 Importance of characteristics in a potential health SNS

4.3 Interactions of attributes

In order to understand what attributes participants associate with medical SNS nine attributes as well as their antonyms were presented within the survey. Participants could tick a box if they agreed that the said attribute would fit a description of a medical SNS. The attributes interesting, useful, practical and insecure were those that were mentioned most frequently (N = 20, 18, 16, 10). Out of 113 selected attributes 87 were of a positive connotation while only 26 had a negative connotation (see Figure 5). Given that each participant could only select maximal three items, 29 selected three items, nine selected two and eight selected only one item. All chi-square tests showed no significant interaction if corrected by Yates for all user characteristics.

Figure 5 Bubble graph of associated attributes. Larger circles mean that more participants associated this attribute with a medical SNS. Distance of circles relates to co-occurrence. If two attributes were selected by the same participant, they are closer together.

5 DISCUSSION AND LIMITATIONS

The results of this study suggest, that German users put high emphasis on data security. Even in the expectation that a health SNS would be useful, interesting and practical, data insecurity plays a dominant role in the attitude towards health SNS. In particular sharing or disclosing health related data or information is not desired. Nonetheless users like to consume professional guidance and benefit from other users sharing their information. This might reflect a typical German perspective, as in Germany data security is regarded as very important.

In order to avoid this conflict other forms of ehealth-technology could be implemented as support for disease stricken patients. Ambient assistant living or video conferencing with your personal doctor could benefit from this attitude.

The sample in this study is rather small and of high education. This might lead to a skewed view especially because all of the participants were of the "lurker" type and not contributing to any health SNS. In future research different ways of increasing trust should be examined. Especially standardized quality seals of data security or government control should be regarded.

The survey was constructed in a rather abstract fashion, as no concrete implementation of a health SNS or prototype was presented or familiar to the participants. This could have lead to filling information gaps with prejudices guided by horror scenarios. In many prior cases acceptance of SNS where higher after usage when the user could reap and feel the benefits of the service provided.

In this study we also assessed locus of control and technical expertise, but both did not yield any significant outcomes. This might also be due to the selective highly educated sample.

Maybe the time has not yet come for German health SNS. Interestingly users would try a service if friends or the personal doctor recommended the service. Health insurance companies should refrain from recommending usage, as they are disregarded as commercial advertising.

ACKNOWLEDGEMENTS

The authors would like to thank all participants of the survey for their support. We also thank Kathrin Pereira for her supportive performance.

This research was funded by the German Federal Ministry of Education and Research and the European Social Fund (ESF) under the funding code 01HH11045.

REFERENCES

Arning, K. & Ziefle, M., 2009. Different Perspectives on Technology Acceptance: The Role of Technology Type and Age. In *Proceedings of the 5th Symposium of the Workgroup Human-Computer Interaction and Usability Engineering of the Austrian Computer Society on HCI and Usability for e-Inclusion*. Linz, Austria: Springer-Verlag, pp. 20–41.

Belleghem Van, S., 2011. Social networks around the world 2010. *SlideShare*. Available at: http://www.slideshare.net/stevenvanbelleghem/social-networks-around-the-world-2010 [Accessed February 26, 2012].

Boyd, D.M. & Ellison, N.B., 2007. Social Network Sites: Definition, History, and Scholarship. *Journal of Computer-Mediated Communication*, 13(1), pp.210–230.

Calero Valdez, A. et al., 2009. Effects of Aging and Domain Knowledge on Usability in Small Screen Devices for Diabetes Patients. In A. Holzinger & K. Miesenberger, eds. *HCI and Usability for e-Inclusion*. Berlin, Heidelberg: Springer Berlin Heidelberg, pp. 366–386.

Calero Valdez, A. et al., 2010. Task performance in mobile and ambient interfaces. Does size matter for usability of electronic diabetes assistants? In *2010 International Conference on Information Society (i-Society)*. 2010 International Conference on Information Society (i-Society). IEEE, pp. 514–521.

Calero Valdez, A., Schaar, A.K. & Ziefle, M., 2012. State of the (net)work address Developing criteria for applying social networking to the work environment. *Work: A Journal of Prevention, Assessment and Rehabilitation*, 41(0), pp.3459–3467.

Davis, F.D., 1989. Perceived Usefulness, Perceived Ease of Use, and User Acceptance of Information Technology. *MIS Quarterly*, 13(3), pp.319–340.

Hartmann, M. et al., 2011. Web 2.0 im Gesundheitswesen – Ein Literature Review zur Aufarbeitung aktueller Forschungsergebnisse zu Health 2.0 Anwendungen. *Wirtschaftinformatik Proceedings 2011*. Available at: http://aisel.aisnet.org/wi2011/111.

Lahlou, S., Langheinrich, M. & Röcker, C., 2005. Privacy and trust issues with invisible computers. *Commun. ACM*, 48(3), pp.59–60.

Moturu, S.T., Liu, H. & Johnson, W.G., 2008. Trust evaluation in health information on the World Wide Web. *Conference Proceedings: ... Annual International Conference of the IEEE Engineering in Medicine and Biology Society. IEEE Engineering in Medicine and Biology Society. Conference*, 2008, pp.1525–1528.

Schaar, A.K. & Ziefle, M., 2011. What Determines Public Perceptions of Implantable Medical Technology: Insights into Cognitive and Affective Factors. In A. Holzinger & K.-M. Simonic, eds. *Information Quality in e-Health*. Berlin, Heidelberg: Springer Berlin Heidelberg, pp. 513–531.

Scheermesser, M. et al., 2008. User acceptance of pervasive computing in healthcare: Main findings of two case studies. In *Pervasive Computing Technologies for Healthcare, 2008. PervasiveHealth 2008. Second International Conference on Pervasive Computing Technologies for Healthcare*. Pervasive Computing Technologies for Healthcare, 2008. PervasiveHealth 2008. Second International Conference on Pervasive Computing Technologies for Healthcare. pp. 205–213.

Venkatesh, V. et al., 2003. User Acceptance of Information Technology: Toward a Unified View. , 27(3), pp.425–478.

Ward, R., 2006. Blogs and wikis. *Business Information Review*, 23(4), pp.235–240.

Wilkowska, W. & Ziefle, M., 2011. User diversity as a challenge for the integration of medical technology into future home environments. In *Human-Centred Design of eHealth Technologies. Concepts, Methods and Applications*. Hersehy, P.A.: IGI Global, pp. 95–126.

Ziefle, M. & Röcker, C., 2010. Acceptance of Pervasive Healthcare Systems: A comparison of different implementation concepts. In *4th ICST Conference on Pervasive Computing Technologies for Healthcare 2010*. ICST Conference on Pervasive Computing Technologies for Healthcare 2010.

Ziefle, M. & Schaar, A.K., 2011. Gender differences in acceptance and attitudes towards an invasive medical stent. , Volume 6(Issue 2).

Ziefle, M. & Schaar, A.K., 2010. Technical Expertise and Its Influence on the Acceptance of Future Medical Technologies: What Is Influencing What to Which Extent? In G. Leitner, M. Hitz, & A. Holzinger, eds. *HCI in Work and Learning, Life and Leisure*. Berlin, Heidelberg: Springer Berlin Heidelberg, pp. 513–529.

Ziefle, M. & Wilkowska, W., 2010. Technology acceptability for medical assistance. In *4th ICST Conference on Pervasive Computing Technologies for Healthcare, 2010 (CD Rom)*. Conference on Pervasive Computing Technologies for Healthcare.

CHAPTER 10

A New Collaborative Tool for Visually Understanding National Health Indicators

Songhua Xu, Brian Jewell, Chad Steed, Jack Schryver

Oak Ridge National Laboratory
Oak Ridge, TN, USA 37831
{xus1, jewellbc, steedca, schryverjc}@ornl.gov

ABSTRACT

We propose a new online collaborative tool for visually understanding national health indicators, which facilitates the full spectrum of investigation of indicators, from an overview of all the correlation coefficients between variables, to investigation of subsets of selected variables, and to individual data element analysis. This tool is publicly accessible at http://cda.ornl.gov/heat/heatmap.html. In this paper, we discuss the key issues regarding the interface design and implementation. We also illustrate how to use our interface for analyzing the health indicator dataset by showing some key system views. In the end, we introduce and discuss some ongoing research efforts extending this work.

Keywords: National health indicators warehouse, visual analytics, heatmap, scatterplot, online collaborative analysis platform

---

This manuscript has been authored by UT-Battelle, LLC, under Contract No. DE-AC05-00OR22725 with the U.S. Department of Energy.

1 INTRODUCTION

The National Center for Health Statistics develops the Health Indicators Warehouse (HIW) website, http://healthindicators.gov/, as a platform for publishing thousands of national health indicators and their values. Some indicators only provide national level data while others provide detailed individual Hospital Referral Regions (HRR), county and state level data for specific demographic groups of the population as characterized by people's age, race, and many additional sociodemographic factors. Some indicators are also dedicated to profiling patients of specific diseases. Overall, these indicators provide a rich source of data for informing the public about the general health care quality and population health conditions in a certain region. These data resources can also support health policy decision making and analysis on various levels, ranging from counties, states, to the federal administration.

Despite the promising value potentially offered by the warehouse, a major problem remains in effectively discovering knowledge from such a comprehensive data source. Some key challenging tasks include: 1) how to examine values of these indicators and comparatively study value differences across multiple states, counties, and HRRs; 2) how to understand value correlations between multiple indicators and detect regions in the country where such correlations behave abnormally; 3) how to understand health condition and healthcare quality, as characterized by multiple indicators, for the same region as well as these indicators' relative standing in the country.

Drawing answers to the above questions can effectively support health policy makers and legislators to change existing healthcare regulations and amend the healthcare laws for providing more affordable healthcare with better care outcome. Unfortunately, existing mainstream business intelligence tools failed to provide user friendly solutions for policy researchers and makers to freely and effectively explore answers for the above questions and many variants of these questions.

2 BACKGROUND AND RELATED WORK

Continuing technological advances have resulted in increasingly complex multivariate data sets which, in turn yield information overload when explored with conventional visual analytics techniques. The ability to collect, model, and store information is growing at a much faster rate than our ability to analyze it. However, the transformation of these vast volumes of data into actionable insights is critical in many domains, such as health care. Without proper techniques, analysts are forced to discard layers of information in order to fit the tools; therefore, new approaches are necessary to turn today's information deluge into opportunity.

One promising solution for this challenge lies in the continued development of techniques in the realm of intelligent user interfaces. The intelligent user interface adapts and learns from the user's interaction with the data to adjust levels of detail and highlight potentially significant associations among sets of interrelated variables. Like the related field of visual analytics, intelligent user interfaces combine the strengths of humans with those of machines. While methods from knowledge discovery, statistics, and machine learning drive automated analytics and augment the display, human capabilities to perceive, compare, and interpret strengthen the iterative process.

Over the years, there have been many approaches to the visual analysis of multivariate data (see Wong and Bergeron, 1997). However, the techniques employed in most operational systems are generally constrained to non-interactive, basic graphics using methods developed over a decade ago; and it is questionable whether these methods can cope the with complex data of today. For example, analysts often rely on simple scatter plots and histograms which require several separate plots or layered plots to study multiple variables in a data set. However, the use of separate plots is not an ideal approach in this type of analysis due to perceptual issues described by Healey et al. (2004) such as the extremely limited memory for information that can be gained from one glance to the next. These issues are illustrated through the so-called change blindness phenomenon (a perceptual issue described by Rensink (2002)) and they are exacerbated when searching for combinations of conditions.

One approach often used by statisticians to overcome this issue is to use the scatterplot matrix (SPLOM), which represents multiple adjacent scatterplots for all the variable comparisons in a single display with a matrix configuration (Wong and Bergeron, 1997); but the static SPLOM requires a large amount of screen space and forming multivariate associations is still challenging.

3 APPROACH

To address the limitations of using existing methods for analyzing the national health indicators, we designed and developed a novel collaborative tool for visually understanding values and relationships of indicators in HIW. In the current work, we have enhanced the traditional SPLOM by providing cues that guide and refine the analyst's exploration of the information space. This approach is akin to the concept of the scented widget described by Willett et al. (2007). Scented widgets are graphical user interface components that are augmented with an embedded visualization to enable efficient navigation in the information space of the data items. The concept arises from the information foraging theory described by Pirolli and Card (1999) which models human information gathering to the food foraging activities of animals. In this model, the concept of information scent is identified as the "user perception of the value, cost, or access path of information sources obtained by proximal cues" (Pirolli and Card, 1999).

Our approach is perhaps closest to that proposed by Friendly (2002), who subsumed his suggestions to visualization of correlation matrix data in color- and shape-coded formats under the name "corrgram." Their study introduced several interesting alternatives to the display of sign and magnitude of correlation coefficients, and novel visualization techniques for summarizing higher-order shape or trend information in scatterplots.

In the current work, the focus is on a particular set of health care indicators from the Health Indicators Warehouse (HIW). The HIW hosts a number of indicators for national, state, and community level statistics and serves as a data repository for the Department of Health and Human Services (HHS) Community Health Data Initiative. We sampled a collection of variables from this repository to provide an effective visual interface to explore and discover significant associations between health indicators. We focused our efforts on a set of CMS indicators from HIW that contain values at the HRR level, augmented by multiple CMS component cost variables downloaded from the Institute of Medicine website http://iom.edu/Activities/HealthServices/GeographicVariation/Data-Resources.aspx.

Our system, which is publicly accessible at http://cda.ornl.gov/heat/heatmap.html, facilitates the full spectrum of investigation of indicators, from an overview of all the correlation coefficients between variables, to investigation of subsets of selected variables, and to individual data element analysis. This capability is realized through a level-of-detail algorithm that includes more detail as the analyst zooms into the display. Initially, the analyst is given an overview of the variable correlations in the form of a heatmap visualization of the correlation matrix. Correlation mining is an important data mining technique due to its usefulness in identifying underlying dependencies between variables. The correlation mining process attempts to estimate the strength of relationships between pairs of variables to facilitate the prediction of one variable based on what is known about another. The linear relationship between two variables *X* and *Y* can be estimated using a single number, *r*, that is called the sample correlation coefficient (Walpole and Myers 1993). Our correlation matrix display uses the Pearson product-moment correlation coefficient to measure the correlation between two variables. Given a series of n measurements of *X* and *Y* written as x_i and y_i where $i = 1, 2, \ldots, n$, r is given by

$$r = \frac{n\sum x_i y_i - (\sum x_i)(\sum y_i)}{\sqrt{[n\sum x_i^2 - (\sum x_i)^2][n\sum y_i^2 - (\sum y_i)^2]}}.$$

r = 1.0 RGB(165, 15, 21)
r = 0.0 RGB(247, 261,255)
r = -1.0 RGB(8, 81, 156)

For each pair of variables in the display, the system computes *r* which results in a correlation matrix. The correlation matrix is a symmetric *n* by *n* matrix where each *i, j* element is equal to the value of *r* between the *i* and *j* variables. The intersection of the variables is represented graphically as a color-filled square.

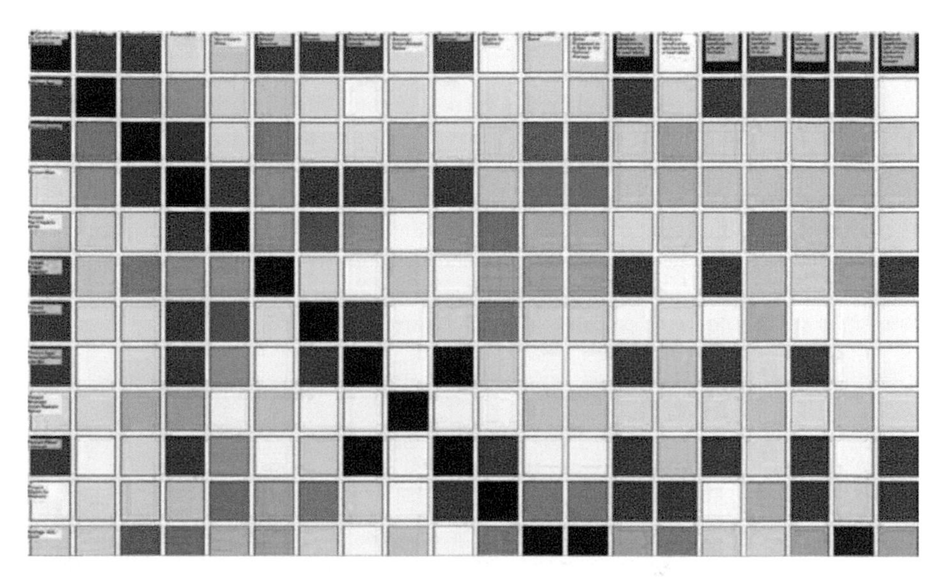

Fig. 1 Top-level overview on pairwise health indicator correlations.

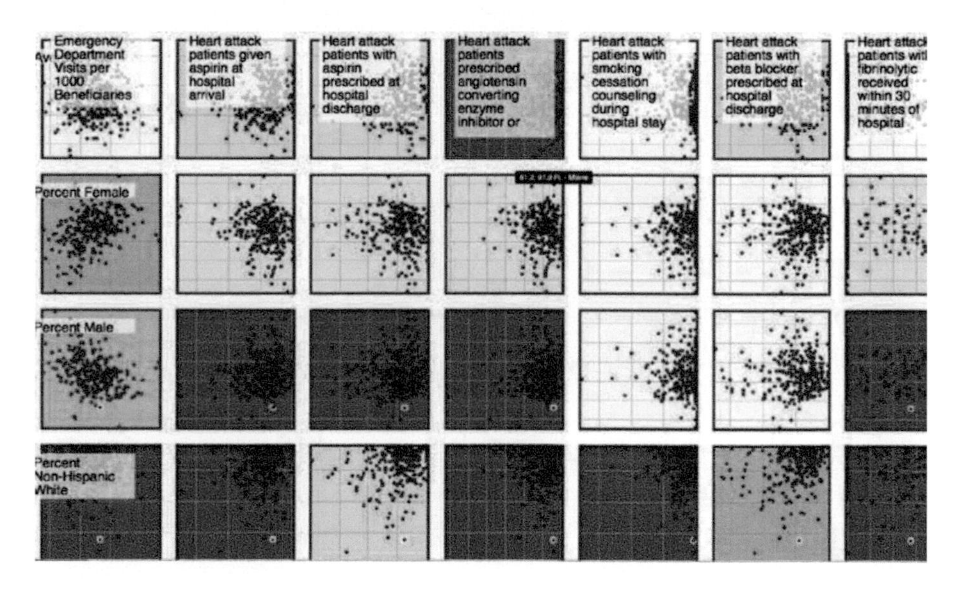

Fig. 2 Zoomed-in view of county-level health indicator correlations.

The colors used to fill the squares are calculated based on the value of r between the variables using the color scale shown on the previous page. This color scale results in shades of blue for negative correlations and red for positive correlations. The color scale maps the saturation of the color to the strength of r so that the strongest correlations are displayed more prominently. Characteristic of correlation matrices, the correlation of a variable with itself is always a perfect positive correlation (1) and the diagonal of the matrix is therefore shown as a series of squares with the highest saturated red.

From the initial correlation matrix, the analyst can zoom into the display using the mouse scroll gesture to focus on a particular set of variables. As the analyst zooms into the display, the number of variables shown decreases. When the number of variables across the row or column dimension of the visible display is reduced to 6 or fewer variables, the display fades in the scatterplots for each variable intersection. That is, the square block for the variable intersections maintains the background fill that is associated with the correlation strength, and the individual points that contribute to the correlation measure are revealed.

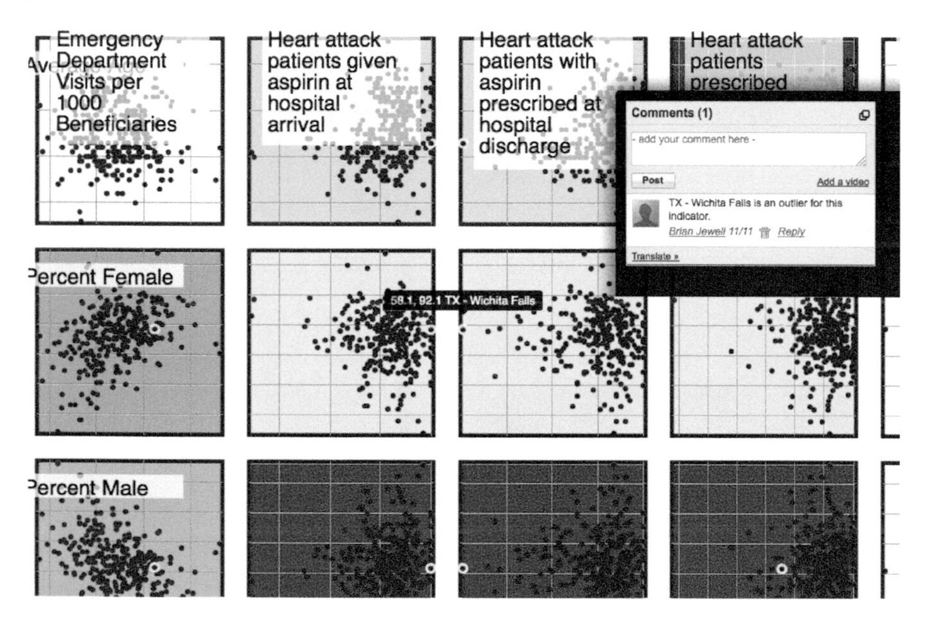

Fig. 3 Sharing analysis results and findings using the comment and tagging feature built into our tool.

When the scatterplots are shown, the user can hover the mouse over individual points to see the associated values and geographic place names for the item. This information is shown as a tooltip in the display. Furthermore, subtle grid lines are shown in the scatterplots to provide reference points for comparing distributions across the SPLOM rows and columns.

Overall, our tool offers the following views to explore indicator values and their relationships: 1) a high-level graphic overview on pairwise correlations between indicator values in the form of a heatmap (Fig. 1), 2) a zoomed-in view of such correlations, where both the national average correlation strength and that of a region's local value are visually presented (Fig. 2), also upon selection of a certain region, a highlighted view on the region's relative standing in the country as characterized by multiple indicators, 3) a tagging view where an analyst can document thoughts and discoveries over a particular indicator for a particular region; since we provide the tool as an online application, a group of analysts can also collaboratively exchange their analytic findings by using our tool as a new collaboration platform for conducting social analysis (Fig. 3).

4 KEY IMPLEMENTATION ISSUES

Given the desire for the visualization to be accessible in a common format by a wide audience for collaboration a web platform was chosen. With further considerations of browser compatibility and interactivity, we chose to implement our application using the Data-Driven Documents (D3) library (Bostock et al., 2011). D3 is a library that allows for the direct manipulation of the standard document object model (DOM). Using D3 we were able to bind data directly to SVG elements in order to construct a web-based visualization with all the desired intractability.

The code in Fig. 4 comes directly from our implementation for creating the high-level heatmap view, and is included to demonstrate how D3 was used. On line 1 and 2, a DOM element is selected and a SVG element is appended. D3 then allows for the setting of element attributes (attr, style, property, html, text) through operators, which wrap the W3C DOM API. When data is bound a hierarchical structure of host elements is created for us to accommodate each array element. Thus, when our field array is bound on line 12, SVG elements for each column of our heatmap are created, and on line 17 and 18 the columns are then selected. Finally, row elements are created for each field in each column and data is appended appropriately creating a DOM based representation of our heatmap. We then append visual svg:rect elements to this representation colored using the data which is bound to their parent elements producing the heatmap which the user sees.

In order to optimize performance, the lower level view elements aren't added to the scenegraph until the users view is limited to a roughly 6x6 matrix. Addition of new elements is then limited to the users viewable area (viewbox). This means that the performance of the visualization is primarily bound to the *O(m*n)* (or *O(n^2)* for a symmetric representation) complexity of the matrix used for the high-level heatmap view. This does in fact affect the performance scaling of our application greatly. Complicating the further optimization is two factors 1) creating and removing DOM elements is expensive and 2) we do not wish to restrict the users ability to zoom/pan across the entire heatmap. A future implementation may be improved by rasterizing this high level view on the server side to increase scalability.

```
01    var svg = d3.select("#chart")
02    .append("svg:svg")
03    .attr("width", size * health.fields.length)
04    .attr("height", size * health.fields.length)
05    .attr("class", "BlRd")
```

```
06        .attr("pointer-events", "all")
07        .call(d3.behavior.zoom()
08        .on("zoom", redraw)).append("svg:g");
09
10        // One column per field.
11        var column = svg.selectAll("g")
12        .data(health.fields)
13        .enter().append("svg:a")
14        .attr("transform", function(d, i) { return "translate(" + i * size +
    ",0)"; });
15
16        // One row per field.
17        var row = column.selectAll("g")
18        .data(cross(health.fields))
19        .enter().append("svg:g")
20        .attr("x", padding / 2)
21        .attr("y", padding / 2)
22        .attr("width", size - padding)
23        .attr("height", size - padding)
24    .attr("transform", function(d, i) { return "translate(0," + i * size + ")";
    });
25
26        row.append("svg:rect")
27        .attr("x", padding / 2)
28        .attr("y", padding / 2)
29        .attr("width", size - padding)
30        .attr("height", size - padding)
31        .style("stroke", rectColor)
```

Fig. 4 – Implementation of high-level heatmap view in d3.

5 DISCUSSION AND FUTURE WORK

In this project, we introduce a new platform for visually analyzing national health indicators. In contrast to the traditional visual analysis interfaces, our new platform offers a mixed heatmap and scatterplot view, allowing end users to hierarchically examine the correlations between multiple indicators, both on the national level and on finer geographical resolutions. Such a multi-resolution view enables policy analysts to comprehensively understand the healthcare performance and cost of a specific region in the context of the national average performance as well as that of its peer locations. Another unique function of our interface is that it allows analysts to share their discoveries through spatially anchoring their comments inside our visual interface onto individual geographic regions. By setting one's comments public, the analyst can invite other peer users in the system to participate in his/her analysis efforts over the region. With such an online comment and discussion forum feature, our visual analytics platform also serves as an online collaborative platform for uncovering hidden, underlying relationships embedded in the national health indicator warehouse through engaging social intelligence in the cloud.

Several conceptual extensions of the tool are being considered for future development. First, we conceive the analyst workflow as proceeding from the highest levels of abstraction to drill-downs into more limited areas of interest. One simple but important option to aid directed drill-down is to permit the analyst/user to select a smaller group of indicators for viewing in a smaller visual area; for example, clicking on multiple rows or columns generates a more compact matrix containing only those indicators of sufficient immediate interest.

Second, we desire to leverage human capability to detect complex color patterns within a visual matrix. The basic idea is to reorder indicators in the basis of similarity with respect to the vector of correlations. Reordering can lead to emergence of color patterns in the matrix and identification of latent factors in the form of variable clusters. Friendly (2000) illustrated the use of reordering techniques for baseball and automobile data; our approach is to use hierarchical clustering as a basis for reordering rows and columns. If the initial order is random, a clustered heatmap usually has the effect of transforming a random color matrix into a highly patterned diagram that illustrates regions of high negative and positive correlation. Although the clustered heatmap has been used to great effect in profiling gene expression data from micro-arrays (e.g., Rajaram and Oono, 2010), we are not aware of any applications to visualization of health care indicators.

We note that the correlation/scatterplot matrix is symmetric, i.e., half the information is redundant. In addition, information displayed in basic scatterplots is exclusively bivariate. In order to utilize available space more efficiently we can extend the capability of the tool to explore relationships between two indicators while controlling for the effects of one or more additional indicators. Mediating factors are commonly required for adequate interpretation of aggregated multivariate health care data. This kind of information can be represented in partial correlation coefficients and partial regression plots. A promising approach introduced by Davison and Sardy (2000) is the partial scatterplot matrix. The authors introduced a mixed partial scatterplot matrix that displays univariate histograms along the diagonal (which typically contains no quantitative information in a conventional scatterplot matrix), and partial regression residuals after adjusting for all other variables along the lower panel.

Finally, although the correlation coefficient is useful for summarizing the dependency between two variables, it is primarily a linear index. There are also many possible nonlinear relationships that can be very informative from a data mining perspective. Examples of noteworthy nonlinear relationships include, for example, the inverted U-function, exponential or logarithmic functions, sinusoidal functions, non-coexistence relationships, and composite relationships such as piecewise linear functions. The latter can arise when two or more distinct groups independently cluster in a scatterplot. Recent advances in data mining are making the goal of characterizing nonlinear relationships increasingly feasible with current computer architectures. For example, the maximal information criterion and related family of MINE statistics for nonlinear characterization (Reshef et al., 2011) are computationally expensive but scalable to visualizations of HIW-sized datasets like those described in this paper. Algorithms of this type can complement more conventional nonlinear visualization techniques such as locally weighted regression (Cleveland and Devlin 1988). We are exploring methods of displaying nonlinear relationship information in readily consumable forms within the framework of a scatterplot matrix.

ACKNOWLEDGMENTS

This work is sponsored by U. S. Centers for Medicare & Medicaid Services (CMS). Songhua Xu performed this research as a Eugene P. Wigner Fellow and staff member at the Oak Ridge National Laboratory.

REFERENCES

Bostock, M., V. Ogievetsky, and J. Heer. 2011. D3: Data-Driven Documents, IEEE Transactions on Visualization and Computer Graphics (Proc. InfoVis):2301-2309.

Cleveland, W. S., and S. J. Devlin. 1988. Locally weighted regression: an approach to regression analysis by local fitting. Journal of the American Statistical Association 83(403): 596-610.

Davison, A. C., and S. Sardy. 2000. The partial scatterplot matrix. Journal of Computational and Graphical Statistics, 9(4): 750-758.

Friendly, M. 2002. Corrgrams: exploratory displays for correlation matrices. The American Statistician 56(4): 316-324.

Healey, C. G., L. Tateosian, J. T. Enns, and M. Remple. 2004. Perceptually-based brush strokes for nonphotorealistic visualization. ACM Transactions on Graphics. 23(1):64–96.

Pirolli, P. and S. K. Card. 1999. Information foraging. Psychological Review 106(4):643–675.

Rensink, R. A. 2002. Change detection. Annual Review of Psychology 53:245–577.

Reshef, D. N., Y. A. Reshef, H. K. Finucane, S. R. Grossman, G. McVean, P. J. Turnbaugh, E. S. Lander, M. Mitzenmacher, P. C. Sabeti. 2011. Detecting novel associations in large data sets. Science, 334: 1518-1524.

Rajaram, S. and Y. Oono. 2010. NeatMap - non-clustering heat map alternatives in R. BMC Bioinformatics, 11:45.

Wong, P. C. and R. D. Bergeron. 1997. 30 years of multidimensional multivariate visualization. In: Scientific Visualization - Overviews, Methodologies, and Techniques, IEEE Computer Society Press, pp 3–33.

Walpole, R. E. and R. H. Myers. 1993. Probability and Statistics for Engineers and Scientists, 5th ed., Prentice Hall, Englewood Cliffs, New Jersey.

Willett, W., J. Heer, and M. Agrawala. 2007. Scented widgets: Improving navigation cues with embedded visualizations. IEEE Transactions on Visualization and Computer Graphics 13(6):1129–1136.

CHAPTER 11

Sociotechnical Considerations in the Delivery of Care to the Frail Elderly at Home

Ken Eason[1] *and Patrick Waterson*[2]

1The Bayswater Institute, 9 Orme Court, London, W2 4RL UK
k.d.eason@lboro.ac.uk
2 Loughborough Design School, Loughborough University, Leics, LE11 3TU, UK
p.waterson@lboro.ac.uk

ABSTRACT

A case study is reported of a local health community that has created a Frail Elderly Pathway with the aim of treating elderly patients in their own homes who might otherwise be admitted to hospital. The development of the Frail Elderly Pathway is a sociotechnical systems challenge that requires a multi-strand programme involving process, organisation and technical developments. The experience of delivering the pathway is reported and demonstrates the difficulty of creating temporary virtual acute care teams with members from different agencies. The degree to which integrated care can be delivered is seen as dependent upon a variety of electronic health systems being in place that enable real-time sharing of patient information. Teleheath equipment has been purchased as part of this development but it is not widely used. The paper concludes that, in an immature service, the practices needed to make rapid and appropriate use of telehealth facilities are not yet in place.

Keywords: Telehealth, frail elderly, sociotechnical systems, virtual wards, e-health systems

1. INTRODUCTION

In the UK as in other countries the treatment of the elderly, who may suffer from multiple chronic conditions, is a national priority. The elderly are often ‘frequent flyers’ in the Accident and Emergency Departments of Hospitals when they suffer falls or other emergencies. It is generally agreed that if they could be treated at home or elsewhere in the community whenever possible it would be better for both the patient and for the health service (Dept. of Health 2008). This paper reports a major effort in a metropolitan borough in the West Midlands of England to implement a Frail Elderly Pathway (FEP): a healthcare process that intervenes when an elderly person in crisis might be admitted to hospital and develops for them an acute care programme that can be delivered to their own home. The aim is to spare the patient the trauma of hospital admission whenever possible and to save the health service the additional costs associated with hospital care. The paper describes the sociotechnical system that is being created to deliver this healthcare pathway and the role that telehealth is playing in this process.

2. AN OBVIOUS APPLICATION OF TELEHEALTH?

Caring for frail elderly people in their own homes who are in crisis means looking after people whose condition might be unstable and changeable. In a hospital setting they could be constantly monitored and an array of specialist services would be near at hand if they were needed. In the creation of the FEP the vision was that the elderly person would be in a ‘virtual ward’ with a similar array of services available if required. Monitoring the condition of the patient on a regular basis would be necessary and the FEP was seen to be an obvious application for telehealth equipment. The aim was to install it in the patient’s home in order that results of tests could be sent to a control centre that could alert medical and other services when they were required. This would be a short-term application of telehealth and it would be removed when the patient no longer needed acute care. It is a different kind of application to the more common provision of telehealth for people with long-term conditions. We know from these applications (McLean et al 2011) that success is dependent on careful matching of the technology to the patient and the context. One question for this research was how difficult it would be to make successful use of telehealth in this short-term acute care setting.

3. THE RESEARCH METHODOLOGY

This case study was undertaken as part of a broader programme to examine the extent to which e-health systems were supporting integrated care in a range of healthcare pathways: the EPICOg project (Electronic Patient Information Crossing Organisational Boundaries) (Eason et al 2011). The e-health systems included any that involved the processing and storing of electronic patient information. The aims of the research programme were:

- to develop a sociotechnical systems map of each pathway that depicted the process of the pathway, the healthcare agencies that delivered each part of the pathway and the e-health systems that provided electronic patient information.
- to study the experience of healthcare staff of the use of e-health systems in order to understand the contribution the systems were making to integrated care.
- to examine the development processes by which healthcare pathway systems were created, including process design, organisational change and technical system development.

The research was conducted in two local health communities in England and studied nine healthcare pathways. The FEP case study was conducted over a two-year period, involved interviews with over 30 healthcare professionals from a range of agencies involved in the pathway and included observational studies of several strands of technical development delivering e-health systems to the pathway.

4. THE SOCIOTECHNICAL SYSTEMS CHALLENGE OF THE FRAIL ELDERLY PATHWAY

Figure 1 below is a simplified sociotechnical mapping of the vision for the Frail Elderly Pathway. The definition of the frail elderly is problematic and there is no accepted national definition in the UK (Sternberg et al 2011). In this local health community they are defined as persons over 75 years of age with multiple conditions. Pragmatically in the pathway they are people over 75 who are in a crisis who can be successfully treated at home rather than being admitted to hospital. If an elderly patient has a crisis the rapid assessment team of the pathway have a short time to assess whether they can put in place an intensive care process that can treat the patient safely in their own home. If this is possible an acute care team is created to deliver an intensive care plan for a period of up to 10 days. When the patient has recovered sufficiently there is a 'stepping down' to a post-acute team who, in time, discharge the patient back to their General Practitioner and any other health and social care arrangements that are appropriate for more long term support.

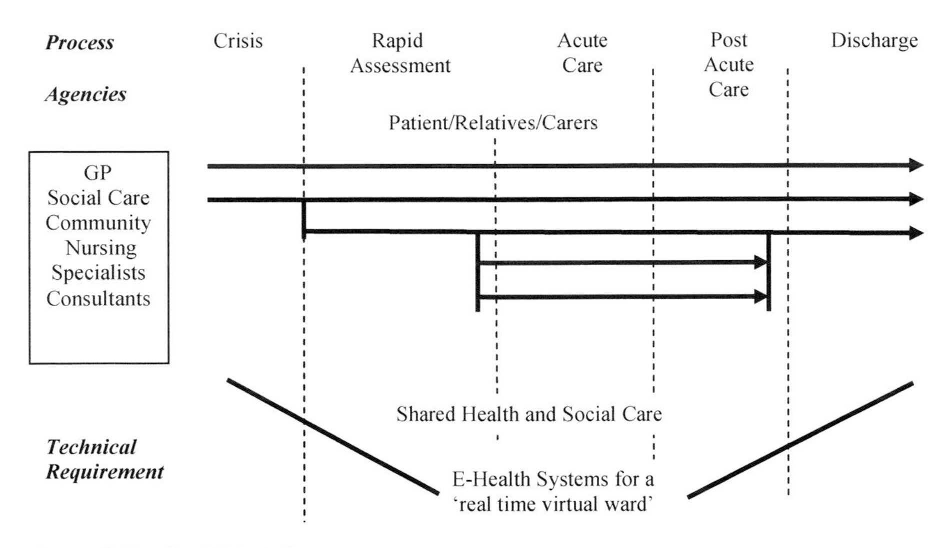

Figure 1: The Frail Elderly Pathway

This pathway is an example of shared care, in which many agencies, from both health care and social care, need to co-operate closely to ensure integrated care is delivered to the patient. Many of the agencies are based in different parts of the local community and in delivering care their staff spend most of their days visiting patients. The team is therefore for most of the time a 'virtual' team who need access to good quality, up-to-date information about the condition of the patient and who is providing what services to the patient if care is to be effectively co-ordinated.

In the plan for the pathway a comprehensive electronic system was envisaged that would ensure all members of the care team had this patient information available to them. At the start of the process there was an existing community nursing team who cared for less sick elderly patients in their own homes. Each of the existing agencies had its own electronic patient administration system (PAS) and most sharing of information between agencies was accomplished through the paper-based SAP (Single Assessment Process) system. This was a collection of agreed forms that each member of the care team completed after a visit to the patient. Each member of staff took a copy of their form back to their office but the complete set was stored at the patient's home so that each visiting member of the team could know what services other members of the team had delivered to the patient. The SAP documentation was also available for the patient, their relatives and carers to consult and add to as they wished.

Because it would be supporting unstable patients in the community a comprehensive, electronically based patient information system was considered to be essential if real-time co-ordination of care by the acute care team was to be achieved. The technical developments envisioned were:

- A 'virtual ward' information system that contained an up-to-date summary of information about each patient on the pathway. The local health community had already developed an electronic portal system that permitted a wide range of users in different agencies to view the patient information held by different agencies and the plan was to create a module in this system for the patients on the Frail Elderly Pathway that would be accessible to all the agencies.
- An e-SAP system. The paper-based SAP system provided the detail of the care delivered to each patient and the plan was to create an electronic version which each visitor to the patient would complete and which would be available to all members of the virtual team not just in the patient's home but wherever they were.
- Shared health and social care information. Whilst health information could be shared across health agencies it was not easy to share information with social services who were responsible for many related services to the patient. The plan was to find ways of providing access to the 'virtual ward' for members of social services and similarly to give healthcare staff access to the relevant records held by social services.
- Laptops for the healthcare team.Before the pathway was implemented the community staff that visited patients only had access to electronic patient information when they were in their offices and only had access to the comprehensive SAP information when they visited the patient. In order that they could access and input new information to e-SAP and to the virtual ward the plan was to issue each team members with a laptop.
- Telehealth equipment. In order that regular information about the condition of the patient could be provided to team members it was planned to install telehealth equipment in the homes of the more unstable patients. The specification for the equipment was that it would be extremely portable so that it could be installed and removed very easily and quickly and that it provided for the sending of a limited range of information, e.g. temperature and blood pressure results, to the pathway control centre. No automatic monitoring was envisaged; it was believed that during the acute phase many visits to the patient would be made by many agencies and, together with help from carers and relatives, it would be possible to maintain a schedule of frequent tests and reports. The aim was that the control centre would alert medical staff immediately results were outside expected limits and that the medical staff would intervene as appropriate.

5. PROGRESS AND CHALLENGES

A multi-strand programme of development work led to the Frail Elderly Pathway (FEP) being implemented in April 2010 and there are currently over 100 patients on the pathway at any one time. The implementation was undertaken before the full system was in place because there was an urgent need to begin showing results in this national priority area. The research team followed the progress of the various strands of development and interviewed the staff of the different agencies about their experiences in delivering the services of the pathway.

The development involved the creation of new processes and many organisational changes. New processes were required, for example, to specify how the rapid assessment process would work alongside the ambulance service and the Accident and Emergency Department that would otherwise be admitting the elderly person to hospital. Processes were also established for each General Practitioner to identify elderly patients at risk so that pre-emptive work could be done to bring them onto the pathway before a major crisis if possible. Since the pathway represented an important way of re-balancing services offered in the community rather than in hospital, monitoring procedures were also put in place to facilitate management judgments of whether the pathway was succeeding in reducing unnecessary admissions to hospital.

The new pathway involved organisational changes for many agencies. The existing community nursing team was strengthened with more specialists in, for example, physiotherapy so they were able to deliver a wider array of services and the service hours were extended to provide 7-day cover. The management was strengthened with the appointment of two co-ordinators one for patients being re-directed from hospital admissions and another for admissions from the community. A new administrative control centre was established and a rapid assessment team created.

Many of the technical developments lagged behind the process and organisational changes. The virtual ward had been created in the portal system and progress had been made in giving Social Services access to this system. The co-ordinators of the FEP had also been given access to information about patients on the pathway held in the Social Services system. Trials had been undertaken of a laptop for use by the rapid assessment team but progress with e-SAP was slow. A limited number of telehealth kits had been purchased but they had only been used in a few instances.

The staff delivering the pathway felt they had made a lot of progress but that there was much still to do to provide a safe and well co-ordinated service for their vulnerable patients. They were managing to look after the patients on the pathway although in many cases they were in the acute phase longer than expected. There was concern that some patients were being referred to the pathway for the wrong reasons, e.g. to provide relatives with respite care, and that this could deprive patients in real need of the service of the opportunity of being cared for in their own homes. There was also concern to get the accurate performance measures for the pathway because the figures were not showing an overall reduction in elderly patients being admitted to hospital. Planned admissions were also being counted in this measure when the pathway could only help avoid some unplanned admissions. The scale and speed of the service had increased and many operational staff were worried that they did not have access to patient information sufficiently quickly to co-ordinate care. They gave examples of cases where a visit had been missed because of an assumption that another agency was undertaking the task. For some staff the need for access to real-time, electronic patient information wherever they were was growing more critical as the scale and speed of the pathway increased. Staff were very much aware that they were devoting a lot of time to entering information into several different systems, for example, into the paper-based SAP documents and then into the electronic system of their own agency when they returned to their office. Any help in reducing double data entry requirements would help to provide more time for patient care.

The relative lack of use of the telehealth equipment was not a big issue for most staff. In most instances the condition of the patients they were treating could be monitored effectively by the members of the acute care team during their regular visits. The problems of deciding which patients could benefit from telehealth and co-ordinating its delivery and assembly in a short time together with training the people available (members of the healthcare team, carers of the patient etc) in its use usually meant that communicating test results was undertaken by traditional means even if it was slower. For the telehealth service to be effective they also needed to be able to get test results from the control centre to medical staff quickly and there was still work to do on that part of the system. Some of the managers of the pathway were also concerned that, once telehealth equipment was installed, patients and their relatives might appreciate the security of having a link to the control centre and may not want it removed when the patient came out of the acute phase. With limited sets of equipment available, the managers had to avoid this kind of outcome. As a result of these factors there had to be a very clear reason before the telehealth equipment was installed. There was a belief that, as the pathway matured and more of the other technical systems were in place, more use would be made of the telehealth facilities but that it would still only be useful in a minority of cases.

6. DISCUSSION.

Three features of this development stand out. First, in order to establish the pathway a multi-strand, multi-agency development programme had to undertaken. The total system needed to deliver the pathway is a socio-technical system of some complexity and changes in the pathway process, the organisation and the technology had to be made in a co-ordinated way. Second, the pathway began operation without the complete system being in place and it has continued to evolve slowly over the period it has been in operation. The pathway began operation when most of the organisational resources were in place but many of the planned technical developments were still on the drawing board. As a result, as the scale and speed of service delivery has increased staff who are still managing with existing ways of sharing information are feeling less able to cope with their workload.

The third feature of this development is a realization on the part of many involved of the sociotechnical systems challenge they have set themselves in creating the pathway. Not only are they seeking to intervene at a critical time when a patient may be taken to hospital but they then have to create a short-term temporary virtual team from a multi-disciplinary, multi-agency pool of staff that will need to co-ordinate their care of vulnerable person for a short time in what may be quickly changing circumstances. This is a difficult context in which to create effective teamwork and the need to up-grade the information support for this teamwork was widely recognized. Starting the pathway without the planned electronic systems in place meant an effective sociotechnical system was not available and the staff were trying to deliver a new service handicapped by old methods of sharing information. It is sometimes reported that front-line staff in organisations resist the use of new technology and wish to continue using tried and trusted paper-based methods. There was no indication of this attitude in this case. On the contrary there was a clamour for laptops, for e-SAP and the virtual ward because staff appreciated they needed rapid access to information to provide a co-ordinated service.

It is striking that, in this context, there was no clamour for the wide application of telehealth equipment even though it was available. Neither the patients, their carers and relatives or the health and social care staff felt the need to push for more telehealth equipment to be installed even though it had been purchased to provide for remote monitoring of patients who needed medical care in just such a setting. We can adduce a number of factors to account for this relative lack of enthusiasm. First, the condition of the patients meant it was only relevant in a number of cases and, even when it could be considered appropriate, the regularity of attendance of members of the acute care team meant sufficient monitoring and reporting could often be done by traditional means. Against the perceived need also needs to be set the perceived difficulty of getting the technology into effective use: the ease of use of the technology in this context was very significant. To be effective it needed to be installed quickly into whatever were the conditions of the patient's home, relevant people had to be recruited and trained to make tests and report the results and procedures had to be in place for the control centre to report abnormal results to the relevant members of medical staff. The practices and procedures of the FEP were gradually maturing but, at the time the research was undertaken, there had been little effort put into developing the practices around the use of the telehealth equipment. In common with some of the other technical developments the procedures for using technical potential had not been developed because the pressure to deliver care to patients was taking priority. It might be that, in time as the pathway becomes more established, the procedures around the use of the telehealth equipment will also become established so that acute care teams will know exactly when and how to install it to good effect.

This case is one example of a situation where telehealth equipment might have value. Unlike the situation where it might be used to help a person with a long-term condition over an extended period, this is a case of using it quickly for a short period. In these circumstances having a mature healthcare delivery sociotechnical system in place that is practiced at making appropriate and effective use of the technology is essential.

7. CONCLUSIONS

The goal of creating a community service that can provide healthcare support to elderly people in their own homes proves to be a major sociotechnical systems challenge. Ensuring that a virtual team can have good access to patient data is a major need and led in this case to the development of a variety of electronic health systems. On this evidence they have to be designed to fit closely with the procedures and practices by which healthcare is delivered if they are to be effective. This is true in particular for the use of telehealth facilities. Wyatt (2012) notes that “the fundamental telehealth fallacy is that technology alone is the solution“. This case study makes the point very clearly. In order to care for vulnerable elderly people in their own homes a complex sociotechnical system has to be put in place that is multi-disciplinary and multi-agency and which can quickly create acute care teams that can operate virtually to deliver what might be a rapidly changing care plan to patients in their own homes. Whilst it is easy to argue that telehealth equipment can play a valuable role in this context, it cannot do so without the well-practiced support of the acute care teams. At present, in this example, the teams are busy developing other aspects of their healthcare delivery and the use of telehealth equipment is of low priority. It may be sometime before they are in a position to realize the full potential of the technology for their patients.

ACKNOWLEDGEMENTS

We gratefully acknowledge the contributions to this project of our co-workers Professor Mike Dent, Dr. Dylan Tutt, Dr. Andrew Thornett and Mr. Phil Hurd. The funding for the EPICOg project was provided by the NIHR Service Delivery and Organisation programme under grant number 08/1803/226. The views and opinions expressed herein are those of the authors and do not necessarily reflect those of the NIHR SDO programme or the Department of Health. We are grateful also for the co-operation and help we have been given by the staff of the Walsall and Northamptonshire Local Health Communities.

REFERENCES

Department of Health (2008) *Delivering Care Closer to Home: Meeting the Challenge* 8th July 2008, London, Department of Health, London

Eason, K., Dent, M., Waterson, P., Tutt, D., Hurd, P., Thornett, A. (2011) *Getting the benefit from electronic patient information that crosses organisational boundaries.* Final report. NIHR Service Delivery and Organisation Programme, Department of Health, London

McLean S., Protti D. and Sheikh A. Telehealthcare for long-term conditions (2011) *BMJ* 342:374-378

Sternberg S A, Wershof Schwartz A, Karunananthan S, Bergman H, Mark Clarfield A. (2011) The identification of frailty: a systematic literature review. *J Am Geriatr Soc.* 59(11):2129-213

Wyatt J. (2012) More work needed on telehealth. Guardian Professional Online 6 January - Healthcare network http://www.guardian.co.uk/healthcare-network/2012/jan/06/more-work-needed-telehealth?INTCMP=SRCH

CHAPTER 12

Explorative Web Development for Using ICF as a Patient Classification Tool with the Rare Disease of ALS

Romy Elze, Christian Zinke, Klaus-Peter Fähnrich

University of Leipzig,
Leipzig, Germany
elze@informatik.uni-leipzig.de

ABSTRACT

The role of standardized, practical patient classifications in medicine is becoming increasingly important. This applies not only to billing systems but also increasingly to the standardized description of the clinical case for treatment purposes. The specific circumstances that play a role in the treatment of ALS patients (patients with Amyotrophic Lateral Sclerosis) require a broad and multidimensional description. Analyses to find an appropriate classification for this purpose have preceded the development work. The International Classification of Functioning, Disability and Health (ICF, (WHO, 2006)) was identified as most suitable for this purpose. The ICF includes more than 1,400 categories assessment schemes, which makes dealing with this classification very difficult. The main issue with the development work presented here was the following: How can the knowledge, which is based on the ICF, be made accessible in an easy and intuitive manner? And how can this accessible knowledge be applied in practice? The Edit-Patient Approach presented here uses web technologies such as sensitive search, word suggestions and the possibilities of visualization to make the ICF classification applicable to a wide group of users. The creation of templates simplifies the reusability of relevant categories and a sharing feature links the various treatment partners.

Keywords: ICF, Web-based Classification Tool, Rare Disease, ALS

1 INTRODUCTION

In the field of patient information systems, describing the health situation of the patient is one of the most important challenges. Existing patient classification systems are primarily focused on the medical billing services. However, the basis for medical treatment decisions is the description respectively classification of the health status of the patient including the symptoms, effects, and environmental variables. The limited consultation time of the doctors requires electronic support systems based on standardized patient descriptions. The utilization and usability of patient classification are crucial in developing intelligent patient information systems.

One of the most useful classifications in the field of rare diseases is the International Classification of Functioning, Disability and Health (ICF (WHO, 2002)).The Edit-Patient Tool was developed as a multidimensional user-friendly web application so that this classification can be used as an electronic description tool. The demand for a higher level of usability of the ICF through software systems (Schaller, 2009) was concrete, exemplary and applied in an exploratory manner in the Edit-Patient Tool.

The concept, implementation and future utilization options of the Edit-Patient Tool are discussed below.

2 PROBLEM DEFINITION

Physicians invest significant amounts of time and effort in communication with patients, especially patients with complex. Amytrophic lateral sclerosis (ALS) is one example of such a complex disease (Mitchell and Borasio, 2007). ALS is a fatal disease with an unpredictable progression. A broad healthcare system and many interacting stakeholders are needed to achieve an adequate quality of life (Eidt, et al., 2009, Kuschel, 2006, Taruscio and Seyoum, 2003). To receive this quality of life means to manage the permanent injury and to decide whether life-prolonging measures should be applied. For ALS patients, it is important to be informed about the individual problem areas. The problem areas concern

1. the physical disability (symptoms),
2. the independent and self-determination act,
3. social and professional environment factors and
4. the financial and living situation.

To provide the affected patients with the problem-specific requires a detailed information base about the patient (Elze, et al., 2011b). The system describing the patient's health state should address the following competency concerns:

1. Using an international standardized classification for symptoms, effects, and environmental variables (which occur in ALS disease), which allows for a multidimensional description

2. Feasible for heterogenic stakeholders
3. User-friendly for the different stakeholders
4. Exchanging patient classification between interacting stakeholders
5. Interoperability of the saved data

3 STATE OF DEVELOPMENT

Detailed analyses of the use case ALS have shown that recording and storage of information about a patient depends on a treatment field (Elze, et al., 2011a). For example, the family doctor uses his patient administration system, patient representatives such as the German Society for Muscle Diseases uses personal conversations about the environment and the conditions of the patient, the speech therapist has a detailed questionnaire for the exact definition of dysphagia (NOD-level concept (Ickenstein, et al., 2009)) and the ALS clinic (Charité Campus

Virchow-Klinikum Berlin, Neurological Clinic) documents the clinical course using standardized ALS assessment questionnaires (ALSFSR (Cedarbaum, et al., 1999)). Between these various sources information about the patient, there is a common subset that is of interest to all those involved in the disease. Such a common view on the patient's health state could realize qualitative and temporal potentials.

In medical context, there are the Electronic Health Record (EHR) on the one hand and patient classifications on the other hand. The Electronic Health Record (EHR) realizes the interagency, patient centered documentation (Haas, 2006). Beyond this, the patient classifications are sector-specific descriptions of the health state of the patient. Consequently, there are patient classification models of rehabilitation such as the Functional Independence Measure-Function Related Groups (Fischer, et al., 2006), which capture the functional independence state of the patient. There is also Form-36 Health Survey (SF-36), which is an indicator for the quality of life and the mental state (Neubauer and Ranneberg, 2005) as well as the International Classification of Functioning, Disability and Health (ICF) (WHO, 2002).

In order to resolve the problem, a decision was made in favor the classification system ICF, whose advantages have already been presented. By means of this system, an exemplary attempt to facilitate a multidimensional patient description was made. The ICF is a classification system developed by the WHO (World Health Organization). It is used to describe the health of humans and other domains related to the issue of health. The ICF is divided into four major areas and focuses on the problem areas of body structures, body functions, activities and participation and environmental factors. Thus the ICF includes most identified problem areas on the one hand and meets the challenge by standardization on the other hand.

However, this classification, with its approximately 1424 categories, is not suitable for routine use (Morita, et al., 2006). The ICF Research Branch in cooperation with the WHO Collaborating Centre for the Family of International Classifications in Germany (at DIMDI) has generated lists of generally agreed upon

ICF categories for certain diseases (Kesselring, et al., 2008). These lists are called ICF Core Sets. An ICF Core Set is a widely accepted terminology but in practice there is as little software support as for the ICF itself. An example for an ICF support tool, which represents only a browser, is the ICF-Browser (WHO, 2012).

To find a level that can be used for the health problems of patients with ALS on the basis of the ICF, we have developed the ICF-based web tool Edit-Patient Tool.

4 EDIT-PATIENT APPROACH

The Edit-Patient Tool uses the ICF classification and measurement of all ICF categories. Through intelligent search mechanisms and the possibility of involving core sets, the ICF web tool makes this classification applicable. The ratings are versioned in time, so the disease can be documented and evaluated. The implementation of patient classification using the ICF offers the possibility of a multidisciplinary description level, which forms the basis for the complex treatment for the disease ALS.

For the development process of the Edit-Patient Tool a bottom-up approach was chosen, which emphasizes the explorative character of the respective application. Medical practices are determined by enormous time pressure and the need for the fast and simple administration of systems which are to assist in the daily routines and operations. Therefore, the following Use Case was assumed among others: A physician wants to generate an ICF-based description template to help him describe a patient comprehensively. Furthermore it should be possible for him to add individual aspects and elements from ICF that ensued from a conversation with the patient. The Edit-Patient Approach allows the physician to generate a template by means of the concise ICF section-tree (see Figure 1). The left side shows an overview of the ICF Library (l). The visualization (v) of body-parts, which were connected to specific ICF elements, helps with navigation.

Figure 1 Edit-Patient Approach according to ICF

A context-sensitive search supports the user through further search proposals. As shown in Figure 1, the search enriches the target word (w) with word suggestions occurring as predecessors (p) and successors (s) in the ICF, and it shows the words containing the target word (c). Furthermore, the hit rate (r) in the ICF section tree shows the percentage of hits matching this category. In the middle frame, the detailed ICF categories appear, including the target word, which is highlighted there. The physician or other user can add (a) the relevant category to a template.

This multiple contextualization of the search allows the use of the Edit-Patient Tool for non-medical personnel and patients as well. The user receives this support not only during the creation of templates, but also when coding individual ICF elements that describe the patient's health state as shown in Figure 2. The relevant ICF categories must be evaluated to give a statement about the extent of the health condition. The extent of the health restriction is expressed in numbers that qualify (q) the category. The user can evaluate the category by using the drop down box (u) or by clicking on the numbers (q). The window on the right visualizes changes during the course of the disease. The red line (B) shows the development over the cumulated group of Body Functions. Meanwhile the blue line shows the changes for the selected category. Clicking on the history points (h) displays the health state at an earlier point in time. The date (d) provides an overview of the temporal dimension.

patient record T. Testuser (201103172)

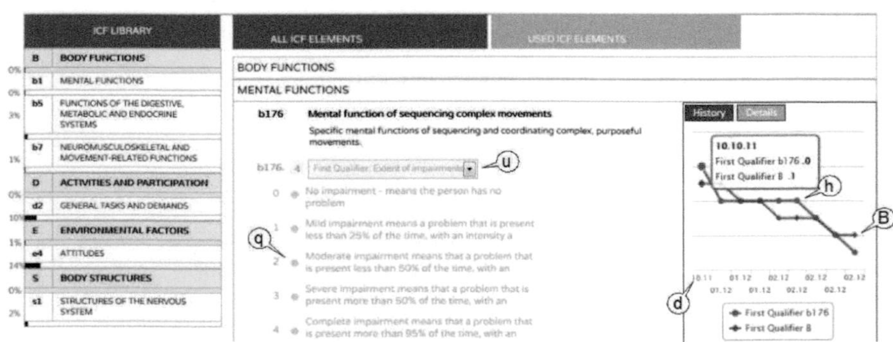

Figure 2 Application of the ICF

The patient description of a concrete patient with the Edit-Patient Approach includes a sharing function. This means that the concrete patient description can be shared and used by different stakeholders. The treatment of the patient's description is stored daily as a history and can be shared with other users if needed. The use of the ICF has been kept as simple as possible so that the patients themselves can use the ICF classification with the Edit-Patient Tool. Users are able to comply, modify, and apply those templates (of the Core Sets or directly from the ICF) they deem the most relevant and suitable in practice. The shift from set to template emphasizes the context of re-use and a sustainable use of the ICF and its Core Sets for specific disease patterns, as shown using the example of ALS.

2 CONCLUSIONS

Finding the targets, a classification that is relevant to the needs of ALS patients and to give a tool the proper support to describe the patient's situation has been prototypically achieved with the Edit-Patient Tool. Based on the problem definition, the classification ICF includes the description of physical disability in the category groups of Body Structures and Body Functions. The description of self-determination is noted in the ICF in the category Activities and Participation, while social factors and the living situation are coded in the category Environmental Factors. The ICF as a knowledge base for the Edit-Patient Approach still corresponds with the requirement to use an international standardized classification for describing the situation of ALS patients. The applicability of the ICF for different stakeholders was realized by means of intuitive functions in the Edit-Patient Tool. In order to allow the parties to interact, the Edit-Patient Tool includes a sharing function for the patient description.

The requirement of interoperability of the saved data must be considered and tested in future work. Finding a suitable exchange data format is particularly important for use in the context of medical information systems such as the Dispedia project (Elze, et al., 2012). Another next step is to test and evaluate the usability of the Edit-Patient Tool in practice in collaboration with relevant partners from the health sector.

REFERENCES

Cedarbaum, J., N. Stambler, E. Malta, C. Fuller, D. Hilt, B. Thurmond and A. Nakanishi 1999. The ALSFRS-R: a revised ALS functional rating scale that incorporates assessments of respiratory function. *Journal of the neurological sciences,* 169.646

Elze, R., M. Böttcher and S. Klingner 2011a. Konzept patientenzentrierter Wissens-modellierung für eine seltene Erkrankung. *In:* G. S. a. D. H. a. E. Ammenwerth (ed.) *eHealth2011 - Health Informatics meets eHealth - von der Wissenschaft zur Anwendung und zurück.* Wien: Österreichische Computer Gesellschaft.

Elze, R., M. Böttcher and S. Klingner 2011b. Konzept patientenzentrierter Wissensmodellierung für eine seltene Erkrankung. *In:* G. S. a. D. H. a. E. Ammenwerth (ed.) *eHealth2011 - Health Informatics meets eHealth - von der Wissenschaft zur Anwendung und zurück.* Wien: Österreichische Computer Gesellschaft.

Elze, R., M. Martin and K.-P. Fähnrich 2012. Dispedia.de – a knowledge model of the service system on the rare disease. *AHFE 2012 and its affiliated conference on human factors and ergonomics in healthcare.* San Francisco California: Purdue University, USA and Tsinghua University, P.R. China (to appear).

Fischer, W., J. Blanco, M. Mäder, P. Zangger, F. M. Conti, L. Babst and B. Huwiler 2006. Das TAR-System und andere Patientenklassifikationssysteme für die Rehabilitation. TAR-Forschungsbericht und Kurzbeschrieb von Systemen aus Deutschland,

Frankreich, Australien und den USA. Wolfertswil: Zentrum für Informatik und wirtschaftliche Medizin.

Haas, P. 2006. *Aspekte der Gesundheitstelematik,* Berlin, Heidelberg, Springer.

Ickenstein, G., A. Hofmayer, B. Lindner-Pfleghar, P. Pluschinski, A. Riecker, A. Schelling and M. Prosiegel 2009. Standardisierung der Diagnostik und Therapie bei Neurogener Oropharyngealer Dysphagie (NOD). *In:* G. z. E. d. U. m. e. T. G. Team) (ed.).

Kesselring, J., M. Coenen, A. Cieza, A. Thompson, N. Kostanjsek and G. Stucki 2008. Developing the ICF Core Sets for multiple sclerosis to specify functioning. *Multiple sclerosis (Houndmills, Basingstoke, England),* 14, 252-4.278

Morita, E., M. Weigl, A. Schuh and G. Stucki 2006. Identification of relevant ICF categories for indication, intervention planning and evaluation of health resort programs: a Delphi exercise. *International journal of biometeorology,* 50, 183-191.994

Neubauer, G. and J. Ranneberg 2005. Ergebnisorientierte Vergütung der neurologischen Rehabilitation. *Abschlussbericht.* München: Unversität der Bundeswehr München.

Schaller, A. 2009. Anwendung der ICF in der stationären Rehabilitation nach Totalendoprothese *Implementierung in Rehabilitationskonzeption und Ergebnismessung.* Köln: dissertation.de.

WHO 2002. ICF The International Classification of Functioning, Disability and Health.621

WHO 2006. International Classification of Diseases (ICD).652

WHO 2012. ICF Browser. WHO. Available: http://apps.who.int/classifications/icfbrowser.

Section III

How to Improve Quality of Ergonomics in Healthcare

CHAPTER 13

Can Inclusive Environmental Design be Achieved in Acute Hospitals?

Sue Hignett

Loughborough University
Loughborough, UK.
S.M.Hignett@lboro.ac.uk

ABSTRACT

The effectiveness of healthcare delivery is determined, in part, by the design of the physical environment and the spatial organisation of work. This paper will consider firstly whether ergonomic input to provide recommendations for work space requirements may restrict patient autonomy and secondly, whether design developments for patient benefit may lead to difficulties in providing clinical care.

The findings from two research studies are used to discuss the impact of physical layout on work systems with respect to staff well-being (space to work), patient care (monitoring) and patient experience (privacy and dignity). Several approaches to design and ward layout are considered, including Harness, Nucleus, AEDET, Planetree and Sengetun. Finally, the involvement of both staff and patients through a participatory ergonomics framework in building design is explored. It is suggested that mapping criteria for user participation in building design briefing with the participatory ergonomics framework may offer potential to improve and enhance patient involvement in hospital design.

Keywords: inclusive design, patient safety, participatory ergonomics

1 INTRODUCTION

Although it is acknowledged that the physical environment has a significant impact on health and safety (including confidentiality, cross infection and travel time), it has been suggested that hospitals have not been designed with the explicit goal of enhancing staff and patient safety through facility design innovations (Reiling et al, 2004).

The findings from two research studies will be used to consider the impact of physical layout on work systems with respect to staff well-being (space to work), patient care (monitoring) and patient experience (privacy and dignity). It is proposed that the patient perspective in the system has received relatively little attention and that combining the theoretical concepts of participatory ergonomics and inclusive briefing for building design may offer a better framework for environmental design in care facilities.

2 DESIGN FOR STAFF: SPACE TO WORK

The space provision in hospitals has been debated since the time of Florence Nightingale when she recommended the construction of a larger ward as a long open space for approximately 30 patient beds to achieve cross-ventilation and nurse efficiency (Gesler et al, 2004). The racetrack concept was introduced after 1945, with multi-bed bays grouped around a central core of utilities (part of the Harness initiative; Francis, 1998). A variation on this theme, the Falkirk layout (1960s) used dispersed nurse stations to support a flexible response to variations of workload. The most widely used design in the UK is Nucleus, a cruciform template with six-bedded bays and single rooms, some with en-suite facilities. The response by architects and stakeholders to Nucleus is that '*it was too prescriptive, tended to stifle creativity and simply failed to address design issues such as location, legibility and sense of place*' (Francis, 1998). An alternative ward layout (bed courtyard) has been developed in Norway to improve monitoring, the '*sengetun*' (Rechel et al, 2009). The sengetuns (courtyards) are groups of 6-8 single rooms are arranged as '*pearls on a string*'. Each has a central fully operational (not dispersed) nursing workstation with sufficient space for documentation, observation, medication and storage.

All these design approaches have been influenced by government legislation and guidance. For example, in the 1990s health and safety law that required '*every room [to] have sufficient floor area, height and space for the purposes of health and safety*' (The Workplace (Health, Safety and Welfare) Regulations 1992, regulation 10). However space planning has continued to be a matter for debate and there have even been legal challenges to hospital plans (BBC, 2004). In 1996, Palmer suggested that Nightingale would now challenge design professionals to create patient bedrooms of a sufficient size to accommodate two caregivers simultaneously, as well as the visitor/patient chair, bedside locker, over-bed tray, straight-backed chair, and washing facilities and allow a trolley, bed or wheelchair to be move in and out of the room.

In the UK an ergonomic database was developed to encourage those involved in hospital design to think in terms of the relationship between a user and a particular component and other components located within a room with respect to critical minimum space requirements (Dept. of Health and the Welsh Office, 1986). Unfortunately user data were not included due to a lack of time and resources (Stanton, 1983) so to address this deficit, Hignett and Lu (2008) investigated spatial requirements for frequent and safety critical tasks in medical, surgical and intensive care wards. Data were collected from field observations with 34 nurses and 58 tasks (Figure 1) and laboratory simulations (n=90 datasets). The results were compared with previous recommendations where a gradual increase in the recommended space can be seen from 6.96m^2 in 1961 (HBN 04) to 10.84m^2 in 2008 (Hignett and Lu, 2008). However, even though the new recommendations were based on empirical data and included user data (staff) they still failed to include patient activities and preferences.

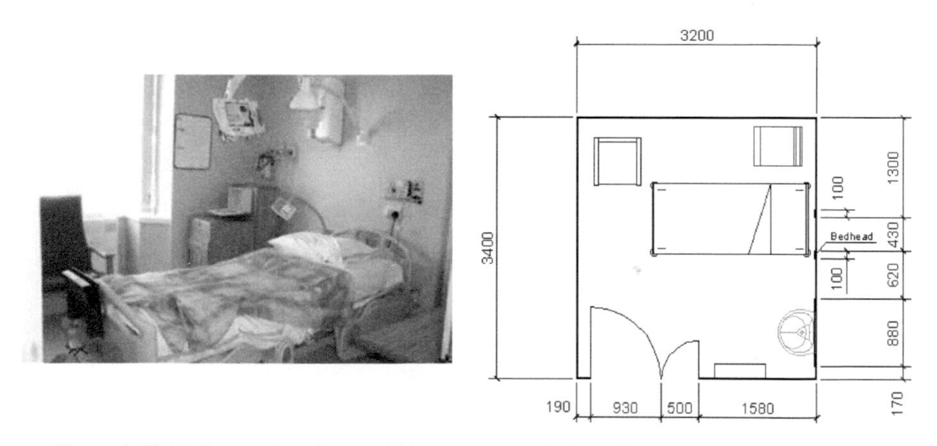

Figure 1 Field observations for spatial layout on medical wards

3 DESIGN FOR PATIENTS: SINGLE ROOMS (PRIVACY)

The bed space becomes the patient's domain during their stay in a care facility; the hub of their clinical and care experience. It provides a private space either as a single room or a cubicle with curtains or screens and will be where they store their personal belongings, receive visitors and interact with staff for many aspects of their treatment and care. Cartledge (2007) reported a divergence in privacy preferences with older people concerned about being in single rooms because they 'don't trust the staff to be there to help if they fall out of bed, and they don't like being alone and may depend on other patients in the room for assurance and safety'. In contrast, younger people are reported to prefer the privacy of a single room with the ability to have treatment and care without moving location. In UK hospitals the percentage of single bedrooms as a proportion of total available beds has increased from 22.6% in

2002/03 to 32.7% in 2009/10 (DH, 2010) as a design development for patient benefit in response to privacy requirements.

Falls in care facilities account for over 33% of reported incidents in the UK with over 70% of reported falls being un-witnessed (Healey et al, 2008). One of the interventions for managing falls is to increase the level of monitoring for at-risk patients (Hignett, 2010). The second research study (Sands et al, 2011) looked at the contributory factors of falls by collecting data in an overnight bedrail audit at 18 UK hospitals (n=1,799 beds). The level of monitoring (observation) was explored by recording bed visibility from nursing stations. The participating hospitals had more than 900 beds (n=4), 500-899 beds (n=8) and less than 499 beds (n=6). Seven were built in the 2000s, 3 were built in the 1980s/90s, 4 were built in the 1970s, and 5 were built in the 1940s and 1880s with refurbishment in the 1970s/80s.

Table 1. Profile of participating hospitals (*, †, ◊ = part of same NHS Trust, with individual hospitals in different towns)

Hospital number. Approx. opening date of current building	No. of beds	No. of single rooms with en-suite	% Beds visible	% Patients described as confused
1. Refurbished from 1940s	152	-	28%	62%
2. 1990s◊	155	7 (5%)	30%	35%
3. No date available†	255	64 (25%)	4%	28%
4. 1988◊	380	38 (10%)	0%	33%
5. 1992*	448	25 (6%)	13%	45%
6. 1970s†	451	58 (13%)	12%	36%
7. 1976*	503	24(5%)	24%	50%
8. 2010	516	40 (8%)	21%	56%
9. 1972	581	2 (<1%)	6%	64%
10. 1980s, refurbished from 1850s†	591	44 (8%)	61%	24%
11. 1970s, refurbished from 1870s†	617	56 (9%)	24%	49%
12. 2010	744	306 (41%)	58%	49%
13. 2007	809	66 (13%)	25%	53%
14. Part 1993 and part refurbished from 1800s	884	74 (8%)	29%	40%
15. 2003	900	-	16%	47%
16. 2010	1000	440 (44%)	20%	39%
17. 2009	1010	317 (31%)	2%	50%
18. 1976	1106	169 (15%)	39%	41%

[1]Dr Foster Hospital Guide (http://www.drfosterhealth.co.uk/hospital-guide/ accessed 18th May 2011). Data Source: England, qualified nursing, midwifery & health visiting staff (full time equivalent) in NHS Hospital and Community Health Services: Staff by main staff groups in England as at 30 September 2009 from the Non Medical Workforce Census. The number of beds at each hospital is published in the General and Acute (available) column of *'Bed availability and occupancy, England - KH03'* return and is for the financial year 2008/09.

It was found that only 23% (0% - 61%, median 24%) of beds were visible from a nursing station (Table 1). Most of the accommodation was provided in 4-6 bed bays, with en-suite single rooms accounting for a median of 10% (range <1% to 44%). Beds that were visible from the nursing station were significantly more likely to be a multi-bed bay with no toilet ($p<0.001$, Phi=0.230). Beds that were not visible from the nursing station were significantly more likely to be located in a single en-suite room ($p<0.001$, Phi=0.230). So there is a need for balance in environmental design to support both safe observation (monitoring) and patient privacy (Essence of Care, 2010).

4 COMPLEXITY OF DESIGN

The evolution of design approaches have been described as changing focus from designers (1960s), to design for healthcare planners, and finally design for service delivery (Glanville, 2006). In the USA, the Planetree philosophy of patient-centred care is gaining momentum for emphasising 'trust, intimacy, dignity, security and confidence, holistic care and treatment, information, participation in decision-making, health promoting physical surroundings, and network support' (Jenso and Haugen, 2005). The involvement of patients in the hospital design process has been discussed for many years. Ronco (1972) offered 3 reasons why patients were not involved, including the need to prioritise functional efficiency before habitability, and the lack of data relating physical environment to behaviour. The challenge of including patient input continues, for example the Achieving Excellence Design Evaluation Toolkit (AEDET) was developed to assess new healthcare buildings for functionality (user, space, access), impact (character and innovation) and build standard (performance, engineering, construction). But its effectiveness has been questioned as it requires stakeholders (including staff and patients) to translate quite complex qualitative judgements about several discrete questions into single scores (Gesler et al, 2004).

Attaianese and Duca (2010) suggested, in a theoretical discussion of human factors in building design that many of the methods (in particular task analysis and participatory methods) in offer benefits for (1) environment design and sustainability, (2) functionality (effectiveness and efficiency) and user satisfaction, (3) accessibility including way finding, emergency response and inclusive design, and (4) value creation through design management (economic). Their model is an excellent step forward but limited in terms of hospital design to physical impairments (mobility, vision, hearing) with only fear of falling as a cognitive impairment. This excludes many of the in-patient population particularly confused patients (dementia and delirium). The difficulties for researchers to deliver empirical data on hospital design may be due to the *'inherent logistic difficulties in performing or interpreting studies in care homes or hospitals associated with population, setting, design, and outcome measurement. Getting consent from or randomising frail, confused, unwell elderly people, who are often in the institution for only a short stay, is challenging*' (Oliver et al. 2007).

5 PARTICIPATORY ERGONOMICS

One theoretical model that offers potential for future patient involvement is Participatory Ergonomics (PE). PE can be very simply described as a concept involving the use of participative techniques and various forms of participation in the [*work*]place (Vink and Wilson, 2003). Wilson (1995) defined participation in ergonomics projects as *'the involvement of people in planning and controlling a significant amount of their own [work] activities, with sufficient knowledge and power to influence both processes and outcomes in order to achieve desirable goals*'. This is being explored in building design, for example Jensen (2011) suggesting that user participation in briefing should be a continuous process before and during the design and construction activities. He reflects that the view of buildings has changed *'from seeing buildings as mainly architectural expressions or passive physical constructions to..... facilities that must support the needs of an organisation'* and describes this as 'inclusive briefing' rather than the traditional model of where users were mainly involved as data sources. This involvement has been mapped onto the dimensions of the Participatory Ergonomics Framework (Table 2).

Table 2. Participatory Ergonomics Framework mapped with selected elements of inclusive design criteria and user building design briefing (modified from Haines et al, 2002; Jensen, 2011)

PEF Dimension (*range and/or scope*)	User participation in briefing
Decision-making (*group to individual*)	The result is acceptance of solutions based on a brief
Mix of participants (*operators to management*)	Concerns all client/user needs in developing facilities
Remit (*problem to solution*)	A guided learning and dialogue process
Role of ergonomics specialist (*initiate and guide to consultation*)	
Involvement (*full direct to representative*)	
Focus (*equipment/job design to strategy)*	A continuous process with changing focus in different phases
Level of influence (*organisation to work group)*	Users actively involved as part of a corporate change process
Requirement (*compulsory to voluntary)*	
Permanence (*on-going to temporary)*	

6 CONCLUSION

Healthcare is a complex system with multiple users of equipment, products and treatment/care environments. Whether inclusive environmental design can be achieved remains to be seen but the indications are positive with building designers exploring human factors methodologies including user-needs analysis, task analysis and participatory ergonomics.

The challenge for clinicians, designers and researchers is to work together using robust high quality research methods to analyse the activities of all the user groups. This should be used at all stages of the briefing (design) process to achieve inclusive facilities that provide both functionality and habitability including autonomy, privacy (where appropriate) and safety.

ACKNOWLEDGMENTS

The author would like to acknowledge all her collaborators in the studies: Jun Lu, Gina Sands, Paula Griffiths, Jane Youde, Frances Healey, Mike Fray, Penny Xanthopoulou.

REFERENCES

Attaianese, E., Duca, G. 2010. Human Factors and ergonomics principles in building design for life and work activities: an applied methodology. *Theoretical Issues in Ergonomic Science.* 13: 2, 187-202

BBC. 2004. *Legal bid on hospital bed spaces.*

http://news.bbc.co.uk/1/hi/health/4013509.stm Accessed 14th February 2012

BS 7000-6. 2005. *Guide to Managing Inclusive Design.* London: British Standards Institution

Cartledge, S. 2007. Planning for Patient-Centred Care. *Hospital & Healthcare* December 25-26.

Department of Health and the Welsh Office. 1986. *Health Building Note 40, Common Activity Spaces Vol. 1 – Example layouts: Common components.* London, HMSO.

Dept. of Health. 2010. Accessed 29 Dec 2010, from *Hospital Estates and Facilities Statistics 2009-10. Main Findings* http://www.dh.gov.uk/prod_consum_dh/groups/dh_digitalassets/documents/digitalasset/dh_120811.pdf

Essence of Care. 2010. *Benchmarks for the Fundamental Aspects of Care. Benchmarks for Safety (Ref. 14641).* Norwich: The Stationary Office.

Francis, S., 1998. A golden record, but is planning past caring? *Hospital Development.* June, 14-18.

Gesler, W., Bell, M, Curtis, S., Hubbard, P., Francis, S. 2004. Therapy by Design; Evaluating the UK Hospital Building Programme *Health and Place*, 10: 117-128.

Glanville, R., 2006, HD/MARU Forum. *Hospital Development.* 12-16.

Haines, H.M., Wilson, J.R., Vink, P. and Koningsveld, E. 2002. Validating a framework for participatory ergonomics. *Ergonomics.* 45: 4, 309-327

Healey F, Scobie S, Oliver D, et al. 2008. Falls in English and Welsh hospitals: a national observational study based on retrospective analysis of 12 months of patient safety incident reports. *Quality and Safety in Healthcare*;17: 424-430.

Hignett, S. Lu, J. 2008, Ensuring bed space is right first time *Health Estate Journal.* February *http://www.healthestatejournal.com/Story.aspx?Story=3395* Accessed 11-11-2011.

Hignett, S. 2010. Technology and Building Design initiatives in interventions to reduce the incidence and injuries of Elderly In-Patient Falls. *Healthcare Environments Research and Design Journal* 3: 4, 62-84

Jensen, P.A. 2011. Inclusive Briefing and User Involvement: Case Study of a Media Centre in Denmark, *Architectural Engineering and Design Management*, 7:1, 38-49

Jenso, M., Haugen, T. 2005 Usability of hospital buildings Is patient focus leading to usability in hospital buildings? *Proceedings of the 11th Joint CIB International Symposium.* Accessed 19 February 2012 http://www.metamorfose.ntnu.no/dok/050708UsabilityHelsinkipaperMJ.pdf

Oliver, D., Connelly, J.B., Victor, C.R.. et al. 2007. Strategies to prevent falls and fractures in hospitals and care homes and effect of cognitive impairment: systematic review and meta-analyses. *BMJ* 334: 82-7.

Palmer, I.S. 1996. What Florence Nightingale would tell us today. *Journal of Healthcare Design.* 8: 19-22.

Reiling, J.G., Knutzen, B.L., Wallen, T.K. et al. 2004. Enhancing the traditional hospital design process: a focus on patient safety. *Joint Commission Journal of Quality and Safety.* 30: 3, 115-124.

Rechel, B., Erskine, J., Dowdeswell, B. et al. 2009. Capital investment for health. Case studies from Europe. *European Observatory on Health Systems and Policies.* 168-9.

Ronco, P. 1972. Human Factors applied to Hospital Patient Care. *Human Factors* 14: 5, 461-470.

Sands, G., Hignett, S., Fray, M. et al. 2011. An overnight bedrail audit in UK hospitals exploring the relationships between environmental factors and patient characteristics. In Albolino, S. et al. (Eds.) *Proceedings of the 3rd Healthcare Ergonomics and Patient Safety Conference.* June 22-24, 2011. Oviedo, Spain. 446-450

Stanton, G. 1983. The Development of Ergonomics Data for Health Building Design Guidance. Ergonomics, 26, 3, 785-801.

Vink, P., Wilson, J.R. 2003. Participatory Ergonomics. *Proceedings of the XVth Triennial Congress of the International Ergonomics Association and The 7th Joint conference of the Ergonomics Society of Korea/Japan Ergonomics Society. 'Ergonomics in the Digital Age'.* August 24-29, 2003. Seoul, Korea

Wilson, J.R. 1995. Ergonomics and participation. *In Evaluation of Human Work: A Practical Ergonomics Methodology* (2nd Ed.) J.R. Wilson & E.N. Corlett (Eds.) London: Taylor & Francis. 1071-1096

Workplace (Health, Safety and Welfare) Regulations. 1992. [S.1.1992 No. 3004] Accessed 14th February 2012 www.legislation.hmso.gov.uk/si/si1992/Uksi_19923004_en_1.htm.

CHAPTER 14

Workshop Concept "Fabrica Medica ®" - Enhancing Collaborative Work in Heterogeneous Design Teams

B. Podtschaske, D. Fuchs, W. Friesdorf

Department Human Factors Engineering and Product Ergonomics
Berlin Institute of Technology
Beatrice.Podtschaske@awb.tu-berlin.de

ABSTRACT

The ergonomic design of work systems improves effectiveness and efficiency of the system by a well-balanced workload. Ergonomic design processes require scientific findings with the know-how of the affected persons must be tied together suitably. Heterogeneous backgrounds of experience and knowledge in such an expert team allow coping with complexity by a comprehensive reflection. However, there is a risk that communication and coordination barriers throw up and prevent collaborative work. Therefore in this study, a workshop concept for optimizing the collaboration work of heterogeneous expert teams in design processes is developed and evaluated. The workshop concept is developed through a concept-based approach. It was exemplary conducted for two complex design tasks and tested thorough a satisfaction survey. First results show that the workshop concept called "Fabrica Medica ®" allows a straight and subject oriented solution finding while considering clinical expertise and relevant stakeholder. The used methods and instruments contribute to the success of the collaborative work process within the heterogeneous expert team. At a scale from 1 ("very good") to 5 ("poor") 90% of the participants valued the workshops with 1("very good") to 2 ("good").

Keywords: complexity, collaborative work process, heterogeneous design teams, interdisciplinary, health care

1 INTERDISCIPLINARY EXPERT COLLABORATION TO SOLVE COMPLEX DESIGN TASKS

The work system performance is depending on problem free interactions of technical, organizational and human factors. To increase human well-being and with it the overall system performance, is the aim of an ergonomic work design (IEA, 2000). Clinical employees need adequate support to be able to operate in this highly complex, emotionally demanding work system. (Friesdorf et al., 2011). It appears that arising design tasks often are very challenging. Therefore ergonomic design processes require scientific findings with the know-how of the affected persons must be tied together suitably. Hence, ISO 6385:2004 recommends an interdisciplinary team for coping with design task. The strength of multidisciplinary teams is to combine knowledge from different disciplines. This guarantees a comprehensive approach that also considers perhaps less clearly evident requirements and contexts of use (ISO 6385:2004).

On the one hand this specialization promises to cope with complex medical tasks. But at the same time, managing collaborative work effectively and efficiently is a significant challenge. Heterogeneous backgrounds of experience and knowledge in such an expert team allow coping with complexity by a comprehensive reflection. However, there is the risk that communication and coordination barriers throw up and prevent collaborative work. Specific methods and instruments contribute to the success of the collaborative work process within the heterogeneous expert team by building up a common understanding and common ground of knowledge (Steinheider, 2000).

ISO standards provide an adequate framework for ergonomic analysis, design and evaluation of work systems. They describe the design of work systems as an iterative and structured process. The guidelines demand the application of supporting procedures, like simulation and group discussion, in the design process. However, their successful application is not concretised further. Even though there are different methods and instruments supporting expert collaboration, there are no integrative concepts to synthesize findings of the current scientific state of the art related to interdisciplinary teamwork.

2 OBJECTIVE

Heterogeneous backgrounds of experience and knowledge in such an expert team allow coping with complexity by a comprehensive reflection. However, there is a risk that communication and coordination barriers arise and prevent collaborative work. Therefore in this study, a workshop concept for optimizing collaboration work of heterogeneous expert teams in design processes is developed and evaluated. Focus is the participatory design of productive work systems (processes, technology, and qualification) in the domains "Hospital" and "Home Care".

3 METHODS

The workshop concept is developed through a concept-based approach. The foundation is formed by a summary of the current scientific state of the art related to interdisciplinary teamwork. The scientific findings were synthesized to a workshop concept and approved on the basis of observation, questioning and documents. The concept was exemplary conducted for two complex design tasks: hospital building process and clinical process benchmarking. The suitability of the workshop concept was also tested thorough a satisfaction survey which was applied to the workshop participants.

4 ENHANCING COLLABORATIVE WORK IN HETEROGENEOUS DESIGN TEAMS

The workshop concept "Fabrica Medica ®" addresses issues in the domains "Hospital" and "Home Care". The focus is on participatory design for productive work systems, with the aim to provide safe and efficient treatment procedures.

4.1 Theoretical Background

The analysis of the current state of the art in cooperation science revealed four particularly relevant models and methods that will be presented in more detail below.

According to the model of interdisciplinary cooperation, the success of expert collaboration depends on the three sub-processes communication, coordination and shared knowledge; it is essential to optimize these sub-processes (Steinheider & Legrady, 2000):

1. Process of Communication: A frictionless exchange of data, information and knowledge between the involved work persons is often possible. In this context, intercultural factors can also play a role. This process can be supported by suitable communication-technical aids.

2. Process of Coordination: Tasks, activities and interactions of involved participants are regulated unambiguously. All individual sub tasks are integrated and harmonized to reach the overall goal. This process can be supported by methods and tools of project management.

3. Process of shared knowledge: Creating a common or shared knowledge supports successful communication and coordination processes. To achieve a shared knowledge, a common understanding of the aims, work content/work tasks and terms/definitions is essential. To support this process, there is a considerable need for further research. Nevertheless, methods such as process visualization and haptic models can be helpful.

If these sub processes are optimized ergonomically, the working system is more likely to succeed.

In interdisciplinary cooperation the process of shared knowledge plays a key role. Only if we succeed in building a shared understanding in the form of shared mental models, heterogeneous knowledge can be linked to new, innovative solutions. This two-step process of sharing knowledge and generating new knowledge is described by the second so-called "bridge model of sharing knowledge" (Bienzeisler et al., 2007).

The joint production of new knowledge and innovative design solutions in the form of a creative and discovery procedure is also called "cooperative modeling" (Hornecker et al., 2001). This is the third model, which is concidered during the development of the workshop trilogy. Sharing knowledge processes and generating solutions are largely supported by haptic models and other tools (Hornecker et al, 2001, Wetzstein et al, 2003). In addition to haptic models items such as visualization, task-related information exchange, reflection techniques, storytelling and simulations are valuable. These tools support innovation processes by

- enabling a high level of participation,
- preserving a natural structure of communication,
- concretizing abstract issues,
- allowing an experience-oriented access,
- supporting a empathic-intuitive problem-solving process.

Finally, aspects of the "systems engineering" method by Haberfellner et al. (1997) are included in the concept. This method is used to solve complex design problems. The "systems engineering" postulates a number of principles, as

- principle of systems thinking,
- procedure principle "from general to detail"
- principle of composing alternative solution,
- principle of phase structure as a macro-logic,
- principle of problem-solving cycle as a micro-logic.

The methodical approach is divided into the procedures "goal seeking", "search for solutions" and "selection". In particular, the step of finding solutions as a creative part of the method is included into the workshop concept. Thus the search for solutions consists of a creative and constructive (synthetic) and a critical-analytical step (analysis). Guiding theme of the problem-solving process is "thinking in alternatives" (Haberfellner et al., 1997).

4.2 The Workshop Trilogy "Fabrica Medica ®“

According to ISO 6385:2004 the design process is understood as an iterative process. The aim is to find innovative solutions to complex design issues. Through a creative recombination of existing knowledge new knowledge is produced. The specific approach and the use of methods and tools help to sensitize the participats to the complexity of the design problem as well as the need for interdisciplinary cooperation. Special attention was paid to the methodological support of sharing knowledge processes to build shared mental models.

The identification of relevant issues is carried out by a team of researchers and practicioners. If a suitable design problem is detected, preparations for implementing a "Fabrica Medica ®" trilogy begin. The preparations include research and analysis of the chosen topic, selection and response of about 30 experts and the selection and adaptation of specific tools. It is especially important to involve more than one person/representative from each stakeholder group. This redundancy is to reduce the effects of personal opinions.

For a period of 6-9 months (incl. preparation and follow-up) three one-day interdisciplinary design workshops are carried out in regular interval. During the first workshop ("Introducing") a common ground of understanding is developed with the help of haptic models, simulations and visualizations. During the second workshop ("Matching") the participants combine their different knowledge to innovative design solutions. Duringthe third workshop ("Agreeing") the participants discuss and evaluate the developed solutions and a consensus of the final solution is intended. Further appropriation of the results is fixed and the workshops series is closed.

During the workshops, group discussions are provided in plenary as well as small group work. The highly innovative character prevents any detailed planning in advance. Therefore the detailed planning of each workshop will take place iteratively and depending on the respective preliminary or interim results. While the trilogy theme selection, viewing perspectives, and design priorities are repeatedly questioned, critically reviewed and adjusted as necessary. With progression of the trilogy the perspective shifts from the design problem to possible alternative solutions until finally for the last session a concrete solution is chosen. This final solution has to be supported by all participants if possible.

The cooperation between the participants, in particular sharing knowledge processes will be supported by using different tools. The tools used include haptic models, visualizations and simulations. These tools are intended to develop a common understanding and thus convey different perspectives and, if any, organizational hierarchies. In particular, haptic models allow an intuitive access to the issue and possible solutions and help to lift creative potential. Additionally the stringent, structured approach accelerates the innovation process.

The 2- to 3-month period between the workshops is used for documentation and further elaboration of the developed results. Continuous consultation with the workshop participants guarantees targeted and high quality results and consensus. A core group of scientists coordinates and facilitates the workflow of "Fabrica Medica ®" trilogy. They assume coordination and moderation tasks therefore the participants are relieved and can focus on the problem-solving process. Overall, the workshop approach is highly participatory. Common reflection of the results and democratic decision-making about their further use, promote the identification, help to build trust and thus utilize existing problem-solving potential.

In one of the applications of "Fabrica Medica ®", a six-level model for clinical process benchmarking was developed. This model was applied to support information transfer in Intensive Care Units. The following figure 1 shows the used haptic model.

Figure 1: Example of a haptic model to support the sharing knowledge process. Source: Authors.

4.3 Questionnaire

The workshop concept has been conducted on two issues so far. The first trilogy was entitled "hospital building process" and addressed the question of whether the planning process of hospitals can be standardized and normed. The second trilogy was entitled "clinical process benchmarking", and aimed to develop a methodology for process benchmarking of clinical work processes.

Data on the constructs of "person", "collaboration" "design" and "evaluation" were collected. These constructs have impact on the "satisfaction" with the overall workshop quality. They were operationalized using a variety of dimensions and attributes. For the 43 attributes corresponding question items were phrased. For the closed Questions 5-point-scaled answers from 1 (very good) to 5 (poor) or from 0 (not true) to 4 (apply completely) were used. All scales stages were marked with a number and named exactly. The table 2 shows the constructs and their operationalization.

The overall workshop quality is sub-divided into "collaboration" and "design" according to the concept of "cooperative modeling". According to the "cooperation model" the construct "collaboration" distinguishs three dimensions "coordination", "communication" and "shared knowledge". The construct "design" based on the "bridge model of shared knowledge" is divided into the two dimensions "generation of (new) knowledge" and "evaluation of (new) knowledge". For the construct "person" attributes such as discipline or expertise of the workshop participant were collected. Using the construct of "evaluation" assessments of the suitability of the workshop approach and the excellence of the workshop are accomplished.

Table 2: Theoretical constructs and their operationalization. Source: Authors.

construct	dimension	attributes	items
person	characteristics	discipline, expertise, experience, commitment, mindfulness	4
collaboration	coordination	moderation,structure, division of work	6
	communication	information transparency, attitude	4
	sharing knowledge	awareness, tools, common ground	6
design	generation of (new) knowledge	know-how, participation	3
	evaluation of (new) knowledge	result, participation	3
evaluation	approach	„interdisciplinarity“, implementation	2
	quality	excellence of workshop	6

4.4 Results of the Survey

4.4.1 Survey Participants

A total of six workshops were conducted. Five workshops were evaluated by a satisfaction survey. A total of 78 workshop participants took part in the survey. The participants belong to the following disciplines:

- "operating level" (physicians and nurses),
- "hospital management" (principal and controlling),
- "planner" (architects and process planners),
- "financier" (health insurance and government),
- "industry" (medical equipment industry and IT industry),
- "Other"

The most represented group with 32% is the hospital staff on the operating level. In each workshop at least four hospital employees were representing the operating level and participated in the survey. The group of "planners", together with the group of "industry" is the second largest discipline with 20% each. Staff from the hospital management with 14% of surveyed participants is involved. Associates of the disciplines of "financier" (6%) and "other" (8%) are represented in each workshop at least once. They constitute the smallest group.

4.4.2 Reliability and Validity of Questionnaire

The quality of the questionnaire is assessed with the three main quality criteria. The objectivity of the questionnaire is satisfactory. The standardized questions ensures high operating objectivity. Similarly, a sufficiently high level of analysis objectivity and interpretation objectivity is assured by requirements for a standardized evaluation, as well as descriptions of the scales and of the survey population. The reliability of the questionnaire is still very satisfactory. To the constructs "collaboration" and "design" associated scales brought most acceptable to good internal consistency (Cronbach's alpha ranges from α = ,662 to α = ,841) and good corrected item-scale-correlation (up to two exceptions ranges from r_{it} = ,04 to r_{it} = ,07). The scales consist of relatively few items, so the values reached a higher importance. Exceptions are, particularly in the scale "shared knowledge", due to the small sample size and the subject-related differences in the workshops and thus high heterogeneity of the items. The validity of the questionnaire is also satisfactory. The development and evaluation of the workshop approach was systematic and theoretically well-founded. In addition, an expert focus group assessed both the workshop concept and questionnaire. The criterion validity was calculated using the Pearson product moment correlation. All scales experienced significantly correlation (correlation are from r = ,614 to r = ,833; p = ,05). A factor analysis of the scales finally confirmed satisfactory construct validity. The results show the consistency, validity and reliability of the developed questionnaire.

4.4.3 Quality of the Workshop Trilogy "Fabrica Medica ®"

The following three questions were selected to represent the workshop' quality:

- How do you rate the workshop overall?
- How do you assess the approach to deal with this question interdisciplinary?
- How successful did you find the selection and composition of the participants for this workshop?

The majority of the participants (90%) estimated the overall workshop quality as "good" up to "very good". These very satisfied participants come from all six groups. The remaining 10% of the less satisfied participants (workshops has been assessed as "satisfactory" or "sufficient") come from all disciplines except the group "financier". The representatives of the discipline "financier" have always been fully satisfied with the workshops (evaluation only "good" or "very good").

The satisfaction with the workshop is potentially influenced by the question whether the topic should be handled with all disciplines. Therefore, the general attitude of the participants was checked whether they consider the approach of the interdisciplinary collaboration of the topic as appropriate. The overwhelming majority of the respondents (N = 66, 12 invalid responses) agree with the approach. 97% rated the interdisciplinary approach to the topic as "very good" or "good." Only two respondents (3%) rated the approach of each topic as "satisfactory".

The selection and composition of the workshop participants may also have an impact on overall satisfaction with the workshops. Accordingly, the opinion of respondents was collected as well. Almost all responding participants (98.4%) assessed the composition and selection of "very good" to "good".

Additional factor of influencing the evaluation of the workshops could be the operational expertise to interdisciplinary collaboration. Previous studies have shown that persons with experience in interdisciplinary collaboration operators are more efficient in problem-solving when problems arise (Steinheider, 2000). Therefore, in the context of a self-assessment survey respondents are also asked about their level of experience in interdisciplinary collaboration. 80% of the respondents assessed the extent of their experience in interdisciplinary collaboration with at least "good". Therefore the workshop participants can mainly be classified as very experienced. Participants from the group "industry" differ here from the overall picture: only 50% reported a "good" to "very good" level of experience. For the other groups there are at least 80% of respondents who have a "good" to "very good" level of experience. In the disciplines of "financier" and "Other" none of the surveyed workshop participants is inexperienced in interdisciplinary collaboration.

A comparison of the two groups "experienced" (N = 63) and "inexperienced" (N = 15) and their satisfaction ratings show slight differences in the assessment. The group of "inexperienced" judged both the workshop as a whole, as well as the approach and the selection and composition of the participants of the workshops ratherto be only "good" than "very good". Two respondents judged the overall quality of the workshop with "sufficient". Inexperience may be associated with lower frustration tolerance and therefore with dissatisfaction. However, these respondents are experienced in interdisciplinary collaboration according to their self-assessment. Thus their level of experience is not the cause for the extent of dissatisfaction. In figure 2, this observation is represented graphically.

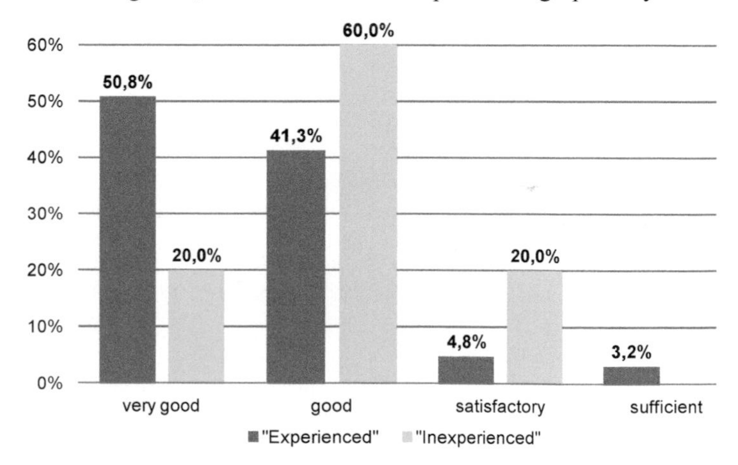

Figure 2: Satisfaction of the workshop participants in accordance to their level of experience in interdisciplinary collaboration. Source: Authors.

CONCLUSION

Overall, the results suggest a high suitability of the workshop concept "Fabrica Medica ®". The survey result shows that the workshop concept supports interdisciplinary design processes of expert teams. The used methods and tools are suitable. Improved collaboration helps to consider multiple perspectives, allowing better results. The survey results show that the workshop concept "Fabrica Medica ®" is suitable to improve interdisciplinary collaboration. In addition, an evaluation of the design results in the future is desirable. The findings should be verified by conducting further design workshops. A workshop trilogy for the design of emergency departments is in progress.

ACKNOWLEDGMENTS

The authors would like to thank the Federal Ministry of Traffic, Construction and Urban Development (BMVBS Bundesministerium für Verkehr, Bau und Stadtentwicklung) and the Representative of the Federal Government for the Federal States in the former East Germany (Beauftragten der Bundesregierung für die neuen Bundesländer Bundesministerium für Inneres BMI) for the research funding. Special thanks go to the participants of the workshop trilogy "hospital building process" as well as "clinical process benchmarking" for their commitment and cooperation (see also www.awb.tu-berlin.de).

"Fabrica Medica ®" is a protected brand of the Institute for Health Care Systems Management Berlin eG (HCMB). For more information please visit www.hcmb.org

REFERENCES

Bienzeisler, B., Kremer, D. and Spath, D. 2007. Wissensintensive Kooperationsprozesse bei der Entwicklung innovativer Produkte. In *Entwicklung und Erprobung innovativer Produkte – Rapid Prototyping,* eds. B. Bertsche and H.-J. Bullinger. Heidelberg: Springer: 70-122.

ISO 6385:2004: Ergonomic principles in the design of work systems. 2nd ed. 2004.

Friesdorf, W., Mendyk, S., Sander, H. und Podtschaske, B. (2011): Arbeitssystem Gesundheitswesen – eine Herausforderung für die Arbeitswissenschaft. In *Mensch, Technik, Organisation - Vernetzung im Produktentstehungs- und –herstellungsprozess*, eds. GfA. Dortmund: Technische Universität Chemnitz: 417- 420.

Haberfellner, R., Daenzer, W. F. 1997. *Systems Engineering. Methodik und Praxis*. Zürich: Verlag Industrielle Organisation.

Hornecker, E., Robben, B. and Bruns, F. W. 2001. Technische Spielräume: Gegenständliche Computerschnittstellen als Werkzeug für erfahrungsorientiertes, kooperatives Modellieren. In *Neue Medien im Arbeitsalltag. Empirische Befunde, Gestaltungskonzepte, Theoretische Perspektiven*, eds. I. Matuschek, A. Henninger and F. Kleemann. Wiesbaden: Westdeutscher Verlag: 193-216.

IEA International Ergonomics Association Council. 2000. What is Ergonomics. Accessed January 18, 2010, http://www.iea.cc/browse.php?contID=what_is_ergonomics.

Steinheider, B. 2000. Cooperation in interdisciplinary R&D Teams. In *Proceedings of the ISATA 2000*, International Symposium on Automotive Technology and Automation, Dublin, Ireland, September 25-27, 2000, Epsom: ISATA Düsseldorf Trade Fair: 125-130.

Wetzstein, A., Dreessen, A., Hoffmann, J. and Hacker, W. 2003. Einsatz und Bewertung von Techniken der Wissensintegration zur Stärkung der Innovationsfähigkeit von Unternehmen. In *Projekt MIK, Projektberichte* (Heft 30), eds. Institut für Psychologie. Dresden: TU-Eigenverlag.

CHAPTER 15

System Ergonomic Analysis, Evaluation and Design of Structures and Work Processes in Health Care

D. Fuchs[1]*., B. Podtschaske*[1]*, W. Koller*[2]*, M. Stahl*[1]*, W. Friesdorf*[1]

[1]Department Human Factors Engineering and Product Ergonomics
Berlin Institute of Technology
Fasanenstr. 1, 10623 Berlin, Germany

[2]Trauma ICU, Klinik für Allgemeine und Chirurgische Intensivmedizin
Innsbruck Medical University
6020 Innsbruck, Austria

ABSTRACT

The health care system as a work system is under increasing pressure: a growing amount of work is facing a lack of personnel resources. Constantly increasing quality requirements from different perspectives as Patients, Cost Carriers and Clinical Management are also challenging factors. There is a need for improving treatment processes by making them comparable and therefore transparent workflows are necessary.

The Department of Human Factors Engineering and Product Ergonomics has developed solutions for process benchmarking of patient care in a three-stage workshop-concept. A 6-layer model of patient treatment for planning and managing treatment processes and controlling resources is developed. The left side of the model reflects the treatment processes at different system layers. Compared to the definition of resource requirements (demand) the provision of resources (supply) is shown on the right side of the model. The connection of both sides is represented by bridges on each system layer. To coordinate and control the resource allocation to the needs of patient treatment processes there are so called bridge managers.

It serves as a basis for improvement projects, better predictability and controllability of treatment processes.

Keywords: processes in health care, process benchmarking, system analysis

1 INTRODUCTION

Today's Medicine is excellent. The latest medical findings are globally exchanged, evaluated and further developed. The survival chances of individual diseases, especially in the industrialized countries have already risen significantly and will continue rise (WHO 2011). Increasing specialization in the medicine, better and innovative medical technology and improved information transparency are further challenging factors to the medical experts. "Over a long period the health care system has developed into a highly differentiated and networked expert system. It is able to cope with unexpected, difficult and time-critical situations (e.g. infarction). Hospitals are a good example how a large number of such different patients can be treated at the same time" (Friesdorf et al GfA 2011). Overall the requirements of quality- and safety-standards of clinical treatment processes are constantly increasing. Additionally economic aspects in patient care gain a higher importance also.

There is a need for improving treatment processes by making them comparable. This requires transparent workflows. Benchmarking is an appropriate method that has been proven successfully in different industry sectors for decades. Of course it cannot simply be transferred but must be adapted to the particular structures of the health care sector.

Medical experts have seen the need for process optimization years ago and have begun to define clinical pathways to standardize the processes as far as possible. This is also to handle increasing complexity due to more specialized medicine and extensive technology applications. But the individuality of a patient leads to complex processes and a directly passing through is no longer possible. A straight visualization of treatments is often difficult. Process planers also neglected the point that not only standard-procedures and treatments exist in the medical field. Within the treatment processes of a single patient there are always individual decisions needed and taken. So that either individual parts of the treatment have to be modified on the patient treatment or parts of the treatment process are even deviating completely from the standard. This is where completely individual treatment processes are constructed.

Therefore, a model to demonstrate and structure clinical treatment processes must be compatible for each of these three types of treatment (1 standardized, 2 modified, 3 completely individualized). The model also needs sufficient flexibility to adapt individual patient treatment (Friesdorf AHFE, 2010). Here the Ergonomics and Human Factors, with its extensive knowledge in process analysis and optimization, can give support.

2 OBJECTIVES

The objective of this paper is to illustrate the development and description of an System Ergonomic Approach for analysis, evaluation and design of structures and work processes in health care. On the basis of medical treatment procedures, the relationships between medical management and the management of personnel and substantial resources are described. System ergonomic methods and tools should be used and transformed to the health care work system.

3 METHODS

The Department of Human Factors Engineering and Product Ergonomics has developed solutions for process benchmarking of patient care in a three-stage workshop-concept (Podtschaske et al. 2011). Approximately 30 experts, including physicians, nurses, process planners, hospital management, medical industry, experts of clinical quality management and ergonomics developed the requirements for a clinical process benchmarking. The whole approach is conducted through the problem solving cycle of systems engineering (Daenzer et al., 1992) also other preliminary work by the department (Carayon & Friesdorf 2006; Friesdorf & Marsolek 2009) is used.
The first workshop "Introducing" is to integrate knowledge of all participants. For this purpose a sub-process of an intensive care unit was simulated and discussed. The second workshop "Matching" is the synthesis of various solutions. In this stage only a small panel of experts (approx. 10 participants) adapted and specified the results. The third workshop "Agreeing" is to find a consensus with all 30 participants again. The individual workshops are carried out at intervals of three months. The time in between is used to elaborate and prepare the intermediate results from each workshop.
For the evaluation of the model key treatment processes of an emergency room are exemplary analyzed and tasks of the management roles are defined.

4 STRUCTURE OF A SYSTEM ERGONOMIC 6-LAYERS MODEL

The main result of the three workshops is a three dimension 6-layer model of patient treatment for planning and managing treatment processes and controlling resources. The left side of the model reflects the treatment processes at different system layers. They define the particular resource requirements to perform the tasks. Compared to the definition of resource requirements (demand) the provision of resources (supply) is shown on the right side of the model, as shown in Figure 1. The connection of both sides is represented by bridges on each system layer. There are so called bridge managers to coordinate and control the resource allocation to the needs of the patient treatment processes.

The distribution of responsibility must be clearly defined. For each layer management roles are defined on the demand as well as on the resource side. At each layer there is the role of a "medical bridge Manger" (MedBrM) on the left side and a "Resource Bridge Manger" (ResBrM) on the right side.

Figure 1 Illustration of the system ergonomic 6 layers Model

Vertically, the model is divided into 6 system layers (E) corresponding to the "natural" or organizational layers of responsibilities. The left side is dividing a treatment task into sub-treatment tasks with increasing granularity: the model as a whole can be viewed as a system classification with a more precisely granulation of the six layers, beginning with the top layer (E1) and ending with the lowest layer (E6).

E1) **Cases**: Classification by case groups (for example DRGs in Germany etc.) considering the overall treatment of an individual patient

E2) **Stages**: Sequences of treatments in structures of different sectors (general practitioner, ambulance, hospital, rehabilitation), the linearity of the treatment process is resolved by the decision modules - such as outpatient or inpatient care of patients

E3) **Stations**: Components of the treatment in different organizational units within one structure, e.g. in hospital: emergency room, operating room, intensive care unit

E4) **Phases**: Sub-tasks within a station- the sequence of the phases of the patient in a particular “Station”

E5) **Modules**: Structuring a treatment “Phase” into diagnosis and patient-specific bundles of individual medical interventions, such as cardiovascular treatment of heart failure

E6) **Arrangements**: All individual medical measures, such as administration of an analog sedation, blood pressure measurement.

The chronological structure of the treatment is illustrated from left to right. Parts of treatments can be done in serial or parallel. Analog to a project management plan it is not possible to jump backwards (loop). From layer 2 to 6 decision making elements are integrated to adapt the treatment process to individual patient.

5 SYSTEM OF MANAGEMENT ROLES IN TREATMENT PROCESSES

The objective of the presented model is to combine the medical tasks with the resource management tasks, connected by bridges on each layer. For each layer management roles are defined on both sides, for managing treatment tasks (demand) on the left side and for the resources management on the right side of the model as already introduced. The 6-Layer model should create the conditions for a optimal communication, coordination and shared-knowledge between all actors involved in the treatment process. These three sub-processes are necessary for successful collaboration (Steinheider 2000).

All bridge managers are responsible for the coordination of the medical treatment to the resources allocation at each layer. It is important to know, that we talk about roles and not individual persons. There can be horizontal, as well as vertical roll-assumptions. This allows the coordination of different process modules within a layer. I.e. the MedBrM on layer 4 (Phases) coordinates various “Phases” of the patient treatment within one particular “Station” (admission, diagnosis, intervention, discharge / transfer, etc.). The ResBrM, which is on the opposite of the MedBrM is responsible for planning, allocating and assigning the required resources. In the Model the medical experts can jump from the MedBrM into the ResBrM-roll of the same layer or from layer 4 MedBrM- to layer 3 MedBrM-roll.

Only if the distribution of responsibility is clearly defined and well established, an efficient and stable planning and controlling of treatment processes is possible. Due to the different granulation of the 6 layers the "bridge managers" are considering different time lines. A long planning interval (concerning the upper layers of the model: E1-3) can consider many patients simultaneously, but only a rough planning (e.g. operating plan) is possible. The focus here is on planning and managing an organizational unit. A short planning interval (concerning the lower layers of the model: E4-6) provides a detailed view of the treatment process, but then only a single patient can be considered (e.g. an intervention of a patient in the emergency room). The focus here is the planning and controlling of a specific treatment process of an individual patient.

A clear and transparent structure of the management responsibilities of the "bridge managers" at all layers ensures an optimal coordination between the different layers. E.g. a transfer of a patient from the emergency room to a peripheral ward needs coordination between two physicians, including the ward managers. Thus a layer of interdepartmental coordination is necessary. The differential term of the bridge manager of each layer is derived from the name of the layer above. For layers 4-6, this means: Bridge manager on layer 4 (Phases) is called "Station Manager" and is responsible for the smooth running of the "Phases" of all patients within a "Station" (e.g. emergency room). The bridge manager on layer 5 is the "Phase Manager" and is responsible for "Modules" within a "Phase" of a particular patient treatment. The bridge manager on layer 6 is the "Module Manager" and is responsible for coordinating the "Arrangements" of one particular "Module".

The system of management roles is evaluated for processes in emergency rooms. First the treatment processes of key patient treatments were structured in layer 4 and 5 and on this basis bridge manager's tasks were derived. Table 1 represents exemplary bridge manager's tasks for E4 and E5 in emergency rooms.

System Layer	Tasks MedBrM Roll	Tasks ResBrM Roll
E4 Phases: Treatment process within an organizational unit	Decision of planned patient admission, Coordination and decision of patient treatment; Analysis/evaluation of treatment plan and patient's status; Management of patient discharge	Overall staff planning, Annual allocation of staff (vacation scheduling), Monthly allocation of staff (shift planning), Short-term allocation of staff ; Equipment requirements planning, Planning equipment purchases and equipment ordering, Planning and coordination of equipment use, Requirements planning, Purchase and coordination of material; Requirements planning of medication; Definition and decision of the unit's medication lists; Drug supply; Planning an decision of hospital admission/transfer, Allocation of physician/nurse to patient or task or room
E5 Modules: Division of treatment phase into diagnostic and patient specific treatment procedures, e.g. analgesic sedation, cardiovascular treatment for cardiac insufficiency	Decision on patient transport, Coordination of patient admission; Planning and performing of initial assessment (anamnesis); Determination of diagnosis- and intervention plan; Decision on tentative diagnosis; Decision on Consultation request resp. further diagnostics; Planning of patient transport to diagnostic unit, Analysis of diagnostic findings; Decision on treatment plan; Preparation of patient information; Preparation of detailed information if necessary	Planning and performing of functional tests of equipment and feedback, Preparation and supply of examination unit (treatment room, examination area/room), Medical equipment and examination couch; Definition of Physicians and Nurse in the shift plan.

6 CONCLUSION

The model is mainly used as a reference point for process analysis, problem screening and target process definition. So far, the model has been used only exemplary for processes of Emergency Rooms and Intensive Care Units and must now be transferred to many other processes in the health care work system. It serves as a basis for improvement projects, better predictability and controllability of the treatment processes. The measurability of the processes under consideration of the individuality of each patient or the individual hospital is guaranteed.

REFERENCES

Carayon, P. & Friesdorf, W. 2006, Human Factors and Ergonomics in Medicine. In: G. Salvendy (Hrsg.): Handbook of Human Factors and Ergonomics. Hoboken: Wiley, 1517–1537.

Daenzer, W.F. & Huber, F. 1992, Systems Engineering. Zürich: Verl. Industrielle Organisation.

Friesdorf, W. & Marsolek, I. 2009, Medicoergonomics – A Human Factors Engineering Approach for the Healthcare Sector. In: C.M. Schlick (Hrsg.), Industrial Engineering and Ergonomics. Visions, Concepts, Methods and Tools. Festschrift in Honor of Professor Holger Luczak. Berlin: 165

Friesdorf, W. 2011, Arbeitssystem Gesundheitswesen – eine Herausforderung für die Arbeitswissenschaft; In: Gesellschaft für Arbeitswissenschaft (Hrsg.), Mensch, Technik, Organisation - Vernetzung im Produktentstehungs- und Herstellungsprozess. Dortmund: GfA Press

Podtschaske, B., Mendyk, S., Sander, H. & Friesdorf, W. 2011, Entwicklung und Evaluation einer Workshop-Reihe zur Verbesserung der interdisziplinären Expertenkooperation im Gestaltungsprozess von ergonomischen Arbeitssystemen. In: Gesellschaft für Arbeitswissenschaft (Hrsg.), Mensch, Technik, Organisation - Vernetzung im Produktentstehungs- und Herstellungsprozess. Dortmund: GfA Press, 207-210.

Steinheider, B. (2000). Cooperation in interdisciplinary R&D Teams, Proceedings of the ISATA 2000, International Symposium on Automotive Technology and Automation (33rd: 2000 Sep: Dublin) (pp. 125-130). Epsom: ISATA-Düsseldorf Trade Fair, Germany.

WHO World Health Organization (2011). Global status report on noncommunicable diseases 2010. Description of the global burden of NCDs, their risk factors and determinants: World Health Organization.

CHAPTER 16

Supporting Structures for Information Transfer in ICUs

G. Yucel Hoge[1,2], *D. Fuchs*[2], *W. Koller*[3], *W. Friesdorf*[2]

[1]Department Industrial Engineering
Istanbul Technical University
Maçka, 34367 Istanbul, Turkey
yucelg@itu.edu.tr

[2]Department Human Factors Engineering and Product Ergonomics
Berlin Institute of Technology
Fasanenstr. 1, 10623 Berlin, Germany
daniela.fuchs@awb.tu-berlin.de, wolfgang.friesdorf@awb.tu-berlin.de

[3]Trauma ICU, Klinik für Allgemeine und Chirurgische Intensivmedizin
Innsbruck Medical University
6020 Innsbruck, Austria
wolfgang.koller@i-med.ac.at

ABSTRACT

Information management in an Intensive Care Unit (ICU) has to guarantee that each staff member knows what he/she has to know in order to make patient treatment effective, safe and efficient. In this study, it is aimed to develop structures to support information transfer between patient treatment and resources in ICUs for typical patient treatment with several days duration, from the admission till the discharge. The six system layer of patient treatment model, which is a proposal to link patient treatment to resources, is used as a basis to improve information transfer structures in an ICU. The proposed information transfer model can support an efficient and safe information transfer between physicians and nurses during a patient treatment in an ICU.

Keywords: Intensive Care Units (ICUs), Information Transfer, Patient Treatment, Resource Management, Bridge Managers (BrMs)

1 INTRODUCTION

Patient treatment in a high dependency environment such as an Intensive Care Unit (ICU) is a very complex and dynamic process. Complex work processes occur due to each patient's health status, unpredictable treatment dynamics, unavoidable ethical problems and increasing fragmentation of the entire processes (Friesdorf and Marsolek, 2007). Thus, acquisition, transferring and documentation of data and information are key elements for the coordination of ICU work processes with many shared tasks and overlapping activities (Rothen, 2010). However, there is no structured way of communicating data and information between shifts and units; there are often simply habits and common practices. Analyzing information transfer such as hand-over processes is a new subject of Human Factors investigations (Donchin, 2010). According to Miller (2010), ICU care coordination involves two distinct information spaces; patient information space and resource information space and optimal care depends on adaptive coordination across these information spaces. However, the mechanisms and support structures associated with transition between these spaces are unclear and future research is needed to explicit the transition be-tween these two spaces of information. Therefore, in this study it is aimed to de-velop structures to support information transfer between patient treatment and re-sources in ICUs.

Table 1 Definition of the 6 Layers and Bridge Managers

Definition of the System Layers	Definition of MedBrM	Definition of ResBrM
Layer 1: Case groups	Medical Bridge Manager Layer 1	Resource Bridge Manager Layer 1
Layer 2: Stages (treatment)	Case manager	Resource Manager Layer 2
Layer 3: Stations (one treatment within a hospital as Organization)	Stage manager	Resource Bridge Manager Layer 3
Layer 4: Phases (of one treatment within a "Station")	Station manager	Station Resource Manager
Layer 5: Modules (of one treatment "phase")	Phase manager	Resource Bridge Manager of one Work place
Layer 6: Interventions (of one treatment "module")	Module Manager	Resource Bridge Manager of Materials and Medications

The 6-layer model of patient treatment which is a proposal to link patient treatment to resources (Friesdorf et al., 2011), is used as a basis to improve an information transfer structure which can guarantee that each staff member knows what he/she has to know in an ICU. The model differentiates six system layers of patient treatment. In this model, each layer contains two sides: the left side represents the process of patient treatment; the right side shows the required resources (see Table 1). Both sides are linked by bridges on which Bridge Managers (BrMs) coordinate patient treatment (medical BrMs) and the provision of resources (resource BrMs). BrMs represent roles which have to be fulfilled by clinical staff. In the model, the implementation of the BrMs` roles promotes coordination between the levels and within levels and thus a stable planning and controlled processes. Defining these roles and identifying information transfer between these roles makes it possible to design a supporting structure for information transfer between patient treatment and resource management in an ICU. To define the required information transfer between BrMs, interviews with head physicians and head nurses of ICU units of two different German Hospitals were conducted.

According to the interview results, the tasks of BrMs, information needs of BrMs and information transfer between BrMs during a typical patient treatment in an ICU are defined. The interview results are merged to build a "To-be" model to support an efficient and safe information transfer between physicians and nurses who fulfill the roles of the BrMs.

2 METHOD

The method of the study is shown in Figure 1. The "To-Be" model for information interchange in an ICU is obtained through the following steps; in the first step, in two different German hospitals "As-Is" models for information interchange in ICUs are obtained through interviews with head nurses and head physicians from ICUs. In the second step, these models were analyzed and combined to build a "To-Be" model with the help of 6-layer model of patient treatment and expert interviews. In the third step, the proposed "To-Be" model is evaluated with interviewees from the two hospitals. In the last step, based on the evaluations the refined version of the aimed "To-be" model is defined.

The interview questions were developed to find out which bridge manager is involved, which information he/she needs, generates and provides during a typical patient treatment. The focus of the interview was defined as a typical patient treatment with several days of duration, from the admission till the discharge. Based on the literature review, main steps of a typical patient treatment are defined as admission, treatment, monitoring, evaluation and discharge. Main tasks of resource management are defined as bed-, staff-, device-, material- and medication-management. The listed questions (see Table 2) are asked for each of the patient treatment steps and for the resource management tasks. Most of the interviewees were familiar with the 6- system layers of patient treatment model since they participated in the model development workshops. For the ones who were not familiar with the model a description of the model and BrMs roles was given before the interviews. Interviews were done in two rounds. In the first round, the questions were asked and the answers are recorded. After the answers were transcribed and analyzed, a second interview with each interviewee was conducted to confirm the results and to complete the missing topics. Two As-Is models of information transfer which were obtained by interviews were used to design a To-Be model with an ICU expert and project members in 4 meetings.

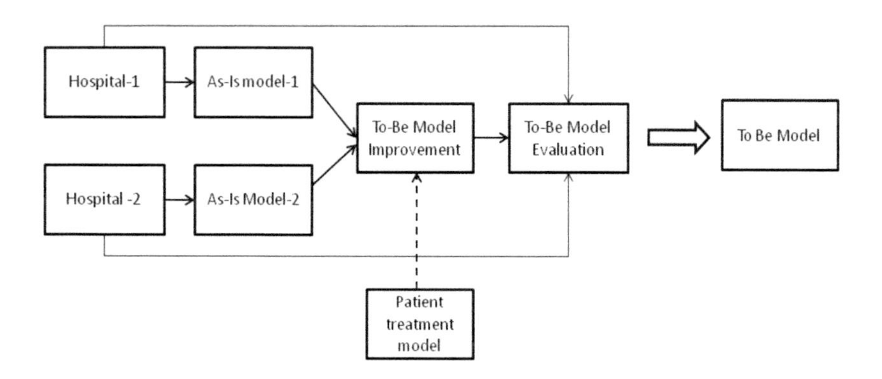

Figure 1 Process of the method of the study

Table 2 Interview questions

Interview questions
What are the main tasks for [patient treatment / resource management]?
According to your opinion, which level does this task belong to?
Who is involved in the task?
In your opinion, does this actor perform a bridge manager role? And why?
What kind of information is needed to perform the tasks?
How is the needed information be provided? (person, paper/electronic health record, white board)
Which information does the actor generate?
To whom does the actor transfer this information?
How is it transferred?

3 RESULTS

Interviews with ICU head physicians and ICU head nurses in two ICUs from two different hospitals were conducted. Both ICUs were post-operative care units. The first ICU has 26 beds. The physician/patient and nurse/patient ratio was: 1/4(5) and 1/2(3) respectively. Occupation rate of the unit is between 95%-97% and each workday 3-5 patients are admitted. The second ICU has 18 beds. The physician/patient and nurse/patient ratio 1/5 and 1/2 (3) and the occupation rate of unit is %100 and each workday 3-5 patients are admitted.

Based on the interviews, As-Is models of information interchange in ICUs were analyzed and reported for each task by the following categories: task level, BrM level, incoming information, information sources, incoming information sources, incoming information transfer tool, out coming information, information receivers and outcoming information transfer tool. According to As-Is models, difficulties are mostly found in the coordination between the layers, so clearly defined and transparent responsibilities of the BrMs role are highly important for an optimized coordination between different layers. Another problem found were undefined information needs for each task. Subsequently the task-actor misses important information, necessary to perform the task. Searching, retrieving and learning all this missing information while performing the task itself, slows down the action and costs substantial work time.

As-Is models were used to develop a To-Be model for information transfer which describes the necessary information for each role. Based on the To-be model, tasks of BrMs, information needs of BrMs (see Appendix 1) and interactions between BrMs were defined during a typical patient treatment in an ICU. Also, the main interactions between actors are grouped and shown in the Figure 2. According to Figure 2, the predominant information flows at the medical (left) side from top to bottom are decisions and orders, from bottom to top reports of treatment and interventions are the most frequent. At the resource (right) side from top to down, information about available resources and allocation of this resources flows. From bottom to top, reports of usage of resources are the main information content. The basic interaction from left to right (medical-side to resource-side) is the demand for patient care. From right to left (resources to medicine) information about available resources and feedback to resource-demands are given. It can be demonstrated by means of the model, that the Medical Bridge manager at level 5 holds the central role of information transfer. Based on the definitions of information flows between bridge managers, it would be possible to build or improve an information management software tool to support bridge managers` tasks.

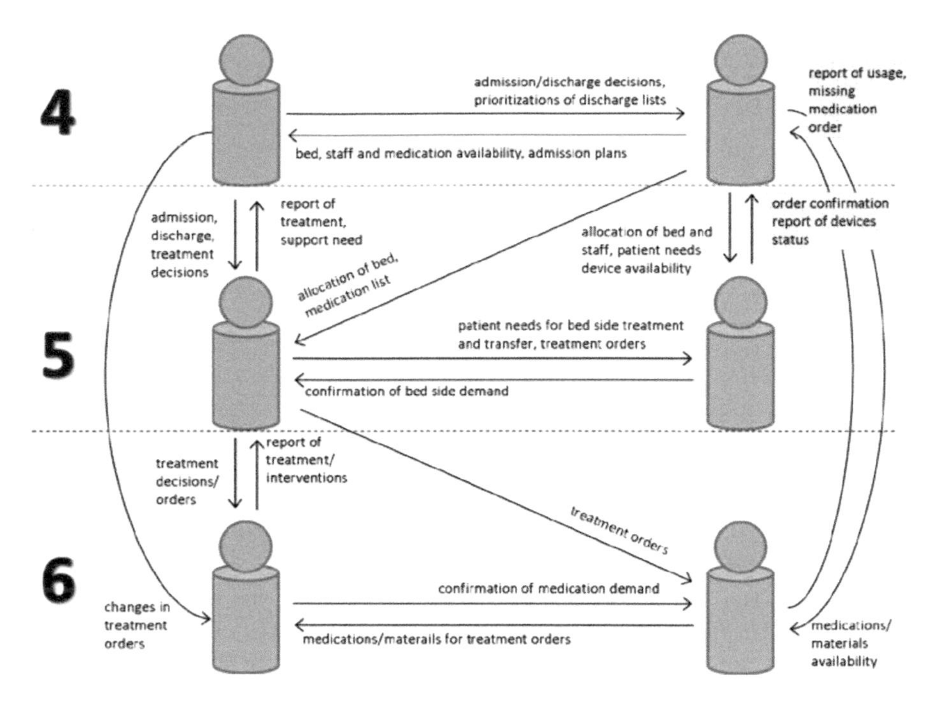

Figure 1 Process of the method of the study

4 CONCLUSIONS

The To-Be Model defines information needs and transfers during a patient treatment in an ICU. It develops the functional roles of Bridge-Managers as crystallizing points in the line of information. The position of the Bridge-Managers in the network of information (horizontally and vertically) can be fixed, and their importance is emphasized. In the future, after validating the model, it seems to be possible to develop checklists and software supporting the information transfer during patient treatment in an ICU could be also developed based on this model.

ACKNOWLEDGMENTS

The authors would like to thank Dr. Dirk Pappert, Dr. Ralf Groessle, Dr. Suanne Tuossaint and all staff from ICU units of Ernst von Bergmann Hospital in Potsdam Vivantes Klinikum in Berlin-Neukölln for their support through the research.

REFERENCES

Donchin, Y. 2010, The role of Ergonomics in Modern Medicine; In: Organization and Management in ICU (Eds: Flaatten, H., Moreno, R.P., Putensen, C. and Rhodes, A.), 251-257, Mwv Medizinisch Wissenschaftliche Verlagsges, Berlin.
Friesdorf, W. and Marsolek, I., 2007, Medico-Ergonomics in ER, OR & ICU – What is state of the art in the design of medical products, SCATA/ESCTAIC Meeting.
Friesdorf, W. 2011, Arbeitssystem Gesundheitswesen – eine Herausforderung für die Arbeitswissenschaft; In: Gesellschaft für Arbeitswissenschaft (Hrsg.), Mensch, Technik, Organisation - Vernetzung im Produktentstehungs- und Herstellungsprozess. Dortmund: GfA Press.
Miller, A., Weinger, B. M., Buerhaus, P. and Dietrich, Mary, S., 2010, Care coordi-nation in Intensive care units: communicating across information spaces, Hu-man Factors, 52 (2), 147-161.
Rothen, H. U. 2010, Organizing the workflow in ICU; In: Organization and Man-agement in ICU (Eds: Flaatten, H., Moreno, R.P., Putensen, C. and Rhodes, A.), 225-239, Mwv Medizinisch Wissenschaftliche, Verlagsges, Berlin.

APPENDIX: TASKS OF BRMS, INFORMATION INPUT AND OUTPUT

MedBrM4		
Input	**Task**	**Output**
ICU bed request patient identification operation /diagnosis preoperative examination findings previous relevant diseases the treatment plan from previous unit the reasons for ICU bed requirement	(planned) ICU admission decision	confirmation of bed request alternative solutions isolation need admission decision patient information
emergent ICU bed Request patient history emergency details	Unscheduled Admission decision	confirmation of emergency bed request emergency patient handover plan bed request to send the new patient to another ICU/IMC
synthesized Report of Patient status data	General Evaluation of the treatment and patient	general judgment of patient course and procedures definition of treatment goals discharge decision treatment change
discharge decision with risk assessment	Managing scheduled Discharge	prioritization of discharge list bed request to next unit

ResBrM4		
Input	**Task**	**Output**
ICU bed request according to surgery plans	weekly bed planning	ICU gross bed assignment plan over capacity situations
confirmation of bed request necessary space for emergencies	daily bed organization for scheduled admission	bed assignment list for the day list of refused patients feedback about admission plan of patient patient specific needs
confirmation of emergency ICU bed request bed availability of ICU	bed organization for unscheduled admission	bed allocation to a Patient patient specific needs (e.g. isolation need, equipment need) no bed availability
business plan of the hospital epidemiological variables job market information	general staff capacity planning	staff request with definition of their qualification and number
staff information future demand	yearly staff planning	yearly staff availability plan internal organizational changes capacity plans
yearly staff availability plan staff information	monthly shift planning	staff schedule shift leaders general information about staff-overload or underload
staff schedule for all ICU staff daily bed booking list unexpected staff shortage information	short term (not planned) staff management underload and overload	singular decision of send staff home staff requirement from other units short term changes about staff scheduling
patient related information nurse availability and qualification	distributing of nurses to patient or procedure	nurses in charge of a specific patient or procedure
patient related information physicians availability and qualification	distributing physician to patient or procedure	physician in charge of a specific patient or procedure
devices' locations devices' returning times to ICU	device movement management	report of devices movements
not enough devices	missing device order	device request
device needs technical feature, price	new device demand	new device request
materials stock level physicians orders material turn over	ordering and stocking materials	materials order

confirmation of orders	Providing materials	documentation of materials arrivals
hospital pharmacy list medication turn over treatment standards	defining ICU standard medication list	standard medication order list
standard medication order list other medication orders	Providing medications	confirmation of medication orders documentation of medications arrivals

MedBrM5		
Input	**Tasks**	**Output**
Pre-notification about patient is ready to be transferred	planning patient transfer	fixed transfer time needs for patient transfer
relevant information happened in the last hours	patient handover (admission)	elements for treatment plan
vital parameters (CV, LV and KI of patient)	first examination	statement of acute patient status decision of necessary procedures
all patient information	defining first treatment plan	patient care goals diagnosis steps of treatment treatment orders
patient information (first response to first treatment, unsolved problems)	establishing/ confirming diagnosis	diagnosis
gap between current diagnosis and new patient findings control points	decision of diagnostic need	diagnostic request (e.g CT, X-ray)
possible appopintment times	planning patient transport to diagnostic	fixed agreement (e.g.transfer time) needs for patient transfer
consultation need (e.g. gap between current diagnosis and new patient findings)	planning consultation visit	consultation meeting time
problem oriented patient information	therapy plan of underlying disease	changes in treatment plan
control points time or event triggered patient data	repeated detailed evaluation of treatment	result of evaluation (e.g. new medications) support of other physicians or head physician synthesized patient status data

all results created by diagnostic tools results from physical examinations special medications other problems of patient	transfer information preparation	comprehensive and structured patient report
patient information request	provide detailed patient information by request	related patient information

ResBrM5		
Input	**Tasks**	**Output**
new patient arrival time special patient needs (e.g. isolation) special equipment needs of patient	bed preparation	bed readiness confirmation
Decision of regular testing and their cycles	regular Device check-ups	fault report or positive report

MedBrM6		
Input	**Tasks**	**Output**
treatment orders	defining needs for interventions	medications/materials for treatment orders
decision of necessary procedures patient care goals diagnosis steps of treatment treatment orders changes in treatment plan result of evaluation (e.g. new medications) confirmation of medications	conducting treatment	report of treatment/interventions
comprehensive and structured patient report	patient handover (discharge)	presentation of patient (relevant factors)

ResBrM6		
Input	**Tasks**	**Output**
medication order	Checking medications	confirmation of medication
		missing medication order
medication order treatment order (e.g. blood pressure)	Providing medications, materials and equipment at the bed side	report of materials/medications usage

CHAPTER 17

Improving Patient Safety through an Ergonomic Technical Solution

M. Stahl[1], *N. Boehning*[2], *L. Kroll*[3], *B. Kujumdshieva-Boehning*[2], *R. Somasundaram*[3], *W. Friesdorf*[1]

[1]Department Human Factors Engineering and Product Ergonomics
Berlin Institute of Technology
Fasanenstr. 1, 10623 Berlin, Germany
[2]iDoc Institute for Telemedicine and Health Communication
Posthofstr. 8, 14467 Potsdam, Germany
[3]Department of Emergency Medicine
Charité - Universitätsmedizin Berlin
Campus Benjamin Franklin
Hindenburgdamm 30, 12200 Berlin, Germany

ABSTRACT

The number of dementia patients that need inpatient hospital care is growing. The unfamiliar environment, strangers, high noise level and the hectic of hospital routine have a negative impact on the patient's cognitive status resulting in "challenging behavior". Dementia patients need additional and specialized care but high workload and limited personnel resources prevent staff to respond adequately to the patient's special needs. Assistive technology might help to address the special needs of dementia patients without increasing the workload of the medical staff. For its successful application a holistic approach to ensure ergonomic quality is necessary. The objective of the present contribution is to introduce such a holistic approach and to illustrate its application during a case study. The results show that the presented approach is suitable for integrating ergonomic principles into the development process of assistive technology. Further development of the approach, i.e. identification and evaluation of suitable methods, is planned and in progress.

Keywords: Assistive Technology, Dementia Care, Medico Ergonomics, Patient Safety

1 SITUATION AND PROBLEM

The number of people with dementia is growing. The Alzheimer Disease International estimates an increase from 35.6 million in 2010 to 65.7 million in 2030 (WAR, 2010). Hence, hospitals have to respond to an increasing number of dementia patients that need inpatient hospital care (Angerhausen, 2008). The hospitalization of patients with dementia leads to several problems for patient as well as hospital staff.

The unfamiliar environment, strangers, high noise level and the hectic of hospital routine have a negative impact on the patient's emotional and cognitive status and may result in "challenging behavior" (Angerhausen, 2008): the patient is not able to actively support the treatment process or may even refuse the treatment. Attempts to wander around (Mavundla, 2000) or even leave the hospital (Müller et al., 2008), screaming (Mavundla, 2000) and aggressive behavior (Wojnar, 2003) are common for dementia patients. This jeopardizes the well being of patient, fellow patients as well as medical staff. Dementia patients need additional and specialized care (Alzheimer's Society, 2009) but high workload and limited personnel resources prevent staff to respond adequately to the patient's special needs.

The question arises how adequate care can be provided for dementia patients without demanding too much resources. One promising possibility is to support dementia patients and medical staff by assistive technology. For its successful application the ergonomic quality plays a key role. The requirements of the future user(s), i.e. dementia patient and medical staff, have to be considered during the development process. This requires adequate methods for identifying the user's requirements and integrating them actively into the product development process. Different methods for user integration are already successfully applied in the design of workplaces or products (e.g. Czaja & Nair, 2006). The medical work system is characterized by its complexity. Existing methods of user integration have to be adapted to address this complexity (Carayon & Friesdorf, 2006; Friesdorf & Marsolek, 2009).

2 OBJECTIVE AND METHODOLOGICAL APPROACH

The objective of this paper is to introduce a holistic approach to ensure ergonomic quality during the design of medical work systems. It can help to integrate ergonomic principles into the development of assistive technology.

The development of the approach is based on the problem solving cycle of the Systems Engineering approach (Haberfellner et al., 1997) and includes the following steps:

1. Identification of existing procedural models on ergonomic design (analysis)
2. Development of holistic approach (synthesis)
3. Identification and integration of domain specific models (analysis)
4. Identification and integration of methods (synthesis)

The approach is applied and evaluated during a case study. The content of this study is the development of an assistive technology to support the care of dementia patients.

3 RESULTS

Holistic approach to ensure ergonomic quality of medical work systems

An initial analysis identified three existing procedural models: The EQUID approach aims to integrate ergonomic knowledge and methods in the design process of products and services (Bruder, 2011). Normative guidelines such as DIN EN ISO 6385: 2004 define a process to integrate ergonomic principles in the design of work systems (DIN, 2004a). When developing medical products special normative guidelines apply: DIN EN 60601-1-6 as well as DIN EN 62366 defines a usability engineering process for the development of medical products (DIN, 2004b; DIN, 2008). The procedural model for organizational development describes a standardized procedure to support changes in organizations (Vahs, 2007). Based on these three models an approach was developed to support ergonomic quality in medical work systems. The approach consists of 3 different levels (figure 1).

The first level defines each development phase and their corresponding tasks. The second level includes existing models to ensure ergonomic quality. The development of assistive technology for medical work systems requires the adaptation of existing models of ergonomic analysis and evaluation. Compared to the industry the medical work system is highly interdisciplinary and interprofessional, resulting in a very complex work system (Carayon & Friesdorf, 2006).

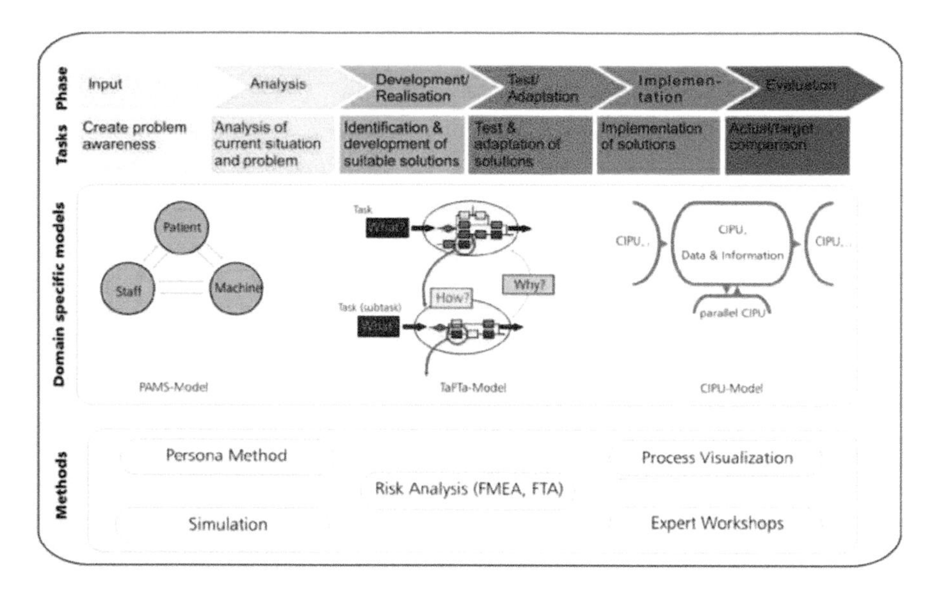

Figure 1: Holistic approach to ensure ergonomic quality in medical work systems

To address this complexity it is necessary to adapt and expand existing models to domain specific models. In the literature this enlargement is named the "medico ergonomic approach" (Friesdorf & Marsolek, 2009). The following domain specific models are integrated into the approach:

The extended Patient-Staff-Machine-Model (PAMS-Model)

In order to ensure ergonomic quality of assistive technology is it necessary to enhance the well-known human-machine interaction model used in the industry. For this purpose the PAMS-model divides the system element "human" into the patient on the one hand and into physician/nurse on the other hand. This differentiation is based on the assumption that the patient is not only a passive working object but also an active co-worker (Friesdorf, 1990).

The recursive hierarchical Task-Process-Task-Model (TaPTa-Model)

The universal approach of the patient-staff-machine-model does not define priority system boundaries; therefore the examination level can be varied corresponding to the investigation focus. Depending on which medical procedure is considered as a work task, a more or less distinguished picture can be detected by decomposition or aggregation of workflows, information flows, material flows etc. This flexible focusing on different levels is described by the Task-Process-Task-Model (Friesdorf & Marsolek, 2009). The TaPTa-Model helps to identify and define tasks and their responsibilities that are necessary to fulfill the task. This serves as a basis for specifying the division of work between human and assistive technology.

The Clinical-Information-Process-Unit-Model (CIPU-Model)

During patient treatment various data and information is produced and used by the involved medical staff. Therefore a careful passing on of this information is necessary and must be adequately supported by technology (Henke et al., 2008). The Clinical-Information-Process-Unit-Model is used to describe these complex information processes by dividing them into units called CIPU. Every CIPU starts with receiving information and ends with transferring information. The CIPU-Model helps to identify and define the information and documentation requirements during the treatment process (Friesdorf et al., 1994).

The third level includes specific methods to help ensuring ergonomic quality during the development process.

Case study: developing assistive technology for dementia patients

The original idea to create a technical solution for dementia patients was developed during a student's project at the Hasso Plattner Institute – School of Design Thinking, University of Potsdam. During this project a cover was developed that helps to soothe the patient by protecting him/her from the disturbing environment. The cover can be used in different wards such as emergency room or geriatric ward. The transformation of this idea into a market-ready product is the objective of a current research project. An interdisciplinary team including physicians (emergency medicine, neurology and geriatrics), nurses, media- and industrial designers, engineers, and ergonomists, is involved in this research project.

The original idea was used as an input during phase 1 of the approach. During the phase 2 a detailed analysis of the context of use was necessary. Domain specific models such as the PAMS-model and TaPTa-model can help to identify requirements by analyzing the potential users and their workflows.

The results of this analysis were discussed and verified during an expert workshop. During the workshop the discussion was focused on the use in a geriatric ward. The following requirements play a central role for cover, media and telemedical application, e.g.:

- Self explanatory, fast and easy to use (even under hectic conditions in the emergency room)
- Quick access in case of emergency
- Fulfillment of hygienic requirements
- No restriction of patient's mobility
- No removable components that can hurt patient or staff
- Music should be hearable for the patient but should not disturb other patients

During phase 3 of the approach a prototype was developed. It consists of a cover that is mounted on the hospital bed (figure 2). To reinforce the soothing effect, a media player to play "relaxing" music complements the cover. Additionally telemedical applications such as a motion, breathing and heart rate sensors are integrated into the system.

Figure 2: Prototype A of the cover mounted on a hospital bed

The prototype has already been evaluated with a group of nurses and non-dementia participants during a simulation. The results show that there is especially a need for improving the mounting of the cover. The development of further prototypes as well as their evaluation is planned to improve the usability of the cover. For assessing potential risks of the cover a risk analysis, e.g. failure mode and effect analysis (FMEA), will be applied. To prove the effectiveness of the cover and the telemedical application a clinical study with dementia patient will be carried out.

4 CONCLUSION AND DISCUSSION

Developing suitable assistive technology that matches the user's requirements is a highly challenging process. Methods on how to integrate the user(s) and their requirements exist but they are insufficient for the complexity of the health care system. Domain specific models and methods are necessary. The holistic approach described above tries to combine domain specific models and methods into a structured process. The approach requires further development, i.e. identification and evaluation of suitable methods for each phase.

The development of a technical solution as described above requires the collaboration of different disciplines such as physicians (emergency medicine, neurology and geriatrics), nurses (emergency nurses, geriatric nurses), media- and industrial designers, engineers and ergonomists. On the one hand the integration of different disciplines helps to consider different requirements that are necessary for developing a technical solution. On the other hand this diversity can impede the collaboration due to different knowledge backgrounds (Bienzeisler et al., 2007). To support interdisciplinary collaboration it is necessary to provide adequate methods. Especially during the beginning of a research project the team must be supported in establishing a shared knowledge. One possibility is the use of personas. Personas are fictitious but specific descriptions that represent real user(s). They can help convey information about the user(s) to the product development team (Cooper, 2004). Especially during the beginning of a project they can support the project team to establish a shared knowledge. The design and application of personas still have to be tested.

5 REFERENCES

Alzheimer's Disease International (2010). Wolrd Alzheimer Report 2010. The Global Impact of Dementia. http://www.alz.co.uk/research/files/WorldAlzheimerReport2010.pdf. Accessed February 20, 2012.

Alzheimer's Society (2009). Counting the cost: Caring for People with Dementia in Hospital Wards. http://alzheimers.org.uk/site/scripts/document_pdf.php?documentID=1199. Accessed February 20, 2012.

Angerhausen, S. (2008). Demenz – eine Nebendiagnose im Akutkrankenhaus oder mehr? Maßnahmen für eine bessere Versorgung Demenzkranker Patienten im Krankenhaus. In: Zeitschrift für Gerontologie und Geriatrie, Vol. 41, No. 6, pp. 460-466.

Bienzeisler, B., Kremer, D., Spath, D. (2007). Wissensintensive Kooperationsprozesse bei der Entwicklung innovativer Produkte. In: Bertsche, B. & Bullinger, H.-J. (Ed.): Entwicklung und Erprobung innovativer Produkte: Rapid Prototyping. Grundlagen, Rahmenbedingungen und Realisierung. Berlin: Springer, pp.70-122.

Bruder, R. (2011). Ergonomische Qualität im Design (EQUID) – ein prozessorientierter Gestaltungsansatz. In: Spanner-Ulmer, B. (Ed.): Mensch, Technik, Organisation - Vernetzung im Produktentstehungs- und - herstellungsprozess. 57. Frühjahrskongress der Gesellschaft für Arbeitswissenschaft. Technische Universität Chemnitz, 23.-25. März 2011, pp. 203-206.

Carayon, P. and Friesdorf, W. (2006). Human Factors and Ergonomics in Medicine. In: Salvendy, G. (Ed.) Handbook of Human Factors and Ergonomics, New Jersey: John Wiley and Son, pp. 1517-1537.

Czaja, S.J. and Nair, N. (2006). Human Factors Engineering and Systems Design. In: Salvendy, G. (Ed.) Handbook of Human Factors and Ergonomics, New Jersey: John Wiley and Son, pp. 32-49.

Cooper, A. (2004). The Inmates Are Running the Asylum: Why High-tech Products Drive Us Crazy and How to Restore the Sanity. Indianapolis: Sams Publishing.

DIN (Deutsches Institut für Normung) (2004a): DIN EN ISO 6385: 2004 – Grundsätze der Ergonomie für die Gestaltung von Arbeitssystemen. Berlin: Beuth.

DIN (Deutsches Institut für Normung) (2004b): DIN EN 60601-1-6: 2004 – Medizinisch elektrische Geräte: Allgemeine Festlegungen für die Sicherheit – Ergänzungsnorm: Gebrauchstauglichkeit. Berlin: Beuth.

DIN (Deutsches Institut für Normung) (2008): DIN EN 62366: 2008 – Medizinprodukte – Anwendung der Ergonomie auf Medizinprodukte. Berlin: Beuth.

Friesdorf, W. (1990). Patient-Arzt-Maschine-System (PAMS). In: Friesdorf, W.; Schwilk, B.; Hähnel, J. (Ed.): Ergonomie in der Intensivmedizin; Mehl-sungen: Bibliomed, pp. 39-46.

Friesdorf, W., Groß-Alltag, F., Konichezky, S., Arndt, K. (1994). Clinical Information Process Units (CIPUs) – a system ergonomic approach to medical information systems. In: Technology and Health Care. Vol. 1, 1994, pp. 265-272.

Friesdorf, W. and Marsolek, I. (2009). Medicoergonomics – A Human Factors Engineering Approach for the Healthcare Sector. In: Schlick, C. M. (Ed.), Industrial Engineering and Ergonomics. Visions, Concepts, Methods and Tools. Festschrift in Honor of Professor Holger Luczak. Berlin, Heidelberg: Springer Verlag, pp. 165-176.

Haberfellner, R., Nagel, P., Becker, M., Büchel, A., von Massow, H. (1997) Systems Engineering. Methodik und Praxis. 9. Aufl., Zürich: Verlag Industrielle Organisation.

Henke, K. D., Friesdorf, W., Bungenstock, J., Podtschaske, B. (2008). Mehr Qualität und Wirtschaftlichkeit im Gesundheitswesen durch genossenschaftliche Kooperationen. Baden-Baden: Nomos.

Mavundla, T.R. (2000). Professional nurses' perception of nursing mentally ill people in a general hospital setting. In: Journal of Advanced Nursing, Vol. 32, Issue 6, pp. 1569–1578.

Müller, E., Dutzi, I., Hestermann, U., Oster, P., Specht-Leible, N., Zieschag, T. (2008). Herausforderung für die Pflege: Menschen mit Demenz im Krankenhaus. In: Pflege & Gesellschaft, Jg. 13, Heft 4, S. 321-336.

Vahs, D. (2007). Organisation. Einführung in die Organisationstheorie und –praxis; 6., überarbeitete und erweiterte Aufl., Stuttgart: Schäffer-Pöschel Verlag.

Wojnar, J. (2003). Demenzpatienten im Krankenhaus. In: Alzheimer Info, Jg. 7, Heft 1, pp. 2–4.

CHAPTER 18

Structures and Processes in Health Care Systems: Lessons Learned from the VA My Health*e*Vet Portal

Kim M. Nazi

Department of Veterans Affairs
Washington DC, USA
Kim.Nazi@VA.gov

ABSTRACT

The use of information technology to improve health care systems and the provision of tools to help patients become more active participants in their care are recognized as important strategies to more effectively address the nation's health care needs. Both Electronic Health Records (EHRs) and Personal Health Records (PHRs) are being implemented to leverage technology to address these needs, however careful attention must be given to the crucial interactions between patients, health care professionals, organizational structures, and processes in order to implement and use technology in ways that are optimal. Using the VA health care system as an exemplar, issues and opportunities for PHRs and PHR systems will be further explored to share some lessons learned and highlight important areas that warrant further research.

Keywords: personal health record, PHR, secure messaging, eHealth

1 HEALTH INFORMATION TECHNOLOGY TO IMPROVE HEALTH CARE: EHRS AND PHRS

In its landmark report *Crossing the Quality Chasm*, the Institute of Medicine (IOM) emphasized the need to redesign health care, calling for the use of health information technology to both improve health care systems and to provide patients with tools that would enable them to more fully participate in their care (IOM, 2001). In the decade since, progress has been made on both fronts. The implementation of Electronic Health Records (EHRs) to improve the efficiency, quality, and safety of health care delivery has been catalyzed by legislation (HITECH, 2009) aimed at incentivizing and supporting organizations and health care providers in the implementation and Meaningful Use of technology (Blumenthal and Tavenner, 2010). The U.S. Department of Veterans Affairs (VA) is credited with creating the first fully electronic enterprise-wide system known as VistA in the early 1980s to improve quality, care coordination, and population health; and to reduce cost (Kizer and Dudley, 2009). Although the nation's progress in implementing EHRs has been modest (Jha, 2010), significant growth is anticipated. While the VA system has demonstrated benefits associated with EHRs, studies of other systems have produced mixed results which several authors attribute to the critical nature of socio-technical factors (Black et al., 2011).

While EHRs represent a health information system managed by the health care delivery system (Mon, 2005), Personal Health Records (PHRs) have emerged as promising tools that can enable health care consumers to manage their personal health information. As Tang and Lansky point out (2005), four of the ten design rules outlined by IOM are directly reliant upon leveraging information technology in new and transformative ways: care based on continuous healing, customization based on patient's needs and values, the patient as the source of control, and shared knowledge with the free flow of information. In this context, however, the mere installation of EHRs will not suffice in engaging patients in the health system. PHRs are needed to provide patients with the shared knowledge and free flow of information intended by the IOM recommendations. Tang and Lansky emphasize that even with the digitization of medical records, the IOM design rules can only be realized if the patient is a full partner with the health care provider, and with the organization providing care. They describe this bridging of the patient-provider health information gap as "the missing link." PHRs have many anticipated benefits, yet further research is needed to design and implement the technology in ways that are most effective and meaningful for patients.

1.2 PHRs AND PATIENT PORTALS

PHRs have been defined as "an Internet-based set of tools that allow people to access and coordinate their lifelong health information and make appropriate parts available to those who need it. PHRs offer an integrated and comprehensive view of health information, including information people generate themselves" (Connecting for Health, 2003). PHRs may be standalone or "tethered' to an organization's EHR system, enabling information from the patient's medical record to populate the PHR (Tang et al., 2006). More recently, a new breed of PHR has emerged in the form of patient portals that combine access to information with electronic services and communication tools such as web-based messaging. Tang et al. assert that the integrated PHR (tethered or interconnected) is the preferred model due to its potential for data-sharing and interoperability. In underscoring the critical importance of a PHR's technical architecture, Tang and colleagues predict that "if they cannot exchange data with other health care systems, PHRs will become 'information islands' that contain subsets of patients' data, isolated from other information about patients, with limited and transient value" (Tang et al., 2006:124).

1.3 THE PHR PARADOX

For the last decade, consumer research has consistently demonstrated strong public interest in utilizing an online PHR. A consumer poll in 2006 revealed that although few adults (2%) had used a PHR, most (64%) would like to (Harris Interactive, 2006). Another public opinion survey commissioned by the Markle Foundation (Connecting for Health, 2006) revealed that 89% of the public would want to look over their medical records if they could, and 65% were interested in accessing their own PHR online. For the Veteran population, data reveals that similar needs exist (Nazi, 2010). Despite consistent reports of high consumer interest in PHRs, adoption remains low, with a more recent survey revealing that only about 7% of the American public is currently using a PHR (CFHC, 2010). PHR users report high satisfaction (Halamka et al., 2008; Nazi, 2010), however, the evidence base about the outcomes of use remains limited (Archer et al., 2011) with persistent calls for further research (Kaelber et al., 2008; Nazi et al., 2010). Given the relatively low rates of PHR adoption, there is continued emphasis on privacy concerns, design shortcomings, cost, and difficulties sharing information across organizations as barriers to broader implementation and adoption (Kahn et al., 2009). Although PHRs have historically been cast as consumer empowerment tools for patients, preliminary findings in the literature also suggest that clinician endorsement may be an important factor in a patient's choice to adopt a PHR (CHCF, 2010; Dunbrack, 2011), and that continued clinician engagement with patient PHR use may be required to achieve and sustain anticipated positive outcomes (Jimison et al., 2008; Wynia et al., 2011). Taken together, these findings point to the paradox of PHRs: despite significant consumer interest and anticipated benefits, adoption and use of PHRs remains low. Notable exceptions are emerging in the form of organizationally sponsored PHR portals which combine PHRs with tools and services that support information sharing and communication.

1.4 PHRS IN THE CONTEXT OF HEALTH CARE

After a decade of PHRs being promoted as independent tools to support a broad notion of consumer empowerment, recognition of the potential for collaborative use of these tools in clinical settings is emerging as an area that holds important potential. This evolution of the PHR paradigm requires not only providing patients with tools that empower them to be active participants in their health care, but also creating an environment that supports and integrates this work within the organizational context of health care delivery; from the patient/physician interaction to the representation of information within the clinical information system. This is reflected in the Chronic Care Model as productive interactions between the "informed activated patient" and the "prepared proactive practice team" (Bodenheimer et al., 2002). In the context of health care, patient use of PHR systems may be an important mechanism for collaboration with the health care team, with potential improvements in patient activation, engagement, safety, and care coordination. As predicted by the Care Transitions Intervention model (Coleman et al., 2006), health care providers and other members of the health care team may play an important role in fostering effective patient use of PHRs. Significant progress is being made at VA to explore this potential, with the implementation of Patient Aligned Care Teams (Klein, 2011) and emphasis on New Models of Care that leverage technology to support patient care.

1.5 STRUCTURES AND PROCESSES IN HEALTH CARE

Healthcare organizations are among the "largest, most complex, technologically rich, and value-infused of any human arrangement" (Lammers et al., 2003: 319). A significant body of literature emphasizes the importance of social and organizational factors which influence the implementation and use of technology (Ball and Gold, 2006; Berg, 2004; Greenhalgh et al., 2004). Existing organizational structures and processes shape technology use, and are reflexively shaped by the ongoing situated use of the technology (Orlikowski, 2000). One important aspect of this dynamic is the impact of the technology on workflow and existing work practices (Rice and Katz, 2006). Health care work is information intensive (Ramaprasad et al., 2009), making information flow and associated communication pathways critical components of the system. To further explore how PHRs can support collaborative health care work, use of PHR features must be considered in the broader clinical context of relevant health care processes (Walker and Carayon, 2009).

2 THE VA MY HEALTHEVET PHR PORTAL

In 2003, VA introduced a web-based PHR to complement traditional services, improve co-managed care, and empower patients and their families to play a more active role in Veterans' health care. The My Health*e*Vet PHR portal (www.myhealth.va.gov) enables Veterans to create and maintain a PHR that includes access to health education information, a comprehensive personal health journal, and electronic services such as online prescription refill requests. VA patients can view a comprehensive summary of their VA medication history, and those who are authenticated can view additional data from the VA EHR including VA appointments, allergies and adverse reactions, chemistry and hematology lab test results, and wellness reminders for preventative care. The VA Blue Button was added to the portal in August 2010, enabling registered users to view, print, or download a single electronic file with all of their available personal health information (www.va.gov/bluebutton). Secure Messaging provides authenticated VA patients with the ability to communicate electronically with their health care team for non-urgent needs, supplementing traditional health care interactions. A triage model is used in which a designated member of the team responds to incoming messages, or assigns action to another member of the team. These tools comprise a PHR portal that is "tethered" in that copies of data from the VA EHR are available to authenticated patients in their portal. Personal health journals also enable Veterans to self-enter information.

The study of health information tools in actual use is important because it enables practical insights (Kaplan, 2001) and can reveal unanticipated consequences that can be both desirable and undesirable. Drawing upon the interdisciplinary science of ergonomics which emphasizes the crucial interactions that occur among the organization, patient, health care professionals, and organizational factors (Harrison et al., 2007), some "lessons learned" are offered based on VA experience with My Health*e*Vet PHR portal. These issues and opportunities can be broadly categorized into four areas: value, structures, processes, and communication.

2.1 VALUE

Organizations should purposefully define the unique PHR features and services that will hold the greatest value for the specific patient population begin served, taking into account constraints that may limit value for certain segments of the population and offering alternatives to reduce the potential for disparities. The definition of 'value' should be based on direct evidence, engaging patients and their family members and caregivers to understand their needs and preferences, and recognizing that these may change over the patient trajectory. Our experience reveals that VA patients place the greatest value on three specific aspects of the PHR: increased access to data from the EHR, transactional services such as online prescription refills that make the accomplishment of tasks more efficient and convenient, and tools such as Secure Messaging that enable improved communication with the health care team. While other PHR tools are perceived to have potential value, their usefulness may be constrained by other factors. Understanding these constraints and developing strategies to address them can remove some of the barriers that would make these features more useful.

One example of this principle in practice is the VA Blue Button that was implemented in August 2010 in order to directly enhance Veteran access to data. An incremental approach was undertaken, beginning with the personal health information that was available via the My Health*e*Vet PHR portal. Early Veteran feedback identified the need to enhance the Blue Button by enabling users to customize their output to meet specific information needs, and by adding additional data classes from the VA EHR. Enhancements were deployed in successive iterations including the ability for users to customize their Blue Button output by date range and/or data class. For example, a Veteran preparing for a VA or non-VA clinic visit may want to create a time based summary of information as an update for their health care team. Additional data classes from the VA EHR were also made available for inclusion in the Blue Button output. Consultations with Human Factors specialists and Cognitive Engineers have supported the development of an enhanced PDF file format to optimize the presentation of information in ways that enhance usability. Feedback from Veterans has validated the value of these enhancements. Additional research is underway to generate further insights about the utility of the Blue Button feature for Veterans, and to evaluate the efficacy of information sharing with both VA and non-VA health care providers. We also anticipate that Veterans will value the ability to authorize sharing of their information with others, and the development of a secure mechanism to transport their information.

2.2 STRUCTURES

To support the effective adoption and use of PHR systems, organizations should examine the physical, social, and organizational structures that are needed to facilitate use. Physical structures can enable access to PHR systems in ways that complement home use, while social structures are important to support meaningful use of the system. Organizational structures can serve as facilitators or inhibitors to use of the system, and these may need to be adapted as actual use provides additional insights. One example of this principle in practice is the provision of computers intended to enable on site access to the PHR portal at a VA facility. Access to My Health*e*Vet at VA facilities was deemed important to ensure accessibility for all patients. To address this need, patient-accessible computers have been deployed at VA facilities across the country. Our experience has revealed that the physical placement of these computers can be important in two distinct ways: ensuring access as part of the flow of the patient visit, and providing adequate resources that support patients in using the system. Feedback from VA staff reveals that placement of patient-accessible computers in close proximity to the clinic area is important to enable patients to access information immediately prior to their clinical encounter. Proximity also enables the "teachable" moment by enabling opportunities for staff to demonstrate use of PHR tools and features to patients. Computers located in other locations may be under-utilized, unless significant resources are available to encourage and support patient use. Some notable examples of the success of separate patient computer labs point to the importance of human resources and social support to facilitate patient's use of the system. Additional research is needed to explore how both models meet these distinct but related patient needs. Another structure that is important to facilitate adoption and use of the PHR system is user training and education. Our experience reveals that users have different preferences and needs for training, and a variety of alternatives should be readily available so that users can choose among them, including self instruction using provided materials posted on the portal, in-person group instruction, and individual instruction. While the provision of user training represents an investment of resources, these investments can ultimately result in more effective use with higher satisfaction. Additional research is currently underway to examine these needs, focusing on both health and computer literacy.

In terms of the collaborative use of PHR systems engaging both patients and members of the health care team, a crucial aspect of structural alignment for health care professionals is integration within the clinical workflow environment. One example of this in practice is the My Health*e*Vet Secure Messaging system, and our experience has shown that "fit" with existing systems is crucial. Health care teams are vocal about where fit has been achieved (i.e., the ability to save a Secure Messaging interaction as a progress note within the VA EHR), and where lack of fit has been problematic (notification messages via the enterprise email system would be of greater value within the clinical workflow system). Alignment with structures facilitates use of the system, while misalignment is not only frustrating for users, but also results in inefficiencies in an already time-constrained system.

2.3 PROCESSES

Organizations should focus on health care processes to identify how use of the PHR system aligns with these processes or points to the need for process redesign. A process-oriented perspective enables the identification of tasks and activities but also places these in the context of other factors that influence adoption and use. One example of this in practice is the examination of implementation processes that are key factors influencing rates of adoption. Efforts should be made to identify and remove process barriers in order to enable new users to easily accomplish account registration and authentication. Our experience has demonstrated that these barriers can be minimized by examining processes and developing strategies to facilitate needed improvements. One example of this has been the decentralization of processes to enable staff within the clinic to assist patients in completing registration and authentication as part of their clinical visit. To further streamline the process of authentication we are exploring the use of an online authentication method as an alternative to the current requirement for In Person Authentication.

Another aspect of this principle is the need to examine the processes that PHR features are intended to support. One example of this principle in practice is the process of medication reconciliation. For VA patients, many of whom are co-managed with health care providers outside the VA system, reconciliation is critical to ensure that the medication list documented in the VA EHR contains the full range of prescriptions and medications that a patient may be taking. The reconciliation process also is important in verifying that actual medication use matches intended use. Tools to enable patient self entry of medications are included in the My Health*e*Vet PHR, and patients can also generate a complete medications list that combines this information with their VA medication history. The VA Blue Button enables VA patients to generate a single electronic copy of this information which can be viewed, printed, and downloaded. The process of medication reconciliation however requires interactive communication with the patient (or their family members or caregivers). Communication is needed to validate, clarify, enable feedback, and ensure the timeliness of updates. We discovered that while patient self entry of medications offered value, this value was constrained until Secure Messaging became available to enable timely information sharing and interactive communication between the patient and their health care team. Secure Messaging has enabled patients to provide timely updates about changes in their medication usage, to correct any observed inaccuracies in their VA medication history, and to enable the interactive dialogue that is needed to ensure understanding and feedback in between periodic face-to-face visits. In this way, PHR system features support the process of medication reconciliation as patients can easily communicate with their health care team to ask questions or provide updates. Members of the health care team can then document these interactions in the VA EHR and make updates to the medication list record. Aligning use of the My Health*e*Vet tools with the process of medication reconciliation was crucial in order for these tools to be most effective.

2.4 COMMUNICATION

Organizations should ensure that patient communication needs are effectively met, avoiding the problems associated with "phone tag." One example of this in practice is the use of web-based messaging as a complement to other tools and resources provided via PHR portal systems. Our experience has been that the addition of Secure Messaging has improved communication between patients and the health care team, increasing patient satisfaction and complementing traditional channels of communication. Secure Messaging has also enabled the flow of information; supporting health care processes by enabling patients to share information in a timely manner and communicate directly with their team. The implementation of Secure Messaging has required careful alignment with the clinical workflow, focusing on both structures and processes. For example, one strategy that was used effectively early in the implementation was to separate the triage process from the technology by using simulation. Teams that were gathered around a table were given a paper to represent a Secure Message, and instructed to "pass the paper" to experience the triage process and identify how it would work most effectively. Although the clinic clerk may have initially been perceived as the best candidate for the role of triage person, the simulation exercise often revealed that having a nurse serve in this role would enable many requests to be handled without further assignment. This example illustrates the critical importance examining information flow within the context of structures and processes.

Another example of this principle in practice is the use of Secure Messaging for process redesign. For example, our experience has demonstrated that the use of Secure Messaging for pre-appointment planning results in the effective exchange of information in advance of the visit, enhancing the quality of the face-to-face encounter by enabling it to be more fully focused on the patient's agenda. Secure Messaging is also being used to accomplish processes that traditionally have necessitated a patient visit or telephone discussion. One example of this is the use of

Secure Messaging with selected patients for diabetes management. Secure Messaging has enabled patient self report of blood glucose readings and insulin titration by the clinical pharmacist. Redesigning the clinical workflow to accommodate such changes requires careful attention to optimize communication, while recognizing the potential for unintended consequences. Further research is underway to examine the impact of Secure Messaging use and the potential for process redesign.

3 CONCLUSIONS

Despite consumer interest in PHRs and widely anticipated benefits, adoption and use remain relatively low. The early design and implementation of PHRs has evolved into PHR portal systems that include access to information from an EHR along with communication via web based messaging. A re-conceptualization of these tools for "consumer empowerment" should recognize the potential to foster patient use in collaboration with clinicians and other members of the health care team. To support this collaboration, careful attention must be given to the crucial interactions between patients, health care professionals, organizational structures, and processes in order to implement and use technology in ways that are optimal. Examples from the VA experience reveal important insights that warrant further attention. The science of ergonomics will provide an important foundation for further exploring the use of PHR systems in the context of structures and processes in health care systems. Additional research is needed to further explore these areas and to identify consequences of use. This work is crucial to realize tangible improvements in patient-centered care.

REFERENCES

Archer, N., U. Fevrier-Thomas, and C. Lokker, et al. 2011. Personal health records: A scoping review. *J Am Med Inform Assoc,* 18(4): 515-522.

Ball, M.J. and J. Gold. 2006. Banking on health: Personal records and information exchange. *J Healthc Inf Manag,* 20(2): 71-83.

Berg, M. 2004. *Health information management: Integrating information technology in health care work.* New York, NY: Routledge Health Management Series.

Black, A.D., J. Car, C. Pagliari, et al. 2011. The impact of eHealth on the quality and safety of health care: a systematic overview. *PLoS Med,* 8(1): e1000387.

Blumenthal, D. and M. Tavenner. 2010. The "meaningful use" regulation for electronic health records. *N Engl J Med,* 363(6): 501-504.

Bodenheimer, T. 2007. "*The science of spread: How innovations in care become the norm.*" California Health Care Foundation.

California HealthCare Foundation. 2010. *National Health IT Consumer Survey.* April 2010. Accessed February 25, 2012, http://www.chcf.org/~/media/Files/PDF/C/ConsumersHealthInfoTechnologyNationalSurvey.pdf

Coleman, E.A., C. Parry, and S. Chalmers, et al. 2006. The care transitions intervention: Results of a randomized control trial. *Arch Intern Med*, 166(17): 1822-1828.

Connecting for Health. 2003. *The Personal Health Working Group: Final Report.* July 2003. Accessed February 25, 2012, http://www.connectingforhealth.org/resources/final_phwg_report1.pdf.

Connecting for Health. 2006. *Survey finds Americans want electronic personal health information to improve own health care*: Markle Foundation, December 6, 2006. Accessed February 25, 2012, http://www.markle.org/downloadable_assets/research_doc_120706.pdf

Dunbrack, L. 2011. *Vendor Assessment: When Will PHR Platforms Gain Consumer Acceptance.* (HI227550). IDC Health Insights.

Greenhalgh, T., G. Robert, and F. Macfarlane, et al. 2004. Diffusion of innovations in service organizations: Systemic review and recommendations. *Milbank Q,* 82(4): 581-629.

Halamka, J.D., K.D. Mandl, and P.C. Tang. 2008. Early experiences with personal health records. *J Am Med Inform Assoc,* 15(1): 1-7.

Harris Interactive (2006). Few patients use of have access to online services for communication with their doctors, but most would like to. *Wall Street Journal Online/Harris Interactive Health-Care Poll,* Accessed February 25, 2012, http://www.harrisinteractive.com/news/allnewsbydate.asp?NewsID=1096

Harrison, M.J., R. Koppel, and S. Bar-Levy. 2007. Unintended consequences of information technologies in health care—An interactive sociotechnical analysis. *J Am Med Inform Assoc,* 14: 542-549.

Health Information Technology for Economic and Clinical Health Act (HITECH), Public Law 111-5, Section 13001 and following, 2009.

Institute of Medicine, Committee on Quality Health Care in America (2000). *To err is human: Building a safer health system.* Washington, DC: National Academy Press.

Jha AK. 2010. A progress report on electronic health records in US hospitals. *Health Affairs*, 29(10).

Jimison, H., P. Gorman, and S. Woods, et al. 2008. Barriers and drivers of health information technology use for the elderly, chronically ill, and underserved. AHRQ Publication No. 09-E004. Rockville, MD.

Kaelber, D.C., A.K. Jha, and D. Johnston, et al. 2008. A research agenda for personal health records (PHRs). *J Am Med Inform Assoc*, 15(6): 729-736.

Kahn, J.S., V. Aulakh, and A. Bosworth. 2009. What it takes: characteristics of the ideal personal health record. *Health Aff (Millwood),* 28(2): 369-376.

Kaplan, B. 2001. Evaluating informatics applications--some alternative approaches: theory, social interactionism, and call for methodological pluralism. *Int J Med Inform,* 64(1): 39-56.

Kizer, K.W., and R.A. Dudley. 2009. Extreme makeover: Transformation of the veterans health care system. *Ann Rev Public Health*, 30: 319-339.

Klein, S. 2011. The Veterans Health Administration: Implementing Patient-Centered Medical Homes in the nation's largest integrated delivery system. The Commonwealth Fund, September 2011. Accessed February 25, 2012, http://www.commonwealthfund.org/Publications/Case-Studies/2011/Sep/VA-Medical-Homes.aspx.

Lammers, J.C., and J.B. Barbour. 2006. An institutional theory of organizational communication. *Commun Theory,* 16: 356-377.

Mon, D. T. 2005. PHR and EHR: What's the difference? *J AHIMA,* 76(10): 60-61.

Nazi, K. (2010). Veterans' voices: Use of the American Customer Satisfaction Index (ACSI) survey to identify My Health*e*Vet Personal Health Record user's characteristics, needs, and preferences. *J Am Med Inform Assoc,* 17(2): 203-211.

Nazi, K., T. Hogan, and T. Wagner, et al. 2010. Embracing a health services research perspective on personal health records: Lessons learned from the VA My HealtheVet system. *J Gen Intern Med,* 25(Suppl 1): 62-67.

Orlikowski, W. 2000. Using technology and constituting structures: A practice lens for studying technology in organizations. *Organ Sci,* 11(4): 404-428.

Ramaprasad, A., S.S. Papagari, and J. Keeler. 2009. eHealth: Transporting information to transform health care. Proceedings of HEALTHINF 2009, L. Azevedo & A. R. Londral, eds., Porto, Portugal: INSTICC Press, pp. 344-350, 2009; UIC College of Business Administration Research Paper No. 09-03. Accessed February 25, 2012, http://ssrn.com/abstract=1328747

Rice, R. E. and J.E. Katz. 2006. Internet use in physician practice and patient interaction. In Murero & Rice (Eds.), *The Internet and health care*: Mahwah, NJ: Erlbaum.

Tang, P. C., J.S. Ash, and D.W. Bates, et al. 2006. Personal health records: Definitions, benefits, and strategies for overcoming barriers to adoption. *J Am Med Inform Assoc,* 13(2): 121-126.

Tang, P. C., and D. Lansky. 2005. The missing link: Bridging the patient-provider health information gap. *Health Aff,* 24(5): 1290-1295.

Walker, J.M., and P. Carayon. 2009. From tasks to processes: The case for changing health information technology to improve health care. *Health Aff,* 28(2): 467-477.

Wynia, M.K., G. Torres, and J. Lemieux. 2011. Many physicians are willing to use patients' electronic personal health records, but doctors differ by location, gender, and practice. *Health Aff,* 30(2): 266-273.

Section IV

Physical Aspects and Risk Factors for Patients and Caregivers

CHAPTER 19

Identifying and Evaluating Risk Factors for Musculoskeletal Disorders in Equine Veterinary Work

Meghan Rogers, David Kaber, Kinley Taylor

Edwards P. Fitts Department of Industrial and Systems Engineering,
North Carolina State University, Raleigh, NC, 27695-7906, USA
meghanrogers23@gmail.com; dbkaber@ncsu.edu; kinley.b.taylor@gmail.com

ABSTRACT

Equine veterinarians are exposed to tasks in day-to-day work that often lead to work-related musculoskeletal disorders (WMSDs). This study involved an ergonomic assessment of veterinary tasks to identify injury risk factors and violations of established criteria for worker protection. Phase 1 involved shadowing equine veterinarians at a university veterinary hospital. Ten tasks were observed and videos were analyzed to identify whole-body risk for WMSDs. Results revealed lameness exams, rectal palpations, positioning and obstetric procedures to pose high risks. Phase 2 involved collecting quantitative data on hand forces and wrist postures to identify violations of validated ergonomic criteria. The Strain Index was used to determine whether wrist postures contributed to a risk of developing distal upper-extremity disorders. Results showed lameness exams, lifting, and performing ultrasound procedures to place veterinarians at high risk for disorders. Comparison of average low-back compression forces with NIOSH spinal compressive force limits revealed no task exceeded these limits. Finally, a measure of job risk was calculated as the ratio of the product of task frequency by job demand divided by worker capacity. The six analyzed tasks were ranked in terms of risk from highest to lowest as: ultrasounds, restraint, lifting, palpations, lameness exams, and injections. Based on the quantitative data analysis, the tasks of ultrasound, lameness exams, and lifting were considered to have the highest priority for ergonomic interventions. Recommendations were made in the form of engineering controls, administrative controls, and personal protective equipment. Examples included raising horses to minimize flexion of the back and using a hoof jack to limit the need for prolonged awkward postures.

Keywords: equine veterinarians, WMSDs, low-back compression, upper extremity disorders

1 INTRODUCTION AND MOTIVATION

Veterinarians' day-to-day work tasks put them at risk of various injuries and disease. Previous studies have shown veterinarians are at risk for cuts, scratches, radiation exposure and a variety of other issues with high prevalence (Jeyaretnam, Jones, & Phillips, 2000; Loomans et al., 2008). In addition, it has been found that veterinarians are prone to work-related musculoskeletal disorders (WMSD). A survey of New Zealand veterinarians reported that over a 12-month period, there was a 100% prevalence for equine and large animal veterinarians (Scuffham et al., 2010). Also, Poole et al. (1999) showed that 68.9% of large animal practitioners had an injury resulting from lifting, as part of work tasks. Furthermore, these prevalence rates are higher than those for their healthcare counterparts, including nurses, doctors and dentists (for human patients). Such high prevalence rates lead to high costs for individual practices in terms of lost work days due to injury, the need to train additional labor on work, etc.

There is a need for research to identify veterinarian tasks that contribute to cumulative trauma injuries based on ergonomic analysis. This study involved an assessment of equine veterinary tasks to identify ergonomics-related risk factors and violations of established ergonomics criteria for worker protection. These objectives were met through the following steps: 1) analysis of typical veterinary tasks for treating large animals; 2) screening of veterinary jobs for ergonomics-related risk factors associated with WMSDs in other domains using both qualitative and quantitative methods; 3) identification of ergonomics guideline violations in current task designs; and 4) recommendations of ergonomic interventions to improve veterinary task safety. The risk assessment focused on the potential for distal upper-extremity and low-back disorders. The study was expected to reveal those tasks contributing most to the overall risk of veterinarian injury.

2 METHODS

There were two phases to this research, including: 1) a risk identification phase; and 2) a detailed ergonomics analysis phase. The outcomes of Phase 1 were used to inform the direction of Phase 2. The overall flow of the study and the outputs are shown in Figure 1 and described in detail below.

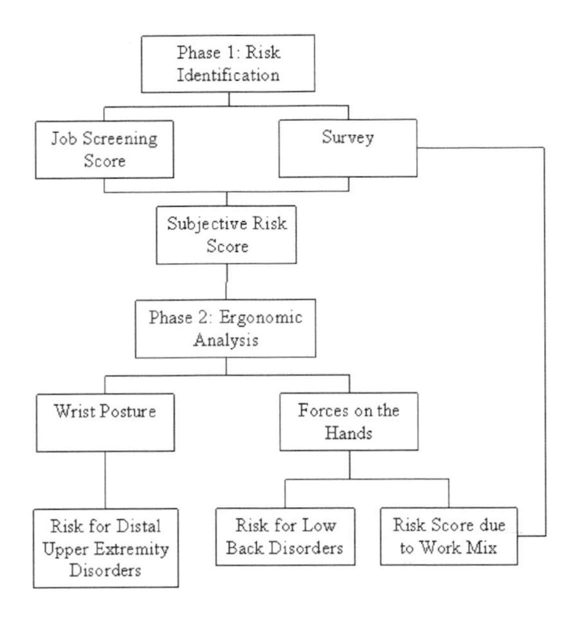

Figure 1. Flow Diagram of Study

2.1 Phase 1

This phase involved work shadowing in order to analyze veterinary task procedures. Two researchers shadowed veterinarians during their daily work activities and used videotape to record typical equine veterinarian tasks. The observed participants were not instructed to perform specific tasks or to manipulate their work in any way. All task observations, except those made on dental work, were completed at the Large Animal Hospital (LAH) of North Carolina State University's (NCSU) College of Veterinary Medicine (CVM). The CVM is a referral hospital; therefore, some of the tasks done at the hospital are not typically performed in private practices. Participants for the observational portion of the study included NCSU LAH staff members as well as one private practitioner. Participants were all knowledgeable in performing large animal veterinarian procedures.

Ten tasks were observed in this study including: ultrasounds, dentistry, lameness exams, lifting, obstetric procedures, positioning, rectal palpations, restraint, and surgery. These tasks were chosen based on previous literature identifying the subactivities as being likely to cause WMSDs. The observations served as a basis for reducing the set of tasks to those meriting further quantitative analysis. Videos of task performance were analyzed to determine subjective whole-body risk for MSDs using a job screening tool that focused on occurrence of extreme posture positions, high force/pressure requirements, and repetitive motions at various joints. The target tasks were assigned "low", "moderate", or "high" subjective ratings for the potential risk factors of extreme posture, high force, and repetitive motion. Two ergonomists made independent ratings for each body part, including the neck, back, and right and left shoulders, arm/elbow, wrist and leg. The risk factor ratings were used to calculate a risk priority level for each body part. A total job score was then determined by summing the weighted scores from the individual body part ratings. The total score was then categorized into one of three ergonomic risk groups: "high", "moderate", and "low".

The equine practitioners at NC State's CVM were also asked to complete a survey combining Mats Hagberg's (1995) discomfort survey with questions about their typical work mix. The discomfort survey asked participants to identify any discomfort they experienced in their work by indicating the anatomical area of discomfort on a chart of the front and back of the body. The nature of the facility in which the study was conducted is such that veterinarians do not perform standardized or consistent sets of tasks each day. Veterinarian workload fluctuates according to the needs of the patients referred to the CVM. Therefore, in order to better understand the performance frequency of the tasks being analyzed, equine veterinarians were surveyed regarding the number of times they performed the 10 target tasks in days, weeks, months or years. To further quantify task and risk factor exposure, the veterinarians were also asked to identify the time spent per day performing the tasks, based on the assumption of an 8 hour work day (Loomans et al., 2008; Moore and Garg, 1995).

Based on the job severity ratings by the ergonomists and frequency of exposure data from the veterinarians, risk scores (Equation 1) were calculated for the 10 observed tasks. The risk scores were used as a basis for determining where to focus the quantitative ergonomics data collection (i.e., what tasks to study in Phase 2). The severity of the task was determined based on the total job score results from the job screening tool. The frequency of the task was measured in terms of how often a veterinarian was exposed to the task per year.

$$\text{Subjective Risk} = \text{Frequency} * \text{Severity} \qquad \textbf{Equation 1}$$

2.2 Phase 2

This phase involved collecting quantitative hand forces and wrist postures and using the data to evaluate violations of validated ergonomic criteria. The tasks and body parts analyzed were informed by the results of Phase 1. Participants were outfitted with electro-goniometers on their right (dominant) wrist and four force sensing resistors (FSR) located at the palm. Participants were asked to perform the six high-risk tasks twice, with the exception of the injection task (due to the nature of the task and safety of animals). Video recordings were also made for data verification and further postural analysis.

The sample size for the quantitative analysis was limited to two veterinarian participants (1 female and 1 male), based on the availability of the staff at the CVM. One was a graduate from NCSU's CVM with 1 year of experience and the other was an equine surgery resident with 3 years of experience.

Wrist posture (flexion, extension, and ulnar deviation) was collected as an input to the Strain Index (Moore and Garg, 1995) measurement technique for identification of risk for distal upper-extremity disorders. Maximum wrist posture (greatest deviation from neutral) for flexion, extension, or ulnar deviation measured by the electro-goniometer was used as the inputs into the Strain Index. The Strain Index was calculated according to Equation 2.

Strain Index (SI) = (Intensity of Exertion Multiplier) x (Duration of Exertion Multiplier) x (Exertions per Minute Multiplier) x (Posture Multiplier) x (Speed of Work Multiplier) x (Duration per Day Multiplier) **Equation 2**

The intensity of exertion multiplier was determined using a modified Borg CR-10 scale (Borg, 1998). The duration of exertion, exertions per minute, and speed of work multipliers were determined from the videotapes of the various tasks. The duration per day multiplier was determined from the results of the veterinarian survey. Finally, the posture multiplier was based on the wrist posture data collected using the electro-goniometer.

The forces exerted between a participant's hand and the task object being used (medical equipment, patient, etc.) were measured using the FSRs. The results were used as inputs into the University of Michigan's 3D Static Strength Prediction Program (3DSSPP; Center for Ergonomics, 2007). Local force maxima were determined by identifying the highest four peaks in force across sensors for each task performance period. These data were averaged to determine the mean maximum force (Rash & Quesada, 2006). This value (in pounds) was input into the static strength-modeling program. The force data was used for an analysis similar to that by Cooper and Ghassemieh (2007), where forces at the hands were used as inputs to the 3DSSPP software to determine compressive forces at the low–back. The resulting compressive forces for the six high-risk veterinarian tasks were compared to the NIOSH recommended spinal compressive force limit of 770 lbs in order to identify those tasks that posed risks for low-back disorders.

Finally, using a typical work mix, the ratio of job demands to worker capabilities was determined for the six tasks identified above. The job demand was determined by the amount of force required to perform the task, as measured with the FSRs. Worker capabilities were determined using the 3DSSPP software by comparison of observed strength with the strength of the 10th percentile of the U.S. female population (The Eastman Kodak Company, 2004). For each task, the forces at the hands were manipulated in 3DSSPP with a manikin posed in the same postures modeled for the low-back compression force analysis. It was found that the observed forces corresponded to the 90 percentile strength capabilities of the female population. This was used as a measure of the worker's capability. Work mix was determined by the number of times a task was performed per year, based on the results of the survey. A risk score taking into account overall equine veterinarian demand was calculated (Equation 3) and the tasks posing the greatest risk for the work mix were identified.

Risk due to Job Mix = (Job Demands/Worker Capabilities) x Work Mix
Equation 3

3 RESULTS

3.1 Phase 1

Results of the survey revealed the most common areas of discomfort for equine veterinarians to be the neck (4 counts) and the hands/wrist. Other identified areas of discomfort included the back (2 counts), ankle (1 counts), and thigh/knees (1 count).

Results of the job screening tool revealed lameness exams to yield the highest aggregate subjective risk score of 33.8 ± 4.3. All of the ten jobs analyzed received an average score above the level of a low risk job (a score of 16 or below), meaning all jobs were considered to be moderate (a score of 17-25) or high risk (a score > 25). In addition to lameness exams, those jobs receiving high risk scores included: patient positioning (M= 28.5, SD = 10.4), obstetric procedures (M= 28.5, SD = 5), and scanning/ultrasound (M= 25, SD = 2.7). Jobs receiving a rating of moderate risk were palpations (M= 23.7, SD = 2.6), injections (M= 20.0, SD = 7.3), restraint/handling (M= 22.3, SD = 3.9), lifting (M= 17.2, SD = 2.8), surgery (M= 23.7, SD = 6.4), and dentistry (M= 24.3, SD = 0.4). Inter-rater reliability across all jobs, based on Spearman's rank correlation coefficient analysis (Quinn & Keough, 2002) was 93.9%, indicating high consistency between raters.

Results of the subjective risk score analysis revealed the order of highest to lowest subjective risk to be: restraint, administering injections, rectal palpations, ultrasound, lifting, lameness exams, surgery, obstetric procedures, dentistry, and positioning. The top six tasks were identified for further analysis based on these results and their contributions to WMSDs reported in the literature.

3.2 Phase 2

3.2.1 *Distal Upper-Extremity Risk Analysis*

Results on raw goniometer data revealed lameness exams to expose the veterinarians to the most extreme wrist postures in terms of flexion (M = 27.39°), extension (M = 42.84°), and ulnar deviation (M = 31.81°). According to the Strain Index, three of the tasks were considered "probably hazardous" and thus posed a high risk for developing distal upper-extremity disorders. These tasks included lameness exams (M = 22.9, SD = 14.5), lifting (M = 21.6, SD = 14.6), and ultrasound (M = 7.2, SD = 6.4). One task, palpations (M =3.1, SD = 1.6), received a score that indicated an "increased risk for an individual." However, this score was just above the criterion value of 3. The final two tasks were rated as being "probably safe," including: injection (M = 0.6, SD = 0.4) and restraint (M = 2.6, SD = 2.1). Inter-rater reliability for application of the Strain Index was 87%, indicating a high degree of agreement among calculations.

3.2.2 *Low-Back Risk Analysis*

Results of the raw force data revealed the tasks of performing ultrasounds (M = 6.41 lbs) and lifting medical equipment (M = 6.13lbs) to require the greatest amounts of force at the hands. Results of the low-back static strength analysis (using 3DSSPP) revealed that none of the analyzed tasks violated the NIOSH low-back compressive force limit of 770 lbs. The tasks of lifting and performing lameness exams both resulted in the highest low-back compressive forces (M = 584.75, SD = 164.31 lbs-f and M= 576.43, SD = 133.30 lbs-f, respectively) compared to all other tasks. Furthermore, the lowest back compression forces were seen in the tasks of restraint (M = 123.38, SD = 68.68 lbs-f) and administering injections (M = 122.0± 14.14 lbs-f).

3.2.3 *Risk due to Job Work Mix*

The ratio of job demands to worker capacity was integrated with the frequency of exposure data for the analyzed tasks. The tasks with the highest ratios represent the greatest risks for the typical work mix, as specified by the participating veterinarians. It was revealed that the six analyzed tasks contributed to worker risk from highest to lowest, as follows: ultrasounds, restraint, lifting, palpations, lameness exams, and injections.

4 RECOMMENDATIONS

Ergonomics interventions were formulated with the intent of mitigating some of the equine veterinarian risk exposures. The recommendations were organized in terms of: (a) interventions at the patient; (b) interventions in the "path" between the patient and the veterinarian (i.e., how the veterinarian interacts with the animal); and (c) interactions at the source of the procedure (i.e., the veterinarian). The interventions were further classified based on the type of control to be applied, including: (a) engineering controls - modifications to the design of the system; (b) administrative controls - modifications to the work process; and (c) PPE - equipment to protect the veterinarian. Proposed interventions for the tasks of lameness exams, ultrasounds, and lifting are discussed below. All recommendations were reviewed for practicality by an equine veterinarian and any caveats to the interventions are identified.

Interventions to mitigate the risks of performing lameness exams should focus on reducing the forces and awkward postures required of veterinarians. In order to reduce the time spent in an awkward posture, an engineering control in the form of a hoof jack could be used to lift the horse's leg. However, this intervention may have limited usefulness if the lameness exam requires flexing the fetlock joint versus just holding the horse's leg passively up in the air. Also, flexion tests, as part of lameness exams, could be conducted on an elevated platform to reduce bending.

The platform should have a sloped ramp in order to be able to trot the horse off immediately after the flexion test. Additionally, administrative controls should be implemented, including: 1) restricting the number of lameness exams performed sequentially, 2) training in postures that increase bending at the knees instead of flexion at the back, or 3) requiring flexion tests to be performed in a seated position. A stool could be practical for a front leg flexion test but probably would not be tall enough to work for a hock flexion test. Also, it is important to ensure the horse is well-behaved to avoid potential for a kicking injury.

Recommendations of engineering controls aimed at mitigating the risks due to lifting include using buckets with larger diameter handles or two handles. Such designs would help with coupling issues and distribution of the load. Administrative controls, such as filling buckets with water at the destination instead of carrying a full bucket, or using a two-person lift when performing heavy lifting and carrying tasks, should be adopted. If a second person is not available, a cart should be used in lieu of carrying heavy objects by hand. Finally, it is important when lifting to bend down at the knees and avoid excessive forward and lateral flexion at the back.

Recommended interventions for conducting ultrasounds focus on the posture of the veterinarian. Engineering controls should be put in place to raise the animal when performing ultrasounds underneath the body or on the lower extremities in order to reduce bending by veterinarians. The horse could be placed in lifted stocks or on a raised platform when conducting these types of tasks. If raising the animal is not possible, administrative controls should be applied, such as requiring veterinarians to sit on a stool (lower the body) or assume a kneeling posture with the aid of knee pads. Such postures are preferable to requiring significant flexion of the back. However, kneeling during ultrasounds limits the veterinarian's ability to move away from the animal quickly in order to avoid a kick. Furthermore, if the area to be probed is high on the animal, the veterinarian should stand on a mini step-ladder to prevent working with his/her arm elevated above shoulder level. Another option when working high on the animal is to sit on a raised stool next to the horse, if it is restrained in stocks. Finally, engineering controls in the form of adjustable monitors or stands and ergonomic probes would help reduce awkward postures in the neck and wrists of veterinarians. Although some of the ultrasound machines used at the CVM have adjustable monitors, the range of adjustability needs to be increased so screens, or portable machines, can be lifted or lowered to approximately 15° below the eye level of a veterinarian during a procedure (Salvendy, 1987).

5 CONCLUSION

After reviewing the previous research and results in the area of MSDs among veterinarians, it can be seen that these types of injuries are often more frequent for this specific occupation, when compared to human healthcare workers. A number of studies have been performed to assess the prevalence rates of WMSDs among large animal veterinarians; however, few studies have provided insight into which tasks maybe causing cumulative traumas as well as the magnitude of the risk factor exposure (e.g., postures and forces). The present study focused on ergonomic evaluations of equine veterinarian tasks to identify risk factors and levels exceeding ergonomic guidelines for injury prevention.

A subjective screening tool identified the tasks of lameness exams, rectal palpations, positioning and obstetric procedures as being of high risk to the whole body. However, when combined with data on the frequency of exposure to these tasks, the six tasks identified as having the highest subjective risk (and therefore requiring further quantitative analysis) were restraint, injections, rectal palpations, lifting, ultrasounds, and lameness exams. Further analysis of these tasks revealed those contributing most to the risk of WMSDs for equine veterinarians, as well as for the specific body parts, including the wrists and low back.

Analysis of the Strain Index revealed lameness exams, lifting, and performing ultrasound procedures to be most hazardous for developing distal upper-extremity disorders. Although none of the average low-back compressive forces observed during these tasks violated the 770lbs criterion identified by NIOSH (1981), the same three tasks (lameness exams, lifting, and ultrasounds) exposed equine veterinarians to the highest low-back forces compared to all other analyzed tasks. Finally, these three tasks were among those posing the highest risk when comparing job frequency, demand, and capacity. Based on these results, it is clear interventions should focus on the tasks of lameness exams, lifting, and ultrasound.

5.1 Limitations and Future Research

Caution should be taken in interpreting and applying the results of the current research due to certain limitations. Due to the availability of equine veterinarians at the CVM and the resources for this study, only two participants were observed during the Phase 2 data collection. Also, as previously mentioned, the types and frequencies of tasks performed at the CVM may not be completely representative of those seen in private practice. Therefore, the results of this study may generalize to certain equine veterinarian operations and practitioners better than others.

Future research in this area should focus on addressing the limitations of the present study. In particular, more participants should be examined in a larger investigation in order to promote the sensitivity and reliability of the quantitative analysis. Furthermore, work should be done to identify differences in risk factors between referral facilities and those seen in private practice.

The current research only focused on the anatomical areas of the wrist and low-back in quantitative ergonomics analysis. Future studies should look into risk factors for other body locations. Previous research has identified MSD prevalence in almost every joint of the body for veterinarian tasks (Schuffham et al., 2010). The rates for the neck and shoulders are also particularly high. Methods of quantitative risk factor exposure for these body parts need to be carefully determined. Finally, any proposed or otherwise identified interventions should be empirically tested to ensure they reduce the risks involved in equine veterinarian work.

ACKNOWLEDGEMENTS

The authors would like to thank NIOSH for their support of this research through a grant to the NC State Industrial and Systems Engineering Ergonomics Lab (No. 2 T42 OH008673-06).

REFERENCES

Borg, G. (1998). *Borg's Perceived Exertion and Pain Scales*. Champaign, IL: Human Kinetics.

Center for Ergonomics, University of Michigan. (2007). 3D Static Strength Prediction Program (Version 5.0.8) [Software]. Retrieved from http://www.engin.umich.edu/dept/ioe/3DSSPP/index.html.

Cooper, G., & Ghassemieh, E. (2007). Risk assessment of patient handling with ambulance stretcher systems (ramp/(winch), easi-loader, tail-lift) using biomechanical failure criteria. *Medical engineering & physics*, *29*(7), 775-87.

Hagberg, M., Silverstein, B., Wells, R., Smith, M. J., Hendrick, H. W., Carayon, P., Perusse, M. (1995). *Work Related Musculoskeletal Disorder (WMSDs): A Reference book for Prevention*. (I. Kuorinka & L. Forcier, Eds.). Taylor & Francis.

Jeyaretnam, J., Jones, H., & Phillips, M. (2000). Disease and injury among veterinarians. *Australian Veterinary Journal*, *78*(9), 625-629.

Loomans, J. B. A, van Weeren-Bitterling, M. S., van Weeren, P. R., & Barneveld, A. (2008). Occupational disability and job satisfaction in the equine veterinary profession: How sustainable is this "tough job" in a changing world? *Equine Veterinary Education*, *20*(11), 597-607.

Moore, J. S., & Garg, A. (1995). The Strain Index: A Proposed Method to Analyze Jobs for Risk of Distal Upper Extremity Disorders. *American Industrial Hygiene Association Journal*, *56*(5), 443-458.

NIOSH (1981). Work Practices Guide/or Manual Lifting, NIOSH Technical Report No. 81-122, US Department of Health and Human Services, National Institute for Occupational Safety and Health. Cincinnati, OH.

Poole, A., Shane, S., Kearney, M., & McConnell, D. (1999). Survey of occupational hazards in large animal practices. *Journal Of The American Veterinary Medical Association*, *215*(10), 1433-1435.

Quinn, G., & Keough, M. J. (2002). *Experimental Design and Data Analysis for Biologists*. Cambridge, U.K: Cambridge University Press.

Rash, G. S., & Quesada, P. M. (2006). Electromyography: Fundamentals. In W. Karwowski (Ed.), International encyclopedia of ergonomics and human factors (Second ed., pp. 3107-3114) Taylor & Francis Group, LLC.

Salvendy, G. (Ed.). (1987). *Handbook of Human Factors*. New York, New York, USA: John Wiley & Sons.

Scuffham, A. M., Legg, S. J., Firth, E. C., & Stevenson, M. A. (2010). Prevalence and risk factors associated with musculoskeletal discomfort in New Zealand veterinarians. *Applied ergonomics*, *41*(3), 444-53.

The Eastman Kodak Company. (2004). Evaluation of Job Demands. *Kodak's Ergonomic Design for People at Work* (2nd ed.). John Wiley & Sons.

CHAPTER 20

An Aspect Designing an Ergonomic Dentist's Room: Survey of Appropriate Room Design among Dentists

Alperen Bal, Ceren Salkin, Eda Bolturk

Istanbul Technical University
Istanbul, TURKEY
abal@itu.edu.tr

ABSTRACT

Nowadays, more and more frequently used the term-"ergonomics" –is getting more important in everywhere -especially in dental care process. Deal with monotonous jobs that means doing the same things almost every day as well as employer and employee-or patients could contributed to face some of the ergonomic problems. In that situation, with a point of dental ergonomics; dentist's working posture, position of the patients, dentist's ability to use the dental equipments(-like dental hand piece, mouth mirror, etc.) dentist's chair and the design of the surgery room (i.e. brightness, the color and the size of the room, noise, direction of equipments and furniture etc.) and also other factors- particularly personal factors like sex, age, weight, history of disease- are all effect dentist's productivity and efficiency while working in his/her surgery room. A good ergonomically-designed surgery room plays an important role in not only increasing work capacity and productivity but also improving patient's satisfaction, comfortableness and dentist's health.

In this case, one of the major points in the dental ergonomics to design a working area in a surgery room was investigated and discussed related with the issues that were mentioned above. At the end of the article, a survey was done to students and lecturers (from faculty of dentistry) who are from some of the universities in Turkey about this topic. As a matter fact it has been consulted one hundred sixty dentists to understand and define the most appropriate design of the surgery room according to their needs, expectations and health that is the most important aspect in this application. The questions were directly related to the work place (surgery room) and data was collected about somatic complaints like pain and grouped according to risk factors. In addition to that, opinions about the ideal work place design were gathered. The survey results were evaluated with software package program (SPSS) and results were interpreted. With these results it has been reached a conclusion and opinions have been suggested about ergonomically-designed surgery room and a comparison has been done between the real situation and ideal situation to make improvements. The intended purpose is to share the conclusion with Istanbul Chamber of Dentists and to change some surgery room for showing the differences to all dentists in Turkey and all over the world.

To sum up, in this study, it is aimed to show how ergonomics is important and how it can change dentist's most important health problems in an easy way. Hopefully this study would be a good example for every surgery room.

Keywords: dental, dentist room, survey, ergonomics, spss

1. INTRODUCTION

Ergonomics is getting more important today's world in everywhere especially in working areas. Deal with monotonous jobs that means doing the same things almost every day as well as employer and employee-or patients or customers could contributed to face some of the ergonomic problems. Every year thousands billion of dollars are spending because of the disorders caused from ergonomic mistakes. According to the Bureau of Labor Statistics, chronic musculoskeletal disorders represent over %60 of occupational disease. A recent study noted that back pain and musculoskeletal pain accounted for over a third of lost productive time in workers afflicted with pain. The cost of this lost productivity can be staggering and is estimated to exceed $60 billion per year. (Hayes M. et al 2009) In this study, we discussed one of the major areas in ergonomics, dentists' ergonomic problems that occur during dental care process and solutions of these problems by arranging an appropriate dentist surgery room. It is such a big problem that nearly 65 percent of participating dentists reported that they are dealing with painful symptoms. (Hayes M. et al 2009) That will cause serious physical disabilities later on. It demonstrates us that dental care process needs important improvements.

It is known that because of unsuitable equipments, improper and poor posture, repetitive movements and vibration, problems could appear especially in the type of upper extremity musculoskeletal disorders like tendonitis, carpal tunnel syndrome (CTS), ulnar and radial nerve entrapment syndromes and various shoulder, neck, back and hand pains. (www.who.int/ occupational_health/ publications/ muscdisorders/en/index.html) Additionally, gender, hormonal changes, immune system situation and anatomic inadequacies causes that kind of disorders. Academic studies indicate that dentists should stand on a "correct" position; extreme and rigid body postures are associated with an increased risk of neck, shoulder, and back musculoskeletal disorders. Treatments, preventions and different ways of appropriate posture are generally discussed in literature but one may ignore the important point, the dentists' environment. According to OSHA, the most efficient and effective way to remedy "ergonomic hazards "causing musculoskeletal strain should be through engineering improvements in the work station-for us- in surgery room. (Simmer-Beck M., 2010).

2. RELATED WORKS AND APPROPRIATE SURGERY ROOM DESIGN PARAMETERS

In this part, we will discuss how the surgery room should be ideally with literature reviews. By putting the ideal ways of the ergonomic surgery room, comparison of the ideal and real situation will be easier.

As a matter of looking at the surgery room, it has become like a site for detached areas of technology; The addition of clinical microscopes, lasers, curing lights, air abrasion techniques, intraoral video, CAD-CAM, root canal apex finders, sonic scalars, radiovisiography, pressure assisted anesthetic devices, separate dental unit water lines and clinical computer systems has the treatment room looking like garage sale. (jobaccess.gov.au/ Advice/ ProductOrSolutionOne/ Pages/ Ergonomic dentalandsurgicalinstruments.aspx) The arrangement of the room will include lots of factors. If one or more factors are ignored, somatic complaints like pain will occur.

It is not a new phenomenon that workers are afflicted from musculoskeletal disorders. In the literature, a 17th century physician and the founder of occupational medicine, Ramazzini indicates that "chair workers" who "become bent, hump-backed, and their heads down," a description that easily matches dentists today. (Hayes M. et al 2009)

If we have a look up to former studies, for dental ergonomic area, dental ergonomic tools and dentist chair are generally discussed. First, before arranging the surgery room, hand tools should be suitable for dentists and tools should reduce the force on hands, and the torquing of the hands by using hand instruments-like gloves-and dentists should prefer light-weighted, balanced and well-sharpened hand tools. For decreasing vibration and heat rising, hand tool material and gloves should be covered with absorbent materials. (www.ada.org/ sections/ education And Careers/pdfs/ergonomics.pdf.) With these ergonomic tools, two of the most important factors cause musculoskeletal disorders, vibration and frictions will less influence the dental process.

Once and for all, dentist chair should have the properties that are written below:

- Foots should step on the floor almost completely. For this, chair should be easily regulatable. While sitting, chair should be reachable enough to all dental instruments in the optimal sight angle.
- Back support should provide optimal °90 -110° sitting angle and again, should be adjustable.

- While sitting, the knee angle should suitable with the optimal back angle. (90°-110° angle.)
- There should be supportive elements for putting their arms or hands while resting. (jobaccess.gov.au/ Advice/ProductOrSolutionOne/Pages/ Ergonomicdentalandsurgicalinstruments.aspx)

With ergonomic dental tools and appropriate dentist chair, musculoskeletal pain will rarely occur and the first two characteristic surgery room design parameters will be achieved.

In foot control dental units, foot control should be designed with a pedal on which the foot is placed either entirely or partially. Placing the whole foot on the pedal causes an unfavorable load which results in the unequal position of the right and left foot which in turn causes an asymmetrical, harmful strain on pelvis and vertebral column. Therefore it is necessary to place the pedal on the floor so it can support the foot. (www.designbyfeel.com/information-for-consideration-in-an-ergonomic-standard-for-dentistry/) Here, the foot angle is the critical point for ergonomic design.

Also dental operating light is another critical point for brightness. A dental operating light must be positioned around the head of the dentist, before and sideward so that the light beam is running parallel to the viewing direction, with a maximal deviation of approximately 15°. In order to achieve this, the reach of the foreside of the dental operating light has to be so great that it can be positioned in the vertical plane of the back of the dentist. Hence, dental operating light should be situated on the ceiling or the wall. Under these circumstances the length of the lamp arms can be shorter so more space at the side of the patient chair is available for the parts of equipment fixed there. (www.designbyfeel.com/information-for-consideration-in-an-ergonomic-standard-for-dentistry/) Operating light should be adjustable with respect to dentist length, sitting position, working height and focus distance. In general room design, light should reach all the sides of the room equally. Moreover, shading and light direction in the room is significant. To prevent the glare, curtains are needed and usually sunrise is adequate for surgery room brightness, and except evenings, additional lamps are not being required. It can be said that, while considering brightness, in addition to the dental care process needs, dentist's age and eyesight are critical points that should be considered while designing. Additionally, to prevent heat rising, air circulation in the room should be provided with air conditioners or dental ventilators.

There is one more important point about dental equipments; dynamic instruments' position. Dynamic instruments such as using for tubing should be positioned within the field of the vision, °30 right and left of the mid -sagital or symmetrical plane of him. Dentist should avoid as far as possible the fatiguing accommodation and adaptation of the eyes outside of the field of vision.

Control panels should be stated at the side of instrument console so dentist should be able to look at the console while sitting and working. (www. ada. org/ sections/ educationAndCareers/pdfs/ergonomics.pdf.) While positioning the screen there will be three important factors that should be considered;

- Looking distance and the height of the letters,
- Height of the screen in relation with the height of the eyes of the dentist,
- Reflection, contrast and type of the screen.

While stating the control screen, the dentist's horizontal eye line and the screen should not exist 60° angle. Turning device should be needed for turning the screen as vertically as possible and if it is needed, a flat screen can be used in relation to a dental unit in connection with its size, so it will produce less heat. Reflection can be avoided by using high frequency neon lights. (www.designbyfeel.com/information-for-consideration-in-an-ergonomic-standard-for-dentistry/)

In the dental process, another design parameter is noise. Noise can be disturbing while dentist is working; it generally arises during using hand instruments with vibrations. It should be absorbed within the design of hand equipments with acoustic materials such as cotton. If it is inadequate, breaks should be arranged frequently. Noise absorbed acoustic wall materials can be used. (www.designbyfeel.com/acquiring-improved-dental-performance-skills/)

Another surgery room design parameter is color. It is a general view that color affects human physiology. Because of that reason, a surgery room should make patients and dentist relax, confident and refreshing. Light colors are suitable for surgery room wall. If it two colors are preferred, two colors can affect each other. Additionally, if two colors are used, there may appear contrast difference which could make somebody tired while looking. (www.eugenol.com/attachments/0007/6874/papers_ergonomic_principles_part1.pdf)

Finally, the last design parameter is the working space. The surgery room should be large enough to fit all the dental equipments, patient chair and dentists' chair, and of course, there should be enough space for the movements of dentist assistant. If the room is too narrow, all the equipments will stand close to each other and dentist will stay in a limited area. Limited area will restrict both the dentist and his assistant. Conversely, if the surgery room is too large and equipments are standing far away to each other, unnecessary movements and undesirable reach will appear. It causes the decreasing of productivity and efficiency. According to the ergonomic requirements in work place, to reduce dispensable motions, time and motion studies should be done to determine the room size. (jobaccess.gov.au/ Advice/ProductOrSolutionOne/Pages/Ergonomicdentalandsurgicalinstruments.aspx)

3. MATERIAL AND METHOD

Questionnaires were modified to suit the requirements of dentists and basic reference for the questionnaire is Prof. Dr. Fatih C.Babalik, "Ergonomics for Engineers" book. Questions were modified from the section "Control of workplace with respect to ergonomics." Questionnaire consist of eight sections: working area, equipments and devices, demand for sense organs, working method, load and strain, working hours, ambient conditions with respect to illumination and climate. Dentist answers measured with five scales: definitely agree, agree, fair, don't agree and definitely don't agree.

Structured questionnaires were mailed to after random sampling dentist from metropolitan teaching hospitals, local hospitals and medical centers. Totally 160 questionnaires were collected from the whole survey.

According to results 40 percent of the dentists care of 11-20 patients per day. Among those dentists equally men and women who are 55 percent is married. The largest age groups were those younger than 30 and those between 41-50 years old. (35 percent) In terms of education 55 percent is under-graduate degree and the rest is graduate and post graduate degree. Answers for dentist chair showed that half of the participants think chair is not sufficient for appropriate movement and to apply necessary force and not have suitable angle to see the working area. Recipients commonly indicated that noise level is disturbing during the treatment and also is making difficult to communicate with patient. (45%) Also 45 percent of recipients chose that generally it is not possible to avoid of overtime working during 35 percent were choosing work hours were not appropriately scheduled in accordance with physiological performance abilities and human wants.

4. RELATED WORKS AND APPROPRIATE SURGERY ROOM DESIGN PARAMETERS

Two hypotheses were set and results were evaluated with SPSS one-way ANOVA.

4.1. The relation between lighting and dazzling

Our first assumption is we have a positive correlation between lighting and dazzling. Variable 1 means lighting and Variable 2 means dazzling. We first investigated minimum, maximum, mean and standard deviation of each variable in SPSS. We can conclude that each variable is waving significantly.

	N	Minimum	Maximum	Mean	Std. Deviation
VAR00001	160	,10	,90	,6100	,22919
Valid N (listwise)	160				

	N	Minimum	Maximum	Mean	Std. Deviation
VAR00002	160	,10	,70	,4500	,19331
Valid N (listwise)	160				

Var 1: lighting Var 2: Dazzling

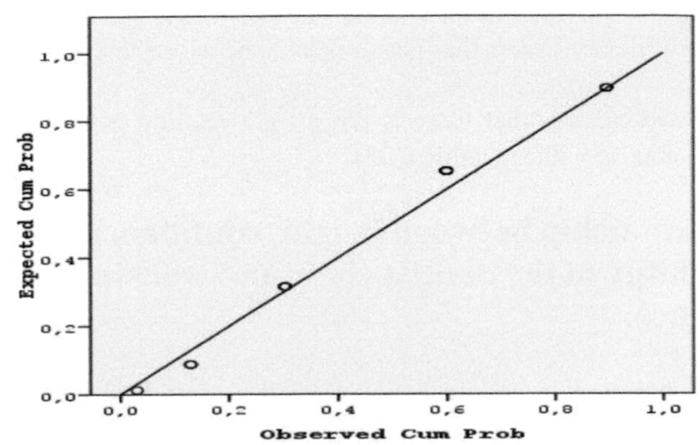

If we look the distribution of lighting, one can easily see the points are close to the line. Thus, lighting values are standing similar to each other. It means the given points seem similar.

VAR00001 * VAR00002 Crosstabulation

Count

		VAR00002				Total
		,10	,30	,50	,70	
VAR00001	,10	0	1	0	0	1
	,30	0	2	1	0	3
	,50	0	3	0	1	4
	,70	1	2	2	3	8
	,90	0	1	1	2	4
Total		1	9	4	6	20

From the cross table, we can see the difference of the each criteria's answers visually.

ANOVA

	Sum of Squares	df	Mean Square	F	Sig.
Regression	,078	1	,078	1,536	,231
Residual	,920	18	,051		
Total	,998	19			

The independent variable is VAR00002.

Coefficients

	Unstandardized Coefficients		Standardized Coefficients	t	Sig.
	B	Std. Error	Beta		
ln(VAR00002)	,129	,104	,280	1,239	,231
(Constant)	,726	,106		6,821	,000

By doing ANOVA test, our aim is to measure variance between two variables. In ANOVA test, there are two critical values: one is F and the other is Sig. If the sig value is less than 0, 05 with 95% confidence interval, we should conclude that there is no significant difference between our two conditions, lighting and dazzling. Here, with 95% confidence interval, 0,231 sig value shows the differences between conditions have a medium relationship if dazzling is assumed as independent variable. F indicates the differences between means. If F is closer to 1, means will be closer. Here, F demonstrates that group means are close to each other with 1.5 value.

As a result, we can say that there is not a tight relation between lighting and dazzling as we suppose with sig value 0,231.

4.2. The relationship between height, confidence and adjustability of the dentist chair and working area confidence.

Our second assumption is we have a positive correlation between height, confidence and adjustability of the dentist chair and working area confidence. Variable 1 means height, confidence and adjustability of the dentist chair and Variable 2 means working area confidence. We first investigated minimum, maximum, mean and standard deviation of each variable in SPSS. We can conclude that each variable differs obviously as we can see the descriptive statistics scheme below.

Descriptive Statistics

	N	Minimum	Maximum	Mean	Std. Deviation
VAR00001	160	,30	,90	,4800	,21423
VAR00002	160	,10	,90	,5700	,23642
Valid N (listwise)	160				

If we look the distribution of height, confidence and adjustability of the dentist chair, one can easily see the points are not close to the line, they are not following the direction of the line. Thus, height, confidence and adjustability of the dentist chair values are not standing similar to each other. It means the given points seem similar.

On the other hand, if we look the distribution of working area confidence, one can easily see the points are close to the line. Thus, working area confidence values are standing similar to each other. It means the given points seem similar.

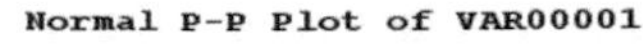

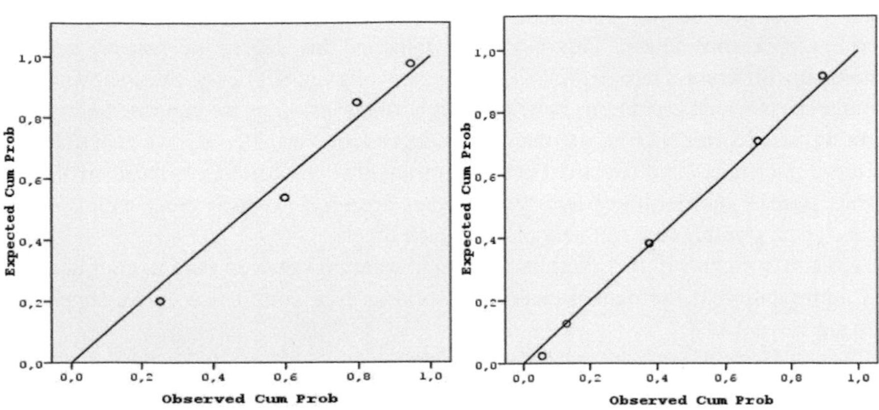

Correlations between height, confidence and adjustability of the dentist chair and working area confidence can be shown with Pearson correlation. Pearson correlation value is 0,237, and this value indicates that there is a medium-tight relation, not as much as tight relation, between height, confidence and adjustability of the dentist chair and working area confidence.

Correlations

		VAR00001	VAR00002
VAR00001	Pearson Correlation	1	,237
	Sig. (2-tailed)		,315
	N	160	160
VAR00002	Pearson Correlation	,237	1
	Sig. (2-tailed)	,315	
	N	160	160

We used sig value to make the variance analysis. As we mentioned before, if we want to the fit the relation distribution between height, confidence and adjustability of the dentist chair and working area confidence, we should consider working area confidence as an independent variable and we can comment logarithmic sig value 0,848 as a strong relation indicator. It means height, confidence and adjustability of the dentist chair differs by the variant values of working area confidence as we can see below.

Model Summary and Parameter Estimates

Dependent Variable: VAR00001

	Model Summary					Parameter Estimates	
Equation	R Square	F	df1	df2	Sig.	Constant	b1
Logarithmic	,002	,038	1	159	,848	,491	,016

The independent variable is VAR00002.

With this information, ANOVA test value, sig 0,117 represent that there is a relation between height, confidence and adjustability of the dentist chair and working area confidence. This sig value tells us the degree of means' being statistically different. Here, with 95% confidence interval, 0,117 sig value shows the differences between conditions have not a tight relationship as we suppose before if working area confidence is assumed as independent variable. As we mentioned before, F indicates the differences between means. If F is closer to 1, means will be closer. Here, F demonstrates that group means are close to each other with 2,207 value. Thus, variables are not very close to each other.
As a result, we can say that there is not a tight relation between height, confidence and adjustability of the dentist chair and working area confidence as we suppose with sig value 0,117.

VAR00001

	Sum of Squares	df	Mean Square	F	Sig.
Between Groups	,323	45	,081	2,207	,117
Within Groups	,549	87	,037		
Total	,872	159			

5. CONCLUSION

Literature research and survey show some results in dentist's ergonomics life. From the survey, we will have a chance to re-design ergonomics in dentist's life which is popular nowadays. By making some cross-correlation between variables in SPSS, we will take a chance to control different design parameters. Before we started to this study, we assume that increasing lightening is fine. But this study demonstrated that increasing lightening create confusion on color perception which is very dangerous about comment diagnosing the diseases by making ANOVA one way test. Also, we can comment about the relationship between adjustability of the height and relaxation of the patient's seat and working area confidence that there is a minor relation between them. Beside as indicated in the questionnaire dentists mostly uncomfortable due to dentist chair. So some improvements are needed for dentist chair. Additionally according to common wisdom it does not seem dentist devices are very uncomfortable however some improvements are needed in terms of sound level and shape of the devices.

Both of literature and survey showed that ergonomics is very important in dentist's life, but significant differences were found about ergonomic surgery room design parameters.

REFERENCES

Fatih C.Babalık, Ergonomics for Engineers , Ankara, Sep-2005

Hayes M, Cockrell D, Smith DR. A systematic review of musculoskeletal disorders among dental professionals. Int J Dent Hyg 2009 Aug; 7(3):159-65

http://www.who.int/occupational_health/publications/muscdisorders/en/index.html<last accessed:05.02.2012>

Simmer-Beck M and Branson BG. An evidence-based review of ergonomic features of dental hygiene instruments.Work, 2010 Jan 1; 35(4):477-85.

http://www.ada.org/sections/educationAndCareers/pdfs/ergonomics.pdf

http://www.designbyfeel.com/information-for-consideration-in-an-ergonomic-standard-for-dentistry

http://www.designbyfeel.com/acquiring-improved-dental-performance-skills

http://jobaccess.gov.au/Advice/ProductOrSolutionOne/Pages/Ergonomicdentalandsurgicalinstruments.aspx

http://www.eugenol.com/attachments/0007/6874/papers_ergonomic_principles_part1.pdf

CHAPTER 21

Musculoskeletal Symptoms, Stress and Work Organizational Aspects in Hospital Nursing Activities

Ariane Silva, Fabiana Foltran, Roberta Moreira, Cristiane Moriguchi, Mariana Batistão, Helenice Coury.

Federal University of São Carlos
São Carlos, Brazil
helenice@ufscar.br

ABSTRACT

Introduction: Nursing has long been associated with musculoskeletal disorders. The problem is commonly attributed to different factors such as stressful and poorly designed jobs. Most studies evaluating the association between symptoms and job stress are based on worker perception. However, in new studies observational methods are needed for greater precision. An observational assessment tool has been developed by NIOSH (WMSD Research Consortium) to assess aspects of work organization that can affect employee production and health, but no reports were found associating collective organizational aspects with individual perception of job demands. Thus, the objective of the present study was to evaluate musculoskeletal symptoms, job demands and organizational aspects in an intensive care unit (ICU) and a general ward of a Brazilian hospital. Method: Seventy-four nursing staff members, age 35.9±10 years, from an ICU (N=30) and a general ward (N=44) were recruited for the study. These 67 females and 7 males represented, respectively, 37% of the ICU and 27% of the general ward staff. They were registered nurses and nursing aides who had worked for at least 6 months in their current job with12h day (N=48) or night (n=26) shifts. The study was approved by the university's ethics committee (CAAE - 1080.0.000.135.10). The professionals completed a questionnaire that included demographic data and a visual analogical scale (VAS) for pain associated with the standardized Nordic musculoskeletal questionnaire, as well as a job stress scale. The collective organizational aspects of the sectors were evaluated with the NIOSH assessment tool. The data were evaluated with the chi-square test and relative risk calculation (odds ratio - OR). Results and Discussion: A high number (85%) of workers reported low back (42%), neck (28%) and shoulder (22%) symptoms. No associations between job stress and body regions or general pain (p=0.8380) were identified, even when considering the three stress scale domains as different entities (demand, control and social support). However, an association was identified between the work sector and job stress (p=0.030), indicating that ICU workerswere2.88 times (IC 1.097-7.607) as likely to present higher levels of stress than general ward workers. Considering the different job stress scale domains, only control was associated with work sector (p=0.030), such that ICU workers were 1.52 times (IC 0.57-3.99) as likely to report lower levels of control than general ward workers. Organizational analysis of the sectors showed that the ICU requires higher levels of attentiveness, specificity, responsibility and skill than the ward, even considering that the ICU sector includes more supervisors (1 supervisor/10 nursing aides) than the ward (1/24). More sensitive observational tools for the evaluation of collective organizational aspects are still necessary to improve the understanding of the complex interaction between pain and stress in the hospital context.

Keywords: physical therapy, nurse, pain

1 INTRODUCTION

Work related musculoskeletal disorders (WRMDs) have severe effects on the health and quality of life of nursing professionals (Gershon et al., 2007). Several factors have been associated with the high prevalence of symptoms, including psychosocial work factors that, in addition to contributing to the development of WRMDs, have the potential to aggravate occupational stress levels (Bongers, Kremer and Laak, 2002; Grosh et al., 2006).Nursing professionals are exposed to a high psychosocial demand at work due to long working hours, shift schedules, and high levels of psychological tension. This is due to the nature of the activity performed, since these professionals work in hospitals in shifts covering 24 hours of activity each day(Grosh et al., 2006; Malinauskienė, Leisyte and Malinauskas, 2009) and are in permanent contact with situations that involve human suffering and death.

The main consequences of exposure to such occupational activity include high levels of absenteeism and sick leave, the development of chronic pain, increased levels of occupational stress, medication expenses, occupational accidents, claims for compensation, changes of profession, early retirement(Pompeii, 2009; Lambert and Clinton, 2010) and decreased quality of patient care (Kawano, 2008).

According to Bongers et al. (1993),in order to better understand the relationship between psychosocial factors and the development of WRMDs, psychosocial factors should be evaluated by subjective methods, to capture the workers' perception about risks, as well as, by observational methods, which allow more precise evaluations.

Karasek and Theorell (1990) demand control model is among the most commonly used tools in the literature for evaluating the psychosocial demands of work that are based on worker self-reporting. The purpose of this model is to identify possible psychosocial stressors and their repercussions on worker health (Alves, 2004). On the other hand, there is a lack of valid and reliable observational tools for evaluating the collective psychosocial stressors of work. One such tool is an observational assessment instrument developed by NIOSH (WRMDs Research Consortium), as reported in Howard et al.(2009), that evaluates the impact of work organization on worker health. This assessment tool includes, among other items, demographics, environmental and organizational aspects of work, and the attention level, responsibility and restrictions imposed in the work structure.

The hospital environment involves a very specific, hierarchical, complex and exhausting work reality, where each work sector entails individualized routines that simultaneously interact with and depend on the others. Therefore, evaluating sectors by means of collective and individual approaches could contribute to a better understanding of the risks involved and to more effective measures for improving the health and quality of life of these professionals.

Therefore, the objective of the present study was to identify the organizational risk factors in two organizationally different hospital sectors and evaluate the prevalence of musculoskeletal symptoms and the stress level among staff in these sectors.

2 MATERIALS AND METHODS

2.1 Subjects

A total of 74 nursing professionals (67 women; age 35.9±10 years)employed at an Brazilian hospital in São Carlos, Brazil, were evaluated. Forty-eight of these individuals worked during the day shift and 26 during the night shift. Forty-four worked in the general ward, and 30 in intensive care units (ICUs), totaling 34% of the hospital's 217 nurses.

2.2 Inclusion criteria

The study included workers who were part of the nursing staff(registered nurses, practical nurses and/or nursing aides) and had direct contact with patients, were regular employees of the hospital and agreed to take part in the study.

2.3 Questionnaire application

The participants answered a questionnaire applied by physiotherapists during the regular work shift of the nursing staff. This questionnaire included personal information, information regarding the work sector, shift, pain evaluation and stress levels. Pain was evaluated in general terms using the Nordic Questionnaire translated and validated into Portuguese (Barros and Alexandre, 2003) and pain intensity was evaluated using the VAS. The stress levels were evaluated using the Job Stress Scale in a version translated and validated into Portuguese (Alves et al., 2004).

2.4 Organizational Analysis of the Job Posts

The organizational demand of the sectors was evaluated using the checklist proposed by Howard et al.(2009). The evaluations were carried out separately by three trained evaluators who reached the final results by consensus.

2.5 Data analysis

The intensity of the pain was measured with the VAS and classified in the4 severity levels proposed by Breivik et al. (2008): none (0); mild (1-3); moderate (4-6) and severe (>7). The symptoms were evaluated as a whole and also specifically by body part.

To classify the domains of the stress scale as high or low the median of a sample of 129 workers was calculated. The individuals who obtained values greater than or equal to the median were classified as having a high stress level; those whose values were less than the median were classified as having a low stress level. The individuals were subsequently reclassified according to the model proposed by Karasek et al. (1981) and placed in the following quadrants: high strain, passive, active and low strain.

The chi-square test was used to verify the association between the presence of WRMD symptoms, work sector and stress. Analysis of the data from the Job Stress Scale considered the domains of the scale as well as the classification used by Urbaneto et al. (2011), which includes the following quadrants: high strain and passive (highest health risk) and low strain and active (lowest risk). The association between occupational exposure (work sectors) and pain symptoms was calculated with the odds ratio in Epi Info version3.5.2.

3 RESULTS

3.1 Musculoskeletal symptoms

A high prevalence of WRMDs was identified among the nursing staff, with85% of the subjects (63 individuals) reporting pain in some body region. The lumbar region was the most frequently reported (42%), followed by the shoulders (32%), neck region (26%) and ankles and feet (22%). The hip and thigh regions (16%), knees (14%), wrists and hands (5%) presented lower prevalence. Moderate or severe pain levels occurred mainly in the lumbar spine (>30%), shoulders and neck (>20%) and feet and ankles (approximately 20%).

3.2 Job Stress Scale

The results from the Job Stress Scale showed that 54% of the evaluated workers (40 individuals) presented high demand levels at work, while51% (38 individuals) presented low control levels and 46% (34 individuals) presented low levels of social support. According to the quadrant model proposed by Karasek et al. (1981), of the 74 evaluated individuals, the work of 26% (19 individuals) was high strain, the work of26% (19 individuals) was passive, the work of 28% (21 individuals) was low strain, and the work of 20% (15 individuals) was considered active.

The work sector classification according to the domains demand, control and social support identified that the work of most of the ICU workers was low control and low social support (Table 1), whereas the ward workers' activities involved high demand (Table 1).

Table 1. Percentage of individuals classified in the domains demand, control and social support by sector.

Stress scale domains		ICU (%)	Ward (%)
Demand	High	50	56.8
	Low	50	43.1
Control	High	33.3	59.0
	Low	66.6	40.9
Social Support	High	43.3	59.0
	Low	56.6	40.9

Table 2 presents the percentages of symptomatic and asymptomatic individuals according to risk classification. There was no association between general pain and stress when classified according to risk ($p=0.83$) or the domains demand ($p=0.40$), control ($p=0.84$) or social support ($p=0.29$).

Table 2.Percentage of symptomatic and asymptomatic individuals according to health risks and stress levels.

Sectors	Risks	Symptomatic (%)	Asymptomatic (%)
ICU	Higher health risk	53.3	13.3
	Lower health risk	20	13.3
Ward	Higher health risk	38.6	2.27
	Lower health risk	52.2	6.8

Considering the symptoms per body region, there was no association between stress and pain in the neck ($p>0.08$), shoulder ($p>0.21$), lumbar spine ($p>0.27$), hip/thigh ($p>0.27$), knee ($p>0.23$), wrist/hand ($p>0.50$) and ankle/feet ($p>0.20$). However, there was an association between work sector and health risk level ($p=0.03$); the ICU workers had a greater chance (OR 2.88; IC 1.097-7.607) of being placed in the health risk quadrants than the ward workers. Among the stress scale domains, only control was associated with the work sector ($p=0.03$); the ICU workers had a greater chance (OR 1.52; IC 0.57-3.99) of exposure to low control work than the ward workers.

3.3 Collective Organizational Aspects of the Work

Some organizational aspects of the ICU and wards were similar. However, while ICU activities were more adequate ergonomically than those in the ward (less lifting and the presence of job rotation and informal breaks), they were also more demanding in terms of organizational aspects such as very high attentiveness demands, responsibility for the safety of others and structural job constraints.

4 DISCUSSION

The results indicated a high prevalence of WRMD symptoms among nursing workers. The highest number of symptoms were reported in the lumbar region, shoulders, neck, ankles and feet. High symptom prevalence for these regions have also been reported in Brazil (Gurgueira and Alexandre, 2003) and other countries (Lagertrom et al.,1995; Daraiseh et al., 2003; Smith et al., 2003). These studies have pointed out that psychosocial, organizational and ergonomic factors are associated with the occurrence of disorders.

Regarding psychosocial aspects, the results of the present study showed that the work of most of the nursing staff involved high demand, low control and low social support, which would indicate high strain and health risk. Among these results, the work of 36% of the ICU workers was classified as high strain and the work of 30% was classified as passive, i.e., conditions associated with low control. Thus, it was verified that a higher number of ICU workers were classified in the higher health risk quadrants .According to Karasek and Theorell (1990), people with high strain jobs present adverse psychological reactions such as fatigue, anxiety, depression and physical illness when continuously exposed. On the other hand, passive jobs are related to reduced learning capacity and apathy (Alves, et al., 2004) and low control jobs are associated with limited autonomy and reduced self-esteem (Urbanetto et al., 2009). A number of studies have reported similar results (Lee et al., 2002; Ha and Park, 2005). The low control levels identified in the present study by analyzing the organizational aspects of the work seem to be related to the nature of the work, particularly in the ICU.

The evaluation of sector organization according to the checklist proposed by Howard et al. (2009) showed that the ICU and ward shared similar organizational characteristics. However, the ICU presented organizational restrictions that led to lower levels of control, higher levels of attention and responsibility for the safety of others and physical confinement compared to the ward. This is because the ICU sector involves complex care that demands a high degree of commitment qualification and, consequently, leads workers to emotional exhaustion (Ferrareze, Ferreira and Carvalho, 2006). Due to its complexity, the ICU also has more supervisors, which reduces general staff autonomy.

Nevertheless, regarding the physical environment, ergonomic work conditions, including lighting, controlled room temperature, and space to carry out procedures, were better in the ICU. Despite that, the work is arduous and involves frequent patient handling. This may explain, at least in part, the high number of WRMD symptoms.

Even though WRMD symptoms were quite prevalent, there was no association with stress. There is no consensus in the available literature regarding this issue, as some studies show a positive association between these outcomes (Lipscomb et al., 2002; Gershon et al., 2007), where as a systematic review found no evidence of association (Bongers et al., 1993).Several different factors, such as the personal characteristics of the participants and the methodological design of the studies may play an important role in explaining the divergent results.

5 LIMITATIONS

Due to the lack of available instruments for evaluating the collective aspects of work, we selected an assessment tool that was not developed for the hospital environment. This caused some difficulties during the application and some incompatibility issues that had to be adjusted by the observers.

Another limitation was the relatively small sample, since our results are based on data from only 34% of the workers in the evaluated sectors.

6 CONCLUSION

There was a significant association between higher levels of health risk and lower levels of control over work with ICU workers. More sensitive observational tools for evaluating collective organizational aspects are still needed in order to improve understanding of the complex interaction between pain and stress in the hospital context.

7 REFERENCES

Alves, M. G. M., D. Chor, E. Faerstein, et al. 2004. Short version of the "job stress scale": a Portuguese-language adaptation. *Revista de Saúde Pública* 38(2),164-171.

Barros, E. N. C. and N. M. C. Alexandre. 2003. Cross-cultural adaptation of the Nordic musculoskeletal questionnaire. *International Nursing Review* 50(2):101-108.

Bongers, P. M., A. M. Kremer, J. Laak. 2002. Are phychosocial factors, risk factors for symptoms and signs of the shoulder, elbow, or hand/wrist?: A review of the epidemiological literature. *Amererican Journal of Industrial Medicine* 41;315-342.

Bongers, P. M., C. R. Winter, A. J. M. Kompier et al. 1993.Psychosocial factors at work and musculoskeletal disease. *Scandinavian Journal of Work Environmental Health*19;297-312.

Breivik, H., P. C. Borchgrevink, S. M. Allen, et al. 2008. Assessment of pain.*British Journal of Anaesthesia* 101(1);17–24.

Daraiseh, N., A. Genaidy, W. Karwowski, et al. 2003. Musculoskeletal outcomes in multiple body regions and work effects among nurses: the effects of stressful and stimulating working conditions. *Ergonomics*46(12);1178-1199.

Ferrareze, M. V. G., V. Ferreira, A. M. P. C. Carvalho. 2006. Perception of Stress Among Critical Care Nurses. *Acta Paulista de Enfermagem*19(3);310-15.

Gershon, R. R. M., P. W. Stone, M. Zeltser, et al. 2007. Organizational climate and nurse health outcomes in the United States: A Systematic Review. *Industrial Health* 45;622-636.

Grosch, J. W., C. C. Caruso, R. R. Rosa, et al. 2006. Long hours of work in the U.S.: Associations with demographic and organizational characteristics, psychosocial working conditions, and health. Amer J Indust Med. 2006;49:943-952.

Gurgueira, G. P., N. M. C. Alexandre, H. R. C. Filho. 2003. Prevalência de sintomas músculoesqueléticos em trabalhadoras de enfermagem. Revista Latino-Americana de Enfermagem 2003;11(5):608-13.

Ha, M. and J. Park. 2005. Shiftwork and metabolic risk factors of cardiovascular disease.*Journal of Occupational Health* 47;89-95.

Howard, N., P. Spielholz, S. Bao, et al. 2009. Reliability of an observational tool asses the organization of work. *International Journal of Industrial Ergonomics* 39;260-266.

Karasek, R. A. 1979. Job demands, job decision latitude, and mental strain: Implications for job redesign. *Administrative Science Quarterly, 24,* 285-308.

Karasek, R. A., D. Baker, F. Marxer, et al. 1981. Job decision latitude, job demands, and cardiovascular disease: A prospective study of Swedish men. *American Journal of Public Health* 71;694-705.

Karasek, R. A. and Theorell, T. 1990. Healthy Work: Stress, Productivity, and the Reconstruction ofWorking Life . New York: Basic Books.

Kawano, Y. 2008. Association of job-related stress factors with psychological and somatic symptoms among Japanese hospital nurses: effect of departmental environment in acute care hospitals. *Journal of Occupational Health* 50;79-85.

Lagerstrom, M., M. Wenemark, M. Hagberg, et al. 1995. Occupational and individual factors related to musculoskeletal symptoms in five body regions among Swedish nursing personnel. *International Archieves of Occupational and Environmental Health* 68;27-35.

Lambert, V. A., E. L. Clinton. 2008. Nurses' workplace stressors and coping strategies. *Indian Journal of Palliative Care* 14(1);38-44.

Lee, S., G. Colditz, L. Berkman, et al. 2002.A prospective study of job strain and coronary heart disease in US women. *International Epidemiological Association* 31;1147-1153.

Lipscomb, J. A., A. M. Trinkoff, J. Geiger-Brown, et al. 2002. Work-schedule characteristics and reported musculoskeletal disorders of registered nurses. *Scandinavian Journal of Work Environmental Health* 28;394-401.

Malinauskiene, V., P. Leisyte, R. Malinauskas. 2009. Psychosocial job characteristics, social support, and sense of coherence as determinants of mental health among nurses. *Medicine* (Kaunas) 45(11);910-917.

Pompeii, L. A., H. J. Lipscomb, A. L. Schoenfisch, et al. 2009. Musculoskeletal Injuries Resulting From Patient Handling Tasks Among Hospital Workers. *American Journal of Industrial Medicine*52;571–578.

Smith, D. R., N. Wei, L. Zhao, et al. 2004. Musculoskeletal complaints and psychosocial risk factors among Chinese hospital nurses. *Occupational Medicine* 54;579-582.

Urbanetto, J. S., P. C. Silva, E. Hoffmeister, et al. 2011. Workplace stress in nursing workers from an emergency hospital: Job Stress Scale analysis. *Revista Latino-Americana de Enfermagem*19(5);1122-31.

CHAPTER 22

Evaluation of Bamboo Charcoal Effect to Comfort Based on Physiological Data

*Ken'ichi Takao *, Noriaki Kuwahara**, Noriyuki Kida***

* H 'in Solution Co., Ltd.
Osaka, Japan
hin_works_k@alpha.ocn.ne.jp

** Kyoto Institute of Technology
Kyoto Japan
nkuwahar@kit.ac.jp
kida@kit.ac.jp

ABSTRACT

Because bamboo charcoal is known to have high porosity, it is widely used in those products which has good qualities for keeping the room environment comfort for example the wall material, the formed panel, and so on. So far, however, there is no research that evaluated the effectiveness of this material to the comfort based on physiological data. In this research, the sympathetic nerve activity level (SNS) and the parasympathetic nerve activity level (PNS) that were calculated by using the heart-rate (HR) data were measured in order to evaluate the bamboo charcoal effect of odor eliminating. Also, the subjective evaluation of the comfort level of the subjects was conducted, and the correlation between the physiological data and the subjective data were examined. The experimental results showed that SNS of the control group (w/o bamboo charcoal) was higher than that of the experiment group (w bamboo charcoal) and PNS of the control group was lower than that of the experiment group even in a situation where the odor measuring device could not detect the odor of the cigarette smoke. It means that subjects of the control group felt the stress from the odor of the cigarette smoke more than those of the experimental group did which was proved and supported by the questionnaire results of the subject.

Keywords: bamboo charcoal, odor eliminating, physiological data

1 INTRODUCTION

Bamboo material has an extraordinary micro-structure: it has a high absorptive capacity after carbonization. Bamboo charcoal can be used to purify water and eliminate organic impurities and smells. There are reports on the functional assessment of the bamboo charcoal for eliminating environmental toxins like formaldehyde (Obata, 2001). Also, it is widely used as the wall material, the formed panel and so on. Such products have good qualities for keeping the room environment comfort. So far, however, there is no research that evaluated the effectiveness of this material to the comfort based on physiological data. In this research, the sympathetic nerve activity level (SNS) and the parasympathetic nerve activity level (PNS) that were calculated by using the heart-rate (HR) data of the subjects were measured in order to evaluate the bamboo charcoal effect of odor eliminating. Also, the subjective evaluation of the comfort level of the subjects was conducted, and the correlation between the physiological data and the subjective data was examined.

2 METHOD

2.1 Bamboo Charcoal Board and Experimetal Setup

The bamboo charcoal board is the building material that is the composite of the bamboo charcoal of mousou bamboo and the paper fiber. It is eco-friendly material because it is biodegradable and doesn't contain any petroleum-based chemical substance and. Figure 1 shows the example of the bamboo charcoal board. The size of the board is 6mm(D)×303mm(H)×606mm(W). As enlarged view of the cross-section of the board presented in Figure 1, bamboo charcoal board is highly porous.

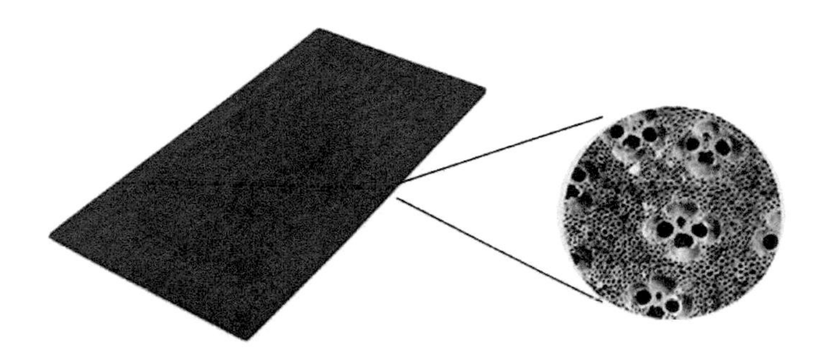

Figure 1 Bamboo charcoal board and close-up of its cross-section

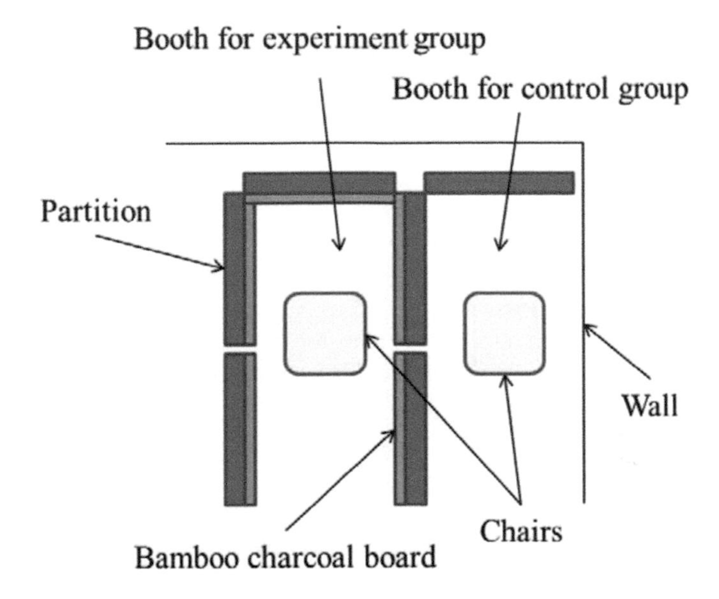

Figure 2 Illustrative example of the experimental setup

Figure3 Scene of the experiment

Figure 2 shows the illustrative example of the experimental setup. Two booths that were enclosed by partitions on three sides were prepared. The size of the partition was 455mm (W) × 910mm (H). The partitions of one of these booths were coated by bamboo charcoal formed panels (experiment group). On the other hand, those of the other booth were uncoated; the surface material was mainly chemical textile (control group). The subjects sat on the chair in either booth facing the back partition as shown in Figure 3.

2.2 Subjects

The number of the subjects is five. All subjects are male. The age of them is from 30's to 60's. Two of them are smokers.

2.3 Experimental Protocol

Figure 4 shows the experimental protocol. This protocol is designed with reference to previous work (Fujibayashi, 2008) on the evaluation of the comfort. After sitting on the chair, Period i) the subjects kept themselves at rest without the olfactory stimulus for 20 minutes; Period ii) Next, the subjects were exposed to the olfactory stimulus of the cigarette smoke for 15 minutes; Period iii) Finally, the subjects kept themselves at rest again with forced ventilation for 20 minutes. From Period i) to Period iii), the HR data of the subjects were kept measuring. The questionnaire survey for asking the subjects their comfort level was also conducted.

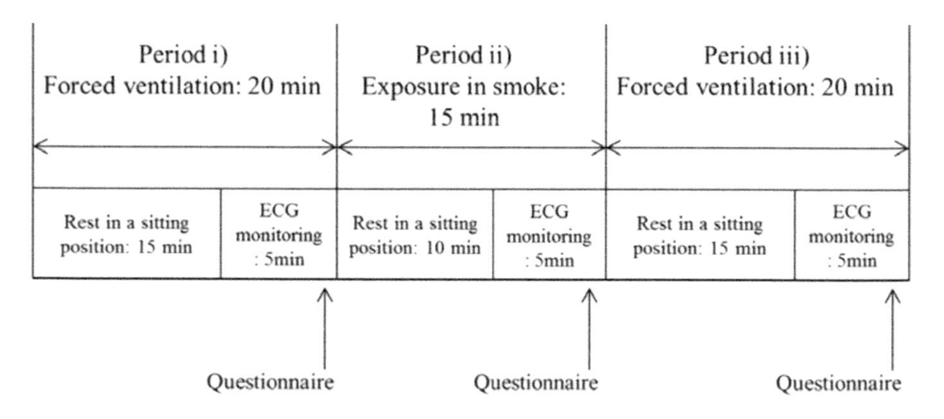

Figure 4 Experimental protocol

When the questionnaire was conducted to the subject, room temperature, humidity, noise, and odor level were measured. Heart rate data was recorded by using RS800-CX of Polar Corp. Room temperature and humidity were recorder by using RS-13 of ESPEC Corp. And noise data was recorded by using DT-8852 of CEM Corp. Additionally, odor level was monitored by using OMX-SR of SHIN-EI Corp.

2.4 Heart Rate Data Analysis

In order to calculate both SNS and the PNS, the HR data of the subjects in the last five minutes of each period of i), ii), and iii) was used as shown in Figure 4. The autonomic nervous system (ANS) activities were measured during the resting condition by means of HR variability power spectral analysis, which enables us to identify separate frequency components; low frequency (LF; 0.04-0.15 Hz) reflects mixed sympathetic (SNS) and parasympathetic nervous system (PNS) activity; high frequency (HF; 0.15-0.4 Hz) mainly associates with PNS activity; total power (TP; 0.04-0.4 Hz), evaluating the overall ANS activity. According to the previous researches (Hayashi, 1994) (Matsumoto, 1999) (Moritani, 1993), SNS index was calculated by the ratio of the LF to the HF, and PNS index was calculated by the ration of the HF to the TP.

3 RESULTS

3.1 Odor, Tempertuare and Humidity

Figure 5 shows the averages of the odor level monitored by OMX-SR in Period i), ii) and iii). OMX-SR measures the odor level of the atmosphere in a relative manner. Therefore, from Figure 5, there was no difference of the odor level between Period i) and Period iii) in both experiment group and control group. However, in Period ii), the odor level were much higher than those in Period i) and Period iii) because of the exposure of the cigarette smoke. The odor level of control group was higher than that of the experiment group, but there was no significant difference.

On the other hand, the average of the temperatures was 23.8 Celsius, and the standard deviation of it was 0.4 Celsius in Period i), ii) and iii). The average of the humidity was 47.5 %, and the standard deviation of it was 11 % in Period i), ii) and iii).

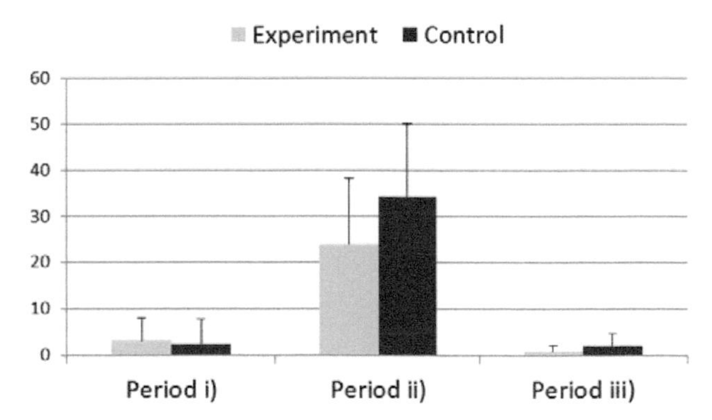

Figure 5 Odor level in the experiment

3.2 Questionnaire

Figure 6 shows the difference of the comfort level between Period i) and Period ii). Positive value means that the comfort level was worse in Period ii) than in Period i). The result shows that the comfort level was apparently worse in Period ii) than in Period i) in both experiment and control groups because of cigarette smoke. Figure 7 shows the difference of the comfort level between Period i) and Period iii).

Positive value means that the comfort level was worse in Period iii) than in Period i). As shown in Figure 7, the comfort level in Period iii) was almost the same as in Period i) in case of experiment group. On the other hand, nevertheless odor level was almost zero in both experiment and control group, the comfort level remained worse in Period iii) in case of control group.

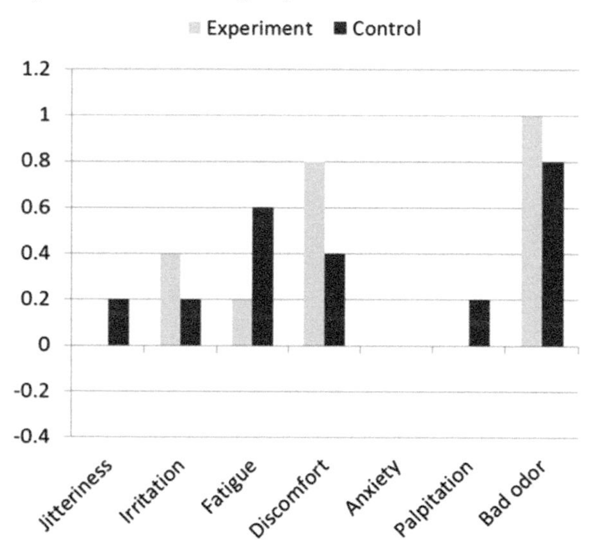

Figure 6 Difference of the comfort level between Period i) and Period ii)

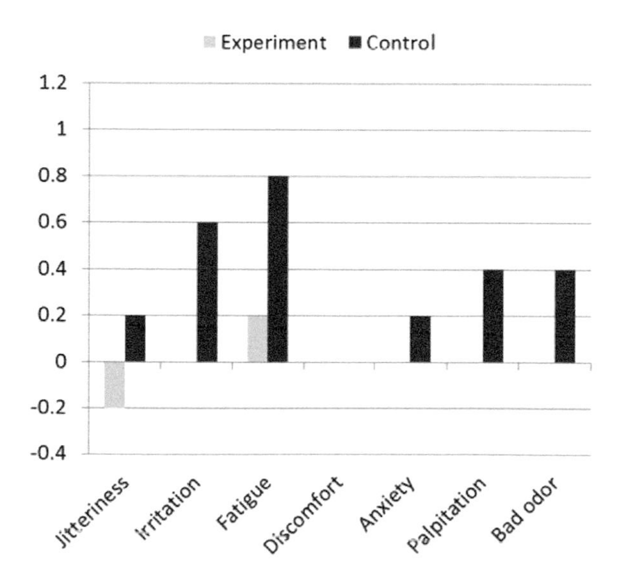

Figure 7 Difference of the comfort level between Period i) and Period iii)

3.3 Heart Rate Data

According to the calculation results shown in Figure 8, SNS became higher from the Period ii) to iii) in case of control group. On the other hand, SNS index kept almost the same level from the Period i) to iii) in case of experimental group. Figure 9 shows that PNS index was lower in case of experimental group than in case of control throughout periods. Higher SNS index means that subject felt more stress (uncomfortable). Also higher PNS index means that subject were relaxed more or felt less stress (comfortable). It corresponds to the questionnaire result as shown in 3.2.

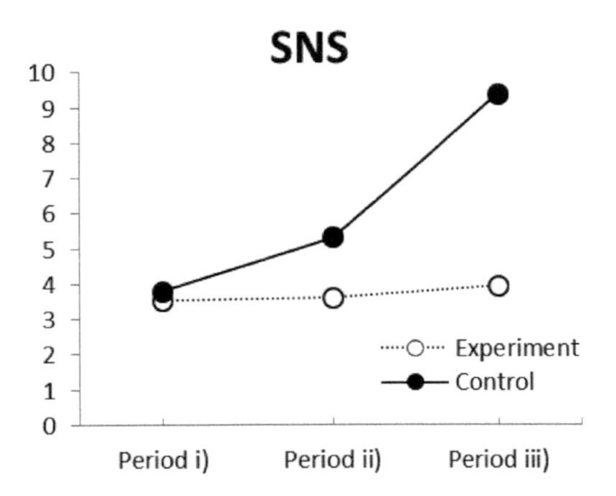

Figure 8 SNS index

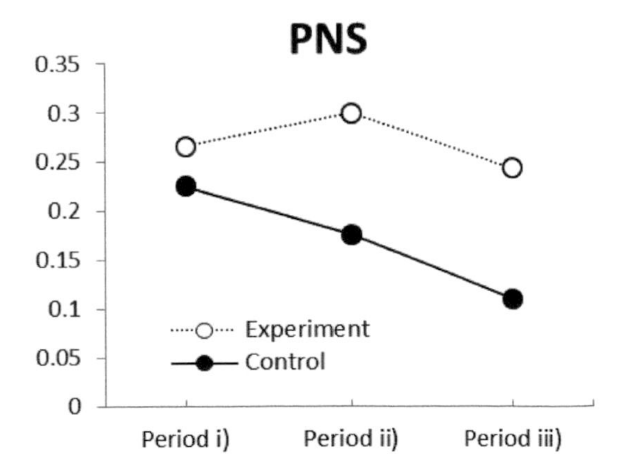

Figure 9 PNS index

4 CONCLUSIONS

In this paper, the sympathetic nerve activity level (SNS) and the parasympathetic nerve activity level (PNS) that were calculated by using the heart-rate (HR) data of the subjects were measured in order to evaluate the bamboo charcoal effect of odor eliminating. The experimental protocol consisted of three periods. In the first period, the baseline data was obtained. In the second period, the subjects of both control group (w/o bamboo charcoal) and experiment group (w bamboo charcoal) were exposed to the cigarette smoke. In the third period, by forced ventilation, the odor level in the third period monitored by OMX-SR became as same as that of the first period (baseline).

According to the results of SNS and PNS, the subjects in the experiment group felt less stress than those of the control group in the third period. Additionally, , the result from the subjective evaluation of the comfort level of the subjects showed the correlation between both SNS and PNS.

These results confirm the effectiveness of the bamboo charcoal for keeping the comfort of a human based on both physiological data and subjective data in the situation where the odor measuring device cannot detect the odor.

ACKNOWLEDGMENTS

The authors would like to acknowledge Professor Hiroyuki Hamada for making this work possible, and also for his encouragement and helpful discussions.

REFERENCES

Fujibayashi, M., Saito, M., Ota, Kaori., et al. Cosmetic facial mask: effects on psycho-physiological relaxation measured by autonomic nervous system activity *Journal of Japanese Society of Psychosomatic Obstetrics and Gynecology* Vol.13, No.1: 86–93, 2008. (in Japanese)

Hayashi. T., Masuda. I., Shinohara. M., et al. Autonomic nerve activity during physical exercise and postural change: investigation by power spectral analysis of heart rate variability. *Japanese Journal of Biochemistry of Exercise* 6: 30-37, 1994.

Matsumoto. T., Miyawaki. T., Ue. H., et al. Autonomic responsiveness to acute cold exposure in obese and non-obese young women. *Int J Obes Relat Metab Disord* 23: 793-800, 1999.

Moritani. T., Hayashi. T., Shinohara. M., et al., Comparison of sympatho-vagal function among diabetic patients, normal controls and endurance athletes by heart rate spectral analysis. *Journal of Sports Medicine and Science* 7: 31-39, 1993.

Obata, T., Matsunaga, K., Kasasaku, K., et al. Study on Adsorption Chemical Reactions of Bamboo Charcoal. *Research Report of Kagoshima prefectural Institute of Industrial Technology* No.15: 35–37, 2001. (in Japanese)

CHAPTER 23

Environmental Audit of UK Hospitals to Design Safer Facilities for Frail and/or Confused Older People

Sue Hignett

Loughborough University
Loughborough, UK.
S.M.Hignett@lboro.ac.uk

ABSTRACT

This paper reports the outputs of 2 pilot studies to develop an environmental audit tool for care facilities using a checklist and a hierarchical task analysis. Ten wards at 6 hospitals (10 wards) were audited using the checklist and 3 wards assisted with the development of the HTA. It was found that the audit tool needed further development to include empirical measurements (lighting, contrast and signage) and international environmental standards. The HTA described a complex hierarchy of tasks and subtasks for the goal of going to the toilet without assistance for a frail and/or confused older person in an unfamiliar environment. There is a need for an in-depth understanding of the person-environment interaction to support patient autonomy (reduce the level of disorientation and distress) and improve safety. A validated environmental audit tool will support the delivery of (and evaluate) system level and design solutions.

Keywords: environmental audit, older people, dementia

1 INTRODUCTION

According to projections the number of severely disabled elderly will grow from about 5.1 million in 1986 to 22.6 million in 2040 (Dept. of Health and Human Services, 2011). As well as physical impairment it is expected that there will be over 2 million cases of clinically diagnosed Alzheimer's Disease (AD) at ages 65 and over by 2050, compared with about 3.8 million in 1990 (Evans et al, 1992). It has been suggested that they may be occupying up to 25% of acute hospital beds at any one time (Alzheimer's Society, 2009).

This paper reports the outputs of 2 pilot studies to develop an evidence-based environmental audit tool for care facilities using a checklist and hierarchical task analysis.

2 ENVIRONMENTAL AUDIT

The first pilot study developed an audit checklist to evaluate the environmental design of acute care settings with a focus on autonomy and independence for a confused patient (dementia/delirium). Three published environmental audit tools (Royal College of Psychiatrists. 2010; Dementia Services Development Centre, 2007; Calkins, 1988) were used as the basis for the tool (Hett, 2011).

The Royal College of Psychiatrists (2010, 2011) National Audit of Dementia Environmental Checklist was used to gather information (by self-report) about aspects of the ward physical environment known to impact on people with dementia (PWDem) in acute, community and mental health hospitals in England and Wales. An expert consensus group reviewed data and ideas from professional recommendations and guidance documents (Housing Corporation, 2004; Day et al, 2000; Department of Health, 2007; Brawley, 2001) to agree a series of checklist items that was used to collect data from 144 hospitals. The full report was published after this pilot study so the findings have been included for discussion.

The Dementia Services Development Centre (2007, 2011) Dementia Design Audit Tool offers a framework for making decisions about the design of spaces for PWDem. It was produced from expert consensus, literature and examples of best practice and then piloted in a range of settings. Calkins (1988) offers an early example of universal design recommendations for dementia based on professional opinion.

The audit checklist had sections for signage, flooring/walls, lighting, layout and monitoring (sight lines), physical support (grab rails), hygiene facilities and design for patient independence. It was evaluated at 6 UK hospitals (acute and longer stay), 3 of which also participated in the Royal College of Psychiatrists audit. This included 1 mental health trust (2 sites/wards) for longer term dementia care with an average stay of 6 weeks, and 3 acute trusts (4 sites) for 3 orthopaedic wards and 5 acute medical/care of elderly wards with an average stay of 3-10 days. The findings for signage, flooring and lighting are discussed in this paper.

2.1 Signage

It was found that the signage varied across the hospitals from only text, to text with a picture or symbol with a range of different (some contrasting) colours (Figure 1). PWDem may have semantic memory impairment resulting in problems recognising words, objects and pictures (Adlington et al, 2009). The use of primary colours (Wijk et al, 1999) and concrete nomenclature (rather than abstract) with supporting diagrams is likely to be helpful. For all older people it is likely that improving contrast, font size and signage placement may increase independent navigation and successful use of hygiene facilities (Namazi and Johnson, 1991a). In a study where patients (with no confusion) were interviewed about contributory factors for falls, difficulties were identified with distance perception, in particular underestimating distances between objects due to the greater size and distance between hospital fixtures compared to domestic environments (Morse *et al.* 1987).

The results from the Royal College of Psychiatrists audit (2011) found that fewer than half of wards (48%) reported that key areas were clearly marked; 56% reported that information (words and pictures) on signs was in clear contrast to the background; and 38% reported that signs/maps were large, bold and distinctive.

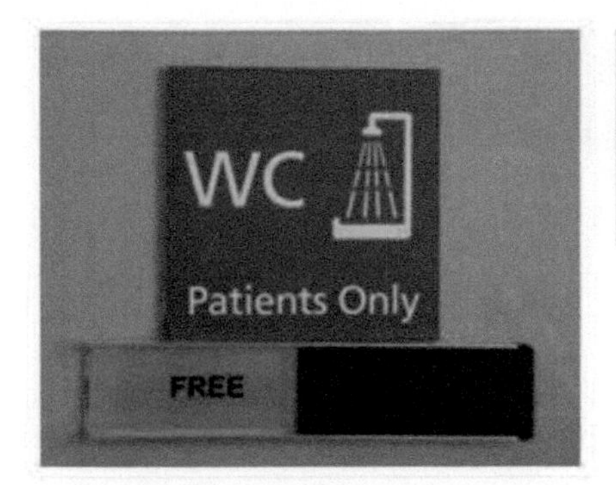

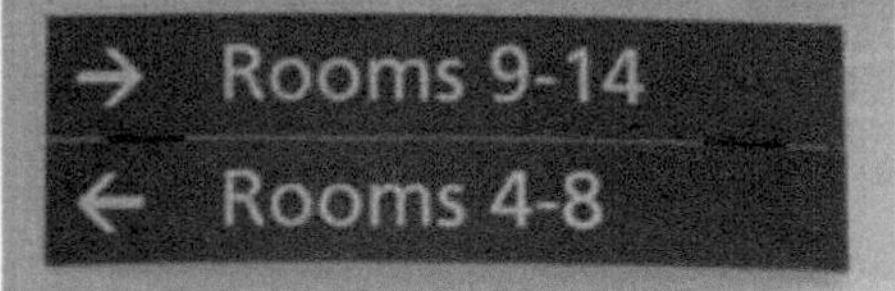

Figure 1 Examples of Signage from Environmental Audit Pilot Study

The pilot study found that the route from the patient bed to the toilet/bathroom varied in terms of the layout (distance and route) and signage. In the multi-bed bays there was usually at least 1 bed without directional information (signage or sightline) to locate the toilet from the bed.

The Royal College of Psychiatrists audit (2011) reported that only 15% used colour schemes to help PWDem find their way around the ward; and only 28% reported that signs to locate the toilet were visible from the patient's bed area/door of room. However, 90% of wards reported that toilet and bathroom doors had signs and 87% had toilet and bathroom doors of a different colour to the walls, with a surprising 38% of wards using toilet paper of a different colour to the wall.

2.2 Flooring

There was a marked difference in flooring between the older (polished, shiny finish) and newer (matt finish) acute wards (Figure 2). As older people may have increased sensitivity to glare a matt (less reflective) floor has been found to be beneficial (Bright et al, 1991; Torrington and Tregenza, 2007).

The Royal College of Psychiatrists (2011) report found that 95% of wards self-reported that floors were plain or subtly patterned; and that 93% (134/144) of floor surfaces were subtly polished rather than high gloss, with 78% reported as non slip.

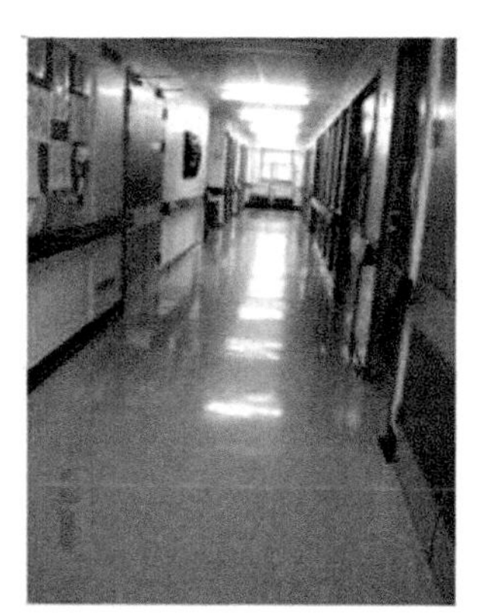
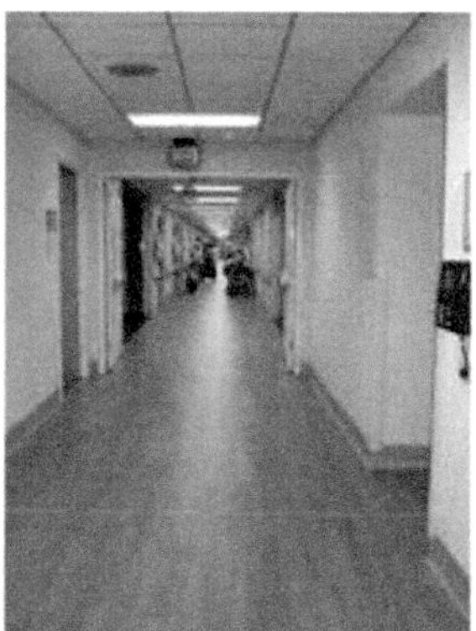
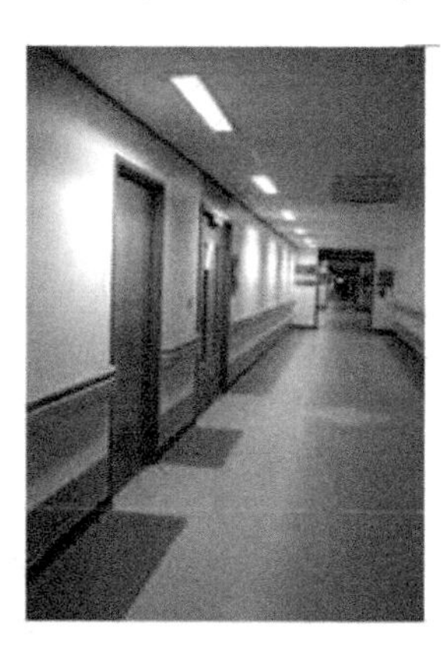

Figure 2. Examples of flooring with a shiny and matt finish, and pattern to indicate doorways

2.3 Lighting

The pilot study found that 80% of the wards had insufficient lighting; this would need to be measured and compared with international standards. One of the newer acute wards had night lighting for the floor around the bed and near the toilet which would be particularly beneficial for people with visual impairment or low vision (Wright et al, 1999).

Serious sight loss affects about two million people and the main causes of sight loss are related to ageing, so most people with sight loss are also aged over 65. Among those over 75, it is estimated that around 100,000 people will have a diagnosis of both dementia and sight loss (Thomas Pocklington, 2010). The Royal College of Psychiatrists audit (2011) did not consider lighting so no comparative data are available.

3 GOING TO THE TOILET: HIERARCHICAL TASK ANALYSIS

Hierarchical task analysis (HTA) is used to provide a more in-depth analysis by dividing a task (goal) into a hierarchy of operations (Kirwan and Ainsworth, 1992). To further develop the environmental audit tool an HTA of a frequent activity (going to the toilet independently) was carried out to identify features which may support or hinder independence and autonomy.

In a previous paper, it was proposed that Maslow's Hierarchy of Needs could be used as a model to explore the motivators for movement and barriers that might be associated with in-patient falls (Hignett and Masud, 2006). The level one needs (physiological) motivation for movement may include bladder and bowel function (to support homeostasis), hunger, thirst and activity. If these needs are not met by caregivers, then the patient will be highly motivated to achieve them independently, resulting in movement from the bed and introducing the risk of falling. This formed the basis for the hierarchical task analysis (HTA) to achieve the goal of going to the toilet independently (Figure 3).

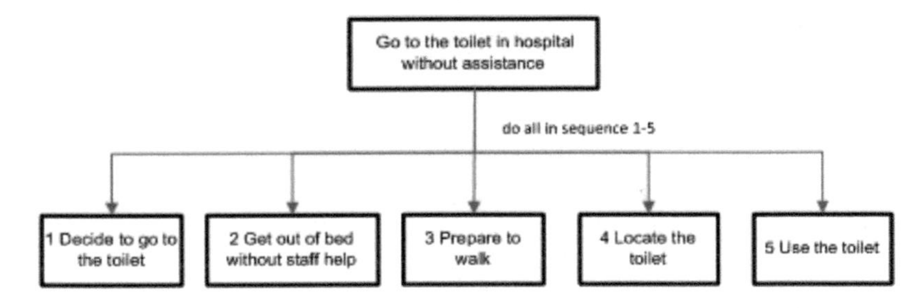

Figure 3. Hierarchical Task Analysis from bed to toilet

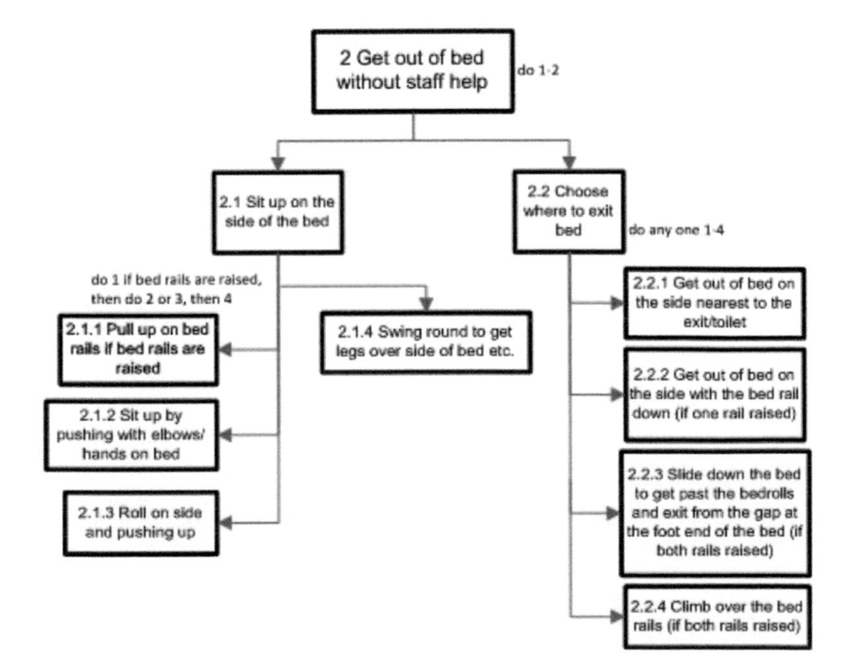

Figure 4. Hierarchical Task Analysis for getting out of bed without staff assistance (Step 2)

One of the first barriers/enablers that will be encountered may be when trying to get out of bed without staff help (Figure 4). The use of bed rails has been discussed since the 1960s, with Fagin and Vita (1965) commenting that *'to many conscious patients, side rails are frightening and imply dangerous illness. To others, side rails are irritating and humiliating because they emphasize the confining aspects of hospitalization.'* Most patients want to retain their independence, in particular with respect to elimination needs (Chung, 2009), for example *'on numerous occasions seriously ill patients climbed over the bed rails to go to the bathroom, thus averting the embarrassment of a soiled bed'* (Parrish and Weil, 1958).

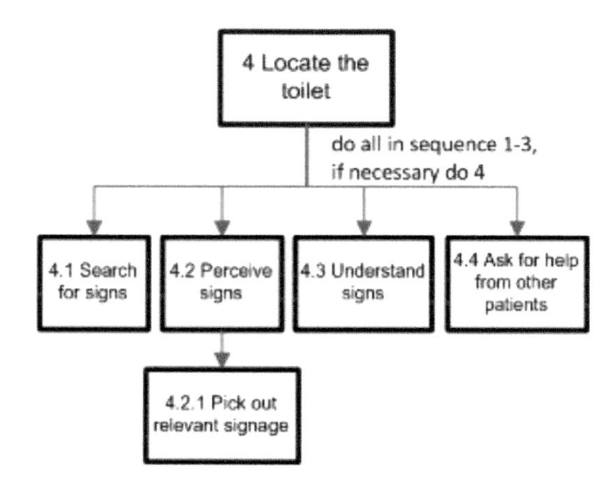

Figure 5. Hierarchical Task Analysis for locating the toilet (Step 4)

The challenge of locating the toilet will be supported or complicated by the characteristics of the pathway between the bed and bathroom. This was found to include the signage (Figure 5), distance to the bathroom, number of directional turns between the two destinations, the availability of support (hand rails), obstacles in the path, bathroom configuration, direction of door swing, and the space within bathroom (Pati et al., 2009). Figure 6 shows two layouts that, it is suggested, have been designed to (a) provide hand rails for patients for physical impairment, giving support (hand rails) between the bed and the toilet along the head wall, and (b) give a direct sight line for patients with cognitive impairment from the bed head to the toilet pan (Namazi and Johnson, 1991b).

4 DISCUSSION AND CONCLUSION

During the development of the pilot tool, it was felt that a major limitation was the lack of environment measures. For example *'the space has good levels of artificial lighting'* (Dementia Services Development Centre, 2007); *'signs/maps are large, bold and distinctive'* (Royal College of Psychiatrists, 2011). Further pilot work is being carried out to continue the development of the environmental audit tool using empirical measurements for lighting, contrast and signage based on ISO recommendations in a nursing home care facility.

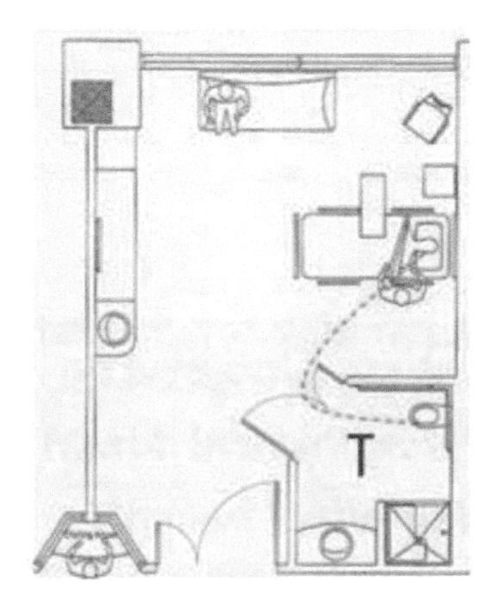

Figure 6. Design for (a) frail patients with hand rails from bed to toilet (Pati et al., 2009) and (b) confused patients with a direct sight line of the toilet (NHS Estates, 1998)

The Thomas Pocklington Trust (2009, 2010) reported that models for dementia care have previously not given sufficient emphasis to broader issues of ageing e.g. visual impairment. This has resulted in design guidance for people with sight loss focusing on maximising independence and with a higher level of detail and more precise specifications than design guidance for people with dementia. There are common principles between the two sets of guidance but there are also potential conflicts, for example memory objects placed along circulation routes to assist in way-finding for PWDem could become hazards and obstacles for people with sight loss. They recommended the development of a sensory model of care practice for dementia care settings. However, staff in acute hospitals may be *'unaware of a dementia condition in their patients…and the admission itself can lead to the worsening of the effects of dementia…due to the disorientation and distress that arise from separation from familiar people, environment and routine'* (Royal College of Psychiatrists, 2011). The HTA identified environmental enablers as well as barriers but the main finding was the complexity of achieving the goal of the simple hygiene task.

It has been estimated that the cost implications of hospital admission for PWDem include a greater risk of medical problems leading to an extended recovery period with an increased length of stay. These costs have been estimated to be in excess of £6 million per year in an average general hospital (National Audit Office, 2007). There is a need for an in-depth understanding of the person-environment interaction to support patient autonomy (reduce the level of disorientation and distress) and improve safety. A validated environmental audit tool will support the delivery of (and evaluate) system level and design solutions.

ACKNOWLEDGMENTS

The author would like to acknowledge the collaborators in both pilot studies: Danielle Hett, Colette Nicholle, Simon Hodder and all the NHS collaborators. She would also like to acknowledge the knowledge and experience gained from the Dementia Design School (Dementia Services Development Centre), University of Stirling.

REFERENCES

Adlington, R.L., Laws, K. R., Gale, T. M. 2009. Visual processing in Alzheimer's Disease: surface detail and colour fail to aid object identification. *Neuropsychologia* . 47, 2575-2583.

Alzheimer's Society, 2009. *Counting the cost: caring for people with dementia on hospital wards*. Alzheimer's Society: Great Britain.

Brawley, E.C. 2001. Environmental design for Alzheimer's disease: A quality of life issue, *Aging & Mental Health*, 5:S1, 79-83

Bright, K., Geoffrey C., Harris, J. 1991. The importance of flooring pattern and finish for people with a visual impairment. *The British Journal of Visual Impairment*, 17:3.

Calkins MP (1988) *Design for Dementia. Planning Environments for the Elderly and the Confused.* Maryland: National Health Publishing.

Chung, H. 2009. *The lived experience of older adults who fall during hospitalization.* PhD Dissertation. College of Nursing, Graduate School of the Texas Woman's University.

Day, K., Carreon, D., Stump, C. 2000. The therapeutic design of environments for people with Dementia: Review of the Empirical Research. *The Gerontologist,* 40:4, 397–416.

Dementia Services Development Centre. 2007. Dementia Design Checklist. https://dementia.stir.ac.uk/files/DementiaDesignChecklist.pdf accessed 11 Nov. 2011.

Dementia Services Development Centre. 2011. Dementia Design Audit Tool (Version 2). Stirling: University of Stirling.

Dept. of Health and Human Services, 2011. Accessed 11 Nov. 2011 http://www.aoa.gov/AoARoot/Aging_Statistics/index.aspx,

Department of Health (2007): *Essence of care – care environment benchmark.* http://www.rcn.org.uk/__data/assets/pdf_file/0003/194709/Essence_of_Care_PDF.pdf Accessed 20 February 2012

Evans, D. A., Scherr, P. A., Cook, N. R. 1992. *The impact of Alzheimer's Disease in the United States population.* In R. M. Suzman, D. P. Willis, & K. G. Manton, (Eds.), *The Oldest Old.* New York, NY: Oxford University Press

Fagin, I.D., Vita, M. 1965. Who? Where? When? How? An Analysis of 868 Inpatient Accidents. *Hospitals*, 39, 60-65.

Hett, D. 2011. *Developing an Environmental Audit Tool that supports autonomy and safety for confused (Dementia/Delirium) Patients in an Acute Setting.* B.Sc. project report. Loughborough Design School, Loughborough University.

Hignett, S. 2010. Technology and Building Design initiatives in interventions to reduce the incidence and injuries of Elderly In-Patient Falls. *Healthcare Environments Research and Design Journal* 3, 4, 62-84

Hignett, S. Masud, T. 2006. A Review of Environmental Hazards associated with In-Patient Falls. *Ergonomics.* 49: 5-6, 605-616.

Housing Corporation (2004) *Neighbourhoods for life: A checklist of recommendations for designing for dementia-friendly outdoor environments.* Accessed 20 February 2012.
http://www.idgo.ac.uk/about_idgo/docs/Neighbourhoods.pdf

Kirwan, B., Ainsworth, L.K. 1992. *A Guide to Task Analysis.* London: Taylor and Francis.

Namazi K., Johnson, B. 1991a. Physical environmental cues to reduce the problems of incontinence in Alzheimer's Disease units. *The American Journal of Alzheimer's Care and Related Disorders and Research,* 6, 6, 22-28.

Namazi, K.H., Johnson, B.D. 1991b. Environmental effects on incontinence problems in the Alzheimer's disease patients. *The American Journal of Alzheimer's Care and Related Disorders & Research,* 6: 6, 16-21.

National Audit Office. 2007. *Improving Services and Support for People with Dementia.* London: The Stationary Office.

NHS Estates 1998 *Health Building Note 4, vol 2 In-Patient Accommodation. European Case Studies* London: The Stationary Office

Parrish, H., Weil, T.P. 1958. Patient Accidents Occurring in Hospitals: Epidemiologic study of 614 accidents. *New York State Journal of Medicine*, 58: 6, 838-846.

Pati, D., Harvey Jr, T.E., Reyers, E., et al. 2009. A multidimensional framework for assessing patient room configurations. *HERD*, 2: 2, 88-111.

Royal College of Psychiatrists. 2010. *National Audit of Dementia (Care in General Hospitals).* www.nationalauditofdementia.org.uk. (accessed 11 Nov 2011)

Royal College of Psychiatrists. 2011. *Report of the National Audit of Dementia Care in General Hospitals 2011* Accessed 19 February 2012. http://www.rcpsych.ac.uk/pdf/NATIONAL%20REPORT%20-%20Full%20Report%200512.pdf

Story, M.F., Mueller, I., Mace, R.L. 1998. *The Universal Design File: Designing for People of All Ages and Abilities.* Raleigh, NC: Centre for Universal Design, North Carolina State University.

Thomas Pocklington Trust. 2009. *People with dementia and Sight Loss: A scoping study of models of care.* Chiswick, London: Thomas Pocklington Trust.

Thomas Pocklington Trust. 2010. *Design guidance for people with dementia and for people with sight loss.* Chiswick, London: Thomas Pocklington Trust.

Torrington, J.M., Tregenza, P.R. 2007. Lighting for people with dementia. *Lighting Resolution Technology.* 39, 1, 81-97.

Wijk, H., Berg. S., Sivik, S., Stehen, B. 1999. Colour discrimination, colour naming and colour preferences among individual with Alzheimer's Disease. *International Journal of Geriatric Psychiatry.* 14, 1000-1005

Wright. M, Hill, S., Cook, G. K. 1999. Office task lighting: a user study of six task lights by five workers with low vision. *British Journal of Visual Impairment.* 17, 117-120.

CHAPTER 24

Extreme Makeover: Ergonomic Challenges and Solutions in a Gastroenterology Clinic

Yeu-Li Yeung[1], *Kelly Monsees*[2]

[1] Ergonomics Division
Duke University and Health System
Durham, NC 27710 USA

[2] Kelly Monsees
GI-Adult Diagnostic Services
Duke University and Health System
Durham, NC 27710 USA

ABSTRACT

Nurses and sterile processing technicians working in a gastroenterology clinic at a major health system in the USA have reported musculoskeletal disorder injuries related to computer workstations, patient handling, sterile processing, and the physical environment. Ergonomic evaluations were conducted due to requests from the nurse manager and employee health. Ergonomic stressors were identified. The design of the tools and the physical environment were contributing factors to some of the identified stressors. Administrative and engineering solutions were recommended to reduce musculoskeletal injuries and to increase patient and staff safety. The engineering solutions included adjustable wall-mounted and mobile computer workstations; various patient handling devices; sit-stand seating and step stools; and other low-tech solutions in sterile processing. Recommended administrative solutions included the rotation of job duties and staff training.

Keywords: ergonomics, gastroenterology, musculoskeletal injuries, nurses, sterile processing

1 INTRODUCTION

This paper describes the ergonomic challenges experienced by the nurses and sterile processing technicians in the gastroenterology clinics at a major health system, and the ergonomic solutions recommended for those challenges. There have been numerous studies of gastroenterologists sustaining musculoskeletal injuries due to risk factors such as the duration and high volume of the case loads (Francis et al 2011; Siegel 2007); awkward and static work postures and force used to control instruments (Shergill et al 2009; Siegel 2007); prolonged standing (Francis et al 2011; Shergill et al 2009; Siegel 2007); lack of adjustability in monitors and the static load of lead aprons coupled with the duration of wear (Siegel 2007; Shergill et al, 2009); as well as trip hazards and head concussion due to equipment and layout of the room (Cappell 2011).

Very little evidence has been found describing musculoskeletal injuries sustained by nurses and technicians working in gastroenterology units. Murty (2010) observed and conducted interviews of nurses working in an endoscopy unit. The study found that the neck was reported by nurses as the part of the body most exposed to high or very high levels of risk, followed by the back. The reported influential risk factor was the duration of exposure. Other reported risk factors were related to the visual demands of the task and the need to observe various monitors for information throughout procedures. In addition to assisting physicians during procedures, nurses may have to assist patients with transfers before and after procedures. Francis et al (2011) examined data on the root cause of patient falls in a high volume endoscopy unit. Patients who received conscious sedation for endoscopic procedures are at high risk for falling due to the cognitive effects of the sedation, potential hypotension induced by the sedatives, and the prolonged fasting state. The study reported that the unit continued to implement thorough falls assessments on patients, and hired additional personnel to assist patients leaving the unit in a wheelchair. Interestingly there was no mention of implementation of engineering controls, such as patient lifts, to prevent patient falls as well as increase the safety of both staff and patients.

2 OBJECTIVE

Musculoskeletal related injuries have been reported in a gastroenterology clinic of a major health system. Ergonomic challenges related to computer workstations, patient handling, sterile processing, and the physical environment were identified. Administrative and engineering solutions were recommended to reduce the risk of injury.

3 METHODS

Observations and interviews were conducted with physicians, nurse managers, nurses and sterile processing technicians. Recommendations were made according to the Occupational Safety & Health Administration (OSHA)'s Occupational Noise Exposure guidelines and ergonomics principles. Equipment trials with various vendors were coordinated, and feedback from staff was collected in order to identify the appropriate engineering solutions. Staff training was provided and follow-up visits were conducted.

4 OBSERVATIONS

4.1 Computer Workstations

Nurses work in three main areas at the clinic: the peri-operative area, procedure room, and recovery area. In all three areas, nurses have to enter patient information in either wall-mounted or mobile computer workstations. The time spent at each type of computer workstation depends on the work area and the procedures. The ergonomic stressors identified were:

- Awkward and static postures of the neck, wrists, and back when using the non-height adjustable wall-mounted and mobile computer workstations as shown in Figures 1 and 2;
- Prolonged periods of standing;
- Contact stress placed on the wrists and the forearms when using the mouse and keyboard at the wall-mounted or mobile workstations as shown in Figures 1 and 2.

Figure 1: A non-height adjustable wall-mounted computer workstation.

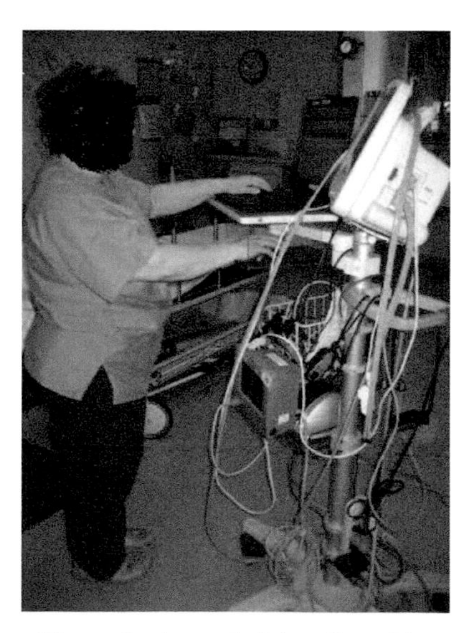

Figure 2: A non-height adjustable mobile computer workstation.

4.2 Patient Handling

Nurses also perform patient handling tasks, which include: occasional transfers between the wheelchair and stretcher; repositioning patients (lateral, prone, and supine); and holding limbs in position during procedures. The ergonomic stressors identified were:

- Force, awkward postures, and contact stress while:
    - Transferring patients between wheelchairs and stretchers;
    - Repositioning patients in lateral, prone, and supine positions during procedures;
    - Holding limbs in position during procedures;
    - stopping a patient from falling;
- Prolonged duration of exposure to the awkward and static postures as well as increased force applied due to repositioning semi-sedated or totally-sedated patients.

4.3 Sterile Processing

Sterile processing technicians work in two main work areas: #1 dirty scopes and #2 clean scopes. At work area #1, there are two sinks where each technician manually washes the dirty scopes. After the scope is washed, it is placed in an ultrasonic disinfector for further disinfection. On average a total of 60 scopes are washed manually each day between two technicians. Each scope takes approximately 15 minutes to wash. As a result, each technician can spend an average of 7.5 hours (or 450 minutes) standing at the sink, washing scopes. At work area #2, one technician further dries the clean scopes with an air nozzle for about one minute per scope, and then hangs them on the hooks that are mounted on the wall. The ergonomic stressors identified were:

- Prolonged standing as well as awkward and static postures of the neck and upper body while looking down at deep sink to wash dirty scopes as shown in Figure 3, and at the counter to air dry the clean scopes with an air nozzle;
- Contact stress placed on the forearms, elbows, and mid-section when leaning against the sink to wash the scopes;
- Awkward shoulder posture as a result of reaching over the counter and above shoulder height to activate and control the air that flows out of the air nozzle. The counter depth is 27" and the handle is mounted at 60.5" from the floor;
- Noise level from the air nozzle peaked at 101.3 dBA when in use, which exceeds an 8-hr Time-Weighted Average of 90 dBA, according to the OSHA and Duke policy;
- Repeatedly reaching above the shoulder height to hang the clean scopes on the hooks a shown in Figure 4.

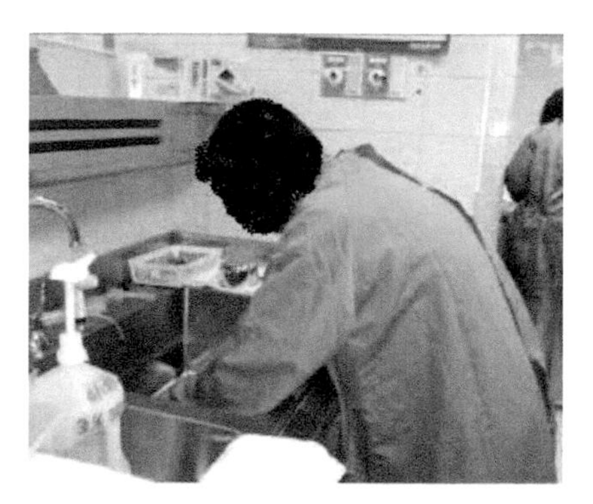

Figure 3: A technician standing at the deep sink to wash the dirty scopes.

Figure 4: A technician hanging up the clean scopes.

4.4 Physical Environment

The physical layout and space of the procedure rooms do not readily accommodate the latest medical technology and equipment used during procedures. The floor space of each room is typically occupied with various types of equipment, tubes, and cords. The ceiling is also occupied with arms that hold the monitors used during procedures. The ergonomic stressors identified were:

- Trip hazards from the suction tubing and other cords as shown in Figure 5
- Low placement of the monitor and the sharp edges of the monitor can cause head concussions as shown in Figure 6

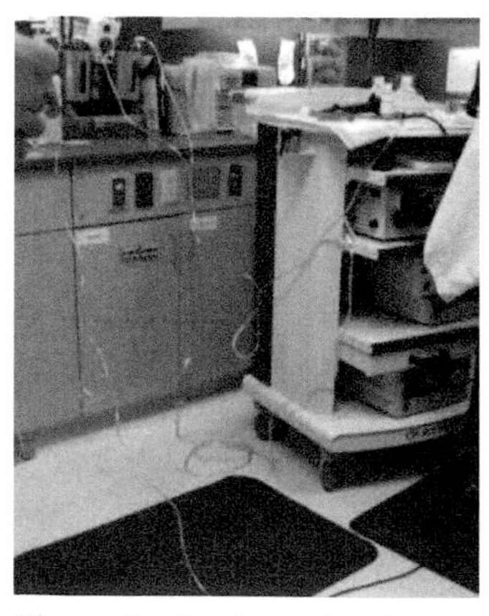

Figure 5: Suction tubes hanging off a cart and on the floor.

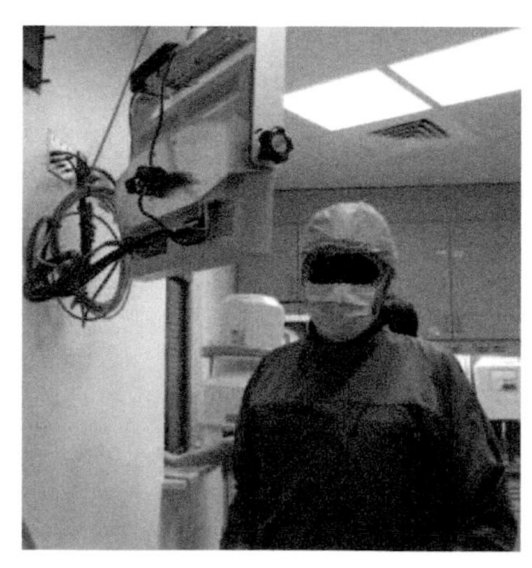

Figure 6: A ceiling-mounted monitor with sharp edges is hanging at a fixed height.

5 RESULTS

5.1 Computer Workstations

After trialing workstations from three different vendors and with feedback from the staff, the appropriate solutions were identified for both the wall-mounted and the mobile computer workstations. All the wall-mounted computer workstations now have separate height and swivel adjustability for the monitor arm and keyboard/mouse tray as shown in Figure 7. All the mobile computer workstations were adapted to allow mounting of the vital sign monitor along with separate height and swivel adjustability for the monitor and keyboard/mouse tray as shown in Figure 8. Both types of workstations can now properly accommodate staff of all height ranges so that they can work more comfortably. Also, a sit-stand stool was recommended to the staff to allow alternating postures as shown in Figure 9.

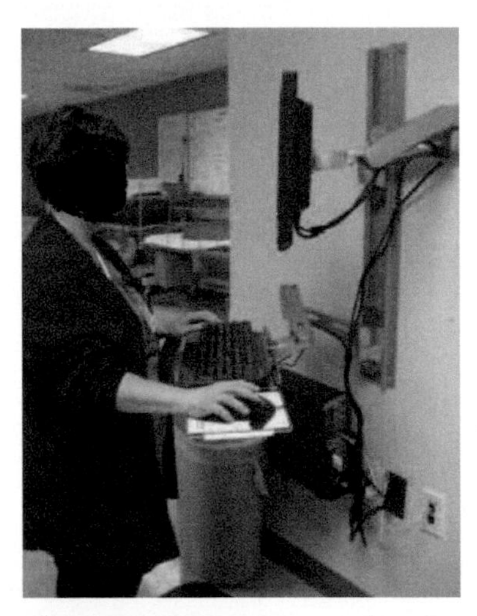

Figure 7: New wall-mounted computer workstation with separate adjustability.

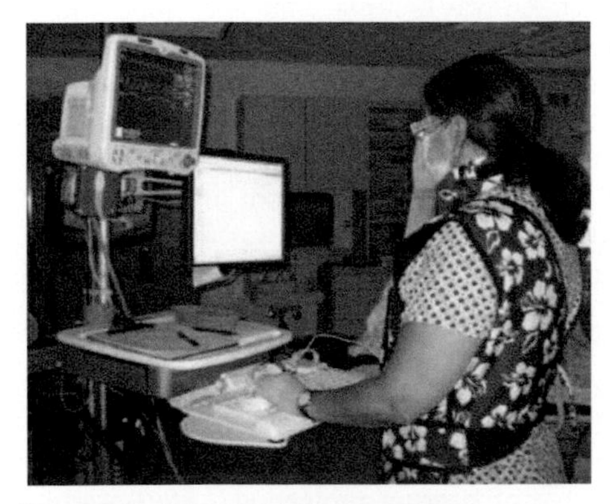

Figure 8: New mobile computer workstation with separate adjustability and vital sign monitor attached.

Figure 9: Sit-stand stool

5.2 Patient Handling

In the out-patient clinic setting, a powered sit-stand lift was made available for the staff to use with patients who require assistance to stand and transfer, and are at risk for falling as shown in Figure 10. This increases safety for patients and staff by reducing potential injuries from manually handling patients and falls. An air-assisted lateral transfer device was trialed for turning patients between prone and supine during procedures. The results were favorable and equipment evaluation is ongoing. For maintaining a patient's limbs in position during procedure, it was recommended to use wedges or bolsters instead of holding limbs manually. It was also recommended that staff should be limited to working in the procedure rooms for fewer number of hours to reduce the duration of exposure to ergonomic risk factors. In addition, it was suggested to management to rotate job duties within the shift. For example, a nurse would work 4 hours in the procedure rooms and 4 hours in the recovery area rather than 8 hours in any one area.

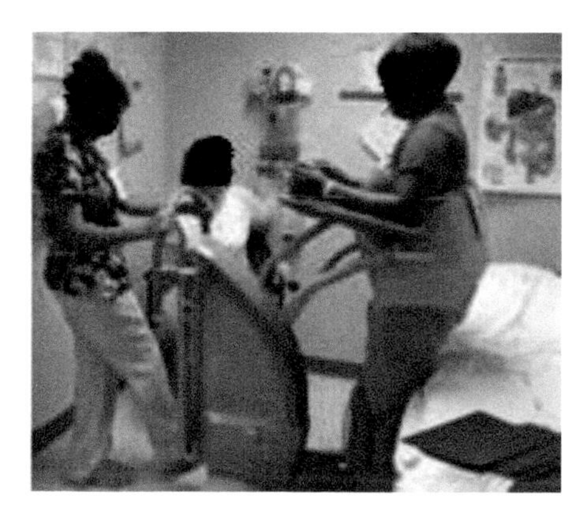

Figure 10: Staff using the powered sit-stand lift.

5.3 Sterile Processing

The long-term engineering solution recommended was to replace the deep sink with a shallower sink and to automate part of the manual disinfecting process to reduce repetition as well as awkward and static postures. Meanwhile, the short-term engineering solution was to bring the dirty scope closer to the staff. The dirty scope is washed in a plastic bin, which is elevated with a riser underneath as shown in Figure 11. The technician can perform this task by sitting, as shown in Figure 12, or standing. The administrative solution was to rotate the tasks of washing and air-drying scopes. Both of these solutions can reduce the frequency and duration of exposure of awkward and static postures.

At the scope drying area, the air nozzle was replaced with an air gun to reduce the noise level to below 85 dBA. Also, staff were provided with protective hearing devices such as ear plugs to reduce noise exposures below 85 dBA and required to take a one-time training course on proper use of hearing protection. Staff were reminded to use a step stool when hanging the scopes, to reduce repetitively reaching above shoulder level.

Figure 11: A dirty scope in a bin elevated with a riser underneath.

Figure 12: A technician washing the dirty scope while seated.

5.4 Physical Environment

The recommendations included:

- Minimize the amount of tubing on the floor and to reduce trip hazards by mounting J-hooks to the ceiling
- Expand the track that holds the monitor horizontally and replace the current monitor arm with one that is height adjustable. This will allow physicians and staff of various heights to view the monitor during procedures and to provide head clearance.
- Pad the sharp edges of the monitor to avoid injury from hitting the head.

6 CONCLUSION

Nurses and sterile processing technicians working in a gastroenterology clinic are at risk of sustaining sustained musculoskeletal injuries similar to gastroenterologists. Ergonomic risk stressors were identified in computer workstations, patient handling, sterile processing, and the physical environment set up. Recommended engineering and administrative controls were implemented to reduce staff injuries and to increase safety for both patients and staff.

REFERENCES

Cappell, M (2011). Accidental occupational injuries to endoscopy personnel in a high-volume endoscopy suite during the last decade: mechanisms, workplace hazards, and proposed remediation. *Dig Dis Sci* 56:479-487.

Francis, D., Prabhakar, S., Bryant-Sendek, D., Larson, M. (2011). Quality improvement project eliminates falls in recovery area of high volume endoscopy unit. *BMJ Qual Saf* 20: 170-173.

Murty, M. (2010). Musculoskeletal disorders in endoscopy nursing. *Gastroenterology Nursing.* 33(5): 354-361.

"Occupational Noise Exposure". Occupational Safety and Health Administration, United States Department of Labor. Accessed February 10, 2012. Available from www.osha.gov/SLTC/noisehearingconservation/

Siegel, J. (2007). Risk of repetitive-use syndromes and musculoskeletal injuries. *Techniques in Gastrointestinal Endoscopy.* 9:200-204.

Shergill, A., McQuaid, K., Rempel, D. (2009). Ergonomics and GI endoscopy. *Gastrointestinal Endoscopy.* 70(1): 145-153.

CHAPTER 25

Upper Extremity Musculoskeletal Disorder Risk in Clinical Gastroenterology

Tamara James[1], *David Tendler*[2]

[1]Ergonomics Division
Duke University and Health System
Durham, NC 27710 USA

[2]Durham Gastroenterology Consultants, PA
Durham, NC 27713 USA

ABSTRACT

This case report describes the evaluation of a colonoscopy procedure using direct observation and video analysis of a gastroenterologist. The purpose of the evaluation was to determine whether colonoscopy-related tasks were contributing to the pain he experiences in his left thumb and forearm as well as both wrists. The Strain Index (SI) was used to analyze a colonoscopy. The SI score for performing a colonoscopy was found to be significantly higher than 7 (the score at which it is believed there is a higher risk of distal upper extremity injury. Recommendations, including those for equipment redesign were provided for reducing the SI score by nearly two-thirds.

Keywords: Ergonomics, Upper Extremity Pain, Musculoskeletal Disorders, Strain Index, Equipment Design

1 INTRODUCTION

This case report describes an evaluation of a gastroenterologist to assess his work tasks and work environment. Work-related upper extremity musculoskeletal disorders have been reported in the past in various medical professions such as ultrasound, dental providers, and surgeons. In recent years, a few studies have reported injuries among gastroenterologists. One case investigated by Cappell (2006) indicated that there is a direct relationship between one gastroenterologist's orthopedic left thumb injury and left thumb flexion required during colonoscopies. Byun, et al. (2008) administered a questionnaire to endoscopists in Korea and found 89.1% reported musculoskeletal pain in at least one anatomic location and 47.3% reported having severe musculoskeletal pain. Shergill, et al. (2009) looked at pinch force and forearm-muscle load during colonoscopies and found right thumb pinch forces exceeded the injury threshold and left forearm-muscle activity was at or exceeded the ACGIH hand activity level (HAL) action limit. However the researchers did not examine left thumb pinch forces. Kuwabara, et al. (2011) sent questionnaires to Japanese endoscopists and non-endoscopist physicians and found the frequency of pain in the hand and wrist and especially the left thumb was significantly higher in endoscopists (17%) than non-endoscopists (6%). Lee, et al. (2012) quantified wrist postures associated with colonoscopies and determined these postures presented a risk for wrist and forearm musculoskeletal injuries. To our knowledge, no studies have determined the risk of injury using a risk assessment tool such as the Strain Index nor have specific recommendations to redesign colonoscopes been presented in previous studies.

2 OBJECTIVE

An ergonomic evaluation was requested by a gastroenterologist to evaluate work tasks and environment. The purpose of the evaluation was to determine whether the colonoscopy task is contributing to the pain experienced in the left thumb and forearm as well as both wrists. The goal was to use a standard risk assessment tool (Strain Index) for distal upper extremity musculoskeletal disorders for this purpose. In addition to determining risk of injury, recommendations for reducing risk were also requested as part of the evaluation.

3 METHODS

A colonoscopy procedure was observed as this is the type reported as the most painful for the gastroenterologist. The gastroeneterologist was a right-handed male, age 44, with 15 years of clinical experience. The equipment he used was a Pentax fiberoptic video colonoscope. Digital photos and videos were taken during the observations. The Strain Index was subsequently used to analyze this task. Analysis was performed using Moore and Garg's Strain Index (1995) which is a semi-quantitative exposure assessment methodology that results in a numerical score (SI score.) This score is believed to correlate with the risk of developing distal upper extremity disorders. The SI score represents six task variables: intensity of exertion, duration of exertion, exertions per minute, hand/wrist posture, speed of work, and duration of task per day.

3.1 Observations

Gastroenterologists use their left thumbs to turn both control dials on colonoscopic heads for horizontal or vertical tip deflection for steering during colonoscopies with no power assistance. The thumb provides all mechanical force for this movement against the resistance of the colon. The frequent repetitive thumb motion with frequent thumb extension and force requirements take a toll on thumb tendons. One case was report of a gastroenterologist with De Quervain's syndrome with pain exacerbation during left thumb flexion during colonoscopies. With increasing volume of colonoscopies due to increased awareness of screening for colon cancer, this type of injury may be on the rise.

The gastroenterologist in this case performs multiple colonoscopies throughout the week. His typical volume of cases for the week is provided in Table 1.

Table 1: Case Volume of Weekly Colonoscopies

Day of the Week	Estimated Colonoscopy Volume
Monday	6-7
Wednesday	12
Thursday	8-9
Friday	8

Ergonomic stressors observed during the colonoscopy were:

- High, prolonged force to grip the colonoscope handle with the left hand and moderate force to move the colonoscopic tip with the left thumb;
- Prolonged gripping and torque required of the right hand to manipulate the colonoscopic shaft;
- Repetitive use of the left thumb to control the dials;
- Extreme awkward postures (hyperextension) of the left thumb;
- Prolonged periods of standing.

These observations are consistent with published studies. The studies also report risk of injury appears to be related to endoscopy volume. Volume reported by Cappell (2006) in his unpublished data found 3 of 10 gastroenterologists surveyed had left thumb pain associated with colonoscopies and those 3 performed >1000 cases per year. One study of surgeons who perform colonoscopies (Liberman, et al., 2005) reported 39% had at least one injury or pain episode and the risk increased to 47% among those who performed more than 30 procedures per week. Left thumb, finger, and hand injuries were attributed to repetitive turning of control dials and prolonged gripping of the colonoscope handle as shown in Figure 1.

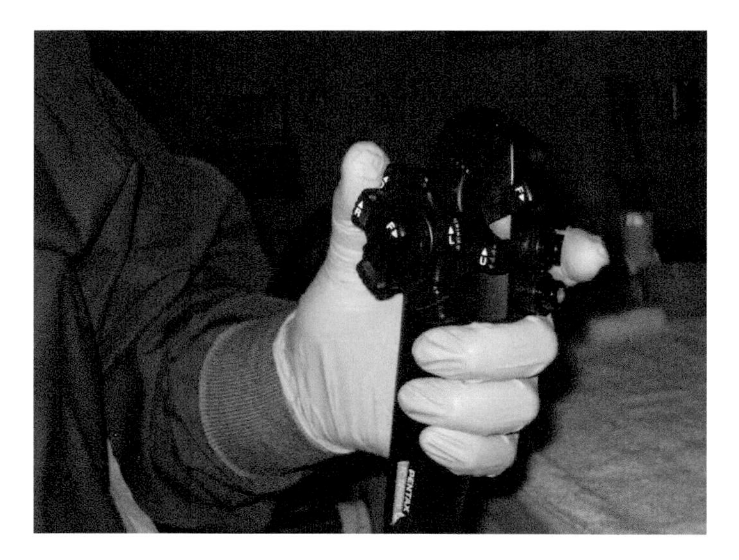

Figure 1: Left Hand Posture Required to Manipulate Colonsocope

4 DATA ANALYSIS

In addition to observing work tasks and identifying ergonomic stressors, the Strain Index was used to further analyze these tasks as a means of quantifying risk of musculoskeletal injury. As previously mentioned, the Strain Index is a semi-quantitative job analysis methodology that results in a numerical score (SI score) that is believed to correlate with the risk of developing distal upper extremity disorders.

The SI score results from measuring and assessing each of six independent task variables: intensity of exertion, duration of exertion, exertions per minute, hand/wrist posture, speed of work, and duration of task per day.

- *Intensity of exertion* is an estimate of the force requirements of a task and reflects the magnitude of muscular effort required to perform the task, such as the amount of effort to manipulate dials on the colonoscope.
- *Duration of exertion* reflects the physiological and biomechanical stresses related to how long exertions are maintained, such as the amount of time required to grip the colonoscope handle.
- *Efforts per minute* is the number of exertions per minute and is an indication of repetitiveness of the task, such as the number of times the dials are manipulated.
- *Hand and wrist posture* refers to the anatomical position of the wrist or hand relative to neutral (0 degrees for extension, flexion, and ulnar deviation.)
- *Speed of work* is an estimate of the perceived pace of the task and how it relates to fatigue.

- *Duration of task per day* reflects the total time a task is performed each day and attempts to include positive effects of job rotation and negative effects of overtime.

5 RESULTS

According to the researchers who developed this methodology, jobs that have a total SI score of 3 or less are considered safe. Jobs that have a score of 7 or greater have a higher risk of injury. Those between 5 and 7 have a low risk of developing a distal upper extremity disorder so a total score less than 7 is ideal. Calculations and determinations for individual task variables are presented in Table 2 for the left hand only assuming a Wednesday schedule which has the highest volume of colonoscopies. The final Strain Index (SI) score is also presented.

Table 2: Task Variable Multipliers for Colonoscopy

Task Variable	Colonscopy Task (Left Hand)	Task Variable Multiplier
Intensity of Exertion	Somewhat Hard	3
Duration of Exertion	100%	3
Efforts per Minute	40.2	3
Hand/Wrist Posture	Good	1
Speed of Work	Fair (91 – 100%)	1
Duration per Day	4-8 hours	1
	Final Strain Index Score	27

6 SUMMARY

As indicated by the final SI score of 27, this Gastroenterologist in this case is at a higher risk of injury. Changes to the equipment and or work schedule were recommended to decrease this risk and to decrease the current pain level. For the Strain Index task variable multipliers shown in Table 1, the way to reduce risk of injury is to reduce one or more multipliers. Ideally the strain index should be 7 or less to minimize risk of injury. This can be achieved through implementing any of the recommendations (or some combination of these) presented in Table 3.

Table 3: Recommended Changes to Reduce SI Score

Multiplier	Recommendation(s)	Potential Final SI Score
Reduce Intensity of Exertion	• Manufacturers should investigate ways to reduce the resistance required to manipulate the dials. Currently the thumb provides all mechanical force without any power assistance	Reduce by two thirds, if resistance was considered "light" rather than somewhat hard
Reduce Duration of Exertion	• Attach a strap to the colonoscope handle to decrease the prolonged grip force required. • Place the handle on top of the patient to reduce the amount of time gripped for prolonged period of time • Investigate the use of a bracket that would attach to stretchers and "hold" the handle	Reduce by half, if grip force is eliminated
Reduce Efforts per Minute	• Manufacturers should investigate ways to reduce repetitive motion of the thumb and to improve the design to promote more neutral thumb postures. A joystick or touchpad type of control would be preferable to the dials. • Occasional use of the right hand to turn the control dial to reduce left thumb strain. • Manufacturers should consider a foot pedal control for suctioning as the forefinger has to extend around the scope and repetitively depress the suctioning button.	Reduce by one third, if alternative controller is used, right hand is used periodically, or foot is used to perform some motions
Reduce Duration Per Day	• Fewer numbers of cases per day would reduce the duration exposure.	Reduce by half if no more than 4 cases per day; reduce by three fourths (which would put this task in the "safe" zone) if no more than 2 cases per day

Colonoscopes were initially designed such that movement of the colonoscopic head was held with the right hand and movement of the tip was controlled by the left hand while a second gastroenterologist held and manipulated the colonscopic shaft with either hand. The equipment control was later modified as a means of eliminating the second gastroenterologist so that now one person is performing the work of two. Furthermore, the left thumb now controls all colonoscopic tip movements that were previously performed by the entire left hand. In addition to the recommendations provided in Table 3, manufacturers would be wise to narrow the distance between the grip and the thumb controlled dials to reduce the strain put on the thumb, which currently has to move repetitively while in an extended position. That may be accomplished by extending the dials out further (bringing them closer to the thumb) and/or by narrowing the neck of the scope, where it is gripped.

The consequences of ignoring the need for colonoscope equipment redesign are significant. Thumb injuries such as DeQuervain's may require surgery and can ultimately require providers to reduce colonoscopy volume or may end their ability to perform these procedures at all. With the current shortage of providers who perform colonoscopies (Chang, 2004) and the increased burden on them, it is critical to make changes to protect them. Reduced volume does not appear to be an option given the supply of providers and growing demand for these procedures. Interventions such as improvements to work practices or posture may be somewhat helpful. However, with the aid of a tool like the Strain Index it is clear that the best way to reduce risk of upper extremity disorders for these providers is through improvements to the equipment design and manufacturers should immediately begin to do so.

REFERENCES

Byun, YH: Procedure-related musculoskeletal symptoms in gastrointestinal endoscopists in Korea. *World Journal of Gastroenterology*. 2008 July 21; 14(27): 4359–4364.

Cappell, MS: Colonoscopist's thumb: DeQuervain's syndrome associated with overuse during endoscopy. *Gastrointestinal Endoscopy*. 64, No 5:841-843, 2009.

Chang, M, Schroy, PC: Endoscopic colorectal cancer screening-can supply meet demand? *Gastroenterology*. 2004; 126:1482-5.

Kuwabara, T, Urabe, Y, Hiyama, T, Tanaka, S, Shimomura, T, Oko, S, Yoshihara, M, Chayama, K: Prevalence and impact of musculoskeletal pain in Japanese gastrointestinal endoscopists: A controlled study. *World Journal of Gastroenterology*. 2011 March 21:17(11): 1488-1493.

Lee, DL, Rempel, D, Barr, AB, Shergill, A: Ergonomics of Colonoscopy: Wrist Postures of Gastroenterologists Performing Routine Colonoscopy. *Proceedings of the Human Factors and Ergonomics Society Annual Meeting 2010*. 54:1205-1209.

Liberman, AS, Shrier, I, Gordon, PH: Injuries sustained by colorectal surgeons performing colonoscopy. *Surgical Endoscopy* (2005) 19: 1606-09.

Moore JS, Garg A: The strain index: a proposed method to analyze jobs for risk of distal upper extremity disorders. *American Industrial Hygiene Association Journal.* 56:443-458, 1995.

National Institute for Occupational Safety and Health (NIOSH,) *Musculoskeletal Disorders and Workplace Factors*, CDC, July 1997

Putz-Anderson V: *Cumulative Trauma Disorders: A Manual for Musculoskeletal Diseases of the Upper Limbs.* Taylor & Francis, Bristol, PA, 1992.

Shergill, AK: Asundi, KR, Barr, A, Shah, JN, Ryan, JC, McQuaid, KR, Rempel, D: Ergonomics and GI Endoscopy. *Gastrointestinal Endoscopy.* 69, No 1:142-146, 2009.

Shergill, AK: McQuaid, KR, Rempel, D: Ergonomics and GI Endoscopy. *Gastrointestinal Endoscopy.* 70, No 1:145-153, 2009.

CHAPTER 26

Rapid Improvement Event to Prevent Patient Falls

Laurie Wolf, Eileen Costantinou, Pat Matt and Liz Schulte

Barnes-Jewish Hospital
Operational Excellence Department
St. Louis, MO, 63110, USA
lwolf@bjc.org

ABSTRACT

Barnes-Jewish Hospital (BJH) is a large urban teaching hospital in St. Louis, Missouri. (9,438 employees, 1,845 physicians, 803 residents/interns/fellows, 1,111 staffed beds, 83,997 Emergency Room visits and 54,733 admissions in 2009). Barnes-Jewish Hospital is part of BJC HealthCare and is affiliated with Washington University School of Medicine. BJC is attempting to reduce the number of patient falls and falls with injury throughout all their hospitals. A Rapid Improvement Event (RIE) was conducted with three oncology divisions at BJH to standardize patient assessment and then to implement appropriate fall interventions. Post fall activities were also defined to help develop an understanding of contributing factors. Data transparency processes were put in place to make fall occurrences more visible to all staff. The project described in this abstract is included as part of a PhD program at Loughborough University and will be building toward a wider scope of work in the future.

Keywords: Fall Prevention, Rapid Improvement Event Process, Patient Safety, Get up and Go Test, Short Portable Mental Status Questionnaire, Fall prevention interventions, Oncology, Patient Falls with Injury

1 OBJECTIVE AND SIGNIFICANCE

The objective of this project was to reduce patient falls and falls with injury on three oncology divisions at BJH. A gap analysis showed that patient assessment of gait and mental status were not being conducted in a consistent manner. It also revealed that if an assessment indicated a specific intervention (such as a bed alarm or low bed), that the intervention may not be put in place until after the patient had fallen. By standardizing assessment and intervention processes, we hope to decrease patient falls and falls with injury by 50% and 30% per 1,000 patient days respectively.

2 METHOD

BJC conducted a consortium with fall prevention experts representing all BJC hospitals to develop a package of minimum standards that would represent the best practice, evidence-based methods to assess patients and select the appropriate interventions. The minimum standard package had three components: assessment, interventions and data transparency.

In order to implement the minimum standard package a three-day rapid improvement event (RIE) was held to establish how the standards would be implemented on three oncology divisions at Barnes-Jewish hospital that had the highest number of falls and injuries from patient falls in BJC. To standardize gait assessments the Get-Up and Go (GUG) test was selected. For mental assessments the Short Portable Mental Status Questionnaire (SPMSQ) was selected.

The following Figure 1 shows that three of the four divisions with the highest number of falls with injury are Oncology divisions. Consequently, these divisions were chosen to participate in the RIE.

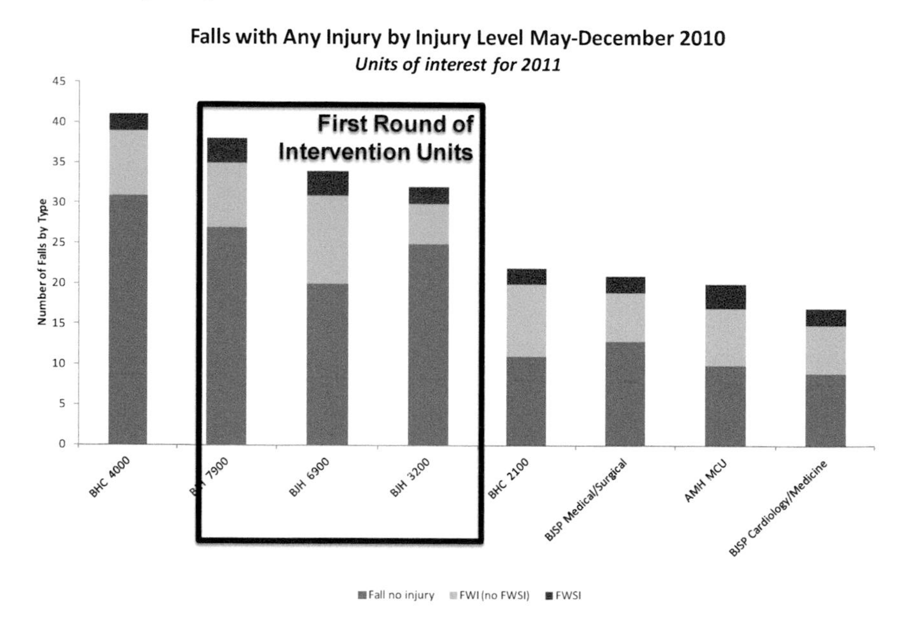

Figure 1. BJC divisions with the highest number of falls and injuries.

Assessment: Since a new Fall Risk Assessment was under development at the time of the RIE, the team decided to enhance the inpatient assessment by trying to standardize a screening technique for gait and mental status. The group developed a standard work document for use of the Get-Up-and-GO (Currie, 2008) and Short Portable Mental Status Questionnaire on all three oncology divisions. A reminder card was developed to highlight steps involved with the assessments. Standard work was also developed for documentation of results.

Intervention: Another subgroup worked on developing an algorithm to ensure the appropriate interventions were selected and implemented based on assessment and clinical expertise. This algorithm was based on a Linking Evaluation and Practice (LEAP) grid that was developed by BJC and the system-wide group of subject matter experts. The LEAP grid recommended interventions according to what deficit was found during the patient assessment in an attempt to mitigate risk associated with individual risk factors and common combinations. For example if a patient was found to have altered elimination, mental confusion and altered gait, a low bed and bed alarm would be recommended. The LEAP grid is based on best-practices interventions according to fall prevention literature (ECRI, 2006).

Data Transparency: The third subgroup established the processes that would be required after a patient fell. Each division was already conducting a post fall huddle but was using different processes and documentation. All agreed on one form with some investigative questions required to be completed within 60 minutes of the fall and the remainder within 48 hours. Another process developed to make information visible about each fall after it occurs. A managing for daily improvement (MDI) board was posted on all divisions where information could be posted such as: reason for getting up when fall occurred, contributing factors (medications, clutter, wet floor, lighting), scoring on assessments, type of interventions that were in place and follow up that occurred after the fall. Information gathered help to develop an Action Plan to resolve issues as they are discovered.

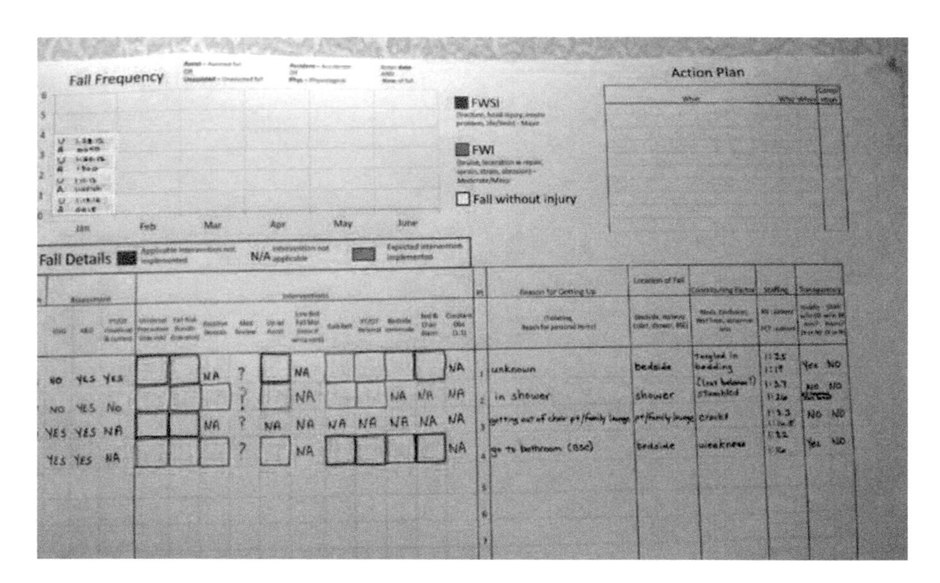

Figure 2. MDI board for posting information after a fall occurred

This board shown in Figure 2 is posted in the hallways so all staff members can see when a fall occurs and what action items are being taken for prevention.

3 RESULTS

Revisions are being made to the fall prevention process as issues are encountered and solutions are agreed upon by the RIE team. For example, the nurses reported resistance from the patients as they repeatedly performed the mental status test on patients with an extended length of stay. The team decided to stop using the SPMSQ and instead standardized the way they asked and rated "Alert and oriented" questions were already incorporated into daily patient assessments.

Figure 3 shows the overall fall rates on the RIE divisions are trending downward, however falls resulting in injury still occur. We plan to continue to pursue opportunity for refinement by participating in a Six Sigma project with the Joint Commission Center for Transforming Healthcare. Seven hospitals across the US are collaborating with the Center to reduce falls with injuries in healthcare.

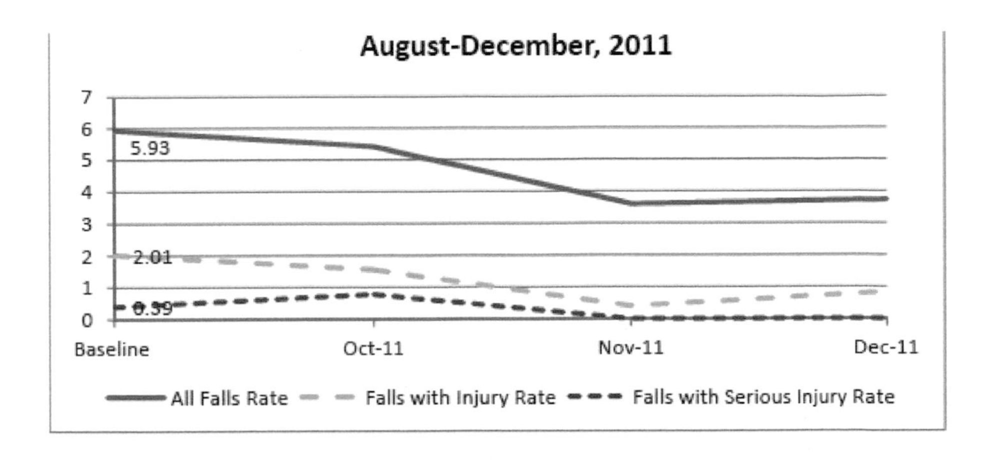

Figure 3. Fall rates comparing baseline at start of RIE in August through December 2011 for Oncology divisions participating in the RIE.

Human Factors and Patient Fall Model

As part of a PhD program at Loughborough University the risk factors of fall prevention have been aligned with the SEIPS model (Systems Engineering Initiative for Patient Safety) model of Human Factors developed at the University of Wisconsin. The SEIPS model recognizes five components in the interdependent nature of a work system where the care provider (1) is performing various tasks (2) using tools and technology (3) in a given environment (4) within an established organization (5). Next this work system performs the process of patient care to achieve patient and hospital outcomes (Carayon, 2007). Intrinsic risk factors were placed outside the work system box because they are associated with the patient and first come into play during assessment under the patient care process. There are additional human factors that align with the entire care giver staff such as: capabilities, limitations mental workload, stress, perception and communication issues. This is an attempt to gain insight into the multi-faceted complexity of patient falls.

4 CONCLUSIONS

Since implementing the minimum standards package on Aug 10, many concerns have been addressed including: lack of availability of low beds, nursing dissatisfaction administering the SPMSQ assessment, concerns with sitters causing possible delay in placement to rehabilitation facilities. These ongoing issues touch the surface of the numerous factors and interactions that make fall prevention so complex. Consequently patient fall prevention must be a continuous improvement process requiring ongoing vigilance from all care givers, hospital staff, patients and families.

ACKNOWLEDGMENTS

The authors would like to acknowledge the participation of the entire RIE team: nurse managers, physical therapy, pharmacy, physicians, patient safety and quality, staff nurses, unit secretaries, patient care techs and clinical engineering. Thank you to the Directors of Oncology and Surgical services for their guidance and support. A special thank you to the extremely dedicated Clinical Nurse Specialists: Cathie Limbaugh, Kathy Rensing and Phyllis Gabbart without whom this project could not sustain.

REFERENCES

Carayon, P. (2007). Handbook of Human Factors and Ergonomics in Health Care and Patient Safety. New-York, NY : CRC Press Taylor & Francis Group

Currie, L. (2008). Fall and Injury Prevention. In R. Hughes (ed.), Patient Safety and Quality: An Evidenced-Based Handbook for Nurses (Chapter 10). Rockville, MD : Agency for Healthcare Research and Quality (AHRQ).

ECRI. Falls Prevention Strategies in Healthcare Settings. ECRI publisher, Plymouth Meeting, PA (2006).

CHAPTER 27

Designing for Patient Safety – Considering a Patient Safety Risk Assessment

Ellen Taylor, AIA, MBA, EDAC, Anjali Joseph, PhD, EDAC, Xiaobo Quan, PhD, EDAC

The Center for Health Design
Concord, CA
etaylor@healthdesign.org

ABSTRACT

Objective: A growing body of research indicates the built environment (light, noise, air quality, room layout, etc.) contributes to adverse outcomes like healthcare associated infections, medication errors, and falls in healthcare settings (Joseph & Rashid 2007; Ulrich et al. 2008). It has become increasingly clear that the problem of patient safety does not lie solely in the hands of clinicians or frontline staff; rather adverse events result from the latent conditions of complex healthcare interactions (Reason 2000). Often, these latent conditions are built into the physical environment.

Significance: Since the release of the 1999 Institute of Medicine (IOM) report (Kohn et al. 2000), 'To Err is Human', patient safety improvements have remained elusive, in spite of numerous interventions (Wachter 2010). Recent studies have demonstrated no significant improvement for a number of healthcare associated conditions including the failure to reduce postoperative, blood stream and catheter associated urinary tract infections (The Agency for Healthcare Research and Quality (AHRQ) 2010). A study of 10 North Carolina Hospitals over 10 years found 25.1 harms per 100 admissions (Landrigan et al. 2010); and a Department of Health and Human Services' Office of the Inspector General's report found that 13.5 percent of hospitalized Medicare patients experienced adverse events and another 13.5 percent experienced temporary harm (Levinson 2010). This impacts the nation's healthcare bill, with 1.5 million errors estimated to contribute $19.5 billion dollars (Society of Actuaries 2010). Perhaps these results reflect an incomplete understanding of the puzzle that quality healthcare represents. A growing body of research suggests that an understanding of the multiple components of the healthcare system (of which the built environment is one component) is crucial for improving patient safety. Given the massive investment anticipated in healthcare facility construction in the next 10 years, there is an urgent need for a well-defined and standard methodology to identify and eliminate built environment conditions that impact patient safety in healthcare facilities.

Methodology: A national seminar funded by the Agency for Healthcare Research and Quality (AHRQ) and the Facilities Guidelines Institute (FGI) brought together multidisciplinary experts to address the role of the built environment in improving patient safety in healthcare facilities. The primary focus of this seminar was on identifying the tools and processes that would support decision making related to patient safety during the healthcare facility design process.

Results: Design teams themselves are often unfamiliar with the built environment impact on patient safety and even less familiar with ways to address concerns during design. While other industries have been able to harness human factors, engineering, and cognitive science that result in the preferred human response and improved safety, no similar method exists for the design of healthcare facilities and renovation projects. Seminar participants felt several commonly used evaluation techniques could be effectively used throughout the design process. A focus is needed, however, on the pre- design phases - strategic planning, master planning, and functional programming, as the decisions made during pre-design significantly impact the design parameters going forward and resulting safety-related project outcomes.

Keywords: Healthcare facility design, patient safety, safe design tool, design activity, patient safety risk assessment, PSRA

1 PATIENT SAFETY AND THE DESIGN PROBLEM

Since the release of the 1999 Institute of Medicine (IOM) report (Kohn et al. 2000), 'To Err is Human', patient safety improvements have remained elusive, in spite of numerous interventions (Wachter 2010). Recent studies have demonstrated no significant improvement for a number of healthcare associated conditions including the failure to reduce postoperative, blood stream and catheter associated urinary tract infections (The Agency for Healthcare Research and Quality (AHRQ) 2010). A study of 10 North Carolina Hospitals over 10 years found 25.1 harms per 100 admissions (Landrigan et al. 2010); and a Department of Health and Human Services' Office of the Inspector General's report found that 13.5 percent of hospitalized Medicare patients experienced adverse events and another 13.5 percent experienced temporary harm (Levinson 2010). This impacts the nation's healthcare bill, with 1.5 million errors estimated to contribute $19.5 billion dollars (Society of Actuaries 2010). Perhaps these results reflect an incomplete understanding of the puzzle that quality healthcare represents. A growing body of research suggests that an understanding of the multiple components of the healthcare system (of which the built environment is one component) is crucial for improving patient safety. Given the massive investment anticipated in healthcare facility construction in the next 10 years, there is an urgent need for a well-defined and standard methodology to identify and eliminate built environment conditions that impact patient in healthcare facilities.

A growing body of research also indicates the built environment (light, noise, air quality, room layout, etc.) contributes to adverse outcomes like healthcare associated infections, medication errors, and falls in healthcare settings (Joseph & Rashid 2007; Ulrich et al. 2008). It has become increasingly clear that the problem of patient safety does not lie solely in the hands of clinicians or frontline staff; rather adverse events result from the latent conditions of complex healthcare interactions (Reason 2000). Often, these latent conditions are built into the physical environment.

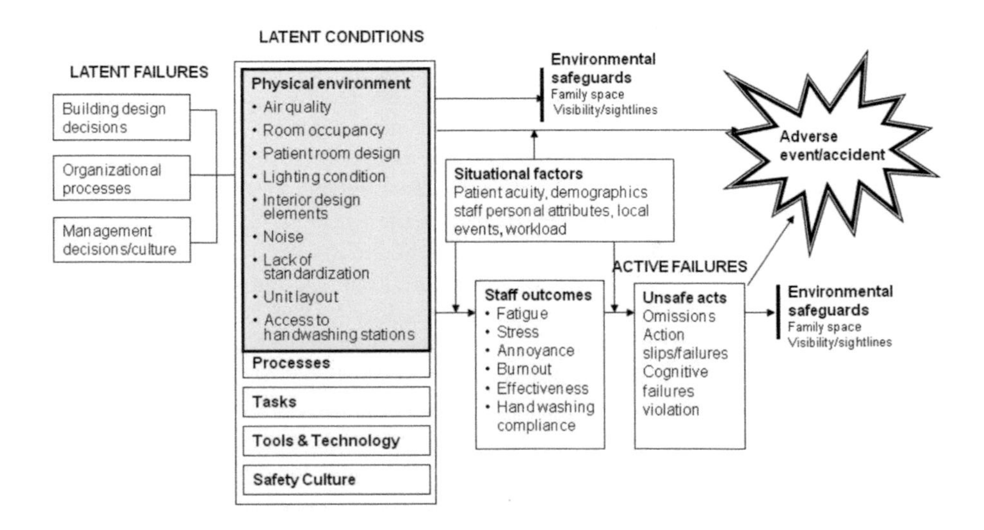

Figure 1: Conceptual model based on Reason's model showing the role of the environment as a latent condition or barrier to adverse events in healthcare settings. Source: (Joseph, 2007)

The conceptual model in Figure 1, (based on Vincent et al. 1998; Reason 2000), shows the role of the physical environment elements as the latent conditions that contribute to patient safety. Often, these latent conditions that adversely impact patient safety are built into the physical environment during the planning, design and construction of healthcare facilities. For example, the location of emergency departments and intensive care units might necessitate the transport of critically ill patients over long distances potentially causing patient complications. Hand-washing sinks located in inconvenient or inaccessible locations might result in poor hand hygiene compliance among physicians and nurses. Nursing unit areas with high levels of traffic and activity can result in distraction and interruption, leading to possibilities of medication errors.

Emerging regulations associated with the passage of the 2010 Patient Protection and Affordable Care Act legislation have changed the landscape of the healthcare community. As part of this legislation, the Centers for Medicare & Medicaid Services (CMS) reduce or prohibit payments to doctors, hospitals and other health care providers for services that result from certain preventable healthcare acquired conditions, including events such as hospital acquired infections (HAIs) and falls. As healthcare organizations make significant investments to replace or renovate aging healthcare infrastructure, the reimbursement realities have honed the focus on patient safety and other reimbursement related factors such as patient satisfaction.

1.1 Origins of the Patient Safety Risk Assessment (PSRA)

Consideration of patient safety is certainly not new. The Hippocratic Oath, "First, do no harm," dates back to the 5th century BC. A paper evaluating the impact of the IOM report on patient safety publications and research awards found the rate of all types of patient safety publications increased from 59 to 164 articles per 100,000 MEDLINE publications following the release of the IOM report, while research increased from 5 to 141 awards per 100,000 federally funded biomedical research awards. It is interesting to note that the most frequent subject of patient safety publications prior to the IOM report was malpractice, while following the report, organizational culture was the most frequent subject (Stelfox et al. 2006). Despite this increase in awareness, hospitals remain a dangerous place to be. An often untapped consideration of patient safety is the facility itself and its impact on outcomes. The focus on the design of the built environment and patient safety has also increased in the past decade, with papers considering varied aspects of noise, air quality, ergonomics, the human factors of work environments, adjacencies, and specific unit types or room types. Unfortunately, while fields such as aviation and other high-risk industries have been able to incorporate human factors, engineering and cognitive science that result in design solutions to facilitate the intended outcome, a similar approach is not widely used in the architectural design process.

Introductory language surrounding a patient safety risk assessment (PSRA) was introduced in the appendix of the FGI *Guidelines for the Design and Construction of Health Care Facilities* in 2010. While this document is used as a code or referenced standard by The Joint Commission, many federal agencies, and 42 states, components of the appendix text are not required, but are included for "consideration." Part of the challenge to include the PSRA as a requirement was the lack of definition on how such an assessment should be conducted.

The national seminar on Designing for Patient Safety was formulated to develop additional detail around existing tools and methodologies to assess risk and design and reduce the related latent conditions during the healthcare facility design process. To accomplish this, a diverse group of stakeholders was invited to participate, including: architects, interior designers, planners, clinicians, hospital administrators, researchers, human factors experts, industrial engineers, guidelines experts, and facility managers. The project focused on two key areas of development: resources and background material for evaluation prior to the seminar (a literature review of design tools for patient safety, a compilation of opinion papers by industry and academic experts, and the draft of a Safe Design Roadmap), and a seminar agenda that would provide a hands-on, highly interactive environment for multiple workgroups to focus on discussion and evaluation of an assigned process or tool. The workgroup sessions were followed by a consensus workshop to identify high-priority activities and documentation that could be included in the PSRA. PSRA activities identified by the group in predesign and design/construction phase were ranked as high priority, medium priority, or low priority to identify the top 5 high-priority activities in the design process.

1.2 Development of Resources and Background Material

The literature review focused on the tools and approaches that were potentially useful for incorporating patient safety in the design process. The goal was to generate a set of tools or methods used to enhance patient safety in the design process that could be discussed and evaluated in the national seminar. A scan of design tools and approaches for patient safety was conducted in the fields of human factors, architecture, engineering, business management, and others, using PubMed, EBSCO, and Internet search engines. Relevant articles, books, or other publications were reviewed, in addition to two compendiums around patient safety published by AHRQ in recent years (Henriksen et al. 2008; Henriksen et al. 2005). Additional tools were recommended by the advisory committee and other experts in the field. The result of this step was a list of 14 design tools and approaches. Following the compilation of this list, further literature searches and reviews were conducted to gather relevant information about the tool or process. The definition, history, and examples of use in healthcare settings; typical process of implementation; limitations; and additional resources were extracted from the literature for each tool/approach. The information for each tool/approach was then synthesized into a brief summary. In the final step, the project team selected seven design tools for workgroup discussion in the national seminar. These included:

- link analysis,
- root cause analysis,
- failure modes and effects analysis (FMEA),
- simulation,
- work sampling,
- balanced scorecard, and
- process analysis.

The selection of tools was based on a set of criteria including the relevance to the facility design process, scope of use, documented effectiveness, and validity of tools. It excluded high-level design approaches or philosophies (e.g., lean). Prior to the event, participants assigned to a workgroup were asked to evaluate a tool by reviewing the prepared summary and relevant research articles, as well as scoring usability, relevance, feasibility, generalizability, and several additional questions.

1.3 Seminar Results: Activites and Tools by Design Phase

The discussions from the workgroups produced rich insights into the activities of designing for patient safety. There was consensus that time and effort was needed to focus on patient safety during the earliest phases of a healthcare facility project (predesign phases of strategic planning, master planning, operational planning, and programming). The decisions made during predesign significantly impact the design parameters and outcomes of the project from a safety perspective. Attendees also noted flaws in the traditional linear design process. They felt design efforts should be an important part of the overall continuous improvement of patient and staff safety in any healthcare organization and work iteratively in small cycles.

The workgroups identified a range of activities that should be undertaken during predesign and design/construction phases to improve patient safety outcomes. Table 2 shows the top high-priority activities identified by most of the attendees.

Table 2: High-Priority Activities in Designing for Patient Safety

Design Phase	High-Priority Activities
Predesign	*Articulate of a clear statement around patient safety* at the start of the healthcare facility design project was of paramount importance as it sets the tone for the activities of the team through the course of the project (at the creation of the vision statement).
	Clearly defining future state operations and planning processes is necessary to help achieve those states prior to starting any design.
	Using *simulation and mock-ups* early in design to visualize key concepts can identify possible built environment latent conditions.
	Using a *process-led design* of defining the care processes should happen in in parallel with a flexible building design.
	Establishing *goals for improvement* requires collecting baseline metrics around key outcomes (e.g., falls, HAIs, medication errors).
Design and construction	*Simulations and mock-ups* were considered the most important activity to identify built environment latent conditions.
	Safety priorities needed to be institutionalized, and the teams needed to have *regular check-ins during all phases* of the project to ensure that safe design is being implemented as envisioned.
	Post-occupancy evaluations were identified as a key to ensure that the building was effective in providing safe care and supporting the staff in conducting their work in a safe and efficient manner.
	Safety reviews would enable the team to review plans and construction documents using a patient safety lens.

Participants felt the design team should be formed as early as possible and include multidisciplinary team members to ensure that patient safety issues were effectively addressed. Suggested team members included clinicians, administrators, facility managers, architects, designers, consultants, human factors specialists, and researchers. There was also recognition that different team members may lead the team effort in different stages. Table 3 lists tools and documentation identified by attendees. The participants felt that conducting a patient safety risk assessment might involve healthcare design teams documenting their findings from using the selected tools, as well as the documentation from other risk assessments (e.g., the Infection Control Risk Assessment (ICRA), Patient Handling and Movement Assessment (PHAMA)). An operational plan that documents key processes in the facility could also be required, along with consideration for caregiver-related safety.

Table 3: Tools and Documentation Recommended by Seminar Attendees

Design Phase	Tools	Documentation
Predesign	Balanced scorecard, Benchmarking, Brainstorming, Case studies, Communication plan, Critical pathway analysis, Failure modes and effects analysis (FMEA), Focus groups, Lean and six sigma, Link analysis, Pareto analysis, Photo journal, Process mapping/analysis, Culture of Safety assessment, simulations/mock-ups , Spaghetti diagrams, Statistics gathering, Task analysis , time motion study	Business case (line-item budget for safety), Documentation of current safety issues and safety opportunities (data + Root Cause Analysis), Measurable goals defined/metrics, Operational plan (flow diagram, narrative), Repetitive room design, Risk management matrix, Strategic plan, Vision/mission statement
Design and construction	Bump analysis, Flow assessment, Failure modes and effects analysis (FMEA), Link analysis, Operational safety risk, assessment , Post-Occupancy Evaluation (POE), Priority matrix in patient safety issues, Safety plan during construction, Safety-related punch list, Safety review, Simulation	Documentation of Evidence-based Design and safety design elements, Construction documents, Risk matrix, Safety plan, POE documentation, Punch list

Balanced scorecard, process analysis (process mapping), and link analysis were ranked high on all key criteria. Simulation and FMEA were also highly ranked, and workgroups felt that these methods were already being used in the facility design process and could be modified to make it feasible for projects of different scopes. The balanced scorecard was suggested as a tool for continuous monitoring of patient safety. Other tools such as process analysis and root cause analysis (RCA) (aggregated data from RCAs being most beneficial) would be critical during predesign and planning. FMEA, simulations, and link analysis could be effective at different design phases and could support decisions at varying levels of detail.

Seminar participants also reviewed the draft Safe Design Roadmap. The document was envisioned as a facility management tool to provide CEOs and their leadership team with a method to capture opportunities to use physical environmental features to improve patient safety outcomes throughout the project life cycle. This was supplemented by a design framework and considerations for safe design based on prior works (Joseph & Rashid 2007; Reiling et al. 2004). Groups felt it either needed less detail for a CEO or more detail for a design team.

1.4 Informing the Guidelines - Proposals

The PSRA aims to direct the attention of healthcare providers and design team to key patient and caregiver safety outcomes in healthcare facilities that are impacted by design. The consensus of the seminar resulted in a proposal to move the PSRA from the appendix of the Guidelines to the body of the text. The proposed modifications to the Guidelines state that the PSRA shall be a multidisciplinary, documented assessment process. The intent is to proactively identify and mitigate the varied latent conditions of the built environment that can lead to patient and caregiver safety adverse events such as infections, falls, errors, and immobility-related or other injuries. The process identifies and takes into account the patient and caregiver population at risk, the nature and scope of the project, and the functional program of the health care facility as well as models of care, operational plans, and performance improvement initiatives within the health care organization. The PSRA determines the potential risk associated with a hazard and identifies proposed built environment solutions to mitigate the related adverse events. The new language in the Guidelines also describes when such an assessment should be conducted and outlines the process for conducting the PSRA - it is critical to focus on safety issues during the early stages of the facility design process (strategic planning, master planning, operational planning and programming).

It was also proposed that the PSRA include caregiver safety considerations (becoming the Patient and Caregiver Safety Risk Assessment – PaCSRA) and serve as an umbrella for several other required assessments (such as the ICRA and PHAMA) as well as several new proposals focusing on specific risk categories, such as falls, medication errors, and security. As approved for the draft manuscript of the Guidelines to be open to public review through 2012, the updated PaCSRA includes the: Infection Control Risk Assessment (ICRA); Patient Handling and Movement Assessment (PHAMA); patient falls prevention risk assessment; medication safety risk assessment; psychiatric patient injury and suicide prevention risk assessment; patient immobility risk assessment; and security risk assessment.

1.5 Next Steps

While the seminar made significant headway in obtaining consensus from a multi-disciplinary group around key issues that should be addressed in a patient safety risk assessment, much work lies ahead to create a comprehensive toolkit that can support the guidelines and healthcare design teams. Patient safety is a primary concern for most healthcare organizations today, and the features in the built environment provide an important but often unconsidered tool in the on-going effort to improve safety outcome. In order for the PSRA toolkit to be accepted and used by healthcare design stakeholders, several tasks will build on work from the prior seminar. These include the:

- development of an online PSRA toolkit that can be used to conduct a proactive PSRA during a healthcare facility design process,
- development whitepapers and guidelines to support the use of the PSRA and to detail the process for implementation in the facility life cycle,
- further development of a Safe Design Roadmap for healthcare CEOs to integrate with the PSRA toolkit, and
- creation of an education platform to promulgate PSRA activities.

When completed, teams will have access to a comprehensive proactive PSRA that will help them carefully consider and eliminate patient safety risks during the healthcare facility design process. This will not only reduce costs of renovation at a later date to address latent conditions but also contribute to safer environments for patients, staff and families. As part of the toolkit, the Safe Design Roadmap will foster communication and will enable CEOs to understand how the facility design process can be made a part of quality improvement and culture change initiatives in their organization.

2 CONCLUSIONS

Patient safety goals must drive the project strategy and pre-design planning. This necessary to ensure certain environmental features are included in the project scope. As the design continues, the focus is on avoiding design hazards and latent conditions. Nearly all of the seven evaluated design tools (link analysis, RCA, FMEA, simulation, work sampling, balanced scorecard, and process analysis) were considered relevant and applicable to this process. Tools highly rated for feasibility included balanced scorecard and process analysis, while the most generalizable tools included balanced scorecard, link analysis, and process analysis. The Safe Design Roadmap was perceived as an overarching structure to facilitate cross-disciplinary communication and decision-making for administrators. Workgroup participants suggested that the format should be revised and customized to fit the needs of CEOs and other team members, with supporting materials such as glossary terms and examples. Future work will continue to refine these findings into a comprehensive toolkit for healthcare facility design teams.

ACKNOWLEDGMENTS

The authors would like to acknowledge the Agency for Healthcare Quality and Research who funded the seminar in part through grant 1R13HS020322-01A1. Acknowledgement is also give to the Facility Guidelines Institute (partial funding), Virtua Health Voorhees, (event host), The Center for Health Design staff (seminar logistics), and the speakers and participants who evaluated processes and tools for inclusion in the final report.

REFERENCES

Henriksen, K. et al., 2005. *Advances in Patient Safety: From Research to Implementation*, Rockville, MD: Agency for Healthcare Research and Quality. Available at: http://www.ahrq.gov/qual/advances2/ [Accessed February 23, 2012].

Henriksen, K. et al., 2008. *Advances in Patient Safety: New Directions and Alternative Approaches*, Rockville, MD: Agency for Healthcare Research and Quality. Available at: http://www.ahrq.gov/qual/advances2/ [Accessed February 23, 2012].

Joseph, A. & Rashid, M., 2007. The architecture of safety: hospital design. *Curr Opin Crit Care*, 13(6), p.714–9.

Kohn, L., Corrigan, J. & Donaldson, M. eds., 2000. *To err is human: Building a safer health system*, Washington, DC: National Academy Press.

Landrigan, C.P. et al., 2010. Temporal Trends in Rates of Patient Harm Resulting from Medical Care. *New England Journal of Medicine*, 363(22), pp.2124–2134.

Levinson, D., 2010. *Adverse Events in Hospitals: National Incidence Among Medicare Beneficiaries*, Washington, DC: US Department of Health and Human Services, Office of the Inspector General.

Reason, J., 2000. Human error: models and management. *BMJ*, 320(7237), pp.768–770.

Reiling, J.G. et al., 2004. Enhancing the traditional hospital design process: a focus on patient safety. *Joint Commission Journal On Quality And Safety*, 30(3), pp.115–124.

Society of Actuaries, 2010. Society of Actuaries Study Finds Medical Errors Annually Cost at Least $19.5 Billion Nationwide. Available at: http://www.soa.org/news-and-publications/newsroom/press-releases/2010-08-09-med-errors.aspx [Accessed October 20, 2010].

Stelfox, H.T. et al., 2006. The "To Err is Human" report and the patient safety literature. *Quality and Safety in Health Care*, 15(3), pp.174 –178.

The Agency for Healthcare Research and Quality (AHRQ), 2010. *National Healthcare Disparities Report*, Rockville, MD: U.S. Department of Health and Human Services, Agency for Healthcare Research and Quality. Available at: http://www.ahrq.gov/qual/nhdr09/nhdr09.pdf.

Ulrich, R.S. et al., 2008. A Review of the Research Literature on Evidence-Based Healthcare Design. *HERD-HEALTH ENVIRONMENTS RESEARCH & DESIGN JOURNAL*, 1(3), pp.61–125.

Vincent, C., Taylor-Adams, S. & Stanhope, N., 1998. Framework for analysing risk and safety in clinical medicine. *BMJ*, 316(7138), pp.1154–1157.

Wachter, R.M., 2010. Patient Safety At Ten: Unmistakable Progress, Troubling Gaps. *Health Affairs*, 29(1), pp.165 –173.

CHAPTER 28

Ergonomic Issues of Computer Use in a Major Healthcare System

Alan Hedge, Tamara James

Cornell University, Dept. Design & Environmental Analysis
Ithaca, NY 14853, U.S.A.
Email address: ah29@cornell.edu
Community and Family Medicine, Occupational and Environmental Medicine
Duke University Medical Center, Durham, NC 27710
Email address: tamara.james@duke.edu

ABSTRACT

The results of a survey of computer use patterns among 180 physicians and 63 nurse practitioners and physician assistants (NP/PAs) within multiple clinics of a major healthcare system in the USA are summarized. Results show that a majority of respondents work with desktop or laptop computers for much of the day and that while clinical computer use had increased during the 12 months prior to the survey the time for face-to-face interactions with patients had decreased. Two thirds of respondents reported daily use of a desk mounted computer yet only 20% said that they often made adjustments to the keyboard and monitor position. Over two thirds of respondents did not use a mobile computer cart and around two thirds did not use a wall-mounted computer at work. Over two thirds of respondents reported not being at all involved in the planning or the design of either their clinical workplace or their computer workstations and only around 5% said that they had an "expert knowledge" of ergonomics. Implications are described.

Keywords: healthcare information technology, computer use, computer ergonomics

1 HEALTH INFORMATION TECHNOLOGY

Health information technology (HIT) systems are computer-based systems designed to improve the ability of medical care providers to better manage patient care as well as lowering medical costs. In 2009 Congress passed the American Recovery and Reinvestment Act which contained the Health Information Technology for Economic and Clinical Health (HITECH) Act that called for an increased implementation of health information technology, and since that time the rate of adoption of HIT has accelerated (Blumenthal, 2010; Hedge et al., 2011).

In the past few years the rate of publication of research on this topic has exploded. Buntin et al. (2011) screened over 4,000 journal articles and reviewed 154 of these that passed the screening criteria. Of these articles 62% (96) reported clear positive benefits of HIT and 30% (46) found results that were predominantly positive. In 2009 it was estimated that around 44% of office-based physicians were using an electronic medical or health records system and that number was predicted to increase rapidly (Hsiao et al., 2009). Furukawa (2011) undertook a cross-sectional analysis of 62,710 patient visits to 2,625 physicians from the 2006-2007 National Ambulatory Medical Care Survey to evaluate the impact of electronic medical records (EMR) systems. The EMRs evaluated included electronic information on demographics, clinical notes, prescription orders, and laboratory and imaging results. Results showed that EMR use was associated with a 7.7% increase in any examination (95% CI = 2.4%, 13.1%); a 5.7% increase in any laboratory test (95% CI = 2.6%, 8.8%); a 4.9% increase in any health education (95% CI = 0.2%,9.6%); 7.1% fewer laboratory tests (95% CI = -14.2%, -0.1%) and 7.3% fewer radiology procedures during pre/post-surgery visits (95% CI= -12.9%, -1.8%).

However, the focus of almost all of the studies is on some measure of the patient care outcome and the health and well-being of the medical care personnel using the system is seldom considered.

In the rush to implement HIT systems the ergonomic risk factors for musculoskeletal injuries and other concerns are seldom considered but we cannot afford to be unknowingly exposing highly trained medical personnel to any increased injury risk. This paper reports the computer use patterns of medical personnel working in private clinics in a major healthcare system.

2 METHODS

2.1 Sample Profile

The 245 research participants were all members of the Private Diagnostic Clinics (PDCs) which is a for-profit faculty practice plan associated with the Duke University School of Medicine and Health System in Durham, North Carolina. PDC members work in clinics on the main Duke campus but several are located throughout the city of Durham as well as a growing number of community-based clinics in central and eastern North Carolina and southern Virginia. The respondents comprised 180 Physicians (13.8% of all PDC MDs) and 63 Advance Practice

Providers consisting of Nurse Practitioners and Physician Assistants (12.6% of all PDC NP/PAs).

2.2 Survey Instrument and Procedure

A web survey comprising 31 questions (a total of 83 items) was electronically administered. An email invitation to participate in the survey was sent to all PDC MDs and NP/PAs, and two-weeks later a follow-up email invitation was sent. The survey instrument and procedure were approved by the IRBs at Cornell University and Duke University School of Medicine.

2.3 Data Analysis

All data were analyzed using a multivariate statistical package (SPSS v19). NPs and PAs perform similar work and consequently data for the NP and PAs were aggregated for analysis to increase the sample size. Patterns of computer use as well as differences in physical complaints were compared for the MDs and the combined NP/PAs. All results are percentages with the actual frequency following in parentheses. The totals of percentages do not equal 100 in all cases because respondents were allowed to make multiple selections for some questions.

3 RESULTS

3.1 Demographics

There was a significant difference in the sample profile between MDs and NP/PAs for gender (X^2 =31.37, df 1, p=0.0001): 52% (93) MDs were male compared to 11.3% (7) NP/PAs. There was no difference between the two groups in the distribution by age 53.3% (96) MDs were <50 years old compared with 55.5% (35) NP/PAs. There was no difference between groups in the distribution of years of clinical experience: 68.9% (124) MDs had 5 or more years of experience compared with.63.5% (40) NP/PAs. Sixty three percent (114) MDs and 57.1% (36) NP/PAs were in hospital-based clinics; 47.8% (86) MDs were in outpatient clinics as were 41.3% (26) NP/PAs; 6.1% (11) MDs and 9.5% (6) NP/PAs were in primary care clinics.

3.2 Technology Use

Table 1 shows the use of computer technology by the sample. A desktop computer was used by more people at work than at home but the reverse was the case for laptop computers. A minority of the sample used a docking station with their laptop, and used a tablet PC/iPad. Twice as many MDs used a Blackberry or mobile device in the clinic compared with the NP/PAs.

There was no significant difference in the daily hours of computer use at work (MDs = 5.1 hours; NP/PAs = 5.8 hours) or at home (MDs = 2.2 hours; NP/PAs = 1.8 hours). Also there was no significant difference in reports of how the amount of daily computer use at work and at home had changed over the previous year: 55.9% (99) MDs and 70.9% (44) NP/PAs said computer use had increased at work; 34.1% (61) MDs and 44.4% (28) NP/PAs said computer use had increased at home. When asked about the comfort of their home computer setup 83.2% (149) MDs and 70.9% (44) NP/PAs said it was 'fairly' or 'very' comfortable.

Table 1 Technology used by MDs and NP/PAs in their workplace and at home

TECHNOLOGY	WORK		HOME	
	MD	NP/PA	MD	NP/PA
% Desktop computer	90.6 (163)	93.7 (59)	54.4 (98)	44.4 (28)
% Laptop/notebook	44.4 (80)	22.2 (14)	88.3 (159)	81 (51)
% Laptop with docking station	19.4 (35)	19.1 (12)	9.4 (17)	4.8 (3)
% Tablet PC/iPad	15.6 (28)	7.9 (5)	35 (63)	22.2 (14)
% Blackberry/other mobile device	69.4 (125)	36.5 (23)	75.6 (136)	54 (34)

Figure 1 shows the self-reported typing proficiency of the sample. A majority of 78.8% (141) MDs and 79.0% (49) NP/PAs said they use both keyboard and mouse about equally each day; 11.2% (20) MDs and 9.7% (6) NP/PAs said they mostly use the keyboard each day, and 10.1% (18) MDs and 11.3% (7) NP/PAs said they mostly use a mouse each day.

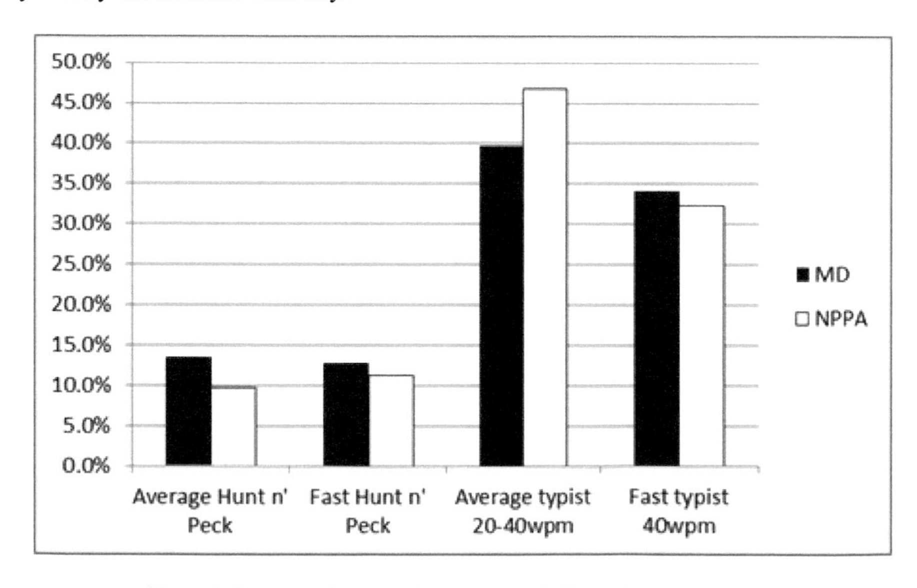

Figure 1 Typing proficiency of the sample of MDs and NP/PAs

3.3 Patient Interactions

Both the MDs and the NP/PAs reported an average of 4.9 hours per day of face-to-face interactions with patients. There was a significant difference in how this has changed during the previous year: 22% (39) MDs reported less time and only 6.7% (12) reported more time in face-to-face patient interactions, whereas 19.4% (12) NP/PAs reported less time but 33.9% (21) reported more time in face-to-face patient interactions (X^2 =30.41, df 4, p=0.0001).

3.4 Clinic computer use

When asked about the frequency of using a wall-mounted computer at work only 19.4% (35) MDs said they used this arrangement at least as frequently as 'some days each week' and 34.9% (22) NP/PAs reported this (X^2 =6.23, df 1, p=0.013). Of those respondents who used a wall-mounted computer at work, 42.8% (15) MDs and 59.1% (13) NP/PAs said that they adjusted the keyboard 'sometime' or 'often', and 57.2% (20) MDs and 52.3% (11) NP/PAs said they adjusted the monitor 'sometime' or 'often'.

For the work desk mounted computer, 82.8% (149) MDs and 87.3% (55) NP/PAs said they used this at least as frequently as 'some days each week'. Of those using a desktop computer, 53.8% (79) MDs and 62.3% (33) NP/PAs said that they 'sometimes' or 'often' made adjustments to their keyboard, and 41.5% (61) MDs and 44.2% (23) NP/PAs said that they 'sometimes' or 'often' made adjustments to their monitor.

Only 6.7% (12) MDs and 8% (5) NP/PAs said that they used a mobile computer cart at least as frequently as 'some days each week. Of those MDs who used a mobile computer cart only 16.7% (2) said that they 'sometime' adjusted the keyboard compared with 80% (4) NP/PAs who said that they 'sometimes' or 'often' adjusted the keyboard, and 25% (3) MDs and 60% (3) of NP/PAs said that they 'sometimes' or 'often' adjusted the monitor.

3.5 Workplace layout

Most examination rooms in the clinics consist of an exam table, a sink, one or more guest chairs, a rolling stool for the clinician, and a fixed or moveable table with desktop computer. The layout of these rooms and placement of the computer can have a major impact on musculoskeletal comfort of health care providers in addition to patient interaction during a clinic visit. Most computers are placed such that clinicians' backs are to the patient as they enter or look up electronic information. With this arrangement in exam rooms, clinicians then twist their torsos or neck in order to periodically ask questions or converse with the patient.

When asked how the current clinical workplace layout affects interactions with patients 43.8% (78) MDs and 47.6% (30) NP/PAs said the layout 'somewhat' or 'definitely' hindered interactions and only 24.7% (44) MDs and 25.4% (16) NP/PAs said it 'somewhat' or 'definitely' helped. Only 18.3% (32) MDs and 11.1% (7) NP/PAs reported involvement in the design of the clinical workplace. In contrast, only 18.9% (34) MDs and 3.2% (2) NP/PAs reported involvement in the design of their computer workstation.

3.6 Ergonomics Knowledge

Surprisingly, only 4.5% (8) MDs and 6.3% (4) NP/PAs said they had 'expert' knowledge of ergonomics, with 58.7% (105) MDs and 57.1% (36) NP/PAs reporting some knowledge of this discipline.

3.7 Computer Work Tasks

When asked what method of processing patient information for medical records they used, 78.9% (133) MDs and 77.8% (49) NP/PAs said direct computer entry. Two-thirds (120) of MDs and 71.4% (45) NP/PAs said they used dictation.

On a daily basis, 98.7% (177) MDs and 95.2% NP/PAs said they used a computer to find or review research articles, and 56.1% (101) MDs and 66.7% (42) NP/PAs did this for less than 1 hour per day. A minority of 13.1% (24) MDs and 23.9% (15) NP/PAs said that they spent 1 or more hours a day entering laboratory results on a computer. Most MDs used email (98.3%: 177) and NP/PAs (98.43%: 62) said that they used email, and the average time for each group was 2.28 hours per day. Almost all MDs (98.9%: 178) and NP/PAs (98.4%: 62) said that they used a computer each day to review patient information/medical records (Figure 2).

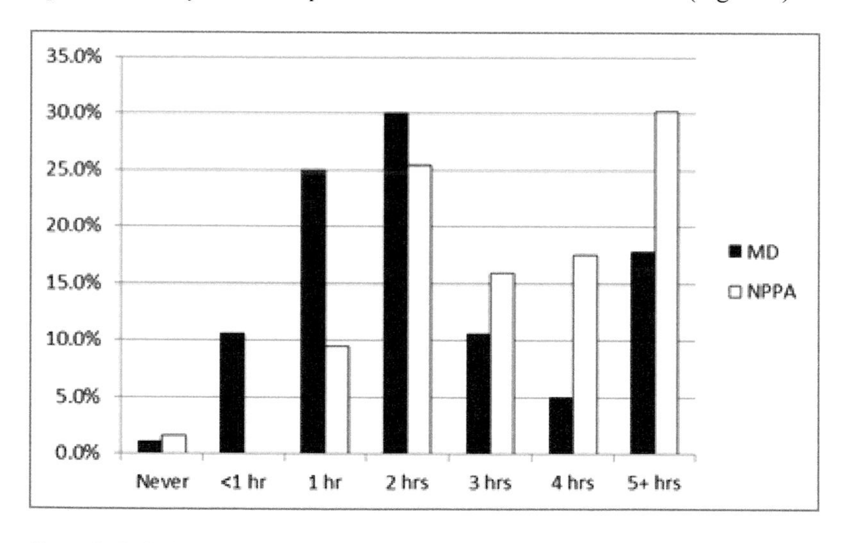

Figure 2 Daily hours spent using a computer to review patient medical records

3.8 Weekly Frequency of Neck Discomfort

Nine out of 10 NP/PAs and 7 out of 10 MDs reported neck discomfort. Figure 3 shows that significantly more NP/PAs than MDs experienced more frequent neck discomfort (X^2 = 13.796, df=4, p=0.008).

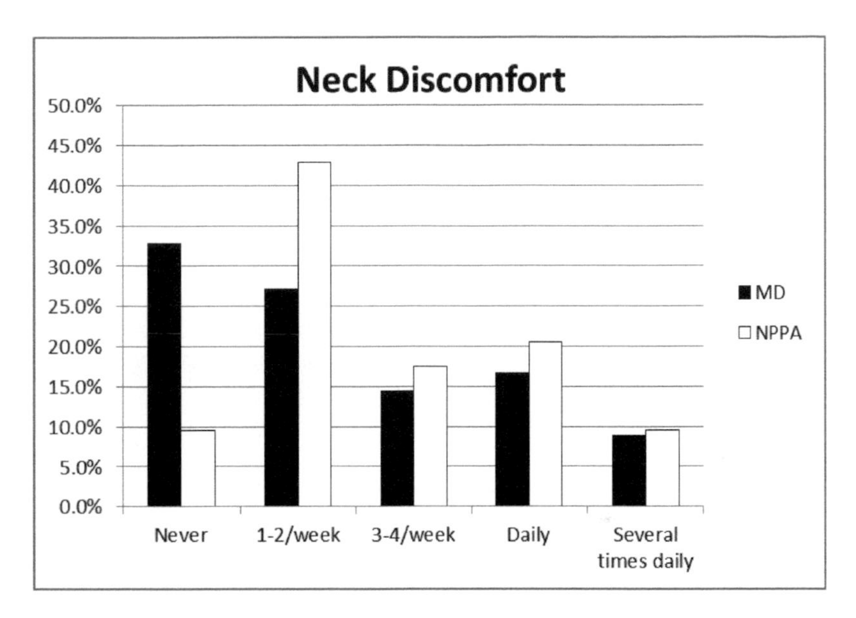

Figure 3 Frequency of neck discomfort in a typical week for the MDs and NP/PAs

3.9 Weekly Frequency of Shoulder Discomfort

Eight out of 10 NP/PAs and 6 out of 10 MDs reported shoulder discomfort. Figure 4 shows that significantly more NP/PAs than MDs experienced more frequent neck discomfort (X^2 = 12.810, df=4, p=0.012).

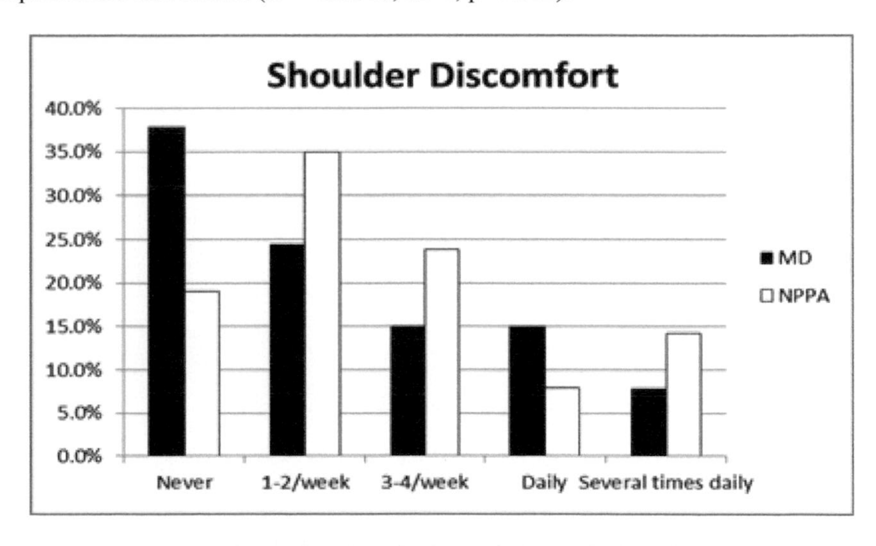

Figure 4 Frequency of shoulder discomfort in a typical week for the MDs and NP/PAs

3.10 Weekly Frequency of Back Discomfort

Fifty-five percent (99) MDs and 69.8% (44) NP/PAs reported upper back discomfort at least once week and 50% (90) MDs and 55.6% (35) NP/PAs reported lower back discomfort at least once week but there was no significant difference between the groups.

3.11 Weekly Frequency of Upper Limb Discomfort

Reports of at least weekly upper limb discomfort are shown in Figure 5. Right hand and right wrist discomfort was twice as prevalent as left wrist or left hand discomfort and also around twice the prevalence of any of the other upper limb segments. There was no significant difference between the groups for any of the upper limb segments.

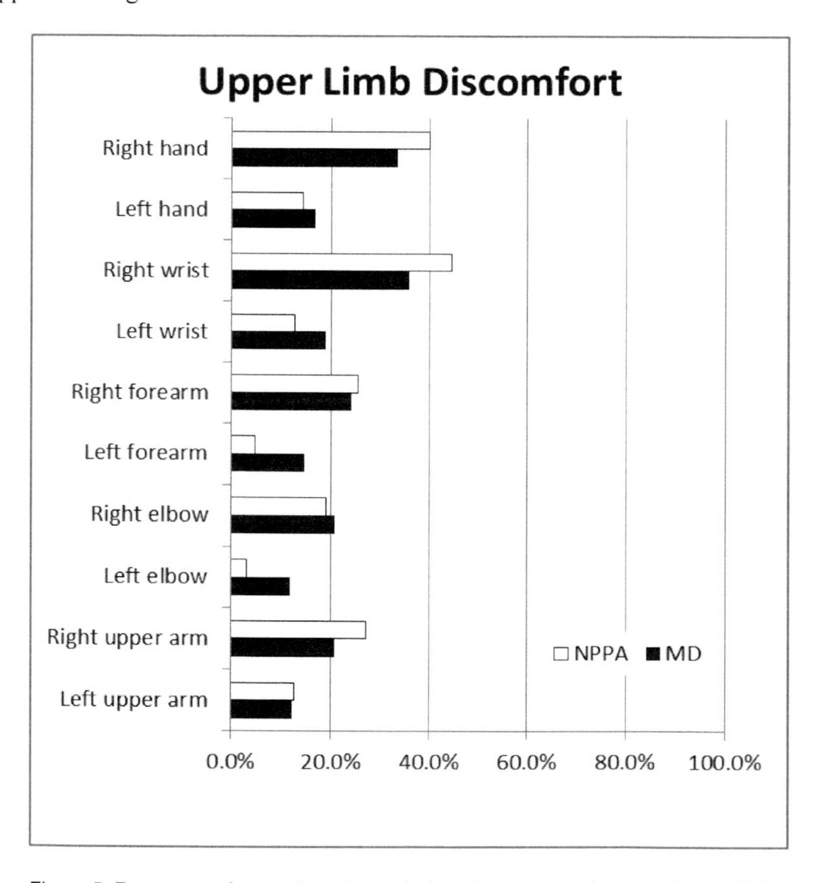

Figure 5 Frequency of upper limb discomfort in a typical week for the MDs and NP/PAs

3.12 Computer work-related injury

There was no significant difference between the groups for reports of computer work-related injuries in the current workplace and 10.6% (19) MDs and 11.1% (7) NP/PAs said that they had been injured. Four MDs and 3 NP/PAs reported chronic neck pain problems associated with computer use; 4 MDs reported wrist problems (2 were carpal tunnel syndrome) and 1 NP/PA reported wrist pain; 3 MDs reported shoulder pain, as did 1 NP/PA; and 2 MDs reported 'tennis elbow' that they attributed to carrying a computer/tablet around.

3.13 Workplace modifications

There was a statistically significant difference between the groups when asked whether or not they had done anything to correct any problems related to pain/discomfort associated with using computers at work, with 43.8% (79) MDs and 44.4% NP/PAs (28) reporting experiencing some musculoskeletal discomfort but not making any workplace changes in response to this; 27.5% (49) MDs and 42.9% (27) NP/PAs said that they had made changes to their workplace because of musculoskeletal discomfort (X^2 = 8.048, df=2, p=0.018). Seventeen MDs and 10 NP/PAs said that they had adjusted or changed their chair, 10 MDs and 7 NP/PAs reported adjusting or changing their keyboard, and 7 MDs and 4 NP/PAs mentioned that they had adjusted or changed their monitor screen position.

4 CONCLUSIONS

This study has confirmed that computer use is a substantial part of the daily work of MDs and NP/PAs working in PDCs in this major health system. A variety of different computing devices are being used both at work and at home, and consequently daily exposures to the potentially adverse effects of poor computer workspace design are quite high. For most of the variables there were statistically significant differences between the MDs and the NP/PAs, and in general the NP/PAs experienced more discomfort associated with computer use. Although around half of the users surveyed reported making some adjustments to keyboards and monitors little attention appears to have been paid to ergonomic principles in designing the workplaces.

The survey results also found that medical personnel had little involvement in the design of their workstations and facilities, and surprisingly few of them had any expert knowledge of ergonomics. With computer placement being a contributor to provider musculoskeletal discomfort and adversely impacting patient interaction, it is critically important that the actual clinicians in addition to ergonomists, have input into the layout of examination rooms and computer work areas. Likewise, ergonomics awareness among HIT users is imperative for injury prevention since so few indicated they had expert knowledge and not all users made adjustments to their work areas.

Given the increasing demands being placed on the U.S. healthcare system as a consequence of factors such as the aging and increasingly unfit patient population, along with a potential shortfall in the number of medical personnel, it is crucial that clinical workplaces effectively support the work of MDs and NP/PAs rather than posing potential injury hazards. To deal with these issues much hope is being pinned to the use of health information technology systems. As the future use of these HIT systems expands, and as the findings of this study confirm, ergonomics can play an important role in the design of clinical workplaces that are efficient and effective, and that protect medical personnel against the risks of musculoskeletal injuries.

ACKNOWLEDGMENTS

The authors would like to acknowledge Paul Newman, Executive Director of the Private Diagnostic Clinics for providing email access to the PDC members and to his assistant, Susan Cole for overseeing distribution of the email invitations.

REFERENCES

Blumenthal D. 2010. Launching HITECH. *New England Journal Medicine*, 362(5):382–385

Buntin, M.B., Matthew F. Burke, M.F., Hoaglin, M.C. and Blumenthal, D. 2011. the benefits of health information technology: a review of the recent literature shows predominantly positive results. *Health Affairs*, 30(3):464-471.

Furukawa, M. F. 2011. Electronic medical records and efficiency and productivity during office visits. *American Journal of Managed Care*, 17(4), 298-303.

Hsiao, C., Beatty, P.C., Hing, E.S. et al. 2009. Electronic medical Record/Electronic Health Record Use by Office-based Physicians: United States, 2008 and Preliminary 2009. (http://www.cdc.gov/nchs/data/hestat/emr_ehr/emr_ehr.htm) Accessed 2/5/12.

Hedge, A. James, T. and Pavlovic, S. 2011. Ergonomics Concerns and the Impact of Healthcare IT, *International Journal of Industrial Ergonomics*, 41(4): 345-351.

CHAPTER 29

Physical and Mental Burden of Caregivers in Providing Bathing Assistance

*YAMAMOTO Akiyoshi**,**, *MATSUYAMA Minoru**, *KAMOI Hiroaki**, *YAMAMOTO Ryousuke**, *KUWAHARA Noriaki***, *KIDA Noriyuki***

*City Estate Co., Ltd.
Osaka, Japan
yamamoto@city-estate.co.jp
matsuyama@city-estate.co.jp
kamoi@city-estate.co.jp

** Kyoto Institute of Technology
Kyoto, Japan
nkuwahar@kit.ac.jp
kida@kit.ac.jp

ABSTRACT

A questionnaire survey regarding the physical and mental load of providing bathing assistance was conducted with a group of paid caregivers. The questionnaire was administered before and after bath time for 11 days with staff (n=34) working at a private nursing home. The following data regarding bathing assistance were collected daily: number of people receiving bathing assistance, average age of bath recipients, and the average nursing care level of people receiving bathing assistance. In addition, participants answered questions about their physical and mental well-being. Participants responded to this part of the questionnaire by using a Likert scale consisting of 5 choices ranging between *never applicable* and *very often applicable*. Through the survey period, a total number of 126 data points were obtained. The results indicated that after bathing an elder, irritation, dullness, and sleepiness significantly increased, whereas vigor significantly decreased . After conducting an analysis of covariance structures, a path diagram was obtained

(GFI=.893, AGFI=.821, RMSEA=.099). Sense of fulfillment was not significantly related to bathing load (.01), but sense of burden was related to bathing load (.20).In addition, the relationship between average nursing care level of the patients and the bathing load was significant (-.42), such that bathing care load was high when conducting assistance for people with low nursing care levels.

Keywords: caregiver, bathing load, analysis of covariance structures

1 INTRODUCTION

Recently, caregiving has become a serious challenge in aging societies. Reasons for this include, because of increasing numbers of older adults with dementia and/or that are bedridden and the extended duration of caregiving due to increases in the lifespan. On the other hand, the conditions of elderly people in need of care are changing rapidly. Care for the elderly can be a heavy physical and mental burden on families and other caregivers. Therefore, it is important to investigate and assess the degree of problems of caregivers and to develop realistic countermeasures to solve those problems.

Some of the main issues faced by caregivers in today's society are a tendency to suffer from low back pain symptoms, cervicobrachial disorder, and depression. In addition, caregivers often complain of daytime sleepiness and arm pain. It is clear from these findings that caregiving can create be extremely difficult and stressful as well as create a high labor load.

Although some reports suggest that the use of nursing equipment may reduce the burden on the lower back, nursing care equipment has not been widely received for many reasons, such as cost, difficulty in use, and effort. In Japan, moreover, as a culture focused on the importance "entertaining" people, there is the idea that nursing has to be done in "the hands of people".

Bathing is essential to maintain cleanliness and is considered an important activity of daily living (ADL). However, bathing assistance constitutes a heavy load, both physical and mental, for caregivers. The degree and impact of this load is not well-understood. A questionnaire survey was conducted regarding the physical and mental load of caregivers when providing bathing assistance. Details of the degree of burden were analyzed.

2 METHODS

The questionnaire research was conducted before and after bath time for 11 days with staff (n = 34, men = 10, women = 24, cf. Table 1) working at a private nursing home. Figures 1 and 2 show the bathroom in the nursing home. Characteristics and care level of the elderly who were nursed in this care-home at the time are presented in Table 2.

The following observable data regarding bathing assistance were collected daily: (1) the number of people receiving bathing assistance, (2) the number of those individuals (from 1) in need of total assistance, (3) the average age of bath recipients, (4) the average height of bath recipients, (5) the average weight of bath recipients, and (6) the average nursing care level of people receiving bathing assistance. Total care burden or load of providing bathing assistance was calculated by combining the average weight of the to-be-bathed people, the number of people provided with bathing assistance, and the average nursing care level of the individuals receiving baths.

Self-rating items were developed based on several questionnaires, including the Survey on Subjective Symptoms (Industrial Fatigue Research Group, 2003) and the Profile of Mood States (POMS). Based on the results of a pilot study, a questionnaire consisting of 7 factors with a total of 22 items was constructed (Table 3). Participants responded to this questionnaire by rating themselves using a Likert scale consisting of 5 choices ranging between 1 (*never applicable*) and 5 (*very often applicable*). Factor analysis of these questions was conducted (see Results section) and two main factors were identified (labeled 'sense of burden' and 'sense of fulfillment'). The sense of burden factor includes physical and mental problems (e.g. anxiety) of caregivers. The sense of fulfillment factor indicates positive emotional aspects (e.g. vigor) of caregivers.

Changes in physical and mental burden scores measured before and after providing bathing assistance were evaluated using a multiple indicator model. The model consisted of the following three concepts: care load, sense of burden, and sense of fulfillment. It was predicted that bathing care load would affect both the sense of burden and the sense of fulfillment of the caregivers. Analysis was conducted using Amos, which is software specifically developed for conducting analysis of covariance and structural equation models.

Figure1. Bathroom of the nursing home

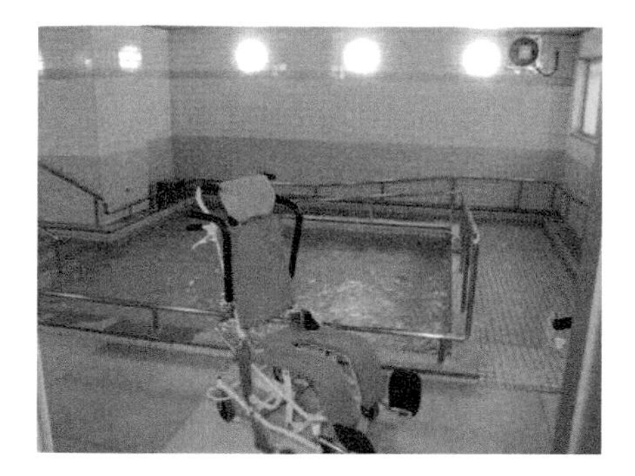

Figure 2. Work environment for bathing in the nursing home

Table 1. Caregiver characteristics

	Number	Age (years)	Height (cm)	Weight (kg)
Men	10	38.0±13.3	171.4±5.6	68.2±8.0
Women	24	38.0±13.5	157.3±6.3	51.7±7.8

Table 2. Characteristics of the elderly receiving nursing care

	Men	Women	Total
Number	20	97	117
Age (years)	86.2	85.7	85.9
Height (cm)	160.0	145.7	152.9
Weight (kg)	52.1	42.9	47.5
Full assistance (n)	6	25	31
Assist need (n)	2	9	11
Care need level 1 (n)	5	22	27
Care need level 2 (n)	4	26	30
Care need level 3 (n)	4	17	21
Care need level 4(n)	3	16	19
Care need level 5(n)	2	6	8

Table 3. Questionnaire items

Factor	Question	Factor	Question
Satisfaction	Feeling a sense of fulfillment	Irritation	Having a headache
	Feeling a sense of relief		Being irritated
Vigor	Being full of energy		Feeling heavy in head
	Being lively		Feeling sick
	Feeling positive	Dullness	Having stiff shoulders
	Feeling good		Feeling heavy in legs
Refreshment	Feeling refreshed		Having a backache

	Feeling calm		Feeling heavy in arms
	Being in high spirits	Sleepiness	Yawning
Anxiety	Feeling depressed		Being sleepy
	Feeling sick		Being not motivated

3 RESULTS

Through the survey period, a total of 126 data points were obtained. The results indicated that the caregivers' irritation, dullness, and sleepiness significantly increased from pre-bathing experience to post-bathing experience, whereas vigor significantly decreased after bathing work (Table 4).

Factor analysis (Principle Factor Method, Promax Rotation) was conducted on the variations before and after providing bathing assistance. Factors with eigenvalues of one or above were extracted yielding two factors that conformed to the model (Table 5). Factor 1 was named "Sense of Burden", and Factor 2 was named "Sense of Fulfillment".

The specifics of the work content of bathing assistance are shown in Table 6. After conducting an analysis of covariance structures, observable variables that did not have significant paths values to bathing load were deleted, and the path diagram shown in Figure 3 was obtained. Fit indexes were relatively high (GFI=.893, AGFI=.821, CFI=.729, RMSEA=.099), but not sufficient.

Table 4. Questionnaire results before and after work

	N	Before work		After work		*F*	
		M	sd	M	sd		
Satisfaction	126	2.4	0.9	2.5	0.9	1.66	
Vigor	126	2.5	0.9	2.3	0.9	2.70	**
Refreshment	126	2.6	1.0	2.6	0.9	0.38	
Irritation	126	1.8	0.8	1.9	1.0	2.58	*
Dullness	126	2.4	1.2	2.7	1.3	4.96	**
Sleepiness	126	2.0	1.0	2.2	1.1	2.43	*
Anxiety	126	1.8	1.0	1.9	1.1	1.00	

*p<.05, **p<.01.

Table 5. Factor analysis

	Factor 1	Factor 2
Sleepiness	.696	.259
Anxiety	.674	-.212
Dullness	.462	-.053
Irritation	.437	-.171

Vigor	.044	.839
Refreshment	-.139	.416
Satisfaction	-.052	.305
Correlation between factors		-.038

Table 6. Work contents

	N	Min.	Max.	Mean	sd
Number of people given assistance	126	1	7	4.0	1.7
Number of people in need of total assistance	126	0	4	1.1	1.0
Average age of assistance recipients (years)	126	71	95	85.8	3.9
Average height of assistance recipients (cm)	126	117	170	148.3	5.3
Average weight of assistance recipients (kg)	126	32	58	43.7	4.8
Average care level of assistance recipients	126	0.8	5.0	2.4	0.8

4 DISCUSSION

An analysis of covariance structures of questionnaire items regarding mental and physical well-being confirmed that the load of providing bathing assistance increased caregivers' sense of burden with the coefficient between bathing load and the sense of burden factor being 0.20. Furthermore, since the coefficients between the sense of burden and both sleepiness and anxiety were particularly high, the sense of burden of caregivers may be expressed in sleepiness and anxiety.

On the other hand, the sense of fulfillment, which is related to positive feelings, was not significantly related to bathing load (.01). Therefore, sense of fulfillment for caregivers in a nursing home may result from factors other than those targeted in the current study. For instance, communication at the work place, the degree of satisfaction with work, techniques for providing bathing assistance were not objectively evaluated in the present study.

The coefficient between bathing load and average nursing care level was significant (-0.42), which suggests that bathing load is higher when conducting assistance for people with lower nursing care levels. One explanation for this may be that people with low nursing care levels might strongly reject or resist being bathed. It is suggested that future studies on this topic should examine this link further and possibly take the degree of dementia into consideration.

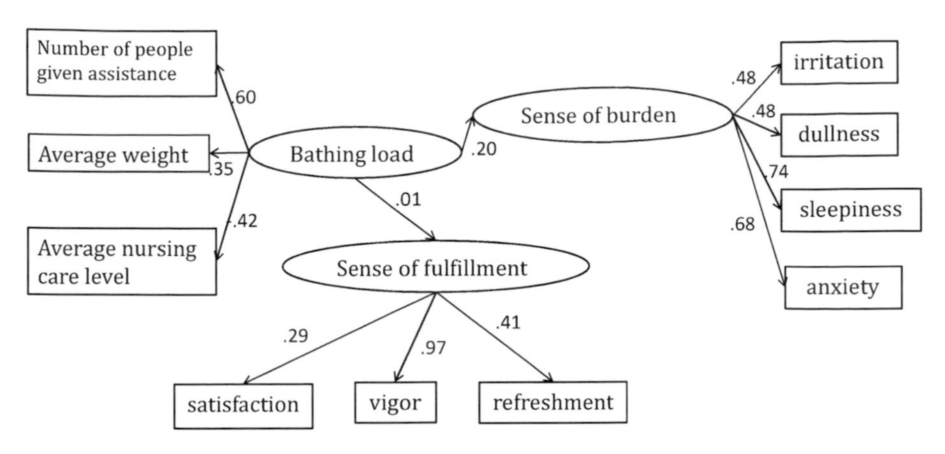

Figure3. Path diagram and results of analysis

ACKNOWLEDGMENTS

The authors would like to acknowledge Professor Hiroyuki Hamada for making this work possible and for his encouragement and helpful discussions.

REFERENCES

Covinsky, K. E., Newcomer, R., Fox, P., Wood, J., Sands, L., Dane, K., & Yaffe, K. (2003). Patient and caregiver characteristics associated with depression in caregivers of patients with dementia. *Journal of General Internal Medicine*, *18*, 1006-1014.

England, M., & Roberts, B. L. (1996). Theoretical and psychometric analysis of caregiver strain. *Research in Nursing & Health*, *19*, 499-510.

Industrial Fatigue Research Group, Survey on Subjective Symptoms, Selected by Japan Society for Occupational Health, 2002

Moens, G., Dohogne, T., Jacques, P., & Van Helshoecht, P. (1993). Back pain and its correlates among workers in family care. *Occupational Medicine (Oxford, England)*, *43*, 78-84.

Nagai, K., Hori, Y., Hoshino, J., Hamamoto, R., Suzuki, Y., Sugiyama, A., & ... Sakakibara, H. (2011). [Subjective physical and mental health characteristics of male family caregivers]. *[Nihon Kōshū Eisei Zasshi] Japanese Journal of Public Health*, *58*(8), 606-616. [In Japanese]

Tomioka, K. (2008). Low back pain among care workers working at newly-built nursing homes for the aged. *Sangyō Eiseigaku Zasshi = Journal Of Occupational Health*, *50*(3), 86-91. [In Japanese]

Tomioka, K., & Matsunaga, I. (2007). [The actual condition of musculoskeletal disorders of workers working at new special nursing homes for the aged in Osaka Prefecture--comparison of musculoskeletal disorder between manager's awareness and care workers' complaints by questionnaire survey]. *Sangyō Eiseigaku Zasshi = Journal Of Occupational Health*, *49*, 216-222. [In Japanese]

CHAPTER 30

Slip, Trip and Fall Prevention for Health Workers in the Operation Room

Kazuhiko Shinohara

School of Health Science, Tokyo University of Technology
Tokyo, JAPAN
kazushin@hs.teu.ac.jp

ABSTRACT

Work-related slip, trip, and fall (STF) incidents can frequently result in serious disabling injuries that impact an employee's ability to do his or her job. Risk assessments of STF problems for health workers in the operation room (OR) were performed in this study. The following hazards for STF incidents in the OR were found: contaminants on the floor, surface irregularities, loose cords and medical tubing, and improper arrangement of medical equipment and step-stools. The restriction of surgeon and scrubbing nurse movement resulting from the operating gowns worn was a major risk for STF incidents. Recently, the amount of cords, cables, and medical equipment required in the OR for endoscopic and computer-aided surgery and computerized physician order entry systems has increased. These circumstances increase the risk of STF incidents in the OR. As countermeasures against these problems, it is essential for the human-machine interface in the OR to be further improved by accounting for OR staff's workflow. Fatigue management of OR staff should also be further discussed and improved.

Keywords: slip, trip, and fall incidents, operation room,

1 INTRODUCTION

Work-related slip, trip, and fall incidents can frequently result in serious disabling injuries that impact an employee's ability to do his or her job. To date, most risk assessments and prevention programs for STF incidents in the healthcare industry have been limited to patients. Surgical practices in the OR)impose heavy physical and mental burden on health workers, but there are few studies examining STF incidents in the OR. In this study, risk assessments of STF incidents were conducted for health workers in the OR.

2 MATERIALS AND METHODS

The objects of this survey were ORs in middle-size municipal hospitals providing general surgery, endoscopic surgery, urology, orthopedic surgery, and ophthalmology and the health workers who work in them. Medical equipments for these surgeries are installed in the OR, alongside a computerized physician order entry system (CPOE) and computerized picture archiving and communication system (PACS). OR workers change their shoes to slippers at the entrance to the OR. Scrubbing with tap water and rubbing with alcoholic disinfectants were used for surgical hand washing. Hazards for STF incidents were surveyed in these ORs.

3 RESULTS

Hazards for STF incidents in the OR were reported as follows.

From the entrance of the OR to the start of the operation:

contaminants on the floor (especially around the washstand for surgical hand washing); improper use of floor mats; surface irregularities; poor drainage pipes; loose cords and medical tubing; and improper arrangement of medical equipment, foot switches, and step-stools. (Fig.1)

Figure 1 Washstand for surgical hand washing (a) and OR door (b)

During the operation:

loose cords and medical tubing; improper arrangement of medical equipment, foot switches, chairs with casters, and step-stools; contaminants on the floor; movements toward medical equipment and computer terminals during surgery; restriction of surgeon and scrubbing nurse movement resulting from the operating gowns and slipper worn; differences in illuminance between the surgical field and the OR; and darkened illumination during endoscopic and microscopic surgery. (Fig 2, Fig 3)

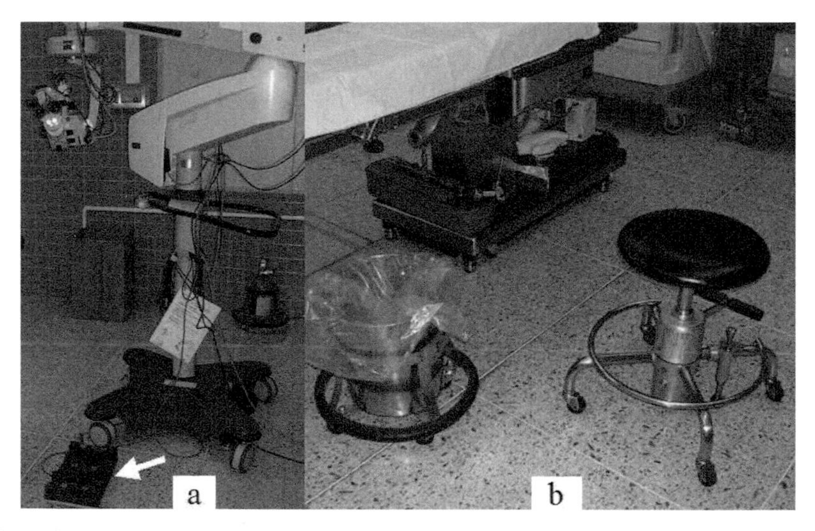

Figure 2 Surgical microscope and foot switches (a),chair with the caster and step-stools (b)

4 DISCUSSION

Recently, the National Institute for Occupational Safety and Health (NIOSH) published guidelines for "Slip, Trip, and Fall Prevention for Health Workers". For health professionals working in the OR particularly, risk management is of some greater importance in that they have heavy mental and physical workloads originating from the need to perform urgent and sterile procedures and involving the use of many medical devices and instruments. Moreover, work-related STF incidents in the OR can frequently result in serious disabling injuries that impact not only OR staff themselves but also the patient undergoing surgery. Among the common hazards for STF in hospitals listed in the NIOSH report are contaminants on the floor, improper use of floor mats, surface irregularities, loose cords and medical tubing, as were found for ORs in the present study. Further hazards peculiar to OR work revealed by this study include improper arrangement of medical equipment, foot switches, and step-stools, which can cause severe STF injuries in the OR. Furthermore, restrictions imposed by the operating gowns on the movements of surgeons and scrubbing nurses increase the risk for STF incidents and so new designs might help reduce STF incidents. Differences in illuminance between the surgical field and the OR as well as mental and physical fatigue including asthenopia among the OR staff were also risk factors.

Recently, the amount of cords, cables, and medical equipment required in the OR for endoscopic and computer-aided surgery, CPOE, and PACS has greatly increased. This increase has also led to the need for surgeons and nurses to make additional movements toward medical equipment and computer terminals during surgery. These circumstances increase the risk of STF incidents in the OR.

As countermeasures against contaminants on the floor around the washstand for surgical hand washing, changes in surgical hand washing from scrubbing with tap water to rubbing with alcoholic was found to be effective. Moreover, replacing the use of OR slippers with rubber shoes is also effective for STF prevention. As countermeasures against other problems, it is essential for human-machine interfaces in the OR to be further improved by accounting for the OR staff's workflow. Moreover, it is necessary to integrate cables and wires, use boom arms that eliminate the need for troublesome cords piling up in the OR, and introduce tablet PC and WIFI environments with electromagnetic compatibility for medical devices. Fatigue management of OR staff should also be further discussed and improved.

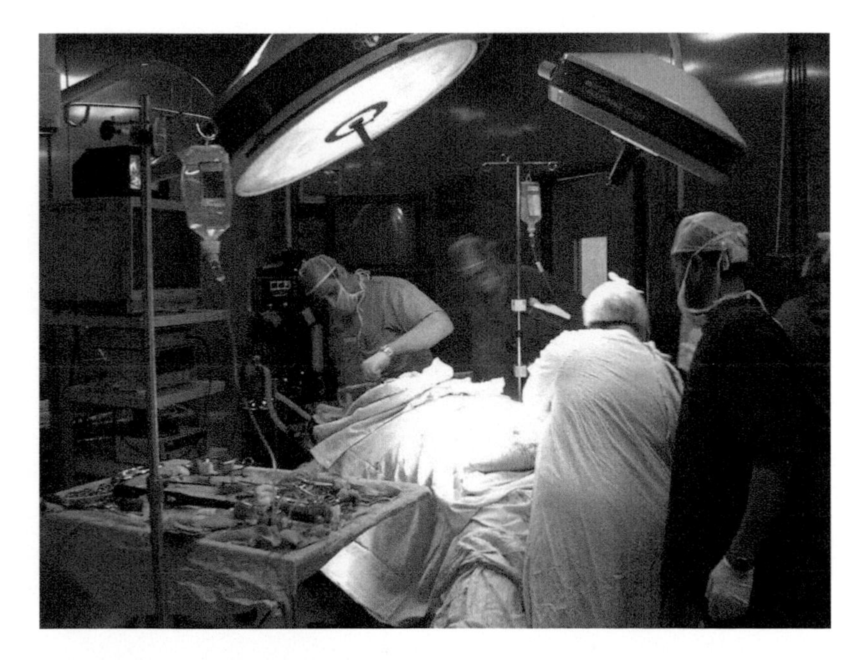

Figure 3 Scenes of endoscopic surgery

5 CONCLUSION

Risk assessments of STF problems for health workers in the OR identified not only conventional STF risk in hospital environments such as contaminants on the floor and surface irregularities, but also hazards peculiar to working in modern ORs. Among the countermeasures needed, human-machine interfaces should be further improved by accounting specifically for OR staff's workflow. Risk management for STF incidents in the OR remains an important issue to safeguard not only the safety of OR staff but also the patients undergoing surgery.

REFERENCES

National Institute for Occupational Safety and Health: 2011. Slip, Trip, and Fall Prevention for Health Workers *DHHS (NIOSH) Publication*. No. 2011–123, 2011

Shinohara, K. 2011. Consideration of the human-computer interface in the operation room in the era of computer aided surgery. *Proceedings of the HCI International 2011*(CD-R), Springer-Verlag,

Section V

Patient Care, Patient Safety and Medical Error

CHAPTER 31

The Study of Drug/Medical Device Recall Data (I) Analyses on Recall Data in Japan

Masaomi Kimura, Hiromi Yoneshige, Michiko Ohkura, Fumito Tsuchiya

Shibaura Institute of Technology
3-7-5, Toyosu, Koto City, Tokyo, Japan
{masaomi,l05127,ohkura}@shibaura-it.ac.jp

International University of Health and Welfare
2600-1, Kita-Kanamaru, Otawara City, Tochigi, JAPAN

ABSTRACT

In Japan, there have been reports of more than 600 recalls per year for drugs, medical devices, cosmetics and other quasi-drugs. The Japanese Ministry of Health, Labor and Welfare defined a classification system referring to that defined by the FDA, which has been used to classify all recall cases since 2000. Nevertheless, a problem in assigning the cases to these classifications has emerged. In this study, we analyzed the recall data on medicines and medical devices disclosed by Japanese authorities to determine the reason for such misclassifications and analyzed them based on a text mining technique, the dependency-link method.

As a result, we found that the confusion of two kinds of extent, namely *the extent of health hazards* and *the extent of the possibility of hazards emerging* could be causes of incorrect classification. We also found that the possibilities were categorized into the possibility of using, possibility of finding a defect, and possibility of taking action against hazards. We also proposed a novel dimension of classification based on these possibilities.

Keywords: Recall, Medicine, Medical equipment, Text mining

1 INTRODUCTION

In Japan, there have been reports of more than 600 recalls per year for drugs, medical devices, cosmetics and other quasi-drugs. The recall reports include cases of varying severity, such as mixtures of foreign substances, misprints on packages and so forth. Obviously, severe recall cases should be recognized by customers in the market. However, if the reporting of the recalls is confused, severe recall cases may be obscured by other, less severe, cases, which happens much more frequently as stated by Heinlich's law.

To prevent such confusion, the Japan Ministry of Health, Labor and Welfare defined a classification system referring to that defined by the FDA and comprising three classes:

Class I: cases where the products possibly cause users severe health hazards or death.

Class II: cases that may involve temporary or medically reversible adverse health consequences.

Class III: cases that are unlikely to cause health hazards.

Since 2000, all recall cases have been classified based on the above classification definitions.

Nevertheless, a problem in assigning the cases to these classifications has emerged. In fact, some recall reports in which the error did not result in any health hazard were assigned to Class II, although they should have been classified into Class III. This situation confuses consumers and may render the classification system meaningless.

Table 1 An example of incorrect assignment of cases to classifications

Attribute Name	Attribute Value (Description)
Product name	Helmitin suppository
Class	Class II
Reason for recall	The results of stability tests of this product reveal the possibility that the amount of allantoin, one of the active ingredients, is below the approved standard six months to one year after production. Accordingly, we decided to recall this product.
Specific health hazards that may appear	Currently, the amount of allantoin is within the approved scope and the amount of other ingredients is not recognized to decrease. *Adverse health consequences are unlikely.* In addition, there has been no report of adverse events attributable to allantoin to date.

In this study, we analyzed the recall data on medicines and medical devices disclosed by Japanese authorities to determine the reason for such misclassifications, and compared the result with the definition for recalls to identify the origin of inconsistency. The target data are descriptive data of the expected health hazards.

We analyzed them based on a text mining technique, the dependency-link method, which provides us with patterns of word dependencies frequently appearing in target sentences.

2 TARGET DATA

We downloaded HTML style recall case data disclosed by the Japan Ministry of Health, Labor and Welfare and the Pharmaceutical and Medical Devices Agency. The collected data consisted of 300 cases of Class I, 2909 cases of Class II and 1473 cases of Class III respectively. The data for each case contained the following items:

- Generic name / Product name
- Target lot number, Quantity and Shipping date
- Manufacturer / Distributer
- Reason for recall
- Specific health hazards that may appear
- Starting date of recall
- Effect and efficacy / Purpose of use
- Other information, such as contact information

In this study, we focused on the item, "Specific health hazards that may appear," in order to determine the origin of the classification error.

We encountered difficulty while extracting recall data from the HTML files. Although we tried to refer to attribute names and tag patterns appearing in the HTML to identify the recall data, orthographic variations of the attribute names and differences in tag structures in the original data emerged. For instance, there were three Japanese expressions used to mean "a specific health hazard that may appear," namely "危惧される具体的な健康被害," "危ぐされる具体的な健康被害" and "危惧される健康被害." In this example, the variations come from the replacement of a Kanji character with a Kana character or the omission of the word "specific." Another example is the placement of blanks in an attribute name for aesthetic reasons (e.g. "販 売 名" instead of "販売名" (Product name)). These varieties tend to appear when the HTML files are created / maintained by hand, which hinders the identification of the data. To promote the use of the disclosed data, it is clear that the data should be processable by computer programs, namely, the data should have standardized structures and attribute names without orthographic variations.

3 METHODS

3.1 Word-link method/Dependency-link method

Since the data in the item, "Specific health hazards that may appear," is descriptive data, it is obvious that they cannot be analyzed by statistical or data mining methods. Instead of these methods, we utilized text mining methods instead, namely the dependency-link method, which extracts the patterns of structures in target sentences.

Let us review these methods briefly (Kimura, 2007).

We start with the assumption that target sentences that should be summarized use common words and have similar structures. To find patterns of structures in such sentences, we employ dependency structure analysis, which is a common technique used in natural language processing and allows us to extract the dependency (modification) relationship between words in a sentence. Under our assumption, we can expect that patterns of dependency structures will emerge. By rearranging these dependencies correctly, we will extract the summarized sentences from the target set of sentences. This is the basic idea behind the dependency-link method, the steps of which are as follows:

1. Extract the dependency relationships between words from each of the target sentences using dependency structure analysis.
2. Create a directed graph whose nodes are the dependency relationships appearing as the result of Step 1. Arcs between the nodes denote that they share a word and co-occur in a sentence. If the word w modifies w' and w' modifies w'' in a sentence, we draw an arc from the dependency relationship $(w - w')$ to $(w' - w'')$.
3. Assign the number of sentences containing the dependency to the corresponding arc. If there are N sentences that have both dependency relationships $(w - w')$ and $(w' - w'')$, we assign N to the arc in the graph at Step 2.
4. Define the threshold T of the number assigned at Step 3, and erase the edges whose assigned number is below the threshold. If a dependency relation is included in sentences appearing more than the threshold, it will remain in the graph.
5. Read out the sentences from the resultant graph.

Figure 1 illustrates an example of the output of the dependency-link method. The arc between “可能性—ある” (possibility – exist) and “ある—こと” (exist—that) exists, since these dependency relations share the word “ある” (exist) and co-occur in target sentences. In this figure, we can read a Japanese sentence whose meaning is “Because the fact that there is a possibility is revealed.”

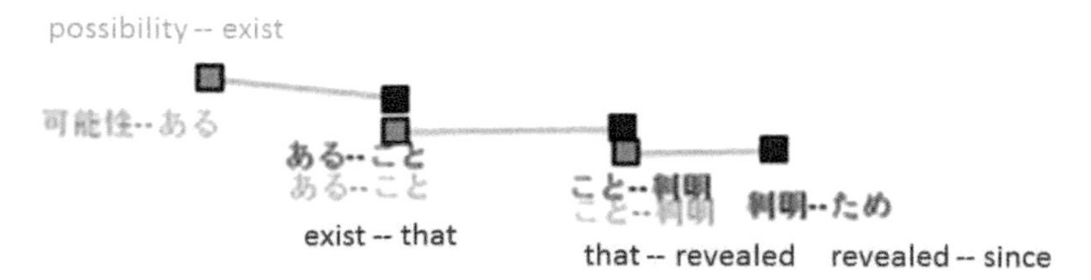

Figure 1 An example of the results of the dependency-link method. We can read a Japanese sentence, “可能性(が)あること(が)判明(した)ため” (Because the fact that there is a possibility is revealed).

3.2 Analysis methods

To see the tendency of descriptions in each class, we applied word-link and dependency-link methods to the data, "Specific health hazards that may appear" with respect to each class. In particular, we focused on comparing the descriptions in Classes II and III, the classification criteria of which are often confused.

Additionally, in order to identify the grounds for manufacturers to judge the class, we extracted the relevant parts of sentences based on keywords, such as "ため" (because), "従って" (therefore) and so forth. Table 2 shows the list of keywords. As shown in this table, we divided them into two categories, since the keywords indicated as "Step 2" are polysemic. For example, though the word "ため" is expected to indicate the part of reasons in our context, it can also indicate the part of purposes. Therefore, we first tried to match the keywords listed in Step 1, and where none of them matched, sought those in Step 2. If multiple keywords matched the words in a target description, we used the one appearing last, because ground statements tend to appear around the end of a description.

Naively, it might be considered that only words depending on these keywords are important. However, the expression to express grounds tends to consist of several words. In this study, we extracted four words in front of the keywords, which are summarized by counting and investigating the pattern of dependency between nouns and verbs.

Table 2 Keywords to extract the ground part in the description

	Keywords
Step 1	従って / したがって /従いまして / したがいまして**(therefore)**
	よって**(thus)**
	以上より / 以上から / このことから / 以上のことから **(as we mentioned)**
	このため / そのため **(because of this)**
Step 2	ので / ため **(because)**
	より **(as)**

4 RESULTS

Figures 2-4 show the result of the dependency-link method applied to the descriptions in each class.

In Fig. 2, we can read phrases or sentences, "in case trouble occurs," "a severe health hazard occurs" and "(We have) not received any health hazard report, because the patients have not used it yet." The last one is a typical phrase appearing in the recall description of blood products. As for blood products, it is usual to stop distribution before they are used. The other phrases suggest that the majority of the description in Class I is related to severe health hazards, which fits the definition of Class I.

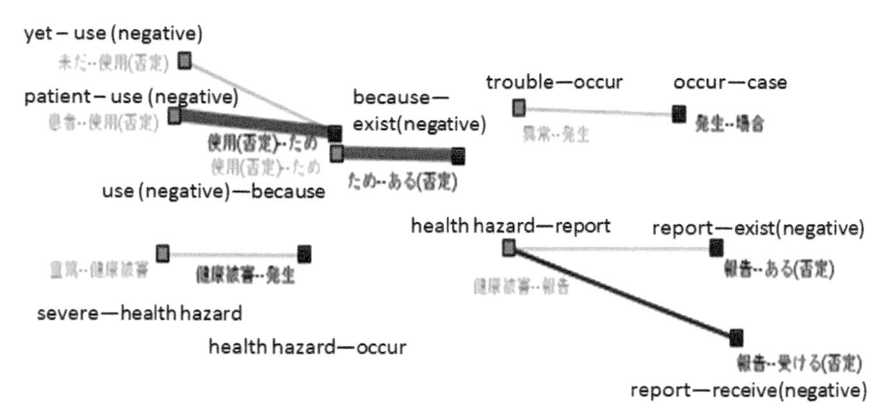

Figure 2 The results of the dependency-link method for the descriptions of Class I (T=9). We can read phrases or sentences, "in case trouble occurs," "a severe health hazard occurs" and "(We have) not received any health hazard report, because the patients have not it used yet."

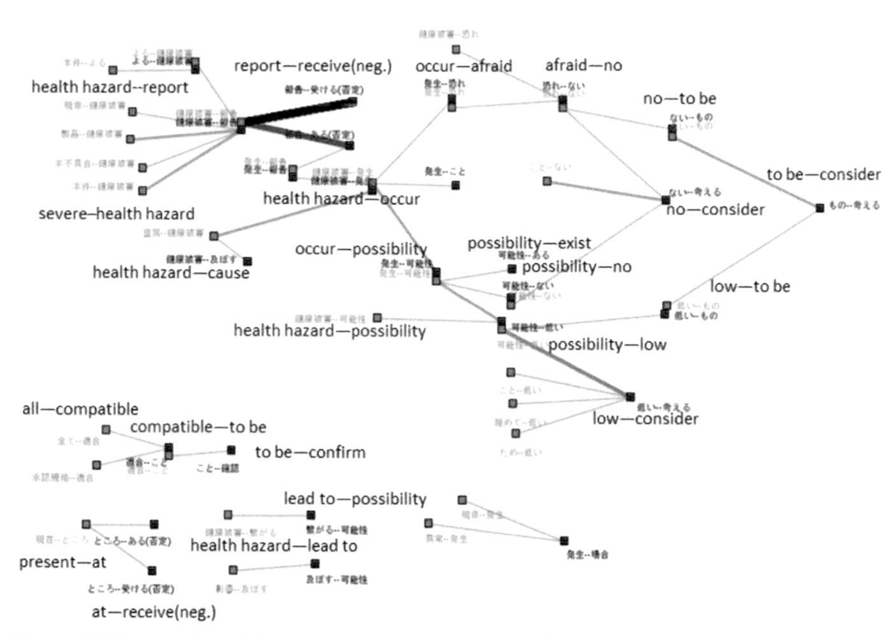

Figure 3 The results of the dependency-link method for the descriptions of Class II (T=59). We can read sentences, not only "We consider that there is no possibility of severe health hazards occurring" but "We consider that there is no or low possibility of health hazards occurring" (the reader should note the omission of the word "severe" in the latter sentence).

Figure 3 shows the result for Class II. While the sentence "We consider that there is no possibility for severe health hazards occurring" is included in this figure, the phrase "We consider that there is no possibility of health hazards occurring" is also described for cases of Class II. The former complies with the definition of Class II. However, the latter states *no* health hazards may occur, which is the definition of Class III. Moreover, we should note that there are also cases involving a *low* possibility of health hazards.

Figure 4 shows the result for Class III. We can read the sentence, "We consider that there is no possibility of health hazards occurring," which complies with the definition of Class III.

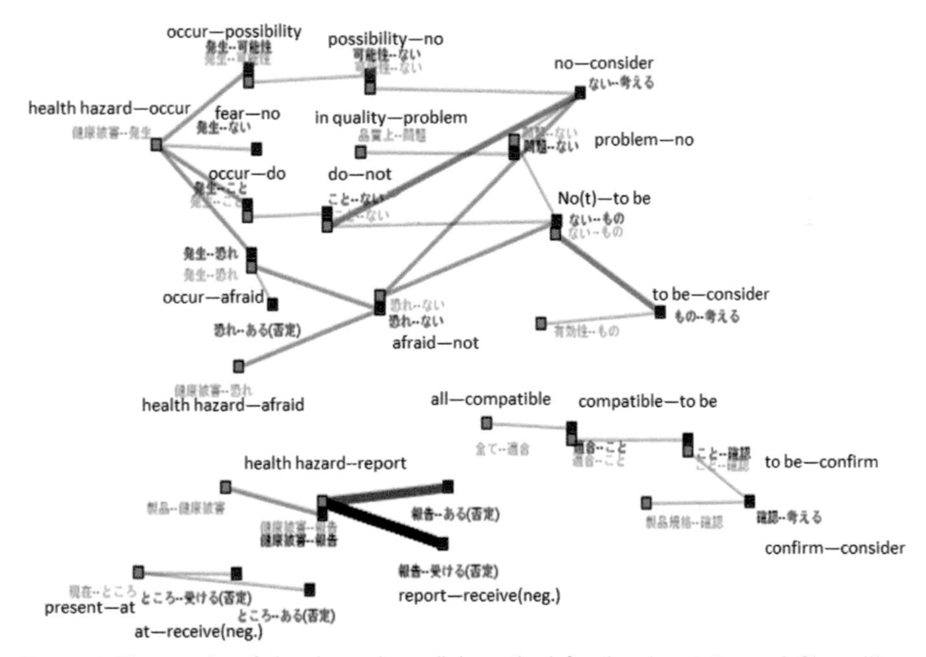

Figure 4 The results of the dependency-link method for the descriptions of Class III (T=50). We can read the sentence, "We consider that there is no possibility of health hazards occurring."

These results show us that some of the descriptions concerning the possibility of health hazards in Class II do not fit its definition. To estimate this quantitatively, we classified the cases in Classes II and III as follows:

- Cases with *no* possibility of *any* health hazards.
- Cases with *low* possibility of *any* health hazards.
- Cases with possibility of *any* health hazards.
- Cases with *no* possibility of *severe* health hazards.
- Cases with *low* possibility of *severe* health hazards.
- Others (including no description of the health hazards)

Table 3 shows the result. As for Class II, the descriptions in 491 cases cite no possibility of health hazards. As we discussed in Fig. 3, these cases should be classified into Class III, while cases with low possibility of health hazards are mostly classified into Class II, which suggests that cases with even low possibility tend to be classified into Class II.

Table 3 shows that the descriptions include information not only about *the extent of health hazards* but also about *the extent of the possibility of hazards emerging*. Originally, the recall classification should be based on the extent of health hazards in the medicine/medical equipment used. The confusion of these two kinds of extent can be regarded as one cause of incorrect classification.

Table 3 The number of cases in Class II/III as classified by the described possibility of health hazards

	Class II	Class III
No possibility of *any* health hazards	**491**	**1050**
Low possibility of *any* health hazards	**495**	**36**
A possibility of *any* health hazards	**4**	**0**
No possibility of *severe* health hazards	**513**	**13**
Low possibility of *severe* health hazards	**216**	**2**
Others	**1226**	**372**
Total	**2909**	**1473**

To identify the grounds used to judge the possibility, we extracted the relevant parts of sentences based on the keywords in Table 2. As a result, we found that the word "できる" (can) frequently appears. This can be regarded as a word which indicates how to handle hazards possibly caused by a recalled product. The word is used in the form "[noun][particle][verb]できる" (e.g. "加湿状態 (humidification state) を確認 (confirm)できる (can)"). We extracted pairs of such nouns and verbs (Fig. 5).

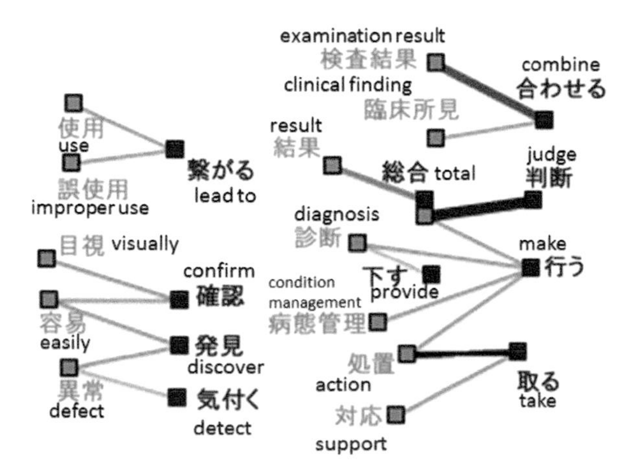

Figure 5 The major noun-verb pairs related to the grounds used to judge the possibility.

The pairs in this figure can be classified into 3 groups as follows:

i. *Possibility of using the recalled product*

The pairs, "使用—つながる" (lead to use) and "誤使用—つながる" (lead to improper use), relate to the use of the recalled product. In fact, referring to original descriptions, we can see sentences such as "Because the defect does not lead to (improper) use, there is no possibility of a hazard occurring." It is obvious, if one does not use the recalled product, the hazard cannot occur and no action is needed. The use of the product is one of the keys to judging the possibility of hazard occurrence.

ii. *Possibility of finding a defect*

We can see pairs, such as "異常—発見/気付く" (discover/detect defects), "検査結果/臨床所見—合わせる" (combine examination results/clinical findings), "診断—行う" (make a diagnosis). This suggests a tendency to report to the manufacturer, since finding a defect early prevents a hazard from becoming severe.

iii. *Possibility of taking action against a hazard*

The pairs, "病態管理—行う" (manage clinical condition), "処置/対応—取る" (take action), suggest the existence of descriptions whereby the action taken against a hazard mitigates the same.

We should notice that these three possibilities are in chronological order. Figure 6 shows the mapping of these possibilities to the players in terms of the product distribution sequence.

This shows that there are three phases: the phase when the recalled products are not distributed to medical institutions and cannot be used by patients (Phase 1), the phase when the product can be used but users can note defects and cease use (Phase 2) and the phase when medical experts or patients take action (Phase 3).

One should remember that the original classification is only based on the extent of health hazards *in case the recall product is used*. However, these results show the need for this additional classification scope.

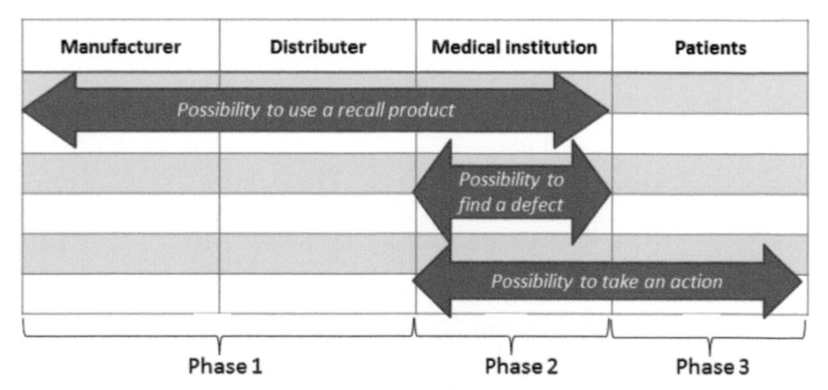

Figure 6 The time-oriented relationships between each possibility and players related to recalls.

5 CONCLUSION

In this study, we analyzed the recall data on medicines and medical devices disclosed by Japanese authorities, and discussed the reason for misclassifications. We focused on descriptive data concerning the expected health hazards, and analyzed them based on a text mining technique, the dependency-link method.

Our results indicate that the descriptions tend to include information not only concerning *the extent of health hazards* but also *the extent of the possibility of hazards emerging*. The confusion of these two kinds of extent can be regarded as one of the causes of incorrect classification. Actually, the possibilities are categorized into three, possibility of using, possibility of finding a defect, and possibility of taking action against hazards.

We proposed a novel classification dimension based on these possibilities. The new classification should prevent unnecessary unease at recalls perceived by consumers.

REFERENCES

Kimura, M, 2009. The Method to Analyze Freely Described Data from Questionnaires. *Journal of Advanced Computational Intelligence and Intelligent Informatics*, 13(3) 268-274.

"Recall information retrieval page," Accessed January 1, 2012, http://www.info.pmda.go.jp/rsearch/html/menu_recall_base.html

"Safety," Accessed January 1, 2012, http://www.fda.gov/safety/recalls/ucm165546.htm

CHAPTER 32

Testing of a Voluntary Patient Safety Reporting System by Think-aloud Protocols: Will a Common Format Help Quality and Efficient Reporting?

Lei Hua, Yang Gong
University of Missouri - Columbia,
MO, USA
gongyang@health.missouri.edu

ABSTRACT

Objective: To conduct think-aloud user testing of a prototypical voluntary patient safety reporting system (VPSRS) that employs a common format of collecting incident details.

Methods: Five subjects were tested on a prototype of VPSRS as they performed a reporting task by think-aloud protocols. The experimenter selected three representative incident cases from the patient fall category for the task, which cover a full scope of questions asked in the common format. An established schema of usability problem categories were applied to code and analyze the transcribed verbatim of the think-aloud reports. A follow-up interview was conducted to supplement subjects' further comments about the prototype.

Results: Subjects made 21 comments on interface problems that were eventually coded into 9 usability problem categories. The most frequently reported problem categories were of "understand system instructions/message" and "expectation on data integrity" from questions of a standardized common reporting format. The subjects also identified usability difficulties in using the system in terms of blocked data entry, system consistency and navigation, layout and screen organization, visibility of system status, graphics and overall ease of use. In addition, the learnability on new interface widgets presented discrepancy between subjects.

Conclusion: The high level of usability problems could affect the effectiveness and efficiency of using the VPSRS, especially on answering the questions in the common format that often require heavy cognitive workload for recalling case details. The think-aloud protocols are able to identify these problems on the prototypical system, and come with power to instruct a better design for ease-of-use and quality reporting according to the protocols.

Keywords: Usability, Think-aloud, Common Formats, Patient Safety

1 INTRODUCTION

A recently published study in 2010 by the Department of Health and Human Services (DHHS) Office of Inspector General (OIG), claims 1 in 7 Medicare patients were injured or killed by health providers as a consequence of medical errors, costing around 4.4 billion taxpayer dollars in 2009 (Levinson, 2010). To manage the errors for life and money saving, a voluntary patient safety reporting system was widely suggested by the Institute of Medicine (IOM) in order to facilitate a culture of incident reporting, effective learning from mistakes, and improvement on patient safety.

Although this approach has been examined to be effective in improving safety across fields of aviation and nuclear facilities, its effectiveness in the medical field is still being challenged by underreporting and low quality of reports (Leape, 1994). The underreporting was estimated in a range from 50% to 96% (Kim and Bates, 2006), and attributed to prevalent human factors in terms of time pressure, fear of punishment and perceived usefulness. From reports voluntarily submitted by a computerized reporting system, e.g. a Patient Safety Network (PSN) – implemented across the University of Missouri Health Care System (UMHC), one of our previous studies identified several severe quality problems: nearly half of the reports (2,735 of 5,654) were removed due to incompleteness or duplication; over 50 typos were identified on date fields that serve as key attributes to locate reports by case reviewers; over 25% of valid reports were mislabeled; and case details were dominantly in narrative format with largely varied levels of quality (Gong, 2009) .

To the best of our knowledge, though a better design of user interface was unable to deal with all of the identified non-technical and technical issues, it is promised to improve effective use of the system if human factors on system safety, ease of use, and perceived usefulness could be properly addressed (Leape, 2002). In this study, we applied think-aloud user testing to identify users' cognitive reflections on these concerns as they interact with a prototypical system to complete incident reports, and look deeply into the strength and weakness of a common standardized reporting format. The findings are helpful to remove users' cognitive difficulties toward a more efficient and effective prototype and pilot more compressive use testing experiments.

2 BACKGROUND

University of Missouri Health Care (UMHC) is one of the most comprehensive health-care networks in Missouri, with approximately 500 staffed beds and 19,000 patient admissions annually, offering the finest primary, secondary and tertiary health-care services. In 2001, UMHC developed a web-based voluntary reporting system, the Patient Safety Network System (PSN), which collects errors, adverse events, and near misses reported from five facilities located in mid Missouri (Kivlahan et al., 2002).

The tested VPSRS prototype applied a technique of two dimensional design as recommended by Nielsen based on usability evaluations of the PSN system (Nielsen, 1994). The prototype remained similar to the layout of the PSN. Some problematic features such as invalid hyperlinks on the PSN navigational bar were fixed, and some new features such as mouse-over tips, autocomplete and shortcut buttons were added. It was narrowed down to one specialty in reporting patient fall incidents with in-depth features for quality improvement of reporting case details. Once we have the development and evaluation of fall reporting functionality successfully done, we would expect to expand prototyping on the other incident categories e.g. lab, treatment, etc. The prototype had its first version in specialty of patient falls in September 2009. Later that year, a heuristic evaluation was employed to inspect induced usability violations in this version (Hua and Gong, 2010). The second version of the prototype had all of the identified severe violations fixed in 2010, and was the object tested in this study.

The biggest difference between interfaces of the PSN and prototype is on the screen of collecting case details. We replaced the original page design and contents with a standardized reporting form in the Common Formats (first official version was released in a PDF format in March 2010) (AHRQ, 2010). It was developed and determined by the Agency for Healthcare Research and Quality (AHRQ) to diminish disparity of tracking, categorizing and describing incidents among the existing VPSRSs, and serve the aggregation at the national level. Having this standardized piece of work implemented in the tested prototype made the study able to identify the problems of the Common Formats and examine its usefulness for reporting quality improvement, prior to being widely adopted in any specialty fields for incident reporting purposes.

For any clinical IT system, usability evaluation is a method for ensuring the system design fits users and their tasks (Norris, 2009). The standard definition of usability is "the extent to which a product can be used by specified users to achieve specified goals with effectiveness, efficiency and satisfaction in a specified context" (ISO9241, 1998). Usability of an IT system can be defined as its ease of use, ease of learning, effectiveness, efficiency, minimal error-proneness and subjective pleasantness (Nielsen, 1994). For VPSRS, usability is the extent to which the end users can easily learn and report an incident completely and accurately in a minimum amount of time without any errors and frustration. Our study is to evaluate this ability of the prototype through testing by real users. Think-aloud protocols were widely used for system testing experiments as they can provide insight into user's task performance and interaction with the tested system for usability enhancement (Ericsson and Simon, 1985).

Voluntary Medical Incident Reporting

Report incidents
New Incident Report
Report Info
Rate Harm Score
General Info
Event Details
Fall
Equipment/device
Event Summary
Custom Layouts
Data Display

Initial Questions

Health Profession*: RN
Anonymous Report
Facility*:
University Hospital (UH)
University Physician Clinics (Clinics)
Ellis Fischel Cancer Center (EFCC)
Columbia Regional Hospital (CRH)
Missouri Rehabilitation Center (MRC)
Patient Involvement*:
Reached Patient (Either harm or no harm events)
Close Call/Near Miss Event (Caught before reaching patient)

Rate Harm Score, 0 ~ 5 (low to high severity)

0 - No clinical changes/no apparent injury to patient, No additional tests ordered or evaluation needed, N
1 - Vital signs monitored, Single test ordered due to event, Brief observation in clinic after event, Minimal
2 - Vital signs changed, More than one test ordered due to event, Moderate level therapy/intervention r
3 - Transfer to higher level of care for observation, hospitalization or surgery, Significant level or unplanne intervention needed due to event, Altered LOC

Patient Information

Medical Record#*: 1596839
First Name*: sasa
Last Name*: lala
Middle Name:
Birthday*: 09/22/1970
Gender*: male female

Event Information

Event Date*: 09/26/2010
Yesterday
Today:09/26/2010 - 14:59:03
Patient Home Unit*: 5 east oncology unit
Attending Physician*: Perry Peggy
Resident/Fellow*: Amaya Paul
Staff involved: add

Event type*

Blood/Blood Products
Medication Reaction
Medication/IVs
Procedure/treatment
Skin Impairment
Equipment/device
Fall
Miscellaneous

Fall Event Details, please answer question one by one.

Fall Details
Was the fall unassisted or assisted?
Unassisted
Was the fall observed?
Yes
Who observed the fall?
Staff
Did the patient sustain an injury as a result of the fall?
Yes
What type of injury was sustained?
Dislocation
Fracture
Intracranial injury
Laceration requiring sutures
Other(PLEASE SPECIFY):
Prior to the fall, what was the patient doing or trying to do?
Toileting-related activities
Prior to the fall, was the patient assessed to be at risk for a fall?
Yes
No

Patient Information (#ext-record-1)[Edit|Save]

Medical Record #: 1596839
Patient Name: lala,sasa
Gender: female
Birthday: 09/22/1970

Event Information[Edit|Save]

Event Happened Date: 09/26/2010
Patient Home Unit: 5 east oncology unit
Harm Score: 0
EventType: Fall

Event Details[Edit]

Patient Fall Details
Was the fall unassisted or assisted?
Unassisted
Was the fall observed?
Yes
Who observed the fall?

Submit
New Report

Figure 1. The collage of user interface screenshots of the prototype

3 METHODS

Subjects

The invitation for subject recruitment was emailed to nursing and health informatics students under the School of Medicine at the University of Missouri. All recipients were in one user group (health students) classified in the PSN for incident reporting. The responders that had no experiences with both the current testing prototype and the PSN system were selected as candidates. Eventually, the top five qualified subjects were selected for the study. Three of them are nursing students and the other two are doctoral students in health informatics.

Procedure

The five subjects were assigned with separated time periods for the testing. The entire on-site session on took around 50 minutes per subject. An experimenter briefly introduced the system and delivered a training video clip (around 4 minutes in length) to subjects about how to manipulate the five screens (as shown in Figure 1) of the testing system for an incident report. The five screens consist of "Initial Questions", "Rating Harm Score", "General Info", "Event Details" and "Event Summary". The required fields on screen are attached a red asterisk as shown in Figure 1. The improved features contrasting to the PSN are: including the Common Formats questions; an auto-complete function to long pull-down lists; and shortcut buttons as event date pickers (Hua and Gong, 2011).

The task given to the subjects was to report three patient fall cases that were written on paper in a fixed order by the system. Each case was selected from a library of 346 patient fall reports and represented respectively one of three case completeness categories defined in a previous study(Gong, 2010). The case details were memorized by the subjects prior starting the report. The subjects were allowed to review patient general information only while reporting as illustrated in Figure 2. All new functions as aforementioned were tested out in this process.

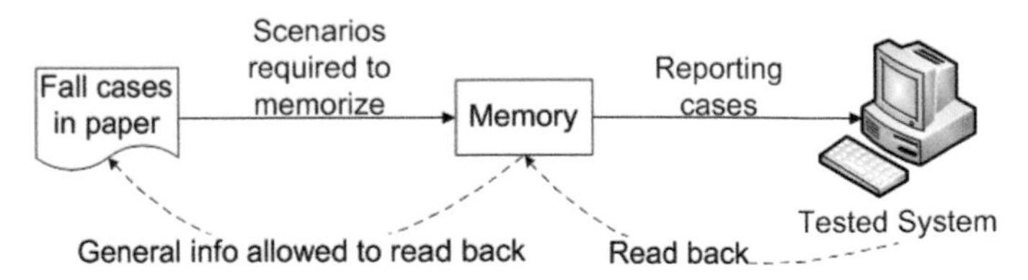

Figure 2. Only generic information of given cases were allowed to be read while starting to report

Subjects were asked to verbalize their thoughts as expressions of think-aloud as they interact with system. The think-aloud verbalization and follow-up interview were audio taped for later transcribing, coding and analysis. The computer screen was recorded as subjects performed the task. During the analysis, the experimenters were able to review all videotapes to scrutinize and annotate usability problems, and roughly estimate the average time expenses on all five screens.

At the end of the session, subjects were interviewed about their overall attitudes to the system. Any further comments or open discussions were encouraged and transcribed as supplement for think-aloud reports. The study is IRB approved and every subject signed a required IRB consent form.

Data Analysis

The analysis of the data was based upon transcribed verbatim of all "think aloud" reports and interviews. All of these reports were annotated and coded to categories that indicate usability issues encountered by subjects during the sessions. For the above steps we adopted the coding schemes described by Kushniruk & Patel (Kushniruk et al., 1996, Kushniruk and Patel, 2004), with a new added category of "expectation on data integrity". Any minor disagreements on annotation and coding were resolved by discussions among team members until reaching a full agreement. The execution times for reporting activities were estimated at the screen level. The typos and inconsistent answers on case detail questions were manually reviewed to interpret identified usability problems in a reporting quality perspective.

4 RESULTS

All five subjects completed the tasks based on the protocols. It showed a good capacity in manipulating user interface of the system to complete three incident reports. The consistency of their answers that relied on recall efforts was varied by question's complexity and sufficiency of contextual information. The follow-up interview reflected overall satisfaction from the subjects on ease of use and easy to learn. Some subjects even expressed their favors to the system and experiment spontaneously right after the testing session. The study, as expected, exhibited multiple critical usability issues through the transcribed reports, which were mainly associated with the common reporting form and the reporting quality.

Table 1. Frequencies of coded problems identified in the transcribed reports

Problems	Subjects					Total	Subjects in problems
	#1	#2	#3	#4	#5		
Understanding instructions/message	3	1	1	3	1	9	5
Expectation on data integrity	2	2	2	2	1	9	5
Comments on overall ease of use	1	1	0	1	0	3	3
Data entry blocked	0	2	0	1	0	3	2
Comments on consistency	0	0	0	1	1	2	2
Comments on navigation	0	0	0	0	2	2	1
Layout/Screen organization	0	1	0	0	0	1	1
Visibility of system status	0	0	0	0	1	1	1
Graphics	0	0	0	1	0	1	1

Each of the reported problems fell into at least one best-matched category under agreement of both the authors, with categorical title and annotations to identify its context. The frequency of coded problems is given for all five subjects in Table 1. The problems that were most frequently coded are "understanding system instructions/message" and "expectation on data integrity" (with all five subjects having both of these problems). The next most coded category is the overall ease of use (three of subjects' comments fitted in) that fills uncovered gaps by the other categories. The other categories also indicate several usability problems with good insight, such as data entry blocked, system consistency and navigation, layout &

screen organization, visibility of completed data and graphics. Although subjects shared a high agreement on the top two most frequent categories, they interpreted differently the extent to which the indicated subjects' training background might have impacts on their perceptions in detail.

The subjects' execution time on all five system screens are given in seconds in Table 2. As can be seen, the subjects reported an incident in an average time of 5 minutes 48 seconds. Two screens, "Rate Harm Score" and "Event Details" highly depended on subjects' memory and took over 50% of reporting time. There were two typos on date fields (patient birthday and event date) identified within the total of fifteen reports.

Table 2. Average times for reporting on screens

Screens	Aver. Time of Reporting	Percent
Initial Questions	35	10%
Harm Score	35	10%
General Questions	102*	29%
Event Details	147	42%
Event Summary	30	9%
Total	348	100%

* The screen on which the typos occurred

5 DISCUSSION

In this study, we adopted think aloud protocols as the major method for usability testing, and supplemented by a post-testing interview. It helped to find fairly distinct sets of usability problems from a previous heuristic evaluation on prior version of the prototype (Hua and Gong, 2010). The transcripts and taped video materials allowed us to examine user-system interaction in a straight-forward and retrospective way to confirm users' operational and cognitive difficulties in an on-going reporting scenario, especially from the common reporting form, in which subjects must recall and infer upon givens and knowledge in memory to accept or reject possible options. As a result, the coded reports served as protocols to scrutinize usability factors that contributed to reporting efficiency and quality, and instruct the optimal solution as correspondences.

To our best knowledge, the study has been the first time to investigate users' reflection in reporting by using the AHRQ Common Formats. It also provided a preliminary outcome of whether the Common Formats would improve reporting quality and efficiency and have new problems incurred.

As the most frequent cognitive difficulties show in the table 1, the problems on "understanding system instructions/message" and "expectation on data integrity" were dominantly manifested when subjects answered questions in this common reporting form. Each of them encountered difficulties in understanding or selecting response items at least once in the session. They complained on unclear or overlapped meaning of terms and phrases applied in options which cost considerable time to determine a proper answer. We believe such language clarity problems impeded accurate answers and prolonged completion times during the testing sessions.

On the other hand, subjects expressed high expectation on information integrity supported by the prototype. All subjects believed it would be extremely helpful, if the prototype was able to ask adequate questions in a hierarchy and collect incident related data completely. During the interview, subject #2 even argued she would not recommend the system to her colleagues just because the level of details upon domain specific questions was inadequate, though the system was overly easy to learn and use. She worried that the insufficient reporting quality would eventually force her to document everything again to anther parallel system e.g. EHR, which was considered as a major barrier to system acceptance (Levinson, 2008).

Average time of a report completion was 5 minutes 48 seconds in this study. Comparing to the existing literature, an incident report with multiple sections combining coded fields and free text entries usually requires around 10 minutes (Mekhjian et al., 2004, Holzmueller et al., 2005, Levtzion-Korach et al., 2009). Although it is hard to justify which reporting system is faster than the other since the context and testing methods differ across these studies, the time span in our study was shorter nonetheless.

There were two typos on date values identified after reviewing the testing case submissions. They were in date fields of patient birthday and event occurrence time, e.g. subject #3 failed to realize the utility of shortcut buttons and did not use them to enter yesterday's date given by author. Instead, she manually transcribed the date value from the paper to the prototype and made a typo on last digit of the year. Consequently, the subject submitted an incident report as if it happened one year ago, and it is impossible to identify such a mistake by merely reviewing reported data. Therefore, we suspect the number of typos (1.5%) previously identified on this field of 2,919 historical incident reports was severely underestimated(Gong, 2009). More importantly, serving as a key attribute for case reviewing, large amounts of typos on this time field would possibly undermine the results of root cause analysis, further impairing the effectiveness of the reporting system. In the newest version of the prototype, we have implemented a date checker to alert reporters if the interval between reporting time and event time is more than seven days (85.7% of 2,919 reports were reported within seven days after occurrence (Gong, 2009)).

The other coded problems were general usability issues about navigation, page layout, and consistency. A few of them were unfixed problems in the last version, and some were newly introduced due to system modification. Overall, none of them were a catastrophic problem to block reporters out or contribute low quality reports.

6 LIMITATIONS AND FUTURE WORK

We conducted a usability test on a computerized prototype of voluntary patient safety reporting with patient fall cases. The findings and discussion presented in this study have their limitations on the specific domain and might not be completely generalizable to other systems development or even the reporting of other incident categories that vary in terms of detection and reporting emphasis.

The recruited subjects, though met the requirement in quantity (5 subjects) for a usability testing experiment (Nielsen, 1994, Nielsen, 2000, Lewis, 2006), were all novices and not representative of all potential voluntary reporters. Moreover, whether thinking aloud has effects on subjects' task performance was not investigated. The differences of subjects, such as background, culture, language proficiency, in particular for their verbal facility, could be contributing factors to this concern. The other option can be to recruit more subjects or have a training of verbalization before the study.

In the future study, we will recruit more subjects from different user groups to consolidate findings and identify highly distinct group types. In fact, a separated experiment was already done to investigate physical and mental efforts of reporters by a retrospective think-aloud method as they interacted with system. The results through these completed and future studies are expected to impel better designs on VPSRSs to improve quality and efficient use while reducing user's efforts.

7 CONCLUSION

The inclusion of a reporting common format is very possible to increase the level of details and efficiency in reporting, but the interface usability problems might lower this probability. The think aloud protocols is a useful method to identify these usability problems with real users in prototype. The results of analysis can benefit the following design and iterative refinement of the target system to fix the problems and improve user performance on the Common Formats and reporting.

ACKNOWLEDGMENTS

This project was in part supported by the Richard Wallace Research Incentive Award at the University of Missouri. Thanks to Mathew Koelling for proofreading of the manuscript.

REFERENCES

Ahrq. 2010. *AHRQ Common Formats Version 1.1: Event Descriptions, Sample Reports, and Forms* [Online]. Available: https://www.psoppc.org/web/patientsafety/version-1.1_documents#fall.

Ericsson, K. A. & Simon, H. A. 1985. *Protocol analysis : verbal reports as data,* Cambridge, Mass., The MIT Press.

Gong, Y. 2009. Data consistency in a voluntary medical incident reporting system. *J Med Syst,* 35, 609-15.

Gong, Y. Terminology in a voluntary medical incident reporting system: a human-centered perspective. Proceedings of the 1st ACM International Health Informatics Symposium, 2010 Arlington, Virginia, USA. 1882996: ACM, 2-7.

Holzmueller, C. G., Pronovost, P. J., Dickman, F., Thompson, D. A., Wu, A. W., Lubomski, L. H., Fahey, M., Steinwachs, D. M., Engineer, L., Jaffrey, A., Morlock, L. L. &

Dorman, T. 2005. Creating the web-based intensive care unit safety reporting system. *Journal of the American Medical Informatics Association,* 12, 130-9.

Hua, L. & Gong, Y. 2010. Developing a user-centered voluntary medical incident reporting system. *Studies in health technology and informatics,* 160, 203-207.

Hua, L. & Gong, Y. Identifying and addressing effectiveness in a user-centered design of voluntary medical incident reporting system. 13th IEEE International Conference on e-Health Networking Applications and Services (Healthcom) 2011. 358-363.

Iso9241 1998. Ergonomic requirements for office work with visual display terminals (VDTs) -- Part 11: Guidance on usability. International Organization for Standardization.

Kim, J. & Bates, D. W. 2006. Results of a survey on medical error reporting systems in Korean hospitals. *Int J Med Inform,* 75, 148-55.

Kivlahan, C., Sangster, W., Nelson, K., Buddenbaum, J. & Lobenstein, K. 2002. Developing a comprehensive electronic adverse event reporting system in an academic health center. *Jt Comm J Qual Improv,* 28, 583-94.

Kushniruk, A. W. & Patel, V. L. 2004. Cognitive and usability engineering methods for the evaluation of clinical information systems. *J Biomed Inform,* 37, 56-76.

Kushniruk, A. W., Patel, V. L., Cimino, J. J. & Barrows, R. A. 1996. Cognitive evaluation of the user interface and vocabulary of an outpatient information system. *Proceedings/AMIA Annual Fall Symposium*, 22-6.

Leape, L. L. 1994. Error in medicine. *JAMA,* 272, 1851-7.

Leape, L. L. 2002. Reporting of Adverse Events. *N Engl J Med,* 347, 1633-1638.

Levinson, D. R. 2008. Adverse events in hospitals: overview of key issues. Washington, DC: US Department of Health and Human Services, Office of the Inspector General

Levinson, D. R. 2010. Adverse events in Hospitals: national Incidence among medicare beneficiaries. Washington, DC: US Department of Health and Human Services, Office of the Inspector General

Levtzion-Korach, O., Alcalai, H., Orav, E. J., Graydon-Baker, E., Keohane, C., Bates, D. W. & Frankel, A. S. 2009. Evaluation of the contributions of an electronic web-based reporting system: enabling action. *Journal of patient safety,* 5, 9-15.

Lewis, J. R. 2006. Sample sizes for usability tests: mostly math, not magic. *interactions,* 13, 29-33.

Mekhjian, H. S., Bentley, T. D., Ahmad, A. & Marsh, G. 2004. Development of a Web-based event reporting system in an academic environment. *Journal of the American Medical Informatics Association,* 11, 11-8.

Nielsen, J. 1994. *Usability engineering,* San Francisco, Calif, Morgan Kaufmann Publishers.

Nielsen, J. 2000. *Why You Only Need to Test with 5 Users* [Online]. Available: http://www.useit.com/alertbox/20000319.html 2011].

Norris, B. 2009. Human factors and safe patient care. *Journal of nursing management,* 17, 203-211.

CHAPTER 33

The Study of Drug/Medical Device Recall Data (II) Analyses on Recall Data in United States

Tomoyuki Nagata, Ryo Okuya, Masaomi Kimura, Michiko Ohkura, Fumito Tsuchiya

Shibaura Institute of Technology
3-7-5, Toyosu, Koto City, Tokyo, Japan
{l08086, ma11044, masaomi, ohkura}@shibaura-it.ac.jp

International University of Health and Welfare
2600-1, Kita-Kanamaru, Otawara City, Tochigi, JAPAN

ABSTRACT

In order to improve patient safety in cases of drug recall, we studied drug recall data disclosed by the Food and Drug Administration (FDA). Since the FDA does not disclose data in the form of a database, we focused on enforcement reports published by the FDA, which are less structured and contain less information. This suggests that it is difficult to identify necessary information from the reports. For this reason, we constructed a database storing drug recall data in the appropriate format. The data sources are enforcement reports and press releases, the latter of which are published by pharmaceutical companies. In this paper, we discuss the problems faced during construction of the database.

Keywords: Recall, Medicine, Text mining, Classification

1 INTRODUCTION

Drugs and medical devices that are found to have defects or problems are recalled. If recalled products have the potential to cause high-risk adverse events, notification needs to made quickly and accurately. For this reason, the

Pharmaceuticals and Medical Devices Agency (PMDA) in Japan and the Food and Drug Administration (FDA) in the U.S. have formularized a recall case classification system that classifies recall cases into the following three classes:

- Class I: cases in which the product can cause serious adverse health consequences or death
- Class II: cases in which the product can cause temporary or medically reversible adverse health consequences
- Class III: cases without risk of adverse events

This makes it easy for us to find high-risk recall information.

Yoneshige et al. pointed out that recall cases disclosed by the PMDA have not been correctly classified into these classes. For example, though the case saying "there is no possibility of adverse events" should be classified into Class III, it was actually classified into Class II. In order to find a solution to this problem, they investigated recall cases of drugs and medical devices disclosed by the PMDA. The reported recall cases are shown in Table 1. Since the recall case classification classifies cases based on the extent of adverse events in use of the recalled product, Yoneshige et al. focused on the sentences describing adverse events that are stored in the column "possible adverse events." They found that the possibilities of adverse events tend to be described more than the specific adverse events themselves. They classified the description of possibilities and found that three possibilities of using a recalled drug can be described, which correspond to the stage of distribution; *the possibility of not being used* before coming to market, *the possibility of medical experts noticing defects* before use and *the possibility of cure* after use. Based on this, they proposed a new classification system by adding the stage of production use to the original system.

The classification system formularized by the PMDA originates in the one defined by the FDA. Fukuda et al. investigated the data in the medical device recall case database disclosed by the FDA, and compared their results with the results of Yoneshige's analyses. They found that the database did not have an attribute that is equivalent to the "possible adverse events" in the PMDA database. Instead of this attribute, they analyzed the information in the attribute "recall reason" and found it also had little information about possible adverse events and the possibilities of use of recalled products. They proposed the attribute that the database should have to store the assessable data for the classification system proposed by Yoneshige.

As we mentioned, the following recall data have already been analyzed in previous studies:

- The recall cases of drugs and medical devices disclosed by the PMDA,
- The recall cases of medical devices disclosed by the FDA.

However, the drug recall case data disclosed by the FDA have not been analyzed. We need to analyze these data, since we expect the results to improve recall case reporting systems and their databases in the PMDA and the FDA.

In fact, the FDA does not disclose the drug recall case data as a database; they provide the data as documents called enforcement reports. Pharmaceutical companies issue press releases on recalled drugs, if they can cause a severe adverse event. In this paper, we first report the analyses to discuss how to extract data to be stored in a drug recall case database. We also evaluate the data in the database by discussing keyword distribution over the classes, under the assumption that the keywords representing the extent of adverse event severity should appear in the sentences in only one class.

Table 1 Attributes in the recall case table disclosed by the PMDA

The names of attributes
Brand name and generic name
Subject lot number, quantity and shipment time
Manufacturers
Recall reason
Possible adverse events
Start date of recall
Efficacy and applications
Other
Contact

Table 2 The table attributes of the medical device recall cases disclosed by the FDA

The names of attributes
Recall Number
Trade Name / Product
Recall Class
Posted Date
Recalling Manufacturer
Reason for Recall

2 TARGET DATA

Our target data is the drug recall data called enforcement reports and press releases disclosed on the FDA website. Enforcement reports are weekly reports that contain information on the action taken for recalled products such as drugs, medical devices, foods and so forth. Press releases are documents issued by pharmaceutical companies that contain recall information related to a product that can cause a severe adverse event.

Table 3 shows data attributes defined in enforcement reports. Although they provide information on recall cases in any class, they only have a small description for each attribute. We should note that there is no attribute to describe the extent of adverse events, which has a relationship to grounds of classification.

By contrast, press releases have no definite attributes in them, since they are composed of sentences. This causes some difficulties in identifying the included information. In addition, press releases generally provide information on recall cases in Class I. This means that they provide data on fewer recall cases than enforcement reports. However, press releases are still useful, since some of them include information that is not included in enforcement reports, such as the extent of adverse events.

Table 3 The data attributes in enforcement reports

Item
PRODUCT
CODE
RECALLING FIRM/MANUFACTURER
REASON
VOLUME OF PRODUCT IN COMMERCE
DISTRIBUTION

First, we defined the attributes of recall information required to compose a database that stores recall case data.

Referring to the attributes of descriptions in enforcement reports, we included basic data such as "product name," "lot number" and "recall reason" in the attributes in the database. We also included "adverse events" and "stage of distribution," since recall cases are classified currently based on the extent of adverse events and should also be classified by the possibility of product use as suggested by previous researches. In order to inform consumers of a product risk, we need the data on "the efficacy of a product" and "Action that consumers should take" to avoid adverse events. Table 4 shows the attributes that the database should have.

Table 4 The attributes of the proposed database

Product name	Lot number	Stage of distribution	Recall reason
Adverse event	Action that consumers should take	Efficacy of product	

Second, we obtained data corresponding to the attributes from enforcement reports and press releases.

Enforcement reports have data on product names, lot numbers and recall reasons. Since enforcement reports are in the form of HTML, we obtained HTML source data from the FDA website and extracted descriptions by pattern matching.

In contrast, press releases are composed of sentences. Obviously, it is not straightforward to identify to which attribute the description belongs. Therefore, we utilized text classification to identify the attribute of the description. We used the Naive Bayes classifier and neural network method as text classification techniques. The Naive Bayes classifier classifies texts based on the word frequency therein. The class $\hat{C}$ to which the text belongs is given by the equation:

$$\hat{C} = \arg\max_{c} P(C) \prod_{i=1}^{n} P(W_i|C) \quad , \qquad (1.)$$

where C is a category of classification, W_i is the words in the text and n is the total number of words in the text. Naive Bayes classifier can be implemented relatively easily, but has high accuracy. However, Naive Bayes classifier does not take the co-occurrence of words into account, because it assumes independence between words.

Thus, we also analyzed the sentences by means of a neural network, which is a simplified method that simulates how the human brain calculates. The function of each neuron is represented by the numerical expression as follows:

$$f = \frac{1}{1+e^{-h}}, \ h = \sum_{k=1}^{K} w_k x_k - \theta \, , \qquad (2.)$$

where x_k denotes an input value, w_k is a weight, and θ is a threshold. A neural network can also be used as a classifier, if we regard the presence (or absence) of the word as input data and the classes as output. In this study, the input data on the neural network is a set of the presence (or absence) of the words in the sentence that appear in each category frequently, and the output is the attribute to which the sentence belongs.

We verified whether these methods correctly identified the class of sentences in press releases. The training data were 35 press releases (1030 sentences) in 2010, and the test data were 20 press releases (649 sentences) in 2011. We compared the results for the test data to the correct classification performed by hand to evaluate the techniques.

Finally, we investigated the problem of drug recall classification in the FDA based on the implemented database. We investigated whether the cases were classified correctly by reference to the description of adverse events. We supposed that similar cases would be classified in the same class, and verified the validity of the classification by examining whether similar cases are actually classified in the same class.

4 RESULTS

Firstly, we classified the sentences in press releases in 2011 into data attributes as a recall drug database by Naive Bayes classifier. During confirmation of the results, we found that proper nouns such as company name appear in some sentences that should be classified into the efficacy of the product and *what consumers should do*. However, these proper nouns were mostly included in statements classified into *other*. This fact suggests that the proper nouns potentially make sentences classified into *other*. Therefore, we excluded proper nouns by tagger, which is one of the functions of NLTK.

Table 5 Classification results by Naive Bayes classifier

Categories	Precision (%)	Recall (%)
Lot Number	14.3	9.1
Recall reason	48.0	52.2

Adverse event	60.7	65.4
Efficacy of product	66.7	30.0
What consumers should do	75.0	27.3
Other	88.0	93.7
Average	58.5	46.3

Table 5 shows the precision and recall of classification results by Naive Bayes classifier. The averages of precision and recall were low. In particular, except for *other*, the values of recall tend to be much less precise. This indicates that the target sentences were not classified correctly.

Table 6 Classification results by neural network

Categories	Precision (%)	Recall (%)
Lot Number	25.0	9.1
Recall reason	72.7	29.6
Adverse event	66.7	53.3
Efficacy of product	75.0	14.3
What consumers should do	86.7	34.2
Other	86.1	97.5
Average	69.0	32.0

Table 6 shows precision and recall of classification results by neural network. We can see that the values of recall were as low as the ones for the classification by Naive Bayes classifier.

From these results, we found that the classifications by both methods do not lead to positive results. Since the methods are based on the appearance of words, this shows that there are no characteristic words that assign press release sentences into the categories. This indicates that it is difficult to automatically classify each sentence in press releases by means of text classification techniques.

Therefore, we manually extracted recall data from descriptions in press releases additional to extracted information from sentences in enforcement reports automatically by a program. By storing these extracted data, we created a drug recall case database.

After implementation of the database, we investigated it to see the current status of recall case classification. Unfortunately, we do not have data corresponding to the attribute "possible adverse events" in the PMDA database. Instead of this attribute, we use the attribute "reason," whose data can contain the information including adverse events. We focused on the key words in the description whose attribute is "reason," and discuss the keyword distributions over the classes, under the assumption that the keywords representing the extent of adverse event severity should appear in the sentences in only one class.

Table 7 The keyword distributions over the classes

Words	Class I	Class II	Class III
Specification	5	48	43
Labeling	6	22	26
Dysfunction	13	0	0
Deviations	1	48	10
Contamination	4	40	1
Sulfoaildenafil	11	0	0
Marketed	45	9	2
NDA/ANDA	45	9	2
Stability	6	60	52
Erectile	13	0	0
Vials	6	25	0
Approved	37	7	1
Assurance	2	30	0
Package	0	3	11
Content	2	5	12
Unapproved	20	7	4
Contamination	2	36	3
Vial	0	11	1
CGMP	1	48	10
Subpotent	3	26	24
Particulate	9	22	2
Assay	1	18	11
Undeclared	16	4	0
Expiration	0	9	24
USP	3	25	22
Impurities/Degradation	1	21	32

Table 7 shows the keywords appearing ten times in the "reason" descriptions. We can easily see that the keywords such as "Dysfunction," "Erectile" and "Sulfoaildenafil" appear in a single class. However the other keywords tend to appear in plural classes. In this table, the keyword "Subpotent" is related to the extent of adverse event severity, since this indicates the weakness of efficacy, which itself cannot cause severe hazards. In this sense, the cases whose descriptions include "Subpotent" should be classified into Class III. Table X shows that some of such cases belong to Class I and Class II. The reporting companies might consider that the weakness of efficacy causes something equivalent to a hazard. However, the disease causes a harmful influence not the weak efficacy. The origin of this gap is the ambiguity of the class definition. The current class definition does not distinguish whether a product itself causes a hazard or does not prevent a hazard.

5 PROPOSAL

Based on the results shown above, we propose improving the drug recall reporting system of the FDA.

- The FDA should implement a drug recall database. In current enforcement reports, the plural cases of different types of products (e.g. medicine, foods, cosmetics) in the different classes are mixed up on a single web page. This mixture hinders effective utilization of the recall data to ensure suitable recall classification. They also lack the attributes, such as possible adverse events and product information, necessary to verify the classification. These attributes should be added to the recall database.
- We should improve the recall classification system. The current classification is based on the extent of hazards caused by product use. However, as mentioned in the section of results, the current class definition does not distinguish whether a product itself causes a hazard or does not prevent a hazard. We should introduce an additional axis to distinguish this difference. In the previous study, Yoneshige et al. proposed an axis representing the possibility of product use. The preferred recall classification should include not only the current axis representing the hazard extent but these additional two axes.

6 CONCLUSION

In order to notify recalled products possibly causing high-risk adverse events, the PMDA and the FDA have formularized a recall case classification system. In order to discuss the status of recall case classification, previous studies have analyzed the recall cases of drugs and medical devices disclosed by the PMDA and the recall cases of medical devices disclosed by the FDA. In this study, we investigated the drug recall case data disclosed by the FDA that have not been analyzed.

Since the FDA does not disclose the drug recall case data as a database, we analyzed enforcement reports disclosed by the FDA and press releases provided by pharmaceutical companies.

In this paper, we first discussed how to extract data to be stored in a drug recall case database. We defined the attributes of the database by referring to the attributes in enforcement reports. We tried identifying sentences in press releases with the attributes by means of text classification techniques, Naïve Bayes method and neural network. The resultant precision was about 60% and recall was about 40%.

This suggests that it is difficult to perform automatic assignment of the sentences to the attributes techniques because of lack of characteristic words peculiar to the attributes in the sentences.

On account of this difficulty, we manually implemented the database. We focused on the key words in the description whose attribute is "reason," and discussed the keyword distributions over the classes, under the assumption that the keywords representing the extent of adverse event severity should appear in the sentences in a single class. We found that the cases whose descriptions include the keyword "Subpotent" belong to any class, though the keyword is related to the extent of adverse event severity. Since keyword "Subpotent" indicates the weakness of efficacy, the cases whose descriptions include it should be classified into Class III. Although reporting companies might consider that the weakness of efficacy causes something equivalent to a hazard, the disease causes a harmful influence not the weak efficacy. We suggest this gap comes from the ambiguity of the class definition.

Based on these findings, we proposed improving the recall case reporting system: the implementation of a recall case database and the introduction of new axes to the recall case classification system.

REFERENCES

Hiromi Yoneshige, Masaomi Kimura, Keita Nabeta, Michiko Ohkura, Fumito Tsuchiya. 2009. A proposal of classification for recalled cases of medicines and medical equipments. *IEICE Technical Report, SSS Safety* 109(177):17-20

Yumi Fukuda, Hirotsugu Ishida, Keita Nabeta, Masaomi Kimura, Michiko Ohkura, Fumito Tsuchiya. 2011. Analysis on medical device recall data submitted to FDA. *Proceedings of the 2011 IEICE General Conference, A-18 Safety*:276-277

"Pharmaceuticals and Medical Devices Agency," Accessed February 24, 2012, http://www.pmda.go.jp/.

"U.S. Food and Drug Administration", Accessed February 24, 2012, http://www.fda.gov/.

CHAPTER 34

The Relation between Task Errors and the Graphic Representation of Information in Medicine Inserts in Brazil

Carla G. Spinillo, Dr; Stephania Padovani, Dr; Cristine Lanzoni, BA

Federal University of Paraná
Curitiba, Brazil
cgspin@gmail.com

ABSTRACT

Medicine administration involves patients' decision-making and cognitive load. In this sense, Patient Information Leaflet (PIL) is the main document for users as it conveys information into tasks. However, most studies have looked at legibility and readability aspects, neglecting the usability of PILs. This paper discusses the results of a study on the effects of graphic presentation of PILs on the simulated use of five medicines, conducted with 60 participants in Brazil, and focusing on their information processing, action and verification errors. The results showed that drawbacks in the graphic presentation of the PILs produced task errors, particularly of dosage verification and manipulation of medicine components. The main conclusion was that the graphic presentation of the Brazilian PILs affected their usability in medicine usage; and therefore may lead treatment of diseases to failure.

Keywords: task error, graphic representation, medicine inserts

1 INTRODUCTION

Patient Information Leaflet (PIL) is a mediator between medicine information and usage. It is a mandatory document for users/patients, which not only inform on the medical aspects of a medicine but also how to use, handle and dispose it after use. The task of taking/using a medicine can be as simple as swallowing a pill, or as complex as preparing an oral suspension medicine. In such cases, patients have to deal with manipulation of objects (e.g., syringes, applicators), measurement of dosage, obtaining the ideal state for medicine consumption, among other tasks. These demand patients to make decisions while undertaking the task of taking/preparing a medicine. In this sense the PIL should enable patients to fulfill their information needs, empowering them to make decisions in health treatment processes. Thus, PILs should not only be understood by patients, but also serve as a support for task performance in medicine usage. In this sense, usability tests should be conducted to verify the effectiveness of PILs with patients, making it possible to access errors on medicine usage, due to drawbacks in the product design (e.g., complex applicators) and/or patients' difficulties in carrying on the tasks. Identifying patients' difficulties and behavior in task performance is particularly relevant, for instance, to decide on the inclusion of warnings in PILs which may prevent errors in medicine administration.

Nevertheless, little has been investigated on usability of PILs with patients, since most studies focus on their readability and legibility aspects (e.g., Andriesen, 2006; Maat and Lentz, 2009; Fuchs, 2010). Despite their contributions to improve PILs', they do not consider the context of medicine usage, and overlook the influence of PILs on task performance. Taking this into account, Waarde (2008) criticized European Union directives for medicine leaflets which is based upon readability testing and design template, neglecting the importance of patients' task performance in medicine administration.

Considering the demand for studies on usability of PILs, this paper discusses the relation between task errors and the graphic representation of information in medicine inserts through a study conducted in Brazil funded by the Ministry of Health. For the purpose of this paper, the context of medicine inserts in Brazil is briefly presented followed by an overview of research on human errors in medicines and on the graphic presentation of PILs, to set the ground. Then, the study carried out in Brazil is presented, highlighting the results on participants' information processing, action and verification errors.

2 MEDICINE INSERTS IN BRAZIL

The production of medicines and their PILs is regulated in Brazil by The Ministry of Health through the National Agency for Sanitary Regulation ANVISA. To commercialize a medicine in Brazil the pharmaceutical industry has to comply with a restrict regulation and go through a number of testing procedures. These involve medicine usage by patients to verify their effectiveness and efficacy as well as the risks they may pose to patients, among other safety measures. However, for the production of the medicine PILs the pharmaceutical industry has only to follow general recommendations regarding mainly the information content of the medicine leaflets, and certain legibility aspects of the text (e.g., type size). As the Brazilian medicine manufacturers do not have to validate their leaflets as part of the medicine licensing process (e.g., readability testing), the PILs regulation in Brazil is only based upon the medical and pharmaceutical views on what and how medicine information should be provided to patients. Thus, it neglects patients' views on and behavior towards medicine administration.

As an attempt to differentiate medicine users, the Brazilian regulation states that the inserts should have separate sections for patients and health professionals. However, the information addressed to patients may not be satisfactorily provided in the medicine inserts, as found in the findings of Gonçalves et al study (2002). They investigated the adequacy of information provided in 168 PILs of 41 medicines regarding their compliance with the ANVISA regulation for PILs and in accordance to specialized literature. They found that 91.4% of the information for patients was incomplete and/or incorrect. They concluded that the lack of regulation and supervision by the government agencies, together with a low level of social control by consumers were responsible for such neglecting situation. Going further, one may wonder what would be the effect of those PILs on the administration of the medicines by patients, and what errors they would lead patients to during task performance. Thus, the adequacy of information in PILs is not only a matter of regulation compliance but also a matter of usability, so as to prevent errors in medicine administration.

3 HUMAN ERRORS IN MEDICINE

According to Reason (1990), error may be defined, in a psychological sense, as a generic term to encompass occasions in which a planned sequence of activities (mental or physical) fails to achieve its desired outcome. Failing to achieve an intended goal may involve both cognitive and motor skills.

Ergonomics and human factor literature comprises a huge series of classification schemes of human error and, still, no universally agreed taxonomy. Within this plurality of approaches it is possible to distinguish three main levels: behavioral, contextual and conceptual. At the behavioral level, errors are classified according to observable features of erroneous behavior (e.g., omission/commission, repetition, injuring). The contextual level goes beyond formal error features and explores assumptions about causality, i. e., what prompts an error to occur at a particular stage of a cognitive or operational sequence. The conceptual level considers the cognitive mechanisms involved in error production. Iida (2005), for example, synthesized the latter approach in a three-category framework where error types are associated to the cognitive structures involved at each stage of the information processing cycle: (a) Perception errors; (b) decision-making errors, and (c) action errors. The latter regards muscular actions, such as incorrect movements and delay in action. Rasmussen (1986) also proposed a cognitive-control framework that is error-oriented. The author distinguishes three levels of performance that correspond to increasing degrees of familiarity with the environment, tools or task: (a) knowledge-based level, (b) rule-based level, and (c) Skill-based level.

Regarding the academic medicine literature, much has been researched on human error identification and prevention. The predominant approach focuses on problems in systems rather than on individual blame (e.g., Kaushaul et al., 2001), since the medical care comprises a complex chain of stakeholders and decision-making processes, which are all subject to error at some point. The specialized literature informs us of recurrent errors on diagnosis/prescription (by physicians or nurses), on medicine dispensing/delivering (when pharmacists and patients are involved) and on medicine usage/administration (e.g., medication preparation, use, storage and discard) in both inpatient and outpatient settings (e.g., Graber et al., 2002; Croskerry, 2003).

In the last decade, several publications addressed different types of medical errors by identifying, classifying and proposing strategies to minimize/prevent their occurrence (e.g., Johnson, 2002). Lane et al (2006) took a step further and applied hierarchical task analysis (HTA) to model drug administration and then systematic human error reduction and prediction approach (SHERPA) to foresee which errors are likely to occur. The authors also put forward design solutions to mitigate the errors encountered. Looking forward to decreasing or preventing medical errors in general, some studies stemmed from previously identified and classified errors and proposed solutions for such problems. Rennie et al. (2007), for instance, developed and tested an illustrated job-aid to be used in rapid diagnostic tests (RDT) in malaria. The results indicate that the provision of clear, simple instructions can reduce errors even in simple diagnostic tests. The authors concluded that preparation of appropriate instructions and training, as well as monitoring of user behavior are essential parts of rapid test implementation. More recently, Sandlin (2008) investigated the factors affecting pediatric medication errors and proposed a series of pediatric specific strategies for medication error reduction. The results showed that children are more prone to medication errors because of the lack of specific packaging and instructions for medication administered to children.

Our review of the literature revealed that medication errors involving health professionals in hospital settings is well researched and documented. In contrast, studies involving patient-committed errors are scarce. Moreover, all the studies reviewed dealt with patient erroneous behavior from another stakeholder's point of view, usually nurses (Meredith et al, 2001; Ellenbecker et al, 2004). For instance, Ellenbecker et al (2004) explored and described medication management for patients receiving services from certified home health care agencies (CHHAs) and nurses. The results showed many medication errors, such as patients taking medications in ways that deviate from the prescribed medication regimen. Results also suggest that patients are experiencing many adverse effects from medication usage errors.

Although the abovementioned studies have contributed to broad the understanding of human errors in medical field, there seems to be a lack of research on the effect of PILs on medicine administration errors, particularly regarding the role played by the graphic presentation of information.

4 GRAPHIC PRESENTATION OF PILS AND MEDICINE USAGE

The graphic presentation of a document has been acknowledged in the literature as a key aspect in message understanding and task performance by readers/users (e.g., Wright, 2003; Ganier, 2001). It regards typographic and pictorial aspects of information, such as visual hierarchy of headings to ease sequencing, word emphasis to aid information searching, space in page layout to promote legibility, and visual instructions to facilitate task performance.

Concerning medicine information for patients, whether in inserts or in labels/packaging, most research has been conducted on the effect of their graphic presentation on comprehension of information and on reading strategy by patients. Sless (2004) in a case study on an OTC-Over the counter medicine found that the employment of design principles in the graphic presentation of information to patients improved testing performance. He tested the actual medicine instructions in the label/package against a redesigned version with 19 participants in Australia. The results of the redesigned version of the label were superior to the original one, proving the contribution of design principles to medicine information to patients. In similar studies, Maat and Lentz (2009) in Holland, and Cossío (2011) in Mexico ratified Sless' findings. Maat and Lentz (2009) tested the effect of graphic presentation of European Union PILs on readability and information searching, and Cossío (2011) investigated how graphic presentation of headings in medicine instructions labels influences patients' information searching strategies. In both studies current leaflets/labels and their redesigned versions (according to information design principles) were tested, and their results also proved that the graphic presentation of PILs positively influenced participants' performances. Cossío (2011) particularly found that the use of color and bold type in headings aided participants to locate information and to navigate within the text.

In Brazil, research on the graphic presentation of information in PILs mainly focuses on their readability and legibility aspects. Fujita (2009) found that graphic presentation of PILs in Brazil did not meet patients reading strategies. She carried out a qualitative study on information searching through verbal protocol with four patients. The results showed that drawbacks in the PILs design, such as poor typographic hierarchy of headings, lack of visual emphasis (bullets, bold type) in lists and in relevant words posed difficulties to participants when searching/finding information within the leaflet. Poor typographic solutions to PILs were also discussed in Paula et al study (2010). They looked at people's attitude towards the reading of PILs and their difficulties in such task. The results showed that more than 80% of participants do read the leaflets despite their difficulties; and that type size was considered the main problem in the reading process, making PILs comprehension harder.

Regarding the use of pictures to communicate medicine information to patients, it is acknowledged in the literature as a facilitator in the information processing and message comprehension (e.g., Silva, 2008; Spinillo, Padovani and Miranda, 2008). In this regard, it is worth mentioning Silva's (2008) analytical and experimental studies on the graphic presentation of warnings in PILs in Brazil. The results showed participants' difficulties in understanding certain pictorial warnings, although the use of pictures together with text was considered to promote comprehension in the warning messages. Based upon the results, he highlighted the need for improvements in the graphic presentation of Brazilian PILs, particularly for warnings, to promote safety in medication administration by patients.

Those research outcomes demonstrate the relevance of the graphic presentation of information in medicines to support comprehension in Brazil and worldwide. Nevertheless, the effect of graphic presentation of PILs on medicine administration by patients remains to be investigated, particularly on task errors. In this regard, an experimental study was carried out in Brazil on this topic, which results are discussed next.

5 A STUDY ON THE EFFECTS OF THE GRAPHIC REPRESENTATION OF INFORMATION IN PILS AND TASK ERRORS IN MEDICINE USAGE IN BRAZIL

The study investigated the effect of PILs on the simulated use of five medicines (vaginal cream, oral inhaler capsule, oral suspension, insulin injection pen, and nasal spray) with 60 participants. To each medicine a HTA was drawn according to its PIL to identify patient decision-making points, steps and task conditions. The testing materials were the medicines, and their PILs. For the insulin injection a sponge was used to simulate the injection procedure; for the oral suspension, a jug of water was provided to allow filling the medicine bottle, and for the vaginal cream, a plastic bottle was provided to allow empting the applicator. Each medicine was tested with 12 participants individually at a time, who were asked to follow the PILs instructions to use the medicines and to verbalize their actions. Tasks were video recorded and semi-structured interviews were conducted after task performance. Participants' task errors were classified into three categories (Table 1): Information processing; Action and Verification errors, which were based upon Barber's (1996) and Rasmussen's (1986) classifications.

Table 1 Classification for errors adopted in the study

Errors		
Information processing errors	**Action errors**	**Verification errors**
Internal (individual repertoire) Pi1 |Wrong/Mistaken assumption *External (insert/package/product)* Pi2|Information was not read/searched Pi 3| Information was incompletely read/searched Pi 4| Wrong information searched	A1|Task/action was not performed A2| Task/action was incompletely performed A3| Task/action was performed in wrong/inappropriate moment A4| Very long or very short Task/action A5| Task/action performed	V 1| Verification not done V2| Verification incompletely done V 3| Verification in a wrong moment V4| Right verification in a wrong object V 5| Wrong verification in right object V6| Wrong verification in wrong object

Pi 5| Information was searched but not found Pi 6| Information was searched and founded but not understood	in a very little or very large amount/quantity A6| Task/action in wrong direction A7| Wrong alignment A8| Right task/action in wrong/mistaken object A9| Right task/action but in a wrong part/component of a right object A10| Wrong task/action in a right object A11| Wrong task/action in a wrong object A12| Selection not done A13| Wrong selection done	V7| Verification in a very little or large amount/quantity

5.1 Results discussion

Task errors occurred in all five medicine administration tasks, reaching a total of 352 errors. The highest scores were in action errors (N= 179) followed by information processing errors (N=121). The medicines that scored highest in action and in information processing errors were the insulin injection (N=68 and N=60) and the oral suspension antibiotics (N=39 and N=32). These results suggest that action errors were tightly related to information processing errors: difficulties in comprehending the medicine information in the PILs had driven participants to failure at some point of the medicine usage. The results seem also to indicate a relation between errors and task complexity in terms of object manipulation, dosage measurement and number of actions to administrate the medicines. The administration of insulin injection and the oral suspension medicines involved using syringes/applicators with numbered scales to set dosage, demanding participants to process specific/detailed information and to perform actions accurately. With respect to the number of actions to administrate the medicines, the insulin injection (N=19), the oral suspension (N=17) and the oral inhale capsule (N=20) medicines presented the highest number of steps in their instructions of use, as previously identified in their HTAs. Performing many steps might also have increased participants' cognitive effort (read and understand), and demanded particular motor skills, resulting in information processing, verification (dosage) and action errors.

Regarding the graphic presentation of the PILs, it seemed to be particularly related to the information processing errors, and then to action and verification errors. The participants' interview responses and their outcomes in task performance suggested that drawbacks in typographic aspects of the text, visual instructions and page layout negatively affected information searching/finding and message comprehension, leading to errors in the medicines' usage. Among these typographic aspects which affected PILs usability, some are worth mentioning as they contradict design principles for legibility and text navigation. They were: the use of small type size for the text; blocked texts (justified text alignment) that make following line breaks difficult; lack/poor hierarchy in headings and warnings; lack of visual separation cues to distinguish different contents; the use of bold type for long texts, affecting legibility; and the absence of horizontal space (e.g. line space) to separate chunks of information, making reading difficult. These results are in consonance with those of Sless (2004); Maatz and Lentz (2009); and Cossío (2011) giving evidence of the importance of typographic aspects to the usability of PILs.

As for the visual instructions, lack of correspondence between text and image, confusing alignment of images, lack of figure-ground contrast; and lack or even misrepresentation of actions conveying the steps, are examples of problems found in the PILs which lead to errors in task performance. Figure 1 shows details of the visual instructions for the oral suspension antibiotics (left) and the oral inhale capsule (right) in which the communication effectiveness of text-image relation is jeopardized.

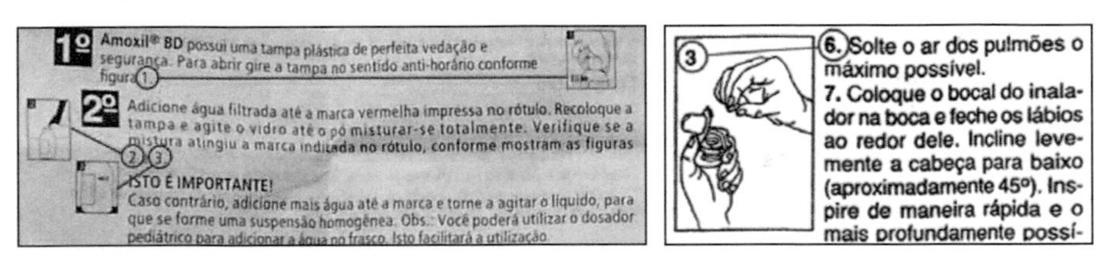

Figure 1 Examples of poor text-image relation in the visual instructions for the oral suspension antibiotics (left) and the oral inhale capsule (right).

It is also worth commenting drawbacks in the visual instructions for the vaginal cream. Ambiguous pictorial representations and lack of emphasis to show details of the task and of the medicine components (applicator and cream tube) lead to misinterpretation of the depictions and task errors. For instance, a participant used the applicator in a wrong/inverted position resulting in action and verification errors (Figure 2a left) even consulting the PIL during task performance. Moreover, depictions of the interior of the woman body (Figure 2b right) may not succeed in communicating instructions to certain patients in Brazil, as they may not familiar to them. Such depictions demand specialized knowledge of the female physiology (visual repertoire). Similar drawbacks in pictorial representation of medicine information to patients were also found in Silva's (2008) and Spinillo, Padovani and Miranda's (2008) previous studies, indicating that they affect not only message comprehension but also task performance.

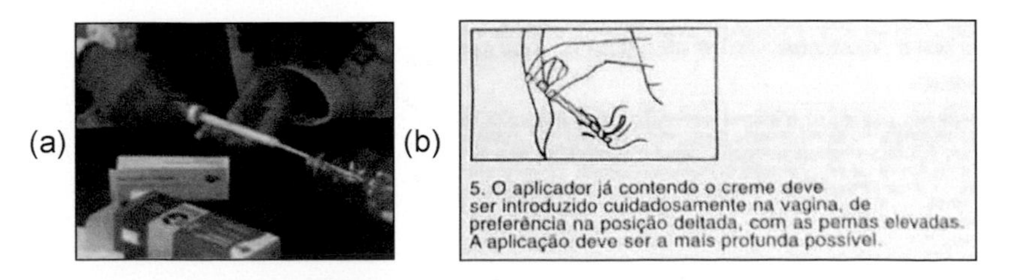

Figure 22 Example of the test and of the visual instructions for the vaginal cream medicine.

6 CONCLUSIONS AND FINAL CONSIDERATIONS

The main conclusions of the results are that: (a) the graphic presentation of PIL affects its usability as well as its readability; and (b) drawbacks in the graphic presentation of PILs lead to task errors particularly of dosage verification and manipulation of medicine components. Patients may then fail in using the medicines and if so, the treatment of diseases will be jeopardized.

It is also possible to conclude that information processing errors and task errors (action and verification) are in direct relation. Therefore, by improving the quality of pictorial and typographic aspects of PILs, patient cognitive load and task performance in medicine usage may also be enhanced.

Finally, based upon the results, it is pertinent to assert that the Brazilian legislation for PILs needs revision, in order to include the graphic aspects of this instructional document to facilitate patients/users' comprehension and the medicine inserts' usability.

ACKNOWLEDGMENTS

We would like to thank the volunteer participants for making this research possible, and the CNPq/Brazilian Ministry of Health for funding this study.

REFERENCES

Andriesen, S., 2006. Readability Testing of PILs – A New 'Must'. EPC. Autum, pp. 42-44. Samedan Ltd Pharmaceutical Publishers.

Barber, C., Stanton, N. A., 1996. Human error identification techniques applied to public technology: predictions compared with observed use. *Applied Ergonomics*, Vol. 27. No 2, pp. 119-131.

Cossío, M. G., 2011. Importance of Headings for the Correct Reading of a Text. Research in Medicine Instructions. Speach at The 5th Information Design Conference. Florianópolis, Brazil: Brazilian Society of Information Design, 2011.

Croskerry, P., 2003. The importance of cognitive errors in diagnosis and strategies to minimize them. *Academic Medicine*, Vol.78, No. 8, August, pp. 775-780.

Ellenbecker, C., Frazier, S, and Verney, S., 2004. Nurses' Observations and Experiences of Problems and Adverse Effects of Medication Management in Home Care. *Geriatric Nursing*, Vol. 25, No 3, pp.164-170.

Fuchs, J., 2010. The Way Forward in Package Insert User Tests From a CRO's Perspective. *Drug Information Journal*, Vol. 44, pp. 119–129.

Fujita, P. T. L., 2009. Análise da apresentação gráfica do conteúdo textual da bula de medicamento na perspectiva de leitura do paciente em contexto de uso. Unpublished Master Dissertation, The Federal University of Paraná, Brazil, Curitiba, 2009.

Ganier, F., 2001. Processing text and pictures in procedural instructions. *Information Design Journal*, No 10, Vol. 2, pp. 143-153.

Gonçalves, S. A., Melo, G., Tokarski, M.H.L. and Barbosa-Branco, A., 2002. Bulas de medicamentos como instrumento de informação técnico-científica. *Revista Saúde Pública*; 36(1):33-9 33. Acessed December 28, 2011, http://www.fsp.usp.br/rsp.

Graber, M., Gordon, R. and Franklin, N., 2002. Reducing Diagnostic Errors in Medicine: What's the Goal? *Academic Medicine*, Vol.77, No.10, October, pp. 981-992.

Iida, I., 2005. O erro humano. In: *Ergonomia: projeto e produção*. São Paulo: Edgard Blucher.

Johnson, C., 2002. The Causes of Human Error in Medicine. *Cognition, Technology & Work*, 4:65–70.

Kaushaul, R., Barker, K., Bates, D., 2001. How Can Information Technology Improve Patient Safety and Reduce Medication Errors in Children's Health Care? *Arch Pediat Adolesc Med*, Vol.155, September, pp. 1002-1007.

Lane, R., Stanton, N. and Harrison, D., 2006. Applying hierarchical task analysis to medication administration errors. *Applied Ergonomics* 37: 669–679.

Maat, H. P. and Lentz L., 2009. Improving the usability of patient information leaflets. *Patient Education Couns*.

Meredith, S. et al., 2001. Possible medication errors in home healthcare patients. *JAGS*, 49: 719-724.

Paula, C. S. et. al., 2009. Análise crítica de bulas sob a perspectiva do usuário de medicamentos. *Visão Acadêmica*, Curitiba, v.10, n.2, Jul. - Dez./2009 - ISSN 1518-5192.

Rasmussen, J., 1986. *Human Error. Information Processing and human-machine interaction*. New York: North Holland, pp.140-169.

Raynor, D. K., Knapp, P., Moody, A. and Young, R., 2005. Patient Information Leaflet: Impact of European regulation on safe and effective use of medicines. *The Pahrmaceutical Journal*. V. 275, pp. 606-611.

Reason, J., 1990. *Human Error*. New York: Cambridge University Press.

Rennie, W. et al., 2007. Minimising human error in malaria rapid diagnosis: clarity of written instructions and health worker performance. *Transactions of the Royal Society of Tropical Medicine and Hygiene*, 101: 9-18.

Sandlin, D., 2008. Pediatric Medication Error Prevention. *Journal of PeriAnesthesia Nursing*, Vol 23, No 4 (August), pp. 279-281.

Silva. C. R. L. Da., 2008. Contribuições da ergonomia cultural para a representação gráfica em advertências de medicamentos. Unpublished Master dissertation. The Fedeal University of Pernambuco, Recife, 2008.

Sless, D., 2004. *Labeling code of practice: designing usable non-prescription medicine labels for consumers*. Acessed December 28, 2011, http://www.communication.org.au/cria_publications.

Spinillo, C. G., Padovani, S. and Miranda, F., 2008. Graphic and information aspects affecting the effectiveness of visual instructions in medicine inserts in Brazil. In: AHFEI- Applied Human Factors and Ergonomics Conference, 2008, *Proceedings of the AHFE International Conference 2008*. Louisville: USA Publishing, 2008. v. 1.

Waarde, K., 2008. Measuring the quality of information in medical package leaflets: harmful or helpful? *Information Design Journal*, 16:3, pp. 216–228.

Waarde, K. Visual information about medicines for patients. In: *Designing Effective Communications: Creating contexts for clarity and meaning*. New York: Allworth Press, 2006. p. 38-50. 2006.

Wright, P., 2003. Criteria and ingredients for successful patient information. *Journal of Audiovisual Media in Medicine*, 26, 1, pp. 6-10. London: Taylor and Francis.

CHAPTER 35

How to Bring your Doctor Home. User-centered Design of Trustworthy Telemedical Consultation Services

Shirley Beul[1,2], *Martina Ziefle*[1], *Eva-Maria Jakobs*[2]

[1]Communication Science, eHealth Group,
[2]Textlinguistics and Technical Communication,
Human-Computer Interaction Center, RWTH Aachen University
Aachen, Germany
{beul, ziefle}@humtec.rwth-aachen.de; {s.beul, e.m.jakobs}@tk.rwth-aachen.de

ABSTRACT

Telemedical consultation services promise solutions in maintaining the area-wide supply of healthcare in times of fast growing patient numbers, though their acceptance is low. This study aims at detecting how a computer mediated medical encounter must be designed to win users' trust in such services for establishing them as serious alternatives to face-to-face consultations. 15 semi-structured scenario-based interviews with older female heart patients were conducted using the Future Care Lab of RWTH Aachen University in Germany as an example. Users identified an intuitively usable interface and the preserving of patient's privacy as important requirements. Data security played a subordinate role. Besides, integrated ICT must remain unimpaired or even improve the quality of the physician-patient interaction. However, users doubt these highly innovative applications in default of own or indirect user experience. Thus, developers must consider users trust decision-making based on analogies to established, similar appearing ICT of the same domain (e.g. health) if possible. Additionally, initial user experience must be provided to reduce patients' irrational beliefs and facilitate a critical evaluation.

Keywords: telemedicine, ehealth, Ambient Assisted Living, telemedical consultation, trust, doctor-patient communication

1 CHALLENGES IN TIMES OF GRAYING SOCIETIES

Within industrialized countries healthcare systems currently alter to cope with the upcoming consequences of the demographic change. The postwar baby booms induced an extremely large cohort of people, who presently begin turning into the "third age", the stage of personal self-fulfillment (Laslett, 1991). Sooner or later, the majority of them will enter the fourth and last phase of the life cycle, the decrepit period characterized by depending on (geriatric) care. Longevity among the elderly, improvements in medical care, and this demographic imbalance lead to an increasing number of old people, who have to be supplied with healthcare in the close future (Leonhardt, Hexamer, and Simanski, 2009). Therefore, a higher demand for medical personnel and rising healthcare expenditures must be considered in prospective health reforms (Beul, Ziefle, and Jakobs, 2011).

One promising solution for this problem is the implementation of information and communication technology (ICT) into healthcare (Ziefle, 2010), which can bridge a geographical distance between health care provider and patient (Miller, 2001). Particularly, telemedical consultation services, designed as real-time audiovisually communication channels, play an important role (Ferguson, 2007). These services enable physician and patient to discuss the latest recorded vital data, talk about current ailments and their treatment, and negotiate the therapy or the next steps in the treatment respectively without traveling (Beul, Ziefle, and Jakobs, 2011). They facilitate the patient to access medical treatment and information independently of time and location. Consequently, patients' resources (e.g. time, money) can be preserved (Miller, 2003). Physicians' human resources can be allocated more efficiently, for instance, by locating them in a telemedical center and distribute their medical advice online (Yousuf et al., 2002).

Crucial for the success of such a consultation service is its design, which is exceedingly significant for the acceptance of users. Before launching using barriers and user requirements on telemedical services, consequently, must be identified to ensure their success (Rogers, 1995; Wilkowska and Ziefle, 2011). Among the user requirements, trust-building features must be specially emphasized because users' confidence is regarded as a major prerequisite for the acceptance of telemedical applications as well as their providers (Katsikas, Lopez, and Pernul, 2008).

2 USERS' TRUST IN TELEMEDICAL CONSULTATION SERVICES

The doctor-patient-relationship is fragile, based by the doctor's promise to treat the patient and the patients' trust in her/him, particularly in her/his medical competence and professional handling of their personal data. Hence, possessing patients' trust is a crucial precondition before they allow insights in their privacy.

Trust is an elusive phenomenon, which is conceptualized and operationalized concerning discipline, approach, and focus of the specific research context (McKnight and Chervany, 1996). However, Wang and Emurian (2005) ascertained four cross-disciplinary characteristics, which are relevant for the on- and the offline context: The trust relationship consists of *trustor* (e.g. patient) and *trustee* (e.g. telemedical consultation service). Trust is developed based on the trustee's ability to act on behalf of the trustor's interest and the degree of trust that the trustor places in the trustee. Furthermore, trust is *vulnerable* and a *subjective matter* varying on several contextual factors. Out of the trustor's perspective, trust arises necessarily in risky situations and is determined by *trustee's actions* in service of the trustor.

Applied to the research subject, prospective users are in the unsecure situation of using telemedical consultation services, which are unfamiliar to patients owing to their high degree of innovation. Actual usage bears subsequently a risk. According to Braczyk, Barthel, and Fuchs (1998), using novel ICT requires users' trust building, for which the system's reliability is determinant. Apart from this, users assess a potential damage involved in the usage. When it is rated as too high, familiar alternatives are chosen. If there are non, trusting the innovation becomes legitimately. Therefore, the system's reliability remains exclusively important. It usually is anticipated based on direct or indirect experiences made by other users, what is impossible due to the weak diffusion of such consultation services. Consequently, crucial assessment criteria are contributed by analogies of technologies appearing comparable to them (investigated for the mobility context by Wirtz, Jakobs, and Beul, 2010). Derived from trust research about web-based systems (e.g. Fruhling and Lee, 2006; Belanger, Hiller, and Smith, 2002), "usability of the user interface", "data security" and "protection of privacy" seem particularly promising as trust-related characteristics of telemedical consultation services. After all, the quality of the supplied medical service must be considered, too.

3 METHODOLOGY

In the following, the design of this study is delineated: The Future Care Lab of RWTH Aachen University, an Ambient Assisted Living prototype, is chosen and described as an example of application. Besides, addressed research questions, the participant selection, and the applied survey method are, finally, explained.

3.1 Research subject: Future Care Lab at RWTH Aachen University

The Future Care Lab is an existing Ambient Assisted Living (AAL) environment developed by the eHealth group of RWTH Aachen University (http://www.ehealth.rwth-aachen.de). This project aims at developing new, integrative models for the design of user-centered healthcare systems and novel concepts of electronic monitoring systems within ambient living environments. In this prototype, user experience is embedded into a spatial context (e.g. living room). This requires designing spaces like the one under examination in a way that they support technology seamlessly through everyday objects (e.g. furniture), but also room components (e.g. floor, walls). Thus technology may be designed to overtake different roles, functionalities and services (assisting and care) (Beul et al., 2010).

The lab is conceptualized and technically realized as a smart living room for exploring patients' "life" at home in order to investigate possible usage barriers and perceived benefits. It is equipped with different medical devices and user interfaces. Figure 1a shows fundamental room components: a wall-sized interactive multi-touch display and a pressure sensitive floor. The multi-touch wall shifts the primary function of the wall as a room component towards an active, graphical in- and output device for human-computer interaction. The floor monitors unobtrusively persons' movement behaviors to detect walking patterns, fall events or other abnormal movement behaviors that would indicate a medical issue of the inhabitant. In case of emergency, help can be called automatically (Leusmann et al., 2011).

The integrated medical biosensors acquire four vital parameters, which are important to monitor cardiac patients: blood pressure, blood coagulation, body temperature, and weight (Klack et al., 2011). Figure 1b illustrates the integration of the devices.

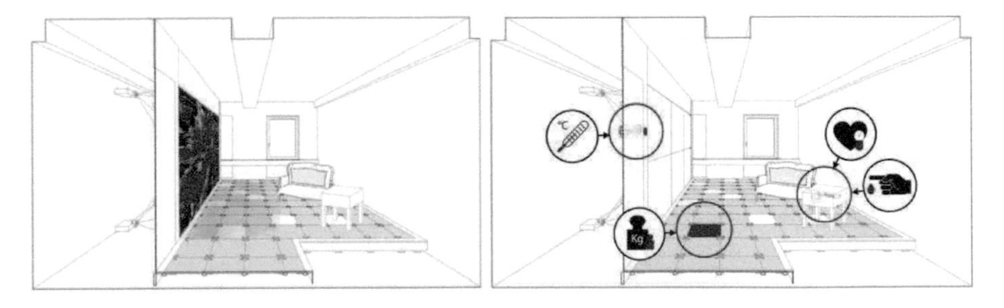

Figure 1a+b Illustration of the Future Care Lab and the integrated medical technology.

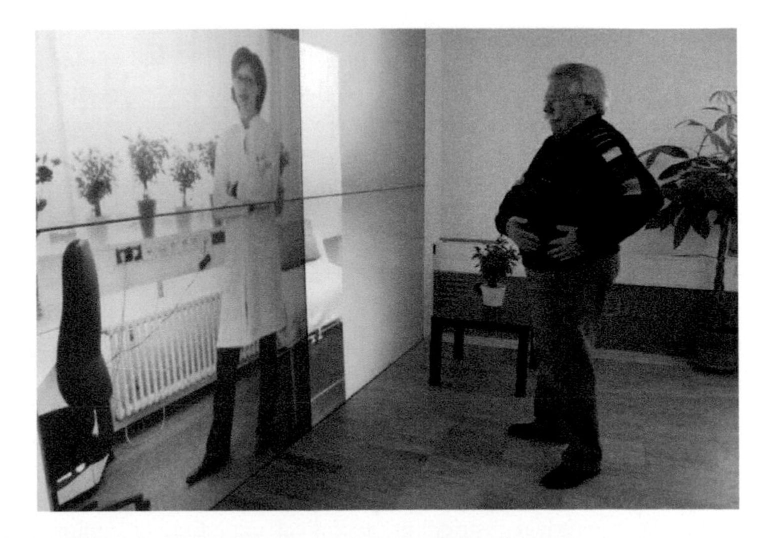

Figure 2 Simulation of a telemedical encounter in the Future Care Lab.

Besides the patient monitoring, the Future Care Lab can be used for doctor-patient communication. A video channel can be implemented to link the doctor's office with the patient's home (Fig. 2). The interactive wall offers the possibility of applying new formats like life-size presentation of interlocutors (Ziefle et al., 2011). Despite the obvious advantages of telemedical consultation services in general and the interactive wall in particular, the success of such a service depends on the integrated system, especially on the users' acceptance and their trust in it.

3.2 Adressed research questions, participant selection and applied survey method

In order to explore how prospective users generally assess telemedical consultation services in the Future Care Lab, research questions of two levels were addressed: (1) As regards content, it is necessary to find out how a computer-mediated medical encounter must be designed to win users' trust in this service that it becomes a serious alternative to face-to-face consultations. Using motives and also barriers are considered as well as using preconditions named by participants. Thus trust-related characteristics are ascertained, which are important for the design of telemedical consultation services or AAL environments respectively. (2) With respect to methodological aspects, it is relevant to find out to which degree complex future technologies can be explained to participants with a scenario. In case, it is too abstract for a lay understanding, further methods have to be developed to forward the evaluation of innovative, but unknown ICT.

The demographic change is also characterized by an imbalance between genders. Currently, a feminization of ageing occurs in several societies (e.g. Tews, 1999). This gender effect notably grows with increasing age. Because of womens' higher life expectancy, they are mostly widows. When they loose their independence, they often are single and rely on geriatric care, wherefore inhabitants in nursing homes are mainly female. Hence, there is a high demand of older women for telemedical solutions. The prevalence for chronic disease, especially cardiac, increases with rising age (Ho et al., 1993).

For this reason, 15 female cardiac patients aged 55+ (M = 57,2) were recruited via advertisements in doctors' offices, pharmacies and senior clubs as participants. Six of these ladies lived on their own, four together with a partner. Only three of them stated to work, the others actually were not working (any more). Concerning physical restrictions, eight out of ten wear glasses. Three users had a hearing impairment.

Regarding technical expertise, the telephone was the most often used information and communication technology for private communication as well as talks with the doctor. Only two participants stated to use emails to communicate with their physician. Secondly, five interviewers each ran one additional interview a) to validate the collected data of the first stage of survey and b) to control possible interviewer effects. No significant difference was found in contrast to the first group with respect to their physical restrictions, technical expertise, or communication with doctors.

Semi-structured interviews based on an interview guide were applied. The core of the guideline was a scenario, in which the interviewee was living in the Future Care Lab and could conduct telemedical consultations through her interactive wall in real-time. The life-size representation format, the option to exchanging medical data like vital data diaries, and the possibility to record the talk were explicitly exposed. Participants named and reviewed pros and cons about this scenario. Finally, they were asked about their opinion about data security in this Ambient Assisted Living system. Verbal data was recorded, transcribed afterwards, and analyzed qualitatively by content.

4 RESULTS

In general, participants showed a positive attitude towards the interactive wall, stated to be convinced about the usefulness of telemedical consultation services, but mentioned to see no need for them. As long as they stay in the stage of the „third age", no questioned women would be willing to use the service. Out of the perspective of participants, living in rural areas far away from the next doctor, being decrepit, or physically immobile are comprehensible using motives.

As apparent advantages pragmatic aspects like saving time, flexibility, and comfort are named. However, the size and aesthetics of the interactive wall are seen as drawbacks because most homes have not been designed yet to integrate this technology. Costs and logistical effort additionally deter. Furthermore, it was assessed as "bothersome" to tidy up the home for doctor's virtual visit.

The non-working interviewees lament the loss of the pressure to see the doctor personally, too, what compels them to dress up and leave the house. Also, cultivating social contacts with other patients and the doctors' receptionists are missed. The still employed women were delighted about timesaving and decreasing infection risk by avoiding contact with other sick people in the waiting room.

Regarding privacy issues, the female interviewees were concerned about third parties who could sneak in their telemedical consultation. These unauthorized persons could participate by hiding in the doctor's office or illegally join the encounter through the interactive wall. The ladies are, especially, afraid of hackers who could enter the doctor-patient talk for voyeuristic motives. Besides, installing cameras in their home environment makes them feel uncomfortable because they refuse to become a "vitreous human". Opinions were divided concerning the recording function of the talk: One female stated to fear speaking imprudently what will be conserved. Others welcome this documentation as an enhancement of legal safeguarding towards the doctor and also as a reminder of doctor's advice.

According to the quality of the telemedical consultation service, the physical absence of the physician is mostly criticized. They assess the remote interaction with him/her as "impersonal", "anonymous", "cold", and "sterile". The ladies state to miss the human touch in this type of medical encounter. Moreover, they believe that a telemedical consultation cannot transfer an authentic, genuine, or natural perception of an interlocutor for all senses because it is realized via an interposed machine. For example, they think that non-verbal signals are not transmitted well, but they are attributed as essential for the understanding between physician and patient. One female mentioned to read a diagnosis' seriousness out of her doctor's mimic, what she thinks is not realizable on the interactive wall. In the opinion of another woman her physician would take patients more seriously when they are present in the same room. What they also doubt is the feasibility of a remote diagnose – without touching the patient. This is why the participants question the effectiveness of the treatment via this ambient environment.

In comparison with the telephone, the interviewees prefer the interactive wall for a talk with their physician. Positively regarded is the telemonitoring: the measurement procedure in the Future Care Lab is evaluated as easier than the handling of conventional medical devices. Apart from this, it facilitates an overview of measured vital data, which allows them a better understanding of their disease and to detect anomalies by themselves.

Surprisingly, data security is of secondary importance. Data abuse is discussed as a serious invasion of privacy in the context of telemedical services. On the contrary, the questioned women appraise their medical as less sensitive information than their financial data. They cannot imagine a data abuse or a damage resulting from it. No interviewee expressed that medical data was more sensitive than other personal data. They usually handle their personal data (e.g. address data) naively, but show a critical attitude towards the topic data security. Protection against spying is rated as technically hardly doable, and that data security mainly depends on the doctor's handling of the data.

The possibility of being stigmatized as ill by the wall was mentioned by one woman, but only for the case of an exclusive using of this ICT for telemedical purposes. Additionally, another lady feared (unnecessarily) noxious radiation of the interactive wall.

The interviewed females formulated a catalogue of using preconditions: In addition, they wish to have a mobile control unit and a voice control for the ICT. They doubt their capability to install or use the interactive wall. Therefore, the interaction with the system should be easy and intuitively to understand ("contact via push of a button").

The communicative procedure in this telemedical consultation service should be designed like using a telephone: One dials a number and the interlocutor confirms to answer. Here, the asked women demand that only the patient is allowed to initiate the talk. The doctor must wait for the patient's confirmation or invitation respectively. Moreover, the physician must be presented authentically and realistically on the wall-sized screen.

Concerning the organization of the service, an initial face-to-face talk must be executed before the first computer-mediated contact. This leads to the conclusion that a foundation of trust must be built in real life before telemedical services can be accepted, and actually used. Besides, these ladies would like to contact the doctor via interactive wall around the clock. Finally, the telemedical consultation should not replace the traditional face-to-face encounter. It must be possible to visit the physician in her/his office.

5 DISCUSSION AND CONCLUSION

In this research, we explored prospective users' acceptance, especially their trust, in a telemedical consultation service integrated in an AAL environment to establishing it as a serious alternative to face-to-face consultations. As expected, one identified user requirement is the usability of the user interface. Another relevant one is the warranty for data security, although participants rate their health data as less sensitive as assumed. A data abuse seems improbable to them, the potential damage involved is assessed as slight. There are three possible explanations for this: (1) The occurring damage is not directly noticeable because it does not affect user's life actively. (2) Users cannot imagine reasons why third parties are interested in stealing their medical data. Obviously, media does not make this as a subject of discussion sufficiently. (3) They believe data security is important, but do not know in fact what this term really stands for. They just know it from the public debate. Thus "data security" in the context of telemedical consultation services must be operationalized in future research.

The invasion of their physical privacy is the most serious using barrier. The interviewees fear manipulation of cameras and intruding of strangers in their telemedical encounter and also in their home. Besides, the volatility of the oral communication is invalidating in the delineated application, which has not been prevailingly refused. Surprisingly, the record function has been evaluated as useful to empower patient rights legally and to enhance information recall. They can also listen again to unclear content, what facilitates a better understanding of doctor's decisions, and treatment instructions, and consequently patient's coping.

The lack of touch during the talk is frequently criticized. Patients miss the physical examination, which is a well-known part of a routine consultation. This loss leads to the fact that a computer-mediated medical encounter is only restrictedly a serious alternative. Performing exclusively talk-based encounters exclusively (e.g. for exchanging medical data) seems, however, an appropriate using motive. Moreover, the quality of transmitted nonverbal communication is doubted. Finally, all questioned ladies prefer the face-to-face talk instead of the telemedical solution because it seems more personal and authentic to them.

To guarantee the success of these services, the computer mediation must remain unimpaired or even improve the quality of the physician-patient interaction because patients derive criteria out of it to assess the medical treatment. Creating authenticity for users is, therefore, in top priority. A supportive function could be a personal definition of display details (e.g. close up on doctor's face), which users need for a good doctor-patient communication assessed out of their individual perspective. Moreover, users could "screen" the doctor's office to ensure that there is no undesirable audience hidden somewhere, which gives them a feel of control.

Concerning the methodology, it was found that using scenarios for explaining older users complex future technologies is challenging. The elderly have difficulties to imagine the ICT abstractly which does not fit in their idea of technology. In consequence, a comprehensive understanding could be facilitated when (1) the scenario emphasizes ICT's functionalities detailed, and (2) interviewers illustrate its design at least. A simulation study is still necessary, in which doctor and patient experience the system in action. Then, users' "perceived usefulness" and also their "perceived ease of use" could finally be understood to a significant degree, as experts of acceptance research demand it (e.g. Wilkowska and Ziefle, 2011).

In conclusion, telemedical consultation services are promising solutions for maintaining the area-wide supply of healthcare in times of the demographic change. However, users doubt these highly innovative applications in default of own or indirect user experience. System developers, therefore, must consider users trust decision-making based on analogies to established, similar appearing ICT of the same domain (e.g. health) if possible. Additionally, first user experience must be published to reduce patients' irrational beliefs and facilitate a critical evaluation.

ACKNOWLEDGMENTS

We have to express our gratitude to our students, especially Verena Bock and Teresa Schmidt, who supported this research by collecting and transcribing the interview data. We also would like to acknowledge our interviewees for answering our questions so patiently. The Excellence Initiative of the German federal and state governments funded this research.

REFERENCES

Bashshur, R. L. 1995. On the Definition and Evaluation of Telemedicine. *Telemedicine* 1(1), 19–30.

Belanger, F., J. S. Hiller, and W. J. Smith 2002. Trustworthiness in electronic commerce: the role of privacy, security, and site attributes. *Strategic Information Systems* 11, 245–270.

Beul, S., L. Klack, and K. Kasugai et al. 2012. Between Innovation and Daily Practice in the Development of AAL Systems: Learning from the experience with today's systems. In. *Electronic Healthcare*, eds. M. Szomszor and P. Kostkova. Third International Conference, eHealth 2010, Casablanca, Morocco. Heidelberg: Springer, 111–118.

Beul, S., M. Ziefle, and E.-M. Jakobs 2011. It's all about the medium. Identifying patients' medial preferences for telemedial consultations. In. *Human-Computer Interaction Information Quality in E-Health.* eds. A. Holzinger and K.-M. Simonic. Berlin, Heidelberg: Springer, 321–336.

Braczyk, H.-J., J. Barthel, and G. Fuchs, et al. 1998. Vertrauensbildung aus soziologischer Sicht-das Beispiel Sicherheit in der Kommunikationstechnik. In. *Mehrseitige Sicherheit in der Kommunikation,* Vol. 2., eds. G. Müller and K.-H. Stapf. Bonn, 119–150.

Fruhling, A. L. and S. M. Lee 2006. The influence of user interface usability on rural consumers' trust of e-health services. *Electronic Healthcare* 2(4), 305–321.

Ho, K. K. L., J. L. Pinsky, and W. B. Kannel, et al. 1993. The Epidemiology of Heart Failure: The Framingham Study. *American College of Cardiology* 22 (4), 6A–13A.

Katsikas, S., J. Lopez, and G. Pernul. 2008. The Challenge for Security and Privacy Services in Distributed Health Settings. In. *eHealth. Combining Health Telematics, Telemedicine, Biomedical Engineering and Bioinformatics to the Edge.* eds. B. Blobel, P. Pharow, and M. Nerlich. Amsterdam: IOS Press, 113–125.

Klack, L., T. Schmitz-Rode, W. Wilkowska, et al. 2011. Integrated Home Monitoring and Compliance Optimization for Patients with Mechanical Circulatory Support Devices (MCSDs). *Annals of Biomedical Engineering* 39(12), 2911–2921.

Laslett, P. 1991. *A fresh map of life. The Emergence of the Third Age.* Cambridge: Harvard University Press.

Leonhardt, S., M. Hexamer, and O. Simanski 2009. Smart Life Support: model-based design and control of life-supporting systems. *Biomedical Engineering (Berlin)* 54 (5), 229–231.

Leusmann, P., C. Möllering, L. Klack, et al. 2011. Your Floor Knows Where You Are: Sensing and Acquisition of Movement Data. *Proceedings of the* 12th IEEE International Conference on Mobile Data Management (MDM2011), 61–66.

McKnight, D. H. and N. L. Chervany. 1996. *The Meanings of Trust.* Charlson School of Management, University of Minnesota.

Miller, E. A. 2001. Telemedicine and doctor-patient communication: an analytical survey of the literature. *Telemedicine and Telecare* 7(1), 1–17.

Miller, E. A. 2003. The technical and interpersonal aspects of telemedicine: effects on doctor-patient communication. *Telemedicine and Telecare* 9, 1–7.

Rogers, E. M. 1995. *Diffusion of Innovations*, 4th ed. New York, NY: The Free Press.

Tews, H. P. 1999. Von der Pyramide zum Pilz. Demographische Veränderungen in der Gesellschaft. In. *Funkkolleg Altern 1. Die vielen Gesichter des Alterns.* eds. A. Niederfranke, G. Naegele, and E. Frahm. Wiesbaden: Westdeutscher Verlag, 137–186.

Wang, Y. D. and H. H. Emurian. 2005. An overview of online trust: Concepts, elements, and implications. *Computers In Human Behavior* 21, 105–125.

Wilkowska, W. and M. Ziefle 2011. Perception of privacy and security for acceptance of E-health technologies: Exploratory analysis for diverse user groups. *Proceedings of the 5th International IEE Conference on Pervasive Computing Technologies for Healthcare*, 593–600.

Wirtz, S., E. M. Jakobs, and S. Beul. 2010. Passenger Information Systems in Media Networks – Patterns, Preferences, Prototypes. In. *Proceedings of the Internaional Professional Communication Conference – Communication in a Self-Service Society* (IPCC 2010). Twente, Netherlands.

Yusof, K., K. H. B. Neoh, M. A. Hashim, et al. 2002. Role of Teleconsultation in Moving the Healthcare System Forward. *Asia-Pacific Journal of Public Health* 14 (1), 29–34.

Ziefle, M. 2010. Potential and pitfalls of age-sensitive technologies in the e-health field. In. *Proceedings of the 1st European Conference on Ergonomics. Ergonomics in and for Europe. Quality of Life: Social, Economic and Ergonomic Challenges for Ageing People at Work.* Brugge, Belgium

Ziefle, M., C. Röcker, Wilkowska, W., et al. 2011. A Multi-Disciplinary Approach to Ambient Assisted Living. In. *E-Health, Assistive Technologies and Applications for Assisted Living: Challenges and Solutions, eds.* C. Röcker and M. Ziefle. Hershey, P.A.: IGI Global, 76–93.

CHAPTER 36

A Human-Centered Design for Supporting Long-term Care Management

Arpita Chandra, Yang Gong

University of Missouri
Columbia, USA
gongyang@missouri.edu

ABSTRACT

This research aims at developing a set of requirements for long-term care EMRs and how the present EMRs can be improved to facilitate decision-making and communication among providers. It also provides feasibility for a holistic and comprehensive view of all available residents' health information through a single interface using a human-centered prototype enabling improved clinical decision-making. TigerPlace, a true aging in place, is equipped with three disparate systems: electronic medical records to store their clinical data, telemedicine system to store vital signs and feel, and a relational database system to store date and time stamped sensor information, that do not communicate with each other. By conducting user task, function, and representation analyses, a human-centered evaluation framework was employed which provided an in-depth description of user characteristics, preferences, systems functionality, and effective representations. The objectives are: 1) identify the available data that need to be used to design the holistic representation, that could explore feasibility of better clinical decision support; 2) analyze the design requirements that could be extended for understanding the requirements of a long-term care EMR; and 3) how some of the ideas can be used to meet the information needs of care providers and facilitate communication among providers.

Keywords: Long-Term Care, Electronic Medical Record, Human Factors, Congestive Heart Failure, Aging in Place, Human-Centered Design

1 INTRODUCTION

Information technology can assist patients with self-management, communication with their doctors and participate in prevention and treatment programs. Information systems are needed in current healthcare settings to assist providers with the care process. Use of electronic health record (EHR) and electronic medical record (EMR) in healthcare systems, hospitals, and clinics are being mandated by healthcare reforms to meet the standards that improve patient-care and interoperability of health information across the healthcare systems. The health information systems are essential in long-term care (LTC). Long-term care refers to the patients' care process for chronic diseases and often multiple chronic conditions, either from home, an institution or acute care setting, including both inpatient and ambulatory care settings. It includes taking support services as needed, i.e. physical therapy, rehabilitation etc. to stay healthy. LTC and chronic disease management has special implication for the elderly population who have co-morbidities and need enhanced avenues to take care of their health and stay functionally (physically and mentally) active. At present, the numbers of elderly population in the United States are on an increase as the baby boomer age and they are the highest users of healthcare.

Aging in Place is a care model for the elderly that has environmental and health supportive services promoting the health of older adults and provide options for advanced senior healthcare research (Rantz, Marek et al. 2005). Aging in Place program uses a combination of supportive, restorative and assistive services to improve the senior's health and well being without the need for a more traditional nursing home care (Rantz, Marek et al. 2005) or skilled nursing facility, unlike TigerPlace, most elderly are not functionally active enough to be able to live independently on their own. They require assisted services from nurses and other skilled healthcare providers to be able to do their daily personal activities i.e., using restroom, getting to bed etc. In both independent living and nursing home setting, the senior residents receive some form of nursing care. The care providers at LTC have a very high-level of information needs in order to provide necessary care to the residents. Care providers require ways to monitor the residents' activities, trends of a disease or condition, improvement or deteriorating conditions and ways to compare available data with historical values for optimal treatment. It is mandatory for the care providers to have access to the resident's health data to compare the residents' health status, to observe trends of functional decline, recommend change of plan of care, and change of prescription etc.

1.2 Background and Significance

TigerPlace is a retirement community which is a true "Aging in Place" (Rantz, Marek et al. 2005), where seniors live independently with the support of technology that addresses their common health problems and health declines i.e., functional decline, at old age (Cowan 1999). At present, TigerPlace is occupied by 34 residents between the age of 70 and 94 years. The number of residents who have chronic illnesses is approximately 90% and co-morbid conditions e.g., heart condition and problems with mobility and cognition is almost 60% (Rantz, Skubic, Miller, & Krampe, 2008). Different forms of assistive technologies have been employed at TigerPlace to address common aging problems associated to their functional decline (Cowan 1999). TigerPlace is equipped with advanced health technology for the senior residents to enable the care providers to monitor the progress of residents' health in the presence of multiple chronic conditions and problems. The major three health technology employed at TigerPlace are: an electronic medical record (EMR) system which stores the medical records of patients from different clinical encounters that assists in diagnosis and treatment of the patients, a telemedicine system that tracks the daily or weekly progress of the residents in the form of questionnaire which is in turn stores the vital signs, chief complaints, symptoms and related information in the EMR system, and a sensor system that monitors the mobility and daily activities of the senior residents, effective in tracking and collecting their activities of daily living (ADLs) and instrumental activities of daily living (IADLs) aimed towards achieving independent living for the residents and detecting risks of injury (Skubic, Alexander et al. 2009). These three systems are independent and have their individual database systems (Aud 2010) providing minimal provisions for communication in real time and prevents the clinicians to see a holistic view of the resident's entire health information present at TigerPlace (Gong and Chandra 2011).

Though it was realized that TigerPlace requires sophisticated technology to monitor the daily activities of the residents, it lacks the support of efficient tools for a holistic view of relevant clinical information for the residents' care e.g., medical administration, medication management, easy retrieval and diagnosis aided by tracking progress on their daily activities, mobility, general health, chronic diseases, signs of functional decline due to aging, etc. (Aud 2010). The vast amount of staggered data in the three different systems and the lack of holistic representation can easily overwhelm care providers. It hinders their ability to understand the information conveyed in each system and how the information may be relevant to a residents' overall health condition and detect trends of functional decline, which might prolong the clinical decision making (Gong and Chandra 2011).

1.3 Needs of long-term care

The key for long-term care patients with co-morbidities and multiple disabilities is nursing documentation for healthcare quality and nursing assessments to evaluate outcomes and risks. The providers in these facilities require improved avenues to deal with the decorating health conditions of the elderly, especially for pain management and pressure-ulcers etc.

Implementing an EHR with templates with pre-formulated assessment variables related to various nursing assessments need to be developed for pressure ulcers to facilitate user-friendly documentation (Gunningberg, Fogelberg-Dahm et al. 2006) and for recording pressure-ulcer prevalence and prevention (Gunningberg, Dahm et al. 2008). The accuracy in pressure-ulcer recording improves in the EHR compared to the paper-based health record as it gives accurate and reliable feedback to the healthcare organization. High priority needs to be given for developing standardized documentation practices for various areas of care in nursing practice (Gunningberg, Dahm et al. 2008; Gunningberg, Fogelberg-Dahm et al. 2009) and evidence-based pressure ulcer prevention should be provided to the nurses (Gunningberg, Fogelberg-Dahm et al. 2009).

Chronic pain is also a common and costly syndrome effecting almost one in every three United States adults. Factors such as shortened length of the medical visit, increased availability of technological approaches to care, and a more informed patient suggest for a new paradigm for chronic pain management (Marceau, Link et al. 2007). Using pain management documentation and its results can be implemented into clinical nursing practice, which requires an unified way of maintaining and recording the documentation to process the information (Asteljoki, Kesanen et al. 2009). Structured and classified documentation can help utilize different reports from electronic patient record in nursing management. The knowledge of pain management in the electronic patient record can be utilized for nursing management (Kesanen, Asteljoki et al. 2009).

A congestive heart failure (CHF) scenario was used to develop a prototype, which may contain the critical conditions of chronic pain management and pressure ulcer assessment as shown in Figure 1.

1.4 Curent status of LTC Systems

The success of the treatment for the elderly depends on repeated modification of the patient's treatment regimen or simply ongoing assistance with applying a static treatment plan, which requires optimal frequency of use or degree of involvement by health professionals. Ultimately, the consumer's perception of benefit, convenience, and integration into daily activities facilitates the successful use of the interactive technologies for the elderly, chronically ill, and the underserved (Jimison, Gorman et al. 2008). Additionally, the care management plans of residents needs to be adjusted often and this change has to be clearly communication back to the patient with tailored recommendations. Present systems have usability problems and are unreliable in nature. The convenience of using the technology has become an important factor. Data entry by providers is often a cumbersome task and the intervention needs to fit into the user's daily routine. It has been noted that the frequency of interactions with a clinician helps in improving the patients satisfaction (Jimison, Gorman et al. 2008).

At present, most information systems in use lack a holistic view to see the relevant or most important health information related to a patient's condition. Personalized information for patients is often not available to the providers that can aide in enhanced clinical decision support. Most systems are not interoperable and do not have the facility to share information or communicate among themselves. The patient information is not portable and the user view of the data is not tailored according to the information or treatment needs of the patient. Many believe that a shared understanding of medical conditions between patients and their health care providers may improve self-care and outcomes (Malik, Giamouzis et al. 2011).

2 PROPOSED GOALS / OBJECTIVES

The primary goal of the research is to identify and develop the design considerations of LTC EMR that might benefit LTC settings. It focuses towards meeting the information needs of the providers for taking better-informed clinical decisions for the elderly. The secondary goal is to evaluate the design consideration in the form of a prototype to find an effective way of representing the data available in a holistic manner at TigerPlace. It should enable the providers to tailor health information according to the disease condition and personalize it according to the needs of the elderly. The design prototype additionally supports understandability and usability of the end user.

3 METHODS

We analyzed the information needs at TigerPlace to understand the overall information needs at LTC settings. Information related to the systems available at TigerPlace was obtained from thorough literature review of the research publications available. Information was also obtained from existing literature about the needs of LTC facilities, which provides the potential to borrow some of the ideas from TigerPlace to fulfill the LTC needs.

Information related to the technology employed at TigerPlace that enables residents to live independently was obtained. Information related to the types of data present in the systems was collected from the data managers, who work with the data collected at intervention levels of the residents. Several field trips to TigerPlace were done and interviews were conducted with the registered nurses and nursing aides to obtain important insight about the flow of clinical information there. The interaction and information enabled understanding and evaluation of the current shortcomings of the system. The research also focused in identifying relationship among the telemedicine data and sensor data for certain medical conditions taking CHF into consideration, which is a common health condition for the elderly residents. It supported exploring an effective holistic health data integrated display for the care providers, interested in monitoring the residents' health status. The analysis of the information needs of care providers at TigerPlace helped in understanding similar needs in other LTC setting and what could be borrowed from the technological perspective and extended to LTC facilities e.g. EMR requirements focused for LTC settings.

A dashboard design prototype was developed using human centered design framework, UFuRT (Zhang, Patel et al. 2002) that provides a holistic view of the user, task, function and representation from the three systems using a single interface for the care providers. Relational data is present in different formats i.e.

numbers, symbols, colors etc. (Zhang 1991; Zhang 1996) and supports the representation effect of the data, which is an important component of the residents' health records in TigerPlace. Numbers are presented in forms of Relational Information Displays (RIDs), which represent the relationship between dimensions. The two basic types of dimensions are: represented dimensions (the dimensions of an original domain in the world) and representing dimensions (the physical dimensions representing the dimensions of the original domain in the world). These two dimensions need to be matched in different scales for accurate representation between the system display and the real world (Zhang 1996). The RIDs represent relations among type of information i.e. a patient's name in the EMR along with his sensor data constructs multiple relations.

4 RESULTS

4.1 Design considerations for effective clinical decision support in LTC EMR

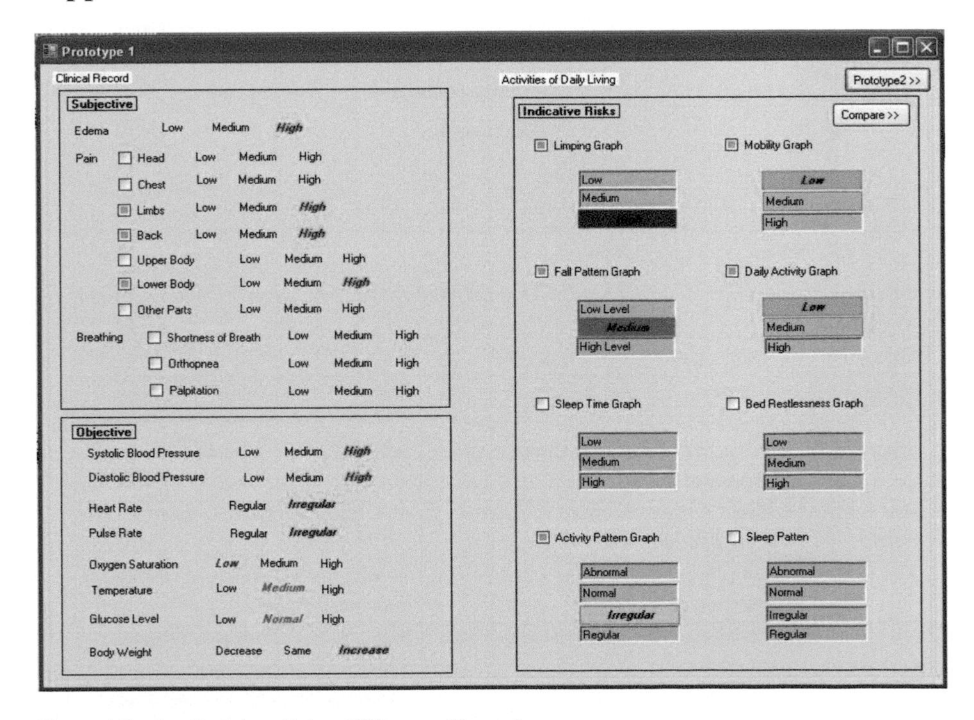

Figure 1 Design Prototype Using CHF as an Example

The designed prototype enables a quick synopsis of a CHF resident at TigerPlace, presented in Figure1. The designed prototype alone cannot assist clinicians with effective care coordination among themselves by just having a view of the prototype. It is essential to provide additional tools that will enable care providers to do easy documentation, care coordination and communication among themselves. These tools should have embedded usability features and adopting providers to use the new system and visualize changes in the health status of the residents. Figure 2 represents the relation of nursing documentation along with the nursing assessment needs (Kane and Kane 2000). It's a high level view of the different nursing assessments at LTC settings and how LTC EMRs can support them.

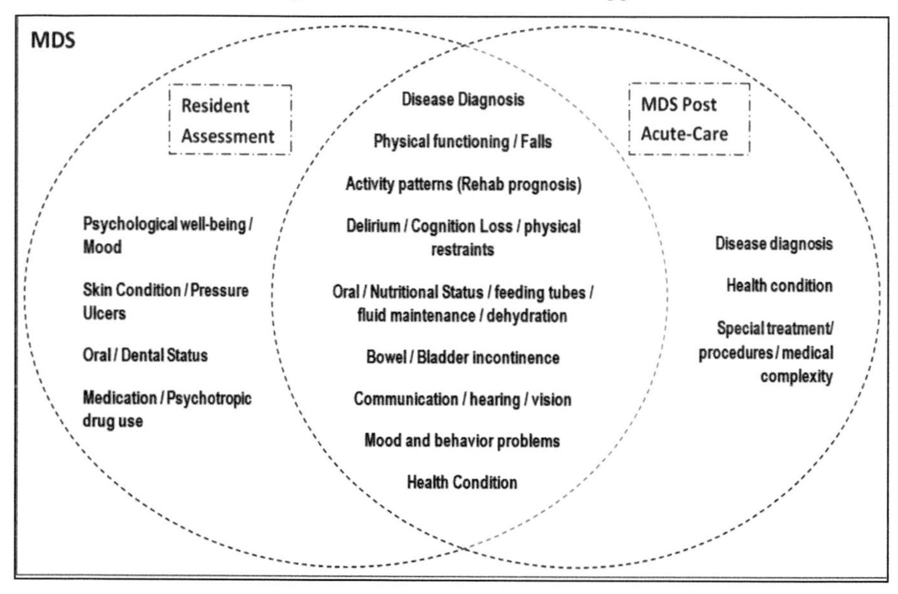

Figure 2 Relation between Different Resident Assessment Needs in a Minimum Data Set (MDS)

Table 1 First Design Considerations

Purpose	Approach	Explanation
Visibility of all health related information	Holistic representation of ADL functions & graphs, nursing notes, progress trends of residents' multiple chronic conditions	Prediction of emergent problems of residents earlier than the actual appearance of the symptoms
Better disease prediction	Availability of ADL functions, vital signs, nursing notes and evidence-based medicine	Aid for better clinical and nursing assessment
User-friendly	Using minimal number of screens and fitting into the workflow of the care provider	Visualizing change of care plans with advent of time for accuracy and quality
More efficient clinical decision-making	Reconciliation of nursing documentation in a single note section	Nursing documentation can be done from any available screens in the prototype
Avoiding information overloaded	Personalization of the resident information according to individual resident needs	Tailoring appropriate health information based on providers' and residents' needs

The first consideration goes to the ability of the tool to be used or embedded with the current EMR/EHR system as shown in Table 1.

The tools cannot function efficiently without the help of a prototypes designed similar to the one designed for the CHF scenario. The second consideration goes to the ability of the tool along with a designed prototype as shown in Table 2.

Table 2 Second Design Considerations

Purpose	Approach	Explanation
Taking nursing documentation while viewing relevant health information	Documenting process and clinical workflow to be structured with the relevant health information e.g. list of problems, vitals & ADLs, etc.	Nursing documentation to be analogous to the information displayed on the screen and once notes saved from any screen should display on the main nursing notes
Features to review documentation	Individual electronic signatures with date and timestamp on notes	Button and additional comment section for amendments on notes by peers
Usability: less is more	Limiting number of screens from 3-5	Easy printing of notes sections
Prioritizing visibility of information	Sort historical health information based on date range, visit, priority etc.	Examples are latest hospitalization report, discharge summary, etc.
Easy visualization	Using colors or numbers for visual effect	Flagging data with colors for prioritization
Easy comparison of information	Prediction of trends and thresholds for ADL functions and vitals	Comparing 2-3 trend graphs over a date range
Information security	Access to health information of users governed by levels of security and confidentiality	Registered nurse getting access to more confidential information compared to physical therapist

5 DISCUSSION

The design considerations listed above are limited to our findings. Since, the elderly residents keep moving to and from acute care to LTC, care providers require information to be exchanged between the two care settings. Health information needs to be made available to providers for better treatment decisions and facilitating interoperability of the same plays a pivotal role in effective clinical decision-making. It is necessary that the health status and related health information of the residents are available to the care providers at the point of service (POS) using mobile technologies. The LTC EMR should facilitate embedding nursing assessment templates for easy clinical documentation. The minimum data-set (MDS) documentation is necessary for the nursing assessments for quality and payments(Resnick, Manard et al. 2009). The documentation of quality of life (QOL) indicators also needs to be supported by the nursing assessments embedded in the LTC EMR (Kane and Kane 2000). The healthcare needs and assessments of residents at LTC are very different and our design considerations focuses distinctly on these needs. Residents at LTC are typically monitored round the clock and even a small decline in their health status may become an emergent condition, as the elderly are frail and more vulnerable to health deterioration compared younger adults. LTC settings also have different quality goals related to patient safety and care providers have to look at different assessments and intervention outcomes for quality reporting and payment for health services. Evaluation is further required on how our design analysis and prototyping can be extended to different kinds of LTC settings e.g., semi-skilled nursing and assistive living. Research is also required to find the information needs of home health and rehabilitation for the senior residents. The design also needs considerations related to the needs of different caregiver user groups at LTC facilities.

6 LIMITATION OF THE STUDY AND FUTURE WORK

Our study focused on the present literature review of the aging in place model e.g., TigerPlace, and current state of LTC EMR. The prototype design and the design considerations for LTC EMR need to be enhanced by detailed interviewing with care providers, which is an ongoing process and may provide important insights into the areas that have yet not been determined. The process of interviewing, transcribing the information and notes to be taken during the interview together will provide requirement specification for the design enhancement. There is a plan to analyze the collected information categorize them further for evaluating them more granularly. Apart from just elaborating the high-level design requirements of LTC EMR, the threats also need to be evaluated and discussed.

ACKNOWLEDGMENTS

The project was supported by Research Enrichment and Dissemination (READ) grant from the Interdisciplinary Center on Aging at the University of Missouri.

REFERENCES

Asteljoki, S., J. Kesanen, et al. (2009). "Documentation of pain management with TT- 2000+ in May 2008 viewed from clinical nursing practice." Studies in Health Technology & Informatics **146**: 503-505.

Aud, M. P., M; Rantz, MJ; Skubic, M; Alexander, G; Koopman, R; Miller, S (2010). "Developing a Comprehensive Electronic Health Record to Enhance Nursing Care Coordination, Use of Technology, and Research." Journal of Gerontological Nursing **36**(1): 13-17.

Cowan, D. T.-S., Alan (1999). "The Role of Assistive Technology in Alternative Models of Care for Older People." Research, HMSO **2**: 325-346.

Gong, Y. and A. Chandra (2011). "Developing an integrated display of health data for aging in place." Human Factors and Ergonomics in Manufacturing & Service Industries: 1-10.

Gunningberg, L., M. F. Dahm, et al. (2008). "Accuracy in the recording of pressure ulcers and prevention after implementing an electronic health record in hospital care." Quality & Safety in Health Care **17**(4): 281-285.

Gunningberg, L., M. Fogelberg-Dahm, et al. (2006). "Nurses' perceptions of feed-back from the electronic patient record for the quality on pressure ulcer care." Studies in Health Technology & Informatics **122**: 850.

Gunningberg, L., M. Fogelberg-Dahm, et al. (2009). "Improved quality and comprehensiveness in nursing documentation of pressure ulcers after implementing an electronic health record in hospital care." Journal of Clinical Nursing **18**(11): 1557-1564.

Jimison, H., P. Gorman, et al. (2008). "Barriers and drivers of health information technology use for the elderly, chronically ill, and underserved." Evidence Report/Technology Assessment(175): 1-1422.

Kane, R. L. and R. A. Kane (2000). "Assessment in long-term care." Annu Rev Public Health **21**: 659-686.

Kesanen, J., S. Asteljoki, et al. (2009). "The utilization of electronic patient record in pain management from the perspective of nursing management." Studies in Health Technology & Informatics **146**: 171-173.

Malik, A. S., G. Giamouzis, et al. (2011). "Patient perception versus medical record entry of health-related conditions among patients with heart failure." American Journal of Cardiology **107**(4): 569-572.

Marceau, L. D., C. Link, et al. (2007). "Electronic diaries as a tool to improve pain management: is there any evidence?" Pain Medicine **8 Suppl 3**: S101-109.

Rantz, M. J., K. D. Marek, et al. (2005). "A technology and nursing collaboration to help older adults age in place." Nurs Outlook **53**(1): 40-45.

Resnick, H. E., B. B. Manard, et al. (2009). "Use of Electronic Information Systems in Nursing Homes: United States, 2004." Journal of the American Medical Informatics Association **16**(2): 179-186.

Skubic, M., G. Alexander, et al. (2009). "A smart home application to eldercare: current status and lessons learned." Technol Health Care **17**(3): 183-201.

Zhang, J. (1991). "The Interaction of Internal and External Representations in a Problem Solving Task." Proceedings of the Thirteenth Annual Conference of Cognitive Science Society.

Zhang, J. (1996). "A representational analysis of relational information displays." International Journal of Human-Computer Study.

Zhang, J., V. L. Patel, et al. (2002). "Designing human-centered distributed information systems." Intelligent Systems, IEEE **17**(5): 42-47.

CHAPTER 37

Reducing Interruptions, Distractions, and Errors in Healthcare

Gary L. Sculli, Amanda M. Fore

Department of Veterans Affairs
National Center for Patient Safety
Ann Arbor, Michigan, USA

ABSTRACT

Nurses represent the largest component of the healthcare workforce and their actions are directly related to patient outcomes. Despite the importance of critical nursing tasks, such as assessment and medication administration, nurses are required to operate in complex, and highly interruptive, work environments. In aviation, sterile cockpit methodology prohibits non-essential activities during critical phases of flight such as, takeoff and landing. These flight regimes are analogous to the task of medication administration in nursing practice. As part of a larger crew resource management program, three nursing units implemented sterile cockpit methodologies while completing specific safety sensitive tasks. Two units implemented this approach during medication administration, and another implemented the methodology during the collection of vital signs and blood glucose readings at the beginning of each shift. Results suggest a decrease in distractions and interruptions, medication errors, and amount of time needed to complete these tasks. Moreover, the sterile cockpit principle led to improvements in team communication and staff morale. The use of crew resource management techniques, including the application of aviation's sterile cockpit rule, can have a significant impact on patient safety.

Keywords: nursing, aviation, distractions, errors

1 CREW RESOURCE MANAGEMENT

Starting in 2010, the Department of Veterans Affairs (VA) National Center for Patient Safety (NCPS) implemented a nursing centered patient safety program on over 20 nursing units across the healthcare system. Over 700 nurses were trained as part of a program called Clinical Crew Resource Management (CCRM) (Sculli, et al., 2011). CCRM applies high reliability tools and behaviors, found in the airline industry, directly to the nursing unit to reduce the risk of patient harm. The program consists of five didactic modules and a practice session using high fidelity simulation (Table 1). Training highlights strategies to build teamwork and open communication, and provides assertive communication algorithms used to safely resolve clinical problems. The program identifies threats to situational awareness and emphasizes countermeasures for managing those threats to optimize clinical decision making. In addition, participating units implement crew resource management (CRM)-based projects, receive monthly consultation, and attend recurrent training sessions one year after the initial training.

Table 1 Components of Clinical Crew Resource Management in the Veterans Health Administration

Program Component	Description
Pre-work & Logistics	(3-4) 30-60 minute phone calls 6-8 weeks before learning session.
Didactic Learning Session	4.5 hour interaction including: crew resource management and safety culture; leadership, followership, and assertive communication; situational awareness; briefings, debriefings, and checklists, reducing distractions and countering fatigue
Clinical Simulation Module	2 hour scenario session using high fidelity human patient simulator
Unit-Based Project Implementation	Unit-based implementation of: briefings and debriefings, clinical checklists, sterile cockpit methodologies (reducing distractions), or situational awareness strategies, fatigue countermeasures
Recurrent Training	4.5 hour session including the interactive didactic learning session and clinical simulation combined

1.2 Crew Resource Management in Nursing

Nurses represent the largest component of the healthcare workforce and perform critical tasks to detect errors and prevent adverse events (Institute of Medicine (IOM), 2004). Nurses provide 24-hour patient care and what they do or fail to do is directly related to patient outcomes (Doran, 2011); yet, nurses are often required to perform critical tasks in complex and highly interruptive work environments.

Medication administration involves a tightly coupled mixture of steps that are interrelated and sequential. Errors, including omissions, during medication delivery can have dire consequences for hospitalized patients. An estimated 770,000 people suffer injury or death annually in hospitals as a result of adverse drug events (IOM, 2004). Likewise, initial patient assessment, which involves the "early in shift" collection of specific physiological information such as vital signs and blood glucose values, is critical in formulating a plan to manage patient care.

Despite the importance of these critical nursing tasks, they are often carried out in environments replete with distractions and interruptions. Distractions and diversions can be environmental, but are also a byproduct of the cultural norms that structure work design. The occurrence and frequency of distractions and interruptions during critical tasks are significantly associated with negative consequences. In a 2010 study, Westbrook et al. reported that each interruption was associated with a 12.1% increase in procedural failures and a 12.7% increase in clinical errors, independent of hospital and nurse characteristics. A study by Hall et al. (2010) showed that 90% of interruptions to nursing practice resulted in negative outcomes such as delays in treatment and loss of concentration and focus.

2 STERILE COCKPIT METHODOLOGY

Sterile cockpit methodology is used on airline flight decks to manage distractions. The principle, derived from aviation statute, applies to all flight operations below 10,000 feet. It states that, "no flight crewmember may engage in, nor may any pilot in command permit, any activity during a critical phase of flight which could distract any flight crewmember from the performance of his or her duties or which could interfere in any way with the proper conduct of those duties". The rule also mandates that airlines cannot require pilots to assume additional duties not directly related to the safe operation of the aircraft during a critical phase of flight. The rule has teeth and is headed by the professional culture.

Implementation of the sterile cockpit principle in healthcare holds promising results for patient safety with regard to tasks that demand concentration and thought work. First applied in medical surgical nursing in 2003 during a study by Teresa Pape, sterile behaviors were shown to markedly decrease distractions and interruptions during medication administration. Pape's multifaceted program included a "Med-Safe" protocol which applied a team approach to medication delivery. The protocol centered on avoidance behaviors used by nurses to deflect distractions and interruptions. For example, medication nurses avoided all non-essential conversation and interaction during medication administration. In addition, support staff intercepted phone calls and other diversions in an effort to shield and protect the medication nurse. The most salient feature of the protocol was the use of a vest worn during the medication delivery period. The vest, displaying the words "Med-Safe Nurse Do Not Disturb", denoted medication delivery as a critical task and served as a visual warning to would be distractors. Sculli and Sine (2011) further describe implementation of the sterile cockpit concept using a plain vest along with removable signs placed on a medication delivery cart only during medication administration times (Figure 1).

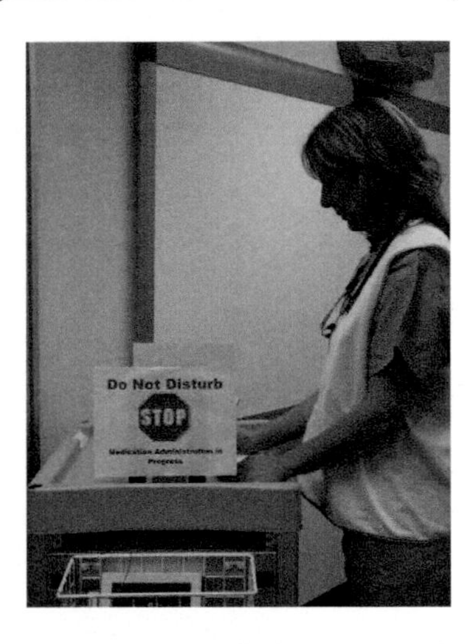

Figure 1 Medication nurse dons yellow vest and displays warning signs during medication delivery

3 IMPLICATIONS IN NURSING

This paper describes three unique applications of sterile cockpit methodologies. Each case describes a unit-based sterile cockpit project implemented at VA medical centers as part of the CCRM program.

3.1 Case 1

Nurses on a 30-bed medical-oncology unit in the Mid-Atlantic VA Healthcare System reported multiple distractions, particularly during peak medication administration times. To decrease the number of distractions and interruptions, this unit implemented sterile cockpit methodologies using "Do Not Disturb" signs and orange vests worn by a medication delivery nurse. In addition, staff members and patients were educated about the project and staff members were asked to protect the medication nurse by intercepting distractions (Fore, et al., in press). Using a self-reported data collection tool, the number and type of interruptions were collected. Furthermore, medication error rates were reported one year prior to, and after, implementation (Fore, et al., in press).

Similar to Pape's studies (2003 & 2005), Fore et al. (in press) showed a decrease in the number of distractions and interruptions when implementing sterile cockpit methodologies. The mean number of interruptions decreased from 4.1 at week 1 to 1.5 at week 11 ($p=.10$); simple linear regression analysis showed a decrease in the mean number of distractions over time ($\beta=.193$, $p=.02$) (Fore et al., in press). The most significant decrease included the mean number of interruptions from hospital staff ($p<.05$) and patients ($p<.05$) (Fore et al., in press). This is likely attributed to education efforts. More importantly, the annual medication error rate, one year prior to and after implementation, decreased by 42.78% ($p<.05$) (Fore et al., in press). These findings have positive implications for patient safety and the reduction of costs associated with adverse drug events.

3.2 Case 2

Embracing CRM strategies and tools, a 36-bed medical-surgical oncology unit in the VA Mid South Healthcare Network implemented sterile cockpit procedures during medication administration. Strategies included the use of a medication administration checklist, "Do Not Disturb" signs and vests, and staff running interference for the medication nurse (McCarthy & Chase, 2011). The medication checklist (McCarthy & Chase, 2011) was designed to ensure the five rights of medication: right medication, right dose, right patient, right time, and right route. The number of distractions and the time required to complete medication administration was assessed using three groups: 1) control group, 2) Certified Nursing Assistant (CNA) accompanying the medication nurse to intercept distractions, and 3) Registered Nurse (RN) and CNA supporting, but not accompanying, the medication nurse.

Running interference or using other staff members to intercept distractions to protect the medication nurse was a critical element in the project. When comparing the three groups, it was found that the group that included a CNA accompanying the medication nurse during medication administration experienced the least number of distractions. Nonetheless, running interference of any kind resulted in a 50% decrease in the number of distractions and interruptions. In addition, efficiencies were also realized. The time it took to complete medication administration decreased by 29% with the CNA accompanying the medication nurse and 19% when the team (CNA and RN) ran interference.

3.3 Case 3

An innovative application of the sterile cockpit rule was implemented on a 36-bed cardiac unit located in the southwest region of the VA healthcare system. Using sterile cockpit methodology, protected time was created for CNAs to collect patient vital signs and blood glucose values at the beginning of each shift (West et al., 2012). RNs and Licensed Vocational Nurses (LVNs) ran interference for CNAs by completing tasks that CNAs were typically required to perform, such as answering call lights, delivering messages, and assisting patients with activities of daily living. Time required to complete patient assessments, medication passes, vital signs, and blood glucose values were tracked (West, et al., 2012). In addition, staff perception of teamwork, communication, and moral was measured using an adaptation of the safety attitudes questionnaire (West, et al., 2012; Sexton, et al., 2006).

The process of applying sterile cockpit methodology during vital sign assessment and glucose monitoring by CNAs resulted in improved efficiency, communication, and staff morale. The most important success included improved patient safety related to the reduction in time needed for CNAs to collect and communicate information critical for clinical decision making (West et al., 2012). The average time required for CNAs to collect patient vital signs and blood glucose levels dropped from 2.5 to 0.5 hours (West, et al., 2012). Since critical patient information was available earlier in the shift, the average time required to complete nursing assessments decreased from 3 to 1 hour (West et al., 2012). Furthermore, staff responses to the safety attitudes questionnaire showed notable improvements in all areas of patient safety culture (West et al., 2012). Statistically significant improvements in staff morale were also seen (West et al., 2012).

4 CONCLUSIONS

Healthcare leadership must recognize that human error is ubiquitous and inevitable, and therefore support the importation of practices used in other safety sensitive industries. CCRM applies high reliability concepts and behaviors in clinical practice to reduce the risk of patient harm while positively influence safety culture. Applying sterile cockpit methodologies on acute care medical-surgical units is a feasible approach to reduce interruptions, distractions, and adverse events in healthcare.

ACKNOWLEDGMENTS

The authors would like to acknowledge those who have participated in and supported Clinical Crew Resource Management, especially the nurse managers and frontline nurses who demonstrate an inexorable commitment to our nation's Veterans.

REFERENCES

Doran, D.M. 2011. *Nursing Outcomes: The State of the Science.* Sudbury MA: Jones & Bartlett Learning.

Fore, A.M., G. Sculli, D. Albee, and J. Neily. in press. Improving patient safety using the sterile cockpit principle during medication administration: a collaborative, unit-based project. *Journal of Nursing Management.*

Hall, L.M., M. Ferguson-Pare, and E. Peter, et al. 2010. Going blank: factors contributing to interruptions to nurses' work and related outcomes. *Journal of Nursing Management* 18: 1040-1047.

Institute of Medicine. 2004. *Keeping Patients Safe: Transforming the Work Environment of Nurses.* Washington, D.C.: National Academy Press.

McCarthy, D. and D. Chase. 2011. Advancing patient safety in the U.S. Department of Veterans Affairs. *Commonwealth Fund* 9: 1-33.

Pape, T.M. 2003. Applying airline safety practices to medication administration. *MEDSURG Nursing* 12(2): 77-93.

Pape T.M., D.M. Guerra, and M. Muzquiz, et al. 2005. Innovative approach to reducing nurses' distractions during medication administration. *Journal of Continuing education in Nursing* 36(3): 108-116.

Sculli, G.L., A.M. Fore and J. Neily, et al. 2011. The case for training veterans administration frontline nurses in crew resource management. *Journal of Nursing Administration* 41(12): 524-530.

Sculli G.L. and D.M. Sine. 2011. *Soaring to Success: Taking Crew Resource Management from the Cockpit to the Nursing Unit.* Danvers, MA: HCPro.

Sexton, J., R. Helmreich, and T. Neilands et al. 2006. The safety attitudes questionnaire: psychometric properties, benchmarking data, and emerging research. *BMC Health Service Research* 6: 44.

West, P., G. Sculli, and A. Fore, et al. 2012. Improving patient safety and optimizing nursing teamwork using crew resource management techniques. *Journal of Nursing Administration* 42(1): 15-20.

Westbrook, J.I., A. Woods, and M.I. Rob, et al. 2010. Association of interruptions with an increased risk and severity of medication administration errors. *Archives of Internal Medicine* 170(8): 683-690.

CHAPTER 38

Identifying and Modeling Perceptions of Risk Factors in Hand Hygiene during Healthcare Operations

Kinley Taylor, David B. Kaber

Edwards P. Fitts Department of Industrial and Systems Engineering,
North Carolina State University, Raleigh, NC, 27695-7906, USA
kinley.b.taylor@gmail.com

ABSTRACT

Hand hygiene is considered to be one of the primary practices for preventing hospital acquired infections (HAIs). However, hygiene compliance remains low in many healthcare facilities. An in-depth analysis of risks to healthcare workers (HCWs) in common tasks was needed to further understand noncompliance issues. Human factors research methods were used to: (1) characterize and compare actual and perceived risks to hygiene among HCWs in clinical tasks; (2) identify individual and workplace predictors of hygiene procedure compliance; and (3) make recommendations for further promoting HCW hygiene and reducing HAIs. A taxonomy of risk factors to hygiene was compiled from the literature and used to assess actual and perceived frequency and severity of risk factor exposure. Sixteen participants, including registered nurses, nurse practitioners and phlebotomists, were recruited for the study. Half were observed in the target tasks and the remainder were surveyed on background and interviewed regarding hygiene risk factor exposure. Observations were made at two clinics and a lab in a single hospital. Deviations from established hygiene procedures and counts of actual risk factor exposure were recorded. Study results revealed extra-individual risks, including equipment and supplies, to have high observed and perceived frequency of occurrence. The highest rated risk factors for severity concerned training opportunities, such as lack of HCW concern for acquiring infections from patients and lack of knowledge of hygiene protocols. Examination of differences among actual and perceived risk ratings revealed the need for training on extra-individual and psychological factors in hygiene. Work location and risk factor exposure were also foundto be predictors of hygiene behavior. It was recommended that healthcare facility administrators focus on work design and HCW training towards preventing noncompliance and reducing HAIs.

Keywords: healthcare, hand hygiene, risk assessment, human factors methods, safety control measures

1 INTRODUCTION

Based on Semmelweis' (1847) seminal research, knowledge of the role of hand hygiene in hospital acquired infections (HAIs) has existed for more than a century. Hand hygiene is considered to be the practice of hand cleaning, with or without antiseptic, or using antiseptic handrub (Centers for Disease Control (CDC), 2002). A lack of hygiene promotes cross-transmission of microorganisms among patients and healthcare workers (HCWs). In general, pathways for exposure to infectious microorganisms include direct contact and airborne, and most commonly occur by the hand of a HCW. Unfortunately, hand hygiene compliance remains low in many healthcare facilities, averaging below 50% (Creedon, 2005; Pittet et al., 2004). The U.S. CDC made initial recommendations regarding hand hygiene over 20 years ago (1975) and has continually updated guidelines for healthcare facilities. Some fundamentals include: (1) decontaminaing the hands before and after contact with a patient; (2) wearing gloves when dealing with patient bodily fluids; and (2) use of alcohol-based hand rub when hands are not visibly soiled. Although such practices are simple, there are many ways hygiene can be impeded and neglected. Most previous hand hygiene studies have focused on HAIs from the patient's perspective. In-depth analyses of hygiene risks for HCWs in common tasks are needed to further understand noncompliance issues. In particular, there is a lack of information on differences among actual risk factors in the workplace and HCW perceived risks as well as individual and workplace factors mediating perceptions.

2 METHODOLOGY

A preliminary descriptive study, with limited sample size, was conducted to characterize actual and perceived risks to hand hygiene among HCWs in two common clinical tasks. There were six steps to this study:

(1) compilation of a taxonomy of risk factors to hygiene based on the literature;
(2) direct observation of HCWs performing the clinical tasks to determine the actual frequency of exposure to hygiene risks;
(3) direct observation of task performance to assess HCW compliance with hygiene protocols (e.g., washing after removing gloves) and time-to-task completion;

(4) a survey of HCW background information;
(5) HCW interviews to capture perceptions of the frequency of risk exposure and severity of outcomes; and
(6) statistical analyses to describe how: (a) actual risk exposure relates to perceptions of risk; (b) predictors of hand hygiene (e.g., professional position, experience, work location, etc.) relate to actual compliance; and (c) predictors of hygiene relate to perceptions of risk exposure.

With respect to the taxonomy of risk factors, multiple sources were used (Barrett & Randle, 2008; CDC, 2002; Hugonnet & Pittet, 2000; Jumma, 2005; Larson et al., 1991; O'Boyle & Henly, 2001; Pessoa-silva, et al., 2005; Pittet, 2001; Pittet et al., 2004). All factors were organized according to Swain's (1976) human performance shaping factor taxonomy (see Figure 1).

Regarding the direct observation of work, the two clinical tasks under study were injection and venipuncture. These tasks were observed at three different facilities in a single hospital, including an Outpatient Clinic, an Occupational Health Clinic and a Phlebotomy Lab. Sixteen participants were recruited for the study, including registered nurses (RNs), nurse practitioners (NPs) and phlebotomists. Half were observed performing the target tasks, including seven females and one male. In making observations on compliance, binary response measures were recorded including: (1) Performs proper hand sanitation before donning gloves?; (2) Dons gloves correctly?; and (3) Performs proper hand sanitation after removal of gloves?

The remaining participants were surveyed as to their background and interviewed regarding risk factors to hygiene. All were female with a mean age of 48 years. The survey also captured their profession, years of experience, frequency of reporting injuries, and frequency of being sick from contact with patients. The interviews posed participants with lists of risks factors included in the taxonomy that had relevance to injection and venipuncture. Participants were asked to identify which factors occurred in performing each task. For each risk factor selected, participants used an adaptation of the Military Standard 882B – Hazard Severity Classification Table to provide ratings of severity of exposures to hygiene risks. The severity scale of the standard was adapted to fit the HCW's range of outcomes from negligible to death. The perceived severity ratings and perceived frequency of exposures were ultimately integrated to calculate a perceived risk.

Regarding the statistical analyses, descriptive statistics (mean and range) were used to identify risk factors with high frequency (actual and perceived) and perceived severity. Correlation analyses were used to determine the degree of association among the actual and perceived responses. Contingency tables were used to assess role of: 1) predictor variables, including gender, work location, task time and actual risk exposure, in compliance rates; as well as 2) the predictors of profession, work location, age and medical experience in perceptions of risk.

2.1 Taxonomy of hygiene risk factors

Figure 1 presents the taxonomy of hand hygiene risk factors developed from the literature review and evaluated in this study. In Swain's taxonomy of human performance factors, the major classes include extra- or intra-individual. Subgroups to external factors include situational characteristics and task or equipment features. Subgroups to internal factors include psychological or physiological stress, "organismic" factors (individual differences), and understanding of job or task methods. We used the same subgroups in our taxonomy of hygiene factors.

Figure 1: Taxonomy of risk factors potentially compromising hand hygiene.

Of the risk factors in the taxonomy, some were observable during injection and venipuncture and some were not. For analysis purposes, the factors were divided into two groups consisting of observable and non-observable risks. In general, the unobservable factors included management policies, HCW beliefs and attitudes, HCW role models, staffing issues, and training and knowledge levels.

2.2 Hypotheses

Based on Parush et al. (2010) research, extra-Individual risk factors, particularly those concerning use of equipment and supplies, were expected to be prevalent in venipuncture and injection. Based on the HAI literature, risk factors with high potential for infection, including lack of hygiene facilities, overloaded clinicians and interference with patient care, were expected to have high ratings of severity and risk by the HCWs. In addition, based on Parush et al. (2010), clinician perceptions of risk factors in injection and venipuncture were expected to correspond to actual risk factors occurring in the environment. Beyond this, previously identified predictors of hand hygiene compliance, including work location and risk factor exposure, were expected to be associated with hand hygiene compliance (Pittet et al., 2004). Predictors of hygiene procedure compliance, such as age, were expected to be associated with perceived severity of outcomes, including how often HCWs reported getting sick by patient contact (Kumar & Burns, 2008).

3 RESULTS

3.1 Actual and perceived frequency of risks

Study results revealed several risk factors with high observed and perceived frequency of occurrence in the tasks. These factors were primarily extra-individual, including equipment and supply issues. For the venipuncture task, the factors with the highest actual frequency were: 1) working during the week versus weekend, posing higher risk of infection exposure; and 2) performing activities with a high risk of cross-transmission. Risk factors with the highest perceived frequency were: 1) lack of soap and paper towels; 2) patient needs take priority over hygiene; and 3) having many requirements for hand hygiene. For the injection task, factors with the highest actual frequency were identical to those for venipuncture, plus "a lack of available skin care promotion and agents." There were no risk factors that were considered to have a higher perceived frequency than others for injection.

3.2 Severity Ratings

Of the risk factors actually observed to occur frequently in tasks, those ranked with the highest severity did not overlap with the risk factors hypothesized as being critical to infections. For example, lack of hygiene facilities/supplies, HCWs being too busy for hygiene, or hygiene practice interference with patient care were all expected to have high severity ratings. However, the highest severity risk factors primarily concerned training issues. In venipuncture, factors included "perceived low risk of acquiring infection from patients", "lack of knowledge of guidelines or protocols", and "forgetfulness or not thinking about hygiene risks." The same factors were perceived to pose high severity in injection along with "performance of activities with higher risk of cross contamination."

With respect to perceived risk scores for the various factors (i.e., the product of frequency × severity), as would be expected, a subset of the factors with high perceived frequency and severity in venipuncture emerged as being most risky. The same was true for injection.

3.3 Comparison of perceived and actual risk exposures

In general, study results revealed differences between risk factors that phlebotomists and RNs/NPs perceived in their environment and what actually existed. For example, "failure to wear gloves" and "failures to address hand hygiene requirements" were observed to occur with high frequency but were not perceived to be prevalent in the work environment. In general, HCW comprehension of the importance of such risk factors appears to underestimate actual importance. Conversely, HCWs appear to overestimate the importance of risks to hygiene, including "inconveniently located or shortage of sinks", "lack of soap and paper towels", and other physiological factors, relative to actual frequency of exposure.

Non-parametric correlation analyses were used to assess the degree of association among the average actual frequency of exposure from observations with the average perceived frequency of occurrence from the interviews. Correlations were also determined for the average actual frequency with the average perceived severity. We found positive assocations of actual frequency with perceived frequency and severity for the extra-individual risk factor groups, including situational characteristics and task and equipment characteristics. The correlations indicated HCWs correctly perceived such factors to be risky in the work environment. However, there were also negative correlations between the actual frequency and perceived frequency and severity for risk factor groups including psychological stresses, intra-Individual organismic factors, and intra-individual job and task instructions. These correlations represented disagreement between HCW perceptions and reality.

3.4 Significant predictors of compliance and perceived risks

Contingency table analyses were used to identify significant predictors of hygiene procedure compliance and perceived severity of risk exposures. Predictors including work location and frequency of risk factor exposure were found to be useful in explaining observed hygiene compliance rates (see Table 1). Whether HCWs worked in one of the clinics (Occupational Health Clinic and Outpatient Clinic) or the Phlebotomy lab significantly predicted the use of gloves in venipuncture ($\chi 2=10.226$, $p=0.006$) and hand santizer after the task ($\chi 2=6.884$, $p=0.032$). There was also a significant difference in actual risk factor exposure ($\chi 2=16.048$, $p=0.0003$) among the clinics and lab. In general, the clinics had lower hygiene compliance and higher risk factor exposure in task performance than the Phlebotomy Lab. Some possible reasons for this finding could be large variations in the tasks performed and less standardized procedures in the clinics.

Table 1: Predictors of hygiene procedure compliance in venipuncture

PREDICTOR \ RESPONSE	Use of Sanitizer Before	Use of Gloves	Use of Sanitizer After	Risk Factor Exposure (Number of Risk Factors ≥4)
Work Location	Not significant (ns)	(M_{OH} = 33.3%; M_{PL} =100%)	(M_{OH} = 100%; M_{PL} =100%)	(M_{OH} = 100%; M_{PL} =0%)
Task Time	ns	ns	ns	ns
Risk Factor Exposure (Number of Risk Factors ≥4)	ns	($M_{<4}$ =100%, $M_{\geq4}$ = 25%)	ns	N/A

(Note: M_{OH} = Occupational Health Clinic mean; M_{PL} = Phlebotomy Lab mean; $M_{<4}$ = Less than four risk factors present mean; $M_{\geq4}$ = Greater than or equal to four risk factors present mean.)

With respect to the perceptions of the HCWs, the predictors of profession, location, age and years of experience were analyzed for effects on the responses of how often HCWs reported getting sick from their patients, how often they reported work injuries, and the total number of risk factors identified as being present during the specific task. Results are presented in Table 2.

Table 2: Predictors of perceived severity of outcomes in venipuncture

PREDICTOR \ RESPONSE	Perceived Sickness Caused by Patients	Report Work Injuries or Illnesses	Risk Factor Exposure (Number of Risk Factors ≥ 4)
Profession	Ns	ns	ns
Work Location	Ns	ns	ns
Age	($M_{\leq30}$=100%; $M_{>30,\leq40}$ =100 %; $M_{>40}$ =0 %)	ns	ns
Years of Medical Experience	ns	ns	ns

(Note: $M_{\leq30}$ = HCWs less than or equal to 30 years old mean; $M_{>30,\leq40}$ = HCWs between 31 and 40 years old mean; $M_{>40}$ = HCWs over 40 years old mean.)

The analysis revealed a significant association between age and how often HCWs reported getting sick ($\chi 2$=7.638, p=0.0219; see Table 2). HCWs of greater age (over 40 years) reported getting sick less than younger HCWs. The venipuncture severity ratings provided no additional insights into the role of the predictor variables.

Contingency table results for the injection task revealed no evidence of effects of the predictor variables - work location, task time and risk factor exposure - on whether HCWs performed hand sanitation before the task, used gloves correctly, followed proper sanitation procedures after the task, or in the level of exposure to risk factors during injection observations. We also did not find HCW profession, location, age or experience to be significant in explaining the perceived severity ratings for risks in injection tasks.

4 DISCUSSION

Our first hypothesis was that risk factors in the extra-individual category of the taxonomy, including equipment and supplies, could be more prevalent than others during venipuncture and injection. This hypothesis was supported in that several risk factors had higher observed and perceived frequency of occurrence in tasks than other factors. In general, these results imply the most prevalent risk factors influencing HCW hygiene are generally out of their control. The factors represent specific aspects of the environment and medical tools, which could be controlled through engineering design or administrative/management interventions.

Our second hypothesis was that those risk factors with high potential for infection were expected to contribute to a high risk factor severity and overall risk score (assuming clinician comprehension). The highest severity risk factors rated in this study primarily concerned training issues and were similar to risk factors the CDC (2002) has identified as receiving limited coverage in current healthcare training programs and contributing to hygiene problems.

With respect to our third hypothesis, the perception of risk factors in injection and venipuncture did not closely correspond to the risk factors actually present in the work environment. This was contrary to conclusions of Parush et al. (2010). Furthermore, through correlation analyses, some risk factor groups revealed negative relations of actual and perceived frequency of occurrence. This suggested that HCWs perceived there to be other risks in the environment, which were not directly observed in the study. Based on the differences between actual and perceived frequency ratings for the two clinical tasks, it was evident that training of HCWs is needed to appropriately calibrate perceptions of risks, including extra-individual and psychological factors, in order to promote hygiene compliance and reduce HAIs.

Our fourth hypothesis was that previously identified predictors of hand hygiene compliance would be associated with actual hygiene compliance among the observed HCWs. In general, work location and risk factor exposure were predictive of compliance with hygiene procedures. Interestingly, increased risk factor exposure had a negative association with correct glove use (i.e., glove use decreased with a greater number of risks present). However, this suggests that high workload and time pressure, for example, may lead to HCWs taking "short cuts" in task performance.

Our last hypothesis was that predictors of hand hygiene compliance were expected to be associated with perceived severity of outcomes of risk factor exposure. We found that age was associated with getting sick less. This contradicts findings of Kumar and Burns (2008) that immunity decreases with age; however, the older HCWs had more experience and may have been more effective at protecting themselves from contracting infections.

5 CONCLUSION

Based on the results of the observational study and HCW interviews, recommendations were formulated for promoting compliance with hygiene guidelines. In general, healthcare administrations, and occupational health and infectious disease departments should focus on work design and training methods towards preventing noncompliance and reducing illnesses of staff and patients. Some example engineering controls include: ensuring hand hygiene facilities are accessible and improving substandard facilties; providing task flow charts in the work environment to maintain standardized procedures; identifying boundaries of work environments to maintain awareness of where hygiene practices should occur; and showing what constitutes a clean work environment. With respect to administrative controls, management should attempt to reduce high HCW workload by preventing understaffing through scheduling and integrate hygiene protocols among required steps for medical tasks. Regarding training, programs need to be developed to address high risk factors, including: skepticism on the value of hygiene, a lack of knowledge of guidelines, HCW disagreements with hygiene recommendations, and perception of no immediate consequence of lack of hygiene. HCWs also need to be trained on warnings regarding HAIs.

This was a preliminary investigation of hand hygience risk factors with restrictions on observations of clinical procedures and, therefore, the number of HCWs involved in the study. A larger investigation is needed for greater statistical reliability in results. One direction for future research is identification of actual individual risk factors to hand hygiene in outpatient clinics and diagnostic labs for which perceived risk may not be concordant.

ACKNOWLEDGMENT

This study was supported by a grant from the U.S. National Institute for Occupational Safety & Health (NIOSH) (No. 2 T42 OH008673-06). The opinions expressed in this paper are those of the authors and do not necessarily reflect the views of NIOSH.

REFERENCES

Barrett, R., & Randle, J., 2008. Hand hygiene practices: nursing students ' perceptions. *Journal of Clinical Nursing, 17*, pp.1851-1857.

Centers for Disease Control & Prevention., 2002. *Guideline for Hand Hygiene in Health-Care Settings: Recommendations of the Healthcare Infection Control Practices Advisory Committee and the HICPAC/SHEA/APIC/IDSA Hand Hygiene Task Force* (pp.1-48). [online] Available at:<http://www.cdc.gov/handhygiene/Guidelines.html>

Creedon, S. A., 2005. Healthcare workers ' hand decontamination practices: compliance with recommended guidelines. *Journal of Advanced Nursing, 51*(3), 208-216.

Hugonnet, S, & Pittet, D., 2000. Hand hygiene-beliefs or science? *Clinical Microbiology and Infection, 6*(7), pp.350-6.

Jumaa, P. A., 2005. Hand hygiene: Simple and complex. *National Journal of Infectious Diseases, 9*, 3-14.

Kumar, R., & Burns, E., 2008. Age-related decline in immunity: Implications for vaccine responsiveness. *Expert Review of Vaccines, 7*(4), 467-469.

Larson, E., McGeer, A., Quraishi, Z. a, Krenzischek, D., Parsons, B. J., Holdford, J., & Hierholzer, W. J., 1991. Effect of an automated sink on handwashing practices and attitudes in high-risk units. *Infection Control and Hospital Epidemiology, 12*(7), 422-8.

O'Boyle, Carol A, Susan J. Henly, E. L., 2001. Understanding adherence to hand hygiene recommendations : The theory of planned behavior. *American Journal of Infection Control, 29*(6), 352-360.

Parush, A., Campbell, C., Ellis, J., Vailancourt, R., Locketr, J., Lebreux, D., Wong, E., et al., 2010. A human factors approach to evaluating intravenous morphine administration in a pediatric surgical unit. In *Proc. of the 2010 AHFE Conference.* Boca Raton, FL: CRC Press.

Pessoa-silva, C. L., Posfay-barbe, K., Pfister, R., Touveneau, S., Perneger, T. V., & Pittet, Didier., 2005. Attitudes and perceptions toward hand hygiene among healthcare workers caring for critically ill neonates, *26*(3), 305-311.

Pittet, D., 2001. Compliance with hand disinfection and its impact on hospital-acquired infections. *American Journal of Infection Control, 48*, S40-S46.

Pittet, S. A., Hugonnet, S., Pessoa-Silva, L., Sauvan, V., & Perneger, T., 2004. Hand hygiene among physicians: Performance, beliefs, and perceptions. *Annals of Internal Medicine, 141*(1), 1-9.

Swain, A. D., 1976. *Sandia Human Factors Program for Weapons Development.* (Tech. Rep.: SAND 76-0326). Albuquerque, NM: Sandia National Laboratories.

Semmelweis, I. F., & Codell, C. K., 1983. The etiology, concept, and prophylaxis of childbed fever. *Journal of Public Health Policy*, 8, 582.

CHAPTER 39

Workflow Characterization in Busy Urban Primary Care and Emergency Room Settings: Implications for Clinical Information Systems Design

Atif Zafar, MD[1]
Mark Lehto, Ph.D.[2]
[1] Indiana University School of Medicine, Regenstrief Institute for Healthcare, Inc.
[2]Purdue University Department of Industrial Engineering

ABSTRACT

Clinical Information Systems have been identified by the Institute of Medicine as a viable solution for mitigating medical errors. In order for these systems to be widely adopted they have to be designed to conform to the workflows within complex care settings. We studied the workflows within two highly complex settings (emergency care and outpatient clinic) within a large metropolitan hospital system. We found inefficiencies that could be improved by the use of clinical information systems. These tended to focus on improved communication among staff, mobile data access and entry, and better situational awareness and task orientation within highly interruptive care settings. We propose some software and hardware design elements that could be adopted to make clinical information systems in these settings more efficient and user-friendly.

1. INTRODUCTION

Inefficient processes can add a significant cost burden to any business endeavor. This is especially true in healthcare where such inefficiencies cost the US healthcare system an estimated $1.2 trillion a year (half of the total annual healthcare budget)[4]. Such inefficiencies have the potential to introduce medical errors that can subsequently result in increased morbidity and death. Indeed, the Institute of Medicine reported in 2001 that medical errors contribute to over 44,000 unnecessary deaths each year[5]. Much of the waste results from care fragmentation, inefficient processes, inefficient or non-existent communication between providers, in-homogeneities in the standards of care across care settings and geographic regions, inequitable access to care, poor patient education and costly, often un-needed, testing and treatment (including duplicate testing)[4]. Solutions to this problem lie in technological innovations that improve communication and coordination of care between healthcare providers as well as enabling decision support at the point of care[5]. Implementing such new technologies, however, necessitates a change in clinical and administrative workflows, something very few individuals and organizations are often willing to engage in unless there is a clear value proposition[6].

In order to understand some of these barriers to adoption of health information technology we first sought to better understand the workflow processes of providers and staff in busy ambulatory and emergency room (ER) settings at a major teaching hospital associated with an academic medical center. We conducted workflow-sampling (time-motion) studies in these environments in order to better understand the inefficiencies and develop solutions to overcome these inefficiencies. During several successive academic years, groups of senior undergraduate students enrolled in an industrial engineering program at Purdue University conducted several separate time motion studies of workflows in the ambulatory and ER environments. We report on the results of these studies in this paper and discuss implications for clinical information system design.

Our methodologies apply standard industrial engineering (IE) principles to the healthcare setting. Industrial Engineering focuses on a "human factors" approach whereby the *user* is at the center of design rather than the process to be developed[3]. Thus it deals with how best can a *given user* (novice or expert) complete *this* task? The various tools available for analysis then deal with a "man-in-the-loop" type approach. Some of the tools include *task analysis* (time-motion observation and expert interview), *simulation* (especially of discrete events), *statistical modeling* (of process chains and fault analysis – also known as statistical process control) and *process optimization techniques*[3].

2. MATERIALS AND METHODS

During the academic years 2006-2009, three groups of four senior undergraduate students enrolled in the IE-431 Senior Design Class at Purdue University's Department of Industrial Engineering convened and studied the workflows within three busy clinical care settings associated with a major academic medical center. Three separate studies were performed, two in the ambulatory care setting and one in the ER.

In 2006 we conducted an initial time-motion study of the workflows in ambulatory site A (Wishard) and then subsequently performed a validation study in 2007 at ambulatory sites A and B (Cottage) to evaluate whether the same groups of tasks are common to other ambulatory internal medicine clinics in the same care system. We then applied the same methodology to site C (ER setting) in 2009.

2.1 Sample Characteristics

For the first study the times of observation were from Feb 20, 2006 to March 27, 2006. A total of 61.21 hours of observation was performed at site A. The times of the observations ranged from the 8 am to 4 pm. Four students observed 5 staff types (check-in clerk, check-out clerk, nurse, nurse's aide and doctor) and 6 actual individuals in these roles.

We then performed a validation study of the workflow types from January 20, 2007 to April 15, 2007. We replicated the study at the same site (A) as in Study #1 but then also extended to another site (Site B). A different group of 4 undergraduate students observed 5 types of staff in the work environments. Approximately 8 hours of observation was performed during this study at various points of the day, both morning and afternoon.

For the third study in the emergency department setting, another four groups of students observed and recorded the activities of several staff types including doctors, nurses and clerks. Each team member shadowed one ER doctor for 3 hours. The study was conducted during March and April of 2009.

2.2 Study Procedures

During the first study, the students initially developed taxonomies to describe the tasks of a medical clinic. This was done by interviewing, observing and timing each of the different staff positions at the clinic. During actual time observation additional elements were added to the taxonomy or clarified as needed. The study itself was composed of 61.21 hours of time and observational data collected at a single clinic.

During the second (validation) study there were a total of two clinics studied. The task codes created by the previous group were used as much as possible in order to help validate the study. When new task codes were need they added to the taxonomy and a notation made as to which clinic utilized those procedures. At the first clinic, each position was studied for three hours. Eight hours were collected at the second clinic, giving two hours for each position except the doctor position. Due to time limitations the amount of data collected led to only the ability to do basic analysis such as percentage of time spent on a task.

For the third study (in the ER), team members shadowed one ER doctor for three hours by observing them every 30 seconds. First, the location (see diagram below) of the doctors was recorded ("O" Observation, "AT" Ambulatory Triage, "FH" Front Hall, "H" Holding, "T" Trauma and "W" Staff Workstation). Second, the category of the work that the doctor is performing was recorded and classified into one of these categories: Patient data, Other computer, Ordering test via computer, Walking, Patient care, Conferring with staff, Ordering test via paper, Idle, Other.

After the data was collected, a time study analysis was performed to determine what consumes most of the ED doctors' time, and in particular, which computer tasks (such as ordering tests, querying patient data) took the longest time to complete. In addition, the team studied the travel patterns and frequencies of the doctors in the physical layout at the hospital.

3. RESULTS

Studies 1 and 2: Time motion evaluation of ambulatory care workflows

The data from the first two studies in the ambulatory setting is summarized in the **Figure 1-5** below. The Appendix lists the workflow codes as depicted in these charts.

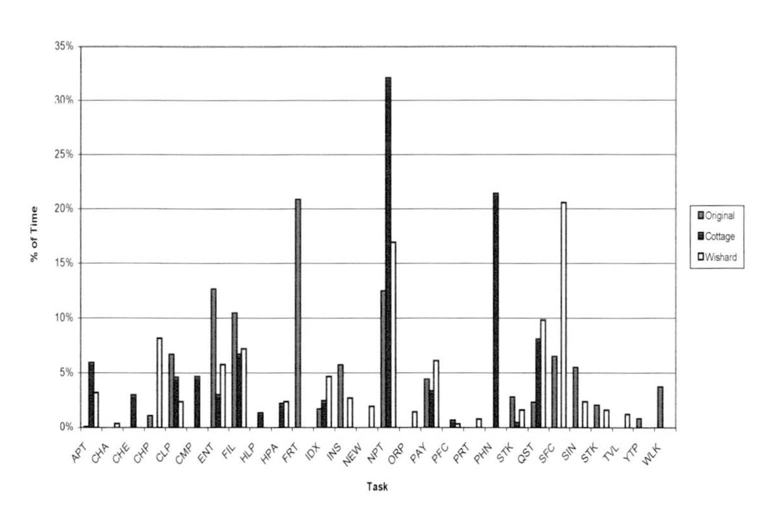
Check-In Nurse
35%
30%
25%
20%
15%
10%
5%
0%
% of Time
Original
Cottage
Wishard
APT
CHA
CHE
CHP
CLP
CMP
ENT
FIL
HLP
HPA
FRT
IDX
INS
NEW
NPT
ORP
PAY
PFC
PRT
PHN
STK
QST
SFC
SIN
STK
TVL
YTP
WLK
Task

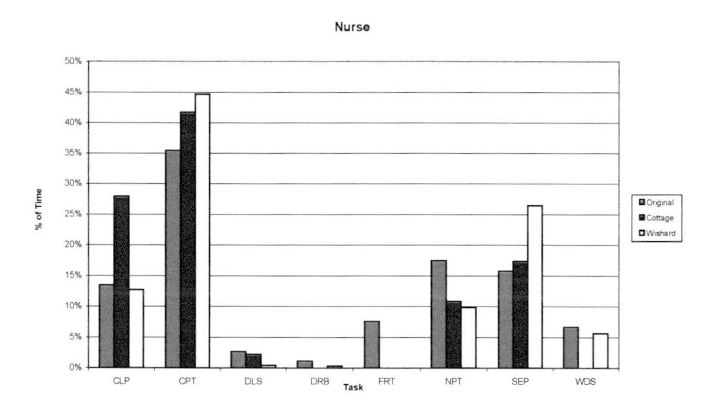
Nurse
50%
45%
40%
35%
30%
25%
20%
15%
10%
5%
0%
% of Time
Original
Cottage
Wishard
CLP
CPT
DLS
DRB
FRT
NPT
SEP
WDS
Task

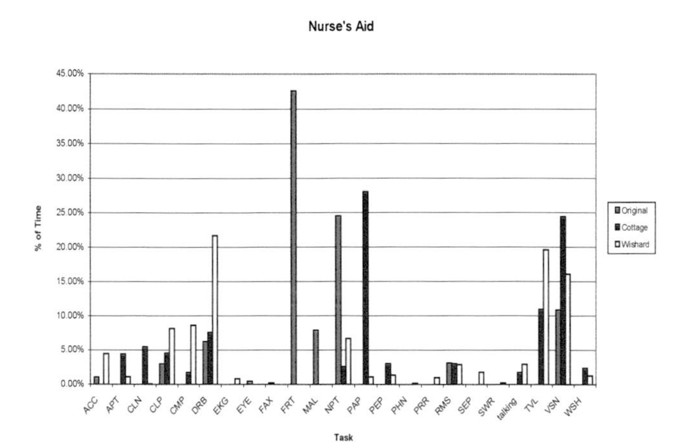
Nurse's Aid
45.00%
40.00%
35.00%
30.00%
25.00%
20.00%
15.00%
10.00%
5.00%
0.00%
% of Time
Original
Cottage
Wishard
ACC
APT
CLN
CLP
CMP
DRB
EKG
EYE
FAX
FRT
MAL
NPT
PAP
PEP
PHN
PRR
RMS
SEP
SWR
talking
TVL
VSN
WSH
Task

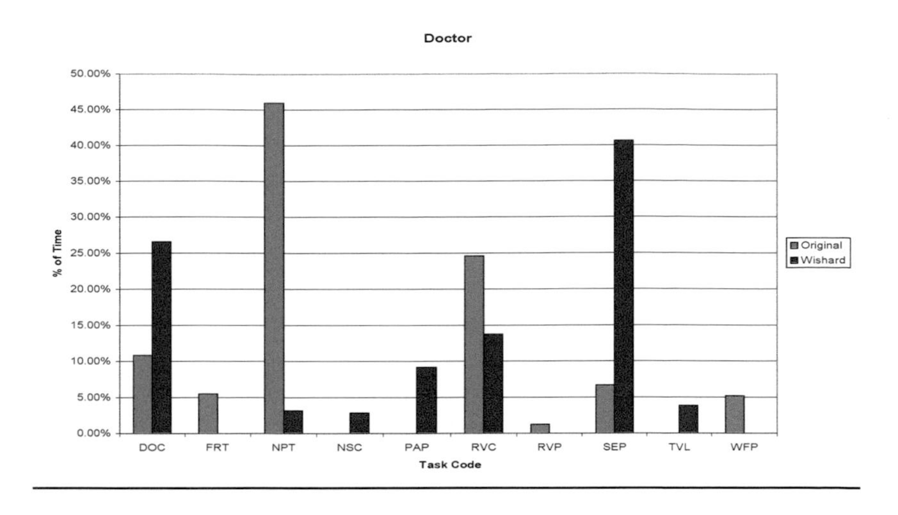

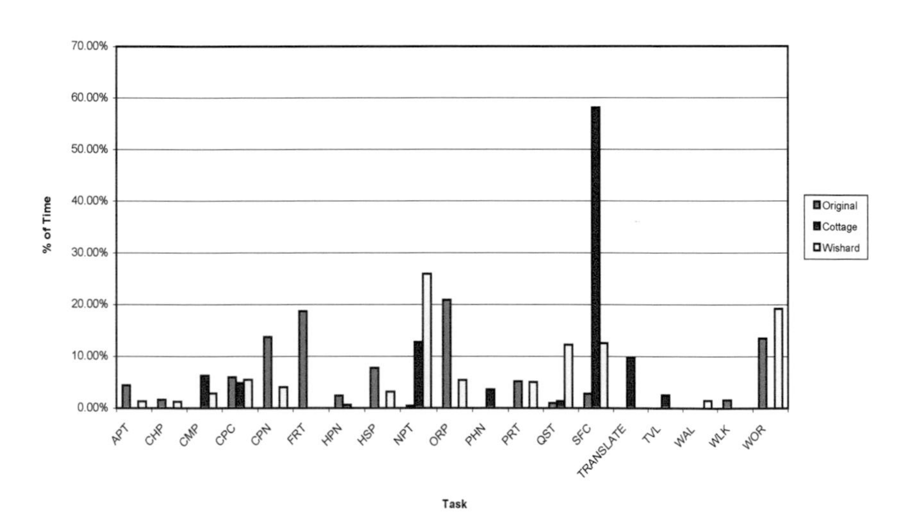

Study 3: Time motion study in the Emergency Room

During the academic year 2008-2009 we collected time-motion data on providers and staff working in the Emergency Room of a major urban teaching hospital. The data is provided below. The first diagram below shows the workflows and the second shows the various "compartments" in the ER and the patterns of travel around in these areas.

Table-1: Workflow in the Emergency Department (ER)

Patient Action	Data Action
Patient arrives to ED	Clerk enters patient data into Gopher, charts created
	Charts placed in area's bins
	Patients assigned to doctors
Patient assigned a bed	Patient's data appears on grease board
Doctor sees the patient	Doctor removes chart from bins, writes notes on chart, writes any orders or prescriptions on chart.
	Places chart in the inbox.
Clerk/nurse retrieves chart from inbox	Clerk or nurse enters order/Rx into WizERD
Clerk or nurse returns chart to sorter	
Doctor retrieves chart from sorter	
Lab work completed, Rx filled	Lab department enters results into Gopher
Doctor queries Gopher, sees lab results	Doctor records lab results onto chart
Doctor sees patient, etc.	

In terms of the actual times spent in each activity we found the following breakdowns:

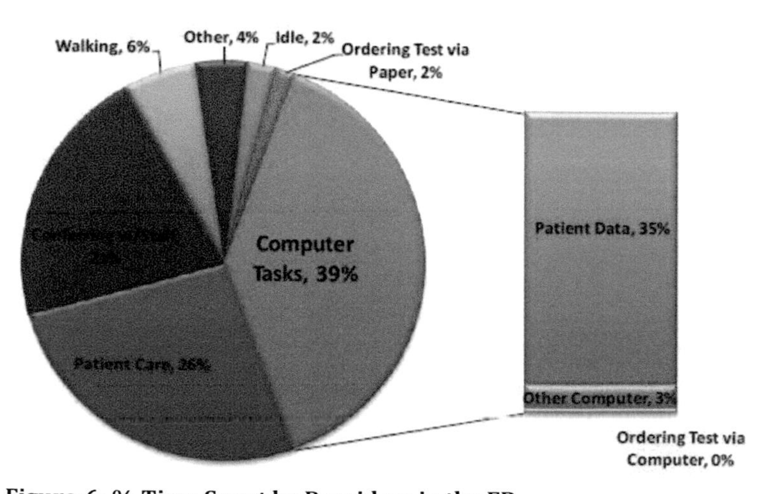

Figure-6: % Time Spent by Providers in the ER

In terms of the workflow in each zone, the following data was obtained:

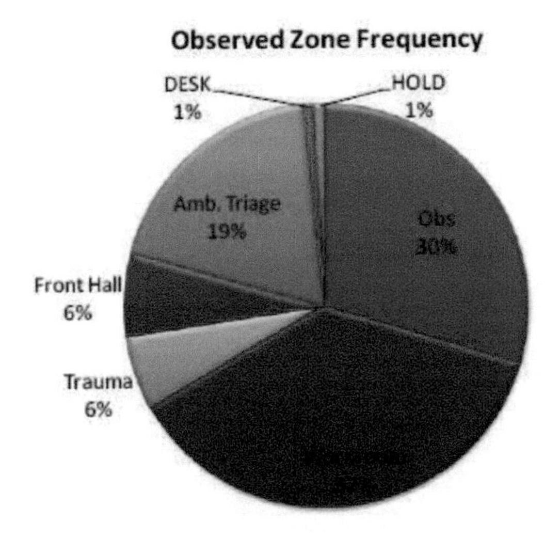

Figure-7: % Time Spent in Each Zone in the ER

4. DISCUSSION

We found that the workflows we studied were highly mobile activities for the provider. Whether in the ambulatory setting or emergency care (ER), providers and staff often moved from room to room for the purposes of actual patient encounters, data lookup or documentation tasks. Precious time was lost during these transition states between rooms that could have been directed towards more productive activities. For example, providers often jotted down information on small pieces of paper or a chart while seeing a patient only to have to later re-enter it into a computer system. Or they forgot what was discussed and had to walk back to a patient location to reaffirm that. Such "transition" activities added cost to the healthcare process and had the potential to introduce errors. As such, mobile computing devices have the potential to alleviate some of these problems by providing instantaneous data lookup at the bedside, ability to record notes and generate orders for tests and medications and provide a function to document encounters.

The time-motion studies in the two environments in this academic medical center showed that although there is significant overlap of the types of activities from clinic to clinic there are some unique activities at each site. There is also some role and responsibility transposition and overlap between staff. Furthermore, lack of communication between check-in clerks and triage nurses often results in a delay in the times for when a patient actually gets placed into an exam room for the provider. This can potentially be improved by more timely communication about when the check-in process has been completed and when a patient is ready for triage (checking vital signs etc.) by nursing staff. Mobile devices or "greaseboard" functions can effectively provide this information. We also found that when the staff goes on lunch breaks round 11:30 that this results in a backlog of check-outs for the afternoon resulting in delayed check-in for the afternoon patients. Quicker check-out procedures that occur in parallel for multiple patients could alleviate some of this delay. Finally, there were significant periods of downtime for some types of staff often due to bottlenecks elsewhere in the system. Simulation studies to depict the bottlenecks and develop creative ways to alleviate them were done and are being reported elsewhere.

From the ER data we can deduce the following: an average of 2 trips per patient is made to look up current medications, past medical records, lab results etc. We know that the average salary of an ER doctor is around $200,000 and the average salary of a nurse is around $60,000. According to our results, 35% of a doctor's time is spent looking up patient data and 6% of doctor's time is spent walking. This equates to $70,000 per doctor per year looking up patient data and $12,000 spent just walking around in the ER! We found that initially very little time is spent walking around the ER but that it accumulates significantly over the course of a year. This is related to response time, as well as to not having computer bedside and not having alerts for test/lab results.

Mobile electronic devices that allow access to electronic patient data and which can be used to enter orders and retrieve test results can play a major role in reducing the "walking time" for providers and staff. The $12000 per year that is spent just walking around has no consequence to patient care. We found that 39% of the time was spent on "computer activities" and better interfaces with more task oriented characteristics could reduce these times. This would reduce the $70,000 per year spent on "looking up patient data"!

5. CONCLUSION

We have performed a workflow analysis in the ambulatory and ER settings of a major urban teaching hospital. Our data suggests that process inefficiencies exist that account for substantial cost incursions. These could be mitigated by better process design and better information tools in a highly mobile setting. Mobile electronic devices have the potential to dramatically reduce the costs associated with these highly mobile workflows and clever interface design strategies could help reduce the "computer interaction" times resulting in improved productivity and better patient care.

REFERENCES

1. J. Marc Overhage, Susan Perkins, William M. Tierney, and Clement J. McDonald, "Controlled Trial of Direct Physician Order Entry: Effects on Physicians' Time Utilization in Ambulatory Primary Care Internal Medicine Practices", J Am Med Inform Assoc. 2001 Jul–Aug; 8(4): 361–371

2. D. W. Bates, D. L. Boyle, and J. M. Teich, "Impact of computerized physician order entry on physician time", Proc Annu Symp Comput Appl Med Care. 1994; 996

3. Lehto, MR, Buck, JR, "Introduction to Human Factors and Ergonomics for Engineers", Lawrence Erlbaum Assoc Inc, 2007

4. Price Waterhouse Cooper Study (http://www.pwc.com/us/en/healthcare/publications/the-price-of-excess.jhtml)

5. http://www.iom.edu/Object.File/Master/4/117/ToErr-8pager.pdf

6. Simon, SR, et.al., "Correlates of Electronic Health Record Adoption in Office Practices: A Statewide Survey", J Am Med Inform Assoc. 2007 Jan–Feb; 14(1): 110–117

APPENDIX: List of Tasks Observed During the Studies

Observer Type	Code	Task
Checkin, Checkout	APT	Make Appt
Checkin, Checkout	CHP	Check Prescription
Checkin, Nurse	CLP	Call Patient
Checkin	ENT	Enter Data
Checkin	FIL	Fill Paperwork
Checkin	FRT	Free Time
Checkin	IDX	Use IDX System
Checkin	INS	Check Insurance
Checkin	NPT	Non-Patient Time
Checkin	PAY	Take Co-Pays
Checkin	PFC	Place Form in Chart
Checkin	QST	Answer Patient Questions
Checkin	SFC	Stuff Charts
Checkin	SIN	Sign In Patients
Checkin	STK	Place Chart on Stack
Checkin	TYP	Ask about Visit Type
Checkin	WLK	Walk-Ins (extra work)
Checkout	CPC	Copying Charts
Checkout	CPN	Copy All Notes
Checkout	HPN	Hand Papers to Nurse
Checkout	HSP	Hand Prescription to Patients
Checkout	ORP	Organize Papers
Checkout	PRT	Print Requisitions
Checkout	QST	Answer Questions
Checkout	WOR	Work on Releasing Patients
Doctor	RVC	Review Computer Records
Doctor	RVP	Review Paper Charts
Doctor	SEP	See the Patient
Doctor	DOC	Document Release Orders
Doctor	WFP	Wait for Printer
Nurses Aide	ACC	Do Accucheck Glucose
Nurses Aide, Nurse	DRW	Draw Blood

Nurses Aide	EYE	Give a Vision Test
Nurses Aide	FAX	Fax Forms
Nurses Aide	MAL	Mail Forms
Nurses Aide	RMS	Get Patients into Rooms
Nurses Aide	VSN	Take Vital Signs
Nurse	DLS	Deliver Charts
Nurse	WDS	Wait for Doctor Signature

CHAPTER 40

Testing the Systems and Error Analysis Bundle for Healthcare using a Prescription Administration Process in a Cardiothoracic Unit, Case Study

Liam Chadwick[1], *Enda F. Fallon*[1] *and Wil J. van der Putten*[2]

[1]Centre for Occupational Health & Safety Engineering and Ergonomics (COHSEE)
College of Engineering and Informatics
National University of Ireland Galway, Galway, Ireland
[2]Department of Medical Physics and Bioengineering
Galway University Hospitals,
Galway, Ireland

ABSTRACT

Health Care Failure Modes and Effects Analysis (HFMEA®) is an established tool for risk assessment in health care. A number of deficiencies have been identified in the method. A new method called Systems and Error Analysis Bundle for Health Care (SEABH, pronounced 'SAVE') was developed to address these deficiencies. SEABH was tested and validated using the Validation Square and supported by application to a number of medical process case studies. One of these case studies, 'Prescription Administration Process in a Cardiothoracic Unit' is reported in this paper in the context of the validation and testing process employed. The case study supported the validity of SEABH with respect to its capacity to address the weaknesses of HFMEA®.

Keywords: Healthcare risk assessment, HFMEA™, validation and testing, prescription administration, SEABH

1 INTRODUCTION

A new method for human error and system failure analysis in health care called Systems and Error Analysis Bundle for Health Care (SEABH, pronounced 'SAVE') was developed to overcome a number of weaknesses of the HFMEA® method (Derosier *et al.*, 2002), as identified by Wetterneck et al. (2006), Habraken et al. (2009) and Chadwick and Fallon (2010, In Press). These include:

- No support for failure mode or human error assessment
- Difficulties associated with using the decision tree
- Dependency on the quality of the facilitator
- Problems using the hazard scoring system
- Difficulty determining corrective measures
- Time & resources required
- Non facilitation of continuous improvement
- Lack of a comprehensive process modeling method
- Lack of consideration of 'Data/ Information' failures or errors

2 SEABH

The SEABH method includes constructs and concepts from a number of existing tools e.g. CREAM, IDEFØ (Hollnagel, 1998, IEEE, 1998) and has been designed to improve the efficiency of the analysis process, while increasing the support information available for use by health care stakeholders. The methodology primarily uses and builds on aspects of the following methods:

- IDEFØ (Process modeling tool)
- Failure Modes and Effects Analysis (FMEA)
- Cognitive Reliability and Error Analysis Method (CREAM)
- Health Care Failure Mode and Effect Analysis HFMEA®
- Irish HSE Risk Assessment Tool

SEABH utilizes a team-based approach similar to FMEA or HFMEA®. IDEFØ models are used to clearly describe the stakeholder functions and task requirements. The SEABH flowchart has been designed to overcome a number of issues identified with the HFMEA® flowchart (Chadwick and Fallon, In Press). Principally, it addresses those related to; prioritization of risk detection, recovery and control measures before formal risk assessment; implied ability to recover from errors once detected; absence of description of controls and detection methods; and confusion surrounding some of the terms used. It reduces the need for the risk assessment of individual process steps through the logic of its flowchart; see Figure 1 a simplified version of the SEABH Flowchart used for illustrative purposes.

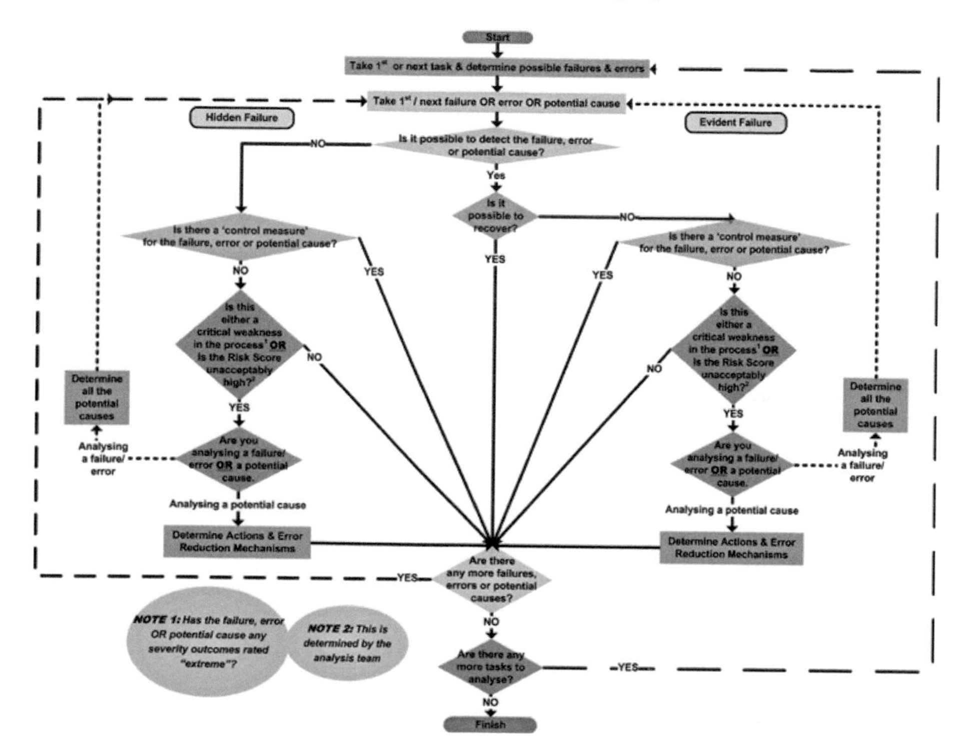

Figure 1. Simplified Version of the SEABH Flowchart

SEABH provides greater support for health care analysts in the identification of potential human errors than previously applied tools. SEABH supports the identification of error related Performance Shaping Factors (PSFs) and Psychological Error Mechanisms (PEMs) based on the CREAM methodology. It directly links the identified human errors to potential causes which in turn support the determination of appropriate control measures. Table 1 shows an extract of the error types and causes for the error category related to 'Interpretation' tasks.

SEABH requires team members to analyze sub-processes or tasks using the FMEA categories presented in IEC (2006). Also, a modified version of the Irish HSE Risk Assessment Tool which includes a new data/information category is used to formally complete the risk assessment (Chadwick and Fallon, In Press).

Table 1 SEABH Error Type and Causes for the Error Category 'Interpretation'

General Error Type	Specific Error Type	Definition/ Explanation	Potential Specific Error Cause	Further Potential Error Causes *(Related Error Category)*
Faulty Diagnosis	Wrong Diagnosis	The diagnosis of the situation or system state is incorrect.	Confusing Symptoms Error in Mental Model Misleading Symptoms Mislearning Multiple Disturbances New Situation Erroneous Analogy	**Cognitive Bias** *(Permanent Person Related Functions)* **Wrong Identification** *(Observation)* **Inadequate Procedure** *(Procedure)*
	Incomplete Diagnosis	The diagnosis of the situation or system state is incomplete		

3 TESTING AND VALIDATION PROCESS

SEABH has been validated using the Validation Square (Pedersen *et al.*, 2000, Seepersad *et al.*, 2006), a prescriptive tool for the validation of new methods, and tested using two distinct theory-testing case studies (Chadwick, 2010). The Validation Square is a framework for guiding the process of validating an engineering design/analysis method which is based on a Relativist / Holistic /Social View of Scientific knowledge and its validation. The approach incorporates inductive reasoning where application of the theory is used to build confidence in the theory's usefulness with respect to its stated purpose. The criterion for validation requires the following: *'the theory was developed to fit some purpose and its validity must prove its usefulness with respect to achieving this purpose.'*

1. Usefulness has two components: Effectiveness and Efficiency
   a. Effectiveness incorporates three elements;
      i. accepting the individual constructs validity
      ii. accepting the constructs consistency
      iii. accepting the appropriateness of the example problems that will be used to verify the performance of the method

b. Efficiency also incorporates three elements
   i. accepting that the method is useful with respect to the initial purpose for some chosen example problems
   ii. accepting that the achieved usefulness is linked to applying the method
   iii. accepting the usefulness of the method is beyond the example problem

The validity of the developed theory (i.e. SEABH) under a Relativistic / Holistic / Social View of Knowledge requires the assessment of the theory using examples or case studies. Evidence to support the validation of the developed SEABH method was provided through its application in a multiple-case study comprising two holistic cases. The first case study, prostate brachytherapy, had not been previously analyzed using similar methods, but which fits the appropriate application criteria and boundaries of the developed theory's (SEABH) use. The second of the two case studies, Prescription Administration Process in a Cardiothoracic Unit, is briefly described here in this paper.

4 CARDIOTHORACIC UNIT STUDIED

The Cardiothoracic Unit studied opened in 2009 and utilizes a Clinical Information System (CIS) with Computerised Physician Order Entry (COPE) and a Clinical Decision Support System (CDSS). It is used for the treatment of cardiothoracic surgical patients before surgery and after surgery following a period in the intensive care unit. The prescription administration process is completed by nursing staff supported by a mobile CIS trolley in 6 wards varying in size from 4 beds to single beds. The patient to PC access ratio varies from 1:1 to 4:1.

The implication of the higher ratio in the unit is that the nurses must move from the patient bedside to another location in the room to access the PC terminal. Additionally, the unit has a mobile PC terminal access point which is mounted on a trolley. Unfortunately, the 'CIS trolley' was not considered by Unit staff to be particularly 'user friendly': staff believed the trolley was difficult to manoeuvre and difficult to fit through the door frames in the Unit. Consequently, it typically remains in a stationary position in the clinical room beside the drug trolley where the prescriptions are prepared for administration by the nurses.

The intended design of the system was that the nurses would use the mobile CIS trolley to check the patient charts and to assist the preparation of the prescriptions. They could then validate ('sign off') drug administration after it had been given to the patient using the CIS trolley or the in-room access terminal. Analysis of the data logs for the CIS terminals in the ward rooms by hospital clinical engineers identified that the in-room access terminals were not significantly being used to check patient charts or to validate the administration of the prescribed drugs.

Additionally and unfortunately, the unit's drug trolley shared similar problems to the CIS trolley, i.e. size and manoeuvrability and consequently remained stationary in the clinical room. The clear implication of these results was that the CIS trolley, the in-room access terminals and the drugs trolley were not being used as specified in the system design. The perception was that patients were at significant risk of being administered the incorrect medication. SEABH was used to determine the extent of this risk. The case studied satisfied the criteria outlined above for testing the validity of SEABH.

5 APPLICATION OF THE METHOD

An IDEFØ model of the Prescription Administration process was developed. The SEABH ('SAVE') flowchart was applied to the clinical activities of this model. It was found that safety checks were prioritised for drugs bound by the Misuse of Drugs Act Ireland (Amendment) (2009), but fewer precautions were given to other drugs, which the patient could be allergic to or could result in contra-indications or complications. Over 60 individual IDEFØ model processes were analyzed. Only one required the determination of its risk score and the identification of potential causes, as the identified errors were neither recoverable nor controlled (A1232 Standard Prescription Checks).

Using the SEABH error taxonomies the following factors were considered to potentially cause and influence the occurrence of the 'Standard Prescription Check' errors:

- Temporary Incapacitation
- Other Priority
- Excessive Demand → Unexpected Tasks/Parallel Tasks
- Distraction → Loss of Orientation
- Habit

The first four of these identified potential causes are related to the time and work pressures placed on the staff when completing their drug rounds and the potential for the nurses to be interrupted during their tasks.

The requirement for nurses to validate a drugs administration on the CIS after it has been given has made the administration process more time consuming than if a hardcopy patient chart was located at the patient bedside. Nurses must now log on to the CIS, open the patient file and prescription chart and validate the drugs administration. A paper based chart needs only a physical signature. This has resulted in the nurses circumventing the intended drug validation on the CIS after administration by validating it in the clinical room after preparation and not after administration.

The CIS has changed the nature of the administration process and has introduced the potential for new errors in the process, e.g. interruptions in the process disrupting the nurses intended activity.

Forty-six of the 60 processes analyzed had errors identified as being detectable and recoverable through the utilisation of checks at different stages in the treatment process or were controlled. This dependency on checking functions as part of the administration process highlights the importance and criticality of their correct completion. Several recommendations were identified regarding potential improvements to the detection methods, recovery mechanisms or control measures for some of the identified failures and errors, as part of a continuous improvement program.

Table 2 below shows a comparison of the potential human error results as determined from the completed SEABH analysis for the step of getting a tablet drug packet from the drug trolley and the equivalent step as analyzed by Lane *et al.* (2006) for a similar prescription administration process using the SHERPA method.

Table 2 Comparison of the SEABH and SHERPA Error Descriptions for the same Process Steps

<table>
<tr><th>Completed SEABH Analysis
(A121621 Get Physical Tablet Packet)</th><th>SHERPA Analysis (Lane et al., 2006) ([Select Tablets] 2.5.2.1.1 Take bottle out of store/drug trolley)</th></tr>
<tr><td>Similar object (Prescription packet) → Wrong I.D. → Incorrect I.D.</td><td rowspan="3">A6 Right operation on wrong object - Take wrong bottle out of store/drug trolley</td></tr>
<tr><td>Similar object (Prescription packet) → False Observation → False Recognition (of the prescription drug name)</td></tr>
<tr><td>Similar object (Prescription name) → Inattention</td></tr>
</table>

It can be seen from the comparison that the SEABH descriptions include information related to the Psychological Error Mechanisms (PEMs) underpinning the potential errors and have identified a greater number of ways in which the task can be performed incorrectly.

6 DISCUSSION AND CONCLUSIONS AND FURTHER WORK

The SEABH 'SAVE' methodology provided a structured and detailed assessment method for the prescription administration process. The analysis identified a number of concerns and potential weaknesses in the process, e.g. the lack of mobility of the CIS system, premature validation of the drugs administered, an over dependency on nursing checks.

SEABH resulted in a better quality of description for the potential failures and errors than other analysis methods support and it directly linked the human errors to potential causes. The SEABH error taxonomies allow for better error descriptions, including contextual information regarding influencing PSFs and PEMs. The use of health care specific risk categories for severity and likelihood of occurrence supports analysts in the completion of risk assessment when required. However, the SEABH Flowchart logic reduces the need for the completion of risk assessment for every process step.

Unfortunately, the benefit of having greater detail in the process and error descriptions can extend the analysis time, but this is balanced in the method by the significantly reduced requirement for the risk assessment of each individual task being analysed.

Overall, the results of the case study provided support for both the external and internal validity of SEABH and its usefulness as a proactive healthcare focused systems and error analysis method. The effectiveness and efficiency of the method was established and the validity of the method was extended beyond its development and initial application in radiation therapy.

Future work is required to further test the efficacy of SEABH in other health care settings and to develop a software support tool for users.

ACKNOWLEDGEMENTS

The authors would like to acknowledge the contribution of the staff of the Cardiothoracic Unit studied in this case study

REFERENCES

Chadwick, L., 2010. A systems engineering and error analysis methodology for health care. Ph.D. Thesis, unpublished. NUI Galway.

Chadwick, L. & Fallon, E.F., 2010. HFMEA™ of a radiotherapy information system - Challenges and recommendations for future studies. *In* V. Duffy (ed.) *Advances in Human Factors and Ergonomics in Healthcare.* Boca Raton, Fl.: CRC Press / Taylor and Francis, Ltd.

Chadwick, L. & Fallon, E.F., In Press. Evaluation and critique of Healthcare Failure Mode and Effect Analysis applied in a radiotherapy case study. *Human Factors and Ergonomics in Manufacturing & Service Industries.*

Derosier, J., Stalhandske, E., Bagian, J.P. & Nudell, T., 2002. Using Health Care Failure Mode and Effect Analysis™: The VA National Center for Patient Safety's Prospective Risk Analysis System. *Journal of Quality Improvement,* 28, 248-267.

Habraken, M.M.P., Van Der Schaaf, T.W., Leistikow, I.P. & Reijnders-Thijssen, P.M.J., 2009. Prospective risk analysis of health care processes: A systematic evaluation of the use of HFMEA™ in Dutch health care. *Ergonomics,* 52, 809 - 819.

Hollnagel, E., 1998. *Cognitive Reliability and Error Analysis Method (CREAM)* Oxford, UK; New York, USA: Elsevier Science Ltd.

Iec, 2006. International Electrotechnical Commission 60812 Analysis techniques for system reliability — Procedure for Failure Mode and Effects Analysis (FMEA). Geneva: International Electrotechnical Commission.

Ieee, 1998. IEEE standard for functional modeling language - syntax and semantics for IDEF0. IEEE Std 1320.1-1998. New York: IEEE.

Lane, R., Stanton, N.A. & Harrison, D., 2006. Applying hierarchical task analysis to medication administration errors. *Applied Ergonomics,* 37, 669-679.

Misuse of Drugs Act (Amendment), 2009. S.I. No. 122 of 2009. Ireland: Irish Statute Book.

Pedersen, K., Emblemsvåg, J., Bailey, R., Allen, J.K. & Mistree, F., 2000. Validating design methods & research: The validation square. *ASME Design Engineering Technical Conferences.* Baltimore, MD: CD-ROM Proceedings of DETC 2000.

Seepersad, C.C., Pedersen, K., Emblemsvåg, J., Bailey, R., Allen, J.K. & Mistree, F., 2006. The Validation Square: How Does One Verify and Validate a Design Method? *In* W. Chen, K. Lewis & L. Schmidt (eds.) *Decision-Based Design: Making Effective Decisions in Product and Systems Design.* New York, NY: ASME Press, 303 - 314.

Wetterneck, T.B., Skibinski, K.A., Roberts, T.L., Kleppin, S.M., Schroeder, M.E., Enloe, M., Rough, S.S., Hundt, A.S. & Carayon, P., 2006. Using failure mode and effects analysis to plan implementation of smart i.v. pump technology. *American Journal of Health-System Pharmacy,* 63, 1528-1538.

Section VI

Medical User Centered Design

CHAPTER 41

Implementation of Usability Engineering Process in a Business Concern

Anja Schultz QIAGEN, Melanie Aust QIAGEN, Dirk Buechel QIAGEN, Nils Neumann QIAGEN, Helge Hoffman QIAGEN

QAIGEN GmbH
Hilden, Germany
anja.schultz@qiagen.de

ABSTRACT

Usability engineering is a powerful process to increase the efficiency of a company and to ensure the success of the business. A professional usability engineering process during product development will be beneficial for the company as well as for the customers. In addition usability engineering is demanded by regulation authorities. For QIAGEN the business aspect as well as the regulatory requirement was the major trigger to establish usability engineering. Time to market, cost and limited resources are obstacle for implementation of new methods. An efficient connection to existing processes and strong support from upper management are important for a successful implementation.

Keywords: Usability Engineering, cost benefit analysis, processes

1 INTRODUCTION

The implementation of new methods and processes is always a cost- benefit calculation for industries. Although usability engineering has many potential advantages, the benefits of better usability are not easily identified or calculated (Rajanen, 2003).

Potential benefits of usability are (Bevan, 2005):

- Reduction of development cost
    - Usability engineering will help to focus on the relevant functionalities (Mauro, 2002)
    - Usability engineering will help to detect problems and reduce or mitigate risks early in the development process (Bias, Mayhew, 1994)
    - Usability engineering will help to reduce the risk of releasing a system which will fail to meet its objectives in use (Landauer,1996)
- Increase product sales as a result of the usability of the product
    - Usability engineering will help to increase revenues (Wixon and Jones, 1995)
- Easy to use systems as benefit for the employers
    - Usability engineering helps to reduce task time for employees (Bias, Mayhew, 1994; Landauer,1996)
    - Usability engineering helps to increase motivation and satisfaction of employees (Bias, Mayhew, 1994; Cooper, 1999)
    - Usability engineering helps to reduce time spent by employees to support colleagues (Gartner Group, 1997)
- Reduction of support and maintenance costs for suppliers and/or employers
    - Usability engineering will help to reduce cost for training (Bias, Mayhew, 1994)

On the other side the costs of usability engineering activities are relatively low because most user centered design techniques are relatively simple and could be scaled to the necessity of the project (Bevan, 2005). The major cost is the working time of the staff who applies the methods.

Starting the discussion at QIAGEN to create a usability engineering program, potential cost and benefits was taking into account.

QIAGEN was founded 1984 as a spin-off at the University of Düsseldorf. Since then the company has continuously developed and operates in over 35 locations worldwide and has more than 3600 employees. QIAGEN provides a broad base of sample and assay technologies. Sample technologies are used to isolate DNA, RNA, and proteins from any biological sample. Assay technologies are then used to make specific target biomolecules, such as the DNA of a specific virus, visible for subsequent analysis. The company develops consumable products and automated solutions for molecular diagnostics laboratories, academic researchers, pharmaceutical and biotechnology companies.

In conjunction with developing products for molecular diagnostics laboratories, international as well as national standards must be fulfilled during a product development process resulting in a medical device.

Incomplete usability of medical devices is a significant patient risk (Brennan 1991) and therefore one major goal of international human factor standards and regulation authorities is to increase the patient safety by reducing use errors. 60% of avoidable harm caused to patient health is caused by use errors and 80% of use errors are generated by insufficient human – machine interfaces (Bleyer 1992;

Schubert 1993). Kohn and Corrigan calculate that between 44.000 and 98.000 persons die in the US due to use errors in hospitals (Kohn 1999).

EN 62366 "Medical devices - Application of usability engineering to medical devices" is a harmonized standard under European Directive 98/79/EC for In vitro diagnostic medical devices and a FDA recognized consensus standard. In addition the FDA distributed in June 2011 a draft guidance "Applying Human Factors and Usability Engineering to Optimize Medical Device Design". These standard and guidelines describe a process for integrating usability engineering into the product design and development to ensure that the needs of the user are appropriately defined and specified and along the company wide established development process verified and validated.

2 PROCESS DEFINITION

To identify the area where professional usability engineering is most required and effective for QIAGEN the company structure and processes were reviewed. This review identified a lot of usability engineering activities, which are often not specifically identified as usability engineering and do not follow a standardized process.

To implement a usability engineering process the following decisions have to be made:

- Define the major scope of usability engineering within the company
- Define the responsibility for the usability engineering activities within the company
- Define a concept how to integrate a usability engineering process into the existing processes of the company

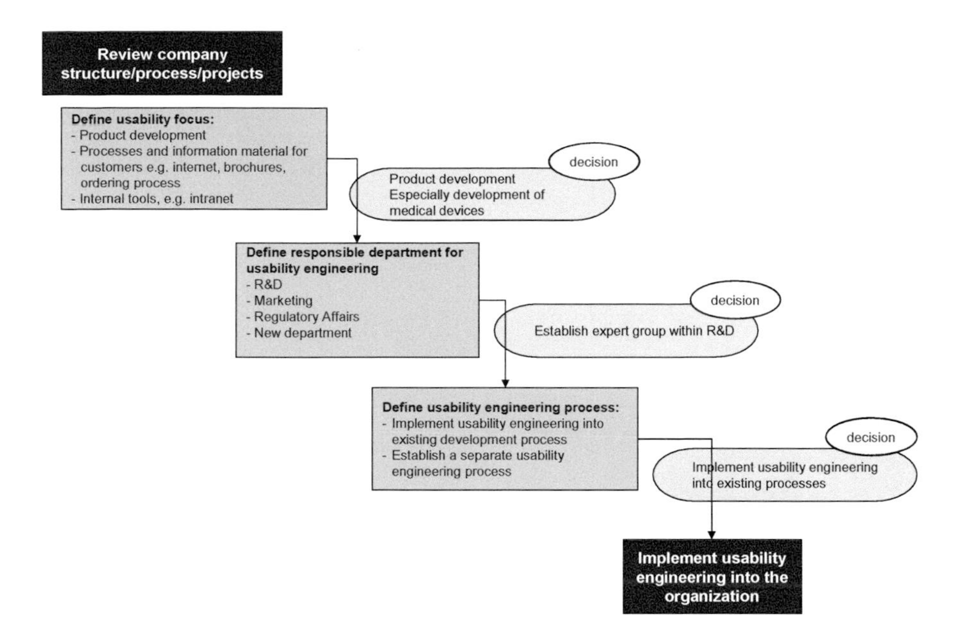

Figure 1 Decision steps for implementation of Usability Engineering

The potential of usability engineering tailored to improve the product quality and relevant product functionality combined with the regulatory requirements was the major trigger to establish usability engineering as part of the product development process.

Product development is done by an interdisciplinary team including marketing, service, production, research and development and regulatory affairs, just to name a few. Usability engineering is a multifaceted process which requires intensive networking within the development team. Nevertheless it is useful to define one discipline to take over this responsibility. At QIAGEN we decided, that the research and development department should take over this responsibility for usability engineering within the product development process. An expert group was instituted to establish a knowledge base, define processes, perform usability engineering during product development and train others.

Performing Usability Engineering within R&D project

- Analyze customer workflow and product context
- Define usability requirements and specifications
- Define interaction design and perform usability verification & validation

Coaching and training of project teams and colleges in terms of UCD

- Support project teams by usability engineering activities
- Perform internal trainings and seminars

Standardize and improve usability engineering

- Define and establish usability engineering process at QIAGEN
- Supplement processes by adding user centered design relevant standards
- Creation of human factor guidelines and master documents

Communicate to user centered design community and continuously improve internal knowledge

- Visit conferences and external seminars
- Liaison with outside human factor experts
- Ongoing educational efforts to improve and update knowledge

Figure 2: The major task of QIAGEN human factors expert group

The first step of QIAGEN user centered design program is to define and establish a usability engineering process for the development of medical devices in the most efficient way. Therefore the existing product development process of the "Total Quality Management" system was analyzed. Different "Standard Operating Procedures" were identified where usability engineering activities are already defined. This confirms the fact that usability engineering and product development are closely linked. We decided not to generate an additional usability engineering process but to implement usability into the existing product development process. Missing usability engineering tasks were implemented into existing "Standard Operating Procedures" for design input, development planning, verification, validation, risk management and market surveillance.

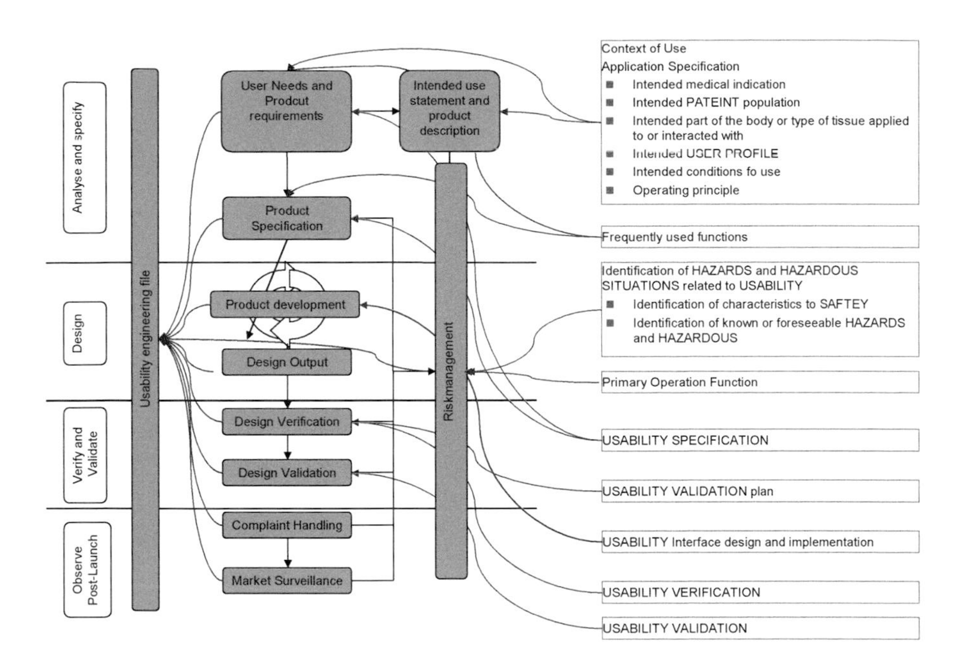

Figure 3 Implementation of usability engineering tasks into the product development process. On the left site are the main steps of product development, in the middle the QIAGEN development process and on the right site the usability engineering task according to EN 62366 "Medical devices - Application of usability engineering to medical devices"

3 CONCLUSIONS

Usability engineering is a powerful process to increase the efficiency of a company and to ensure the success of the business. A professional usability engineering process during product development will be beneficial for the company as well as for the customers. QIAGEN implemented the usability engineering process in 2010 and follows this process for the development of medical devices. Key lessons learned and findings in this organizational effort are listed below:

- Implementation of user centered design activities requires strong support from upper management
- User centered design methods will significantly increase the quality of the product
- User centered design methods will help the project team to focus on the relevant functionality
- The most efficient starting point for User centered design activities is as early in the project phase as possible
- User centered design activities do not extend the project run time or effort if human factor design colleagues are closely linked to the development team.

- The importance of understanding human factors for management and development staff is most clearly illustrated when watching users struggle with usability issues during real case studies

ACKNOWLEDGMENTS

The authors would like to acknowledge colleges discussing and implementing the usability engineering process, especially Andreas Schaefer for the support from upper management and Anke Homann-Wischinski as process owner of the design development process.
Many thank to Courtney Casciano for reviewing this paper.

REFERENCES

Bevan N. 2005. Cost benefits framework and case studies. In Bias, R.G. & Mayhew, D.J. (2005) Cost-Justifying Usability: An Update for the Internet Age.

Bleyer, S.: Medizinisch-technische Zwischenfälle in Krankenhäusern und ihre Verhinderung. In:Anna, O.; Hartung, C. (Hrsg.): Mitteilungen des Instituts für Biomedizinische Technik und Krankenhaustechnik der Medizinischen Hochschule Hannover (1992)

Bias, G. and Mayhew, D. (Eds) (1994). Cost-Justifying Usability. Academic Press, NewYork.

Brennan, T.A.; Leape, L.L.; Laird, N.M.: Incidence of Adverse Events and Negligence in Hospitalized Patients – Results of the Harvard Medical Practice Study (1991). New England Journal of Medicine 324, S. 370-376

Compita (1997) Process Professional Assessment. http://www.processprof.com

Cooper, A. (1999). The inmates are running the asylum: Why high-tech products drive us crazy and how to restore the sanity. Indianapolis, Indiana: SAMS

Draft Guidance for Industry and Food and Drug Administration Staff Applying Human Factors and Usability Engineering to Optimize Medical Device Design DRAFT GUIDANCE This guidance document is being distributed for comment purposes only. Document issued on: June 22, 2011

European Committee for Electrotechnical Standardization, 2008 EN 62366 "Medical devices - Application of usability engineering to medical devices"

Gartner Group (1997). Cited in Gibbs (1997).

Kohn, L.; Corrigan, J. (Eds.): Building a Saver Health System. Institute of Medicine (IOM), Committee on Quality of Health Care in America. Washington, D.C.: National Academy Press (1999)

Landaur (1995) The Trouble with Computers. MIT Press

Mauro, C.L. (2002) Professional usability testing and return on investment as it applies to user interface design for web-based products and services.

Rajanen, M. 2003. Usability Cost-Benefit Models – Different Approaches to Usability Benefit Analysis. *Proceedings of 26th Information Systems Research Seminar In Scandinavia (IRIS26), Haikko, Finland*

Schubert, M.: FMEA – Fehlermöglichkeits- und Einflussanalyse, DGQ- Schrift, 1993. Frankfurt: Deutsche Gesellschaft für Qualität e.V. (1993), S. 7ff.

Wixon, D. and Jones, S. (1995). Usability for fun and profit: A case study of the re-design of the VAX RALLY. In: Human-Computer Interface Design: Success Stories, Emerging Methods, and Real-World Context. Marianne Rudisill, Clayton Lewis, Peter G. Polson, Tim McKay (eds) Morgan Kaufmann Publishers.

CHAPTER 42

Systematic Application of Usability Engineering in Medical Device Development

Dirk Büchel[1], Stephanie Hacker[1], Tom Paschenda[2], Axel Raeker[1], Sascha Stoll[1], Janina Schaper[1]

[1]QIAGEN GmbH, [2]Capgemini Deutschland GmbH
Germany
dirk.buechel@qiagen.com

ABSTRACT

International standardization and regulation authorities commit medical device manufacturers to perform usability engineering. Implementation of usability engineering is defined in various standards and guidelines. However it seems that many manufacturers and engineers consider this approach as an additional cumbersome compliance of standards. Tight budgets, deadlines and a lack of understanding of usability engineering lead to little or no implementation of this process. The systematic application and the benefit of usability engineering in medical device development are shown with a practical example of software and graphical user interface (GUI) redesign of an automated in-vitro-diagnostic device.

Keywords: Usability Engineering; Medical Device; Usability Verification; Usability Validation; EN IEC 62366; QIAsymphony

1 REGULATORY BASIS FOR USABILITY ENGINEERING

Usability engineering is obligatory for medical device manufacturers in Europe and the United States. As harmonized standards of the Medical Device Directive (Council Directive 93/42/EEC concerning medical devices) and the In-vitro

Diagnostic Directive (Council Directive 98/79/EC concerning In-vitro-diagnostic medical devices) the standards EN 60601-1-6 and EN 62366 oblige manufacturers to conduct and document a usability engineering process and risk management according to the standards for CE conformity marking in the European Economic Area.

In the United States usability engineering plays an important role in fulfilling the FDA's design control requirements of the Quality System Regulation, 21 CFR Part 820. To assist manufacturers in this process the FDA has published guidance documents. Specifically the guidance document *Applying Human Factors and Usability Engineering to Optimize Medical Device Design* provides recommendations for medical device design optimization through usability engineering. Additionally the FDA has officially recognized the two European standards for usability engineering of medical devices.

2 THE QIASYMPHONY

The QIAGEN QIAsymphony (Figure 1) is an automation platform for molecular testing and diagnostics that is used in the fields of molecular diagnostics and biomedical research, human identity testing, veterinary applications, biosecurity research and life science applications, such as genomics and proteomics research.

The QIAsymphony Sample Preparation (SP) performs fully automated purification of nucleic acids or proteins using magnetic-particle technology. The QIAsymphony Assay Setup (AS) performs fully automated assay setup using a 4-channel pipetting system. The combined QIAsymphony SP/AS instrument can be operated using a built-in touchscreen located in front of the QIAsymphony SP.

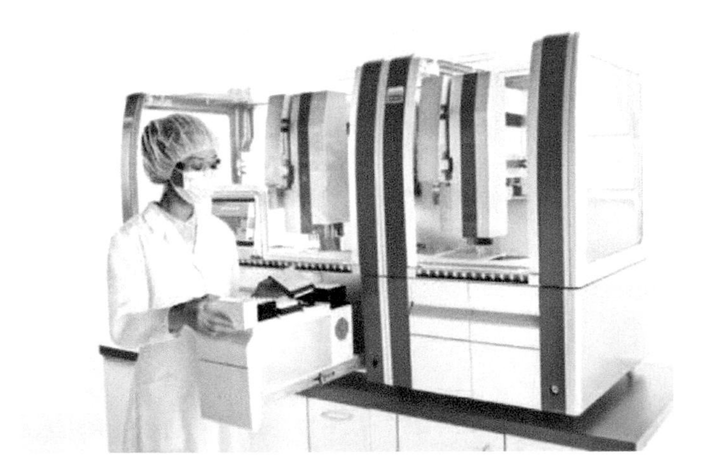

Figure 1 QIAGEN QIAsymphony SP/AS instruments

With a dedicated range of QIAsymphony Kits, the QIAsymphony SP enables sample preparation of DNA, RNA, bacterial and viral nucleic acids from a wide range of starting materials. The QIAsymphony AS extends the capabilities of the QIAsymphony SP by integrating automated PCR assay setup which, in combination with the Rotor-Gene Q, QIAGEN real-time and end-point polymerase chain reaction (PCR) kits, enables to complete an automated PCR workflow.

QIAsymphony Software (SW) Version 4 (V4) which is addressed by this usability engineering study is the first software that supports the feature *Integrated Run* by an automatic eluate transfer. This feature combines SP and AS into one integrated workflow. Prior to this, the eluate rack had to be transferred manually from SP to AS in order to process SP eluates on the AS. With the introduction of the *Integrated Run* this three-step process (processing on SP, transfer, processing on AS) is reduced to a one step workflow. The advantage is that manual operations are minimized, thus eliminating potential contaminations with the effect of improving the results′ reliance. Figure 2 illustrates the minimization of user operations and hands on time with the integrated run compared with the *Independent Run.*

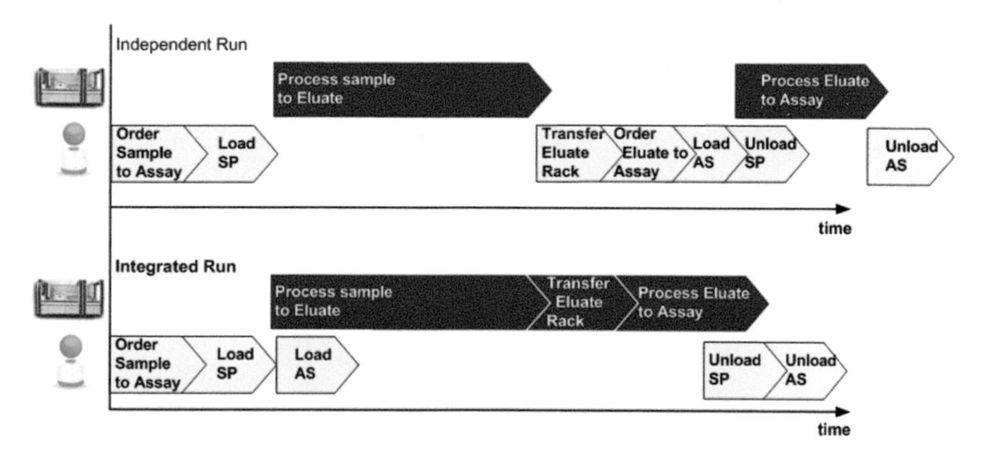

Figure 2 Illustration of how user operations and hands on time with the workflow *Integrated Run* of the QIAsymphony Software V4 are minimized compared with the workflow *Independent Run*

3 METHODOLOGY FOR USABILITY ENGINEERING QIASYMPHONY SW V4

The usability engineering approach in the described project followed the usability engineering process for medical devices, described in DIN EN IEC 62366. A comparable process to the one described in the DIN EN 62366 can be found in the FDA guidance document for usability engineering to optimize medical devices. Table 1 illustrates the similarities of these two processes by comparing the required chapters of the usability engineering documentation. In the following the major steps of the usability engineering process used for the development are described with examples.

Table 1 Comparison of the Chapters of the Usability Engineering Report required for the CE and the FDA regulated market

DIN EN IEC 62366	FDA Guideline
1. Application specification	1. Intended device users, uses, use environments, and training
2. Frequently used functions	4. User task selection, characterization and prioritization
3. Identification of hazards and hazardous situations related to usability	3. Summary of known use problems 4. User task selection, characterization and prioritization
4. Primary operating functions	4. User task selection, characterization and prioritization
5. Usability specification	2. Device user interface
6. Usability validation plan	5. Summary of formative evaluations
7. User interface design and implementation	2. Device user interface
8. Usability verification	5. Summary of formative evaluations
9. Usability validation	6. Validation testing
10. Accompanying document	No separate section, is part of the device
11. Training and materials for training	1. Intended device users, uses, use environments, and training
	7. Conclusion

I. Application specification:
The first task of the usability engineering process was the specification of the application. This specification contained the intended use and the intended indication as well as the intended context of use (user profile, condition of use, use environment, operating principle and system description). In this phase the system analyst and the usability engineer analyzed and described the context of the current system and the additional requirements by the context change emerged by the new features of the software.

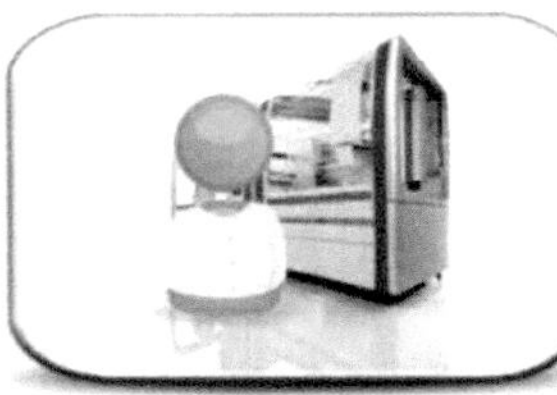

Figure 3 Example from the Context of Use: Interaction Concept

II. Frequently used functions:
Based on the user needs and the application specification the frequently used functions were analyzed and defined. Therefore a task analysis of the current customer workflows was used to define iterations of use scenarios and use cases. With the help of the use cases the concepts of the new workflow *Integrated Run* was described and validated for the first time.

III. Identification of hazards and hazardous situations related to usability:

In addition to the "instrument" risk analysis (EN ISO 14971), an "operator" risk assessment was also performed to identify the specific risks the user may encounter when operating the system. Therefore the method of an expert walkthrough was used. With the help of the workflow description followed by high-fidelity prototypes, results of usability tests and software versions, interdisciplinary experts (usability, system, biologic, assay, software, risk management) analyzed the user tasks step by step regarding user errors and use related hazards. Central questions like “Are any use errors or use-related hazards possible?”, “How might they occur?”, “What are possible consequences?” and “What is a possible mitigation?” supported the expert review.

IV. Primary Operating Functions:

The frequently used functions and the functions related to safety resulted in the primary operating functions. These functions were primarily considered in the usability engineering process.

V. Usability Specification:

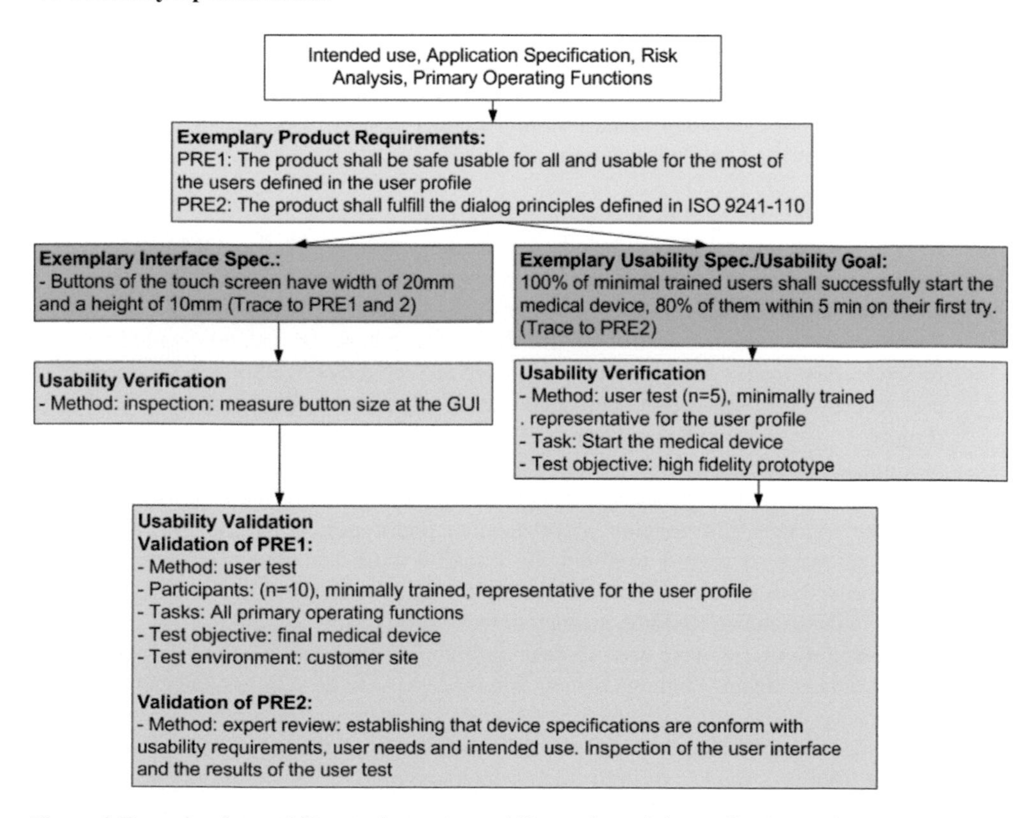

Figure 4 Examples for usability requirements, usability goals and the verification and validation activities

Findings of the context of use, the user needs, the use cases, and the primary operating functions gave basic input to software and usability requirements. The overall usability requirement was an *acceptable residual risk as defined in ISO 14971, associated with usability* (EN 62366) of the QIAsymphony Software V4. This were divided to specific requirements derived from: use scenarios, general usability requirements from standards (e.g. multi-part standard ISO 9241), requirements resulting from the user profile, requirements resulting from the condition of use, user interface requirements for the primary operating functions and requirements from use scenarios that are relevant to safety. In total 16 usability specific requirements were added to the software requirements.

Testable requirements for the usability specification were expressed as interface specifications and usability goals. Usability goals specify the target quality of use in a written form with measurable criteria (performance and user satisfaction). The way to measure whether a device meets its usability goals is to conduct a usability evaluation with the usability goals as the acceptance criteria. For the QIAsymphony SW V4 in total 36 usability goals were defined in cooperation with use experts and other team members, such as marketing, system analyst, product manager and the software developer.

VI. Usability Verification and Validation Plan:

The usability verification and validation plan was part of the product validation and verification plan. It specified the general and minimal concept of the usability engineering, the evaluation phases during product development, the methods used for verification, and validation and the acceptance criteria for the usability goals.

VII. User Interface Design and Implementation:

The procedure to develop and design the interface of the QIAsymphony Software V4 was a design and development process in an iterative fashion. Usability engineering, including usability evaluation, began early in the development process and was continued through the whole design and development lifecycle. First results for a usable device were delivered by developing task models, use scenarios and use cases. This phase was supported by the first low fidelity prototypes like drawings and simple interaction mock ups. Even with these preliminary results, usability evaluations could be conducted. The findings delivered important information to improve, refine and adapt the interaction and interface concepts and develop more detailed prototypes. After the low fidelity prototypes with no or just minimal functionality more complex prototypes were developed to provide a more realistic feeling for the product use. On this basis, valid results of the usability could be evaluated. Evaluation results of simple click models for specific user tasks were used to develop high fidelity interactive prototypes that were evaluated again. Findings during this process were directly included into the issue tracker of the software development team and were subsequently addressed in the design. Figure 5 demonstrates the different phases of the usability engineering starting from task models to high fidelity interactive prototypes (Adobe Flash GUI concepts) and finally first software versions.

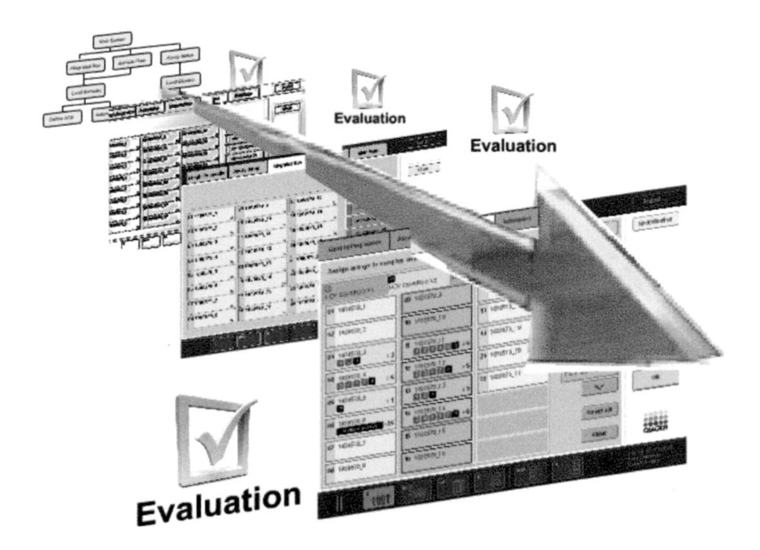

Figure 5 Different phases of the interface design and development with evaluation phases. Starting from task models and simple drawings to interactive prototypes and first SW versions.

VIII. Usability Verification:

The usability verification confirmed that the specific functional and operational requirements of the usability specification and the interface specification of the QIAsymphony SW V4 have been met.

Formative evaluations (Bevan, 2008) were conducted to determine if design concepts meet the acceptance criteria of the usability goals and to identify use-related hazards. These evaluations derive information from the device in different stages of the product development and included various user tests. In the case of unsuccessful usability verification, the interaction concepts would be optimized or redesigned and evaluated again. Formative evaluations have been carried out starting with simple prototypes in an iterative way until the QIAsymphony SW V4 was considered to be usable and safe to a level that met the acceptance criteria of the usability goals.

For this evaluation the usability engineering methods expert review, cognitive walkthrough and user test have been used. In total five evaluation steps involving 18 device users in different stages of the software development have been carried out.

IX. Usability Validation:

After successful formative evaluation the subsequent summative evaluation (Bevan, 2008) for validation should result in good usability and few or no usability concerns. Finally usability validation (Figure 6) demonstrated that the intended users can safely use the software in the intended context of use with the minimum quality of use defined within the usability goals.

The way of measuring if the QIAsymphony SW V4 meets its usability goals was to conduct the usability evaluation methods user test and expert review with the usability goals as the acceptance criteria. In two user tests with seven customer users from four different countries the QIAsymphony SW V4 showed that it met the usability goals and shows no signs of unacceptable risks due to the use of the new features of the software. Usability issues raised by design validation or issues that could not be addressed during the software design were assessed in the risk management. Furthermore the findings delivered useful information for the requirements for the next SW versions and the training process.

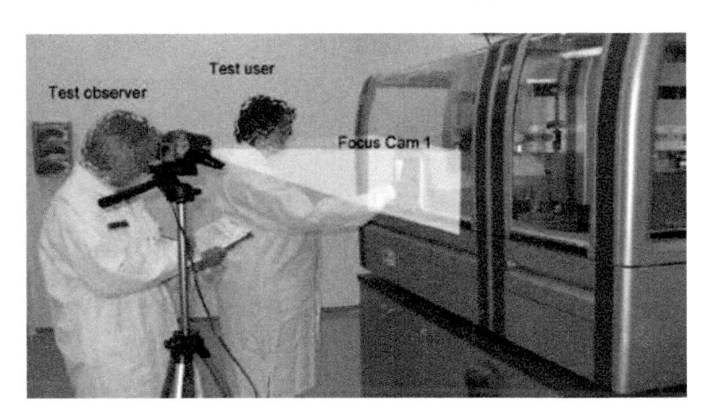

Figure 6 Usability test for validation.

X. Risk assessment:

During the iterative development process usability issues were continuously assessed regarding risk and usability (Figure 7) by an interdisciplinary expert group. A method that combines usability and risk assessment approaches (Büchel and Matern, 2010) was used. When formative evaluations identified unacceptable, use-related risks, they were followed by risk mitigation and device modification. In the summative evaluation no unacceptable use-related risks were identified. The test results showed no usability issues with effect to safety that could be eliminated or reduced through further modification of the design.

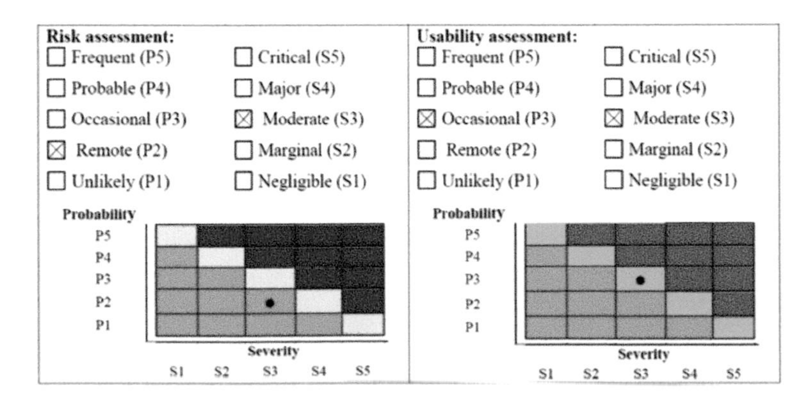

Figure 7 Method to assess use related risk and usability. Example for one usability issue.

4 RESULTS

Essential for an effective and purposeful usability engineering process was the integration into the manufacturer's global design processes. Based on this, product development started with analyzing the context of use, the current customer's workflows, and the requirements for use. This has already shown that significant improvements were made by creating a more efficient workflow and adapting it to the customer's needs. By using the iterative characteristic of usability engineering to design the user interface, the evaluation steps showed successful implementations and usability weaknesses with room for improvements. By the flexible adaption of the simple prototypes, improved concepts were created very quickly and tested again. Formative studies that involve users were useful for identifying problems that were not identified or sufficiently understood using analytical methods. With the help of the outcome of the usability evaluation the software was successfully designed. At the end of the development the usability of the final software was validated with further user tests. Overall five user test phases have been conducted with a total of 23 users.

The final validation showed that effectiveness and efficiency have been drastically improved (Figure 8). For a typical process setup the time for user interaction was reduced by 71% and hands on time during processing by the device was totally avoided. Additional risks were detected and reduced by usability evaluation methods and risk management. These risks would have been left undetected by traditional risk management alone. User ratings for the software were positive regarding satisfaction with the usability of the software.

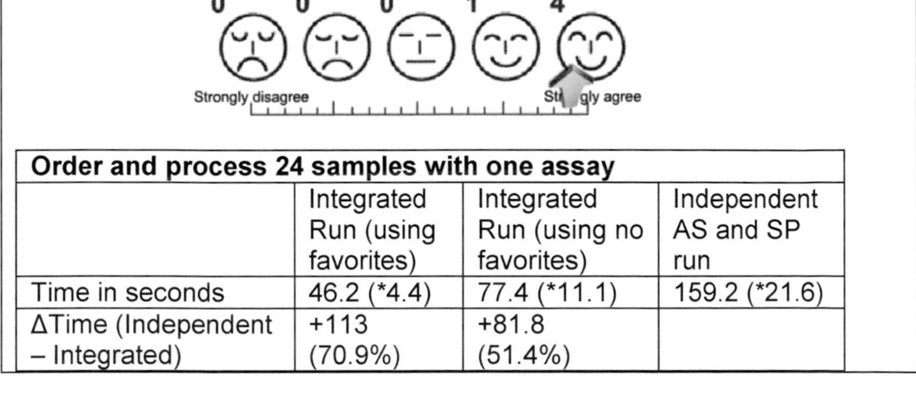

Order and process 24 samples with one assay			
	Integrated Run (using favorites)	Integrated Run (using no favorites)	Independent AS and SP run
Time in seconds	46.2 (*4.4)	77.4 (*11.1)	159.2 (*21.6)
ΔTime (Independent – Integrated)	+113 (70.9%)	+81.8 (51.4%)	

Figure 8 Results of the user test regarding the efficiency and user satisfaction of the *Integrated Run* (*Standard Deviations)

4 CONCLUSIONS

Usability is a key criterion of medical device quality, especially for its user. The best way to achieve a usable device is to start the evaluation and implementation of usability engineering techniques at the very early conception of a new project. Introduction of standardized usability engineering to the company's processes has just started and thus all steps have not been employed yet in a perfect manner. However the results showed that a usability improvement was achieved by early and systematic application of usability engineering. This demonstrates the powerful impact usability engineering can have in any setting and at any stage of development. It is important to start with usability evaluation early in the design process when the methods can be addressed more easily and less expensively. By analyzing the context of use and the user requirements, along with the integration of the evaluation phases in the product development, a quality medical device can be created that is effective, efficient, satisfying and safe.

ACKNOWLEDGMENTS

The authors would like to acknowledge all users that gave feedback during the usability engineering process and the whole development team of the QIAsymphony SW V4. In addition many thanks to Courtney Casciano and Manuela James for reviewing this paper

REFERENCES

21 CFR Part 820: Code of Federal Regulations - Title 21 - Chapter I—Food and Drug Administration Department of Health and Human Services Subchapter H--Medical Devices – Part 820: Quality System Regulations

Bevan N.: Classifying and selecting UX and usability measures; COST294-MAUSE Workshop: Meaningful Measures: Valid Useful User Experience Measurement (2008)

Büchel D., Matern U.: Standardized Evaluation Process of Usability Properties. In Duffy V.: Advances in Human Factors and Ergonomics in Healthcare pp. 296-306 (2010)

Council Directive 93/42/EEC concerning medical devices (1993)

Council Directive 98/79/EC concerning In-vitro-diagnostic medical devices (1998)

EN IEC 60601-1-6:2010 - Medical electrical equipment -- Part 1-6: General requirements for basic safety and essential performance - Collateral standard: Usability

EN IEC 62366:2008 Medical devices - Application of usability engineering to medical devices

EN ISO 14971:2009 Medical devices - Application of risk management to medical devices

Guidance for Industry and Food and Drug Administration Staff - Applying Human Factors and Usability Engineering to Optimize Medical Device Design (2011) (Draft)

EN ISO 9241: Ergonomics of Human System Interaction

CHAPTER 43

Systematical Improvement of an Anesthesia Workstation Considering Physical and Mental Workload

Stefan Pfeffer, Thomas Maier

Research and Teaching Department Industrial Design Engineering
Institute for Engineering Design and Industrial Design
University of Stuttgart, Germany
stefan.pfeffer@iktd.uni-stuttgart.de
thomas.maier@iktd.uni-stuttgart.de

ABSTRACT

In the field of Engineering Design and Industrial Design Engineering the fundamental question of analyzing, assessing and designing usability for human-machine-interaction continuously raises in every thinkable domain. Especially in information-intensive human-machine-systems it is essential not to overload the user and therefore prevent him from operating errors. The present article shows an innovative way for a systematical improvement of an information-intensive human-machine-system in the high reliability organization of human medicine – the anesthesia workstation. The iFlow-analysis is a methodical approach by which this information-intensive human-machine-system can be abstracted and visualized by the information flows between human and machine. This allows an event-related assessment of physical and mental workload. Subsequently, design interventions can be deducted which lead to a better performance for the anesthetist and a lower load.

Keywords: physical and mental workload assessment, iFlow-analysis

1 INFORMATION-INTENSIVE HUMAN MACHINE INTERACTION

Information-intensive human-machine-systems (HMS) are characterized by a high frequency of various information flows (information density) (Bullinger, 1994). In such HMS it is possible that the user is overloaded with information what might lead to operating errors. In contrary, when the information density is low (information-extensive), the user is misled to get into a state of monotony. This might also cause operating errors because of attention deficits. An anesthesia workstation is both information-intensive and information-extensive. In a routine surgical intervention the phases of anesthetic induction and termination are more information-intensive while the intra-operative phase is rather information-extensive (Pfeffer and Maier, 2011a). In the action control loop of the anesthetist there is a lot of information (visual and auditory) to process. Additionally, the anesthetist dynamically has to make decisions and prioritize actions (see Figure 1).

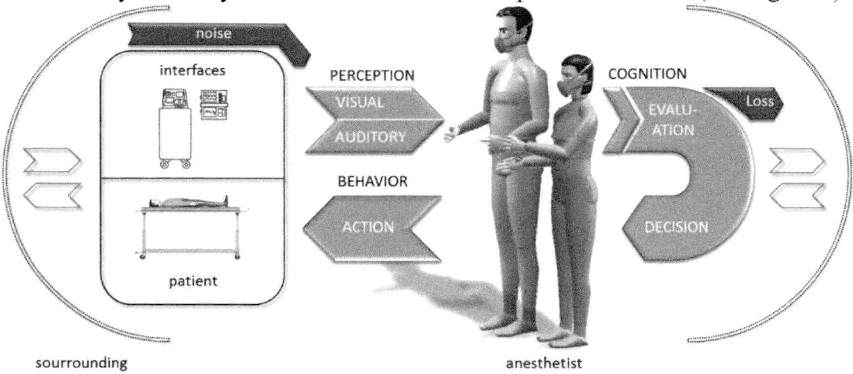

Figure 1 Anesthesia workstation – a complex human-machine-system

Accordingly, the analysis, assessment and design of an anesthesia workstation are very complex but also essential because every interaction of the anesthetist is crucial to the healthiness of the patient. The aim of improving the HMS is to generate a situation-adapted system for the anesthetist in order to submit safe and efficient interactions. Concerning this, it is important not to overload or to bore the anesthetist so that he gets into the state of flow (Csikszentmihaly, 2008). The state of flow is described by the ideal way regarding interaction demands and user skills respectively user capacity.

2 THE IFLOW-ANALYSIS

The improvement of an anesthesia workstation mentioned above can be realized by using the so called iFlow-analysis (information flows), a systematic approach of the Institute for Engineering Design and Industrial Design. It also shows the application areas of the Institute: beginning with the analysis and evaluation using tools and methods from the human factors and ergonomics science to the point of designing optimized human-machine-interfaces. The iFlow-analysis combines the observation of time-related events and their subsequent evaluation concerning physical and mental workload of the operator. On the one hand iFlow matters the afferent and efferent information flows and on the other hand aims an optimally adapted situation of "flow" (Csikszentmihaly, 2008) to the user (Pfeffer and Maier 2010a). The systematic approach consists of 7 main steps that are listed below (Pfeffer and Maier 2010b):

1. **Planning**
   a. Determination of observation aim
   b. Selection of interaction parameters
2. **Preparation**
   a. Gathering of boundary conditions
   b. Choice of recording technics
   c. Determination of layout and setting
3. **Field observation**
   a. Data collection
   b. Live video tagging
4. **Data processing**
   a. Abstraction of information flows
   b. Extraction of features
   c. Generation of iFlow chart
5. **Assessment of physical load**
6. **Assessment of mental load**
7. **Consolidation and deduction of design interventions.**

In a case study the iFlow-analysis was used to improve an anesthesia workstation in a German hospital. The surgical interventions (SI) were a laparoscopic adrenalectomy (SI 1) and a lymph nodes extirpation (SI 2). The intraoperative phase took about 3 hours. Table 1 shows the workflow of the process modules.

The idea or vision of standardization in an anesthesia workstation is daring. In fact, every workstation looks different because of different boundary conditions, devices (type and fabricator), and even arrangements in the operation room (OR). Nevertheless, we chose for the case study a for many other surgical interventions representative arrangement of devices. All the important devices of an anesthesia workstation were used by the anesthetist:

- Respirator (stationary) with gas monitoring
- Infusion pumps x 4
- Patient monitoring system
- Documentation station
- Infusion heater and heat pump
- Infusion bag and roller clamp
- CO_2 insufflator
- Telephone and roll-fronted cabinets x 2.

Table 1 **Workflow of the observed surgical intervention**

Process module	Specification	Phase of anesthesia	SI Phase
1	Connecting respirator and infusion pumps	Anesthetic guidance	Intraoperative
2	Patient positioning		
3	Start intervention 1 (SI 1)		
4	Hand over / break		
5	Documentation of patient recording		
6	End intervention 1		
7	Patient repositioning		
8	Start intervention 2 (SI 2)		
9	Start anesthetic termination	Anesthetic termination	
10	End intervention 2		
11	Disconnecting stationary infusions pumps and respirator		

In step 1 of the iFlow-analysis, i.e. the planning of the observation, it is necessary to select the interaction parameters to be evaluated. The focus was on the anesthetist and his interactions with both the patient and all the devices listed above. This is also described by Friesdorf as the "patient-medical practioner-machine-system" (Friesdorf, 1994). Figure 2 shows this model adapted to our case study and the composition of the three interfaces I1-I3, the human-machine (patient to machine and anesthetist to machine) and the human-human (patient to anesthetist) interfaces whereas the interface between patient and machine is taken for granted and the interfaces anesthetist-machine and anesthetist-patient have been analysed. This structure is also indicated by the general iFlow chart shown in Figure 4.

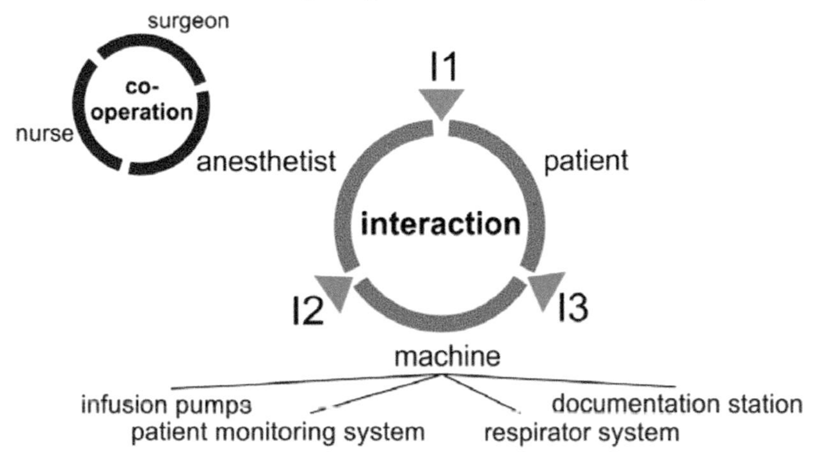

FIGURE 2 Patient-anesthetist-machine-system

Surrounded interaction was also analysed, so the cooperation anesthetist-surgeon and anesthetist-nurse via verbal communication which can be seen as environmental influences. Another step belonging to the planning phase is the choice of interaction parameters from a comprehensive checklist (step 1b). Figure 3 shows the catalogue for the operator (anesthetist).

anesthetist					
perception					visual
					auditory
					tactile
					kinesthetic
					thermal
					olfactory
					gustatory
cognition					plan
					decide
					...
behavior	extremities	left	upper		arm
					hand
					thumb
					index finger
					middle finger
					ringfinger
					little finger
			lower		leg
					foot
					base
					heel
		right	upper		arm
					hand
					thumb
					index finger
					middle finger
					ringfinger
					little finger
			lower		leg
					foot
					base
					heel
					verbal
					facial expression
					gestures
movements					head
					body
posture					sitting
					standing
					kneeling

FIGURE 3 iFlow – Selection of interaction parameters

In step 2 the setting of the OR had to be tested. The area for the anesthetist in the OR was very small – about 5m^2 including all the devices. The field observation of the surgical intervention (step 3) was done via video and audio recording using two cameras for macro- and micro ergonomics. For the subsequent compilation of the iFlow chart it is necessary to have an unimpeded view on the acting person. To get better data of the gaze interactions of the actor it is beneficial to attach an eye tracking system what however was not possible in this case.

Step 4 considers the processing of the video and audio data into the iFlow chart which is a very time-extensive step because all the interactions have to be abstracted as time dependant events and carried over into the iFlow chart by hand. But the generation of the iFlow chart is the essential step for an efficient assessment of physical and mental workload. By the visualization in an abstract form of information; both types of workload can be deducted in a pragmatic way. Figure 4 shows an extracted part of the SI1 and illustrates the physical load in this situation (movement and posture – red time segments) which can be explained by the suboptimal position of the patient monitoring system in relation to the patient itself.

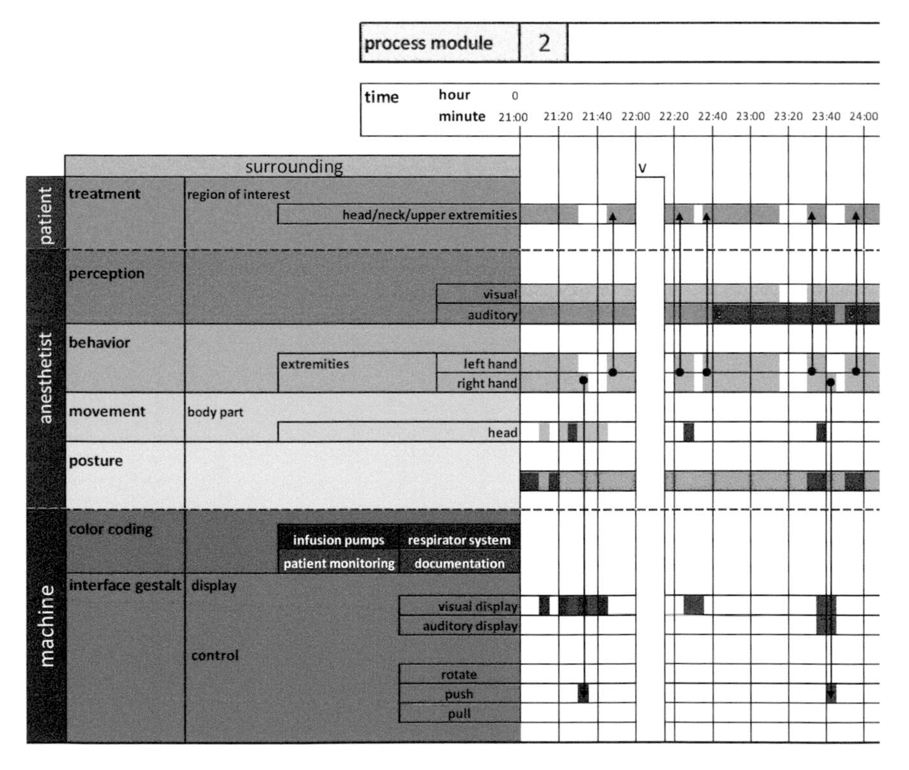

FIGURE 4 iFlow chart – excerpt

The anesthetist simultaneously has to work at the patient´s head and visually checking the monitoring system because of incoming auditory alarms (auditory overload – red time segments). So, in this situation not only the physical but also the informatory or mental load can be qualitative evaluated as high. The abstracted form of the observed interactions and causal coherences allows to deviate several key data to evaluate the whole process of use and to get a focus on the consolidation and deduction of design interventions (step 7). For the example in figure 5 this could be data respective to the frequency of gazing linked with the number of suboptimal head and neck extensions as shown in table 2.

Table 2 Key data – example for physical load evaluation

Anesthetic guidance & termination	Gaze interaction		Head movements	
			light	heavy
	quantity	percentage [%]	quantity	quantity
Infusion pumps	2	6,5	2	0
Respirator system	10	32,3	10	0
Patient monitoring	19	61,3	0	19
Documentation station	0	0	0	0
TOTAL	31	100	12	19

3 RESULTS

The assessment of physical load (step 5) brings out some interesting findings to the dynamic positions of the anesthetist (sit down, stand up, walk, etc.). Due to the concept of workload margin (Hancock, 1988) the predominant factors of physical load are duration, frequency and gravity of interactions. These factors have been analyzed in step 5 with special focus. For example, the frequent gaze changes (up to 200 times per hour) effected a high physical load because of the suboptimal arrangement of the devices (especially head and neck intensive movements).

The assessment of mental load (step 6) considers those sections which are characterized by a high afferent information density or a high information processing rate (many different contents) most of all. This leads to competitive tasks and a high density of efferent information flows (multitasking). In this case the anesthetist is no more attentive to the situation. Operating errors may follow or important signals may remain unnoticed. In literature the "meantime between failures" has been quoted with 30 seconds in stress situations. The study showed a high auditory load additional to the main channel of vision. Figure 5 shows the results of the visual interaction. It is obvious that the patient monitoring system is the most important device for the anesthetist (21,1 % of total time). Similarly to the detection of information-intensive episodes, the information-extensive ones can be extracted of the iFlow chart. Here, the design of the interfaces ideally bridges this gap of attention, so that the awareness of the anesthetist to the situation remains. In our case study the operation procedures were routine tasks for the anesthetist. This was shown in a comparatively long duration of acting with the documentation station (up to 64,2 % of total time of device interactions).

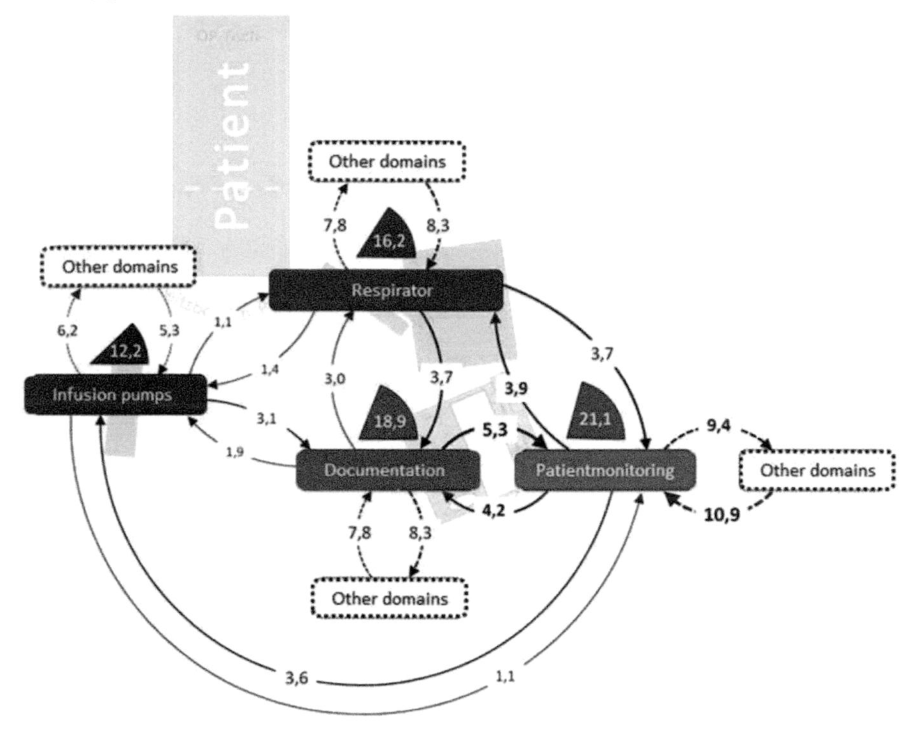

FIGURE 5 visual interaction – example for mental load evaluation

4 CONCLUSIONS

In our case study, only two operation procedures were analyzed. Therefore, the iFlow-analysis shaped up as an effective method to get to improvement potentials. For a significant conclusion more cases have to be observed. The aim is to get to a standardized "cockpit" for the anesthetist. The setting and arrangement of devices in every OR is different – this is in fact a risk factor for the patient safety. Standardization does not mean to simplify the workstation and devices because "simplicity is not the answer" as Donald Norman mentioned (Norman, 2011). The steadily growing capability of the devices makes the human–machine-interaction even more complex. But this complexity is not bad at all because of the useful new options brought to the anesthetist. So, the aim cannot be defined as reducing complexity but as making complexity usable. This is what we call "Usaplexity" (Pfeffer and Maier, 2010a).

Nevertheless our case study already shows several potentials of optimization. The consolidation and deduction of design interventions (step 7) can be divided into two parts. The first part refers to the optimization of the physical load which can be realized on a short-term basis. The rearrangement of the devices due to their frequency and pathways of change for example leads to less movement of the head and neck and therefore to less physical load. The rearrangement should pursuit a cockpit-related composition of the devices.

The second part of optimization aims at the reduction of mental load. These design interventions are on a long-term basis. The iFlow-analysis showed a high auditory load not only for the anesthetist but also for the remaining OR team. In this case of overload the situations of high afferent information density could be optimized for example by using the free channel of haptics by a tactile display. Such tactile impulses could resume acoustic alarms. The expected benefits of these tactile displays which are already used in the aviation domain are a faster classification of alarms without using the channel of vision and thus less gaze interactions and unergonomical movements. There have been made some interesting findings in the basic research at the Institute for Engineering Design and Industrial Design which showed a more efficient performance and quicker reaction on altering visual stimuli whereas a lower demand (e.g. measurement) occurred in action with a tactile display (Pfeffer and Maier, 2011b).

In cases of monotony, the capabilities of the devices could be enriched by optional information. Therefore, the interface should adapt to the situation and accordingly offer a spectrum of opportunities for action (see Figure 6). Similarly, in cases of high load (efferent and afferent information flows), the anesthetist should be able to change into an interface mode with only the core functions and information to lower the information density. This adaption to the situation represents a completely new challenge for the interface design.

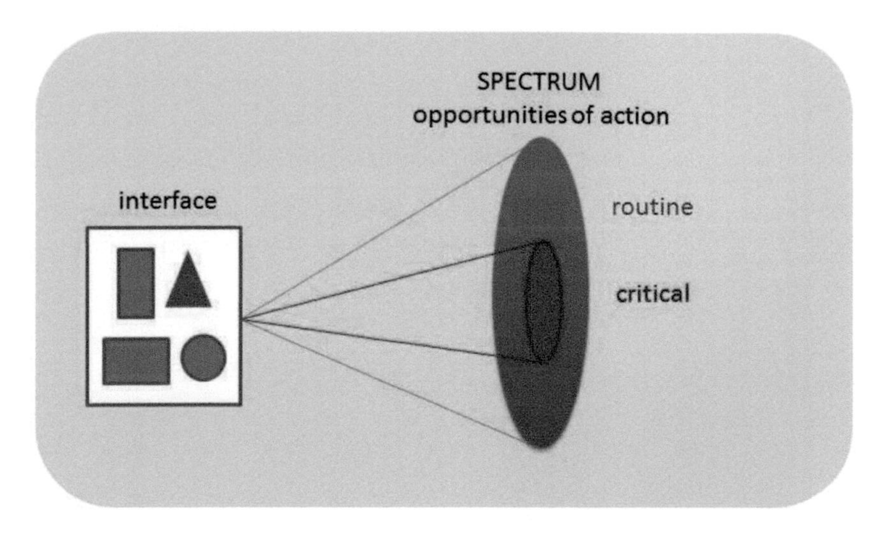

FIGURE 6 Adaptive interface offer

REFERENCES

Bullinger, H.-J. 1994. *Ergonomie. Produkt- und Arbeitsplatzgestaltung*. Stuttgart, Germany: B.G. Teubner Verlag.

Csikszentmihaly, M. 1998. *Optimal experience: psychological studies of flow in consciousness*. Cambridge: Cambridge University Press.

Friesdorf, W. 1994. *Systemergonomische Gestaltung intensivmedizinischer Arbeitsplätze*. Ulm, Germany: Universitaet Ulm.

Hancock, P. A. and Meshkati, N. 1988. *Human mental workload*. Advances in psychology, 52, Amsterdam: North-Holland.

Norman, D. "Simplicity is not the answer," Accessed February 26, 2012, http://www.jnd.org/dn.mss/simplicity_is_not_the_answer.html.

Pfeffer, S. and Maier, T. 2010a. *Usaplexity-Usability in komplexen Mensch-Maschine-Systemen*. In. Useware 2010. Grundlagen-Methoden-Technologien. VDI-Berichte 2099. Duesseldorf, Germany: VDI-Verlag.

Pfeffer, S. and Maier, T. 2010b. *Die Mensch-Maschine-Schnittstelle im Technischen Design - Ein entscheidender Beitrag zur Kostenoptimierung?* In. Design - Kosten und Nutzen. 4. Symposium Technisches Design. Dresden, Germany: Technische Universitaet Dresden.

Pfeffer, S. and Maier, T. 2011a. *Tactile Displays for patientmonitoring in anesthesia*. In. 22nd Annual Meeting ESCTAIC. 12-15th Oct 2011. Erlangen, Germany: Universitaet Erlangen.

Pfeffer S. and Maier, T. 2011b. *iFlow - ein methodischer Ansatz zur Estimation des Workload*. In. 9. Berliner Werkstatt Mensch-Maschine-Systeme. Berlin, Germany: Technische Universitaet Berlin.

CHAPTER 44

The UCD of the Touch Screen Graphical Interface on the Domestic Unit of an In-home Telemonitoring System

Cinzia Dinardo

Consorzio CETMA
c/o Cittadella della Ricerca
S.S.7 Km.706+030, 72100 Brindisi
cinzia.dinardo@cetma.it
+39 0831 449603

ABSTRACT

Consorzio CETMA, on the occasion of the project H@H Hospital at Home for the tweaking of a telemonitoring system for in-home patients by means of a remote control junction box accessible via web, carried out a research to implement the touch screen interface of the domestic unit of this system. The objective was to obtain a user friendly interface, putting the user at the centre of the activities. The quality of the ratio between user and used mean is determined by the ergonomics level. The most important requirement to determine this level is security, followed by adaptability, usability, comfort, pleasantness, comprehensibility, and so on.

The "perceptibility principle" has been used during the design process in order to obtain a high level of the system usability. According to this principle, every information and command necessary for the product utilization should be clearly perceptible. Thanks to a plain hierachy of commands, to a deep knowledge of the different user typologies who will interact with the system and to a clear definition of tasks and scenarios, a graphical interface has been designed trying to make it aesthetically pleasing. The designing of touch screen buttons has generated a hierarchy of very intuitive commands in order to minimize user errors.

Keywords: Telemedicine, e-Health, Ageing, User Interfaces, Touch Monitor, Interaction Design.

1 INTRODUCTION

The design of interactive product should be based not only on the graphic appearance of the interface but mostly on its usability. A really good graphical interface is not only essential in functionality but also well organized in every graphical element, all the functionality should be placed as result of a precise hierarchy perceptive. Furthermore the touch screen technology involves all the problems related to visual perception when not supported by tactile perception, just to remind that "touch screen buttons" have no tactile feed-back. In order to obtain a really high level of usability the project of the system has been developed following the "Perceptibility Principle" that suggests to highlight, and make easy to perceive, all the information that is necessary for product utilization.

2 PERCEPTIBILITY AND VISIBILITY PRINCIPLE

One of the most important principle in the user-centered design is that of perceptibility, very similar to the Norman principle of visibility.

According to this principle every information and every command necessary for the product utilization should be always clearly perceptible, without any extraneous or redundant information.

To satisfy this principle two activities have been executed:

- Assignment of the command hierarchy or of the necessary information;
- Performance of the perceptive hierarchy.

Determining the necessary infomation and commands for the product utilization depends on a good knowledge of the users and on a clear definition of tasks and scenarios. That is why designing for the perceptibility requires some preliminary analysises, which are indispensable to reduce the distance between designer's mental model and the users' one.

The used criteria to assign the hierarchy of the necessary information have been:

- target definition;
- requisites analysis;
- task analysis;
- scenarios;
- use of conventions;
- screen legibility;
- buttons dimension.

2.2 Target definition

Related users analysis is helpful in defining product requisites on the basis of their needs. This analysis enabled the definition of the performances that touch screen interface should ensure to meet requirements, obtaining a calibrated interaction man-product in the context of use and in the user typology. Therefore the analysis focused on the real users' needs and expectations, besides the users typology.

The interface of the system domestic unit will be usable by four different users typologies:

- user – "Patient";
- user – "Patient relative";
- user – "Healthcare assistant";
- user – "Maintainer/Fitter".

2.3 Requisites analysis

After having defined users typologies and their needs, we have defined the functions that will be manageable through the interface.
Every user typology will be able to access different functions, two of which will be available to the first three user typologies, "EMERGENCY CALL" function and "LOG OUT" function.

User "Patient" will be able:

1. to forward an emergency call;
2. to regulate and to exclude audio volume;
3. to refuse or to end a Standard call;
4. to accept or to forward a Standard call;
5. to activate/deactivate oxygen supply (O2):
6. to activate/deactivate Vacuum;
7. to log-out/log-in;
8. to display the Remote Healthcare Assistant;
9. to display one's own picture (framing of the video camera on the monitor).

User "Patient's Relative" will be able:

1. to forward an emergency call;
2. to log-in/log-out;
3. to display patient's vital parameters;
4. to mute alarms;

User "Healthcare Assistant" will display, by default, the patient's screen "Vital Parameters"; through the menu bar, made up of different elements, the user will be able to access his related 5 screens.

User "Healthcare Assistant" will be able:

1. to stop specific time alarm;
2. to stop specific alarm always;
3. to stop heart beats beep;

4. to deactivate specific alarm;
5. avviare measure session (PV);
6. to stop measure session (also single).
7. to display medical records
   a. to display notes (if notes have never been displayed, they are highlighted as pending);
   b. to insert notes;
   c. to display an exam (if report has never been displayed, it is highlighted as pending);
   d. to report an exam
   e. to display reports (if reports have never been displayed, they are highlighted as pending);
8. to display vital parameters values (ECG graphic + SPO2 + most recent values)
   a. Zoom detail on patient's vital parameters
   b. Trends analysis
   c. Equipments
9. to activate/deactivate equipments
   a. Activation/Deactivation Oxygen
   b. Activation/Deactivation Vacuum
10. to make the different typologies of call
    a. Standard call (Note: audio call must be able to remain active also when the other available screens are visualized)
    b. Emergency call
    c. Volume regulation
11. to log-out/log-in from all the screens

User "Maintainer/Fitter" will display by default "manutention" screen and will be able:

1. to display errors technical details;
2. first Intallation;
3. re-installation (download Domestic Unit Configuration);
4. auto-control test;
5. ECG calibration.

All the functions have been defined on the basis of real users skills, that can vary from a specific competence possessed by operators, acquired through short learning courses, to a lower middle competence possessed by some patients' typologies (mainly elderly patients).

2.4 Task analysis

Task analysis consists in the study of the tasks that every user typology must execute, in terms of actions and cognitive processes, to reach a specific objective. For this reason we analyzed and evaluated tasks that all users should have executed through the use of the interface. Through these analysises, buttons, fields and functions executed by every object in the interface have been designed, trying to guarantee comprehensibility of use from every identified user. This aspect can

generate in the users the facility with which object is perceived and recognized for the function that executes , but also a certain grade of pleasantness towards the interface itself.
After having analysed tasks that every user typology should have execute through the interface, a hierarchy of commands (or of the necessary information) has been identified.

2.5 Screen legibility and psychology

In designing the interface SW, usable through a touch screen, the top priority is screen legibility.

2.5.1 Colours

Complementary colours are particular, for they are in maximum contrast, so when they are drawn they create a flickering effect that makes focalization very difficult and strains sight, making difficult the reading. Here is an example:

Table 1 Colour contrasts

Red on green	Yellow on purple	Blue on orange

It is clear how much it is important, in order to reach text legibility, to choose well the colour of text and background. A research carried out by a University of Texas on the legibility based on colours of text and background has reported the following results:

Table 2 Colours legibility on backgrounds

The most legible	The less legible
White on black	Green on red
Grey on black	
White on blue	
Black on white	
Yellow on black	
Blue on white	

Three of the most legible combinations use black as the background. However black is a color that recalls strong negative feelings on users' minds, therefore in the graphical implementation of the interface, positive colours have been associated to the black colour of the background, so that on the whole sight it do not express negative feelings to the users.

Figure 1 "VideoCall" Screen

Recent studies prove that colours affect users' feelings and climates. In choosing the different colour combinations, we considered feelings and climates that sprang from different colours.

The following table associates the main colours to the feelings or to the concepts: obviously people culture distorts colour meaning (for example, in Japan, white is the colour of mourning):

Table 3 Colours legibility on backgrounds

Colour	Positive feelings	Negative feelings	Notes
Red:	Passion, energy, love, power, action	Blood, war, danger, contentiousness	It is one of the "salient" colours, namely a colour that seems to draw nigh the observer.
Green:	Nature, spring, fertility, money, perseverance, readiness	Inexperience, envy	Used as a symbol of "all clear" (i.e. traffic light); it is considered as a "young" colour.

Yellow:	Sun, summer, gold, optimism, helpfulness, altruism	Disease, cowardice, hazard	Particularly visible, also in limit situations (i.e. Fog lights).
Blue:	Stability, calm, unity, satisfaction, harmony, fidelity, sky, sea	Depression, conservatorism	Very used colour by companies for its positive characteristics.
White:	Snow, purity, innocence, peace	Cold, hospital, sterile	
Grey:	Intelligence, dignity, maturity	Shadow, boredom, depression	
Black:	Power, formality, depth, style	Wickedness, fear, death, anonymity	

2.6 Touch screen buttons dimensions

Dimensional and morphological aspects of manual commands concern space necessary to the fingers actions to push buttons. The usability of manual commands and in particular of buttons depends on the dimensions and on the distance among buttons, that is on the movement space allowed to the fingers, and on the possibility to recognize buttons and their position through tactile contact, characteristic which is absent in touch screen buttons. Tactile recognizability of the elements permits their identifiability also with a low visual attention level and, moreover, the possibility of comprehend the outcome of the executed action. The lowering of the buttons represents a feedback through which the user can evaluate if the finger pressure turned out well.

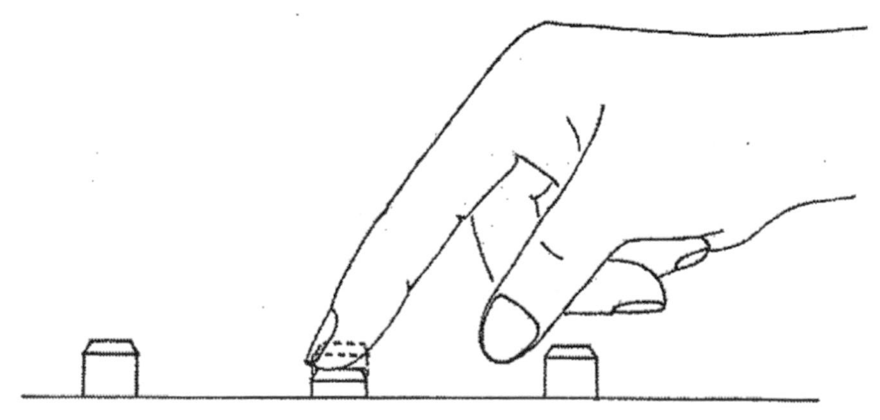

Figure 2 Finger pressure

Table 4 Table captions should be placed above the table

Control device	Operating method	Distances in mm		Recommended dimensions
		Minimum	Suggested	
Button	Finger pressure	20	50	Min.10 – Max 25mm (touch screen)

The problem of a touch screen interface is that, due to the parallax, there is a sensitive gap between what user sees and what user actually clicks.

For this reason the buttons should cover a wide area and should be distant from each other.

Touch screen buttons have been designed by means of a simple and clear graphic art to obtain a high level of recognizability. Moreover, in order to avoid the problems connected with light reflexes on the screen, high chromatic contrasts have been used.

2.6.1. Foreseeing the presence of a feedback

The *feedback* represents the information of return, the system connection in response to the action that the user executed on the interface; it helps in reporting the system status and the outcome of the user's action, including error messages. This characteristic is important to an interface which is usable through a touch screen monitor.

In physical buttons there is a purely tactile feedback: through his fingers the user perceives the pressure and release sensations, and this does not happen when he uses touch screen buttons. For this reason it is very important to provide a double feedback, a visual feedback and a sonic one.

But by when could we expect to receive the feedback?

One tenth of a second is approximately the time to give the user the sensation that the system reacted outright:

1. 1 second is approximately the time to show the reseults of user action without interrupt his reasoning, although he will notice the delay in the system reply;
2. 10 seconds is the time to hold the user attention focused on the dialogue;
3. Over 10 seconds: generally, the user starts another activity while the computer is working: if we want to hold his attention, system reply of his action should be shown in maximum 10 seconds.
4.

Hence we found that a device is easy to use when there is visibility of the possibile actions, when control boards use natural correlations and when there is information of return that informs user about the system status.

REFERENCES

ANCESCHI, G., (a cura di), Il progetto delle interfacce. Oggetto colloquiali e protesi virtuali, Domus Academy, Milano, 1992-1993.

ANDREOLI TERESA, Psy.D. Psychology Assistant Cognitive Disorders Among the Elderly, Lic #PSB 31633, BCIA-C #3949 Fellow *Brain Therapy Center* www.brain-injury-therapy.com/articles/dementia.htm

AA.VV. Progetto Lettura Agevolata www2.comune.venezia.it/letturagevolata

AA. VV. Tecnologie informatiche e utenza debole. La progettazione ergonomica dei siti web e delle postazioni di lavoro per i disabili. *Il Sole 24 Ore Pirola* (2002) - Codice EAN: 9788832447552

BAECKER, R., & Buxton, W., Readings in Human-Computer Interaction: A Multidisciplinary Approach, Morgan-Kaufmann Publishers, Los Altos, CA, 1987.

BAILEY, R., Usability Testing versus Heuristic Evaluation: A Head-to-Head Comparison. Proceedings of the Human Factors Society 36th Annual Meeting, 409-413, New York, 1992.

BIAS, R., and MAYHEW, D. J., Cost-Justifying Usability, AP Professional, Cambridge, MA, 1994.

Hyperlabs - empowering users www.hyperlabs.net.

RASKIN J. “Interfacce a misura d’uomo”, *Apogeo 2003*

CHAPTER 45

User-centered Requirements for Health Information Technology: Case Study in Cardiac Surgery

Raquel Santos[1], *José Fragata*[2]

[1] Human Kinetics Faculty - Technical University of Lisbon
Estrada da Costa, 1495-688 Cruz Quebrada – PORTUGAL
rsantos@fmh.utl.pt

[2] Hospital Santa Marta – New University of Lisbon
Rua de Santa Marta, 1160-024 Lisboa – PORTUGAL
jigfragata@gmail.com

ABSTRACT

Despite the widespread use of technology and information technology in health care there are many problems and flaws registered by several studies, due mainly to the lack of consideration of human factors in its development and implementation. Technology in health care must be based on users and organizations needs, besides the centered concern with its technological power. To accomplish that purpose healthcare must, at first, be aware of their real needs because this context has still a significant lack of knowledge of their complexity. The basis to achieve that objective is the analysis of work system as well as the consideration of usability principles in technology that will permit a better interaction, rendering the system more efficient and safe. The present study is an example of a work system analysis in a pediatric cardiac surgery room.

Keywords: user requirements, human factors, cardiac surgery, efficiency, patient safety

1 INTRODUCTION

Nowadays in health care the relationship between clinicians and patients or between health care professionals themselves is mediated by technology. In hospitals, technology is present all over the place in medical or administrative tasks and its use doesn't stop growing. Technology is used to perform an increasing number of tasks and even in tasks where it is already used the speed of introduction of new or updated models is enormous.

Stakeholders view technology like the panacea that solves the majority of problems and improves the systems' performance. That could explain why the tremendous costs that underlie technology are viewed as a good investment even when there isn't many data available about the return on investment resulting of technology implementation. But in this era of health care reforms, pushed by severe financial constraints, cost-effectiveness must be performed with accuracy considering all influencing factors, and it is inacceptable that wastes resulting from inadequate interactions continue to proliferate.

The advantages brought by technology to health care, namely in prevention and treatment, could not be questionable. Nevertheless, it is undeniable that there exists a notion of waste or at least the idea of a large room for improvement in this process. The problems and flaws registered by several studies, namely regarding information technology (IT), are the proof. The study of Koppel et al. (2005) was one of the forerunners raising the awareness for the flaws of IT showing 22 types of medication error risks in a computer provider order entry (CPOE).

One of the main causes for these problems registered with technology in general and IT in particular is the lack of consideration of human factors issues in the development and implementation of IT. The problems and flaws result from the use of technologies by humans and not from their technical failures. Additionally, health care is a complex context with a great variety of professions and skills, where tasks and its control is divided by several professionals, the information needed to perform the tasks is distributed among several supports and actors, and where safety is still a problem. However, contrary to other complex systems like aviation, health care hasn't still systematically integrated the user-centered design to optimize its system. Thereby that should be an urgent priority of managers' agenda.

2 THE PROBLEM

The problem in many current IT systems designs is that critical clinical applications were designed using the same premises used in the development of previous hospital information systems products, considering primarily the logistic process and incorporating simplistic idealized models of work processes (usually sequential linear processes), therefore ignoring a user-centered and safety oriented design approach (Wears & Berg, 2005).

Moreover, most of the products are designed without consideration of basic human computer interface ergonomic principles. These principles intend to facilitate the three phases of human cognition (perception, cognition and action) when interacting with technology. Usable information systems are essential to promote efficiency and prevent errors in healthcare. But many of the IT actually in use present usability problems being common the existence of error use risks that endangers patient safety.

Additionally, those products are sometimes implemented in highly complex work environments without anticipation of their profound impact on the work or of possible conflicts with existing policies. The introduction of technology changes, for instance, the work processes, the way tasks are performed, the communications needed, the cooperation between professionals, the geography where information is transmitted or the control level on the task. The fact that the system is not adapted to the workflow and needs of professionals may cause workarounds to compensate that inadequacy, mitigating the expected improved efficiency and safety of technology introduction. Thus, not surprisingly, the installation of those complex products may generate unexpected consequences, most of them with negative effects on the healthcare professionals work or on patient outcomes (Ash et al., 2004).

Finally, there are also several work situations where IT is not yet used or isn't potentiated and optimized to support, for instance, clinicians' non-technical skills. Health care is a complex socio-technical system and therefore only a system's approach considering both people, tasks and physical and technical environment can produce recommendations to re-engineer the work system or specify user and organization requirements to develop or updating IT.

In the field, health care institutions and companies commercializing the systems usually address these human factors difficulties by "training the users" on the new technology in order to accommodate the new system workflow and business process. However, training is not enough. It is also necessary a process reengineering to systematize the procedures between professionals avoiding at the same time the adoption of less efficient or more dangerous procedures.

In fact, the image that emerges from this scenario is the almost passive role of different actors involved. Manufacturers provide the late technology with all possible functionalities that professionals might need, managers acquire and implement these tools expecting the improvement of patient safety and organizations' efficiency, and health care professionals watch the arrival of these new devices and must adapt to them. It is assumed by all that the specified technology is the best solution and no other efforts are needed.

This passivity must be replaced by a proactive approach. The design of health care processes and technologies must be adapted to the needs, capacities and limitations of individuals, teams and organizations. Therefore, the basis to succeed is having a complete knowledge of the work system. In other words we must model it and analyze its activity, namely, the users behavior, the environment in which tasks are performed, communications and cooperation modes, identify tasks that cause interruptions in interactions and consider the information processing with high levels of fatigue and stress. Only this way the potential of technology could be totally availed avoiding wastes and negative returns on investments.

3 OBJECTIVES

The study here described presents a work system analysis in a pediatric cardiac surgery operating room (figure 1), regarding specifically social non-technical skills communication, teamwork and leadership (Santos et al., 2012). Based on the user-centered requirements the objective was to produce recommendations for improvement, considering that communication failures are the leading cause of inadvertent patient harm. A specific objective was also to identify emerging needs that could be supported by IT, in order to augment the systems' resilience.

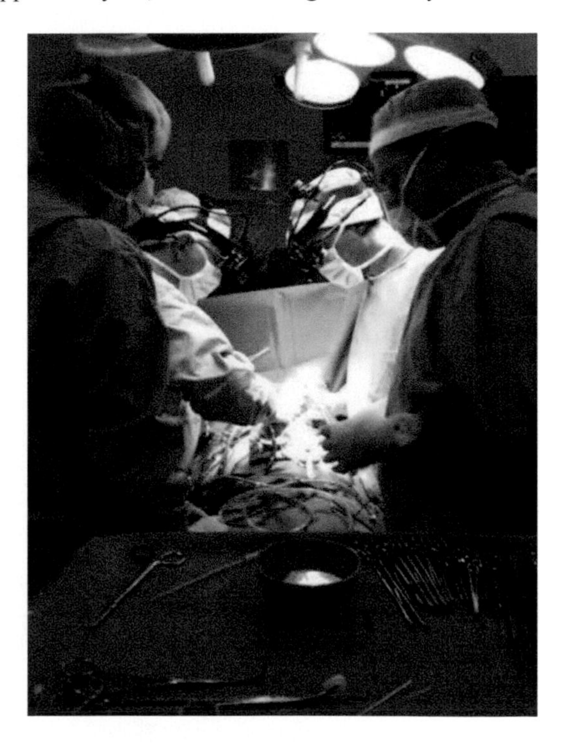

Figure 1 Pediatric cardiac surgery

4 METHODS

For this study 10 pediatric cardiac surgeries were observed, video and tape recorded by a human factors specialist. The team for each procedure was composed by 3 surgeons (S1, S2 and S3), 1 anesthetist (A), 1 perfusionist (P), 1 scrub nurse (SN) 1 circulating nurse (CN) and 1 anesthetist nurse (AN), distributed in the OR as shown in figure 2. The content and effects of relevant communication events to the surgical procedure were systematically described. Utterances were chosen as the unit of analysis and these communications were classified by frequency, direction, type, content and pattern. Disturbing factors to communication were also analyzed.

The communication analysis has also supported the characterization of teamwork and leadership. Despite the little consensus among researchers regarding teamwork and the factors that comprise it, Salas et al. (2004) reviewed team literature and proposed five core factors of teamwork (team leadership, mutual performance monitoring, back-up behavior, adaptability and team orientation) that were used to characterize this variable.

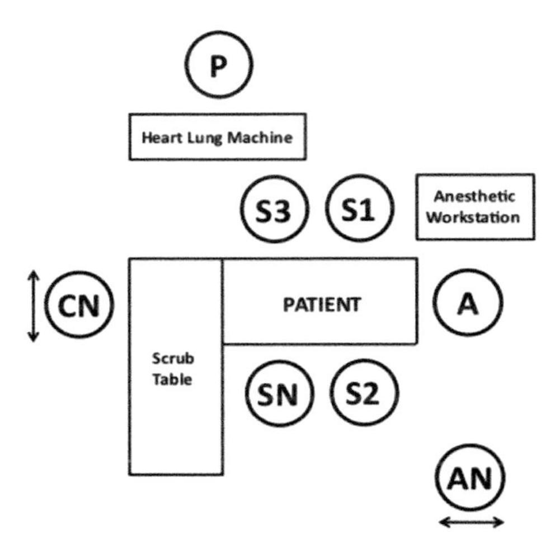

Figure 2 Team positioning in the cardiac operating theatre

5 RESULTS

The main results characterizing non-technical skills were as following (Santos et al., 2012). Twenty-one clinicians have participated in the 10 surgeries (7 surgeons, 3 anestetists, 3 perfusionists and 8 nurses) that produced a total of 10167 communications, with an average of 1017±170.9 per procedure that last on average 136.2±19.5 min (intra-operative phase only). Almost all communications were directed to a specific element of the team. Only 0.7% was directed to all team members, sent mainly by S1. The frequency of communication was maximal between S1 and SN (16% of all), followed by S1-S2 (13.8%) and S1-P (12.4%). Communications S1-A were no more than 5%. Direction of communication also varied. S1 and S2 were mainly senders to SN (13.5% and 8.4% respectively), S1 was mainly sender to S2 (10.5%), but the relation S1-P was more balanced (6.6% - 5.8%). There's clearly the tendency to medical professions (surgeons and anestetists) to be mainly senders, and non-medical professions (perfusionists and nurses) to be mainly receivers. The communications analyzed were categorized in six major types varied from request, question, answer, statement, information and explanation, being different for distinct staff roles. S1-SN involved mainly requests, between S1-S2 statements prevailed in both senses, S1-P involved mainly requests and P-S1 mainly answers. Concerning the content of communications it was observed that between surgeons' staff and SN instrument request dominated, between surgeons the speech is dominated by technical aspects of the procedure and between S1-P the content was about bypass and cardioplegia management. Clearly the communications between S1-P were the more critical if a fail occurred. Finally, communication patterns varied, being closed-loop (with feedback and double check) only between S1 and P, being mostly open, among other team members. In general it was verified an evident lack of systematization and homogeneity of communication between different team members in time and in form of content. This means that there's a potential to information loss.

The main disturbing factors were noise and interruptions. Noise was due to three main sources: teamwork (with team cross conversations), equipment (material unpacking, saw and suction functioning, strokes on the perfusion machine, equipment alarms, etc.) and environment related (high frequency of opening/closing surgery room doors) provoking also many interruptions on surgical activity.

Regarding the factors defined by Salas et al. (2004) for teamwork, the signs of monitoring each other's work (mutual performance monitoring) are low. Sometimes it happens, above all with perfusionists and circulating nurses that in some phases of surgery have more time without action and are more available to do it. Present even in a more reduced percentage are behaviors of providing feedback to improve performance or assist in performing a task and complete the task for another (back-up behavior). However, there are more signs of adaptability of individual team members based on information gathered from the task environment with compensatory behaviors and reallocation of resources (mainly in nurses).

Concerning leadership behavior it was clearly noted that situation awareness, essential to an effective teamwork, is deficient mainly in nurses and especially in perfusionists. One contributing factor may be that leading surgeon rarely gives information about the procedure development to other team members.

5.1 Emerging needs - Clues for optimization

Besides the characterization above-mentioned that have permitted to understand the information flows in the surgery room, other important needs have emerged (Santos et al., 2012). Communication patterns have shown a lack of systematization that could be ameliorated with training to improve the knowledge and coordination of effort among surgeons, anesthetists, perfusionists and nursing staff.

In addiction the relation between S1 and P was very sensitive and must be treated with caution. There is no layer of defense for this interaction if anything goes wrong and a misunderstanding occurs. Each one of these professionals has access to information that the other does not. S1 has visual access to the surgical field, whereas P has visual access to the various displays and controls of the heart-lung machine and other equipment not visible or accessible to S1. Successful execution of cardioplegia management and defibrillation tasks requires effective integration of this information, and P is the element that has less access to live information because he's placed behind S1. The synchronization of their critical actions lies exclusively on the verbal communication established and that explains the strategy of more systematic and structured exchanges developed by these professionals. The discomfort with this constraint was physically expressed by P that, when possible, stood up next to S1 to see the procedure development, and likewise S1 turned back to talk to P when the task permitted.

These facts stressed the need to look for means or equipment to mitigate the potential of information loss and cognitive charge and to improve situation awareness, in order to augment system's resilience. Thus two solutions are going to be tested (Santos et al., 2012).

In the specific communication channel S1-P, but also beneficiating other team members situation awareness, information technology (IT), namely a screen monitor, could play a major role showing images of the procedure progress and additional information needed to be shared (for instance, the main patient characteristics such as age and weight), providing redundancy to a possible communication failure. Furthermore, a better awareness of the surgery development could help teamwork (very deficient in our study) allowing also P and other professionals to anticipate some actions, improving efficiency and decreasing bypass time, so harmful to the patient.

Another additional solution is the change of perfusionist positioning regarding leading surgeon (actually in its back) putting him in front of S1. However, this displacement implies the positioning change of SN to the right-hand side of S1, breaking well-established activity routines. In addition, other variables such as the viewing angle of SN regarding S1 technical actions (that can be diminished) and the increase in movement paths to provide devices to S1 (raising physical fatigue in SN) must be tested, in addiction to other consequences that should also be analyzed and simulated regarding their effects on teamwork.

Furthermore, the disturbing factors detected were almost totally dependent on organizational parameters that can be enormously improved. In a first step, the high frequency of doors opening during surgery must be reduced through new rules for non-technical factors and better organization for technical issues. Nevertheless, the attention dedicated to these variables cannot be resumed only to those registered in the sharp end of the system (OR) but must also be to those at the blunt end, at higher levels of management.

6 CONCLUSIONS

Research into actual practice at the sharp end of health care will provide the basis to understand how IT can support clinical practice. Technology in health care must be based on users and organizations needs besides the centered concern with its technological power. To accomplish that purpose, and similarly to other sectors, healthcare must be aware of their real needs. It's necessary an adequate understanding of what clinicians do and especially how they do it. Until now, this context has a lack of knowledge of their specificity and complexity. That knowledge is essential in order to assist developers to design ongoing adaptation and be more demanding with manufacturers to fulfill their needs. The human factors specialist is the essential element to bridge the stakeholders, analyzing the needs and formalizing the requirements for design purposes.

Like defended by Ferlie & Shortell (2001) organizations must promote a culture that supports learning through the analysis of socio-technical system to produce knowledge. With such an approach, based on a careful process analysis, it will be possible to retain the good practices and detect failures. To overtake these failures two perspectives are usually employed. On the one hand, the processes standardization and the introduction of technology. The efficient use of information and IT (with a better adaptation to the work and the user) and potentiating its use in situations not yet explored, will render the system more safe and efficient. On the other hand, it is essential the active participation of clinicians. If it's true that humans are responsible for the active failures that can contribute to losses and injuries, it is also true that they regularly catch and correct their own and others' errors. Much of the resilience that has been demonstrated by health care systems has been initiated by its human element. For that purpose non-technical skills like communication, teamwork and leadership are essential and must be developed more systematically. This cyclic process, which should be integrated in a more comprehensive quality improvement program, will feed the system with continuous information that should be integrated to provide increased efficiency and safety.

Until now health care has devoted more attention to their own science rather than support issues. But in such times of huge challenges, health care must know its specificities more profoundly and call other disciplines to support its progress in a more sustainable and efficient way.

REFERENCES

Ash, J., Berg, M., Coeira, E. 2004. Some unintended consequences of information technology in healthcare: the nature of patient care information system-related errors. J Am Med Inform Assoc. 11, 10-12.

Ferlie, E., Shortell, S. 2001. Improving the quality of health care in the United Kingdom and the United States: a framework for change. Milbank Quarterly, 79(2), 281-315.

Koppel R., Metlay J., Cohen A., Abaluck B., Localio A., Kimmel S., Strom B. 2005. Role of computerized physician order entry systems in facilitating medication errors. JAMA, Vol 293: No. 10, 1197-1203.

Santos R., Bakero L., Franco P., Alves C., Fragata I., Fragata J. 2012. Characterization of non-technical skills in paedriatic cardiac surgery: communication patterns. European Journal of Cardio-Thoracic Surgery. doi: 10.1093/ejcts/ezs068.

Salas E., Burke C., Stagl K. 2004. Developing teams and team leaders: strategies and principles. In: D. Day, S. Zaccaro & S. Halpin (eds). Leader Development for Transforming Organizations: Growing Leaders for Tomorrow, 325–55.

Wears, R., Berg, M. 2005. Computer technology and clinical work: still waiting for Godot. JAMA, 293, 1261-3.

CHAPTER 46

Design, Usability and Staff Training – What Is More Important?

Matern U.

Medical Faculty
Tuebingen, Germany
ulrich.matern@uni-tuebingen.de

ABSTRACT

Medical devices may lead to patient trauma and time consuming clinical processes. The reason can be a lack of usability and / or untrained personal. We wanted to know which topic is more important to improve patient safety and efficiency. Therefore we performed a usability test with a radio frequency device and evaluated the "teaching" method of the company compared to "no teaching" and "teaching and training". The results showed that the medical device has some minor design deficiencies, but there is no standardized teaching and training concept which could avoid problems caused by the design.

To fulfill international standards for certification of medical devices the manufacturer has the opportunity to improve usability and thereby reduce the effort for teaching and training.

Keywords: ergonomics, usability, medical device, teaching, training

1 INTRODUCTION

Modern medical device are difficult to manipulate. E.g. 70% of German surgeons and 50% of nursing staff have difficulties to manipulate these intuitively and correct. 59% of these surgeons and 40% of the nursing staff declare that they feel not being trained sufficiently in the usage of the devices. (Matern, 2006)

International standards and FDI regulations have been established to reduce these numbers and thereby, increase patient safety (DIN, 2008). These standards should encourage manufacturers to establish a usability engineering process for all medical devices. Additionally IEC 62366 demands data about evaluated teaching and training concepts (duration, re-training).

The aim of the studies described here was to evaluate

- the usability of a common medical device and
- the training necessary to manipulate this device properly.

2 Material and Methods

For the two tests we evaluated a common Radio Frequency Device used for surgical blood control and tissue cutting.

First study – usability test (Büchel, 2007)

In a standardized usability test setting 16 volunteers tested the device regarding: Overall usability, functional design, safety, adequacy for professionals, adequacy for rare use and adequacy for untrained use. Furthermore, usability was evaluated with regard to: start-up, setting values, choosing programs, saving programs and managing error notifications. Measures for usability were, according to the standards mentioned above, "effectiveness", "efficiency" and "user satisfaction".

Second study – Evaluation of different Training concepts (Geiselhart, in progress)

28 OR nursing students were randomized into three equal groups. The control group didn't get any explanation in the usage of the device. The second group got a typical theoretical introduction concerning the manipulation of the device, while the third group got an additional hand's on course by the salesman of the company. After two weeks the students were asked during a usability test scenario to manipulate the medical device.

3 RESULTS

First study – usability test (Büchel, 2007)

Several usability problems were found. Due to these usability problems some volunteers needed prolonged procedure time up to 10 minutes while other did not succeed in any way.

Second study – Evaluation of different training concepts (Geiselhart, in progress)

One salesman of the company taught the volunteers; the groups were informed about different theoretical and practical content. A statistical difference for the manipulation properties of the three groups could not be measured. Single students of each group were able to manipulate the device without any problem while others had huge problems not depending to one of the three teaching methods.

4 CONCLUSIONS

Medical devices are complex machines and most often have to be tuned in difficult medical situations.

The usability of the evaluated medical devices can be improved.

Because of the complexity, there will always be the necessity of additional training.

Standardized teaching and training concepts shall be developed and evaluated by the companies, to follow the international standards for certification.

It can be proposed that optimizing the usability of medical products as well as establishing of evaluated teaching and training concepts will increase patient safety.

The manufacturer has the opportunity to improve usability and thereby reduce the effort and costs for teaching and training.

ACKNOWLEDGMENTS

The authors would like to acknowledge the volunteers and the salesman of the company who all supported this work for free, because the study was not being paid by someone.

REFERENCES

Büchel, D., T. Baumann, U. Matern 2007. Usability of radio-frequency devices in surgery. In: HCI and Usability for Medicine and Healthcare, Ed. A.Holzinger. Springer, Berlin Heidelberg: 97-104; ISBN 0302-9743

DIN EN IEC 62366. Medizinprodukte – Anwendung der Gebrauchstauglichkeit auf Medizinprodukte. Beuth 2008

Geiselhart, I., U. Matern, (in progress) Training effects on safety of device manipulation

Matern U., S. Koneczny, M. Scherrer, et al. 2006. Arbeitsbedingungen und Sicherheit am Arbeitsplatz OP. Dtsch Ärztebl 103(47): A 3187–92

CHAPTER 47

Implementation of a VA Patient-centered Prescription Label

Keith W. Trettin, Erin Narus

VA National Center for Patient Safety
Ann Arbor, Michigan 48106
Keith.Trettin@va.gov

ABSTRACT

The VA National Center for Patient Safety (NCPS) evaluated the impact of culture, age, and education on the understanding of a proposed evidence-based VA prescription label. The evaluation was completed through a structured in-person focus group questionnaire process. Participants were recruited from various regions around the United States to ensure broad representation of various cultural and ethnic groups. The study confirmed that a new standardized VA prescription label is preferred by Veterans and improves Veteran comprehension of prescription label information.

Key Words: prescription label, literacy, standardization

1 PRESCRIPTION LABEL DESIGN AND ITS EFFECT ON VETERAN PRESCRIPTION ADHERENCE

Is Veteran adherence to a prescribed drug regimen influenced by how we ask them to take their medications on a prescription label? When 446 Veterans were asked how many tablets per day would they take when given a prescription with the directions to "Take one tablet daily with meals" only 42 % of the respondents identified the correct answer. When the participants were asked the same question, but with the directions changed to use the word food rather than meals, Ex "Take one tablet daily with food" 81% of the study participants were able to respond correctly. The Department of Veterans Affairs (VA) National Center for Patient Safety has received numerous reports of medication mishaps caused by a lack of understanding by Veterans on how to accurately adhere to the medication regimen as prescribed by their physician. To enhance medication adherence and minimize associated medication mishaps, the VA sponsored this study to elicit the relationship between prescription label literacy and medication adherence, as well as determine if Veterans have a preference on how the VA presents prescription label information to them.

The VA provides services to over 21 million Veterans. Of these Veterans, 4.6 million receive prescriptions through the VA. The VA dispensed 136 million prescriptions at a cost of 3.2 billion dollars in FY 2010. On average, the number of prescriptions filled by the VA is increasing by 4% annually. Seventy seven percent of the VA prescription dispensed are through the seven VA consolidated Mail Order Pharmacies. Local VA Medical Centers or Community Based Outpatient Clinics dispense the remaining prescriptions.

The VA implements many clinical and cost effective strategies to assure the best possible clinical outcome at the lowest possible cost. One cost minimization strategy requires significant prescription label comprehension by the Veteran. A VA cost strategy is to minimize the number of medication strengths stocked by VA pharmacies. In doing so, the pharmacies may reduce inventory and take advantage of pricing which is the same regardless of medication strength. To take advantage of the flat pricing Veterans are requested to take multiple dosage units or split a dosage unit in half to achieve a desired dose. This strategy complicates adherence to a drug regimen and is reliant on Veteran comprehension of how to take their medications.

The VA National Center for Patient Safety (NCPS) evaluated the current Veteran comprehension of VA prescription labels to determine what can be done to increase the understanding of the prescription label information and solicit Veteran preferences for prescription label formats.

The VA National Center for Patient Safety supports evaluation of patient safety incidents through the root cause analysis process. Each VA Medical Center (VAMC) has a patient safety manager who evaluates local patient safety incidents. Using the root cause analysis process they prioritize patient safety incidents, determine root causes, and develop interventions to minimize the incident from reoccurring. This information is then captured nationally by imputing the data into the NCPS data base called SPOT. An analysis of SPOT data indicated there were 644 patient safety incidents between 2000- 2011 associated with misinterpretation of prescription labels by Veterans. For example, a prescription written for glyburide 10mg twice a day before meals was dispensed with 5mg tablets and labeled as **"Glyburide 5mg tablets take two tablets by mouth twice a day (half an hour before a meal)"** . The Veteran misunderstood the directions and took his glyburide before every meal three times a day resulting in hypoglycemia. In addition, 1229 cases were identified in which the Veteran misunderstood how to take a partial tablet to achieve the desired dose. For example, a Veteran would take a whole tablet when given a prescription that read, "Take one-half tablet daily". The VA has not standardized the verbiage directing a Veterans to split a dosage unit or a template for a National VA prescription label.

Many other organizations have recognized the need for prescription labels to be more patient centric and understandable by a wide variety of users, independent of social classes, education level and race/ethnic background.(1-6) These organizations include United States Pharmacopeia (USP), the Institute for Safe Medication Practices (ISMP), the National Association of Boards of Pharmacy (NABP), California State Board of Pharmacy and the American College of Physicians Foundation. The Agency for Health Research and Quality (AHRQ) also identified an association between low literacy and poor health outcomes. AHRQ suggests that only 12% of adults have literacy levels that allow them to interpret conventional prescription labels correctly. The National Priorities Partnership identified that improving patient medication adherence is a $100+ Billion dollar savings opportunity.

Study Design

Four hundred and forty six Veterans representing 11 difference VA care sites and 667 VA pharmacy staff Nationwide participated in the study. The demographics of the Veteran participants was similar to those of the average VA prescription benefit user.

Table 1 Study Participant Demographics

	VA Rx Users	Study Participants
Age	50% > 65 Yr	49%>60 yr
Gender	93% Male	84% Male
High School +	77.7%	97%
Ethnicity	80% White 13% Black 3% Hispanic <1% Asian	45% White 34% Black 14% Hispanic 2% Asian 6% American Indian/Alaska Native

The study participants were queried on their comprehension and location preference of each of the information elements found on a typical VA prescription label. (Figure 1) Veterans participated in face to face group sessions and pharmacy staff participated via an internet survey. Sixty six percent of the participants indicated they were satisfied with their current prescription label.

Figure 1

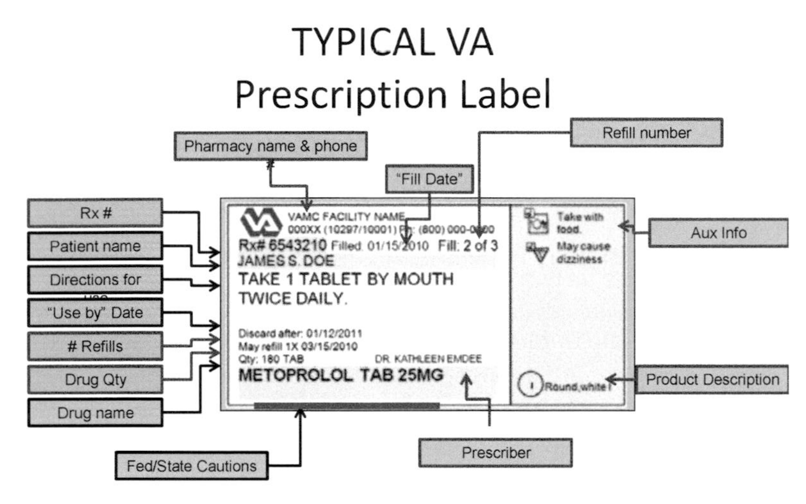

66% of respondents were satisfied with their current label!

1.2 STUDY RESULTS VETERANS DO NOT READ THEIR LABEL EVERY TIME THEY TAKE THEIR MEDICATION.

Only 56% of Veterans surveyed confirmed their name on the prescription label and 55% confirmed the directions prior to each use.

Figure 2.

% VETERANS REFERENCING INFORMATION ON A VA PRESCRIPTION LABEL				
	Never	Once	Sometimes	Every time
Drug name	1%	10%	17%	72%
Instructions	1%	17%	27%	55%
Veteran's name	4%	25%	15%	56%
Doctor's name	8%	28%	29%	35%

1.3 THE WORDS WE USE ON A PRESCRIPTION LABEL DO MAKE A DIFFERENCE IN VETERAN COMPREHENSION!

The word "meal" is found to be confusing to both Veterans and pharmacy staff. Veterans were tested on their comprehension of how many tablets to take daily when presented with several different prescription label directing them to take their medication at different intervals and incorporating the words "meals", "food", twice daily or morning and evening. (Figure 3) In each test the Veteran comprehension was best when labels did not contain the words food or meals, followed by labels incorporating the word "food" and finally the least understood was labels incorporating the word "meals" . Seventy percent of pharmacy staff indicated they preferred using the word "food". The use of the words food or meals in the directions for use is normally redundant because it will also be on the label in the auxiliary information section

Figure 3. % of Veterans with Correct Comprehension of label instructions

LABEL DIRECTIONS	% OF VETERANS
Take 1 tablet by mouth in the morning and evening.	89%
Take 1 tablet by mouth twice a daily.	89%
Take 1 tablet my mouth twice daily with food.	85%
Take 1 tablet by mouth once daily with food.	81%
Take 1 tablet by mouth once daily with meals.	75%
Take 1 tablet by mouth twice daily with meals.	75%
Take 1 tablet by mouth daily with meals.	42%

1.4 WHAT DOES "MEALS" MEAN TO A VETERAN?

Veteran and pharmacy staff interpretation of the word "meals" is quite variable. Veterans and pharmacy staff were presented with the label "TAKE 1 TABLET BY MOUTH TWICE DAILY WITH MEALS". Eighty three (83) % of Veterans responded correctly that they would take the medication with two meals a day, 4% of the Veterans indicated they would take the medication three or four times a day with every meal and 12% would take the medication with only one meal. When Pharmacy staff were asked the same question, 92% answered correctly, 6% answered either 3 or 4 times a day and 2% answered with one meal only.

1.5 DOES IT MATTER HOW WE ASK A VETERAN TO SPLIT THEIR MEDICATION DOSAGE FORM?

The VA dispensed over 8 million prescriptions to 2.9 million Veterans requiring the splitting of a tablet in FY 2011. There is no standardization on how the VA requests the Veteran to split a dosage form. The VA used 18,000 different possible label permutations to direct a Veteran to take a split dosage form. A combination of numeric fractions, verbiage and mg strengths were tested. The correct response rate was from 80-88% depending on the requested methodology.

A combination of using a fraction in conjunction with verbiage provided the best comprehension by the largest number of Veterans surveyed. Education level makes a difference in how a Veteran comprehends how to take a partial tablet. Figure 4. Veterans with an education level of some college or more were more likely than those with a high school education or less to respond correctly to each format.

Figure 4. Veteran comprehension by Education Level

Label Directions	Some College	HS Education or less
Take ½ (one half) tablet by mouth in the morning and the evening	92%	78%
Take one-half tablet by mouth in the morning and in the evening	93%	74%
Take one-half (12.5MG) tablet by mouth in the morning and in the evening	88%	71%
Take ½ tablet by mouth in the morning and in the evening	86%	66%

1.6 VETERANS DO HAVE A PREFERENCE ON HOW THEIR PRESCRIPTION LABELS SHOULD LOOK.

When asked what elements of the label are most important to them, Veterans identified the following label elements, highlighted below, as the most important to them.

WHAT VETERANS IDENTIFIED AS MOST IMPORTANT

Figure 5

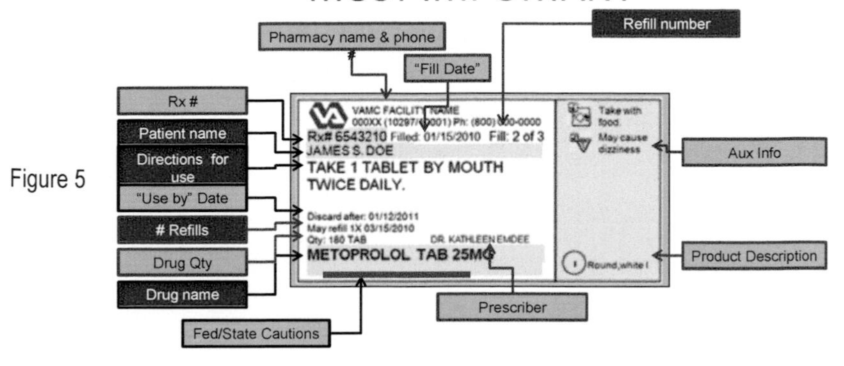

The most important elements identified by Veterans are scattered throughout the label making it difficult to find. Veterans also identified that they preferred bolded type, larger type font, selective highlights of the preferred elements vs entire label highlighted. These results were consistent with the recommendations made by the American Federation of the Blind and results from the California Board of Pharmacy survey.

When asked to compare a typical VA label to a label with the medication name at the top, Veterans had no preference and pharmacy staff preferred (A)

Figure 6

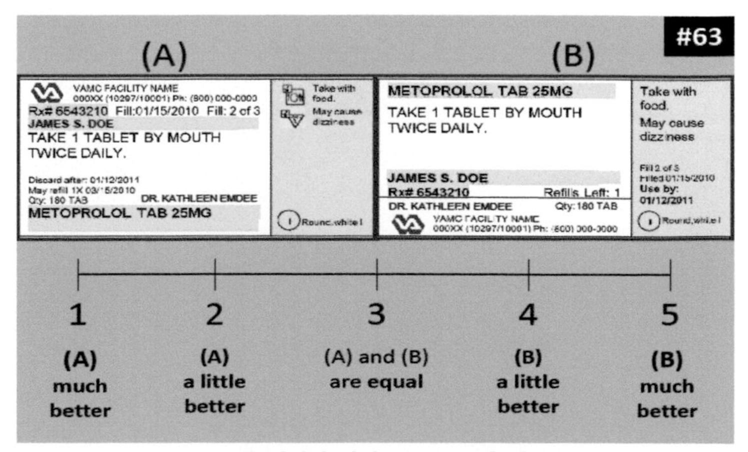

Which label do you prefer?

When asked to compare the typical VA prescription label to a label with the

Veteran name at the top, both Veterans and pharmacy staff preferred (B)

Figure 7

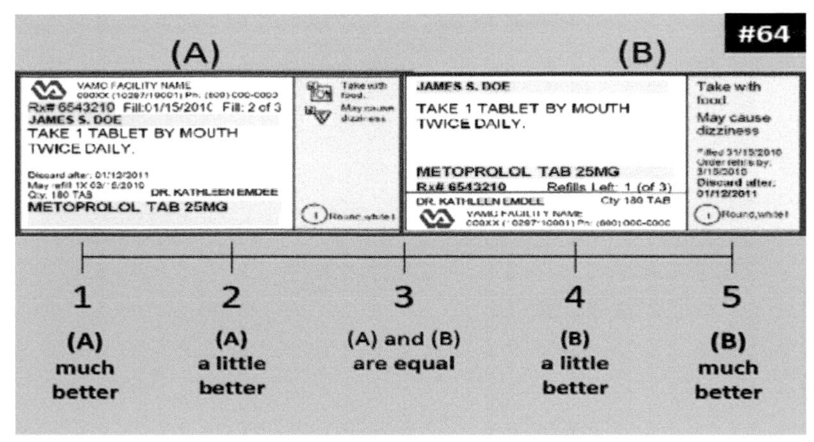

When both groups were asked to compare a label with the Veteran name vs medication name at the top, both Veterans and pharmacy staff preferred the name at the top (A).

Figure 8

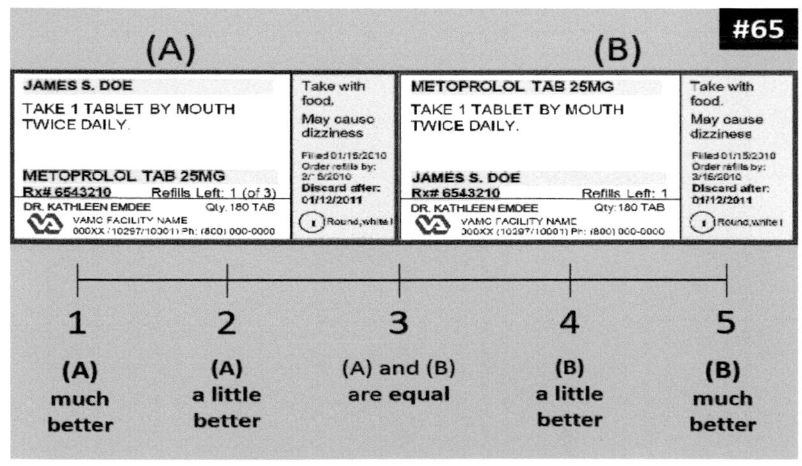

1.7 PATIENT CENTRIC LABEL

Based on the results of the study, 64% of Veterans and 62% of VA pharmacy staff preferred a VA patient centric label with the Veteran name at the top, bolded directions in 12 font, and specific highlighted elements. Figure 10

Figure 9

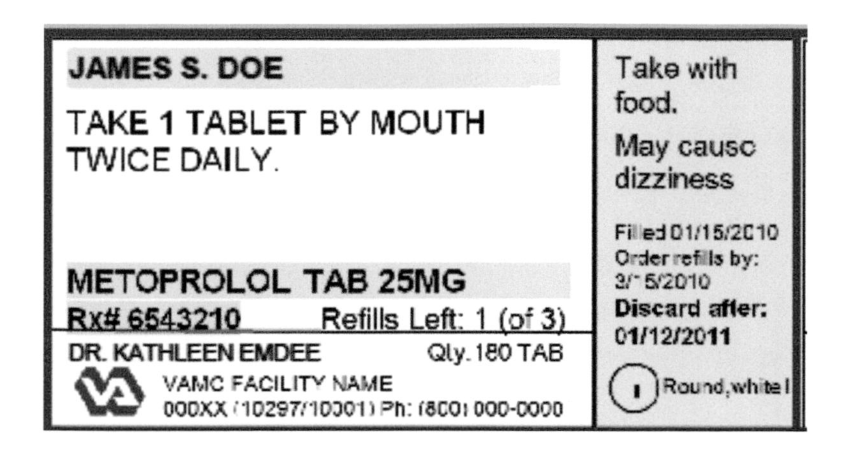

2 CONCLUSIONS

The words used on prescription bottle labels make a difference in the comprehension of the directions by a Veteran. Comprehension of the label influences the ability to adhere to the medication regimen prescribed by their caregiver. Veterans and VA pharmacy staff have differences on what they believe are the most important elements of a prescription label but both prefer a patient centric label as in Figure 10. The VA Medical Advisory Panel and VISN Pharmacy Executive Committee have approved adoption of the patient centric label as the standard for all VA Medical Centers and clinics. In addition to the prescription label format, standardizing the directions to take a partial dosage form by indicating a fraction and written directions has been approved.

REFERENCES

American College of Physicians Foundation (ACPF) 2007" *Improving Prescription Drug Container labeling in the United States"* http://www.acpfoundation.org/files/medlabel/acpfwhitepaper.pdf

NABP "*Report of the Task Force on Uniform Prescription Labeling Requirements."* Dec 2008
http://www.nabp.net/ftpfiles/NABP01/08TF_Uniform_Presc_Labeling_Req.pdf

USP,"*Standardizing Medication Label, Confusing Patients Less"* 2008
http://www.nap.edu/catalog.php?record_id=12077

NABP *Model State Act*, Aug 2009
http://www.nabp.net/index.html?target=/annualmeeting/CTFR.asp&

California State Board of Pharmacy, CA Senate Bill 472 required Board to promulgate regulations for a standardized, patient-centered Rx label. http://www.pharmacy.ca.gov/meetings/agendas/2009/09_jul_bd_pubed.pdf

American Federation for the Blind
http://www.afb.org/Section.asp?SectionID=3&TopicID=329&DocumentID=406

Section VII

Human Modeling and Patient Users of Medical Devices

CHAPTER 48

A Case Study for the Inclusion of Adolescents in Medical Device Design

Alexandra.R. Lang, Jennifer L. Martin, Sarah Sharples and John A. Crowe

Multidisciplinary Assessment of Technology Centre for Healthcare (MATCH)
The University of Nottingham
Nottingham, United Kingdom
Alexandra.Lang@nottingham.ac.uk

ABSTRACT

Adolescent users of medical devices are currently overlooked in the design process and as a consequence may not be using devices which adequately meet their specific requirements. The omission of adolescent user requirements capture in medical device development may contribute to poor adherence and use of the technology by this user population. This paper describes the findings from an interview study with adolescent patient users of a medical device, exploring their satisfaction with it and examining their requirements and preferences. It discusses the results of this study and presents the principle requirements and findings elicited from the adolescent participants. The information in this paper suggests that the inclusion of adolescent users of medical technologies can be useful informants throughout the design process to improve the quality of patient use devices and any subsequent health outcomes.

Keywords: adolescents, young people, user requirements, inclusive design, medical devices, acapella®.

1 ADOLESCENTS IN RESEARCH

Adolescents as a specific user population are often overlooked in the design and development of medical devices (Geljins, Killilea *et al.* 2005). Increasing numbers of young people are living with chronic conditions and are responsible for managing their treatment and monitoring regimes (Viner & Chambers 2000; Perrin, Bloom *et al.* 2007) and medical devices often play an important role in these processes. The design of these devices therefore have a direct impact on the everyday lives of adolescent medical device users, their transition from child to adult and importantly their adherence to regular and correct use of the device. This means that the understanding that medical device developers have of this population will influence the uptake and long term use of devices, which will in turn influence health outcomes.

This paper details the findings of a qualitative study with adolescent users of a medical device; the acapella®, evaluating this technology with respect to their needs as a specific user population. The acapella® case study demonstrates an example medical device which does not adequately meet adolescent user needs (Lang, Martin *et al.* 2010).

1.1 Case Study Background

The acapella® device is used in the airway clearance treatment regime for people with cystic fibrosis (CF), a chronic condition which affects 1 in 2500 babies born (Orenstein 2004). In addition to a highly complex medication regime people with CF have to carry out daily routines of physiotherapy to clear mucus from their lungs, commonly known as chest/ airways clearance. It is well documented that despite the demonstrated links between compliance to recommended physiotherapy and positive health outcomes (Orenstein 2004) these daily practices are met with resistance by patients and adherence is often poor (Pendleton & David 2000; Lask 1994).

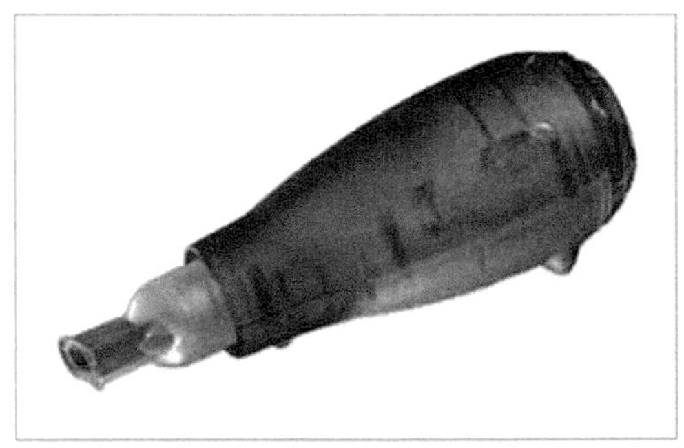

Figure 1. The acapella®

There are several different devices which can be utilised by CF patients in their airway clearance routines (Marks 2007). Figure 1 displays the current design of the acapella®, a device which was identified by healthy adolescent proxies as an example device which does not meet adolescent user needs (Lang, Martin *et al.* 2010).The device works by users breathing out through the acapella®, the airflow works with the internal mechanism to produce vibrations and resistance. The resistance helps to get air behind the mucus secretions, vibrations then help to loosen and move the mucus (NHS 2011). This has to be done at least twice a day for approximately 20 minutes for patients to get benefit from their physiotherapy sessions.

2 STUDY PROTOCOL

Twenty adolescent patients with cystic fibrosis were recruited to participate in an interview study where the aim was for young users to evaluate their satisfaction with the acapella® device. The study took place in paediatric and adult CF clinics within the Nottingham University Hospitals Trust in the UK.

2.1 Ethics

Due to the age of the population being recruited and their condition (cystic fibrosis) there were many ethical considerations which had to be considered prior to and during the study. These special considerations included the appropriate use of informed consent for adolescents 16 years and older and informed assent for those who were under 16 years old, in addition to issues such as confidentiality and appropriateness of topics. With regards to the CF condition, the risk of cross infection and hygiene considerations for carrying out research in clinical environments were also important issues. Further details about this study and the practicalities of involving young people in research in hospitals can be found in Lang, Martin *et al.* (2011). Due to the fact that the study would be carried out on hospital property and participants recruited through National Health Service (NHS) databases, ethical approval was obtained from the NHS NRES system in addition to obtaining an NHS Honorary Contract for research.

2.3 Recruitment

Adolescent participants were recruited by clinical staff members of the CF clinics in two hospital trusts. The recruitment process agreed through the ethical submission meant that potential participants could be recruited through the clinic registers when the adolescents were a) attending their regular clinic appointments or b) if they were an inpatient on the ward and were healthy enough to participate. Participants and their guardians (for those under 16 years old) were given two week's notice to consider whether or not they wanted to participate. Twenty adolescents took part in the study, ranging from 11 years old to 20, with a mean age of 16.65 years. 13 males and 7 females decided to take part in the study. It was a requirement of the recruited participants that they had experience of using the acapella® device. At the time of the interview, 3 of the interviewees had abandoned the device whilst others were using it in conjunction with other forms of physiotherapy.

2.2 Study implementation

The development of the interview schedule, resources and questions were designed to specifically engage the young population being sampled.

The interview began with a brief background history from the participants and identification of devices which they had experience of using. Following this the interviews progressed into an evaluation of the current design of the acapella®. Participants were encouraged to consider their specific needs and to identify positive and negative aspects of the device. The next part of the interview schedule focussed on discussing how the acapella® could be improved for adolescent users. This was facilitated by the use of three vignette images which presented design modifications of the acapella®. The design concepts were explained to the patients and the interview continued through their assessment of the suggested changes in the vignettes. The design process and use of these images is described in Keane, Lang, *et al.* (2012). The adolescent participants were encouraged to critique the original device and the design concepts based on their own experiences, preferences and priorities associated with device use. Throughout the interview participants were assured that there were no 'wrong' answers and that both positive and negative feedback about the acapella® and the redesigns were equally valuable'. Following the evaluation, there was a series of questions about the interview method and the participant's experiences in other research, with the aim of understanding adolescent preferences with regard to participation and methods. The interview concluded with an opportunity for the participants to ask any questions about the study or provide additional feedback.

3 RESULTS

The analysis of the interview data utilised a grounded theory approach so that emergent themes could establish the important issues about adolescent user requirements of medical devices. The following section has broken down this analysis to report the principal findings of this study. Bold words highlight the subthemes which contribute to three emergent categories:

Personal Themes – includes the more visceral elements associated with medical device use and the participants' experiences with the acapella®.

Micro Themes – device specific findings and what adolescents require from their medical items, specifically the acapella®. The concepts within this overarching theme are specific statements of need or responses to design features and the development of solutions.

Macro Themes – broader issues associated with medical device use by the participants.

3.1 Personal Themes and Adherence

Throughout the interviews, the participants expressed a desire to be self determining, **Independent** and **Informed**. Adolescent CF patients appear to want more opportunities to be autonomous in their management of their condition by promoting characteristics of **Control** and **Independence** – *"I just wish they'd trust me to do more on my own"*. This supports existing literature by emphasising the importance placed on adolescent assimilation of independence (McDonagh 2000; Michaud, *et al.* 2007). Currently the transition process from paediatric to adult specialist centres encourages this concept however it is evident that more could be done to empower the adolescent patients through the design of medical devices.

One way of achieving the self reliance which adolescents require is through the inclusion of functions which enable the user to **Monitor** their physiotherapy routines, single sessions and over a long period of time. During the interviews adolescent participants responded positively to this design features and discussed suggested ways in which it would benefit their treatment *"I think if there were a way of recording stuff then I'd be more inclined to use something like that, just so I know, it's better for me, if it's working"*. This facility may also have a positive impact on relations with parents and carers and also clinical staff *"it would be good [to be able to monitor], because then your parents would know you were definitely doing it properly"*. Additionally it can be inferred from the views of the adolescent users that **Adherence** of use may benefit from this kind of function.

Motivation and Incentive appear to be key elements of adolescent medical device requirements. Airway clearance in CF does not always offer a physical relief or obvious, immediate health benefit to the user/ patient. As such there is little stimulus to entice the user into adhering to frequent and regular physiotherapy, an opinion which was reported by the clinical staff who were consulted prior to and after the interviews. The identification of this need can be responded to in device development through incorporation of features which aid **Motivation** to regular and correct use.

3.2 Micro Themes

The theme of **Interaction** and **Engagement** appears to underpin many of the other facets of device design. Current approaches to CF physiotherapy are deemed to be monotonous and repetitive *"I find it boring and find I'm looking at the time all the time, every 5 minutes and that"*. As a result it can have a detrimental effect on **Adherence** of use of the device. To improve this process participants need to be better **Engaged** with the task and to achieve this, the acapella® device would benefit from improved **Interactivity**. Adolescent CF patients appear to want to be more informed about their physiotherapy routine and their use of devices within the care and management of their condition. **Feedback** mechanisms which communicate immediate and short term information to the user could assist the relationship between the user and their device, potentially leading to improved **Clinical Effectiveness** and **Adherence**. Suggestions for improved **Interactivity** ranged from simple sensors, alarms and lights to more intricate gaming concepts. *"On the website there could be games when you have to jump and stuff but you'd have to do that by blowing in your acapella®. It could be racing and it could be sensor activated so when you move it to the left and blow it moves it to the left and when you move it the right and blow it moves to the right".*

It is evident that a variety of **Feedback** functions could be utilised to improve the design of the acapella® and contribute to levels of **Interaction** between the user and their device. From this finding it would appear that some medical devices could benefit from appraisal through the Nielson Heuristics of Use (1994). Further work is required to determine which modes of **Feedback** would be best suited to this specific device and significantly which would be the preferred options by an adolescent user population. For example *"I think a beep because you might not notice a light if you're watching TV, it might just blend in, but if you hear a beep that would be better"* whilst other participants stated preference for a light. Long term awareness of the impact of physiotherapy regimes is a requirement of both patients and carers. Clinical appointments once every three months do not appear to provide frequent enough information and **Monitoring** opportunities. A desired feature of the device is to fill this void and offer a facility enabling habitual **Logging** of the condition and treatment - *"I like the idea of graphs and stuff. That's an incentive for me"*. This can provide early warning for CF crises and also promote compliance of use. The interviews have revealed that improved **Interaction** and **Engagement** may be achievable through the development of **Additional Technology** for use with the acapella®.

Based on the review of the data surrounding the theme of **Clinical effectiveness** it appears that there is a requirement for the device to assist the efficacy of the physiotherapy routine. The themes which emerged have highlighted that apart from the physical vibrations felt during physiotherapy there is no provision for the user to be aware of their technique of use or the therapeutic effects derived from use. The data shows that despite the young age range of the participants their priority for their devices is that they are **Clinically Effective**. However several features of the device were perceived to be unsatisfactory and lacking in this element. Where their needs are not met adolescents are quick to use their experiences with other technologies to provide **Improvement Suggestions** - *"I guess if it could be linked up to the computer and I could see a bit more what was going on then I might take a bit more time over it (physiotherapy).... it's supposed to be doing you good but you can't always tell".* Another clinical need which emerged from the data is the difficulty of users in **Counting Sets** (number of breaths during physiotherapy session) and the impact this has on the physiotherapy task *"Another thing is how many times you do it. Sometime I can't remember if I've done it 3 times or 4. If it came up with how many times I'd done it that would be helpful as it would make me do it better".* This important aspect of the physiotherapy routine has thus far not been identified by physiotherapy device developers as requiring attention in the design of these devices. Again this highlights the need for the device to facilitate the user achieving regular and correct use.

The findings in this section are highly correlated to the themes of **Interaction** and **Feedback** which were discussed previously. It is evident that the integration of systems to address these issues could have potential benefits for **Adherence** of use and **Clinical Effectiveness**.

The **Aesthetic** qualities of a device are not a priority for all adolescent users. However some adolescents do express an interest in how their device looks and believe that the current acapella® does not meet their needs. The personal preferences with regard to **Shape** of the acapella®; some participants felt it was too big whilst others liked the current size, would be difficult to overcome with a mass produced medical item. However options for **Customisation** of the device, might afford the user more **Acceptance** and satisfaction. Any reduction in the **Size** of the acapella® would have to originate from advancements in technology which either enable miniaturisation of the mechanical components or a realignment of the interior of the device, without compromising the **Clinical Effectiveness**.

There are sensitivities about the **Age Appropriateness** of device design. Adolescents are aware of some devices looking 'too young' or 'too old' for their age range and as such state a requirement for designs which are more inclusive of them. This is particularly important with regards to the existing literature which examines adolescent personal identity and the need to not be perceived as children (McDonagh 2000). *"They've got some that are just too young and are for kids....but sometimes they are too adult and boring and they don't get it right in the middle".* There does appear to be a market for **Customisation** of a device as the participants aged 11-16 years were unanimously positive about this aspect of device design, *"If I could design something and make it look how I wanted, I would".* It is worth noting that the personalisation of the device as suggested by the participants was not limited to the physical appearance but also takes into account suggestions for the **Feedback** applications.

From the data it is evident that the **Practical** elements of device design are valued by adolescent users A particular example provided by the participants was the task of cleaning the acapella®, which was considered to be time-consuming and a **Nuisance** to remember and carry out. If the design of the device can overcome this through simpler mechanical arrangement and **Feedback** about frequency of cleaning then this may be less of an issue for users. From the information provided in the interviews, **Materials** used in the current design of the acapella® could be modified to provide a device which is more robust and less likely to break if it has an impact. **Portability** of the devices is also often an issue. This could be alleviated through a reduction in the **Size** of the device and through the provision of an acapella® specific **Bag**. *"I don't know why the acapella® hasn't, it would be good for if it was in your pocket. It would also be good if it was a bit smaller and had a bag that you could clip onto your belt and then 'off you go'...that would be good. It could be a bit like a wallet chain, for security but mostly so you didn't drop or lose it".*

3.3 Macro Themes

There is a wide range of macro issues associated with medical device use by adolescents, many of which involve **Contextual, Social** and **Environmental** aspects of use. This interview study has exposed some themes, all of which require further investigation for better understanding. **Location** of device use is related to **Privacy**, **Acceptance** and time constraints. The device design should take these issues into account so that users are afforded **Choice** of where they can use their device in the knowledge this may have a direct impact on frequency and duration of use. **Independence** has been identified as an important construct of medical device use, whereby the adolescent as they grow older has a widening sphere of use as they encounter new environments on their own. *"I'm not hiding it I couldn't really when I was in the scouts, and not worried about being different or it being different".*

Information provision is an important aspect of device uptake and continued adoption. It is evident that this it is valued by adolescents and that there are two areas where their views are directed. There is a focus on improved **Training** for new users of the acapella® *"if someone puts it in front of you without telling you much, they just expect you to know...that's what happened with me but if they tell you more it's better for being able to use it, I didn't get told very much"*. It was also highlighted that continued education and **Training** with devices is an unmet need as there are situations where current users forget **Breathing Techniques** and this can have a detrimental effect on **Clinical Effectiveness**. A more universal requirement and holistic goal is the improvement of **Information within Society** regarding CF. Better understanding of chronic conditions and the **Acceptance** and **Awareness** of medical device use within the general population is an important issue for those with or associated with a condition.

4 DISCUSSION

It is evident from this study that adolescent users of the acapella® require an improved level of interaction from their device, an important element of this is feedback during device use and long term monitoring. The results suggest that meeting this requirement could improve their technique of device use whilst also making them more aware of the positive impact that physiotherapy might have on their respiratory health on the long term. Another important consideration of device design is the way in which a device may either help or hinder care transition from the adult carer to the adolescent user, thus facilitating the promotion of good health behaviours into adulthood. Currently adolescent user needs are not captured in the design of medical products, but this interview study has shown that they can be constructive in the design process by articulating preference and suggesting ways to develop products which may have benefit to theirs and other user groups.

Figure 2 displays a Venn diagram which illustrates the three categories under which device design can be considered and the interactions which ultimately impact regular and correct use of a medical device.

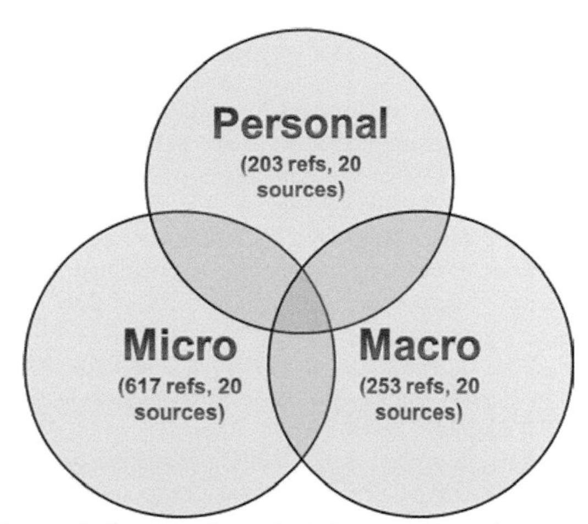

Figure 2. Interaction of adolescent data themes

This demonstrates how adherence to device use is a multifarious construct. In practice a wide range of variables will impact on an adolescent user's decision to comply with the recommended use of their device. Medical device manufacturers need to be aware of this and work with users and clinical staff to account for the variables and design in ways to combat the issues. Further work is therefore required to assess the value of adolescent contributions to healthcare and medical design in both research practises and in industrial applications.

The acapella® case study has provided an example of user requirements elicitation from adolescent users of a medical device. It is evident that the current design of the acapella® does not meet adolescent user requirements and as a result clinical effectiveness and adherence of use are negatively affected. The study provides evidence to suggest that adolescents are both capable and willing to be involved in research, especially when they perceive that their contribution is relevant to them or applicable to real world issues. If user requirements capture for adolescents can be achieved in a manner which provides them a voice in the development of products used by them, then the medical device industry and others could benefit from the inclusion of these young consumers in design and development processes.

ACKNOWLEDGMENTS

This work has been funded by the MATCH Programme (EPSRC Grant GR/S29874/01). The authors would like to thank the patients and staff at the Paediatric and Adult CF clinics within the Nottingham University Hospitals Trust.

REFERENCES

Geljins, A. C., Killelea. B., Vitale, M., Mankad, V. & Moskowitz, A. (2005). The Dynamics of Pediatric Device Innovation: Putting Evidence in Context. In Field, M. J. & Tilson, H. (Eds.), Safe Medical Devices for Children (pp. 302-326). Washington DC: THE NATIONAL ACADEMIES PRESS.

Keane, D., Lang A. R., Craven, M, *et al.* (2012) *The Use of Vignettes for Conducting Healthcare Research.* Proceedings of the 4th International Conference on Applied Human Factors and Ergonomics in Healthcare. 21-25 July 2012, San Francisco, California.

Lang, A. R., Martin, J. L., Sharples, S.et al. (2010) A Qualitative Assessment of Medical Device Design by Healthy Adolescents. Advances in Human factors and Ergonomics in Healthcare. CRC Press, pp. 539-548.

Lang, A. R., Martin, J. L., Sharples, S.et al. (2011) 'Patients/ patience in research'. The challenges of interviewing adolescent medical device users in hospital clinics – the acapella® Case Study. Contemporary Ergonomics and Human Factors 2011. Taylor and Francis Group, pp. 336-343.

Lask, B. (1994). Non-adherence to treatment in Cystic Fibrosis. Journal of the Royal Society of Medicine, 87(21), 25-27.

Marks, J. H. (2007). Airway clearance devices in cystic fibrosis. Paediatric Respiratory Reviews, 8(1), 17-23.

McDonagh, J. E. (2000). The adolescent challenge. Nephrol. Dial. Transplant., 15(11), 1761-1765.

Michaud, P. A., Suris, J. C. & Viner, R. (2007). The Adolescent with a Chronic Condition: Epidemiology, developmental issues and health care provision (WHO discussion papers on adolescents). Geneva Switzerland: World Health Organisation.

Nielsen, J. (1994). Heuristic evaluation. In Nielsen, J. & Mack, R. L. (Eds.), Usability Inspection Methods. New York: John Wiley & Sons Ltd.

NHS (2011). Using your Acapella: A patients guide. Cambridge, Papworth Hospital NHS Foundation Trust.

Orenstein, D. M. (2004). Cystic Fibrosis. A Guide for Patient and Family (3rd Edition.) pg xi, 189-196, 235-252. Philadelphia PA: Lippencott Williams and Wilkins.

Pendleton, D. A. & David, T. J. (2000). The compliance conundrum in cystic fibrosis. Journal of the Royal Society of Medicine, 93(Supp 38), 9-13.

Perrin, J. M., Bloom, S. R. & Gortmaker, S. L. (2007). The Increase of Childhood Chronic Conditions in the United States. JAMA, 297(24), 2755-2759.

Viner, R. & Chambers, T. L. (2000). The scope of adolescent medicine. Journal of the Royal College of Physicians of London, 34(1), 12-15.

CHAPTER 49

The Use of Vignettes for Conducting Healthcare Research

David Keane, Alexandra R. Lang, Michael P. Craven and Sarah Sharples

Multidisciplinary Assessment of Technology Centre for Healthcare (MATCH)
The University of Nottingham
Nottingham, United Kingdom
eexdk1@nottingham.ac.uk

ABSTRACT

Improving healthcare requires engaging with clinicians and patients in order to better understand their needs and expectations. Research methods that are used to conduct healthcare research are selected on the basis of study objectives and practicalities (i.e. finances, resources, time frame, etc.). The methodology of vignettes for conducting healthcare research consists of short descriptive scenarios and/or images to engage participants in hypothetical situations in which their emotional, psychological and sociological responses can be measured. They enable participants to feel comfortable in divulging responses that may be of a sensitive nature and provide insight to situations where they may have little or no experience. Two studies that used vignettes to conduct healthcare research will be discussed in terms of their objectives, conceptualisations, designs and developments, implementations, and outcomes. Validity of the vignettes as a methodology for conducting healthcare research will also be discussed.

Keywords: vignette, user needs, adolescent, medical device, diagnostic procedure, information provision

1 INRODUCTION TO VIGNETTES

Alexander and Becker (1978) define vignettes as 'short descriptions of a person or social situation which contain precise references to what are thought to be the most important factors in the decision-making or judgement-making process of respondents'. They are used by the World Health Organization (Murray, Özaltin, and Tandon, et al., 2003; and Salomon, Tandon, and Murray, et al., 2003) in the form of anchoring vignettes for cross population comparability where participants respond to appropriately designed vignettes using the same response scales as used to evaluate their self-assessed health. Responses to the vignettes can then be used to adjust different health expectations and ratings across different populations.

When attempting to gain a better understanding of participant perceptions and preferences of hypothetical situations where they may have little or no experience, content specific vignettes are used. They introduce participants to a scenario in which a character and/or artefact is to be the focus and explore the interacting dimensions of these on participant responses. They are particularly useful when characters and/or artefacts are manipulated in a controlled experiment to examine the effects of specific variables.

The design of vignettes is an elaborate process with the aim of obtaining appropriate data in response to study objectives. They require participants to submerge themselves into hypothetical situations and as such these situations need to be realistic, relevant, structured, comprehensible, consistent and concise. Vignettes require systematic variation and should be neutrally framed as to not influence participant responses. The following two studies discuss the design and implementation of vignettes to meet study objectives. Outcomes of the studies will also be discussed as will validity of the vignettes as a research methodology.

2 STUDY 1: USER NEEDS OF ADOLESCENTS WITH RESPECT TO MEDICAL DEVICES

The first study examined user needs of adolescents with respect to medical devices and involved three vignettes displaying modified versions of an acapella® for evaluation by adolescent users. The acapella® is a medical device that is used every day in the treatment of cystic fibrosis (CF) for airway clearance physiotherapy (Lannefors, Button, and McIlwaine, 2004). It is well documented that this aspect of treatment is poorly adhered to (Pendleton and David, 2000) and the aim of the research was to better meet the needs of the adolescent user so that the design of the device itself is not a barrier to regular physiotherapy.

Figure 1 The acapella®

2.1 Study design and development

To produce vignettes with appropriately designed modified versions of the acapella® a specification was developed based on data from proxies who had evaluated the device in a previous study (Lang, Martin, and Sharples, et al., 2010). Input from clinical experts was also used. Two design students from The University of Nottingham were employed to interpret the specification and produce new concepts of the acapella®. They also had access to the device, images of the device and abridged versions of qualitative data from the previous study.

Generating and scoping ideas was the first stage of designing new concepts of the acapella®. During this stage ideas were sketched and were then discussed with the researchers. Following constructive feedback and recommendations three ideas were selected to be developed. Development of the ideas involved several modifications and meetings to discuss the modifications. Once the ideas were developed to high standard conceptualisations they were rendered as computer aided design (CAD) outputs. The CAD of one of the acapella® concepts as used in its vignette is displayed in Figure 2. Annotations of the modifications made from the original acapella® are also included, as they were in the other two vignettes.

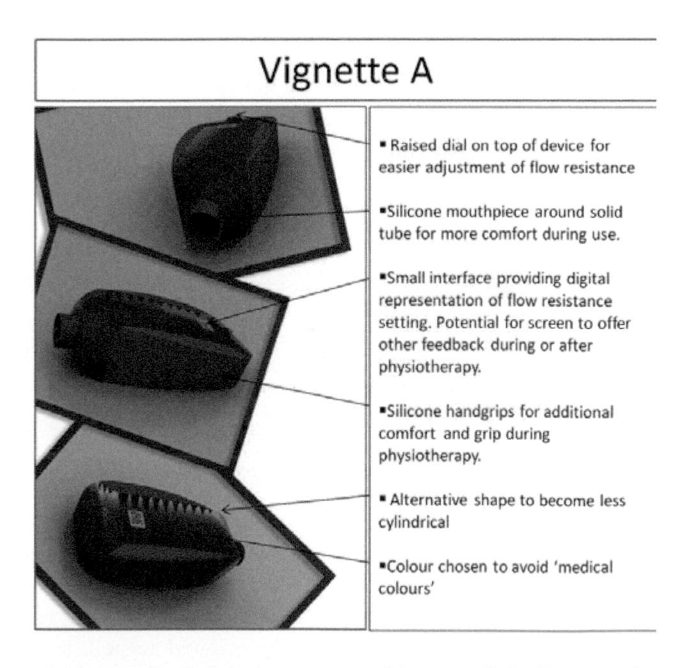

Figure 2 Vignette A – first design concept of the acapella® with annotations

2.2 Study implementation

Vignettes were presented to adolescents with CF during in-depth interviews; 20 participants took part in the study and were aged between 11-20 years.

Consideration of participant young ages, potential vulnerabilities and sensitivities were essential both in securing ethical approval and during interviews.

Interviews began with a brief introduction to the research project and study objectives. Participants were then asked to comment on the acapella® with reference to their own experiences, detailing positive and negative aspects of the device. Following this vignettes were introduced and details of modifications made to the acapella® were explained and questions were invited. Participants were informed that the acapella® concepts were based on information provided by healthy adolescent proxies. Participants were then encouraged to critique the concepts and to disclose positive and negative features of them. Participants were also encouraged to discuss their own preferences and ideas. Throughout this process participants were invited to illustrate their ideas on drawing boards. In the concluding stages of interviews participants were questioned about their involvement in the study and the efficacy of the vignettes.

2.3 Study outcomes

It became evident that the use of vignettes in interviews was a novel and stimulating addition to the study for participants and tended away from traditional studies that they are normally recruited for. None of the participants had ever been asked to comment on their medical devices and previous interview experiences hadn't used additional resources such as vignettes to support the interviewing process. General feedback regarding the use of vignettes was that they were a source of inspiration, provided a reference for discussion and made the interviews more interesting. Comments included:

"I might not have been able to say very much without these giving me ideas; it was nice to be asked".

"It just helps you think a bit more, otherwise you're just looking at the thing you use every day and it's hard to think about changing it".

"I think they were helpful; some of the ideas were ok and I didn't like a few but the pictures helped me to think about it".

The initial stages of interviews were quite limited at times and participants were reluctant to critique the acapella®. The introduction of vignettes stimulated the interviewing process and participants became more active in their evaluations of the device. They raised comparisons between the three acapella® concepts, and expressed their preferences between them and the original acapella®. The vignettes helped participants to focus on different features of the device and to consider their effectiveness of use and/or on personal satisfaction. They also expanded participant assessments of the acapella® from a personal perspective to consider environments of use and social acceptance, in line with changing lifestyles and new experiences as these users transition from child to adult.

With respect to practicalities of using vignettes there were limitations to their utilisation. This was due to limited desk space and displaying vignettes was not always possible. Where participants were asked to contribute their ideas using drawing boards difficulties arose due to room layouts, room ambience (interviews were conducted in hospitals) and confidence in drawing abilities. Though one participant, a college design student, found the creative nature of the study enthusing and became further involved in the research project after the study.

3 STUDY 2: THE EFFECTS OF DIFFERENT TYPES OF DIAGNOSTIC PROCEDURE AND INFORMATION PROVISION PREFERENCES

Exploring the effects of different types of diagnostic procedure and information provision preferences were the objectives of the second study. The study was designed with the intention of gaining a better understanding of user needs with respect to diagnostic procedures that may be encountered in clinical pathways and the role of information provision when encountering such procedures.

3.1 Study design and development

Vignettes were designed to create a diagnostic medical scenario where participants could engage with that scenario and reflect on it as if they were a patient. To create such a medical scenario required two attributes: 1) a symptomatic patient, and 2) a diagnostic procedure to investigate the cause of the patient's symptoms. A number of vignettes were conceptualised and the Map of Medicine (2012) was used to formulate the vignettes so that symptoms and diagnostic procedures were based on clinical evidence. Developed vignettes that were used in the study were based around three sets of condition based symptoms (coronary, gastroenterological and musculoskeletal) and three different types of diagnostic procedure (blood test, imaging procedure and invasive procedure). Using a factorial design nine vignettes were produced (3 × 3).The specific diagnostic procedures used in the nine vignettes are described in Table 1.

Table 1 Diagnostic procedures

	Blood test	Imaging procedure	Invasive procedure
Coronary	Blood test	X-ray	Endomyocardial biopsy
Gastroenterological	Blood test	CT scan	Colonoscopy
Musculoskeletal	Blood test	X-ray	Arthroscopy

Though the blood test type of diagnostic procedure is an invasive procedure it is relatively minimal and provided a basis in establishing findings in the other two types of diagnostic procedure as it is well-known and the same for each set of condition based symptoms. The vignettes were cross referenced against each other to ensure that they were consistently structured and the only variations in them were the generated systematic variation of symptoms and diagnostic procedures, the two independent variables. All other content maintained equivalent stature though phrasing of sentences was varied.

The vignettes were piloted, as was a questionnaire that was developed to obtain participant responses. The pilot of the vignettes and questionnaire provided valuable feedback and recommendations. This was an important process as responses from participants would be dependent on the comprehensiveness of the vignettes and their ability to remain focused on them and the questionnaire. This was to avoid satisficing from participants and Krosnick (1991) identified three conditions that contributed to satisficing: 1) task difficulty, 2) respondent ability, and 3) respondent motivation. Comprehensiveness and conciseness of the vignettes and the length of time it took to complete the questionnaire were important specifications in their designs, especially as participants were to complete three questionnaires in response to three vignettes.

3.2 Study implementation

72 participants from The University of Nottingham took part in the study and the majority were aged between 18-23 years. They were presented with one vignette at a time and encountered each set of condition based symptoms and type of diagnostic procedure only once. Vignettes were distributed in pre-set sequences in which all permutations were encountered uniformly by participants. 211 questionnaires were completed overall, which was five short of the maximum of 216. Participants were introduced to each vignette with the following statement:

'You are presented with a vignette about a patient who is experiencing a number of symptoms and has a test to further understand the reason for their symptoms. Imagine that you are that patient, experiencing those symptoms and that you are having a test to further understand the reason for your symptoms'.

The term 'vignette' has been used in this statement but this term was well understood before the start of the study as participants were informed of what a vignette was. The term 'test' is also used instead of 'diagnostic procedure' as it was regarded as more generic and therefore suitable for each vignette. Incidentally, the invasive procedure for the coronary based symptoms, endomyocardial biopsy, was replaced with 'heart biopsy' in keeping with the requirements of producing comprehensible vignettes.

Each vignette included a brief explanation about the diagnostic procedure that was being used to investigate the cause of the patient's symptoms, which included what the diagnostic procedure was and what it required in terms of patient involvement (i.e. what happens to the patient during the diagnostic procedure). Images of the specific diagnostic procedures were also included.

Reading a vignette and completing a questionnaire took approximately 8 minutes, which was in conjunction with how long participants were informed their participation in the study would be. Participants were instructed that they could refer back to their vignette when completing a questionnaire and they could also ask questions if they needed any assistance in understanding the vignettes and questions in the questionnaire. As well as being of assistance, the study was carried out under supervision as another attempt to prevent satisficing.

The questionnaire included a combination of closed and open-ended questions in order to gain quantitative data for statistical analysis and qualitative data for thematic analysis respectively. The questionnaire was constituted by three fractions: 1) participant preferences for pre-diagnosis information, 2) participant perceptions of diagnostic procedures, and 3) participant preferences for post-diagnosis information. These were developed in response to the study objectives. Pre-diagnosis information refers to information that is provided to patients 'to inform them about their diagnostic options and to promote engagement with clinicians in the decision-making process', and post-diagnosis information refers to 'an outcome or result given to patients from an investigative procedure or test respectively' (Keane, Craven, and Sharples, 2011).

3.3 Study outcomes

The study produced interesting findings that met the objectives of the study. There was high regard for information provision in the pre-diagnosis stage and this was especially important to participants when responding to a vignette with an invasive procedure. This was demonstrated in findings for both statistical and thematic analysis. With respect to thematic analysis, themes that emerged included the *comprehensiveness of or requirement for information provision* about a diagnostic procedure, and *preparing* for possible clinical pathways in the event of a positive, negative or inconclusive result.

Perceptions about the different types of diagnostic procedure produced varied and intriguing results. There were high ratings for accuracies and confidence in the diagnostic procedures to further understand the reason for symptoms described in the vignettes. A major factor contributing to such high ratings was *trust* in clinicians and/or clinical practice, which was a major theme. There were also statistically significant effects of the type of diagnostic procedure on participants' level of apprehension, level of embarrassment and perceived likelihood of proceeding with a diagnostic procedure. These were all effects of the invasive procedures and were contributed by factors such as *physical involvement* and possible *sensations*, which were also major themes. Though there was a significant effect of the type of diagnostic procedure on participants' perceived likelihood of proceeding with a diagnostic procedure, ratings were high to proceed with all procedures and thematic analysis showed that this was because of a need to *understand and improve health*.

Information provision preferences in the post-diagnosis stage were as highly regarded as they were in the pre-diagnosis stage. The majority of participants preferred to receive a detailed interpretation of a diagnostic procedure result and if it was possible to receive the result during or immediately after the diagnostic procedure. The use of new media, such as a mobile phone application, were explored to investigate whether participants would prefer to receive results through such a medium compared to traditional media, such as visiting a general practitioner or specialist clinician. Quantitative data resoundingly found a preference for traditional media and a major theme for such a preference was level of *detail* to be included in the result and the ability to ask questions.

4 VALIDITY OF VIGNETTES

Both studies benefited from the use of vignettes in engaging and eliciting valuable data from young demographics. With respect to the first study, vignettes provided a useful resource to support the interviewing process. They were an alternative approach to focus participant attention and to enable them to articulate their views and opinions. This was especially evident as maturity and competency levels vary considerably between ages 11-20 years and all participants were able to make a valid contribution. With respect to the second study, participant responses were true reflections of their perceptions and preferences rather than of past experiences. This does limit the generalisation of their responses, which Ogden, Daniells, and Barnett (2009) report of in a vignette study exploring choice in healthcare. However, it does provide an insight into user needs and expectations of potential patients. Of course, what has not been examined is the extent to which participant responses to the vignettes would be the same as their responses should they experience similar scenarios in the future. This is a trade-off, although this is a limit, as it avoids bias that may result from past experiences.

Design of vignettes and environments they are used in are important considerations in order for them to be fully utilised and to optimise participant responses. Vignettes in the first study did not have the same descriptive narrative as used in the second study as participants brought their own experiences of CF and user experience of the acapella® to the study. Annotations used in the vignettes were key in discussing the important modified features of the acapella®. These vignettes were constrained by the environment they were used in though and on some occasions the study conductor had to hold the vignettes on display to participants. This made conducting interviews difficult at times. In the second study, as participants were completing three questionnaires in response to three vignettes, it was important that the questionnaire and vignettes were appropriately designed and structured in order for participants to remain focus and avoid satisficing, and to obtain adequate and valid data. The use of open-ended questions in both studies was valuable as they encouraged participants to disclose reasoning for their responses and thus uphold them as 'truth', which is discussed by Torres (2009). Wilks (2004) also discusses the qualitative value of vignettes.

The number of vignettes in both studies was adequate and appropriate in order for study objectives to be met. With respect to the second study, the factorial design used to produce nine vignettes and the distribution of them in pre-set sequences in which all permutations were encountered uniformly by participants meant that the independent variables were sufficiently explored.

Though vignettes in the first study were designed to engage young interviewees and to provide a foundation for them to create and/or expand on their views and opinions, they also provided a basis for demonstrating the purpose of the research project to clinicians. Clinicians gained a better understanding of medical device usage and adherence, and the effects of user needs on these. Parents of the adolescent participants were also encouraged by the use of vignettes.

The use of vignettes in the second study was appropriate due to its exploratory nature. A number of diagnostic procedures were investigated, which would have been difficult to achieve in real life considering ethical requirements needed. The time it would take and access to medical devices would also make it difficult.

5 CONCLUSIONS

Both studies involved young participants and the use of vignettes enabled research where they could make a valid contribution. The vignettes allowed participants to feel comfortable expressing their opinions and that they need not conform to impression-management biases (Alexander and Becker, 1978; and Torres, 2009). Vignettes as a methodology also compensated for participants' lack of product design and healthcare experience for study 1 and study 2 respectively, which Barter and Renold (2000) regard as one of the strengths of this methodology. Responses to vignettes are limited by their hypothetical nature, however, when appropriately designed and implemented they provide participants an opportunity to explore and contribute to research where they may otherwise be unable to, and thus are able to make a valid contribution to healthcare research.

ACKNOWLEDGMENTS

This work has been funded by the MATCH Programme (EPSRC Grant GR/S29874/01).

REFERENCES

Alexander, C.S. and Becker, H.J. 1978. The use of vignettes in survey research. *Public Opinion Quarterly* 42 (1): 93-104.

Barter, C. and Renold, E. 2000. 'I wanna tell you a story': exploring the application of vignettes in qualitative research with children and young people. *International Journal of Social Research Methodology* 3 (4): 307-323.

Keane, D., Craven, M.P., and Sharples, S. 2011. Information provision and decision aids for diagnosis in clinical pathways. In. *Healthcare Systems Ergonomics and Patient Safety*

2011, eds. Albolino, S., Bagnara, S., and Bellandi, T., et al. Boca Raton: CRC Press, 36 (abstract – full paper in accompanying CD).

Krosnick, J.A. 1991. Response strategies for coping with the cognitive demands of attitude measures in surveys. *Applied Cognitive Psychology* 5 (3): 213-236.

Lang, A.R., Martin, J.L., and Sharples, S., et al. 2010. A qualitative assessment of medical device design by healthy adolescents. In. *Advances in Human factors and Ergonomics in Healthcare*, ed. Duffy, V. Boca Raton: CRC Press, 539-548.

Lannefors, L., Button, B.M., and McIlwaine, M. 2004. Physiotherapy in infants and young children with cystic fibrosis: current practice and future developments. *Journal of the Royal Society of Medicine* 97 (supplement No. 44): 8-25.

Map of Medicine. Accessed February 28, 2012, http://eng.mapofmedicine.com.

Murray, C.J.L., Özaltin, E., and Tandon, A., et al. 2003. Empirical evaluation of the anchoring vignette approach in health surveys. In. *Health Systems Performance Assessment: Debates, Methods and Empiricism*, eds. Murray, C.J.L. and Evans, D.B. Geneva: World Health Organization, 369-399.

Ogden, J., Daniells, E., and Barnett, J. 2009. When is choice a good thing? An experimental study of the impact of choice on patient outcomes. *Psychology, Health & Medicine* 14 (1): 34-47.

Pendleton, D.A. and David, T.J. 2000. The compliance conundrum in cystic fibrosis. *Journal of the Royal Society of Medicine* 93 (supplement No. 38), 9-13.

Salomon, J.A., Tandon, A., and Murray, C.J.L., et al. 2003. Unpacking health perceptions using anchoring vignettes. In. *Health Systems Performance Assessment: Debates, Methods and Empiricism*, eds. Murray, C.J.L. and Evans, D.B. Geneva: World Health Organization, 401-407.

Torres, S. 2009. Vignette methodology and culture-relevance: lessons learned through a project on successful aging with Iranian immigrants to Sweden. *Journal of Cross-Cultural Gerontology* 24 (1): 93-114.

Wilks, T. 2004. The use of vignettes in qualitative research into social work values. *Qualitative Social Work* 3 (1): 78-87.

CHAPTER 50

A Comparative Study of the Effectiveness of Leaflets and Videos to Educate Low-literacy Mothers to Take Care of Their New-born Babies

Yah-Ling Hung1, Shuangyu Li 2,

1 Dept. of Communication arts, Fu-Jen Catholic University, Taiwan
School of Design, University of Leeds, UK / sdylh@leeds.ac.uk
2 Clinical Communication Unit, King's College London, UK
shuangyu.li@kcl.ac.uk

ABSTRACT

Low health literacy has been associated with poor outcomes in health care. Recent research suggests that good health promotional media can help to reduce the literacy barrier and enhance health outcome. However, ways to compare the effectiveness of various health promotional media in educating low-literacy patients are still at an early stage. Immigrant populations are vulnerable to serious health disparities, and language barriers may further exacerbate their limited health literacy in accessing health care information. This research considers a specific group within Vietnamese immigrant mothers in Taipei, most of whom have low levels of education. The purpose of this study is to compare the educational effectiveness of leaflets and videos which are designed to deliver knowledge to immigrant mothers about taking care of their new-born babies, thus establishing useful criteria for low-literacy patients to evaluate health promotion media. The findings show that the leaflet group performed significantly better than the video group. Compared with dynamic videos, the advantages of static leaflets are easy to read repeatedly, a clear column type, and user's own control. On the contrary, the disadvantages of videos are information is shown too quickly, narration is difficult to understand, and without user's own control. Since the volume of health promotional media is continually increasing, the need to evaluate its effectiveness becomes more imminent. Future research could examine other evaluation criteria for newly-emerging health care media, such as the internet, mobile devices, and telemedicine for low-literacy patients.

Keywords: Health Communication, Health Literacy, Health Promotional Media

1. INTRODUCTION

1.1 RESEARCH BACKGROUND

Low health literacy has been associated with poor outcomes in health care, which include higher health care costs, a worse assessment of one's own degree of health, and less success in managing chronic diseases (National Institutes of Health, 2002). Recent research suggests that good health promotional media can help to reduce the literacy barrier and enhance health outcome, they can help modify attitudes, shape positive behaviours, and improve patients' self-prevention (Andersen et al., 2008; Atkinson, 2009; Choi & Bakken, 2010). Choosing appropriate health promotion media to meet the needs of specific audience is critical for successful health communication. However, ways to compare the effectiveness of various health promotional media in educating low-literacy patients are still at an early stage.

Immigrant populations are vulnerable to serious health disparities, and language barriers may further exacerbate their limited health literacy in accessing health care information. By the end of 2011, a total of 42,590 new immigrants were registered on the census of Taipei city, and their number of new-born babies was 12,221, which accounted for 6.74% of the total number of new babies born in Taipei. This research considers a specific group within this population, namely, immigrant mothers who have come from China, Vietnam, Indonesia, Thailand, and the Philippines, most of whom have low levels of education, and have married into poor families. In order to help these low-literacy mothers to take care of their new-born babies, health promotional materials are produced in their own languages. However, there are no reliable criteria to evaluate the effectiveness of these materials.

1.2 Purpose of this Study

Apparently, the theoretical foundations of health communication appear to address relatively innumerable sets of goals and concerns relating to public health. However, studies of the comparative analysis of the educational effectiveness of various health promotional media for low-literacy patients are comparatively rare.

The purpose of this study is to compare the educational effectiveness of leaflets and videos which are designed to deliver knowledge to low-literacy mothers about taking care of their new-born babies, thus establishing useful criteria for low-literacy patients to evaluate health promotion media. Even though leaflet is relatively inexpensive to create and provides valuable information for populations, the benefit of its one-size-fits-all approach varies from person to person. On the contrary, video which combine texts, images, audio, and video, seem to be welcome, but whether or not their versatile presentations affect low-literacy patients' health-related behaviour is not yet known. Considering the above-mentioned issues of concern, the primary research objectives of this paper are described below:

- To survey the current strategies, methodologies, and tools to evaluate health promotional media for low-literacy users.
- To compare the educational effectiveness of leaflet and video to deliver children's health care knowledge to low-literacy parents.
- To set up evaluation guidelines of health educational materials for low-literacy patients.

2. Literature Review

2.1 Evaluation of Health Communication

Past studies related to health communication cover diverse disciplines ranging from health science, health education, information technology, and health care application. However, how can behavioural outcomes be improved in the right way through health communication intervention? Neuhauser & Kreps (2010) provided the following guidance:(1) Health communication is more effective when it reaches people on an emotional, as well as rational, level; (2) Health communication is more effective when it is related to people's social or life contexts; (3) A combination of the effectiveness of interpersonal communication and the reach of mass media communication is needed to change popular behaviour; (4) Tailored communication is more effective than generic messages; (5) Interactive communication is more effective than one-way communication. Moreover, Wurzbach (2004) mentioned four methods of programme design and evaluation in community health education and promotion: (1) Using Formative Evaluation to maximise the chance of programme success before the community activity starts; (2)Using Process Evaluation to examine the procedures and tasks involved in implementing a programme; (3) Using an Outcome Evaluation to obtain descriptive data on a project and to document short-term results; (4) Using an Impact Evaluation to focus on the long–range results of the programme and changes or improvements in the health status as a result.

2.2 Problem of Health Literacy

Health literacy is the degree to which individuals have the capacity to obtain, process, and understand basic health information and services needed to make appropriate health decisions and follow instructions for treatment (American Medical Association Foundation, 2008). It is important to examine health literacy because low health literacy has been associated with poor health-related outcomes, which include hospitalisation rates, poor adherence to prescribed treatment and self-care regimens, increased medication or treatment errors, failure to seek preventive care, lack of skills needed to navigate the health care system, disproportionately high rates of diseases and mortality, and the increased use of emergency rooms for primary care (Heather L. Bankson, 2009; Choi & Bakken ,2010).

It is essential to train health care providers to deliver care which is sensitive to the needs of diverse individuals with varying degrees of health literacy. For example, Howard (2005) used a Short Test of Functional Health Literacy to measure whether or not 3260 non-institutionalised elderly persons had the capacity to obtain, process, and understand basic health information and services needed to make appropriate health decisions. The results showed that persons with inadequate health literacy incur higher medical costs and use an inefficient mix of services. Moreover, Trifiletti et al. (2006) developed and evaluated injury prevention materials for people with low literacy skills, and explained literacy and comprehension abilities to a sample of parents from the paediatric emergency department. They suggested that health professionals should consider the reading ability, reading level, content, and design of materials for low literacy visitors.

2.3 Leaftlet Media VS Video Media

Leaflet with illustrations can be used to facilitate patients' understanding of health information. Kripalani et.al. (2007) indicated that patients with low health literacy have difficulty in understanding prescription drug labels and other medication instructions. He described the development, implementation, and preliminary evaluation of an illustrated medication schedule that depicted a patient's daily medication regimen using pill images and icons. The results showed that almost all patients considered an illustrated medication schedule to be a useful and easily understood tool to assist with medication management. On the other hand, Video-based information networks will be an important communicating platform for creating new regional-level virtual communities to further regional invigoration. Lee & Owens (2004) found that video-assisted information of informed consent decreased anxiety levels and improved satisfaction scales in emergent patients undergoing central venous catheter, and improved their recall of information compared with verbal informed consent procedures. In fact, virtual communication continues to be increasing in many IT applications.

2.4 Health Promotional Media for Low-literacy Patients

Low-literacy populations have different ways of accessing information and different preferences of media presentation, and choosing. For example, Kutner et al. (2006) revealed that low-literate adults preferred visual or audio-based health information rather than written text-based sources, such as newspapers, magazines, books or brochures. Burnham (2008) suggested that video tapes and oral communication were recommended for those who felt below the high school reading level. Andersen et al. (2008) found that low-literacy patients preferred images of real people to cartoons and illustrations and disliked the over-use of religious imagery, particularly an inclusion of an image of the Virgin Mary. They recommend developing a dynamic content presentation to provide age-tailored content and to slow users down such as adding sound and animation in the future. Cassell et al. (2010) indicated that the internet had the potential to reach those with low computer and reading literacy levels with a wide array of visual and audible cues. Because of their user-friendly interfaces, such as touch screens, voice recognition, and hand-held remote controls can be used instead of keyboards to interact with programmes on the internet. Based on the above observation, low-literacy patients prefer visual elements instead of text elements, prefer images of real people instead of virtual cartoons, and prefer simple user-control instead of complicated interactivity. It is undeniable that improving the knowledge of preventative measures of the low-literacy population could reduce social risk and cost.

2.5 Chapter summary

What do the authors recommend?

- The bulk of the research concerning health communication covers diverse disciplines ranging from health science, health education, health care applications, information technology, and mass communication.
- Low health literacy has been associated with poor health-related outcomes, health care providers should be sensitive to meet the needs of diverse individuals with varying degrees of health literacy.
- Health promotional media is an effective communication channel to offer an innovative professional care system which provides a more accurate, accessible and applicable educational platform for patients in a diversified society.

What have studies not found and what do further studies need to implement?

- How to compare the educational effectiveness of different type of media to deliver health care knowledge for low-literacy users?

- What are the factors of health promotional media which affect users' satisfaction and help them learn best.
- What are the applicable testing methods to compare the educational effectiveness of health promotional media for low-literacy users.

3. Methods

3.1 Research Subject

By the end of 2011, a total of 42,590 new immigrants were registered on the census of Taipei city, and their number of new-born babies was 12,221, which accounted for 6.74% of the total number of new babies born in Taipei. To assist the new immigrant mothers, in year 2005, Taipei City government set up two New Immigrants' Halls located in the Nangang and Wanhwa districts. Not only do the Halls offers cultural training courses but they also provide newspapers, magazines, and the internet in Vietnamese, Thai, Indonesian, and English for the new immigrants to get news from their home countries. Among those new immigrants, nationality from Vietnam is the majority, most of whom have low levels of education. Therefore, this study recruited 60 participants from this group as the research subjects. Furthermore, existing health promotional media, such as leaflets and videos to educate immigrant mothers to take care of their new-born babies, were collected from the National Health Insurance Bureau. One leaflet and one video bearing the same health care information were sampled out for comparison against each other. In addition, the Pre-Post knowledge testing and usability evaluation for this research were also conducted at the new immigrants' hall.

3.2 Pre-Post Knowledge Testing

Secondly, pre-post knowledge testing will be applied to understand the effectiveness between different media groups in this study. At first, 60 recruited subjects were randomly placed into two groups to explore the educational effectiveness of leaflet and video separately. The leaftlet group was presented with the static leaflet, whereas the video group was presented with the dynamic video. Both groups were asked to fill out the Pre-Post testing questionnaire before and after viewing the specified media presentation. Thus, this research will measure their memory recall of the presented materials.

3.3 Usability Evaluation Questionnaire

Thirdly, a usability evaluation questionnaire was follow-up to assess the usability of the participants. Thus, to identify the factors of health promotional media which affect low health literacy user satisfaction. The IBM had provided evaluation criteria which measure user satisfaction with computer system usability, such as system usefulness, information quality, and interface quality guidelines (Lewis, J.R., 1995). Therefore, each participant was asked to fill out a usability evaluation questionnaire on the presented media. The questionnaire consisted of 9 items of evaluation criteria involving the quality of the information, the quality of the presentation, and the operating quality of the usability testing media in question.

3.4 Questionnaire Design

All the questions in the questionnaire were created at 5th grade readability level or lower, and also made available in Vietnamese. Besides, simplifying the text, strictly limiting the content, keeping the content short and focused, writing the text the way we talk, using plain language, using short sentences, writing in the active voice, and avoiding long strings of nouns will be the strategies for the low-literacy text (Choi & Bakken, 2010). In order to investigate the potential processing problem, and estimate the average duration of viewing the media and responding the questions of the participant, a pilot test was conducted with 5 participants to assess the wording, sequence and clarity of questions and instructions in the questionnaire. After the data had been collected, some statistical analyses were conducted via the SPSS for windows Version 17.0.

As for the Pre-Post knowledge testing questionnaire, get one right answer on each question to get one score, and the higher the score, the better the knowledge a participant has about the healthcare of new-born baby. The rating scale for measuring the performance of the post-knowledge testing was as follows: 1-4=bad, 5-8=ok, 9-12=good, 13-15=excellent, Items=15. As for the usability evaluation questionnaire, the five-point Likert Scale was used for every question, with higher scores indicating the higher appropriateness. The rating scale for the measuring the appropriateness of the questionnaire was scored from 1= disagree very much, 2=disagree, 3=no opinion, 4=agree, 5=agree very much.

3.5 Data Analysis

3.5.1 Pre-Post Knowledge Testing

Table 1 The result of Pair T-test (Video)

Pre-test	7.70 (2.60)
Post-test	7.97 (2.88)
Difference	0.27
t-value	1.55

Note：1. 7.7 is the Mean. 2.60 is the standard deviation.

The results showed that the difference between the pre-test and the post-test is not significant (t=1.55, p>.05). Besides, the rating scale for measuring the performance of the post-knowledge testing was 5-7 which means the performance of the participant is OK.

Table 2. The result of Pair T-test (Leaflet)

Pre-test	7.97 (2.67)
Post-test	9.90 (3.75)
Difference	1.93
t-value	7.37***

Note：1. 7.97 is the Mean. 2.67 is the standard deviation.

2.*** p < .001

The results showed that there is a significant difference between the pre-test and the post-test (t=7.37, p<.001). Besides, the rating scale for measuring the performance of the post-knowledge testing was 8-10 which means the performance of the participant is good.

3.5.2 Usability Evaluation Questionnaire

Category		5	4	3	2	1	Average
Information Quality	Clear	18/60= 30%	12/6= 20%	27/6= 45%	3/60 =5%		225/60 =3.75
	User's need	18/60= 30%	21/6= 35%	18/6= 30%	3/60 =5%		234/60 =3.9
	Understand able	12/60= 20%	12/6= 20%	27/6= 45%	3/60 =5%	6/6 =10%	201/60 =3.35
Presentation Quality	Attractive appearance	15/60= 25%	9/60= 15%	33/6= 55%	3/60 =5%		216/60 =3.6
	Layout design		39/6= 65%	15/6= 25%	3/60 =5%	3/60 =5%	210/60 =3.5
	Cognitive Load	9/60= 15%	45/6= 75%	6/60= 10%			243/60 =4.05
Operating Quality	Accessing	12/60= 25%	39/6= 65%	6/60= 10%			219/60 =3.65
	Comfortable	12/60= 20%	42/6= 70%	6/60= 10%			246/60 =4.1
	Efficient	12/60= 20%	39/6= 65%	3/60= 5%	6/60= 10%		237/60 =3.95

The results reflected in Table 3 indicated that, in terms of the quality of information, 'Users' need' was the most welcome factor, with the average frequency of preference reaching 3.9. However, the factor of 'understandable' is uncommon since its average frequency of preference below 3.5. Besides, in terms of the quality of presentation, 'cognitive load' was the winner, with the average frequency of preference reaching 4.05.Then 'Attractive appearance' and 'Layout design' were also needed factors, since their average frequency of preference also reaching 3.6 and 3.5 respectively. In addition, in terms of the quality of operating, 'Comfortable' and 'Efficient' were the most welcome factors, with the average frequency of preference reaching 4.1 and 3.95 respectively. Nevertheless, the factor of 'accessing' was not popular, with the average frequency of preference reaching 3.65. Basically, all of the above factors were evenly dispersed across leaflets and video media. Participants' average ratings were all nearly 4.0, which indicated that all of the immigrant mothers agreed with the appropriateness of the evaluation criteria.

4. Discussion and Conclusion

The findings showed that the leaflet group performed significantly better than the video group, even though videos can somewhat reduce the anxiety levels and enhance the interest of low-literacy users, they may also split their attention and increase their cognitive load so that they lose the key message. Moreover, participants with a basic idea of caring for a new-born baby felt that they could improve their healthcare knowledge by reading leaflets, whereas participants without a basic concept of a new-born baby thought that they could increase their healthcare knowledge by watching a video. Compared with dynamic videos, the advantages of static leaflets are easy to read repeatedly, a clear column type, and user's own control. On the contrary, the disadvantages of videos are information is shown too quickly, difficult to understand narration, and without user's own control.

The findings also show that the criteria used by low-literacy patients to evaluate health promotion media can be analyzed in terms of the quality of information, presentation, and operation. However, health information should be written in plain language, and accompanied by factual examples to meet the users' need. The visual design should match patients' cultural background, and include iconic images of their native countries to reduce the cognitive load. The operational process should be compassionate, and consist of a comfortable and efficient means of assessment to increase usage. Since the volume of health promotional media is continually increasing, the need to evaluate its effectiveness becomes more imminent. Future research could examine other evaluation criteria for newly-emerging health care media, such as the internet, mobile devices, and telemedicine for low-literacy patients.

5. References

American Medical Association Foundation (2008). Health literacy [Internet].Boston, MA: The Foundation; 2008 Available: <http://www.ama-assn.org/ ama/pub/ca tegory /8115.html >Accessed:150811.

Andersen, P. ;Andersen, S.; Youngblood, E.; Colmenares, E.(2008). Health education kiosk for low-literacy patients served by community-based clinics,2008 IEEE International Symposium on Technology and Society, pp. 1-9.

Atkinson, NL, Saperstein SL, Pleis J. (2009). Using the internet for health-related activities : findings from a national probability sample. J. Med Internet Res 2009;11 (1):e4.

Burnham, Erica (2008). Libraries as Partners in Health Literacy. Journal of Consumer Health on the Internet. Available: <http:// tandfprod.Literatumonline.com/Loi/wchi20 > Accessed: 25 July,2011.

Cassell, Michael.; Jackson, Christine; Cheuvront, Brian.(2010). Health Communication on the Internet: An Effective Channel for Health Behaviour Change? In Krep, Gary L. (Eds.) (pp.17-42). Health Communication. v. 4. Health Communication and new information technologies (eHealth). London: SAGE Publications Ltd.

Choi, Jeungok & Bakken, Suzanne. (2010). Web-based education for low-literate parents in Neonatal Intensive Care Unit: Development of a website and heuristic evaluation and usability testing. International Journal of Medical Informatics, Volume 79, Issue 8, pp. 565-575.

Heather L. Bankson (2009). Health Literacy: an exploratory bibliometric analysis,1997-2007. J Med Libr Assoc 97(2) April 2009.

Howard,D.H., Gazmararian, J., Parker, R.M. (2005). The impact of low health literacy on the medical costs of Medicare managed care enrolees, The American Journal of Medicine, Volume 118, Issue 4, April 2005, pp. 371-377.

Kripalani,S., Robertson, R., Love-Ghaffari, M.H., Henderson, L.E., Praska, J., Strawder, A., Katz, M.G. Jacobson, T.A. (2007) Development of an illustrated medication schedule as a low-literacy patient education tool, Patient Education and Counseling, Volume 66, Issue 3, June 2007, pp.368-377.

Kutner, J.M.; Greenberg, E.;.Jin, Y; Paulsen, C. (2008). The health Literacy of America'sAdults: Results from the 2003 National Assessment of Adult Literacy (NCES 2006-483).U.S. Department of Education, National Center for education, Washington, DC,2006.

Lee,W.W.,& Owens,D.L.(2004). Multimedia-Based Instructional Design: Computer-based Training, Web-based Training, Distance Broadcast Training, Performance-based Solutions. San Francisco : Pfeiffer Liao.

National Institutes of Health (2002). National Institute on Deafness and Other Communication Disorders. "Researcher Strengthens Health, Literacy Link: why Johnny is Sick."

Neuhauser, L. & Kreps, G. L. (2008). Online cancer communication: Meeting the literacy, cultural and linguistic needs of diverse audiences. Patient Education and Counseling, Volume 71, Issue 3, June 2008, pp. 365-377.

Trifiletti, L.B., Shields, W.C., McDonald, E.M., Walker, A.R., Gielen, A.C. (2006). Development of injury prevention materials for people with low literacy skills, Patient Education and Counseling, Volume 64, Issues 1-3, December 2006, pp. 119-127.

Wurzbach, M.E.(2004). Community health education and promotion: a guide to program design and evaluation(2nd ed.).Sudbury, Mass: Jones and Bartlett.

CHAPTER 51

Improving the Patient Pathway in Prosthetic Rehabilitation

Grace Smalley, Laurence Clift

Loughborough Design School
Loughborough University, UK
g.j.smalley@lboro.ac.uk

ABSTRACT

The rate of incidence of vascular disease and associated limb amputation can be anticipated to mirror the reported epidemic of obesity in the Western world. This presents a growing challenge to the provision of prosthetic services. There are currently 43 Disablement Services Centers (DSC) in the UK National Health Service (NHS), each offering prosthetic rehabilitation. A previous study has revealed that there are no guidelines for patient rehabilitation pathways therefore centers are required to produce their own. This has resulted in a situation where the information given to patients is unique to each center. The level of service provision across the UK appears to vary greatly with very little known about the patient experience and whether the service being provided is fulfilling patient's needs and expectations. In order to ascertain patient's satisfaction with the service being provided, an extensive user centered questionnaire survey was undertaken which covered nine main topics. The survey was distributed via the internet and of the 105 participant responses, 96 were usable for analysis. Investigation of the results using SPSS and thematic analysis showed that satisfaction with service provision and prosthesis satisfaction are very closely linked. Participants felt that counseling should be available at every DSC and patient volunteer visitors were considered to be beneficial. The information given to participants by DSCs was primarily verbal with 41% of participants not receiving any information on the ward prior to their amputation. Patients place a high importance on having a spare limb and 71% of those that do not receive one stated that they had not had the reasons for this fully explained to them. Recommendations include standardization of information given to patients with timely support at critical times in the patient pathway.

Keywords: obesity, prosthetic rehabilitation, service provision, diabetes

1 INTRODUCTION

There are currently over 1 billion overweight adults and at least 300 million adults identified as being clinically obese worldwide (World Health Organization, 2003). Research has found that between 1980 and 2002 the prevalence of obesity in Great Britain had almost tripled to approximately 24% of the population (Rennie & Jebb, 2005). There is a strong connection between obesity and Type 2 diabetes with research showing that approximately 80% of people with Type 2 diabetes are obese (Flier, 1994). Between 2006 and 2007 Diabetes accounted for 32% of all amputations in the UK (NASDAB, 2009), which is two and a half times greater than the number reported between 1997 and 1998 (NASDAB, 1999). The increase in prevalence of amputation due to Type 2 diabetes and other vascular conditions is creating challenges for the provision of prosthetic services across the UK.

There are many academic papers that deal with outcome measures following amputation (Herbert et al, 2009), however there are very few which focus solely on the patient experience of the prosthetic service (Legro et al, 1999) and consequently the value offered by the service and patient's acceptance. The Queensland government has set up a client service evaluation of their prosthetic service which is applicable to all providers of prosthetic care (Queensland Artificial Limb Service). The evaluation provides feedback on the range of services patients may encounter and highlights areas of concern. This initiative does not currently exist in the UK and individual centers are responsible for obtaining feedback from patients in the form they believe to be most appropriate.

This paper offers exploration of patient satisfaction with the service provided by the UK National Health Service DSC they attend, with analysis of nine main variables identified in a previous preliminary study.

2 NHS PROSTHETIC SERVICE PROVISION

A preliminary study of 12 NHS DSCs across England was conducted in order to ascertain the current NHS service provision of prosthetic limbs. It was found that very few centers provided a comparable level of service and much of the provision was decided upon by the staff working at individual centres and the budget allocated to them. At the time the study took place there were no NHS guidelines for patient rehabilitation pathways for centres to follow, leaving individual centres to create their own. This introduces great inconsistency in the service provided to patients across the country as centres often employ unique rehabilitation pathways. There did not appear to be any national standardized Key Performance Indicators (KPIs) for the service provided by the NHS therefore creating inherent difficulties in maintaining quality due to the lack of outcome measures. Counseling was recognized by most centres as being an important part of rehabilitation however it was not available at every center. A patient volunteer visitor service was mentioned by very few of the centers visited, however the centers that did provide the service expressed that the service could be of great use to patients. It was also found that there was little consistency with regards the amount and detail of information given to patients as individual centers created their own unique leaflets or packs due to a lack of NHS guidelines. Components provided to patients by centers also varied across the country with some centres able to provide more technologically advanced components than others. The most consistent area was that of spare limbs, in that almost all centres have had to introduce very strict criteria for their prescription due to budget cuts. It was recognized from this work that the differences between centers could have profound effects on patients and the success of their rehabilitation therefore a study was conducted in order to ascertain how the issues outlined by the preliminary study affect the patient experience.

3 USER CENTRED STUDY

The preliminary study highlighted the potential issues with NHS provision and indicated the need for a patient centered evaluation to ascertain how these potential issues may be affecting patients. The purpose of this study was to explore the patient perspective of the service they received and ascertain which areas of the service provision are areas of concern for patients.

3.1 Method

A questionnaire structure was designed for patients of the UK NHS using an online survey tool, using the information gained from the preliminary study. It was decided that the questions should form a broad overview of the service in order to identify the areas of concern for patients. Due to the nature of the questionnaire, questions were primarily quantitative with a small number of open ended questions for greater detail where it was deemed necessary. Questions were kept as simple as possible to avoid ambiguity and participant confusion. The study was conducted completely independently of the NHS to encourage patients to answer the questionnaire as honestly as possible. The most effective technique for obtaining participants was through the internet despite more direct alternatives being tried.

3.1.1 Sampling

In order to avoid NHS involvement it was necessary to use self-selection sampling (Saunders et al., 2007) as it was not possible to obtain participants through contact with the DSCs. Limbless charities were asked for assistance in obtaining participants and an article was placed in 'Step Forward' magazine, published by the Limbless Association, asking for participants. Requests for participants were also published on the Limbless Association website as well as a number of Facebook groups for amputees. A link to the online survey was posted on these websites and in the Facebook groups and also emailed to respondents from the magazine article. Inclusion criteria were explained to participants before they entered the survey as only UK NHS patients were sought. There was no coercion involved as interested parties were able to follow the link of their own accord.

3.2 Results

The results were downloaded from the online survey tool and a preliminary scan was completed to detect any anomalous entries. There were 105 data files, only 96 of which were usable due to duplicates or the entire questionnaire being blank. IBM® SPSS® statistics software was used for statistical analysis and the open ended questions were analyzed thematically. Of the 96 participants 58% were male and 42% female. Age of participants ranged from 24 to 82 with a mean age of 54. Ninety percent of respondents had a lower limb amputation with 57% being a below knee amputation. The most common reason for amputation was trauma, with 51% of participants reporting having a traumatic amputation.

It was initially anticipated that there could be a connection between satisfaction with a prosthesis and satisfaction with service provision. Of the 78 patients that answered both questions 60% expressed satisfaction with the service provision and their prosthesis. Fifteen percent expressed dissatisfaction with both the service provision and their prosthesis and 13% expressed satisfaction with the service provision but dissatisfaction with their prosthesis. The reasons given for the participant's dissatisfaction with their prosthesis were thematically analyzed and showed that 92% of those dissatisfied with both the service and their prosthesis stated that the reason for their dissatisfaction was poor socket fit. The reasons given by those participants satisfied with the service but dissatisfied with their prosthesis were centered on wanting better quality components or better cosmesis, with only one participant mentioning socket fit and stating that this was due to the level of their amputation. When asked if their prosthesis fulfilled their expectations, 31% of the 86 participants that answered the question stated that theirs did not. The reasons for this were thematically analyzed and showed that in 54% of cases, the prosthesis did not fulfill expectations due to the limiting affect it had on their activity levels.

In the initial study, most DSCs visited stated that they believe counseling is a very important part of rehabilitation. When asked if counseling is available at their center of the 87 participants that answered the question, 51% answered Yes, 19% answered No and 30% answered they did not know. Of the 38 participants that received counseling 94% stated that they feel the service should be available at every DSC. When asked at what stage was counseling most useful, 26% of participants stated that it was most useful a year or more after amputation. The participants that did not receive counseling were asked if they would use the service if it were available, to which 52% stated that they would. The same participants were then asked to explain at what time of their rehabilitation would they have found counseling most beneficial. These answers were thematically analyzed and showed that 48% would have liked counseling post- amputation and 35% pre- and post- amputation.

Patient volunteer visitors are active at some, but not all, DSCs, therefore there was value in asking participants whether they felt the service is worthwhile. When asked if they had been visited by a volunteer visitor 20% answered Yes, 79% answered No and 1% answered that they did not know. Of those participants that had a visit form a volunteer, 89% stated that they found the experience beneficial. The participants that did not receive a visit from a volunteer were asked if they would have liked this service, to which 71% answered Yes. Ninety-three percent of all participants stated that they would consider becoming a patient volunteer visitor.

All participants were asked to state the level of information they received at five different stages both pre- and post- amputation. These results are shown in Figure 1. It is clear that the majority of information supplied by DSC staff is verbal with very little written information being provided. Figure 1 also illustrates that 41% of participants received no information on the ward before their amputation. Figure 2 illustrates the breakdown of the written information received by participants at the different stages. The most frequently used form of written information was a leaflet with very few participants receiving booklets or more than one leaflet.

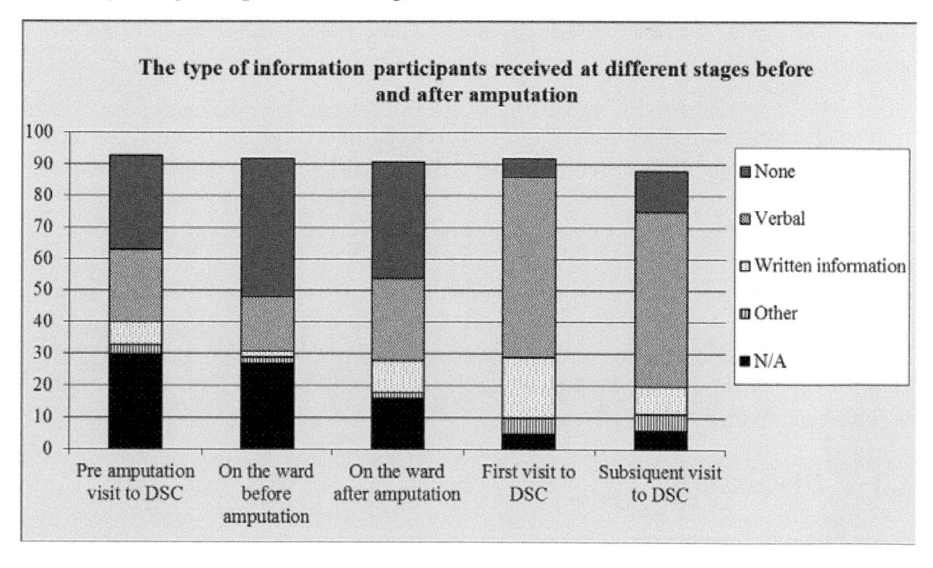

Figure 1 The type of information participants received at different stages before and after their amputation.

The current budget restrictions DSCs face mean that patients are not always prescribed with the most technologically advanced components available. Forty percent of the 87 participants that answered the question stated that they are aware of other components they feel would benefit them, 50% of which stated some, if not all of their information had come from research on the internet. When asked if they would be willing to contribute money to obtain a component they desired, 50% of the 82 participants that answered the question said Yes, with 18% stating they did not know whether they would or not.

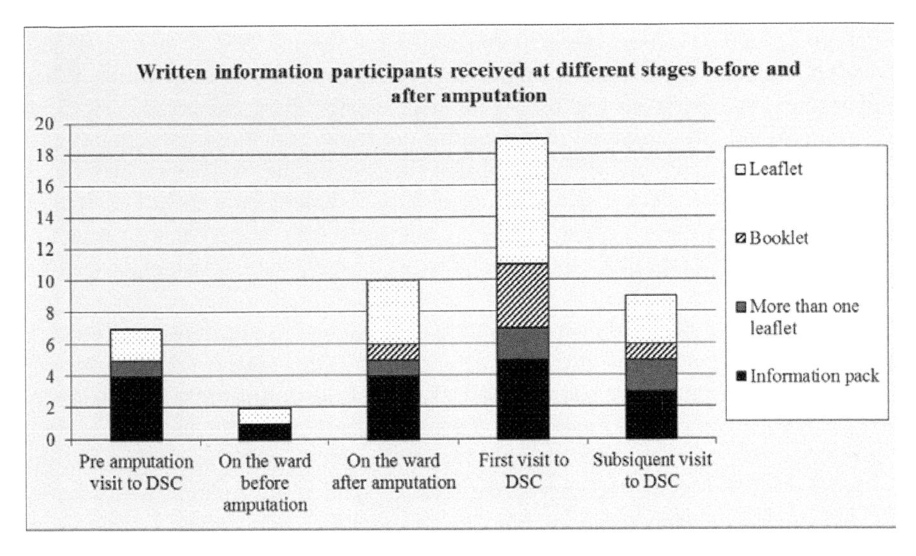

Figure 2 The different written information received by participants at different stages before and after their amputation.

The changing spare limb policy was of particular interest as most centers visited during the preliminary study were no longer giving spare limbs as standard practice. Of the 88 participants that answered the question, 66% receive a spare limb with 87% of them stating that receiving a spare limb gave them peace of mind. Those participants that do not receive a spare limb were asked whether they were happy with this situation, to which 71% answered No. Seventy-one percent of participants also stated that the reasons for not receiving a limb were not properly explained and 77% stated that having a spare limb would have a positive impact on their daily life. Eighty-nine percent of all participants stated that having a spare limb is, or would be, important to their quality of life.

3.3 Discussion

Marquis et al (1983) state that patient satisfaction plays an important role in retaining relationships between patients and healthcare providers. Following amputation patients will require some form of rehabilitation and assistive technology for the rest of their lives, making the relationship between them and their DSC a very important one. In their study of issues of importance for amputees, Legro et al (1999) reported that patients listed having a good prosthetist as being important to having a good life. Legro et al (1999) also reported that the most important function of the prosthesis was to enable walking and the most important characteristic of the prosthesis was fit. The results from the study reported in this paper indicate that satisfaction with the service being provided is closely linked with satisfaction with the prosthesis. This is reflected in the large number of participants that stated that they were happy with both the service and their limb.

The figures quoted are very similar to those of Nicholas et al (1999) who found that 59% of 94 patients were satisfied or somewhat satisfied with the comfort of their prosthesis. Participants of the study reported in this paper that were dissatisfied with their prosthesis, were clearly divided by the reasons for their dissatisfaction. Those that were satisfied with the service but not their prosthesis were mainly dissatisfied with the components in their prosthesis and the aesthetic qualities with only one participant mentioning fit. This patient also mentioned that the problems with fit were down to the level of amputation. It appears that these patients accept the limitations of the NHS with one participant stating "It is a compromise as it is not cosmetically as good as it could be, but it is as good as it is going to get given the system we operate in." These patients do not appear to associate their dissatisfaction with their prosthesis with the service they receive. In complete contrast, those participants that stated they were dissatisfied with both the service and the limb had issues with fit in 92% of cases. When asked if their limb fulfilled their expectations, 92% of these participants answered 'No' giving the reason that it limits their activities in 100% of cases. Further research is required to ascertain whether these patients could be fitted with an appropriate prosthesis or their expectations have not been properly managed, causing their dissatisfaction.

There have been many studies on the psychological effects of amputation (Bhuvaneswar et al, 2007) with a particular study by Callaghan and Condie (2003) finding that there is a "stronger relationship between mental health and quality of life than between physical health and quality of life." Price and Fisher (2002) state that the benefits of counseling indicate that the opportunity to access the service should be available to all interested patients. The results showing that 30% of the participants in this study did not know whether counseling was available at their DSC is an indicator that counseling is not being discussed with every patient. Counseling is regarded highly by those that had access to the service with 94% stating that it should be available at every DSC with 26% stating that the service was most useful over one year after amputation. These results concur with Price and Fisher's (2002) studies which indicated that emotional problems become apparent 6-24 months after surgery. Of those that had not had access to the service 52% stated that they would use the service if it became available which indicates that their psychological needs are not being met by their current service provision. Thirty-five percent of these participants stated that they would have liked counseling pre-amputation which corresponds to Butler et al (1992) and Sherman's (1997) work which indicates that appropriate preparation for surgery alleviates stress and eases rehabilitation. Desmond and MacLachlan (2002) state that development of realistic rehabilitation expectations should be included in this counseling intervention prior to surgery. It has been proven that emotional disclosure of stressful life experiences can have both physical and psychologic benefits and reduce the number of visits to clinicians (Pennebaker 1997). It is therefore in the interest of the NHS to provide counseling for patients as proper expectation management and emotional support could reduce the number of repeat appointments with patients who are dissatisfied with their prosthesis and therefore save clinicians time and the service money.

The advantages of patient volunteer visitors have been well documented with Frogatt and Mawby (1981) stating that an experienced amputee can play an important role in offering advice of a practical and emotional nature to new patients. Briggs (2006) states that meeting and talking with other amputees is important which corresponds with the finding that 71% of patients who did not have a patient volunteer visitor would have liked one and 89% of patients who did have a visitor found it beneficial. Patients are clearly aware of the important role a volunteer visitor can play in the rehabilitation of patients as 93% of participants stated that they would consider taking on this role themselves.

There are very few papers on the importance of patient information, however Mortimer et al (2004) state that minimum standards for information relating to phantom limb pain should be introduced. With so few papers surrounding this subject and the findings that content, mode of delivery and co-ordination of information surrounding phantom limb pain needed improvement, it can be surmised that information relating to other aspects of amputee rehabilitation will be in need of similar attention. The results in this study only relate to the amount of information given to patients, not the content. However it is clear that most information given to patients is verbal with very little written information, consequently this is highly transient and patients have nothing to reference at a later date. In a study of patients about to receive cataract surgery Shukla et al (2012) found that information sheets and videotape presentations were the optimum forms of information when describing the risks, benefits and treatment alternatives. This work is not directly related to amputation, however it provides evidence that verbal information can be inadequate when describing important aspects of surgery to patients. Forty-one percent of patients did not receive any information while they were on the ward prior to their amputation which indicates that the NHS is not preparing all patients for the life changing operation they are about to undergo. Further research is required to ascertain the effects of this lack of information on patients and whether anxiety is increased due to fear of the unknown.

Technical advances are constantly being made in the prosthetics industry with new component information being easily accessible on the internet. It is not surprising then that 40% of participants are aware of components that may benefit them and that 50% of these found the information on the internet. Further research is required to ascertain how this lack of components they desire is affecting patients however it is clear from the results that components are something that the majority of patients are willing to invest their own money in. There is growing media coverage of the Paralympics in the UK which can increase the expectations of new amputees due to the athletes being pictured wearing specialist carbon fiber running blades. Due to this, patients may be disappointed with the service they are being provided by the NHS as they are aware of prostheses which are far superior to those that they have been prescribed.

Due to the recent change in policy, there is no literature pertaining to the advantages of spare limbs for prosthetic patients. These results clearly show that, for participants, having a spare limb is important and that it gives them peace of mind. A very high percentage of participants that do not receive a spare limb are unhappy with this situation which clearly indicates that patients place high importance on these limbs. This is confirmed by 77% of these participants stating that they believe having a spare limb would have a positive impact on their daily life. The figures show that 75% of participants that were unhappy about not receiving a spare limb did not have the reasons for this fully explained to them. Further research is required to ascertain whether having the reasons for not receiving a spare limb being fully explained reduces the importance of the spare limb to the patients.

4 CONCLUSIONS

It is clear from this work that further research is needed in all areas of service provision covered. The majority of participants that took part in this study had traumatic amputations, which is at odds with the statistics in NASDAB (1999 and 2009). Further research is required within the NHS in order to fully represent patients with amputations due to other causes. It is particularly troubling that information provision is so poor and inconsistent and that the reasons for not receiving a spare limb are not being fully explained. It is believed that simply providing the correct information at the appropriate time could alleviate anxiety among patients and allow for a smooth entry into rehabilitation. Information regarding components and spare limbs is also important as patients need to be made aware of what they can realistically expect to receive on the limited NHS budget. It is also clear that centers should be working to provide support for patients in the form of counseling or patient volunteer visitors due to the perceived benefits of these services. It is felt that further research is needed on the connection between socket fit and provision of service, with particular attention paid to whether expectation management and counseling could be used to improve this. It is clear that patients associate bad service provision with an ill-fitting socket and therefore more needs to be done in order to rectify problems surrounding fit. It is hoped that more research will be conducted, within the NHS, from the user perspective in order to gain a better insight into NHS service provision and improvements that need to be made in order to improve the patient experience.

The recommendations from this work include a more informed process of expectation management as well as ongoing and timely support at critical points in the patient pathway. These can be provided by improving and standardizing the information provided to patients, offering counseling at key points and improving the access to patient volunteer visitors who offer peer support. These interventions, which attract little cost, would significantly improve the patient experience and reduce the demand, and hence cost, placed upon the clinical services.

REFERENCES

Bhuvaneswar, C. G., L. A. Epstein and T. A. Stern. 2007. Reactions to amputation: Recognition and treatment. Journal of Clinical Psychiatry, 9(4): 303-308.

Briggs, W. 2006. The mental health problems and needs of older people following lower-limb amputation. *Clinical Gerontology*, 16: 155-163.

Butler D. J., N. W. Turkal and J. J. Seidl. 1992. Amputation: Preoperative psychological preparation. *Journal of American Board of Family Practice*, 5: 69–73.

Callaghan B. G. and M. E. Condie. 2003. A post-discharge quality-of-life outcome measure for lower-limb amputees. *Clinical Rehabilitation*, 17: 858–864.

Desmond, D. and M. MacLachlan. 2002. Psychological issues in prosthetic and orthotic practice: a 25 year review of psychology in Prosthetics and Orthotics International. *Prosthetics and Orthotics International*, 26: 182-188

Flier, J. S. 1994. Obesity. In: *Joslin's diabetes mellitus*. eds. R. C. Kahn and G. C. Weir. 13th ed. Philadelphia: Lea & Febiger.

Froggatt, D. and R. Mawby. 1981. Surviving an amputation. *Social Science and Medicine*, 15: 123–128.

Herbert, J. S., D. L. Wolfe, W. C. Miller, A. B. Deathe, M. Devlin and L. Pallaveshi. 2009. Outcome measures in amputation rehabilitation: ICF body functions. *Disability and rehabilitation*. 31(19): 1541-1554.

Legro, M. W., G. Reiber, M. Delguila, M. J. Ajax, D. A. Boone, J. A. Larsen, D. G. Smith and B. Sangeorzan. 1999. Issues of importance reported by persons with lower limb amputations and prostheses. *Journal of Rehabilitation, Research and Development*, 36(3): 155-163.

Marquis, M. S., A. Ross Davies and J. E. Ware. 1983. Patient satisfaction and change in the medical care provider: A longitudinal study. *Medical Care*, 21: 821-829.

National Amputee Statististical Database, National Health Services (NASDAB). 1999. *The Amputee Statistical Database for the United Kingdom*.

National Amputee Statististical Database, National Health Services (NASDAB). 2009. *The Amputee Statistical Database for the United Kingdom*.

Nicholas, J. J., L. R. Robinson, R. Schulz, C. Blair, R. Aliota and G. Hairston. 1993. Problems experienced and perceived by prosthetic patients. *Journal of Prosthetics and Orthotics*, 5(1): 16-19.

Pennebaker J. W. 1997. Writing about emotional experiences as a therapeutic process. *Psychological Science*, 8: 162-166.

Price M. and K. Fisher. 2002. How does counselling help people with amputations? *Journal of Prosthetics and Orthotics*, 14: 102–106.

Queensland Artificial Limb Service. 2011. *Client Service Evaluation – QUAL*.

Rennie, K. L. and S. A. Jebb. 2005. Prevalence of obesity on Great Britain. *Obesity Reviews*, 6(1): 11-12.

Sherman, R. A. 1997. *Phantom Pain*. New York: Plenum Press.

Shukla, A. N., M. K. Daly and P. Legutko. 2012. Informed consent for cataract surgery: Patient understanding of verbal, written, and videotaped information. *Journal of Cataract & Refractive Surgery*, 38(1): 80-84.

Waites B. and A. Zigmond. 1999. Psychological impact of amputation. In: *Therapy for Amputees*. eds. B. Engstorm and C. Van de Ven C. 3rd ed. Edinburgh: Churchill Livingstone.

World Health Organisation Global strategy on diet, physical activity and health. *Fact sheet* 311, March 2011.

CHAPTER 52

Digital Graffiti: An Augmented Reality Solution for Environment Marking

John Sausman, Kenny Sharma, Alexei Samoylov, Susan Harkness Regli, Polly Tremoulet, and Kathleen Stibler

Lockheed Martin Advanced Technology Laboratories
Cherry Hill, NJ, USA
{john.r.sausman, kenny.sharma, alexei.samoylov,
susan.regli, polly.d.tremoulet, kathleen.m.stibler}@lmco.com

ABSTRACT

The ability to mark a physical environment with meaningful information is critical to disaster relief efforts. However, current field marking methods, such as chalk, spray paint, etc., are not adequate because emergency relief workers have no control over the markings after creation, valuable information is not captured (e.g., who made the marking, when was it made, how long is it valid), and there is no access management of the information to prevent unauthorized personnel from reading, altering, or even removing the markings. We have created a prototype system called Digital Graffiti that employs advanced augmented reality and mobile interaction techniques to allow a user to annotate real world objects with virtual markings. This paper describes multiple advances made while creating a working Digital Graffiti prototype. We describe the system and user interaction design (including a gesture glove and heads-up display), a usage scenario, and our technical approach for input recognition and marking georegistration.

Keywords: augmented reality, mixed reality, georegistration, environment marking, multimodal display, first responder

1 INTRODUCTION

First responders typically employ marking systems to indicate which buildings have been evacuated and searched for survivors in need of aid. Adding information to the environment as part of emergency response provides critical situational awareness information to other disaster relief workers. However, current field marking methods, such as chalk, spray paint, etc., are not adequate. Users have no control over the markings after creation, valuable information may not be captured (e.g., who made the marking, when was it made, how long is it valid), and there is no access management to prevent unauthorized personnel from reading, altering, or even removing the markings.

We have created a prototype system called Digital Graffiti that employs advanced augmented reality techniques to allow a user to annotate real world objects with virtual markings. Users can create, view, and edit virtual markings on real world objects such as buildings or doors. They describe or "mark" the object through the use of natural language and natural gestures as if they had a traditional marking device. The resulting georegistered marking is associated with the object and can be viewed and updated by users with proper access privileges as the situation changes. Our prototype provides the functionality disaster relief workers have come to expect from marking technology while addressing the shortcomings of current techniques.

This paper summarizes our Digital Graffiti research effort. First we review the work done to design the user interaction with the system. Then we describe a scenario in which the prototype is used to mark an environment. Finally, we discuss our technical approach for accurately recognizing user input and georegistering of markings in the physical environment.

2 BACKGROUND

We based our Concept of Operations and use cases on a disaster response effort where emergency relief workers mark the environment during urban search and rescue efforts following the method proscribed in the Federal Emergency Management Agency (FEMA) National Urban Search and Rescue (US&R) Response System Field Operations Guide [FEMA, 2003]. A quadrant symbol is created to present information about the search that was conducted within that structure. Figure 1 provides an example from the Hurricane Katrina post-disaster response; these markings were created with spray paint. Markings are often made using use tape, chalk, pen, or whatever is available in the environment.

Collection systems used in a disaster response context must be mobile. Since the information markers left by disaster relief workers are linked to particular geographic locations, they lend themselves to Augmented Reality (AR) displays. However, a mobile interaction paradigm for creating, editing, and viewing the markers is required. The typical desktop paradigm (windows, icons, menus, and pointer) is often not possible in the environment.

By providing a system interface for entering information, our Digital Graffiti technology can help standardize disaster relief markings; in addition, our prototype captures the information in a digital format that allows it to pass fluidly through time, space, and across levels of authority.

Figure 1 Hurricane Katrina Disaster Response Markings

Other AR interaction work has focused on creating interfaces on available surfaces [Mistry, Maes 2009] [Henderson and S. Feiner, 2010] or using AR for remote collaboration [Lukosch et al., 2012]. Our Digital Graffiti effort focused on using AR in a non-constrained, open environment; we sought to create the simplest possible user interaction, mimicking the existing process as closely as possible. Digital Graffiti markers are meant to build situational awareness of an environment over a period of time across many users as they move through and interact with the environment.

When persisted, Digital Graffiti markers will be available for all those who pass by their locations in the future. Trends in information can be observed over time. Information contained in virtual markers are not at risk of being lost if the environment is altered due to natural or human forces. Virtual markers also allow information to be quickly and easily transferred to remote teammates. They enable disaster relief workers who are coordinating several teammates over a large area to quickly view and assimilate information because the information is natively electronic and does not have to be accepted and entered manually.

3 USER INTERACTION

The Digital Graffiti prototype is designed to support three use cases: creating a new marker, finding and viewing an existing marker, and editing or updating an existing marker.

Based on previous research in AR display design, displays must balance the user's need to maintain situation awareness with the need to focus on specific locations to create meaningful markers. We selected static placement of several heads-up display (HUD) components to ensure this balance (Figure 2). The HUD displays a two dimensional overhead map in the lower left corner that includes markers and icons used to label items in the user's vicinity and visual representations of the user's orientation and field of view.

The user's absolute position is shown at the top center of the display. Status icons in the upper right corner include GPS, network, and battery status. Although not continually used, the lower right portion of the display is reserved to display dynamic commands and alerts. These components were positioned to avoid cluttering the primary viewing area in the center of the display, and show only a select set of crucial information.

To minimize distracting emergency relief workers from their other tasks, we needed a simple yet powerful interaction paradigm for creating and editing virtual markers.

Figure 2 Digital Graffiti HUD Static Components

We explored a multimodal interaction paradigm that combined spoken language and gestural inputs. Glove input required only a few simple actions rather than complex arm and hand movements. The spoken language input design was based on a simple grammar designed to provide control of the system. For this prototype, the gesture input was sufficient for user input so spoken language was only used for system output to the user.

Our Digital Graffiti on-screen menu presentation and navigation design includes a four-component pie menu (north, east, south, west) located in the middle of the display (Figure 3). Each quadrant of the menu is displayed as an icon and selection alternatives include no more than three layers of options into the menu. A "back" option is included on most menus.

The selected menu item is highlighted with a green square. Rotating or flicking the wrist up, down, left, or right selects another item on the display. Item selection is entered through a button press on a wrist-mounted keyboard. A second button press is available to cancel or clear the menu at any time.

Figure 3 Digital Graffiti First-level Pie Menu

3.1 Creating a Marker

The on-display pie menu allows the user to create a marker through a single keyboard button press. The user selects the type of marker to create and navigates through the marker creation menus to add metadata about the marker. Guided wizard assists the user in marking observations of locations, vehicles, people, or events. These menus were focused on the selected item and often included more complex items; for example, the vehicle color selector included six options.

The user may add additional information to any marker. Templated shape creation and freeform drawing are available for all marker types. Before the marker is stored, the system takes a still image from the video with the marker to record the environment when the marker was created. This provides additional information to future users about the situation when the marker was created. The completed marker is stored in the system and transmitted to the server for future access.

3.2 Viewing a Marker

Viewing a marker is simple: Iconified markers are presented in the overhead map as well as in the central part of the display. A multi-step filtering process prevents cluttering of icons in both sections of the display and provides alerts for markers of interest. The processing component adds the icons to the video from the camera that is displayed back to the user through the goggles.

3.3 Editing a Marker

Editing a marker combines the viewing of a marker with a direct interaction. The user approaches a marker and selects it using the selection button input. The resulting pie menu is customized based on the type of marker selected. The user can view information associated with the marker such as photos or maps and update information as needed. During all of these actions, or anytime a user interacts with a marker, the metadata about the interaction is automatically captured by the system. User information, time, location, and change summary are all persisted with the marker so the lifecycle of the marker can be understood.

3.4 Usage Scenario

We developed a full mobile scenario of a search operation to exercise all of these actions using the prototype system. In the scenario, the user conducts a directed search to find a missing person. Commands received through the AR display and spoken language interface initiate a search for a particular person at a last known location. The user goes to the location of the last-known location marker, browses through images of the person and the location using the Digital Graffiti system, and updates the marker to denote the person was not at that location at that time. The user also adds a marker to indicate there is a vehicle that appears to have been abandoned in the location, following on-screen instructions to enter model, color, and location. The user centers the video on the vehicle and the system records an image of the vehicle. Next, the user receives an update to search a second location and follows directions to the marker location that are provided by Digital Graffiti. The user identifies the missing person at that location and creates a new marker to denote where the person was found.

4 WEARABLE PROTOTYPE

The Digital Graffiti wearable prototype was built in parallel with developing the user interaction to provide a platform for evaluating individual concepts and technology components.

The AR display was presented to the user via prototype Wrap™ 920AR video see-through AR eyewear from Vuzix Corporation. The processing was performed on a Dell D820 laptop running the Digital Graffiti software. An external compass was added to the AR goggles to provide head orientation tracking information. An updated commercial version of the goggles now includes an integrated compass.

On-display system navigation and information manipulation was performed by the user with a single glove. The glove included a single tilt compensated compass located on the back of the hand. The movement, specifically the tilt of the hand, was measured to allow the user to navigate through the menus of the system. Use of two buttons from a wrist-mounted keyboard enabled final selection or canceling while navigating the menus.

The system included a commercial GPS, the Garmin GPSMAP 78 with an external antenna. When the user creates a marker, the system combines the GPS location and head tracking sensor information to place the marker at a point in space adjacent to the user's position. As the user views the marker, the system constantly processes the GPS and head tracking sensor information to decide what should be presented into the user's display.

The components were packaged in a backpack (Figure 4). Most components connect to the laptop through USB so a separate USB hub was included in the

Figure 4 Digital Graffiti Wearable Prototype

backpack. The goal was to provide a dynamic platform for testing the interaction designs, not to meet ruggedness requirements.

5 DESIGN CHALLENGES AND SOLUTIONS

A primary research goal was to develop an interaction paradigm for a mobile AR based system. Desktop interaction paradigms do not transfer well to smaller platforms, especially when full user attention is not available. This made early development of a working prototype platform critical to project success because interaction solutions that seemed powerful on a development desktop or laptop were found clunky and useless on a mobile system.

5.1 Marker Creation

The primary interaction challenge for Digital Graffiti was user interaction with geo-rectified markers. We explored several options for marker creation. The first, based on the "graffiti" analogy, used free-form drawing with the hand: tracking of a hand-mounted AR tag would follow the movement the user's hand as they "drew" on an object in front of them. This solution worked well in development and in specific test cases on the mobile prototype but was difficult to use for general control and in open space because it required users to block their own face with their hands, obstructing the view of the environment.

The second solution was to track an accelerometer mounted to the back of the hand as the user drew. This provided "hands-down" input, but integrating the sensor data to create a drawing proved difficult. The resulting image was often too different from the intended image to be acceptable.

We reevaluated the requirements of the marking process and devised a simpler gesture control for marker creation. Templated drawing enabled the user to create any manner of complex drawings by choosing from a number of shapes and rotating their hand left, right, up, and down to increase or decrease the boundaries of the shape. Multiple shapes could be placed on the same marker, providing an easy yet powerful drawing capability. We created guided interactions for creating specific markers such as locations, vehicles, people, or events; these "marker wizards" required a simple interaction to navigate through and select menu items.

This simpler "gesture grammar" reduced the requirements for the glove gesture input and the complexity of the system while increasing recognition accuracy and decreasing training time. We implemented the new gesture grammar with a simple glove-mounted tilt sensor that needed to register only four unique movements.

5.2 Speech and Gesture

Digital Graffiti is designed for use in a highly mobile scenario. To maximize mobility we explored using spoken language interaction where appropriate. For example, system-generated spoken language relays large, wordy pieces of information such as text or instructions so users could keep their eyes on their surroundings in a challenging search and rescue environment. We also developed a simple spoken language grammar for control of the system (e.g., toggling on and off components of the display, increasing and decreasing the brightness, and cycling through several levels of "declutter" of the screen).

We designed a multimodal input system that incorporated glove gesture input and spoken language grammar. For example, the user could specify control of the "decluttering" mechanism and cycles through options using gesture rather than repeating the vocal command. After evaluating the multimodal input system and the spoken language interaction design, we determined that the simple gesture interaction paradigm was sufficient for the prototype system.

5.3 Menu Presentation

The challenge of menu design was to keep the center of the field of view clear while presenting options when necessary. We explored bottom and side bar menus, but eventually settled on the pie menu layout to enable focus on options when the user requested it and complete clearing of menus when not in use. We limited the depth of the menu structure because of the difficulty of presenting nested menus in the simple display environment.

5.4 Remote Icon Placement

To address the challenge of remote icon placement we investigated the use of a laser range finder but determined that it was too complicated for the streamlined CONOPS and would have created an undesirable system dependency. To avoid this complexity we implemented a unique simple trigonometric solution for placement of a marker within ten feet of the user. To place the icon using this process, the user points at the base of the marker location (e.g., the bottom of a vehicle) and the system calculates the distance from the user's location using the tilt sensor on the hand.

6 FUTURE WORK

During the Digital Graffiti effort, we identified several areas for future work. We developed our Digital Graffiti metrics knowing the prototype would use low-cost, COTS hardware such as the compass and GPS. A production system will require research into how to meet stricter requirements for accuracy and update frequency of these sensors. As a step towards this goal, we developed several concepts for meeting these requirements without traditional sensors to minimize price and system complexity.

Research is also needed to most effectively integrate functionality into one easy-to-use piece of hardware. For prototype development, we met several user input requirements with various pieces of hardware based on availability. For example, we used a wrist-mounted keyboard for input of only two buttons and a full glove for one small sensor. Mobile user input design for a production system would need to streamline input mechanisms.

Research into other applications should include expansion of capabilities enabled by the underlying Digital Graffiti technology. For example, our camera-based augmented reality system is able to respond to markings or tags in the environment. AR Tags could be used as placeholders to cue the display of dynamic information such as a virtual "you are here" a map or a virtual clock. Using the AR display, the system could present a synchronized timer to help remote teams coordinate across distances without radios. And given that the prototype system is capable of tracking hand movements, the system could allow gesture-based manipulation of on-display icons such as image browsing. This feature would include the ability to use hand gestures to select and drag images shown on the heads-up display.

We also identified several potential domains for an AR marking system, such as trauma injury site assessment and triage. In a triage assessment, users need to share geolocated, human-generated information in real-time. Digital Graffiti could enable heads-up operation while providing medical care and allow new teams arriving at the site to quickly come up to speed with a glance over the area without repeating the assessment or distracting someone from their current activity.

REFERENCES

Stephan Lukosch, Ronald Poelman, Oytun Akman and Pieter Jonker, 2012. "A Novel Gesture-based Interface for Crime Scene Investigation in Mediated Reality," ACM Conference on Computer Supported Cooperative Work Workshop on Exploring Collaboration in Challenging Environments: From the car to the factory and beyond, Seattle, Washington.

Steven Henderson and Steven Feiner, 2010. "Opportunistic Tangible User Interfaces for Augmented Reality," *IEEE Transactions on Visualization and Computer Graphics*, January/February, vol. 16, no. 1, pp. 4-16.

P. Mistry and P. Maes, 2009. Sixth Sense—A Wearable Gestural Interface. In the Proceedings of SIGGRAPH Asia, Sketch. Yokohama, Japan.

U.S. Federal Emergency Management Agency, 2003. *National Urban Search and Rescue (US&R) Response System Field Operations Guide*. Retrieved from http://www.uscg.mil/hq/cg5/cg534/nsarc/Nat.UrbanSAR_fog25sept2003.pdf

CHAPTER 53

Brain Activation Pattern during Passive Exercise With and Without Active Exercise

*Takahiro Hitomi**, *Hiroshi Hagiwara***

*Graduate School of Science and Engineering
**College of Information Science and Engineering
Ritsumeikan University, Shiga, Japan
ci012071@ed.ritsumei.ac.jp

ABSTRACT

Improving the quality of live (QOL) of elderly individuals is becoming more important for aging societies worldwide. For elderly individuals, who tend to be less physically active, improving both motor and cognitive functions is necessary to improve their QOL. The periodic passive-sway exercise, based on traditional horse-riding therapy in Europe, has recently gained attention for improving physical activity among the elderly because it is a lower load than aerobic exercises like walking, jogging, and so forth. Studies using an exercise machine that simulates horse-riding (JOBA, Panasonic) reported that motor function improved and that adding a cognitive task simultaneously with riding the JOBA also improved brain activation. Moreover, periodic passive-sway exercise is advantageous for elderly people because the seated position is safe and generates low loads on the legs, waist, and hips. In the present study, we measured the brain activation patterns during periodic passive-sway exercise using a newly developed exercise machine. Specifically, brain activity in the frontal cortex was measured using multichannel near infrared spectroscopy (NIRS) during periodic passive-sway exercise (treadle exercise) and jogging with the arms (jarming) using the new exercise machine. The brain activities of 12 healthy subjects were measured using NIRS for a 16-min period, comprising 1 min of rest, 14 min of exercise and 1 min of rest. During the 14-min exercise period, 1-min sets of treadle exercise and treadle exercise with jarming were repeated seven times. All subjects gave informed consent prior to participation. NIRS probes were placed to focus measurements on the activation in the frontal association and motor areas located in the frontal lobe. The oxygenated hemoglobin (oxy-Hb) density in the frontal association area drastically decreased immediately after the start of exercise, but then gradually recovered. The oxy-Hb density increased during treadle exercise, but did not increase during treadle exercise with jarming. In contrast, the oxy-Hb density in the motor area increased immediately after the start of exercise and then gradually decreased. No clear differences in motor area activation between treadle exercise with and without jarming were detected. The hemodynamic changes detected in the brain suggest that passive-sway exercise using the newly developed exercise machine activates the frontal association area. This effect could help improve cognitive function, because the frontal association area is strongly associated with cognitive function. Furthermore, a decrease in oxy-Hb density is likely related to increases in the stress from the exercise, since oxy-Hb density decreased immediately after the start of exercise and was continually decreased throughout treadle exercise with jarming.

Keywords: exercise, f-NIRS, brain hemodynamics

1 INTRODUCTION

In recent years, developed countries have seen a rapid increase in the aging of society. In an aging society, improving the quality of life (QOL) of elderly individuals is important in order to reduce the burden on society, such as medical insurance and pensions. Ideally, as many elderly individuals as possible should live independently. Maintaining physical function is indispensable for improving the QOL of elderly people. In daily life, many situations require motility function and cognitive function, such as walking to avoid people and structures in a crowded area. Therefore, not only healthy body condition but also healthy brain function is essential for a satisfying life without unnecessary difficulty, and maintaining brain function is thus important in improving the QOL of elderly people.

In a previous study of a 6-month exercise intervention in 124 healthy elderly subjects (65-75 years old), those who received aerobic training (walking) showed substantial improvements in performance on tasks requiring executive control compared with anaerobically trained (stretching and toning) subjects. This result suggested that aerobic exercise could have an inhibitory effect on decline in cognitive function. The periodic passive-sway exercise, based on traditional horse-riding therapy in Europe, has recently gained attention based on studies using an exercise machine that simulates horseback riding (JOBA, Panasonic). These studies reported that motor function improved and that adding a cognitive task simultaneously with riding the JOBA machine also improved brain activation. This exercise is a lower muscle load than aerobic exercises like walking or jogging, and it can be conducted safely in small spaces. Periodic passive-sway exercise is advantageous for elderly people because the seated position is safe and generates low loads on the legs, waist and hips. In the present study, we defined periodic passive-sway exercise as a swaying motion that is generated externally, like riding a horse. It displaces the center of gravity, automatically causing muscle activity for postural maintenance. We develop a new periodic passive-sway exercise machine and report our findings from investigating the blood flow in the brain during exercise. We clarify the brain hemodynamic effects of this exercise. Additionally, we investigate the effects on brain hemodynamics when other exercise is incorporated.

2 EQUIPMENT

2.1 Exercise machine

For this experiment, we developed a new periodic passive-sway exercise machine. The machine includes a seat, pedals, and sticks. A motor causes the seat to sway from left to right. There are springs under the pedals. The sticks are movable backward and forward. The swaying motion of the seat displaces the subject's center of gravity, and therefore the subject automatically depresses the pedals for postural maintenance. This sequence of actions is called "treadle" exercise and it is the base exercise induced by this new machine. As an additional option, this machine can combine the treadle exercise with a "jarming exercise" using sticks. Jarming is an active exercise of the arms and upper body using movable sticks, and it is synchronized with treadle exercise.

The previously used JOBA exercise, similar to horseback riding, requires postural maintenance corresponding to the sway from an external force. For treadle exercise using the newly developed machine, the swaying motion of the seat induces the subject to depress the pedals for postural maintenance. Therefore, this exercise focuses on human walking, and the subject can conduct movement of the entire foot (treadle exercise), as well as using the arms and upper body (jarming exercise).

Because this exercise focuses on human walking, it may have potential for application to rehabilitation for elderly and disabled people who have difficulty walking.

2.2 f-NIRS

Near-infrared spectroscopy (NIRS), using near-infrared light corresponding to the light region with a high transmittance in vivo (wavelength range of 780 ~ 900 nm; called the window of living body), is a device based on the modified Lambert-Beer law to measure the relative changes in hemoglobin concentration in vivo. There are great advantages compared to other brain function measurement machine such as functional magnetic resonance imaging (f-MRI) and single photon emission tomography (SPECT). In addition to lower operating cost and high safety level for the human body, there is a high degree of freedom in the measurement conditions. If the light source and light detector surface can be stabilized, it is possible to measure brain hemodynamic during an actual motion task, as in our study.

In our study, we used a NIRStation (OMM-3000, Shimadzu Corporation, Kyoto, Japan) to measure brain hemodynamic during exercise using our new machine. Indicators measured by NIRS are change in oxygenated hemoglobin (oxy-Hb) concentration, change in deoxygenated hemoglobin (deoxy-Hb) concentration, and change in total hemoglobin (total-HB) concentration. Measurement sites with NIRS are shown in Figure 1; we focused on the frontal lobe in this study. We analyzed the 19ch (left frontal association cortex), which is involved in cognition and judgment, and 22ch (primary motor cortex), which is located caudal to the frontal lobe and is related to locomotion.

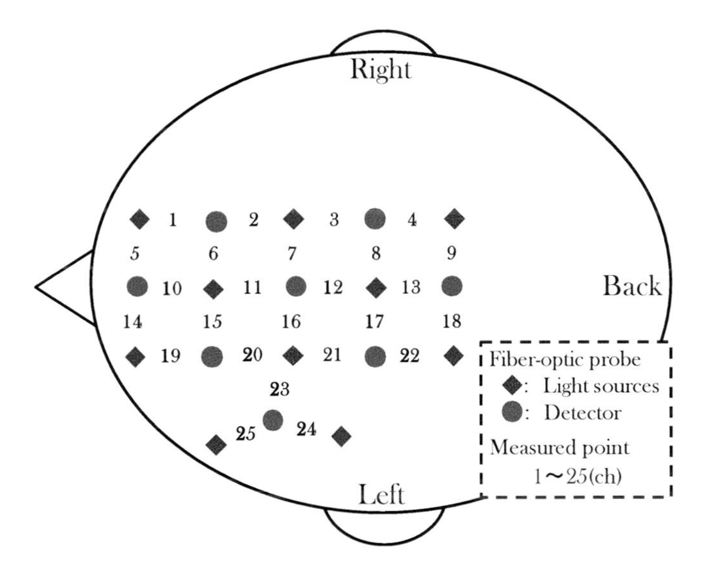

Figure 1 Schematic for location of the fiber-optic probes and measured points. Diamonds indicate light sources and circles indicate detectors. 1~25(ch) are indicated measured points. We selected 19ch (frontal association cortex related to cognition and judgment) and 22ch (primary motor cortex related to locomotion) for analysis.

3 METHOD

The subject was seated on the machine wearing a head-mounted f-NIRS probe. We measured brain hemodynamic during exercise. The periodic passive-sway exercise was initiated following 1 minute of rest with the eyes open. The exercise task lasted 14 minutes, followed by 1 minute of rest, for a total measurement time of 16 minutes. Details of the experimental protocol are shown in Figure 2. The 14 minutes of exercise consisted of 7 repeats of a set of 1 minute treadle exercise and 1 minute treadle exercise plus jarming. Instructions to switch movement were given verbally. The swaying speed of the seat was kept constant at a level 5 which is chosen mid-range level from 1-9.

Subjects were 12 healthy university students (male, 10; female, 2; age range, 21-23 years). All subjects gave informed consent prior to participation, and were fully accustomed to the periodic passive-sway exercise for the experiment. In order to minimize the impact of noise caused by the shaking of the subject's head during measurement, subjects were instructed to unify the line of sight by looking at a fixation point positioned in front of them and to keep their head as still as possible. In addition, subjects all wore short-sleeved T-shirts and shorts and were barefoot. The experimental environment was set at a constant temperature of 24°C in the laboratory.

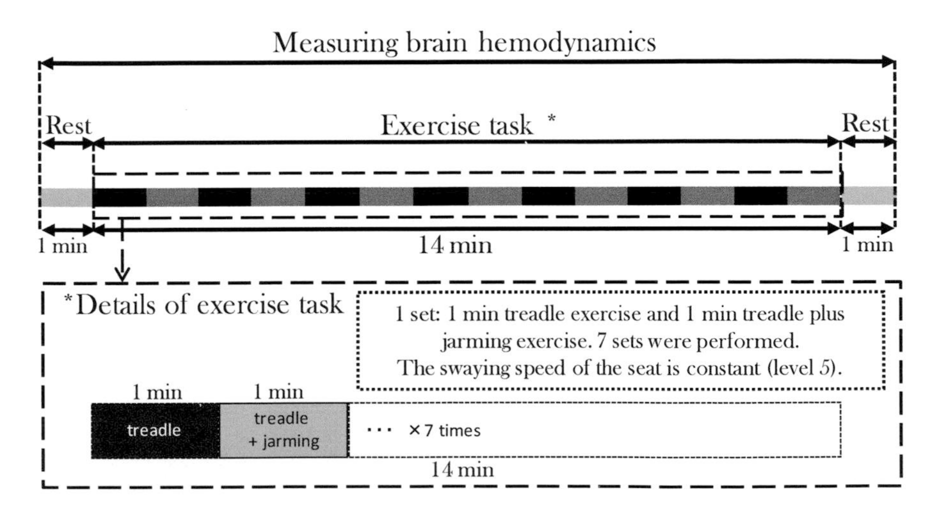

Figure 2 Experimental protocol. *Details of exercise task: 7 repeats of a set of 1 min treadle exercise and 1 min treadle exercise plus jarming. The speed of the swaying seat is constant (level 5).

4 ANALYSIS

Previous experiments using PET have shown that regional cerebral blood flow rises 50% during neural activity, and that the rate of oxygen consumption rises an additional 5%. In other words, more oxygen needed by brain tissue is fed to the site when neural activity increases. Therefore, it is believed that deoxy-Hb concentration decreases with increasing concentration of oxy-Hb in the capillaries. In addition, animal experiments have shown that changes in oxy-Hb correlated best with changes in regional cerebral blood flow. In this study, we selected the more pronounced oxy-Hb parameters for analysis among the three indicators obtained by measuring NIRS (change in oxy-Hb, deoxy-Hb and total-Hb concentration). The data were processed with the following steps: raw waveform measurements obtained NIRS were filtered with a 0.1 Hz low-pass filter to remove high-frequency components and then normalized to each subject. We determined noise contamination by visual examination and excluded data that had excessive noise.

5 RESULT

Figure 3 shows two regional averages of changes in oxy-Hb density during rest before exercise, during treadle exercise, during treadle exercise plus jarming, and during rest after exercise. Oxy-Hb density in the frontal association cortex (19ch) during exercise was reduced as compared to pre-exercise rest. In contrast, oxy-Hb density in primary motor cortex (22ch) was increased during exercise as compared to the baseline resting state. In both channels, adding the jarming exercise, which uses the arms and upper body, decreased changes in oxy-Hb density.

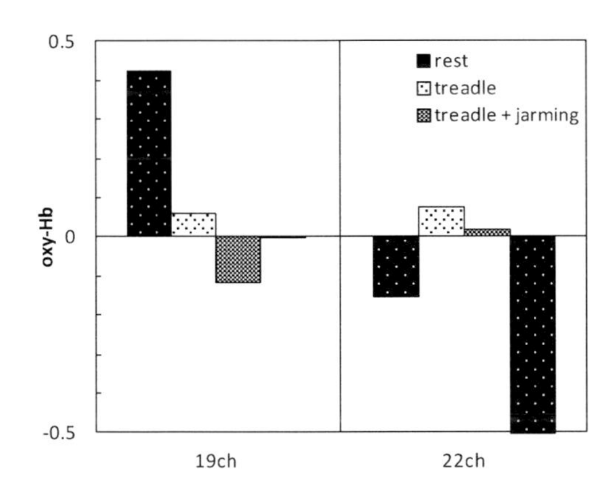

Figure 3 Average changes in oxy-Hb density in 4 conditions: resting before exercise, treadle exercise, treadle exercise plus jarming, and resting after exercise. 19ch: left frontal association cortex; 22ch: left primary motor cortex.

Figure 4 shows the time course of oxy-Hb density changes using regional 1-minute averages in each condition. This demonstrates that oxy-Hb density in the frontal association cortex (19ch) drastically decreased immediately after the start of exercise, but then gradually recovered. The oxy-Hb density in primary motor cortex (22ch) drastically increased immediately after the start of exercise but then gradually decreased, in apparent opposition to 19ch. However, adding jarming to the treadle exercise affects both channels similarly by causing oxy-Hb density to decrease.

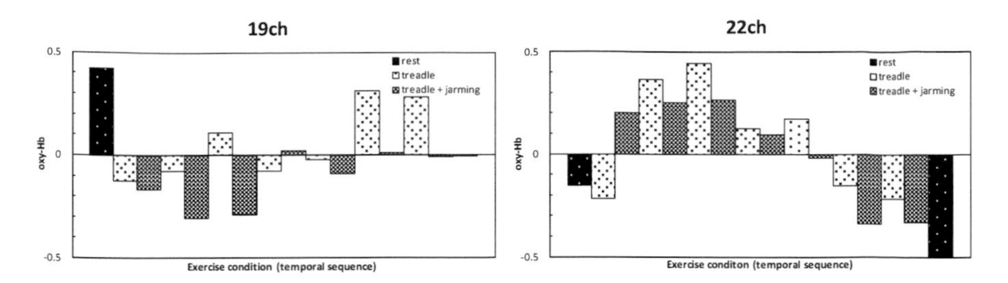

Figure 4 One-minute averages of oxy-Hb density for each condition. 19ch: left frontal association cortex; 22ch: left primary motor cortex.

In Figures 5 and 6, the average oxy-Hb density waveforms from all subjects are shown. These graphs clearly show the oxy-Hb density behavior in relation to the temporal sequence. In the frontal association cortex, oxy-Hb density decreased when jarming was added during treadle exercise. These changes were most apparent in the latter half of the exercise task.

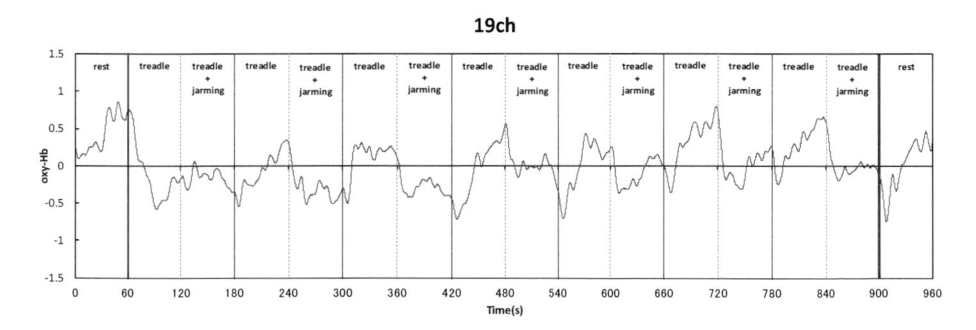

Figure 5 Average oxy-Hb waveform across all subjects in 19ch (left frontal association cortex).

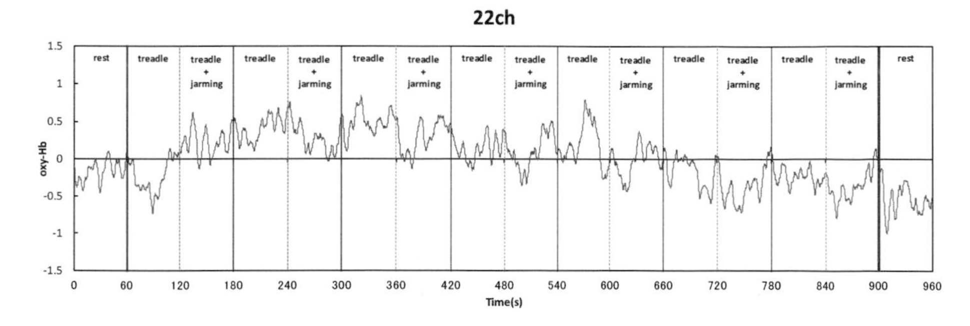

Figure 6 Average oxy-Hb waveform across all subjects in 22ch (left primary motor cortex).

The average waveforms for each exercise condition are shown in Figures 7 and 8. In the frontal association cortex, the different trend between the behaviors of oxy-Hb density change is evident as a function of exercise condition, as mentioned above. In the frontal association cortex, both exercise states reduced oxy-Hb density at the beginning of the tasks. Oxy-Hb density increased by degrees during treadle exercise along, but it did not increase during treadle exercise plus jarming. In addition, there was no difference in oxy-Hb density behavior between treadle exercise and treadle exercise plus jarming in the primary motor cortex.

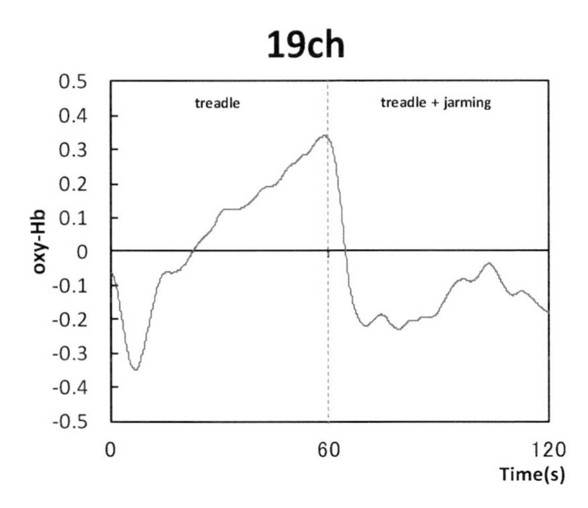

Figure 7 Average oxy-Hb waveform for each exercise condition in 19ch (left frontal association cortex).

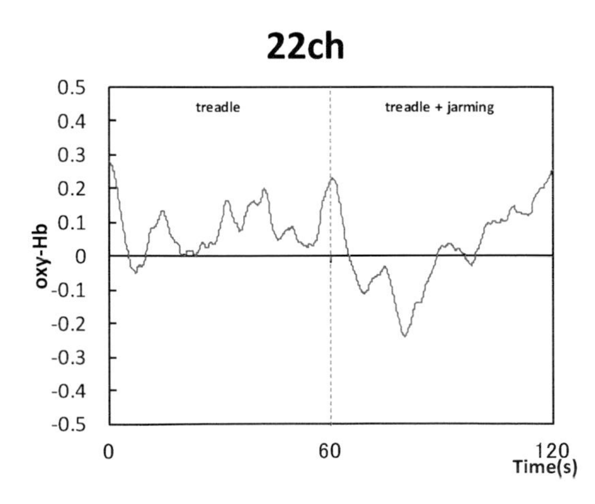

Figure 8 Average oxy-Hb waveform for each exercise condition in 22ch (left primary motor cortex).

6 DISCUSSION

In a previous f-NIRS study using the JOBA exercise format, it was reported that a decrease in oxy-Hb density was found in the frontal association area soon after the start of exercise, followed by an increase in density by degrees. In our study, we confirmed a decrease in oxy-Hb density in the frontal association area during exercise. Notably, a substantial decrease was seen soon after the start of exercise. Additionally, an increase in density was confirmed by degrees. These results agree precisely with the previous study. In the frontal association cortex, the trend of decreasing oxy-Hb density was confirmed during treadle exercise plus jarming compared with treadle only exercise. When the exercise task begins, stress is increasing. The exercise stress during the treadle exercise with jarming represents a greater increase than treadle exercise alone because it includes locomotion of the arms and upper body. Therefore, we suggest that a decrease in oxy-Hb density is related to an increase in exercise stress, probably for locomotion controlled by locomotor regions such as the cerebellum, subthalamic nuclei, and the brainstem. These regions are located deep in the brain and thus cannot be measured by NIRS.

In addition, both exercise states showed reductions in oxy-Hb density at the beginning of the condition (Figure 7). However, oxy-Hb density increased during treadle exercise alone, and did not increase during treadle exercise plus jarming. This suggests that passive exercise increases oxy-Hb density, but adding active exercise inhibits this increase.

For the motor cortex, oxy-Hb density increased during exercise compared to the resting state; however, a decrease in oxy-Hb density during treadle exercise plus jarming compared with treadle exercise alone was confirmed. Because our measurement point of the primary motor cortex was located at the hip representation in the brain map of the motor cortex, the results might have been different if our measurement point covered the arm representation instead.

7 CONCLUSION

Given the results of this study, we suggest that the decrease in oxy-Hb density in the frontal lobe reflects the increased stress of the exercise. In addition, we found that stimulation with periodic passive-sway exercise causes an increase in oxy-Hb density in frontal association cortex. Therefore, we can infer that periodic passive-sway exercise has an effect on frontal association cortex activation.

ACKNOWLEDGMENTS

In this study, received a science research grant (Grant-in-Aid for Scientific Research (C), No. 22500415) from the independent administrative institution, the Japan Society for the Promotion of Science.

REFERENCES

Fox PT, Raichle ME. 1986, Focal physiological uncoupling of cerebral blood flow and oxidative metabolism during somatosensory stimulation in human subjects, *Proc Natl Acad Sci USA 83*: 1140-1144.

Fukao A, Nakano T, Imai B, Michimori A, Hagiwara H. 2009, Quantitative assessment in physiological effects of horseback-riding simulator using near-infrared spectroscopy, *Proceedings of the 24th Symposium on Biological and Physiological Engineering*: 137-140.

Hoshi Y, Kobayashi Y, Tanura M. 2001, Interpretation of near-infrared spectroscopy signals: A study with a newly developed perfused rat brain model, *J Appl Physiol_90*: 1657-1662.

Kramer F, Hahn S, Cohen N. J, Banich M.T, McAuley E, Harrison CR, Chason J, Vakil E, Bardell L, Bouileau RA, Colcombe A. 1999, Aging, fitness and neurocognitive function, *Nature 400*: 418-419.

Kramer F. 2006, Exercise, cognition, and the aging brain, *J. Appl. Physiol 101*: 1237-1242.

Nakano T, Mihara I, Inui K, Imai B, Michimori A, Hagiwara H. 2010, Brain activation and improvement of cognitive performance by exercise on horse-back riding simulator, *Panasonic Technical Journal 56 (1)*: 11-18.

CHAPTER 54

Spine Compression Characteristics Associated with Back Pain Symptoms of Healthcare Workers

Uwe Reischl[*], *Ravindra S. Goonetilleke*[**], *Christine Swoboda*, [***]

[*]Boise State University, Boise, Idaho, USA
ureischl@boisestate.edu
[**]Hong Kong University of Science and Technology, HONG KONG
[***] West Valley Medical Center, Caldwell, Idaho, USA

ABSTRACT

8-hour work induced spine compression and subsequent recovery in response to a 15-minute rest period was studied. Eighteen employees at a regional medical center participated in the study. Seven of the subjects reported experiencing chronic back pain during the previous six months while eleven subjects reported being symptom free. Eight-hour work induced spine compression observed for the symptomatic employees was found to be significantly higher than the spine compression observed for the asymptomatic employees. The results revealed that spine compression and recovery characteristics can be used to differentiate chronic back pain employees from asymptomatic employees. A diagnostic model was created yielding a positive predictive value of 75% and a negative predictive value of 90%. Use of such a method to identify employees potentially at increased risk of work related chronic back pain is suggested.

Keywords: Back pain diagnostics, work induced stature changes, spine compression and recovery

1 INTRODUCTION

Back pain disorders are a leading cause of lost workdays in the industrialized world and represent one of the most costly occupational health problems today. Reducing the risk of back-pain caused or exacerbated by conditions in the workplace continues to be an important service provided by ergonomists and human factors engineers worldwide. Risk of work related back disorders are known to be associated with physical work, personal health factors, and the psychosocial dynamics of a workplace. However, the exact source of back pain is often difficult to identify. Persistent back pain is likely to be caused by structural damage or degeneration in one or more motion segments of the spine. It is believed that a number of factors can contribute to structural damage of the spine. These include constant forward bending, twisting, and repetitive heavy lifting, carrying and pushing objects, asymmetric trunk loading, and inappropriate posture during the day (Eklund & Corlett, 1984; Garbutt, Boocock & Reilly, 1990; Beynon & Reilly, 2001). It has been suggested that back disorders might be viewed as a progression of events in which repetitive and constant loading applied on the spinal column represents the beginning stage of a degeneration process that ultimately results in back pain (Marras et al., 2001).

The purpose of this study was to measure stature compression in employees performing routine 8-hour work tasks and to determine whether the measured changes in stature height could be associated with chronic back pain. A comparison between employees with chronic back pain and employees without symptoms was undertaken to determine whether or not differences in stature compression between the two groups could be observed and whether such differences in stature compression could be used to identify a person's risk of chronic back pain.

It is known that the spine and its associated muscles and ligaments promote the basic mechanical function of load-bearing, mobility, control, and housing of the nervous system. The compressive loading on the spine is primarily the result of trunk muscle forces, the antagonistic muscle forces, and the external load forces. The association between back pain, external workload, and vertebral disc compression has been reported in the literature. Changes in body stature as a measure of disc compression for various daily activities have been found to be a predictor of back pain when the spine is loaded (Flor & Turk, 1989; Flor, Birbaumer & Turk, 1990).

2 METHODS AND PROCEDURES

A precision stadiometer system described previously (Reischl & Weinsheimer, 1995; Reischl & Weinsheimer, 2004) was used in this study to measure changes in the stature of employees during 8-hour work shifts. Measurements were carried out in the standing position while the subjects leaned against a vertical reference frame equipped with a precision scale oriented towards a video camera. A landmark "dot" placed on the skin at the vertebra prominens (C7) was used as the measurement reference point. This landmark is stable in response to any movement of the head. To control a test subject's posture, four points of contact with the reference frame were maintained: Contact of the occiput with an adjustable head-rest; contact of the mid-thoracic vertebra and the sacrum with the vertical frame; contact of the calcaneus with a heel reference block. Moving the video camera up or down, the image of the landmark reference "dot" on the skin of C7 could be centered on the crosshairs which were also projected onto the screen of the monitor. The position (i.e., height) of the reference "dot" was superimposed onto the precision reference scale. A measurement resolution of 0.01cm was achieved. Each stature measurements could be made in 30 seconds.

Eighteen persons employed at a regional medical center participated in this study as volunteer subjects. These subjects were recruited through an inter-departmental memorandum describing the goal and scope of the study. Seven of the volunteer subjects who were selected for this study had previously reported experiencing chronic back pain. Eleven of the subjects reported being symptom free. The symptomatic or asymptomatic status was determined based on a standardized medical questionnaire, a standardized pain questionnaire, and a physical exam conducted by the occupational health nurse. All subjects were healthy, i.e., free of any disability or disease. The characteristics of the "pain-free" group are summarized in Table 1. The characteristics of the back-pain group are summarized in Table 2. The two groups were similar in height, age, and weight where the average age of the symptom free group was 48.4 years, the average height was 67.1 inches, and the average weight was 179.1 lbs. while the average age of the "back-pain" group was 47.5 years, average height was 66.0 inches, and the average weight was 165.7 lbs. There were eight females and three males in the "pain free" group. There were eight females and no males in the "back pain" group. Both groups included a combination of administrators, nurses and technicians.

TABLE 1: Characteristics of participants reporting no back pain symptoms

ID Number	Pain (Yes / No)	Gender (M / F)	Age (years)	Height (in)	Weight (lbs)	Job Classification
207	No	F	30	63	120	Admin
210	No	F	52	68	194	Admin
216	No	F	31	65	197	Admin
217	No	M	52	68	225	Tech
218	No	M	54	72	230	Tech
219	No	M	64	69	150	Admin
221	No	F	55	63	155	Nurse
222	No	F	35	63	98	Nurse
223	No	F	46	62	148	Admin
224	No	F	48	67	168	Nurse
232	No	F	56	66	138	Nurse
Average			47.5	66.0	165.7	

TABLE 2: Characteristics of subjects reporting back pain symptoms

ID Number	Pain (Yes / No)	Gender (M / F)	Age (Years)	Height (in)	Weight (lbs)	Job Classification
203	Yes	M	48	70	183	Admin
205	Yes	F	50	67	150	Admin
211	Yes	M	52	72	245	Tech
212	Yes	M	58	66	180	Pharm
225	Yes	F	49	64	140	Admin
226	Yes	F	53	66	170	Nurse
227	Yes	F	29	65	186	Admin
228	Yes	M	52	71	231	Tech
Average			43.2	67.6	185.6	

Each subject was measured three times during an 8-hour work period. The first measurement was obtained prior to shift begin (7:30 am), the second measurement was taken at the end of the work shift (4:30 pm), and the third measurement was taken immediately after completing a 15-minute rest period after completing the second stature measurement.

3 RESULTS AND ANALYSIS

Table 3 summarizes the stature compression and subsequent 15-minute recovery values obtained for the eleven pain-free volunteer subjects. Table 4 summarizes the stature compression and subsequent 15-minute stature recovery values obtained for the seven back-pain volunteer subjects.

The average 8-hour stature compression value for the eleven pain-free subjects was 0.34 centimeter while the average15-minute recovery value was 0.44 centimeter. This 15-minute recovery value represents a recovery level of 129%. The average 8-hour stature compression value for the seven back-pain subjects was 0.80 centimeter while the average 15-minute recovery value was 0.49 centimeter. The 15-minute recovery for this group, therefore, represents a recovery level of only 61%.

The inter-vertebral disc appears to be the most likely tissue to be undergoing sufficient height change to produce the stature changes observed in this study. Whether this occurs because of anular fiber strain, volume change, i.e. fluid loss, or a combination of both, is not known. The overall range of stature compression measured for the eighteen volunteer subjects are consistent with values reported in previous studies of athletes and nurses (Beynon & Reilly, 2001; Garbutt, Boocock, & Reilly, 1990). The significantly greater stature compression observed for the subjects who suffer back pain, in comparison to the asymptomatic subjects, suggests that additional spinal loading occurred during the work shift. Since relevant variables such as gender, age, height, weight, work load, work rate, and work environment were similar for both groups; it can be assumed that additional spinal loading may be intrinsic to back pain. Persons with back pain may guard against additional pain by using more muscles than they need to (Marras et al., 2001). Also, the increased spine compression may also be explained by the increased static muscle tension that is often associated with the chronic back pain syndrome (Flor & Turk, 1989; Flor, Birbaumer & Turk, 1990). It is important to note that spine compression and recovery is an on-going, non-linear, process that occurs over a 24-hour period. Therefore, obtaining the stature measurements for all of the subjects at the same time period during the day, i.e., 7:30 am, 4:30 pm, and 4:45 pm, was able to control for this non-linear diurnal pattern.

TABLE 3: 8-hour spine compression and associated 15-minute recovery values for subjects reporting no back pain symptoms

ID Number	Pain Yes / No	8-hour Compression (cm)	15-minute Recovery (cm)	100% Recovery (+ / -)
207	No	0.45	0.49	+
210	No	0.32	0.40	+
216	No	0.36	0.30	-
217	No	0.28	0.31	+
218	No	0.22	0.55	+
219	No	0.41	0.45	+
221	No	0.50	0.46	-
222	No	0.17	0.47	+
223	No	0.60	0.65	+
224	No	0.38	0.50	+
232	No	0.06	0.21	+
Average		0.34 (cm)	0.44 (cm)	129 (%)

TABLE 4: 8-hour spine compression and associated 15-minute recovery values for subjects reporting back pain symptoms

ID Number	Pain Yes / No	8-hours Compression (cm)	15-minute Recovery (cm)	100% Recovery (+ / -)
203	Yes	0.62	0.41	-
205	Yes	0.60	0.58	-
211	Yes	1.61	0.34	-
212	Yes	0.28	0.62	+
225	Yes	0.90	0.69	-
226	Yes	0.94	0.58	-
227	Yes	0.73	0.20	-
Average		0.80 (cm)	0.49 (cm)	61 (%)

As shown in Tables 3 and 4, the average spine compression and recovery values for the back-pain group were higher than the spine compression observed for the pain-free group, i.e. 0.80 cm vs. .034 cm. These values illustrate a substantial difference between the two groups. The values for the individual subjects were highly variable and no two subjects in either group exhibited the same values. Therefore, it is necessary to "standardize" these values to allow a comparison. This is achieved using each subject's 15-minute recovery value as a "control" or reference. Dividing the 15-minute recovery by the 8-hour compression value results in a subject relative recovery status. As shown in Tables 3 and 4, the pain-free group exhibited a recovery of 129% while the back- pain group exhibited a recovery of 61%.

Using the 15-minute recovery value to "normalize" each subject's 8-hour compression allows a comparison between the back-pain group and the pain-free group. For example, using a 100% recovery value as the "reference", the data summarized in Table 3 show that 9 out of 11 subjects in the pain-free group exceeded the reference value. However, Table 4 shows that only 1 subject out of 7 in the back pain group met or exceeded this reference value. Using these results, a "diagnostic" model can be established as illustrated in Table 5.

Using the results from the 19 volunteer test subjects and applying the recovery criteria to a diagnostic 2x2 table as illustrated in Table 5, a positive predictive value of 75% and a negative predictive value of 90% is achieved. This shows that by using the spine compression and recovery data only, 75% of the healthcare workers who participated in this study suffering back pain would be identified correctly, while 90% of the asymptomatic employees would be identified correctly.

TABLE 5: 2 x 2 Diagnostic model for stature recovery

	Pain-Free Diagnosis 15-minute Recovery More than 100%	Back-Pain Diagnosis 15-minute Recovery Less than 100%	Total
Subjects with Back Pain	1	6	7
Subjects without Back Pain	9	2	11
Total	10	8	18

The results of this study, therefore, suggest that it should now possible to develop an objective diagnostic that can help identify employees at increased risk of work related back pain. The measurement system used in this study provides a tool for occupational health professionals to conduct evaluations of work associated chronic back pain and to potentially test the efficacy of ergonomic interventions in reducing back pain symptoms.

4 CONCLUSIONS

The results of this pilot study suggest that it is now possible to develop an objective diagnostic strategy to help identify employees at increased risk of work related back pain. The measurement system used in this study provides a practical tool for occupational health professionals to conduct evaluations of work associated chronic back pain and to test the efficacy of ergonomic interventions in reducing back pain symptoms.

HUMAN SUBJECTS

The research protocol and use of human subjects in this study were reviewed and approved by the Institutional Review Board of the Office of Research Compliance at Boise State University. Approval # 193-MED11-008.

REFERENCES

Beynon, C., & Reilly, T. (2001) Spinal Shrinkage during a seated break and standing break during simulated nursing tasks, *Applied Ergonomics,* 32, 617-622.

Eklund, J.A.E., & Corlett, N. (1984) Shrinkage as a measure of the effect of load on the spine, *Spine,* 9, 189-194.

Flor, H. & Turk, D.C. (1989) Psychophysiology of Chronic Pain: Do chronic pain patients exhibit symptom - specific psychophysiological responses? *Psychological Bulletin,* 105, 215-259.

Flor, H., Birbaumer, N. & Turk, D.C. (1990) The Physiology of chronic pain. *Advanced Behavior Research Therapy,* 12, 47-84.

Garbutt, G., Boocock, M.G. & Reilly, T. (1990) Running speed and spinal shrinkage in runners with and without low back pain, *Med. Sci. Sports Exercises,* 22, 769-772.

Marras, W.S., Davis, K.G., Ferguson, S.A., Lucas, B.R., & Gupta, P. (2001) Spine Loading Characteristics of Patients with Low Back Pain Compared with Asymptomatic Individuals, *Spine,* 26, 2566-2577 .

Reischl, U. & Weinsheimer, W. (2004) Work-Related Spinal Compression and Recovery Profiles for Healthcare and Industrial Employees. *Proceedings of the 2nd International Ergonomics Conference*. Zagreb, Croatia. Budimir Mijovic, Editor. 9-15.

Reischl, U. & Weinsheimer, W. (1995) Back-Pain Prevention - Spinal Recovery Characteristics. *Proceedings of the 35th Congress of the German Occupational and Environmental Medicine Association,* R. Schiele, Editor, 381-385, (In German).

CHAPTER 55

3D Vision Gaze with Simultaneous Measurements of Accommodation and Convergence among Young and Middle-aged Subjects

Masaru Miyao[1], *Akira Hasegawa*[1], *Satoshi Hasegawa*[2], *Tomoki Shiomi*[1], *Hiroki Hori*[1], *Hiroki Takada*[3]

[1)]Nagoya University
Nagoya, Japan
miyao@nagoya-u.jp
[2)] Nagoya-Bunri University
[3)] University of Fukui

ABSTRACT

The aim was to compare fixation distances between accommodation and convergence in young and middle-aged subjects while they viewed 2D and 3D video clips. Measurements were made using an original machine, and 2D and 3D video clips were presented using a liquid crystal shutter system. We developed an original machine by combining WAM-5500® and EMR-9® to perform the measurements. As results, subjects' accommodation and convergence were found to change the diopter value periodically when viewing 3D images. When subjects are young, accommodative power while viewing 3D images is similar to the distance of convergence, while subjects are middle-aged, their accommodation is weak. It is generally explained to the public that, "During 3D vision, while accommodation is fixed on the display that shows the 3D image, convergence is at the stereoimage". According to the present findings, however, such explanations are mistaken.

Keywords: 3D, stereoscopic vision, accommodation, convergence

1 IS ACCOMMODATION FIXED ON THE DISPLAY THAT SHOWS THE 3D IMAGE?

It is generally explained to the public that, "During stereoscopic vision, accommodation and convergence are mismatched and this is the main reason for the visual fatigue caused by 3D. During stereoscopic vision, while accommodation is fixed on the display that shows the 3D image, convergence of left and right eyes crosses at the location of the stereoimage". According to the findings presented in our previous report (Miyao, 1996), however, such explanations are mistaken. However, our research has not been recognized in the world. This may be because the experimental evidence obtained in our previous studies, where we did not measure accommodation and convergence simultaneously, was not strong enough to convince people. We therefore developed a new device that can simultaneously measure accommodation and convergence.

2 MATERIAL AND METHOD

The subjects used in this study were six healthy, young men and women in their twenties and four middle-aged subjects in their forties to fifties. We obtained informed consent from all the subjects and the approval from the Ethical Review Board of the Graduate School of Information Science at Nagoya University.

We placed an LCD monitor facing the subjects at a distance of 1 m from them. We presented either a 2D or a 3D video clip on the monitor; in both images, a spherical object moved forward and backward with a cycle of 10 s (Fig. 1).

The spherical object appeared as a 3D video clip located at a virtual distance of 1 m (i.e., the location of the LCD monitor) and moved toward the subjects to a virtual distance of 0.35 m in front of them. We asked the subjects to gaze at the center of the spherical object for 40 s and measured their lens accommodation and convergence distance during that time. The 3D video clip was presented using a liquid crystal shutter system. The 2D video clip was presented using a liquid crystal shutter system.

We developed an original machine by combining WAM-5500® and EMR-9® to perform the measurements. WAM-5500 (Fig. 2) is an auto refractometer (Grand Seiko Co., Ltd.) that can measure accommodative power under natural conditions for the case in which both eyes are open. It can continuously record accommodative focus distance at a rate of 5 Hz. EMR-9 (Fig. 3) is an eye mark recorder (NAC Image Tech. Inc.) that can measure the convergence distance using the pupillary/corneal reflex method.

We used a liquid crystal shutter system combined with the respective bin-ocular vision systems to present 2D and 3D video clips. The experimental environment is shown in Fig. 4.

The video clips we used in the experiment are trade-marked as Power 3D® (Olympus Visual Communications, Corp.). Power 3D is an image creation technique that combines near and far views in a virtual space and has multiple sets of virtual displays whose positions can be adjusted. Power 3D presents a video clip that is similar to a natural image.

Figure 1 Spherical oject vdeo clips.

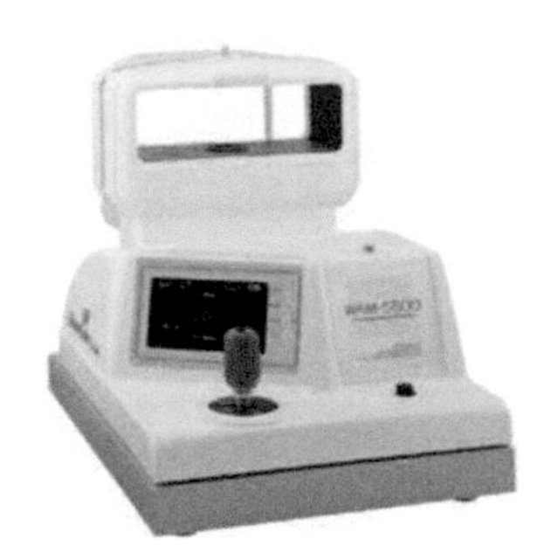

Figure 2 Auto ref/keratometer WAM-5500. (Grand Seiko Co. Ltd.)

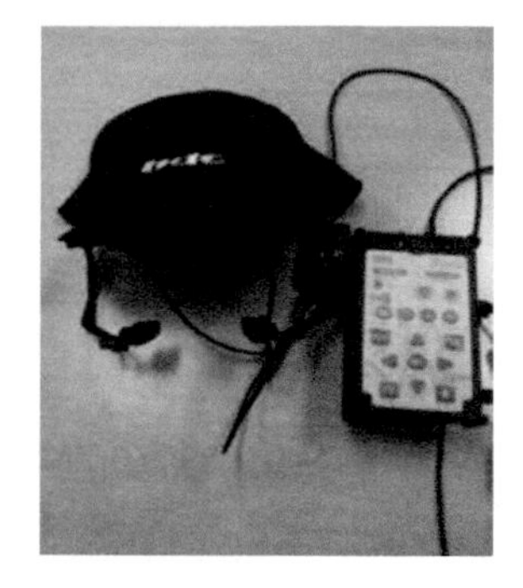

Figure 3 EMR-9 (NAC Image Technology Inc.)

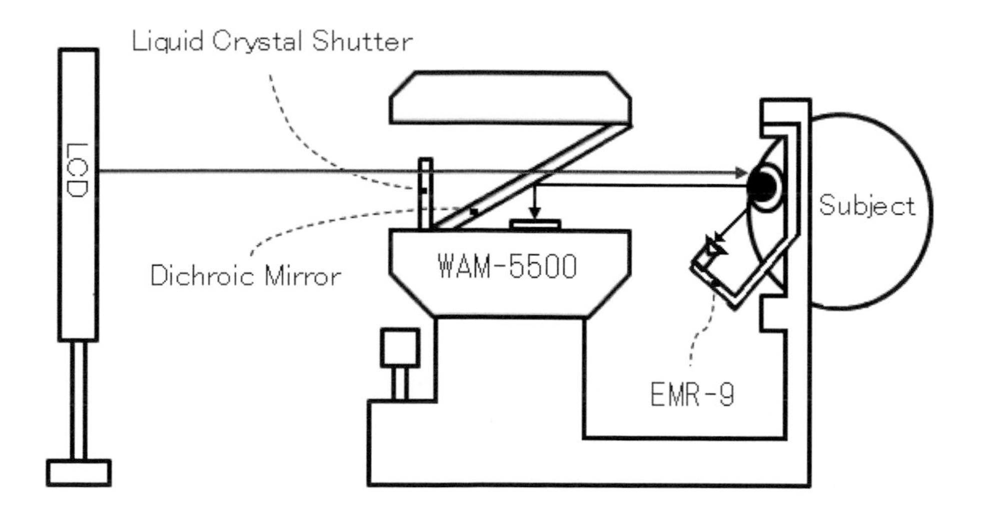

Figure 4 Overview of the experiment.

3 RESULTS IN MEASUREMENT

The measurements for the subjects showed roughly similar results in each age group. For 3D vision, the results for Subject A (23-year-old male wearing soft contact lenses) and Subject B (45-ysear-old male, emmetropia) are shown in Figs. 3-4, as examples.

When Subject A (23-year-old male wearing soft contact lenses) and Subject B (45-year-old male, emmetropia) viewed the 3D video clips presented, accommodation of the young subject varied between 1 diopter (1 m) and 2.5 diopters (40 cm), whereas convergence varied between 1 diopter (1 m) and 2.7 diopters (37 cm). Accommodation of the middle-aged subject varied between 1 diopter (1 m) and 2 diopter (50 cm), whereas convergence varied between 1 diopter (1 m) and 5 diopters (20cm). The changes in the respective values were fluctuating synchronously with a cycle of 10 s.

When the subject was viewing the 2D video clip, the diopter values for both accommodation and convergence remained almost constant at around 1 diopter (1 m). There is not much quantitative difference in the fixation distances between accommodation and convergence when the subject views either the 2D or 3D video clip.

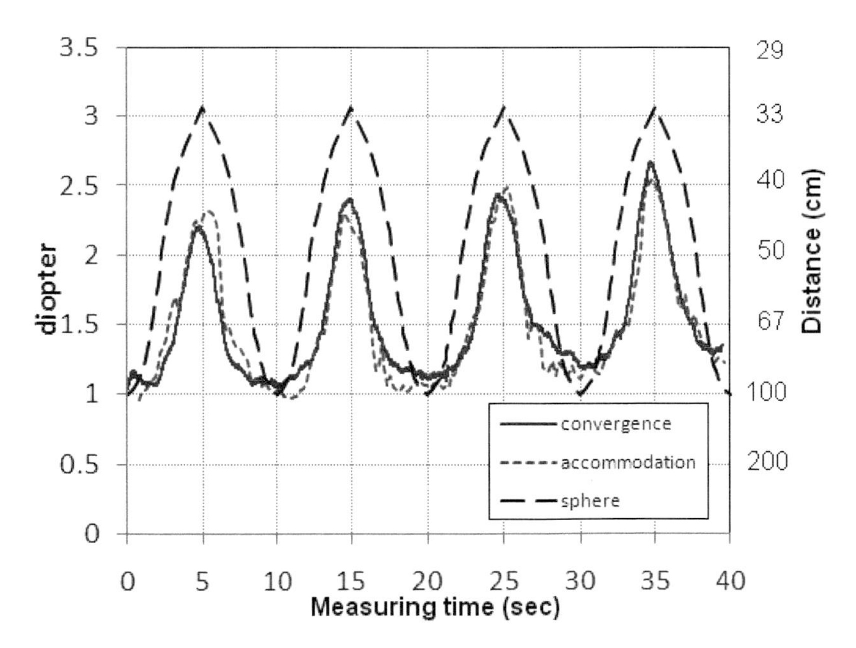

Figure 5 Accommodation and convergence. (Subject A: 23-year-old male)

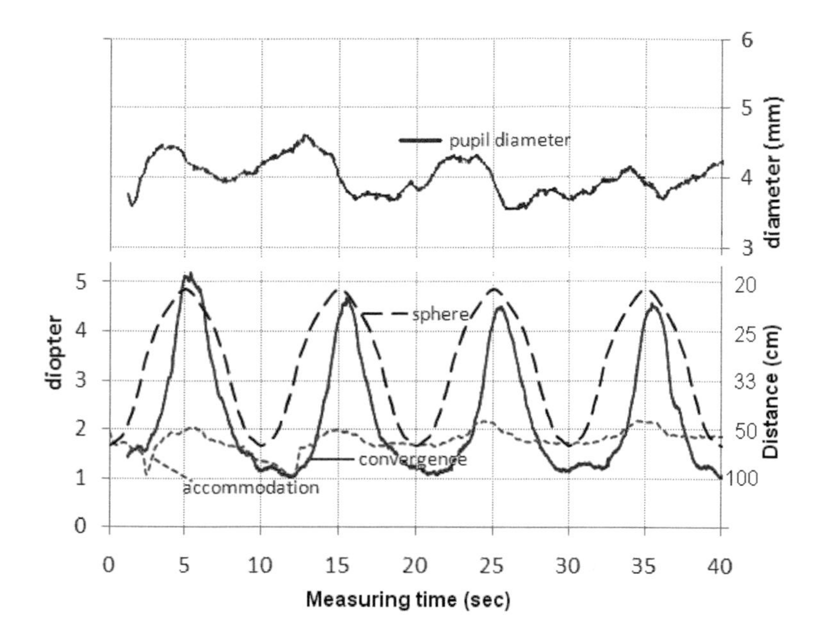

Figure 6 Accommodation, convergence and pupil diameter of middle-aged subject (Subject B: 45-year-old male)

4 DISCUSSION

In this study we used 2D and 3D images with a virtual stereoscopic view. The influences of age and visual functions on stereoscopic recognition were analyzed.

Accommodation and convergence change similarly in young subjects at a constant cycle of 10 seconds, synchronously with to the movement of the 3D image. In the middle-aged subjects, accommodation showed less change than convergence.

Wann et al. (Wann et al., 1995) stated that within a virtual reality system, the eyes of a subject must maintain accommodation at the fixed LCD screen, despite the presence of disparity cues that necessitate convergence eye movements to capture the virtual scene. Moreover, Hong and Sheng (Hong & Sheng, 2010) stated that the natural coupling of eye accommodation and convergence while viewing a real-world scene is broken when viewing stereoscopic displays. Depth information from focus cues is impoetant to know mechanism of 3D vision (Watt, S.J., et al. 2005).

We simultaneously measured accommodation and convergence for subjects viewing 2D and 3D video clips. In young subjects, the difference in the eye functions for accommodation and convergence is equally small in the cases of the observation of both 2D and 3D video clips. In contrast, the middle-aged subjects showed weak accommodation for 3D video clips. They can view 3D images stereoscopically with weak accommodation power. They are supplemented with a deepened depth of field from pupil contraction. Thus, it is thought that with contraction of pupil diameter, images with left-right parallax can be perceived. On the other hand, pupil contraction implies that a decreased amount of light enters the retina. Therefore, middle-aged people might perceive things as being darker than younger people do.

5 CONCLUSION

We compared fixation distances between accommodation and convergence in young and middle-aged subjects while they viewed 2D and 3D video clips. Measurements were made using an original machine, and 2D and 3D video clips were presented using a liquid crystal shutter system. As results, subjects' accommodation and convergence were found to change the diopter value periodically when viewing 3D images. When subjects are young, accommodative power while viewing 3D images is similar to the distance of convergence, while subjects are middle-aged, their accommodation is weak.

ACKNOWLEDGMENTS

The authors would like to thank late Dr. Lawrence W. Stark of Neurology and Telerobotics Units, University of California at Berkeley for his excellent research design teaching. They would also like to acknowledge Olympus Visual Communications, Inc., for their beautiful 3D video clips.

REFERENCES

Hong, H. and Sheng L. Correct focus cues in stereoscopic displays improve 3D depth perception. July 2010. In *SPIE Newsroom*, doi: 10.1117/2.1201007.003109. SPIE.

Miyao, M., Ishihara, S., Saito, S., Kondo, T., Sakakibara, H., and Toyoshima, H. 1996. Visual accommodation and subject performance during a stereographic object task using liquid crystal shutters. *Ergonomics,* 39(11): 1294-1309.

Miyao, et al. 2011. Comparison of 2D and 3D Vision Gaze with Simultaneous Measurements of Accommodation and Convergence among Young and Middle-Aged Subjects. *Ten'th International Meeting on Information Display*, KINTEX, Seoul, Korea.

Shiomi, T, et.al. 2012. Simultaneous measurement of lens accommodation and convergence to real objects, *Forma* (in press).

Sugiura, A., Yamamoto, T., Miyao, M., Takada, H. 2011. Effect of strategic accommodation training by wide stereoscopic movie presentation on myopic young people of visual acuity and asthenopia. *Displays*, 32(4): 219-224.

Ukai, K., and Howarth, P.A. 2008. Visual fatigue caused by viewing stereoscopic motion images: Background, theories, and observations. *Displays*, 29: 106-116.

Wann, J. Rushton, S. and Mon-Williams, M. 1995. Natural Problems for Stereoscopic Depth Perception in Virtual Environments. *Vision Research*, 35(19): 2731-2736.

Watt, S.J., et al. 2005. Focus cues affect perceived depth. *Journal of Vision*. 5: 834–862.

Yano, S., Emoto M., Mitsuhashi T. 2004. Two factors in visual fatigue caused by stereoscopic HDTV images. *Displays*, 25: 141-150.

CHAPTER 56

Assessment of a Virtual Reality-based Haptic Simulation for Motor Skill Training

Michael Clamann[1], *Guk-Ho Gil*[1], *David Kaber*[1], *Biwen Zhu*[1], *Manida Swangnetr*[3], *Wooram Jeon*[1], *Yu Zhang*[1], *Xiaofeng Qin*[1], *Wenqi Ma*[1], *Larry A. Tupler*[2] *and Yuan-Shin Lee*[1]

[1] Edwards P. Fitts Department of Industrial and Systems Engineering, North Carolina State University, Raleigh, NC, 27695-7906, USA
mpclaman@ncsu.edu

[2] Department of Psychiatry, Duke University School of Medicine, Durham, NC 27710, USA

[3] Back, Neck and Other Joint Pain Research Group, Department of Production Technology, Khon Kaen University, Khon Kaen 40002, Thailand

ABSTRACT

This study investigated the utility of types and formats of presentation of haptic and visual interfaces as part of a virtual reality (VR)-based simulator for training psychomotor skill development. The overarching goal of the research is to advance the state of VR-based haptic simulation design for motor rehabilitation applications. The haptic-control interface functions and visual aids were designed, prototyped and tested for use in the context of a Block Design (BD) reconstruction task, based on the Wechsler Abbreviated Scale for Intelligence. Testing was to identify features that might accelerate motor learning. An experiment was conducted in which healthy subjects were trained in the BD task, and observations were made on non-dominant hand performance to simulate minor motor impairment. Training effects were measured with the Rey-Osterreith Complex Figure reproduction test before and after three BD training sessions. Participants were assigned to one of three groups, including performance of the native BD task, use of a basic VR simulation of the task, or an augmented VR simulation with additional visual and haptic aiding. Results revealed a significant improvement in post-test performance over pre-test for the augmented VR training. In general, the study supports integrating haptic control in VR for psychomotor skill training. It also provides useful information for future haptic VR simulation design.

Keywords: Haptic simulation, virtual reality, neuropsychological tests, motor skill training

1 INTRODUCTION

This study was an initial investigation of healthy subject visuomotor control training in native and virtual reality (VR) forms of standardized psychomotor tasks. The objective was to make observations on the degree of skill development that can be achieved with a VR-based haptic simulation, by comparison with native task formats, and to identify unique VR design features that might serve to accelerate psychomotor learning. The overarching goal of this research is to design and prototype custom VR and haptic interfaces for persons seeking recovery from minor Traumatic Brain injury (mTBI) with motor control implications. The study is part of a multi-year research program to conclude with testing of a pathological population.

The VR designs for the present study involved drawing and pattern assembly tasks, as they include fundamental motions and sequences necessary in many work or daily living activities. The Rey-Osterreith Complex Figure (ROCF; Osterreith, 1944; Rey, 1941) reproduction task was used to represent an occupational task and measure training effects following native and simulation-based therapy trials. The ROCF requires participants to reproduce a single picture composed of 18 shapes and patterns (see Figure 1). Scoring is performed by evaluating the components of the figure, referred to as units, on a scale from 0 to 2 in terms of accuracy and placement. The sum of the scores for the 18 units in calculated for a total score between 0 and 36.

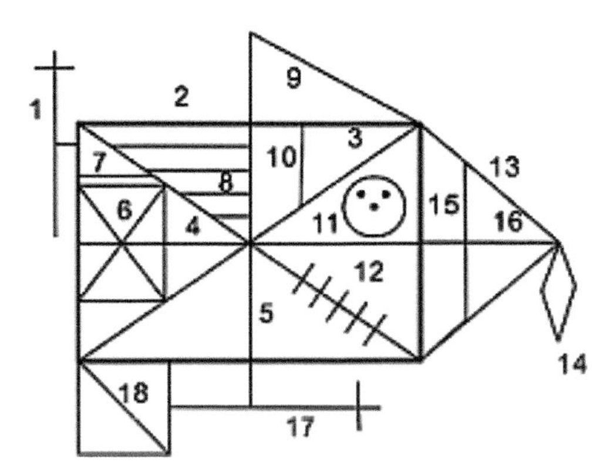

Figure 1. ROCF with unit markings.

With respect to pattern assembly, a VR model of the block design (BD) subtest from the Wechsler Abbreviated Scale for Intelligence (WASI: The Psychological Corporation, 1999) was developed for psychomotor training. The BD task is typically used as a diagnostic tool for evaluating visuospatial and motor skills by requiring participants to build replicas of multiple block patterns. Participants are given a set of nine identical red and white blocks printed with either solid or cross-sectional patterns on each side (see Figure 2 for image of native BD task materials). They are asked to replicate complex designs shown in a series of test cards. Scoring is based on speed and accuracy.

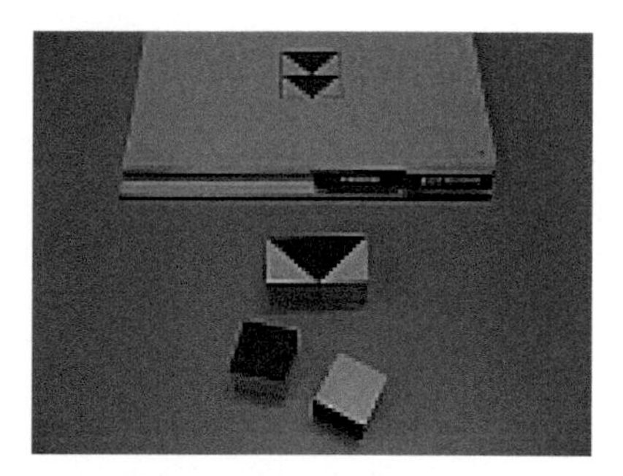

Figure 2. Native BD task materials.

The VR version of the BD task was designed to incorporate augmented haptic feedback, including resistive and assistive forces in motion path guidance, which was reinforced by visual aiding. Such aids were considered applicable for the application of motor skill training. Since mTBI often occurs with both attentional and motor implications (Barnes, 1999), we explored visual features complementary to the haptic features of the simulation.

Hoffman et al. (2003) stated that successful BD stimulus construction requires the observer to:

1) visually encode parts of the model and their spatial organization;
2) select among candidate blocks to determine which to use;
3) check against the model to determine the block's location;
4) place the block in the work area; and then
5) recheck to be sure that the block has been placed in the correct location.

A correction may occur if the checking process reveals that the block has been incorrectly placed.

Lee et al (2009) stated that the process involved in copying an image, such as the ROCF, could be very similar to the BD task process described by Hoffman et al (2003). Both tests require participants to be able to perceptually analyze the parts of an array and then replicate the global configuration. ROCF and BD performance have also been shown to be moderately correlated (Spreen and Strauss, 1998). For these reasons, we expected differences in performance on the native vs. VR versions of the BD task, during training, to be revealed through ROCF test scores.

2 METHODS

2.1 Participants

Twenty-four participants between the ages of 18 and 44 were recruited for this study. All subjects were required to have 20/20 or corrected to normal vision, and all were required to exhibit right-hand dominance. Right-hand dominance was identified through a demographic questionnaire and confirmed using the Edinburgh Handedness Inventory (Oldfield, 1971). Both the questionnaire and inventory were administered with electronic forms prior to a participant visiting the lab.

2.2 Apparatus

The software and hardware used in the experiment included a VR simulation of the BD subtest from the WASI and a subset of tests based on the ROCF. The ROCF was performed using a VR adaptation of the task (Li et al., 2010) presented with a custom workstation featuring a flat-screen monitor mounted in a tabletop and a SensAble Technologies PHANTOM Omni® Haptic Device (see Figure 3 (a)). The Omni includes a boom-mounted stylus that offers 6 degrees of freedom (DOF) for movement and 3 DOF in force feedback. To perform the ROCF, subjects used the Omni to draw the complex figure elements directly on the horizontally oriented monitor. The interface recorded participant performance data automatically, which was one of the primary reasons for using the VR version of the ROCF test. The VR interface for the BD task (see Figure 3 (b)) was presented on a PC integrated with a stereoscopic display using a NVIDIA® 3D Vision™ Kit, including 3D goggles and an emitter. A second Omni haptic device was used as the haptic control interface for the BD task.

(a) ROCF VR workstation (b) BD VR workstation

Figure 3. ROCF and BD workstations.

2.3 Experiment Procedure

The experiment was designed to simulate multiple occupational therapy sessions to promote psychomotor skill development. The duration of the motor skill training was established through pilot testing prior to data collection. Each participant completed four sessions totaling approximately 5 hours, as follows:

- Session 1: Orientation, baseline testing and training (about 2 hours)
- Session 2: Training (about 1 hour)
- Session 3: Training (about 1 hour)
- Session 4: Final testing (about 1 hour)

All participants were required to use their non-dominant hand during the pre- and post-tests and training trials in order to simulate a minor motor impairment; that is, the experiment provided training to participants for their left hand. Previous studies have simulated motor impairment characteristics, such as those associated with mTBI, by using the non-dominant hand in motor skill training (Ozcan et al., 2004; Sainburg and Kalakanis, 2000; Yamashita, 2010). The non-dominant hand requirement was also intended to disadvantage participant task performance in order to promote sensitivity to the training conditions.

All participants were permitted to train with the device prior to baseline testing until they felt comfortable. This step was intended to reduce the potential impact of haptic device usability issues on the results of the study. The device familiarization was followed by a baseline non-dominant hand performance evaluation using two ROCF copy tests presented through the VR workstation interface. During psychomotor skill training, participants were required to complete 8 WASI BD trials. This training occurred with one of three representations of the task, either as 1) a physical task (the native form), (2) a Basic VR task, or (3) an Augmented VR task, including the visual and haptic aiding. Each participant was assigned to one task representation or condition yielding a total of 8 participants per condition. In the last experiment session, participants were retested using the VR-based ROCF copy tests to identify any changes from baseline performance. For those subjects assigned to either of the VR conditions, additional haptic device familiarization was provided prior to the training sessions.

2.4 Experiment conditions

2.4.1 Condition 1: Native Task

The control condition required the participant to train using the Native WASI BD task. Administration of this condition followed the standardized WASI protocol (The Psychological Corporation, 1999).

2.4.2 Condition 2: Basic VR

The Basic VR condition developed for this study featured a visually realistic VR simulation of the WASI BD subtask. The virtual environment (VE) features included a virtual tabletop divided into two parts: a work area and a display area. The display area presented the stimulus figure depicting the BD to be replicated by a participant. The work area was used for block assembly (the design reconstruction). The work area and blocks were presented at approximately 70% of actual size to allow the stimulus figure and work area to be viewed on a 21-inch stereo monitor. The Basic VR task layout is shown in Figure 4 (a).

The Omni haptic device was used to manipulate a cursor (a small blue sphere) appearing on the display. Participants could touch the cursor to a block and press a button to "grab" the block. Blocks could then be lifted from the table surface and rotated along any axis using the stylus. Some haptic features were included in the Basic VR to represent the blocks and the table. Blocks on the work surface were presented as solid objects and participants could feel a solid resistance when the blocks touched the work surface.

2.4.3 Condition 3: Augmented VR

The Augmented VR condition was identical to the Basic VR condition except for the visual display and haptic enhancements designed to assist participants. In this version of the simulation, all BDs were constructed with the aid of a target grid, which appeared as a 2x2 or 3x3 collection of squares, depending on the design stimulus (see Figure 4 (b)). Participants were also provided with assistive and corrective haptic forces during design assembly. If a participant attempted to place a block in an incorrect orientation in the grid, a resistive force equal and opposite to the pressure applied at the stylus prevented the block from touching the table. If a participant moved a block in the correct orientation over the grid, the participant would feel the block being pulled to the matching portion of the grid, by a snap force (see SensAble Technologies OpenHaptics® Toolkit for snap force details; SensAble Technologies, 2008).

The target grid also provided some simple visual decision aiding to reinforce the haptic cues. When a block was rotated so its top surface matched a part of the design that had not yet been assembled in the grid, one or more squares (as applicable) changed color to green to recommend placement of the active block (also see Figure 4 (b), lower right corner of grid). Beyond these visual features, an additional block image was presented in the upper right portion of the display (see Figure 4 (b)). If a participant attempted to place a block in the wrong part of the grid, the image showed the correct orientation for the top of the block.

(a) Basic VR condition

(b) Augmented VR condition

Figure 4. Basic and Augmented Haptic VR simulation conditions.

2.5 Variables

The independent variable for the experiment was the form of the BD training task presented to participants with the three levels of: 1) Native BD, serving as a control; (2) the Basic VR condition; and (3) the Augmented VR condition with enhanced haptic control and visual displays. Dependent variables included the percent improvement in ROCF copy performance. This response measure was calculated using pre- and post-training test scores in the following formula:

Percent improvement = (Post-test Score – Pre-test Score) / Pre-test Score x 100

Subjective confidence ratings of ROCF accuracy were also collected during the first and final sessions. More details on the response measures are provided below in the results section.

2.6 Hypotheses

In general, we hypothesized that participant ROCF test scores would improve as a result of completing any of the three training conditions (Hypothesis (H)1). Basic VR training was not expected to exceed traditional physical task training in improving participant motor skills due to the fact that no additional perceptuo-motor aiding was provided as part of that simulation (H2). The Augmented VR condition was expected to lead to improved performance over the Native form of the BD task due to the visual and haptic enhancements of the VR simulation beyond the constraints of the traditional training (H3). The visual enhancements, including the design grid highlighting and correct block orientation image were expected to assist participants with Steps 1-3 in Hoffman et al. (2003) list for BD completion. The resistive and attractive haptic forces were expected to aid in Steps 4-5 of the BD completion process.

3 RESULTS

The pre- and post-test data were analyzed to identify differences among the three conditions. Data on two of the participants were excluded from further statistical analyses due to problems in data collection occurring during post-testing. Table 1 presents the means and standard errors summarizing the differences between the pre- and post-test scores. In general, findings were mixed and all analyses are described below.

Table 1 Change in pre- and post-test scores across conditions

Condition	N	Pre-Test	Post-Test	% Improvement
Augmented	7	26.93	29.71	10.32
Basic	8	26.56	28.63	7.79
Native	7	28.93	27.07	-6.43

Studentized t-tests were used to compare the pre- and post-ROCF scores for each condition. Only Augmented VR participants showed a significant improvement in scores as a result of training ($t(22)=2.28$, $p=0.018$). On this basis, additional t-tests were conducted to compare Augmented VR pre- and post-scores for individual ROCF units (see Figure 1 for ROCF unit identifiers). Results revealed a significant improvement in scores for Unit 9 (a triangle in the upper right corner of the figure). In addition, all participants showed limited improvement in drawing Unit 12 (five parallel short lines) and performed nearly perfectly on other units (2-8 and 16). Post-hoc analysis revealed Augmented VR training to produce greater improvements in ROCF scores than the Native BD task. This finding was in line with hypothesis (H3).

An ANCOVA model was applied to confidence ratings on ROCF copy trials in order to identify differences in perceived demand and confidence before and after training. Results revealed no significant effects due to training condition.

4 DISCUSSION

With respect to the hypotheses that all participant test scores would increase as a result of training (H1), only the Augmented VR condition resulted in a significant improvement in ROCF performance. Results did, however, support the expectation that the haptic and visual and enhancements to the VR simulation would lead to superior test performance. The findings indicated that the Augmented VR condition supported greater ROCF improvements as compared to the Native task and Basic VR conditions. Therefore, there is evidence that the Augmented VR facilitated a greater improvement in psychomotor skills over the Native BD and Basic VR conditions. Finally, the Basic VR training results also supported our second hypothesis (H2), as there was no significant improvement over the Native BD task. However, this may be due, in part, to the limited sample size for the study.

It is important to note that the Basic and Augmented VR training conditions used the same control device as the ROCF task (i.e., the Omni). This means that participants in the VR conditions received 3 hours of additional training with the stylus than participants performing the Native task. It is therefore reasonable to assume that the VR participants would have some advantage over the Native task participants when performing the ROCF post-test. However, the more important finding of this study is the advantage of the Augmented VR over the Basic VR in terms of performance. Results clearly demonstrated the advantages of the visual (design grid highlighting and correct block orientation image) and haptic (resistive and attractive forces) aiding as part of the Augmented VR simulation on post-test motor skill performance.

5 CONCLUSIONS

This study demonstrated the utility of VR-based haptic simulations for training motor skills to address physical and cognitive limitations. This was an initial investigation of non-dominant manual performance by healthy participants to provide insight into the degree of skill development achievable with VR and as a basis for the design of future VR simulations to support rehabilitation of motor skills in mTBI patients. The study also identified unique visual and haptic features of VR simulation design (facilitating visual pattern recognition and object manipulation and placement) that appear to support efficient and effective motor skill development.

With respect to limitations of this research, although parallels were drawn between physical and cognitive characteristics of non-dominant hand performance and motor planning and control implications of mTBI, there is a need to test an actual pathological population using the present VR technology. A follow-on investigation has been planned to involve current mTBI patients at the Durham Veteran's Administration Medical Center in order to assess the effectiveness of enhanced VR motor training simulator designs for addressing motor planning and control disabilities.

ACKNOWLEDGMENTS

This research was supported by a grant from the National Science Foundation (NSF) (No. IIS-0905505) to North Carolina State University. The technical monitor was Ephraim Glinert. The views and opinions expressed on all pages are those of the authors and do not necessarily reflect the views of the NSF.

REFERENCES

Barnes, M. 1999. Rehabilitation after traumatic brain injury. *British medical bulletin*, 55(4), 927.

Hoffman, J. E., Landau, B., and Pagani, B. 2003. Spatial breakdown in spatial construction: evidence from eye fixations in children with Williams syndrome. *Cognitive Psychology*, 46(3), 260-301.

Lee, J., Okamura, A. M., and Landau, B. 2009. Haptics as an aid to copying for people with Williams Syndrome. In *Proceedings of the World Haptics 2009 - Third Joint EuroHaptics conference and Symposium on Haptic Interfaces for Virtual Environment and Teleoperator Systems* (pp. 356-361). IEEE Computer Society.

Li, Y., Kaber, D. B., Tupler, L. and Lee, Y-S. 2010, Haptic-based virtual environment design and modeling of motor skill assessment for brain injury patients. In *Proceedings of the Computer-Aided Design & Applications Conference* (CAD'10). Dubai: cadanda.com

Oldfield, R. C. 1971. The assessment and analysis of handedness: The Edinburgh inventory. *Neuropsychologia*, 9, 97-114.

Osterrieth, P. A. 1944. Le test de copie d'une figure complexe [The complex figure copy test]. *Archives de Psychologie*, 30, 206-356.

Ozcan, A., Tulum, Z., Pinar, L., and Baskurt, F. 2004. Comparison of pressure pain threshold, grip strength, dexterity and touch pressure of dominant and non-dominant hands within and between right-and left-handed subjects. *Journal of Medical Science*, 19(6), 874-878.

The Psychological Corporation. 1999. *Wechsler Abbreviated Scale of Intelligence*. SanAntonio, TX: Author.

Rey, A. 1941. L'examen psychologique dans le cas d'encephalopathie traumatique. *Archives de Psychologie [Psychological examination of traumatic encephalopathy]*, 28, 286-340.

Sainburg, R., and Kalakanis, D. 2000. Differences in control of limb dynamics during dominant and nondominant arm reaching. *Journal of Neurophysiology*, 83(5), 2661.

SensAble Technologies. 2009. *OpenHaptics® Toolkit version 3.0 Programmer's Guide*. Woburn, MA.

Spreen, O., Strauss, E. 1998. *A Compendium of neuropsychological tests: Administration, norms, and commentary*. (2nd ed.). NY. Oxford University Press.

Yamashita, H. 2010. Right-and Left-hand Performance on the Rey-Osterrieth Complex Figure: A Preliminary Study in Non-clinical Sample of Right Handed People. *Archives of clinical neuropsychology: the official journal of the National Academy of Neuropsychologists*.

Section VIII

Measures and Validation in Healthcare

CHAPTER 57

Development of Moderators and Measuring Criteria in Healthcare TAM research

Byung Cheol Lee[1] *and Vincent G. Duffy*[1,2,3]

[1]School of Industrial Engineering;
[2]Agricultural & Biological Engineering
Purdue University
[3]Bauman Moscow State Technical University
Faculty of Engineering Business & Management

ABSTRACT

To understand care practitioners' perception for newly implemented healthcare IT systems and pursue successful implementation, Technology Acceptance Model (TAM) has been widely applied in healthcare domain. However, TAM has several theoretical limitations and some application problems. Accordingly, this study suggests some moderators of TAM's main constructs to investigate new perspective of TAM research on healthcare IT system adoption and provides recommendations to smooth and implementation healthcare IT systems.

Some studies indicated that unclear definitions of main constructs of TAM limited its effective and diverse applications. Also, it needs to adopt unique domain oriented features to be successful in healthcare domain. Especially, different perspective of user structure is a key to understand healthcare IT system implementation. According to the interview results of care practitioner focus group, it is found that the perceived ease of use (PEOU) is as much important as the perceived usefulness (PU) in the healthcare context. It is proposed that the moderators of PU can be performance and productivity and those of PEOU can be system environment and usability. Additionally, as recommendations for successful implementation, smooth integration with existing systems and appropriate downtime strategies are suggested.

Keywords: Healthcare IT systems, TAM, System implementation, Moderator

1 INTRODUCTION

From the Institute of Medicine (IOM) was published, interest in healthcare information technology (HIT) systems has been remarkably increased and encouraged for many healthcare facilities to adopt various types of systems (Kohn et al., 1999). While some systems achieved intended outcomes such as increase of care quality, reduction of medical/medication errors, cost reduction and decrease of length of stay in the hospital, others just produced serious, blatant complaints from users and are considered as just time and cost devourers (Doolan & Bates, 2002; Han et al., 2005; Koppel et al., 2005; Weiner et al., 1999). Among various approaches to understand such discrepancy, Technology Acceptance Model (TAM) can be considered as one of the promising in the healthcare IT system research area.

However, since TAM is originated in business applications, it has several drawbacks and limitations in application in healthcare discipline. Moreover, a number of studies using TAM indicated that TAM or other TAM descendent models cannot be matched enough with the healthcare structure and need the customization and modification of its frame (Holden & Karsh, 2010; Yarbrough & Smith, 2007). In spite of such disparity, many healthcare studies using TAM merely added some external variables to match with healthcare context, which many scholars criticized as the patchwork type of approach, and it is not enough to provide contextualized conclusions.

Accordingly, the objectives of this study are (1) to suggest new perspective of TAM research on healthcare IT system adoption, (2) to investigate some moderators of TAM main constructs based on current healthcare practitioners' interview about new IT system implementation, (3) identify missing areas being scrutinized for smooth and fitted implementation of healthcare IT systems.

1.1 Technology acceptance and adoption models

TAM was suggested to explain the relationship between user satisfaction and new technology implementation and encouraged to proliferate information system researches on new technology or system adoption by individual level (Davis, 1989; Venkatesh, Davis, & Morris, 2007). It suggested that user decisions about the acceptance of new technology or systems are dependent on two critical constructs: perceived usefulness (PU) and perceived ease-of-use (PEOU) (Davis, 1989, 1993). The relationship between major constructs is shown in Figure 1.

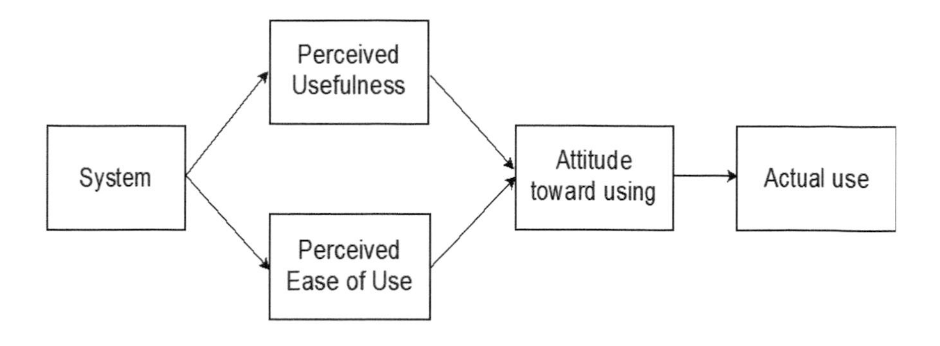

Figure 1 Technology Acceptance Model (Davis, 1989)

The TAM related research has developed into diverse directions. The main areas of post-TAM research can be divided into two directions: the replication of TAM with other technologies and the comparison TAM with other attitude theory models (Lee, Kozar, & Larsen, 2003). These two research directions proved that TAM might predict users' new technology acceptance with a simple, easier approach (Igbaria, Zinatelli, Cragg, & Cavaye, 1997). Recently, TAM has been external variables. For example, TAM II is suggested that PU includes subjective norms, job relevance, image, quality and result demonstrability, and PEOU incorporates self-efficacy, anxiety, playfulness, enjoyment and usability (Venkatesh & Davis, 2000). Even more lately, Venkatesh (2003) introduced the Unified Theory of Acceptance and Use of Technology (UTAUT) which reviewed and tested several factors in four extended TAM studies (Venkatesh et al., 2003a).

1.2 Limitation of TAM

TAM is not the perfect solution for all technology acceptance problems and has several limitations. First, PU and PEOU are ambiguously defined. Such uncertainty leads the relationship between them and other impacting factors (Holden & Karsh, 2010; Karsh, 2004; Yarbrough & Smith, 2007). Second, TAM did not consider boundary conditions such as culture, gender, task, and information system type (Adams, Nelson, & Todd, 1992). Not only depending on boundary conditions, but also simply adding too many external variables weakens TAM's theoretical foundation. Another limitation of TAM is the model's simple and theory oriented constructs and these makes difficult to put into practice (Bagozzi, 2007; Benbasat, Barki, & Montréal, 2005). TAM has developed from the researcher's perspective, but not the user's perspective. So, the focus was not to improve the acceptance rate or achieve smooth implementation but to find some factors that might influence the technology acceptance (Venkatesh, Morris, Davis, & Davis, 2003b; Wang, Wan, Burke, Bazzoli, & Lin, 2005). Due to these shortcomings, several studies blamed the excessive focus on narrow perceptual constructs result in neglecting some of important user behaviors (Bagozzi, 2007).

1.3 TAM Application in Healthcare Domain

TAM has contributed to investigate diverse influencing factors in HIT system implementation and adoption processes. Van Schaik (2002) asserted that TAM can be enough to evaluate the acceptance of information system in healthcare domain by investigating the effects of a portable computerized postural assessment system. Particularly, the results showed that the significant relationship between PU and intention to use is important in the adoption of HIT systems. Chang (2004) applied modified TAM questionnaires to access the perception of a PDA support system in an emergency department. Even though the study did not provide a specific relationship between the constructs, it showed user perceptions about the system and effects on practitioners' work process. Another study applied TAM into general internet-based health applications and it confirmed that job relevance and results demonstrability were more influenced to PU than social factors such as the subjective norm or the image.

TAM was designed for an individual's voluntary use of a technology (Davis, 1989). However, the adoption of most HIT systems is decided by institutional levels and implemented in mandatory environment. Additionally, TAM was developed based on the adoption of simple business information systems such as e-mail or text editor (Davis, 1989). Yet, HIT systems are much more complex in the functions and the implementation process.

2 Method

The interviews and focus group discussion were conducted at an urban teaching hospital in April 2009. During the time of interviews, the hospital was implementing Bar Code Medication Administration (BCMA) system and user-training sessions were held for the potential users in the hospital. Among the users, five nurses and seven pharmacists joined in the focus group discussion and follow-up interviews were carried out. The participants' work experiences from two to 35 years. Interviews were mainly focused on the effect of new IT system implementation on medication administration and communication process. The detailed topics included current IT system environments and back-up processes, communication features and problems in the medication administration process, and expectations and concerns about new IT system implementation. Interview results were recorded and transcribed and content analysis was undertaken to identify key concepts.

3 Results and Discussion

3.1 Some distinctive issues for smooth system implementation

First, the many interviewees indicated that the incomplete integration between existing systems and a new system can generate the resistance for the system adoption. Since several types of existing HIT systems already have been operated in many healthcare facilities, integration problems between a newly implemented system and existing systems frequently occurs. As more new and different types of IT systems are implemented, integration balance between systems will be into a significant barrier to successful system implementation. A study reported that the mortality rate of patients increased due to a new HIT system implementation and the main cause was estimated to the unexpected delays of care process by the incongruous implementation process (Han et al., 2005). Such incomplete system integration also can increase physical and mental workload of the practitioners.

Secondly, the system downtime is another main barrier to successful system adoption. Even in the well-maintained system environment, some downtimes are necessary for system maintenance or data update. Additionally, unexpected system failure occasionally happens. During the downtime, it is not that transition to backup system can produce a lot of miscommunication and mishandling patient data. Since most backup of paper-based format, transition to different system format possibly increases the risk of unintended side effects such as medical/medication errors or adverse drug events.

Another distinctive result from the interviews is that the care practitioners' perception of a new IT system on the patient safety is not uniform. Some expected that the new system would affect nothing on the patient safety and their daily work because they perceive that the new system seems to be duplicated with a previous system and they are reluctant to the new system. On the other hand, others thought the new system would not only improve the work procedures and the patient safety but also positively influence efficient communication processes and cooperation because the new system reduces a lot of the waste of time in the medication data handling and documentation processes.

3.2 Moderators and measuring criteria of TAM constructs

Based on the analysis on the interview results, the moderators and the practical measuring criteria of TAM constructs are shown in table 1.

Table 1 Moderators and measuring criteria in Healthcare TAM

TAM Construct	Moderator	Measuring criteria	Beneficiary
Perceived usefulness (PU)	Performance	Frequency of ADE*, Medical/Medication error, Mortality rate	Patient
	Productivity	Documentation time/quality, Task-fit, Workload, Effective communication	Care practitioner
Perceived easy of use (PEOU)	System environment	Down time strategies, Integration with existing systems	
	Usability	Interface design, Learnability, Memorability, System satisfaction	

The new moderators are necessary to interpret and specify the two TAM constructs because the concept of PU and PEOU can be ambiguously applied in healthcare context. Several studies already maintained that the application of TAM is not appropriate on healthcare domain and needs customized variables and barometers of the healthcare domain properties (Holden & Karsh, 2010; Yarbrough & Smith, 2007). Dansky (1999) also suggested the several important factors to measure PU such as the patient care and the clinical work. Additionally, his study incorporated two different sets of care providers' expectations to assess PU: (1) the system contribution to improve the patient care and (2) the system contribution to improve the work productivity. Thus, the performance of HIT system and its impact on productivity of the care providers can be regarded as two main moderators of PU. As shown on the table, the system performance can be measured by the number of Adverse Drug Events (ADE), medical/medication errors, or mortality rates and these measuring criteria are directly associated with the patient safety (Han et al., 2005; Koppel et al., 2005).

Another aspect is the productivity of care processes. Many interviewees commented that effective communications, workloads, task-fit, time consumption and task redundancy are main evaluating criteria successful system implementation. Even though interviewees did not directly mention, it is easily inferred that many of them are apathetic on the effects of new system performance on the patient safety or they blindly accept the care quality improvement. It is mainly because most system implementation is not decided by the care providers themselves and they put on more values on their additional workloads to learn new system procedures than on the work process improvement.

A few studies are supportive care practitioners' the new system implementation. A study showed that the majority of physicians agree that computers can significantly improve the quality of care and the computerized prescriptions are necessary, yet almost half did not intend to do so (Massachusetts Medical Society, 2004). The main reasons for the non-intention to use were computer systems are too expensive and not secure, systems require high operating costs, and data entry takes too much time. By the same token, Sviokla (1996) differentiates between "efficiency technology" and "transformational technology". According to his definition, "efficiency technology" means the technology reinforces current ways of working, and "transformational technology" means the technology changes the nature of work. Although many healthcare IT systems have the common characteristics of transformational technology, the care practitioners perceive them as efficiency technology. It is probable that efficiency factors affect day-to-day practices, but quality factors in an indirect way and on a long-term basis. Thus, productivity can be the most important moderator of PU with the perspective of the care practitioners.

As moderators of perceived ease-of-use (PEOU), care practitioners acknowledged system environment and usability as main components. While usability of system interface has been frequently discussed in Human Computer Interaction (HCI) domain, few studies have been conducted on system implementation environment. As many interviewees commented, easy transferability of clinical data with back-up system and integration with existing system significantly affect smooth system functions and increase the satisfaction on the newly implemented HIT system. Additionally, optimized strategies for required downtime also reduce the unnecessary effort of practitioners to maintain the care quality.

While most healthcare system research has been focused on the impact of care quality, the influence on practitioners' productivity needs more investigations. Particularly, workload due to system change may affect not only the successful adoption of new system but also the frequency and level of medical/medication errors. Consequently, longitudinal study on healthcare practitioners' cognitive workload of system implementation may provide meaningful results. Another promising future research area is the effect of new system implementation on the communication process. As some interviewees mentioned, the communication within and between departments accounts for considerable amount of working time and the consequences of miscommunication such as distraction or interruption are directly related to medical/medication errors. It is believed that these further studies can demonstrate the relationship between the successful implementation of new HIT system and practitioners' workload and the frequency of errors.

4 Conclusion

This study suggests moderators and measuring criteria of two TAM main constructs: perceived usefulness (PU) and perceived ease-of-use (PEOU), based on theoretical inference and qualitative analysis focus group interviews. The result can be contributed to develop the modified and contextualized TAM framework in healthcare domain. According to interview data with care practitioners, a new framework reflect the healthcare domain's unique user structure and healthcare oriented moderators and measuring criteria. While functional performance would be HIT system, its advantages are not mostly beneficial to the care providers. The productivity factors, such as workload, task-fit, and time allocation, are more meaningful for their satisfaction on the new system implementation and these factors as a critical role in the evaluation of implementation. Such customized moderators and measuring criteria in healthcare context help to understand the new HIT system acceptance/adoption mechanism and provide a solution to remove or reduce the acceptance/adoption barriers.

REFERENCES

Adams, D. A., Nelson, R. R., & Todd, P. A. (1992). Perceived usefulness, ease of use, and usage of information technology: a replication. *MIS quarterly*, *16*(2), 227–247.

Bagozzi, R. P. (2007). The Legacy of the Technology Acceptance Model and a Proposal for a Paradigm Shift. *Journal of the Association for Information Systems*, *8*(4), 244–254.

Benbasat, I., Barki, H., & Montréal, H. (2005). *Quo vadis TAM*. HEC Montréal, Chaire de recherche du Canada en implantation et gestion des technologies de l'information.

Chang, P., Hsu, Y. S., Tzeng, Y. M., Sang, Y. Y., Hou, I., Kao, W. F., & others. (2004). The development of intelligent, triage-based, mass-gathering emergency medical service PDA support systems. *Journal of Nursing Research*, *12*(3), 227–235.

Chismar, W. G., & Willey-Patton, S. (2003). Does the Extended Technology Acceptance Model Apply to Physicians. *Proceedings of the 36th Hawaii International Conference on System Sciences*.

Dansky, K. H., Gamm, L. D., Vasey, J. J., & Barsukiewicz, C. K. (1999). Electronic medical records: are physicians ready? *Journal of healthcare management/American College of Healthcare Executives*, *44*(6), 440–454.

Davis, F. D. (1989). Perceived usefulness, perceived ease of use, and user acceptance of information technology. *MIS quarterly*, *13*, 319–340.

Davis, F. D. (1993). User acceptance of information technology: system characteristics, user perceptions and behavioral impacts. *International Journal of Man-Machine Studies*, *38*, 475–487.

Doolan, D. F., & Bates, D. W. (2002). Computerized physician order entry systems in hospitals: mandates and incentives. *Health Affairs*, *21*(4), 180–188.

Han, Y. Y., Carcillo, J. A., Venkataraman, S. T., Clark, R. S. B., Watson, R. S., Nguyen, T. C., Bayir, H., et al. (2005). Unexpected increased mortality after implementation of a commercially sold computerized physician order entry system. *Pediatrics*, *116*(6), 1506–1512.

Holden, R. J., & Karsh, B. T. (2010). The technology acceptance model: its past and its future in health care. *Journal of biomedical informatics*, *43*(1), 159–172.

Igbaria, M., Zinatelli, N., Cragg, P., & Cavaye, A. L. M. (1997). Personal computing acceptance factors in small firms: a structural equation model. *MIS quarterly*, *21*(3), 279–305.

Karsh, B. (2004). Beyond usability: designing effective technology implementation systems to promote patient safety. *Quality and Safety in Health Care*, *13*(5), 388–394.

Koppel, R., Metlay, J. P., Cohen, A., Abaluck, B., Localio, A. R., Kimmel, S. E., & Strom, B. L. (2005). Role of computerized physician order entry systems in facilitating medication errors. *JAMA: the journal of the American Medical Association*, *293*(10), 1197–1203.

Lee, Y., Kozar, K. A., & Larsen, K. R. T. (2003). The Technology Acceptance Model: Past, Present, and Future. *The Communications of the Association for Information Systems*, *12*(50), 752-780.

Massachusetts Medical Society. (2004). Computers in Clinical Practice Study. Retrieved November 11, 2008, from http://www.massmed.org/pages/120203pr_ hongkong.asp.

Sviokla, J. J. (1996). Knowledge workers and radically new technology. *Sloan Management Review*, *37*(4), 25–40.

Van Schaik, P., Bettany-Saltikov, J., & Warren, J. (2002). Clinical acceptance of a low-cost portable system for postural assessment. *Behaviour & Information Technology*, *21*(1), 47–57.

Venkatesh, V., & Davis, F. D. (2000). A theoretical extension of the technology acceptance model: Four longitudinal field studies. *Management science*, *46*(2), 186–204.

Venkatesh, V., Davis, F., & Morris, M. G. (2007). Dead or alive? The development, trajectory and future of technology adoption research. *Journal of the Association for Information Systems*, *8*(4), 267–286.

Venkatesh, V., Morris, M. G., Davis, G. B., & Davis, F. D. (2003a). User acceptance of information technology: Toward a unified view. *MIS quarterly*, 425–478.

Venkatesh, V., Morris, M. G., Davis, G. B., & Davis, F. D. (2003b). User acceptance of information technology: Toward a unified view. *MIS quarterly*, *27*(3), 425–478.

Wang, B. B., Wan, T. T. H., Burke, D. E., Bazzoli, G. J., & Lin, B. Y. J. (2005). Factors influencing health information system adoption in American hospitals. *Health Care Management Review*, *30*(1), 44.

Weiner, M., Gress, T., Thiemann, D. R., Jenckes, M., Reel, S. L., Mandell, S. F., & Bass, E. B. (1999). Contrasting views of physicians and nurses about an inpatient computer-based provider order-entry system. *Journal of the American Medical Informatics Association*, *6*(3), 234–244.

Yarbrough, A. K., & Smith, T. B. (2007). Technology acceptance among physicians. *Medical Care Research and Review*, *64*(6), 650–672.

CHAPTER 58

Value-motivational Structure of Individuals in Their Relationship with Post-traumatic Functional Reliability in Particular Occupations

Lazebnaya E.O., Bessonova J.V.

Institute of Psychology, Russian Academy of Sciences
Moscow, Russia
leopsy@mail.ru, farandi@mail.ru

ABSTRACT

Personal value-motivational structure has an impact on the level of post-traumatic functional reliability and professional success in particular occupations.

The differences between the obtained groups on level of posttraumatic disorders of rescuers appeared to be significant both on index PSI, and on index GSI of SCL-90-R. Significant differences in rescuer's level of the system posttraumatic stress disorders of professional etiology were found out while analyzing obtained data. Those differences are negative considering the estimation and prognosis of their functional reliability in posttraumatic period.

It is confirmed for the first time in an empirical research that the role of value-motivational structures of subjective-personal level of psychological regulation of behavior is prevalent in the success of overcoming the consequences of mental trauma by persons of dangerous occupations. Rescuers without expressed posttraumatic problems are more harmonious. They stand out for higher balance of the components of both levels of functioning of the value-setting behavior:

cognitive semantic level ideal values and their level of motivational influence on subject's behavior in different living situations.

The study also states that the lack of dominance in the subjects' ideal knowledge about the value of structures reflecting an expressed individualistic focus of a person, contributes to success of overcoming traumatic psychological problems of persons of dangerous occupations.

The study confirmed the hypothesis that there is no direct relationship between the expert evaluation of the success of persons of dangerous occupations by external criteria of its efficiency and by posttraumatic stress disorder symptoms which are the main factor of their functional reliability.

The study also confirms earlier conclusions made by the results of a longitudinal research of the peculiarities of posttraumatic stress adjustment of war veterans (Lazebnaya, 1999; Lazebnaya, Zelenova, 2007). It is proved that the leading role in the process of mental regulation of posttraumatic adaptation activity of a person solving complex and dangerous professional tasks is played by the results of subjective estimation of the state, reflecting subject's potential capabilities to maintain his functional reliability. Besides, the analysis of obtained data allowed seeing the connection between the structures of the subject's value-motivational system, the level of posttraumatic stress disorder and the intensity of the processes of psychological "burning out" of the rescuers overcoming the problems of the professional work (Bessonova, Lazebnaya, 2011).

Keywords: values, post-traumatic stress, rescue team

The special place among the problems generated by labor activity of a person is occupied with the problems connected with experience of traumatic psychological stress of a professional etiology (Lazebnaya et al., 1997; Tarabrina et al., 2001; Ursano et al., 1996).

In particular, one of the significant social impacts of professional psychic trauma may reduce the functional reliability of the subject in hazardous conditions.

The decrease of functional reliability of the subjects having particular occupations may occur as one of the socially significant consequences of psychic trauma (Bodrov, Orlov, 1998).

However, its value and motivational determinants still remain unexplored, that has defined the purposes and problems of our research.

The purpose of the study: examination of value-motivational structure of individuals in their relationship with post-traumatic functional reliability and professional success in particular occupations.

Subjects: professional rescuers, males, 99 persons.

Research methods:

1. Schwartz's Value Inventory (Schwartz, S. H. and Bilsky, W., 1992, 1994).
2. Special psychometric scale for screening estimation of posttraumatic stress disorders of persons of dangerous occupations "SAPSAN" (Lazebnaya et al., 2009).

3. The Symptom Checklist 90 Revised (SCL-90-R) developed by L.R. Derogatis (Degoratis, 1983).
4. Burnout Inventory (Orel, 2005).

During the analysis of the obtained data all samples were divided on a median of Posttraumatic State Index (PSI) of special psychometric scale for screening estimation of posttraumatic stress disorders, which made 85.5 points. The level of PSI in adjusted rescuers group (49 subjects) was 74.7 points, and among maladjusted rescuers (50 subjects) – 102.8 points. There were no difference between these groups in the age and the work experience.

Significant differences in rescuer's level of the system posttraumatic stress disorders of professional etiology were found out while analyzing obtained data. The differences between the obtained groups on level of posttraumatic disorders of rescuers appeared to be significant both on index PSI and on index GSI of SCL-90-R questionnaire (see Fig. 1). Those differences are negative considering the estimation and prognosis of their functional reliability in posttraumatic period.

Thus the level of positive Spearmen correlation between GSI and PSI indexes made 0.696 ($p \leq 0.0001$).

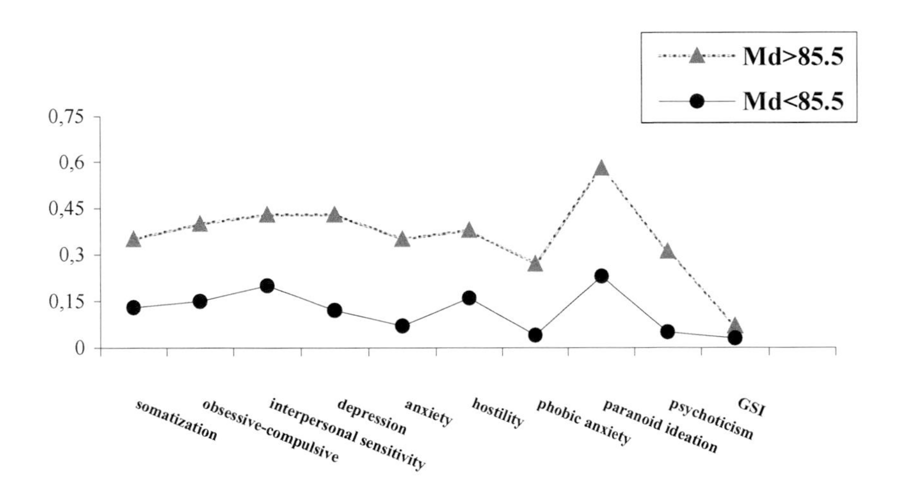

Fig. 1. SCL-90-R profiles in adjusted and maladjusted posttraumatic rescuers groups.

A comparative analysis, dealing with the differences in value-motivational structures of adjustment and maladjustment rescuers groups, was created. Results are shown in tables 1 and 2.

Table 1. **Values as convictions in adjusted and maladjusted posttraumatic rescuers groups**

Med PSI = 85, 5	Maladjusted rescuers group Med > 85,5 n = 50		Adjusted rescuers group Med < 85,5 n = 49		p ≤
Values:	*M*	*SD*	*M*	*SD*	
Conformity	4,66	0,82	4,44	1,08	0,06
Tradition	4,52	0,94	3,88	0,82	0,001
Benevolence	4,75	0,81	4,48	0,72	0,08
Universalism	4,67	0,81	4,98	0,32	0,01
Self-direction	4,66	0,64	4,37	0,58	0,05
Stimulation	4,46	0,71	4,05	0,62	0,05
Hedonism	3,67	1,27	3,17	0,56	0,01
Achievement	5,09	0,81	4,29	0,99	0,0001
Power	4,54	0,83	3,29	0,90	0,0001
Security	4,89	0,85	5,14	0,52	0,08

The value of achievement is the most dominant in the structure of values, beliefs and ideals as convictions (the first part of the Schwartz's Inventory) in maladjusted rescuers group, according to the analysis. The implementation of such values in behavior allows the subject to meet their own needs in the social approval and acceptance. The values of security and benevolence are the next in decreasing order in group of rescuers with severe post-traumatic problems. The hierarchy of values as convictions has a different structure in the group of rescuers without problems owing to traumatic psychological stress of professional etiology. The first rank in that adjusted rescuers group takes the value of security, and the second position are the value of universalism. Both groups of values and related motivational purposes expressed separate personal interests as well as interests of different social groups and communities which the person was included in. The value of benevolence takes the third rank in adjusted rescuers group also.

Comparative analysis of the results of the first part of Schwartz's Value Inventory showed that the adjusted rescuers group has significantly higher expression of such values as universalism and safety (at the level of statistical trends). Other values (such as existential constructs, personal ideals, particularly reflecting the motivational purposes as expressed individualistic orientation) are significantly less pronounced in that group in contrast to the rescuers with post-traumatic problems.

However, the analysis of the data of the second part of the Schwartz's Value Inventory showed the difference between values as convictions and values as behavioral regulators. Although the values of achievement, safety and benevolence are takes principal places in the structure of values in maladjustment posttraumatic rescuers groups, they prefer to regulate their activity in accordance with the values of benevolence, self-direction (self-regulation) and universalism in daily life (see Table 2).

Table 2. Values as behavioral regulators in adjusted and maladjusted posttraumatic rescuers groups.

Med PSI = 85, 5	**Maladjusted rescuers group Med > 85,5 n = 50**		**Adjusted rescuers group Med < 85,5 n = 49**		p≤
Values:	*M*	*SD*	*M*	*SD*	
Conformity	2,39	0,59	2,57	1,03	
Tradition	2,07	0,81	2,20	1,03	
Benevolence	2,77	0,64	3,49	0,66	0,0001
Universalism	2,71	0,61	3,58	0,34	0,0001
Self-direction	2,72	0,34	3,55	0,39	0,0001
Stimulation	2,36	0,81	3,37	1,16	0,0001
Hedonism	2,19	0,57	1,99	0,39	0,06
Achievement	2,49	0,97	2,69	1,20	
Power	1,83	0,46	2,15	0,75	0,01
Security	2,55	0,70	3,22	0,64	0,0001

Besides, the data of survey in adjusted rescuers group doesn't have a complete coincidence of the results of the first and second parts of the Schwartz's Inventory, though both functional levels of their system of values have better coherence.

Rescuers without post-traumatic problems are based on such values as convictions, security, universalism and benevolence. And these values of universalism and benevolence are manifested in their daily life behavior additionally based on the value of self-direction while facing their needs. As for the outsiders in a rating of values as convictions in adjusted rescuers group, we have found the complete concordance of ideals and activity.

Finally, we have found significant differences between adjusted and maladjusted rescuers group in the comparative analysis of value's hierarchy. Those differences are detected for such motivational purposes as benevolence, universalism, self-direction, hedonism, stimulation, power and security. All these values are more expressed and manifested in activity more highly and specifically. But the motivational value of hedonism is the special case - adjusted rescuers have less of its level as compared with maladjusted group.

So, it is confirmed for the first time in an empirical research that the role of value-motivational structures of subjective-personal level of psychological regulation of behavior is prevalent in the success of overcoming the consequences of mental trauma by persons of dangerous occupations. Rescuers without expressed posttraumatic problems are more harmonious. They stand out for higher balance of the components of both levels of functioning of the value-setting behavior: cognitive semantic level ideal values and their level of motivational influence on subject's behavior in different daily situations.

The study also states that the lack of dominance in the subjects' ideal knowledge about the value of structures reflecting an expressed individualistic focus of a person, contributes to success of overcoming traumatic psychological problems of persons of dangerous occupations.

Besides, the study confirmed the additional hypothesis that there is no direct relationship between the expert evaluation of the success of persons of dangerous occupations by external criteria of its efficiency and by posttraumatic stress disorder symptoms which are the main factor of their functional reliability.

Our experts - the commanders of units (2 pers.) - have estimated the level of work success and compliance of the subject's profession status with professional standards. The evaluation criteria consisted of the requirements of job descriptions and personal experiences of the experts. The evaluation procedure contained the parameters of a diagnostic interview and was conducted by the expert-psychologist. A 3-point scale was used.

As a result, each participant obtained an integral quantity rank, which reflects the opinion of the commander about their professional success and quality of implementation of duties: 1 - low success level, 2 – satisfactory one, and 3 – high level. The low level (rate 1) was assessed for 4 persons only, 64 rescuers were evaluated by experts as having satisfactory level (rate 2), and 31 persons obtained high level.

The subjects of the study were divided into groups of "successful" and "unsuccessful" on the median of the expert assessment of the work success level (Me=2 points) for a comparative analysis of rescuers. These groups with various professional successes hadn't differences by main demographic features, as well as by level of post-traumatic disorders (it was defined by the PS -index of the SAPSAN scale).

These results confirmed the research hypothesis that there wasn't direct relation between the assessment of the success by external criteria and features of the subject's status reflecting the level of functional reliability. In fact, while evaluating the work success the experts were guided by procedural and quantitative measures. Experts decided about work success of rescuers but they didn't take into consideration the functional status, the "cost" of the work and successful achievement of another professional purposes which were associated with maintaining the work ability itself, the security, for example (Bessonova, 2003).

This conclusion is also confirmed by the results obtained by the SCL-90-R, which allows estimating the level of general psychopathological distress, subjectively experienced as a concern about their condition (see Table 3).

Table 3. Symptom Scales SCL-90-R in Successful and Unsuccessful rescuers groups.

Symptom Scales SCL-90-R	Unsuccessful rescuers group n = 68		Successful rescuers group n = 31	
	M	SD	M	SD
Somatization (**)	0,18	0,23	0,37	0,32
Obsessive-Compulsive	0,26	0,26	0,30	0,30
Interpersonal Sensitivity (*tend)	0,28	0,23	0,39	0,31
Depression	0,25	0,29	0,33	0,32
Anxiety (*)	0,16	0,23	0,32	0,39
Hostility	0,22	0,29	0,38	0,43
Phobic Anxiety	0,13	0,25	0,21	0,32
Paranoid Ideation	0,35	0,37	0,54	0,52
Psychoticism (*)	0,11	0,28	0,35	0,56
Global Severity Index (GSI) (*)	0,04	0,04	0,07	0,07

** $p \leq 0{,}01$; * $p \leq 0{,}05$; * tend – significant tendency

The most successful rescuers felt more significant concern about their own status that is shown by comparative analysis both for clinical scales and for the total Global Severity Index (GSI). Therefore, this indicates a high "cost" of activity including work activity and stress status of the functional systems which provide, in particular, psychic regulation of their activity.

One purpose of the study was identified features of the structure of rescuer's values and its dependence on the level of burnout. We used Organizational Burnout Inventory OPV-SP (Orel, 2005, Rukavishnikov, 2001) for measure of burnout. This tool allows to assess the "total index of mental burnout " (IPV) as well as a three partial symptoms of burnout according to Maslach's (1996) model: emotional exhaustion, depersonalization and job inefficacy.

The results of comparative analysis are presented in Tables 4 and 5. All samples were divided on a median of IVP index (Me = 4 sten scores) into the 2 groups. The first group included the rescuers without symptoms of burnout (n = 69 pers.). The second group are consisted of the persons with burnout symptoms (n = 31 pers.). There were no differences between these groups in the main indicators of the demographic status.

Persons without burnout have more pronounced level of such convictions values as security, benevolence and universalism. Above it was shown that the same structure of values as convictions is typical for the group of adjusted rescuers, who haven't any symptoms of PTSD. Values of self-direction and achievement are more appropriate for employees with the problems related to burnout. In addition, this group has significantly higher a level of values of power and hedonism.

Table 4. Values as convictions in depending on the level of burnout

Med IVP=4 sten scores	Without burnout group Med < 4 n = 69		With burnout group Med > 4 n = 31		p ≤
Values:	*M*	*SD*	*M*	*SD*	
Conformity	5,16	1,12	4,75	1,54	0,05
Tradition	4,57	1,00	4,31	1,52	0,36
Benevolence	5,67	0,72	4,82	1,92	0,05
Universalism	5,24	0,81	4,62	1,57	0,03
Self-direction	5,31	1,18	5,42	1,13	0,26
Stimulation	4,47	1,32	4,67	1,63	0,59
Hedonism	3,93	1,87	4,16	1,66	0,18
Achievement	5,14	0,92	4,97	1,00	0,05
Power	3,67	1,21	4,18	1,30	0,04
Security	5,75	0,74	4,96	1,46	0,01

Table 5. Values as behavioral regulators in depending on the level of burnout

Med IVP=4 sten scores	Without burnout group Med < 4 n = 69		With burnout group Med > 4 n = 31		p ≤
Values:	*M*	*SD*	*M*	*SD*	
Conformity	2,22	0,70	2,34	1,01	0,83
Tradition	2,17	0,81	1,35	1,00	0,02
Benevolence	2,83	0,92	2,55	1,43	0,24
Universalism	2,71	0,67	2,04	1,10	0,04
Self-direction	3,02	0,80	3,25	0,72	0,35
Stimulation	2,27	0,96	2,39	1,05	0,61
Hedonism	2,19	1,19	2,47	1,03	0,05
Achievement	1,65	0,69	1,80	1,28	0,71
Power	1,36	0,74	1,89	0,94	0,05
Security	2,61	0,76	2,07	0,96	0,05

The absence of symptoms of burnout was closely associated with high level of values of universalism, security, and traditions in behavior too. Whereas significantly predominance of values of power and hedonism as behavioral regulators is the factor that promoted the development of burnout symptoms.

The study also confirms earlier conclusions made by the results of a longitudinal research of the peculiarities of posttraumatic stress adjustment of war veterans (Lazebnaya, 1999; Lazebnaya, Zelenova, 2007). It is proved that the leading role in the process of mental regulation of posttraumatic adaptation activity of a person solving complex and dangerous professional tasks is played by the results of subjective estimation of the state, reflecting subject's potential capabilities to maintain his functional reliability. Besides, the analysis of obtained data allowed seeing the connection between the structures of the subject's value-motivational system, the level of posttraumatic stress disorder and the intensity of the processes of psychological "burning out" of the rescuers overcoming the problems of the professional work (Bessonova, Lazebnaya, 2011).

REFERENCES

Bessonova, J.V. (2003). Formation of professional motivation of rescuers. *The PhD in psychology dissertation author's abstract.* Moscow, Institute of Psychology RAS (manuscript).

Bessonova, J.V., Lazebnaya, E.O. (2011). Value-motivational preconditions of adverse functional states of the subject of work. *Stress, burning out and cooping in a modern context* (eds. Zhuravlev A.L., Sergeenko E.A.). Moscow, Institute of Psychology RAS, p.p. 371-385

Bodrov V.A., Orlov B.Y. (1998). Psychology and reliability: man in the technical control systems. Moscow, Institute of Psychology RAS.

Degoratis L.R. (1983). SCL-90-R Manual. Clinical Psychometric Research, Maryland.

Lazebnaya E.O. (1999). Overcoming of psychological consequences of combat traumatic stress by Afghanistan War veterans. *The Russian Fund of Humanitarian Researches bulletin*, N 4, p.p. 185-191.

Lazebnaya E.O., Zelenova M.E. (2007). Subject and situational determinants of success of posttraumatic re-adjustment of combatants. *Psychology of adjustment and social environment: modern approaches, problems, prospects* (eds. Zhuravlev A.L., Dikaya L.G.). Moscow, Institute of Psychology RAS, p.p. 576-589.

Lazebnaya E.O., Zelenova M.E., Tarabrina N.V., Lasko N.B., Orr S.P., Pitman R.K. (1997). The empirical study of traumatic exposure among Russian veterans of the Afghanistan War. *13-th annual meeting of the International Society for Traumatic Stress Studies (proceedings)*. Montreal, Canada, 1997.

Lazebnaya E.O., Zelenova M.E., Zakharov A.V. (2009). Special psychometric scale for screening estimation of posttraumatic stress disorder symptoms of persons of dangerous occupations. *Problems of fundamental and applied psychology of professional functioning* (issue N 2, eds. Bodrov V.A., Zhuravlev A.L.). Moscow, Institute of Psychology RAS, p.p. 338-364.

Maslach, C., Jackson, S.E, & Leiter, M.P. (1996) MBI: The Maslach Burnout Inventory: Manual. Palo Alto: Consulting Psychologists Press,

Orel, V.E. (2005). Burnout Syndrome. Moscow, Institute of Psychology RAS.

Rukavishnikov, A.A. (2001). Personality determinants and organizational factors of the genesis of psychic burnout in teachers. The PhD in psychology dissertation author's abstract. Yaroslavl. Yaroslavl State Univ.

Schwartz, S.H. (1992). Universals in the content and structure of values: Theoretical advances and empirical tests in 20 countries. *Advances in Experimental Social Psychology*, M. Zanna, San Diego: Academic Press

Schwartz, S.H. (1994). Beyond individualism/collectivism: New dimensions of values. *Individualism and Collectivism: Theory Application and Methods.* U. Kim, H.C. Triandis, C. Kagitcibasi, S.C. Choi and G. Yoon, Newbury Park, CA: Sage.

Tarabrina N.V., Lazebnaya E.O., Zelenova M.E. (2001). Psychological Characteristics of Post-traumatic Stress States in Workers Dealing with the Consequences of the Chernobyl Accident. *J. of Russian and East European Psychology*, vol. 39, No 3, p.p. 29 – 42.

Ursano, R. J., Grieger, T. A., McCarroll, J. E. (1996). Prevention of Posttraumatic Stress: Consultation, Training and early Treatment. *Traumatic stress: the effects of overwhelming experience on mind, body, and society* (eds. B. Van der Kolk, A. McFarlane, L. Weisaeth). N.Y. - L.: The Guilford Press, p.p. 441 – 462.

Van der Kolk, B.A., McFarlane, A.C. (1996). The black hole of trauma. *Traumatic stress: the effects of overwhelming experience on mind, body, and society* (eds. B. Van der Kolk, A. McFarlane, L. Weisaeth). N.Y.- L.: The Guilford Press, p.p. 3 - 23.

CHAPTER 59

Integrated Gesture Recognition Based Interface for People with Upper Extremity Mobility Impairments

Hairong Jiang, Bradley S. Duerstock, Juan P. Wachs

Purdue University
West Lafayette, USA
jiang115@purdue.edu
bsd@purdue.edu
jpwachs@purdue.edu

ABSTRACT

Gestures are of particular interest as a HCI modality for navigation because people already use gestures habitually to indicate directions. It only takes a user to learn few customized gestures for a given navigational task, as opposed to other technologies that require changing hardware components and lengthy procedures. We propose an integrated gesture recognition based interface for people with upper extremity mobility impairments to control a service robot. The following procedure was followed to construct the suggested system. Firstly, quadriplegics ranked a set of gestures using a Borg scale. This led to a number of principles for developing a gesture lexicon. Secondly, a particle filter method was used to recognize hands and represent a generalized model for hand motion based on its temporal trajectories. Finally, a CONDENSATION method was employed to classify the hand trajectories into different classes (commands) used, in turn, to control an actuated device-a robot. A validation experiment to control a service robot to negotiate obstacles in a controlled environment was conducted and results were reported.

Keywords: Borg scale, gesture recognition, particle filter, CONDENSATION

1 INTRODUCTION

Carrying out an independent and autonomous life is deemed a basic need for people with mobility impairments (Cooper, Rninger, and Spaeth, 2006). According to the 2009 US Census Bureau News, 3.3 million people who are 15 or older use a wheelchair. Another 10.2 million use an ambulatory aid such as a cane, crutches or walker and 11 million disabled people need personal assistance with everyday activities. For adults 65 and older over 40% are reported to have some form of disability (U.S. Census Bureau News, 2009). Thus, the demand for more innovative assistive technology (AT) development is needed by a large elderly population and those with movement impairments.

The rapid development of mobile distributed computing systems with effective-human-computer interfaces (HCI) is gaining more popularity (Jacko, 2011). Advanced HCI systems for people with mobility impairments, such as voice, facial and hand gesture based control have been developed where each modality acted alone for the control, or they were combined as multimodal interfaces (Moon, Lee, and Ryu, 2003). Such HCI systems have been used to control or operate wellness monitoring, caregiver assistance, in-home medical alert systems for elder care and intelligent wheelchairs (Scherer, Sax, and Vanbiervliet, 2005; Nguyen, Chahir, and Molina, 2010; Reale, Liu, and Yin, 2011). The use of hand gestures for navigation is an attractive alternative to otherwise cumbersome interfaces, such as joysticks, sip-and puff systems, and tongue controls (Huo and Ghovanloo, 2009). Gesture-based HCI have become popular because they are ergonomic and can be designed to meet individual's particular requirements. Gesture comes naturally to people and is a basic form for individuals to communicate with each other. While not every individual can use gestures, for those who are able to move their hands and upper arms to some degree, gesture-based HCI can be seen as an extremely promising alternative or complement to existing interface techniques.

2 RELATED WORK

There exists a number of works incorporating hand gestures through interfaces to control wheelchair, mobile robot and some other home devices. A hand-gesture based wheelchair navigation system was designed which tracked the hand in real time by using the combination of particle filtering and mean shift method (Shan, Wei, and Tan, 2004). However, it only took into account one hand tracking and only four gestures to control the wheelchair. An intelligent wheelchair control system based on hand gesture recognition was proposed, in which static hand gestures were recognized by employing Haar-like feature detector (Zhang, Zhang, and Luo, 2011). The commands used to control the wheelchair were generated by locating the static gesture in different regions in a window for each frame in the video sequence. The main problem with static hand gestures based control is that the users have to hold their hands in a fixed position for a period of time, which is exhausting when upper extremity impairments exist. A robotic smart house interface was designed to assist people with movement disabilities based on hand gesture, voice and body movements (Park, Bien, and Lee, 2007). The proposed system used ceiling-mounted CCD cameras to observe and recognize the user's hand gestures and used these gestures to control robotic systems and other home-installed devices. But in their system, only pointing gestures were considered. A hand-gesture based control interface for navigating a car-robot was introduced in (Wu, Su, and Wang, 2010). They adopted the dynamic time warping algorithm to classify hand trajectories. Only six commands were considered for car-robot navigation. A natural interaction framework for programming a mobile robot with gestures was developed using two low-cost small mobile robots available on the market (Uribe, Alves, and Rosario, 2011). A comparison between a joystick, wiimote and kinect based interfaces was made in the paper. However, the preliminary gestures selected for controlling the mobile robot required the user to hold their hands over their body for a period of time which made them feel tired during the experiments.

3 SYSTEM ARCHITECTURE

The architecture of this system is illustrated in Figure 1. First, a gesture lexicon was constructed. Subjects with upper extremity impairments were interviewed and a Borg scale (Borg, 1982) rankings were collected to set the guidelines for gestures selection. Detailed description and analysis is shown in sections 4 and 7. After data analysis, eight dynamic gestures were selected to be the components of the gesture lexicon. The hand gesture recognition system includes four parts: foreground segmentation, color-based detection, tracking, and trajectories recognition. A detailed description of the system is shown in sections 5 and 7.

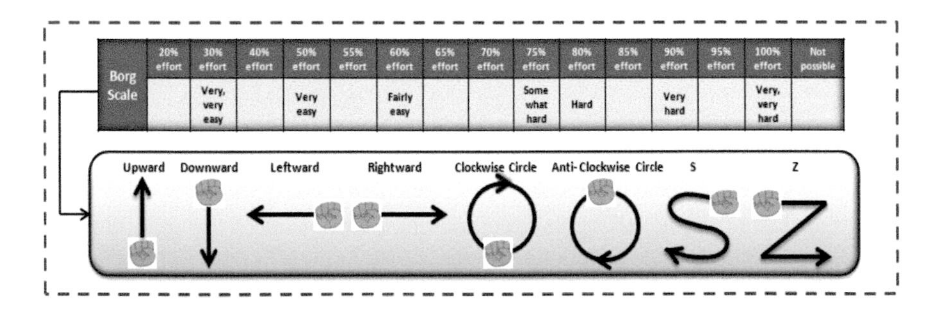

(a) Gesture lexicon

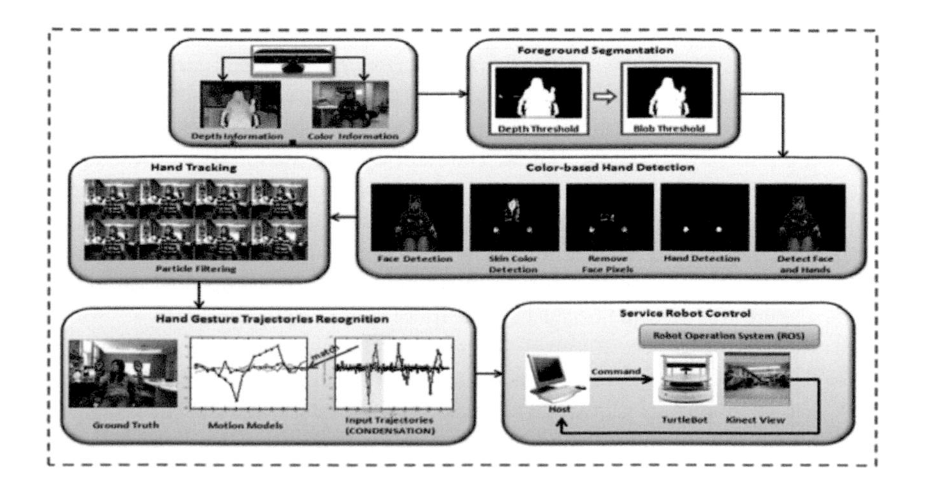

(b) Gesture recognition

Figure 1 System overview

4 GESTURE LEXICON

When developing a gesture-based HCI for persons with upper extremity mobility impairments, the first step is to design a "gesture lexicon". The gesture lexicon is a set of gestures that was determined from quadriplegic subjects using Borg scale metrics. The procedure for designing this lexicon involved first selecting a subset of gestures. This could be accomplished through technology-based approach or human-based approach (Cassell, 1998). The technology-based approach aims to select a set of gestures that can be easily recognized and classified by the system. The biggest problem for the technology-based approach is that gestures selected by this method are not user-centered, which may be difficult to relate to functions or difficult to perform, particularly for persons with upper limb mobility impairments. The human-based approach proposes to construct the gesture lexicon based on studying how people (who are the potential users) interact with each other. In our research, the target population is users with mobility impairments. In such a scenario, it is mandatory that the gestures will be ergonomic (easy to perform). To achieve this goal the human-based approach will be used for constructing the gesture lexicon.

First, a preliminary lexicon including both dynamic and static gestures were shown and demonstrated through a video presentation to subjects with upper extremity impairments and they ranked these gestures employing the Borg scale. The dynamic gestures are hand and upper arm movements by which hand trajectories are generated. The static gestures are postures involving fixed hand and upper limb positions, but not incorporating sophisticated finger variation. Originally, the preliminary gesture lexicon is constructed with 50 dynamic gestures and 15 static gestures. Since the number of gestures that a user can remember is very limited, we limited the number of gestures. The reduced gesture lexicon includes 30 dynamic gestures and 10 static gestures. The gestures were reduced according to the following principles:

(1) Similarity (Proctor and Zandt, 2008): if the trajectory of one gesture has very few differences from another gesture or has many common elements with another gesture, one of the gestures can be deleted. One example is the clockwise circle gesture and the P gesture. Both of their trajectories first go up and then go down as shown by Figure 2(a). The clockwise circle gesture was selected because it was more symmetric.

(2) Redundancy (Yee, 2009): if the trajectory of one gesture contains part of the trajectory of another, this gesture is a redundant gesture of the other. For example: as shown by Figure 2(b) the start gesture (the left one) trajectory contains a Z gesture trajectory. Arbitrary the Z gesture was selected.

(3) Minimize memory load (Nielsen, 1992): reduce gestures that are hard to remember, i.e. the sum gesture as shown by Figure 2(c).

After reduction, the preliminary gesture lexicon includes single hand dynamic gestures in both vertical plane and horizontal plane, two hand gestures in both vertical plane and horizontal plane (as shown in Figure 2(d)) and static gestures.

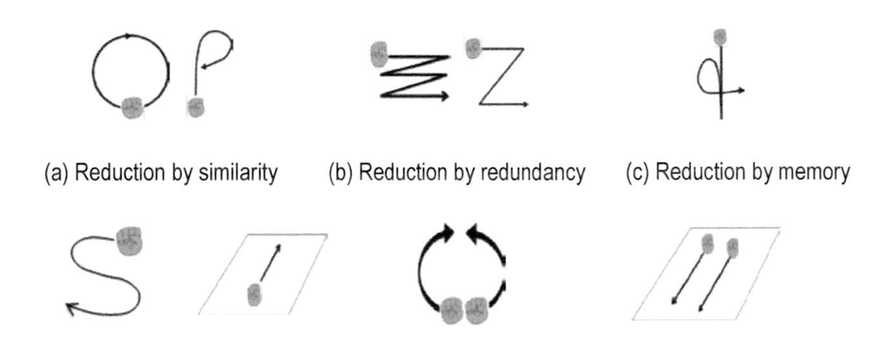

(d) Preliminary gesture lexicon components

Figure 2 Gesture lexicon

5 GESTURE RECOGNITION

The gestures in the lexicon were recognized by the system and the commands corresponding to each gesture were sent to service robot. Four procedures were followed to achieve the gesture recognition. Firstly, the whole human body will be treated as the foreground and segmented from the background (as shown by Figure 3). Secondly, the hands are detected by using face detection and skin color histogram model. The results are shown as in Figure 4. A particle filter method was subsequently used to recognize the hands and represent human hand motion as temporal trajectories (as shown in Figure 5). Finally, dynamic velocity motion models for the gestures in the lexicon were constructed and CONDENSATION-based trajectory recognition method was used to classify the hand trajectories into different classes.

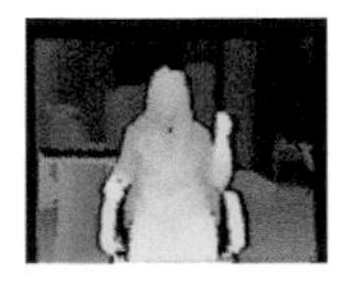  

(a)Depth Image (b) Depth threshold mask (c) Foreground mask

Figure 3 Foreground segmentation

  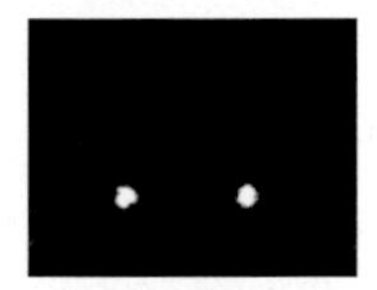 

(a)Face Detection (b) Skin color detection (c) Hand extraction (d) Localization

Figure 4 Face and hand detection

Figure 5 Face and hand tracking

6 SERVICE ROBOT CONTROL

The service robot controlled by the gestures recognition system is the TurtleBot™ robot from Willow garage® (as shown in Figure 6), which includes a mobile base, a Kinect 3D sensor and a netbook with robot operation system (ROS) on a linux environment. TurtleBot was controlled through gesture commands. Two modes were used to control TurtleBot: discrete mode and continuous mode. In discrete mode, the robot moves every time that a command is issued, otherwise it stays still. While in the continuous mode, the robot responds to a given command, until the stop command is issued. To switch between these two modes one distinctive gesture is used.

Figure 6 Turtlebot robot

7 EXPERIMENTS

7.1 Gesture lexicon construction

One female with hemi-media in her left arm and two male subjects with upper extremity impairments were interviewed for data collection. One of the male subjects was a Cervical-4/5 quadriplegic, whose left arm had less movement than his right arm, which performed most of the gestures but no hand movement. During evaluation, both dynamic and static gestures were ranked by each subject from 20% effort to not possible according to the Borg scale. The 15-point Borg scale was chosen because it is more sensitive to the variation of effort subjects spent on each gesture. The subject was asked to perform each gesture, to describe the limitations and physical stress they experience, and then to rank the effort needed to perform the gesture on the Borg scale. Once the subjects finished all the 16 single hand dynamic gestures, 14 two hand dynamic gestures and 10 static gestures, their answers were summarized in a data sheet to help us develop the guidelines for hand movements and postures.

According to the scored rankings on the Borg scale, eight dynamic gestures were selected as the optimal candidates for the gesture lexicon. They are upward, Downward, Rightward, Leftward, Z, Clockwise Circle, Counter Clock Circle, and S gesture in vertical plane (as shown in Figure 1(a)). The effort required to perform the selected dynamic gestures, discarded dynamic gestures, static gestures and two hand dynamic gestures are shown by the histogram in Figure 7. From the histogram, it can be seen that the selected dynamic gestures required the least effort, on average.

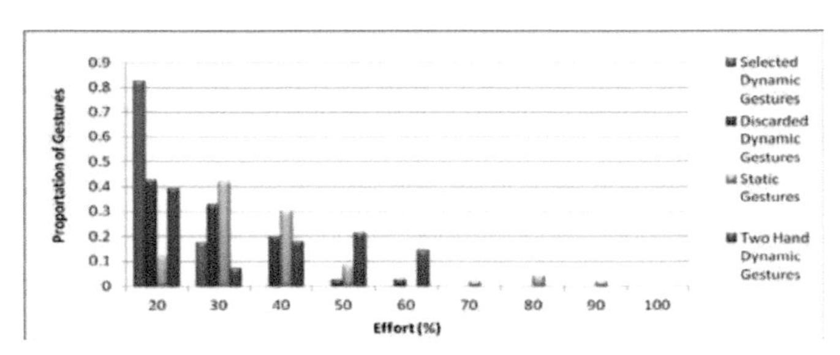

Figure 7 Comparison between selected and discarded gesture groups

ANOVA (Analysis of Variance) and T-test results were given to prove that there was significance difference between the different groups in terms of the effort required to perform the gestures. The mean for the selected dynamic gesture population is μ_1, for the discarded dynamic gesture population is μ_2, for the discarded static gesture population is μ_3 and for the two hands dynamic gesture population is μ_4. The hypotheses were that H_0 states that there is no significant difference among the effort of the selected dynamic, discarded dynamic, static and two hand dynamic gesture population, while H_1 states that there is significant difference among them. The significance level α is set to 0.05. The hypotheses tested are:

$$H_0: \mu_1 = \mu_2 = \mu_3 = \mu_4 \quad H_1: \mu_1 \neq \mu_2 \neq \mu_3 \neq \mu_4$$

P value found was 8.73E-09, which was less than the significance level 0.05. The null hypothesis was rejected, which means there was significant difference among the effort of the selected dynamic, discarded dynamic, static and two hand dynamic gesture populations. Also, it was found that there was significant difference in the effort required between the selected dynamic gestures and the other three gesture groups by applying t-test to each of the two populations. By choosing the null hypothesis as H_{10} and the alternative hypothesis as H_{11}:

$$H_{10}: \mu_1 = \mu_i \quad H_{11}: \mu_1 \neq \mu_i$$

where j = 2, 3, 4. The p value found were 8.19E-05, 2.61E-09 and 3.15E-05, which were less than the significance level 0.05. These indicated that there were significances of effort between the selected dynamic gestures and the other three gesture groups.

7.2 Heuristics

According to (Kortum, 2008), and the scored rankings on the Borg scale, the following heuristics were found to guide the lexicon design process.

(1) Select gestures that do not strain the muscles.
(2) Select gestures that do not require much outward elbow joint extension.
(3) Select gestures that do not require much outward shoulder joint extension.
(4) Select gestures that avoid outer positions.
(5) Select dynamic gestures instead of static gestures.
(6) Select vertical plane gestures where hands' extension is avoided.
(7) Relaxed neutral position is in the middle between outer positions.
(8) Select gestures that do not require wrist joint extension caused by hand rotation.

7.3 Gesture recognition and robot control

The eight-gesture lexicon for the system was tested by nine users, which resulted in a recognition accuracy of about 90.36% on average for all the gestures. The Turtlebot robot was controlled to deliver instruments from place A to B. Figure 8 shows a sequence of the Kinect view from the Turtlebot. The map for the lab and the trajectories of the robot for both discrete mode control (red solid line) and continuous mode control (blue dash line) are shown in Figure 9.

Figure 8 Turtlebot robot kinect view

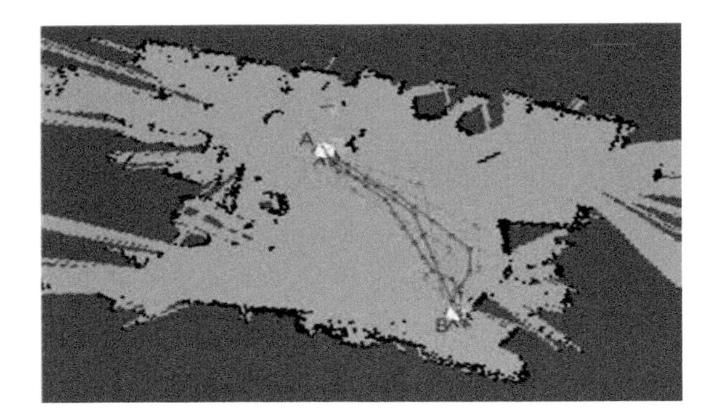

Figure 9 Map of the lab and the robot trajectories

8 CONCLUSIONS

In this paper, a hand gesture lexicon was designed for people with upper extremity mobility impairments. It was shown to require low physical effort based on subjective rankings made on a Borg scale. The heuristics for selecting appropriate gesture lexicons was outlined considering physical and ergonomic constraints. It was found that it is better to select dynamic vertical plane gestures instead of static gestures. In addition, the gestures within the lexicon must not require much outward wrist, elbow or shoulder joint extension. A gesture-based recognition system utilizing this eight-gesture lexicon was tested by nine users and resulted in a recognition accuracy of about 90.36%. The lexicon was validated in a task involving a service robot, which was controlled in a lab environment to deliver laboratory instruments. Two of the gestures in the lexicon were not always recognized due to the fact that the motion models used for those gestures were similar to other gestures in the lexicon. Future work includes developing more robust recognition algorithms based on Bayesian belief networks.

ACKNOWLEDGMENTS

This work was partially funded by the National Institutes of Health through the NIH Director's Pathfinder Award to Promote Diversity in the Scientific Workforce, grant number DP4-GM096842-01.

REFERENCES

Baudel, T. and Beaudoin-lafon, M.. 1993. Charade: remote control of objects using free-hand gestures. *Communications of the ACM* 36: 28-35.

Borg, G. 1982. Psychophysical bases of perceived exertion. *Medicine & Science in Sports & Exercise* 14(5): 377-383.

Cassell, J.. 1998. A Framework For Gesture Generation and Interpretation. *Computer Vision in Human-Machine Interaction, New York: Cambridge University Press* 191-215.

Cooper, R.A., Rninger, M.L. and Spaeth, D.M., et al. 2006. Engineering better wheelchairs to enhance community participation. *IEEE Transaction n Neural Systems and Rehabilitation Engineering* 14(4): 438-455.

Huo, X. and Ghovanloo, M.. 2009. Using Unconstrained Tongue Motion as an Alternative Control Mechanism for Wheeled Mobility. *IEEE Transaction on Biomedical Engineering* 56(6): 1719-1726.

Jacko, J.A.. 2011. Human-Computer Interaction Design and Development Approaches. *14th HCI International Conference* 169-180.

Kortum, P.. 2008. HCI Beyond the GUI: Design for Haptic, Speech, Olfactory, and Other Nontraditional Inferences. *P. Kortum, ed., Morgan Kaufmann* 75-106.

Moon, I., Lee, M. and Ryu, J., et al. 2003. Intelligent Robotic Wheelchair with EMG-, Gesture-, and Voice-based Interfaces. *Proceedings of the 2003 IEEE International Conference on Intelligent Robots and Systems* 3453-3458.

Nguyen, B.L., Chahir, Y. and Molina, M., et al. 2010. Eye Gaze Tracking with Free Head Movement using a single camera. *Proceedings of the Symposium on Information and Communication Technology.*

Nielsen, J.. 1992. The usability engineering life cycle. *IEEE Computer* 25(3): 12-22.

Park, K.H., Bien, Z. and Lee, J.J., et al. 2007. Robotic smart house to assist people with movement disabilities. *Auton Robot* 22:183-198.

Proctor, R.W. and Zandt, T.V.. 2008. *Human Factors in Simple and Complex Systems.*

Reale, M., Liu, P. and Yin. L.J.. 2011. Using eye gaze, head pose and facial expression for personalized non-player character interaction. *IEEE Computer Society Conference on Computer Vision and Pattern Recognition Workshops* 13-18.

Scherer, M.J., Sax, C. and Vanbiervliet, A., et al. 2005. Predictors of assistive technology use: The importance of personal and psychosocial factors. *Disability and Rehabilitation* 27(21): 1321-1331.

Shan, C.F., Wei, Y.C. and Tan, T.N., et al. 2004. Real time hand tracking by combining particle filtering and mean shift. *Sixth IEEE International Conference on Automatic Face and Gesture Recognition* 669- 674.

Suma, E.A., Lange, B. and Rizzo, A., et al. 2011. FAAST: The Flexible Action and Articulated Skeleton Toolkit. *IEEE Virtual Reality*. 247-248.

Uribe, A., Alves, S., Rosario, J.M., et al. 2011. Mobile robotic teleoperation using gesture-based human interfaces. *IEEE Colombian Conference on Automatic Control and Industrial Applications (LARC).* 1-6.

U.S. Census Bureau, [Online]. Available: http://www.census.gov/hhes/www/disability/overview.html.

Wu, XH., Su, MC., Wang, PC.. 2010, A Hand-Gesture-Based Control Interface for a Car-Robot. *IEEE International Conference on Intelligent Robots and Systems.*

Yee, W.. 2009. Potential Limitations of Multi-touch Gesture Vocabulary: Differentiation, Adoption, Fatigue. *Human-Computer Interaction, Part II, HCI* 291-300.

Zhang, Y., Zhang, J. and Luo, Y.. 2011. A Novel Intelligent Wheelchair Control System Based On Hand Gesture Recognition. *Proceedings of the IEEE International Conference on Complex Medical Engineering*

Zimmerman, T., Lanier, J. and Blanchard, C., et al. 1987. A Hand Gesture Interface Device. *In Proceedings of CHI 87 and GI, ACM* 189-192.

CHAPTER 60

Using Computerized Technician Competency Validation to Improve Reusable Medical Equipment Reprocessing System Reliability

Kai Yang[a], Nancy Lightner[b], Serge Yee [c], Mahtab Jahanbani Fard[a], and Will Jordan[b]

[a] Department of Industrial & Systems Engineering, Wayne State University, Detroit, MI 48202, USA
Kai.Yang@wayne.edu
Mahtab.Jahanbanifard@wayne.edu
[b] Department of Veterans Affairs, Veterans Engineering Resource Center, VA-Center for Applied Systems Engineering, Indianapolis, IN, 46222, USA
Will.Jordan@va.gov
Nancy.Lightner@va.gov
[c]Department of Veterans Affairs, Veterans Engineering Resource Center, VA-Center for Applied Systems Engineering, Detroit, MI, 48201, USA
Serge.Yee@va.gov

ABSTRACT

Flexible fiberoptic endoscopes are sophisticated reusable medical equipment. Reprocessing steps depend on the endoscope make and model and require technician proficiency to perform correctly. The Interactive Visual Navigator[1] (IVN™™) ensures that a technician has undergone a competency check prior to displaying reprocessing direction for the endoscope model and prevents the technician from reprocessing endoscopes if their competency check is not current. IVN™ also accepts, stores and displays the history of instruction revisions and ensures the use of the current version. Using these methods, IVN™ assists in the reduction of human errors and coordinates continuous improvement in the reliability and safety of reusable medical equipment reprocessing.

Keywords: proficiency, competency, reusable medical equipment, endoscope, reprocessing

1 INTRODUCTION

Patient safety is an urgent national issue for healthcare today. Morbidity and mortality resulting from human error among hospitalized patients has risen throughout the United States, and heightens concern about professional competency (Kohn et.al, 2000). Technicians, nurses and other health care professionals are under increased pressure to provide safe and effective care. One area of concern affecting patient safety is the use of unclean reusable medical equipment (RME), such as flexible fiberoptic endoscopes (FFE), delivered from the Sterile Processing Service (SPS). Endoscopes are complex pieces of equipment used to identify and evaluate the function of internal organs and cavities. They are inserted into a patient during colonoscopies, endoscopic ultrasounds, endoscopic retrograde cholangiopancreatography and other procedures and are used in over 20 million gastroenterology (GI) endoscopic procedures annually in the United States (Everhart, 2008). It is estimated that infections associated with GI endoscopy occur at a rate of 1 in almost 2 million procedures (Schembre, 2000) and that the source of these infections is either failure to follow established reprocessing guidelines, or the use of defective equipment (ASGE Quality Assurance in Endoscopy Committee, 2011).

Reprocessing involves several steps generally consisting of precleaning, leak testing, manual cleaning, high-level disinfection/sterilization, and rinsing and drying (with alcohol flush) (Pennsylvania Patient Safety Advisory, 2010). Hundreds of endoscope models and other RME devices exist, and the specific processes within each general step vary, depending on the device type and model. This creates a challenging reprocessing environment that necessitates proficient and competent SPS Technicians.

Between January 1 and September 30, 2010, the Veterans Affairs (VA) Office

---

[1] IVN™ is the property of the Department of Veterans Affairs and Wayne State University. Any use of the content presented in this paper without the express written consent of the IVN™ Program Manager is strictly prohibited.

of Inspector General reviewed 45 VA facilities for compliance with VA standards for RME practices and identified six areas for improvement (House Committee on Veterans Affairs, 2011). The Interactive Visual Navigator (IVN™™) system was developed by the Veterans Health Administration in collaboration with Wayne State University to address three of these six areas. These areas are: standard operating procedures (SOPs) are current, consistent with manufacturers' instructions, and located within the reprocessing areas; employees consistently follow SOPs, supervisors monitor compliance, and annual training and competency assessments are completed and documented; and processes for consistent internal oversight of RME activities are established to ensure senior management involvement (House Committee on Veterans Affairs, 2011).

This paper describes the processes and reports available in IVN™ to address improvement in the annual training and competency assessment completion and documentation area identified as problematic within the VA.

2 BACKGROUND

Insufficient decontamination of RME presents a risk for widespread infection to patient populations. The risk of improper reprocessing of endoscopes has prompted organizations around the world to develop guidelines and recommendations in an effort to reduce associated infection (ASGE Quality Assurance in Endoscopy Committee, 2011, O'Brien, 2009, Rey, et al., 2005, Rutala, et. al., 2008, Society of Gastroenterology Nurses and Associates, Inc., 2007, WHO, 2004). The guidelines consistently emphasis 1) staff training, 2) following manufacturers' instructions, and 3) proper storage of clean, sterilized and disinfected equipment. IVN™ was conceived as a tool to implement a standardized method of following established guidelines and especially those specified by the VA House Committee.

In every organization training is crucial and in healthcare, proper training ensures technicians and other workers possess the basic skills necessary to effectively perform their functions. Several types of training have proven effective in reducing human error, including initial skill training (conducted in classroom and supplemented with on-the-job experience), refresher training (minimize worker errors and reduce the potential for a worker's skills to deteriorate) and management systems training (ensure healthcare workers can readily identify and follow relevant management systems) (Rooney et al., 2002).

Competence (or competency) is the ability of an individual to perform a job properly (Boyatzis 1982). A competency is a set of defined behaviors that provide a structured guide enabling the identification, evaluation and development of the behaviors in individual employees. Competence is defined in the context of particular knowledge (understanding fact and procedure), traits (personality characteristics that a person to behave or respond in a certain way), skills (capacity to perform specific actions), and abilities (attributes inherited or acquired through previous experience and brings to a new task) (Landy 1985). A combination of these parameters improves task execution. Regardless of training, competency grows through experience and the ability of an individual to learn and adapt. Measuring competence is possible using a variety of methods: tests, simulations, supervisory performance appraisals and records of performance (Kak, et al., 2001). Job simulations and job samples are reportedly the best predictors of job performance, while written tests are probably the weakest predictor (Spencer et al., 1994).

3 APPROACH

IVN™ is a networked system that presents reprocessing instructions for endoscope models as well as storing data related to technician performance and competency assessment and the date the instructions for an endoscope model were effective. IVN™ instructions are displayed on a 17-inch hermetically sealed touch screen computer monitor attached with a wall-mounted adjustable arm, and an optional mouse and keyboard. Each technician and manager is assigned a unique login identifier which they use to access IVN™ processes. To initiate the display of reprocessing instructions, technicians enter the final four digits of the endoscope's serial number, and a list of endoscopes matching that number appears. For each of the scopes on the list, IVN™ checks against the technician's competency profile. If an endoscope is a model that the technician has passed a competency check within the time specified (one year is the default), IVN™ allows access to the respective Work Instruction Module (WIM). If the competency check has lapsed, IVN™ presents contact information for the SPS Manager who will perform the competency check. Figure 1 contains a screen shot of the competency check and message to the technician. To proceed in reprocessing the endoscope with serial number 33333333, the technician selects the blue arrow in the "Start iNAV" column.

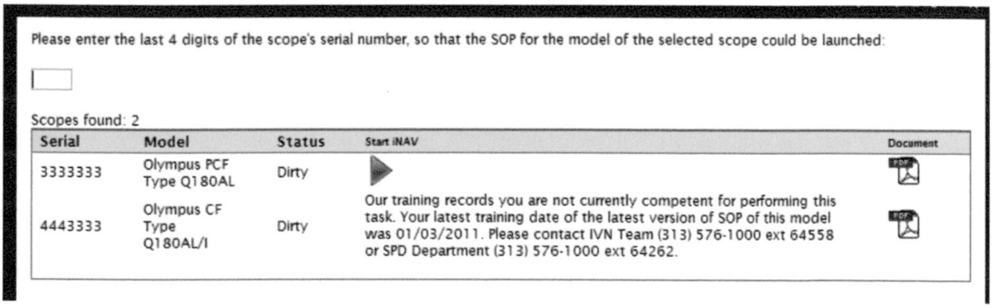

Please enter the last 4 digits of the scope's serial number, so that the SOP for the model of the selected scope could be launched:

Scopes found: 2

Serial	Model	Status	Start iNAV	Document
3333333	Olympus PCF Type Q180AL	Dirty		
4443333	Olympus CF Type Q180AL/I	Dirty	Our training records you are not currently competent for performing this task. Your latest training date of the latest version of SOP of this model was 01/03/2011. Please contact IVN Team (313) 576-1000 ext 64558 or SPD Department (313) 576-1000 ext 64262.	

Figure 1: Screen shot of display after the last four digits of an endoscope serial number entered.

Within this context, competency refers to whether the technician has completed training within an acceptable time frame. IVN™ disables access to an endoscope reprocessing procedure that has a lapse in competency check until the SPS Manager records completion of a competency check by the technician for that model. Figure 2 contains a screen shot of a report on the competency of a specific technician. This report is available to the SPS Manager and the technician. This view indicates checks that have or are about to expire, by endoscope type, manufacturer and model number. The House Committee on Veterans Affairs (2011) indication that annual training and competency assessments are completed and documented motivated the once-a-year assessment requirement in IVN™.

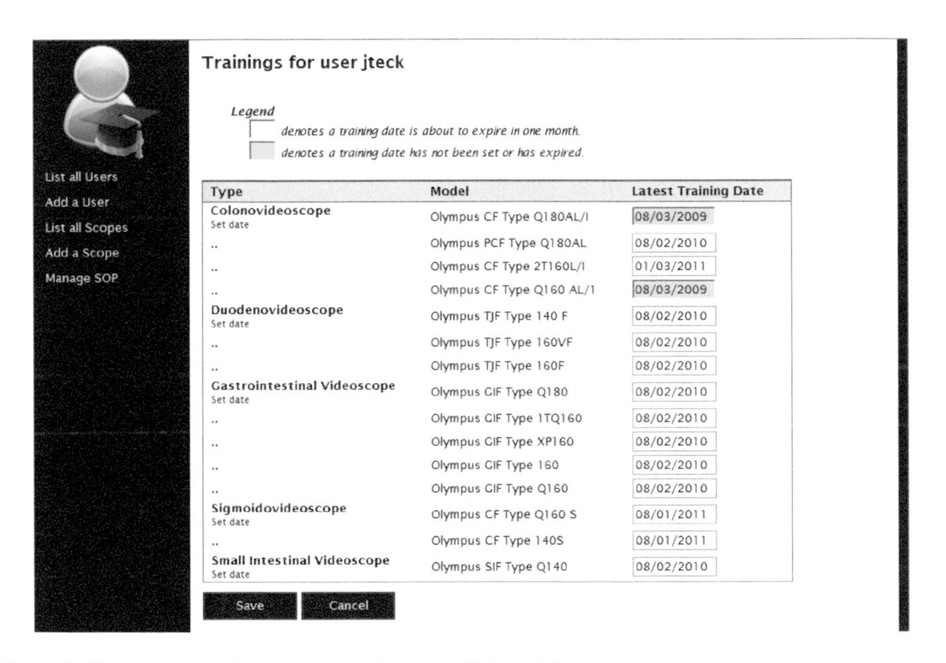

Figure 2: Manager view of competencies for a specific technician.

If an SPS Manager would like to check on the competencies of a technician, the report shown in Figure 3 is available through IVN™. This report is useful for determining whether a technician can reprocess all equipment needed on a given day. Reporting competencies by technician supports personnel scheduling decisions as well as training needs.

BIRT Report Viewer

Showi Run report 1 of 1 — Go to page:

VAMC Detroit — Page 1

Trainings for user Johnny Teck

Model	Last Training Date
Olympus GIF Type Q180	08/02/2010
Olympus CF Type 2T160L/I	01/03/2011
Olympus PCF Type Q180AL	08/02/2010
Olympus CF Type Q180AL/I	08/03/2009
Olympus CF Type Q160 AL/1	08/03/2009
Olympus TJF Type 160F	08/02/2010
Olympus SIF Type Q140	08/02/2010
Olympus GIF Type 1TQ160	08/02/2010
Olympus TJF Type 140 F	08/02/2010
Olympus CF Type Q160 S	08/01/2011
Olympus GIF Type XP160	08/02/2010
Olympus GIF Type 160	08/02/2010
Olympus TJF Type 160VF	08/02/2010
Olympus GIF Type Q160	08/02/2010
Olympus CF Type 140S	08/01/2011

Total Trainings:15

Created on: May 17, 2011 7:43 AM

Figure 3: View of last training dates for a specific technician.

At a facility-specified interval prior to expiration of a competency check, a report is displayed on the screen when a technician logs in to IVN™. This alerts the technician to the need for a competency check within the specified period of time. See Figure 4 for a sample report that indicates competency expiration dates on two Olympus models.

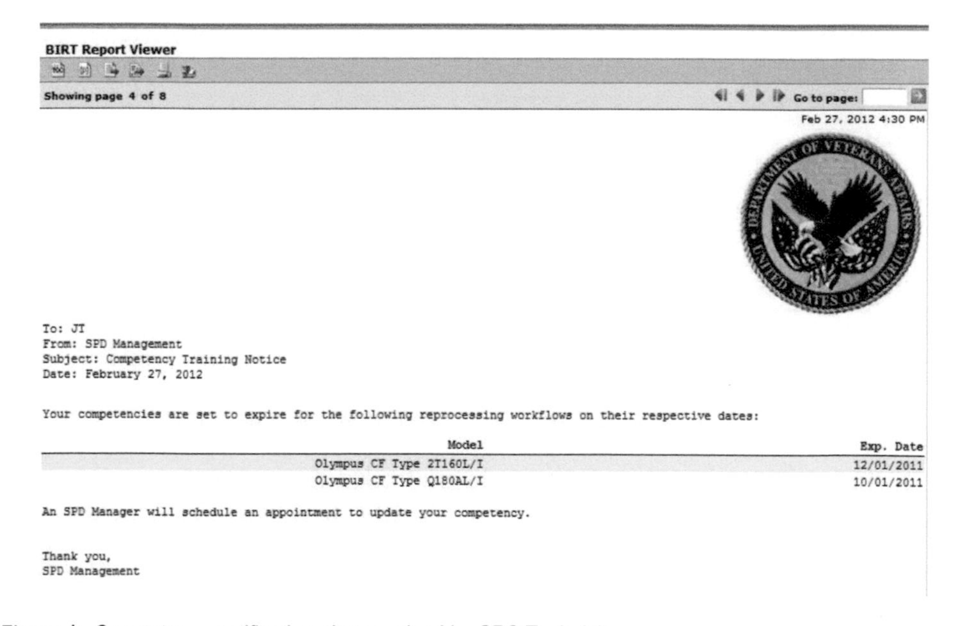

BIRT Report Viewer

Showing page 4 of 8 — Go to page:

Feb 27, 2012 4:30 PM

To: JT
From: SPD Management
Subject: Competency Training Notice
Date: February 27, 2012

Your competencies are set to expire for the following reprocessing workflows on their respective dates:

Model	Exp. Date
Olympus CF Type 2T160L/I	12/01/2011
Olympus CF Type Q180AL/I	10/01/2011

An SPD Manager will schedule an appointment to update your competency.

Thank you,
SPD Management

Figure 4: Competency notification alert received by SPS Technician.

IVN™ also reflects the VA's requirement for regular review of the step-by-step instructions presented to the SPS Technicians. Figure 5 shows an example IVN™ screen that allows an SPS Manager to select a review date to store for this check.

IVN™ tracks the date competency checks were performed, but does not enforce the type or style of competency check. What comprises a competency check is at the discretion of the facility. In any educational or professional setting, making good decisions about competency is difficult. Certification decisions in general are particularly complex and organizations must create procedures to ensure that appropriate decisions about the knowledge and skills of individuals are made. The relationship between the competence and certification of the SPS technician and the reliability of reprocessed equipments is a relatively new area of inquiry.

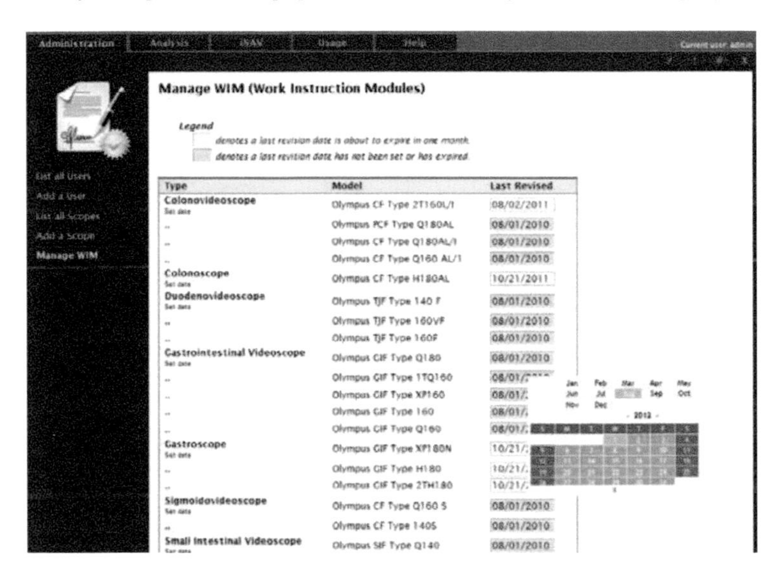

Figure 5: IVN™ screen to update the revision date of reprocessing instructions.

In healthcare and other organization, the term "gold standard" describes practices that are accepted as the best and most effective for a particular problem, disease, or intervention (Kak, 2001). Best practices used in other industries can apply to healthcare; however currently no gold standard for measuring the health care provider's competence exists. The most effective and feasible approach might depend on the situation. In general competence is measured using a variety of different methods, such as written tests, computerized tests, records of performance, simulations with anatomic models, other simulations, job samples, and supervisory performance appraisals (Kak, 2001). Each of these methods has advantages and disadvantages.

A relationship exists between competency and proficiency. The more competent a person, the less they are likely to make errors. IVN™ stores completion time for each step of reprocessing, which managers might use to determine performance measures as well as technician competency. Comparing recorded time with minimum time highlights technicians that are not taking appropriate time to reprocess. It may also indicate faulty work instructions.

Because manufacturers revise reprocessing instructions, IVN™ keeps a history of revision dates and notifies both technicians and SPS Managers when a review is needed. Instructions expire after one year. At eleven months following the last revision date; IVN™ displays a warning on the screen to technicians reprocessing affected endoscope models. In addition, IVN™ notifies the SPS Manager by color-coding on the Work Instruction Module report shown in Figure 4. This feature enables a regular update of reprocessing instructions, which are subject to revisions from manufacturers as well as feedback from reprocessing practices.

4 SUMMARY AND CONCLUSION

FFEs are used in over 20 million procedures annually in the United States. Decontamination and sterilization of this equipment is necessary between procedures to reduce the risk of infection to the next patient. Because reprocessing of each model follows potentially different instructions, technicians are challenged to properly reprocess FFEs. The IVN™ system, developed by the Department of Veterans Affairs and Wayne State University, enforces a competency check for each endoscope model used by each technician before displaying reprocessing instructions for the model. IVN™ also documents revision histories of the instructions used for reprocessing and enforces instruction modification verification. This combination of checks is intended to improve the proficiency and reliability of RME reprocessing to reduce the risk of subsequent patient infection.

ACKNOWLEDGMENTS

The authors gratefully acknowledge the support of the Veterans Health Administration.

REFERENCES

ASGE Quality Assurance in Endoscopy Committee (2011). Multisociety guideline on reprocessing flexible gastrointestinal endoscopes: 2011, *Gastrointestinal Endoscopy, 73(6)*, 1075-1084.

Boyatzis, R.E. 1982. *The Competent Manager: A Model for Effective Performance*. New York. Wiley.

Everhart, JE. (2008). The burden of digestive disease in the United States. *NIH publication no. 09-6443.* Washington (DC): U.S. Department of Health and Human Services.

House Committee on Veterans Affairs. (May 3, 2011). Witness Testimony of Hon. Robert A. Petzel, M.D., Under Secretary for Health, Veterans Health Administration, U.S. Department of Veterans Affairs. Retrieved February 9, 2012 from http://veterans.house.gov/prepared-statement/prepared-statement-hon-robert-petzel-md-under-secretary-health-veterans-health-0

Kak, N., B. Burkhalter, and M. Cooper. 2001. Measuring the competence of healthcare providers. *Operations Research Issue Paper* 2(1). Bethesda, MD: Published for the U.S. Agency for International Development (USAID) by the Quality Assurance (QA) Project.

Kohn, LT, Corrigan JM, Donaldson MS, editors. (2000). To err is human: building a safer health system. A report of the Committee on Quality of Health Care in America, Institute of Medicine. Washington, DC: National Academy Press.

Landy, F.J. (1985). *Psychology of Work Behavior*. (3rd ed.). Homewood, IL: Dorsey Press.

O'Brien, V. (2009). Controlling the process: legislation and guidance regulating the decontamination of medical devices, *Journal of Perioperative Practice, 19(12),* 428-432.

Pennsylvania Patient Safety Advisory. (2010). The Dirt on Flexible Endoscope Reprocessing, *Pa Patient Saf Advis, 7(4),* 135-40. Retrieved February 9, 2012 from http://patientsafetyauthority.org/ADVISORIES/AdvisoryLibrary/2010/dec7(4)/Pages/135.aspx#bm9

Rey, J.F., Bjorkman, D., Duforest-Rey, D. Axon, A., Saenz, R., Fried, M.. et al., (2005). *WGO practice guideline endoscopy disinfection, world gastroenterology organization (WGO).* Retrieved February 8, 2012 from http://www.worldgastroenterology.org/assets/downloads/en/pdf/guidelines/09_endoscope_disinfection_en.pdf

Rooney, J., N. Lee, V. Heuvel, and D. K. Lorenzo. (2002). Reduce Human Error: How to analyze near misses and sentinel events, determine root causes and implement corrective actions. *Healthcare Quality*, 27-36.

Rutala, W.A., Weber, D.J. and Healthcare Infection Control Practices Advisory Committee. (2008). *Guideline for disinfection and sterilization in health care facilities, 2008.* Atlanta, GA: Centers for Disease Control and Prevention.

Schembre DB. (2000). Infectious complications associated with gastrointestinal endoscopy, *Gastrointest Endosc Clin N Am, 10(2),* 215-32.

Society of Gastroenterology Nurses and Associates, Inc. (2007). Standards of Infection Control in Reprocessing of Flexible Gastrointestinal Endoscopes. Chicago, Illinois. Retrieved February 9, 2011 from http://www.sgna.org/Portals/0/Education/Practice%20Guidelines/InfectionControlStandard.pdf.

Spencer, L.M., McClelland, D.C., and Spencer, S.M. (1994). *Competency assessment methods*. Boston: Hay/McBer Research Press.

World Health Organization (WHO). (2004). Practical guidelines for infection control in health care facilities. Retrieved February 9, 2010 from http://www.searo.who.int/LinkFiles/publications_PracticalguidelinSEAROpub-41.pdf

CHAPTER 61

Validity Verification of Coloring Recreation Taking Place at Pay Nursing Home

*KAWABATA Shinichiro**, *NASU Maki***
YAMAMOTO Akiyoshi*** *KIDA Yoshiyuki**
*KUWAHARA Noriaki*HAMADA Hiroyuki**

*Kyoto institute of technology
Kyoto Japan
hhamada@kit.ac.jp

** soliton corporation CO. LTD.
Kyoto Japan
arser3@gmail.com

***City Estate Co., Ltd.
Osaka, Japan
yamamoto@city-estate.co.jp

ABSTRACT

In Japan aging has proceed at a rapid rate, because of a great extension of the average life span by the improvement of the living standard. In 1970 it became aging society and rushed into the aged society in 1994. When this tendency continues, one person in four people comes to enter the age of senior citizen in 2015.

The numbers of dementias patients are also increasing and it has become 1.7 million people in year 2005. It is forecasted to increase up to 2.5 million in the year 2015. Therefore, the coloring which is assumed to be effective to activate the brain and to prevent dementia was paid to attention.

When starting coloring, people needs to observe the original picture carefully. At this time, lobus occipitalis that take charge of the sight work. Moreover, to understand the original picture accurately, the temporal lobe that takes charge of the memory works to refer from the memory the shape and the color sow in the past. The parietal lobe cooperates when the balance of the entire picture is gripped. As written above coloring has the effect to activate a widespread area of the brain.

In this study coloring was taking place at the pay nursing home for the aged tenant as part of the recreation, and the influence to the tenant was analyzed. As an early stage of this experiment we tried to verify which writing equipment can give more effective influence to the brain activity during coloring.

Keywords: brush, coloring recreation, dementia prevention

1 INTRODUCTION

Aging has becoming a serious problem in wide area of the world, as of October 1, 2010, the elderly population aged 65 and over became 2.9 million people to be the highest ever in Japan. Moreover the proportion of the population of the total population over the age of 65 was also recorded the highest of 23.1%. When this tendency continues, one person in four people comes to enter the age of senior citizen in 2015. The stimulus from the outside might decrease when the senior citizen moves in the hospital or the nursing facilities such as pay nursing home, which can lead to the appearance of dementia and the progress of the needing care degree as a result. Therefore, various measures for dementia prevention are taken place in many places.

In Japan, transcribing a sutra is performed from ancient times for mental concentration, nevertheless there are difficulties for the aged person to transcribing a sutra because writing a Chinese character is very delicate work. Accordingly, coloring which is more easily carried out was paid to attention. In this study coloring was taking place at the pay home for the aged tenant as part of the recreation, and the influence to the tenant was analyzed.

2 EXPERIMENT

2.1 Changes of brain activity by difference of writing equipment

As an early stage of this experiment verification of optimal writing equipment during coloring, which can give more effective influence to the brain activity was carried out. The experiment was carried out with four different writing equipment`s such as crayon pastel (SAKURA COLOR PRODUCTS CORP), colored pencil (MITUBISHI PENCIL CO., LTD.), felt-tipped marker (Too Corporation.), and color brush pen (soliton corporation CO. LTD.). Mind Set (Neuro Sky Inc.) was used for the brain activity measurement. The brain activity during coloring in each case was measured.

Mind set enables to measure brain wave information on the alpha wave, beta wave, theta wave, and delta wave, then analyzed by an original algorithm and calculated as attraction (concentrated level) and meditation (relaxation level).

Seven postgraduates cooperated in this experiment as a test subjest. To make experimental conditions impartial, each writing equipment was measured in the same time zone of a different day using the same laboratory where isolated from external sounds. After having installed the Mind set, a test subject rested for three minutes to record the brain wave at the rest situation, then start coloring for five minutes.

The reason for having experimented on postgraduates instead of senior citizen was because it is necessary to wear the headgear for the measurement of the brain activity. We have judged that a mental and physical load would be too large for the senior citizen. Fig 1 shows the measurement scenery.

Fig.1 Measurement scenery of brain activity measuring.

2.2 Coloring recreation at pay nursing home

The coloring was taken place at the pay nursing home as part of the recreation for 56 tenants. The frequency of the recreation carried out was 2~3 times a week, and each recreation was about 1.5 hours. The frequency of the fall accident and the number of the nurse call (sensor mat type) before and after the recreation was recorded.

3 RESULTS AND DISCUSSIONS

3.1 Changes of brain activity by difference of writing equipment

The result of the attention growth rate from at rest situation on each writing equipment`s is shown in Fig.2.

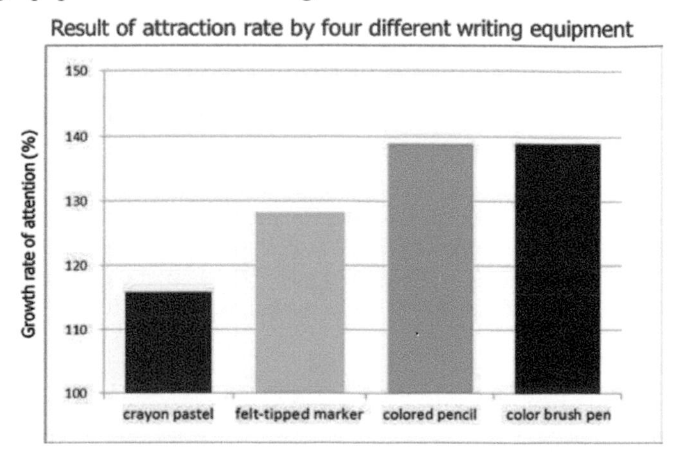

Fig.2 Attention growth rate from at rest.

As for degree of concentration, in the colored pencil and the color brush pen, 39% expansion was seen from the rest situation. That is to say when colored pencil and the color brush pen are used for coloring, degree of concentration increases more and it can expect an effect to brain activation.

The result of the meditation growth rate from at rest situation on each writing equipment`s is shown in Fig.3.

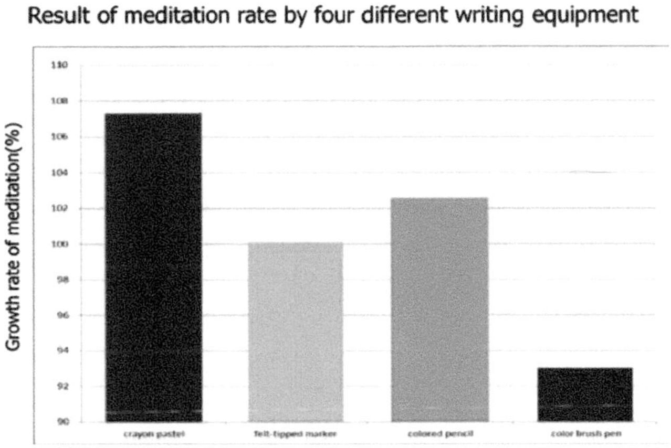

Result of meditation rate by four different writing equipment

Growth rate of meditation(%)

Fig.3 Meditation growth rate from at rest.

The degree of meditation brought a result in which only color brush Pen decreases in number with 93% of a ratio at the time of coloring experiment. This is to consider that color brush Pen has the softest tip in four writing equipment`s, therefor not only a motion of the direction of XY axis, test subject needed to pay attention also in the up-and-down direction of the Z-axis causing meditation rate to decrease during coloring. From this result, color brush Pen is presumed as the most appropriate writing equipment during the coloring recreation taken place at pay nursing home.

3.2 Fall accident number and nurse call frequency investigation

The number of the fall accident per month before and after the recreation is shown in Fig. 4 and the frequency of the nurse call is shown in Fig.5. The averages of fall accident per month at pay home facilities decreased to 4.6 times from 10.7 times which in the percentage by 57% decrease. When thc frequency of the recreation including coloring increased, the frequency of the nurse call decreased to average of 832 from 1469. The decreasing percentage was 35%.

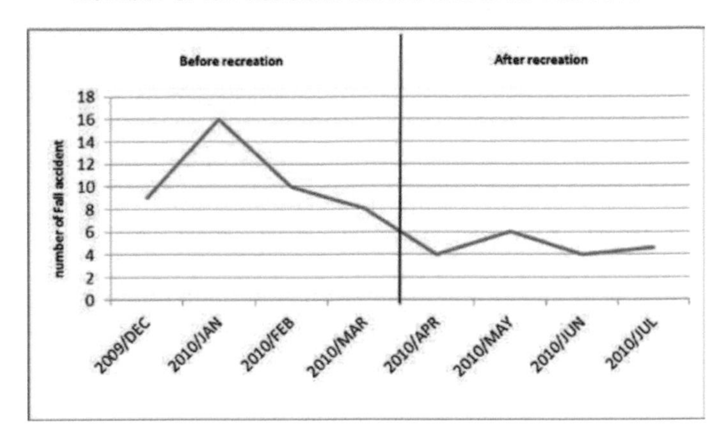

Fig.4 The number of fall accident before and after recreation.

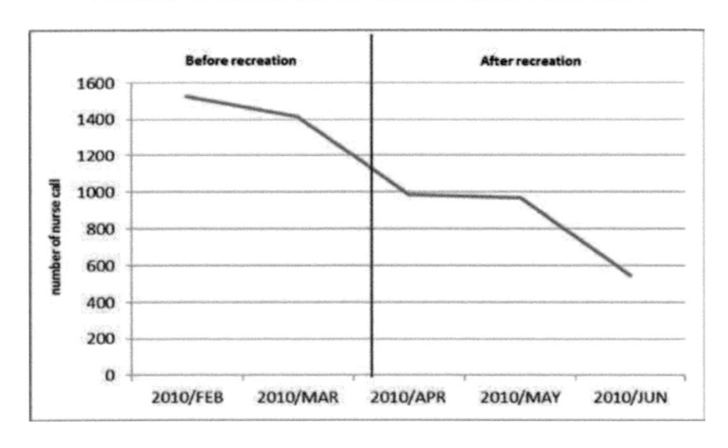

Fig.5 The number of nurse call accident before and after recreation.

There are usually a lot of nurse calls at nighttime when the helper's round is fewer compare to day time. When you increase the frequency of the recreation including the coloring, the frequency of the nurse call decreased. It is suggested the possibilities of the brain and the body received stimulation and produce fatigue which might have led to enough and refreshing sleep at night. Additionally, though it is a result of only the woman, research results are reported that the fall accident risk is higher to an aged woman with short sleeping time. The result shows that good quality sleep was urged by the recreation including coloring recreation, and the possibility of causing a decrease of the fall accident was suggested.

4 CONCLUSIONS

In this study, it was suggested that the coloring showed effect to improvement in quality of the sleep and the fall accident prevention of the senior citizen. Coloring is easily done compared to painting and descriptions, also regardless to the needing care degree, more over a care worker's burden can also be reduced. As the result shows, coloring is suggested as an activity that should be taken as one of the recreations at the pay nursing home.

ACKNOWLEDGMENTS

The authors would like to acknowledge Professor Hiroyuki Hamada, Kuwahara Noriaki, Yoshiyuki Kida for making this work possible and for their encouragement and helpful discussions.

REFERENCES

Hiroaki TANAKA, et al. (2009): Effect that "memories coloring paper" gives to slight dementia patient acknowledgment function, psychology function, and side of daily life Journal of rehabilitation and health sciences 7, 39-42.

Japanese brain Health society (2006): brain Health news No15.

Katie L. Stone, Sonia Ancoli-Israel, Terri Blackwell, Kristine E. Ensrud, Jane A. Cauley, Susan Redline, Teresa A. Hillier, Jennifer Schneider, David Claman, Steven R. Cummings(2008), Actigraphy-Measured Sleep Characteristics and Risk of Falls in Older Women, Archives of Internal Medicine（168: 1768-1）

The Management and Coordination Agency(2005) white paper on aging society.

The Ministry of Health(2003), Labour and Welfare nursing care for elderly people society Nursing care for elderly people in 2015.

CHAPTER 62

Design Validation of Ergonomics through Immersive Interaction Based on Haptic Devices and Virtual Reality

Cinzia Dinardo

Consorzio CETMA
c/o Cittadella della Ricerca
S.S.7 Km.706+030, 72100 Brindisi
cinzia.dinardo@cetma.it
+39 0831 449603

Luca Rizzi

Consorzio CETMA
c/o Cittadella della Ricerca
S.S.7 Km.706+030, 72100 Brindisi
Luca.rizzi@cetma.it
+39 0831 449607

ABSTRACT

The H@H Hospital at Home is a research and development project of Industrial Design and Information Technology departments of CETMA consortium. CETMA by designing "Web Based Measuring System for Health Monitoring at Home", has developed an innovative method for design validation of ergonomics through immersive interaction based on haptic devices and virtual reality.

The main aim of the research activity has been to test and validate the human-machine interaction of the designed device. The consortium has designed a custom protocol able to allow an easy use of 3D model into virtual reality scenery to speed up product design.

CETMA Virtual Reality Center (CVRC) is an internal projection room that allows to reproduce objects and products into immersive virtual reality environment in CAVE configuration (Cave Automatic Virtual Environment), where projectors are directed to three walls of a room-sized cube. Ergonomics validations have been carried out into this virtual environment.

Traditional full-size prototype model has been used to compare results of virtual design validation with standard method. A mockup made of inexpensive materials has been integrated into 2D wall projected scenery and tested with the same validation of ergonomic protocol. Results have been compared in term of time, costs and quality of ergonomic measures.

This work shows how ergonomics validation in immersive environment can greatly improve and speed up product design. Actually industrial design is mostly base on 3D CAD software, and 3D models can be easily viewed with Immersive Virtual Reality devices. It also highlights how test results can only be qualitative and couldn't be transformed into quantitative results useful for comparison with ergonomic defined parameters.

The immersive virtual validation can be a valuable tool for design, but should be always matched with final validation on works-like prototype, that is necessary to better simulate physical interactions.

CETMA purpose is to develop new research lines extending its know-how into ergonomics validation based on virtual models. CETMA would increase this line of research developing new virtual sceneries to analyze and test different products using the CVRC facilities.

Thanks to the research project the consortium aims to offer new services for product development and design. Standard activities could be supported with this new services based on immersive virtual reality.

The research project has been developed through the following steps:

- Defining a new protocol for virtual based ergonomics validation;
- Defining of an operating mode adapted for a specific product;
- Validation Testing and data collecting;
- Data comparison with traditional validating methods.

Keywords: Telemedicine, e-Health, Ageing, Design validation, Virtual reality, Immersive Interaction.

1 INTRODUCTION

The H@H Hospital at Home is a research and development project of Industrial Design and Information Technology departments of CETMA consortium. CETMA by designing "Web Based Measuring System for Health Monitoring at Home", has developed an innovative method for design validation of ergonomics through immersive interaction based on haptic devices and virtual reality.

The main aim of the research activity has been to test and validate the human-machine interaction of the designed device. The consortium has designed a custom protocol able to allow an easy use of 3D model into virtual reality scenery to speed up product design.

CETMA Virtual Reality Center (CVRC) is an internal projection room that allows to reproduce objects and products into immersive virtual reality environment in CAVE configuration (Cave Automatic Virtual Environment), where projectors are directed to three walls of a room-sized cube. Ergonomics validations have been carried out into this virtual environment.

2 DEFINING A NEW PROTOCOL FOR VIRTUAL BASED ERGONOMICS VALIDATION

The system analyzed with the new validation protocol is a medical device similar to a bedside table on which a movable screen with a balanced arm have been assembled. The system is designed for in-home use.

Four typologies of interactions have been defined: self monitor positioning, interaction with touch screen monitor but also interaction with medical gas valves and use of the posterior handle for transportation.

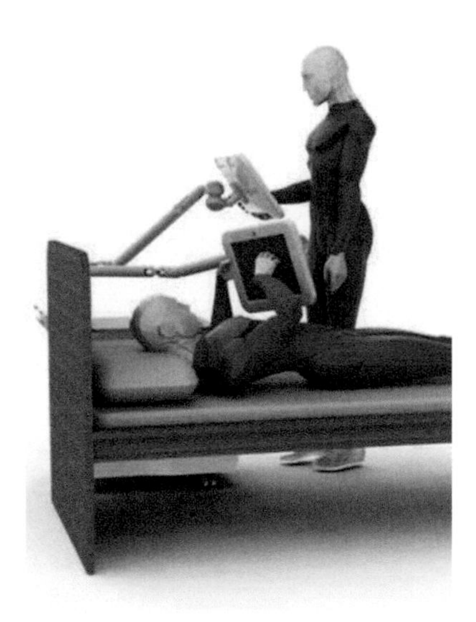

Figura 1 - – Using Positions

Interactions that the designer would simulate should be defined at first and also the group of user that will test the system should be highlighted early (Figure 1). For the virtual simulation project the lower number of degree of freedom should be defined on the previous analysis. Different virtual sets were developed to simulate different interactions and each movable part of the device can determine a different set of variable. Actually arm's handling is the most important interaction that we would reproduce and test for this device.

A large number of users were selected to cover the wider range of representative percentile for system's end-user. A tutor have followed the testing phase recording ergonomic values and parameters, asks user's perception and emotion during device use.

2.2 Interactions

The system is designed with 3 different product customizations for the market. In the present research we decide to analyze the most complex and full accessorized configuration, named H@H – Full. This device allows onsite Medical Air production with high oxygen percentage. The system integrates also a Medical Vacuum valve and in its full customization shows the most complex and interesting architecture from an interaction point of view.

1. Bedside table
2. Movable arm
3. Fully-integrated video system (touch monitor + webcam + audio speaker and microphone)
4. Opening and sensors holder
5. Sensor's connection front-panel
6. Medical Gases
7. Air inlet
8. Wheels
9. Handle and towel holder
10. Maintenance door
11. Aeration grids – Inlet
12. Aeration grids – forced outlet

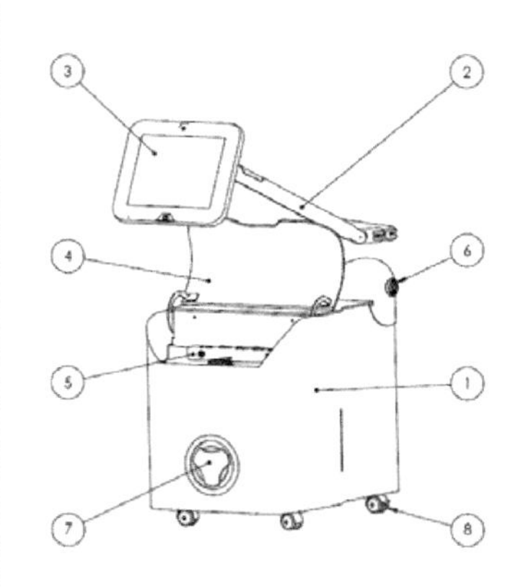

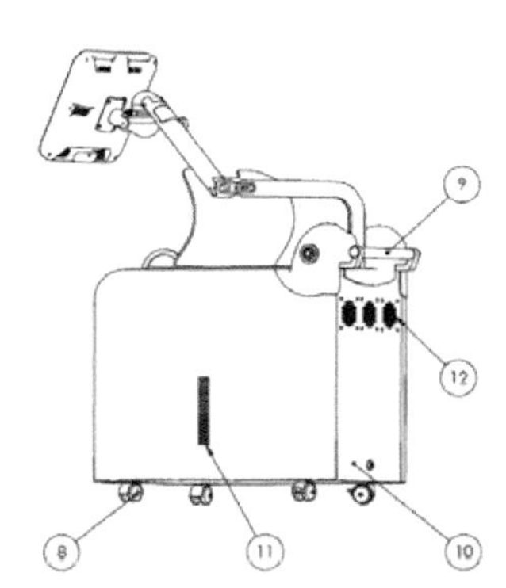

Figure 2 –H@H Full system

Main interactions for a standard player are connected with use of the following parts: 3 (and 2), 4, 6 and 9. Especially parts 4, 6 and 9 are fixed and will not required movement development. The monitor 3 and its arm 2 are adjustable parts and their movement should be reproducible also in the virtual simulation. Interaction will be guaranteed with specific haptic devices. User REACHABILITY to device's components will be simulated only for part 3 and 2.

2.3 Target User definition

The device could be used both in hospital or at home. The system will be placed near patient's bed in the same position of standard bedside table.

Figura 3 - In-home use

Four different user can mainly interact with the device:

- User – Patient;
- User - Patient's relative;
- User - Health Operator;
- User – Installer/maintenance.

Patient will use the system lay on bed in supine position or seated while all the other users mainly use it in standing position.

2.4 Ergonomics evaluation test

CETMA consortium has one of virtual theatre more in the van of European survey. CETMA Virtual Reality Center (CVRC) is an internal projection room that allows to reproduce objects and products into immersive virtual reality environment in CAVE configuration (Cave Automatic Virtual Environment), where projectors are directed to three walls of a room-sized cube. Ergonomics validations have been carried out into this virtual environment allowing the interaction between user and a 3D model of the device showed in stereoscopic real scale.

3 TESTING AND DATA RECORDING

Ergonomics evaluations take place into the CVRC with a representative group of user and reproducing the previously identified application scenery. In-home set was designed to test interaction with user different by height and sex and simulating lie-down, sitting and standing use as showed in figure 4.

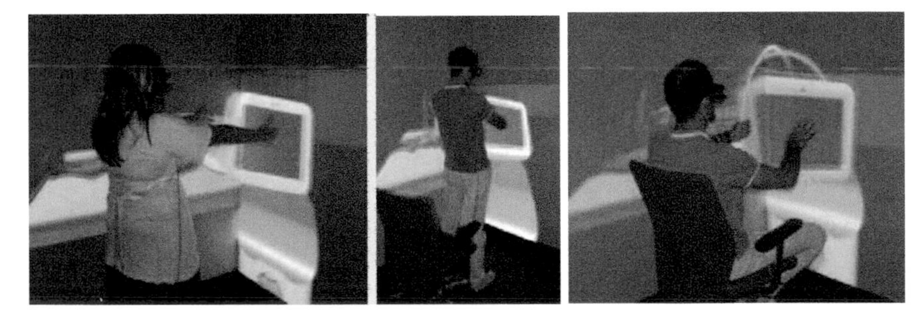

Figura 4 - Touch monitor interaction test in standing and sitting positions

CETMA Consortium Industrial Design Department design the evaluation protocol and collect all the data. Information Technology Department design and develops 3D virtual sets.

3.1 Representative Percentile Definition

User for validating test were chosen paying attention to include the fifth percentile women and the ninety-fifth man that in Italy corresponds to 151 cm (163 for 5° man) e to 185 cm height (172 cm for 95° woman).

Thus the system was tested with the following user selected inside CETMA employees, especially to cover homogeneously the previous mentioned border user.

Table 1 Table Representative Percentile

User	Name	Height [cm]	Representative Percentile
1	Francesco Chionna	190	98° Man
2	Luca Rizzi	183	95° Man
3	Umberto Fioretti	172	40° Man
4	Ubaldo Spina	168	20° Man
5	Domenica Suma	165	70° Woman
6	Giovanni Giodice	164	5° Man
7	Simona Sambati	154	10° Woman

3.2 Task analysis and testing

A team of expert researchers defines the total number of tasks that each user should play to verify the use of monitor's arm and to confirm the correct touch screen reachability.

Specific sets were designed and developed to improve the level of context perception and to better interact with the system.

Furthermore a list of question were designed to introduce and to drive the user using this device. A final questionnaire is then used to interrogate each candidate. Final questions were useful to better understand problems and troubles happened using the system (Figure 5)

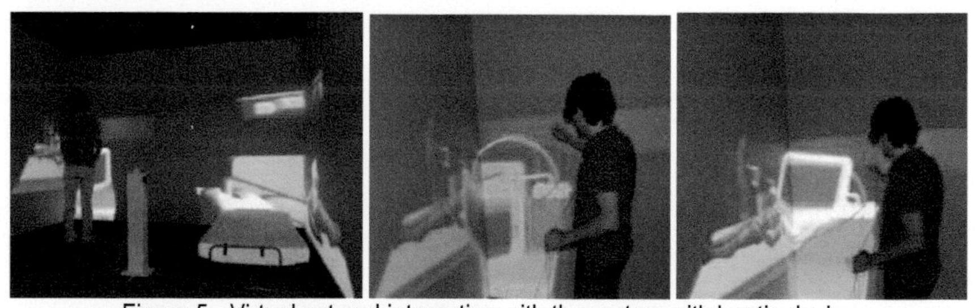

Figura 5 - Virtual set and interaction with the system with haptic device

3.3 Data recording

User's behaviours were recorded during tasks execution and troubles while using the device were highlighted for further investigation.

3.4 Data Analysis

Table 2 Table Data Analysis

User	Height [cm]	Standing up reachability	Lying down reachability	Siting down reachability
1	190	Comfortable	High Comfortable	Low Comfortable
2	183	Very High Comfortable	High Comfortable	Very High Comfortable
3	172	Very High Comfortable	High Comfortable	Very High Comfortable
4	168	Very High Comfortable	High Comfortable	Very High Comfortable
5	165	High Comfortable	Low Comfortable	High Comfortable
6	164	High Comfortable	Low Comfortable	High Comfortable
7	154	Very High Comfortable	High Comfortable	Low Comfortable

3.5 Data recording

Usability test confirms design analysis made during the development of the system, thanks to this investigation we were particularly able to demonstrate onsite that every kind of user could place the arm and then the monitor in the correct position. User never place their body in simply uncomfortable or ergonomically incorrect positions. User easily accepts the sensation to be inside a virtual set, and thanks to a real look of the device we were also able to appreciate design and functional features and propose possible improvement for the system. Importing of 3D models and testing analysis required about 4 days instead of the standard time of 10 days required to make a 1:1 scale mock-up to do traditional test previously showed (Figure 6). Test patient position that lay on bed wasn't possible.

A team of expert researchers defines the total number of tasks that each user should play to verify the use of monitor's arm and to confirm the correct touch screen reachability.

Figure 6 - Traditional Validation with paperboard mock-up

CONCLUSION

The developed protocol shows its applicability also to other products of different typologies. Further development of this validation process could integrate other haptic device like devices able to give a feed-back to the user. For example the interaction with touch monitor could be simply realized with standard virtual-gloves.

REFERENCES

ANDREOLI TERESA, Psy.D. Psychology Assistant Cognitive Disorders Among the Elderly, Lic #PSB 31633, BCIA-C #3949 Fellow *Brain Therapy Center* www.brain-injury-therapy.com/articles/dementia.htm

VV. Tecnologie informatiche e utenza debole. La progettazione ergonomica dei siti web e delle postazioni di lavoro per i disabili. *Il Sole 24 Ore Pirola* (2002) - Codice EAN: 9788832447552

BAECKER, R., & Buxton, W., Readings in Human-Computer Interaction: A Multidisciplinary Approach, Morgan-Kaufmann Publishers, Los Altos, CA, 1987.

BAILEY, R., Usability Testing versus Heuristic Evaluation: A Head-to-Head Comparison. Proceedings of the Human Factors Society 36th Annual Meeting, 409-413, New York, 1992.

BIAS, R., and MAYHEW, D. J., Cost-Justifying Usability, AP Professional, Cambridge, MA, 1994.

Hyperlabs - empowering users; www.hyperlabs.net.

Section IX

Medical Devices and Special Populations

CHAPTER 63

Anthropometric and Scoliosis Survey for Children with Physical and Mental Disabilities

Chao-Yin Wu [1,2] *Te-Hung Chen* [1] *Mao-Jiun Wang* [1]

[1] Department of Industrial Engineering and Engineering Management,
National Tsing Hua University

[2] Department of Rehabilitation Medicine,
Mackay Memorial Hospital, Hsinchu

bessie520@gmail.com

ABSTRACT

Anthropometric data are important references for ergonomics design. Most of the anthropometric data for children are collected from children without disabilities. It may not be applicable for children with disabilities. Scoliosis are often observed in children with disabilities. A study of anthropometric data and scoliosis screening for children with physical and mental disabilities were conducted in Hsinchu, Taiwan. One hundred and four children including 72 boys and 32 girls, aged from 7 to 12 years, participated in this study. Three-dimensional coordinate measurement probe and digital tape were used to measure body dimensions in height, length and girth. Adam's forward bending test was used to screen whether there is a positive sign of scoliosis. The results of forward bending test showed that the percentage with positive sign for boys was 33% and for girls was 28%. In addition, high percentage of positive scoliosis screening was found in children with physical disability (75%). In anthropometric data, non-scoliosis and scoliosis children were similar in heights, lengths, and girths. Furthermore, we also compared our data to a previous study for normal school children in the same geographic area, and found that the anthropometric data for the disability group were smaller than those of the normal group. These results provide important information in designs for children with special needs.

Keywords: Anthropometry, scoliosis, children, physical disability, mental disability

1 INTRODUCTION

Anthropometry is a study of human physical variations by measuring body dimensions, and it is considered as a foundation for ergonomics. In school children anthropometric data applications, several studies have focused on school furniture design which can help school children to adopt a better posture in learning (Jeong and Park, 1990; Ray et al., 1995). Large anthropometric surveys for school children have been conducted in Taiwan (Wang et al., 2002; Chung et al., 2007). Body Mass Index (BMI) and various body dimensions such as arm length, shoulder height, or waist girth were found to be significantly different between these two studies. The authors suggested that these differences might be caused by the different geographic distribution of the subjects. Overall, these data provide important information for different design applications for the normal children with regular size. Children with physical or mental disabilities, however, are often below the norm in terms of functions and abilities. Hence, many designs for ordinary people are not fit to children with disabilities as pointed out by Kroemer (2006) in his book "Extra-Ordinary" Ergonomics. Thus, the needs of measuring anthropometric data for children with disabilities are very obvious.

Scoliosis screening in school children is recommended for early detection of spinal deformities (Scoliosis Research Society, 2007). Morais, et al. (1985) reported the prevalence of scoliosis among school children aged 8 to 15 years was 4.2%, 5.2% among girls and 3.2% among boys. Children with disabilities are often recognized by having poor posture and may lack of self-awareness as well as the ability to correct their posture. The prevalence of scoliosis is expected to be higher in children with physical disabilities compared with other disabilities such as mental or blind. Scoliosis was found in nearly 25% of children with cerebral palsy (Thomson and Banta, 2001). Adam's forward bending test (FBT) is the most commonly used method in school screening; any asymmetry in the contours of the back may suggest a rotational deformity of the spine (Janicki and Alman, 2007).

The purpose of this study is to collect the anthropometric data and evaluate the prevalence of positive scoliosis among children with physical and mental disabilities.

2 METHOD

2.1 Subjects

A total of 104 children (72 males and 32 females) under the special education program in a Hsingchu city elementary school, Taiwan were recruited. The disability children who were not able to perform the forward bending test were excluded for data collection. The means and standard deviations of the demographic data of each age and gender group are presented in Table 1 and Table 2. Parents of the participants were informed about the study procedure and had signed a consent form prior to data collection.

Table 1 Means and standard deviations of height, weight and BMI for boys

Age	Boys						
	N	Height		Weight		BMI	
		Mean	SD	Mean	SD	Mean	SD
7	18	117.69	5.48	21.27	3.86	15.5	1.6
8	13	126.22	6.81	26.42	5.58	16.4	2.0
9	22	127.80	6.69	30.57	10.66	17.9	7.6
10	6	139.22	7.61	37.83	13.93	19.3	6.0
11	9	136.16	5.15	34.28	7.78	18.5	3.9
12	4	159.75	9.18	63.50	17.78	24.9	10.8
Total	72	128.91	11.90	30.39	13.02	17.5	5.6

Table 2 Means and standard deviations of height, weight and BMI for girls

Age	Girls						
	N	Height		Weight		BMI	
		Mean	SD	Mean	SD	Mean	SD
7	5	114.52	4.64	19.50	3.72	14.9	3.0
8	3	131.60	0.71	24.50	0.71	7.2	10.1
9	9	126.91	7.02	28.06	7.28	17.2	3.1
10	4	131.90	11.52	37.88	17.82	21.2	7.9
11	7	133.67	16.00	32.21	13.17	17.4	3.2
12	4	140.38	4.61	31.88	5.81	16.1	2.0
Total	32	128.97	11.75	29.15	10.79	16.6	5.1

2.2 Instrumentation

A 3D coordinate measurement system (Faro Technologies Inc.) was used to measure the 3D coordinates of the 30 markers. The description of the marker positions are shown in Table 3. A digital tape was used to measure curve length and circumference dimensions. The accuracy of the 3D coordinate measurement system and digital tape was about 0.01 mm and 0.1mm respectively. Adam's forward bending test (FBT) was used to screen whether there is positive sign of scoliosis among the school children.

2.3 Procedure

For the procedure of taking measurements, subjects were asked to wear thin clothes and barefoot to ensure the measurement accuracy. They were asked to perform the FBT and examined by an experienced physical therapist. Then, the therapist put the red dot markers on subjects' body landmark positions as listed in Table 3. Anthropometric measurements followed the definition of ISO 7250 standards. Data about length, height, girth, and breadth were measured by two trained research assistants using the FARO and the digital tape.

Table 3 Marker's positions on body surface

Item	Position	Item	Position
1	Vertex	16	Left lateral malleolus
2	Left bitragion	17	Right lateral malleolus
3	Right bitragion	18	Cervical point of 7th
4	Left acromion	19	Suprasternale
5	Right acromion	20	Substernale
6	Left olecranon	21	Navel
7	Right olecranon	22	The maximum anterior protrusion of abdomen
8	Left unlar styloid	23	Gluteal furrow
9	Right unlar styloid	24	Left medial condyle of tibia
10	Left anterior suprailiospinale	25	Right medial condyle of tibia
11	Right anterior suprailiospinale	26	Left medial malleolus
12	Left trochanter	27	Right medial malleolus
13	Right trochanter	28	Left osscaphoideum
14	Left midpoint of kneecap	29	Right osscaphoideum
15	Right midpoint of kneecap	30	Left popliteal

3 RESULTS

The body height and weight increased with advancing age (Table 1 and Table 2). The boys of age 12 and the girls of age 10 had greater variations in weight. The results of the scoliosis screening are shown in Table 4. The percentage of positive

FBT result in boys was 33%, and in girls was 28%. The percentages of scoliosis were also classified by the types of disabilities as shown in Table 5. Seventy-five percent of the physically disabled children and 60% of the children with genetic abnormality were screened as scoliosis. None of the children with Autism was screened with scoliosis. Anthropometric measurements including vertical heights, lengths, and girth are shown in Table 6. The data between non-scoliosis (negative screen result) and scoliosis (positive screen result) group were compared. No significant differences were found between the two groups.

Table 4 Numbers and percentage of positive FBT in boys and girls

Age	Boys			Girls		
	Non-Scoliosis (n)	Scoliosis (n)	Percentage of scoliosis	Non-Scoliosis (n)	Scoliosis (n)	Percentage of scoliosis
7	10	8	44%	3	2	40%
8	8	5	38%	3	0	0%
9	19	3	14%	8	1	11%
10	4	2	33%	1	3	75%
11	5	4	44%	5	2	29%
12	2	2	50%	3	1	25%
Total	48	24	33%	21	9	28%

Table 5 Numbers and percentage of positive FBT in different type of disabilities

Type of disabilities	Non-scoliosis	Scoliosis	Percentage of Scoliosis
Physical disability	2	6	75 %
Mental disability	19	10	34 %
Genetic abnormality	4	6	60 %
Autism	9	0	0 %
Multiple disabilities	10	6	38 %
Others (deaf or blind)	3	3	50 %
No diagnosis	24	2	8 %

4 DISCUSSION

The percentage of positive Adam's forward bending test was relatively high as comparing to the previous studies of the normally developed children. Karachalios et al. (1999) reported that the prevalence of having positive FBT for school children of 8 to 16 years old was 5.8% (156 out of 2700). Kapoor et al. (2008) reported that 2.8% children (11-13 years) were positive for scoliosis test. In this study, 75% children with physical disability had positive FBT. People with physical disability normally have neuromuscular or musculoskeletal impairments that limits the physical ability to a certain degree. Musculoskeletal impairments may cause an imbalance of paraspinal muscle strength, and neuromuscular impairment (i.e. cerebral palsy) may cause abnormal muscle tone. These might contribute to cause scoliosis in children. Hence, it is not surprising to note that children with physical disability had higher prevalence of positive FBT. Saito et al. (1998) reported that 68% of the people with spastic cerebral palsy had scoliosis with a curve of at least 10° Cobb angle, which is a scoliosis index identified from x-ray. Although we did not specify the type of genetic abnormality consisted in our subjects, the scoliosis prevalence among children with genetic abnormality was similar to Shim, et al (2010) who reported a 63.9% of scoliosis in children with Prader-Willi syndrome, a rare genetic disorder. For the accuracy of Adam's forward bending test, inconsistent findings were reported. Karachalios, et al. (1999) reported the sensitivity and specificity of FBT was 84.37% and 93.44%, respectively. However, they also argue that the FBT should not be considered a safe diagnostic criterion for scoliosis detection because of its high false- negative rate. It is recommended to have further medical check-up if there is positive FBT.

For the anthropometric measurements, the results were compared with a previous study by Chung et al. (2007) (Table 7) which measured the elementary school children within the same geographic area, and the number of school children were 1024, aged from 5 to 14 years. All of the vertical height measurements and length of arm, forearm, leg, and shin for the school children in this study were significantly shorter than those of the previous study. This might be due to some development delay for the school children in this study. In addition, the girth of buttock, thigh and calf were also found to be smaller for the children in this study as comparing to the previous one. Chung et al (2007) did report an overweight issue found in their subjects.

5 CONCLUSION

The anthropometric data for children with physical and/or mental disabilities were collected in this study. Smaller body sizes comparing to the normally developed children were found. These findings provide useful information for designing school furniture and apparatus to enhance a healthy learning environment. The high prevalence of scoliosis found in the school children with disabilities bring special caution to parents, school teachers and therapists who involved in the special education system.

Table 6 The comparison of means and standard deviations of the measurements between Non-Scoliosis and Scoliosis children

Measurements		Non-Scoliosis (NS) n=71		Scoliosis (PS) n=33		T-test
		Mean	SD	Mean	SD	P-Value
Vertical Height	Stature	129.62	10.42	127.45	14.41	n.s.
	Left bitragion	116.14	10.26	113.91	14.17	n.s.
	Left shoulder	104.16	9.42	102.13	12.55	n.s.
	Left Olecranon	78.78	7.21	76.91	10.09	n.s.
	Left iliocristale	70.19	7.25	68.36	8.82	n.s.
	Left trochanterion	63.19	6.56	61.74	7.94	n.s.
	Left mid-knee cap	34.42	3.53	31.62	3.99	n.s.
	Left ankle	5.36	0.77	5.31	0.85	n.s.
	7th-cervical	107.39	9.83	104.90	13.12	n.s.
	Suprasternale	103.12	9.51	100.75	12.44	n.s.
	Substernale	91.14	8.66	88.73	11.52	n.s.
	Navel	76.10	7.32	74.74	9.76	n.s.
	abdomen extension	81.61	7.97	79.66	10.42	n.s.
Length	Left arm	42.84	4.35	41.97	4.57	n.s.
	Left upper-arm	26.23	2.90	25.90	2.90	n.s.
	Left forearm	17.51	2.28	17.06	2.35	n.s.
	Left leg length	65.38	6.99	63.11	8.44	n.s.
	Left thigh length	38.27	4.14	36.84	5.19	n.s.
	Left shin length	27.66	3.35	26.73	3.47	n.s.
Girth	Chest	65.37	9.21	64.68	10.59	n.s.
	Weist	61.03	11.36	59.69	10.50	n.s.
	Buttock	70.81	10.37	70.19	12.75	n.s.
	Left lower thigh	32.12	4.70	32.19	5.73	n.s.
	Left calf	27.42	3.65	26.85	4.58	n.s.
	Left biceps, relaxed	19.80	3.53	19.74	3.68	n.s.

*:p<0.05 , **:p<0.01 , ***:p<0.001 , n.s.:non-significant P>0.05

Table 7 Comparison of anthropometric measurements between current study and Chung et al. (2007)

Measurements		Current study N=104		Chung et al., (2007) N=1024		T-test
		Mean	SD	Mean	SD	P-Value
Vertical Height	Stature	128.93	11.80	134.19	11.74	***
	Left bitragion	115.43	11.61	120.41	11.64	***
	Left shoulder	103.52	10.49	107.93	10.40	***
	Left Olecranon	78.19	8.22	80.88	7.89	***
	Left iliocristale	69.61	7.78	73.10	74.74	***
	Left trochanterion	62.73	7.02	68.34	7.39	***
	Left mid-knee cap	32.17	3.68	35.35	3.76	***
	Left ankle	5.34	0.79	4.79	0.80	***
	7th-cervical	106.60	10.97	104.48	10.77	**
	Suprasternale	102.37	10.52	106.53	10.32	***
	Substernale	90.37	9.67	95.47	9.54	***
	Navel	75.67	8.15	77.93	8.01	**
	abdomen extension	80.99	8.81	77.36	9.35	***
Length	Left arm	42.57	4.42	45.58	4.96	***
	Left upper-arm	26.13	2.89	26.40	2.91	n.s.
	Left forearm	17.37	2.30	19.19	2.27	***
	Left leg length	64.66	7.51	68.36	7.34	***
	Left thigh length	37.82	4.52	37.95	4.14	n.s.
	Left shin length	27.36	3.40	30.61	3.40	***
Girth	Chest	65.14	9.62	66.18	7.95	n.s.
	Weist	60.60	11.06	60.75	8.65	n.s.
	Buttock	70.62	11.12	73.25	9.03	**
	Left lower thigh	32.21	5.02	33.83	5.31	**
	Left calf	27.24	3.95	29.40	3.61	***
	Left biceps, relaxed	19.78	3.56	20.34	2.91	n.s.
Other	Weight	30.01	12.34	32.3	11.3	
	BMI	17.25	5.41	17.7	3.2	

*:p<0.05 , **:p<0.01 , ***:p<0.001 , n.s.:non-significant P>0.05

REFERENCES

Chung, M.J., J.P. Chen, and T.H. Chen, et al. 2007. The study of anthropometric data for school children in Taiwan. *Proceeding of the 8th Asian Pacific Industrial Engineering and Management Systems*, Taiwan.

ISO 7250 1996. Basic human body measurements for technological design.

Janicki, J.A. and B. Alman, 2007. Scoliosis: Review of diagnosis and treatment. *Paediatric Child Health.* 12, 771-776.

Jeong, B.Y. and K.S. Park. 1990. Sex difference in anthropometry for school furniture design. *Ergonomics*, 33: 1511-1521.

Kapoor, M., S. G. Laham, and J. R. Sawyer, 2008 Children at risk indentified in an urban scoliosis school screening program: a new model. *Journal of Pediatric Orthopaedics B*, 17, 281-287.

Karachalios T, J. Sofianos, and N. Roidis, et al. 1999. Ten-year follow-up evaluation of a school screening program for scoliosis. *Spine* 24, 2318-2324.

Kroemer K.H.E. 2006 "Extra-ordinary" ergonomics: how to accommodate small and big persons, the disabled and elderly, expectant mothers, and children. CRC press.

Morais, T., M. Bernier, and F. Turcotte, 1985. Age- and sex-specific prevalence of scoliosis and the value of school screening programs. *American Journal of Public Health*, 75; 1377- 1380.

Ray, G.G., S. Ghosh, and V. Atreya. 1995. An anthropometric survey of Indian schoolchildren aged 3-5 years. *Applied Ergonomics*, 26: 67-72.

Saito, N., S. Ebara, and K. Ohotsuka, et al. 1998. Natural history of scoliosis in spastic cerebral palsy. *Lancet*, 351, 1687-1692.

Shim, JS., S. H. Lee, and S.W Seo, et al. 2010. The musculoskeletal manifestations of Prader-Willi syndrome. *Journal of Pediatric Orthopedics*. 390-395.

Scoliosis Research Society (2007) http://www.srs.org/professionals/advocacy_and_public_policy/SRS-AAOS_position_statement.htm

Thomson, J.D. and J. V. Banta, 2001. Scoliosis in cerebral palsy: An overview and recent results. Journal of Pediatric Orthopaedics, 10; 6-9.

Wang, M.J., E.M. Wang, and Y.C. Lin. 2002. The anthropometric database for children and young adults in Taiwan. *Applied Ergonomics*, 33: 583-585.

CHAPTER 64

Human Factors Evaluation of Medical Equipment Reprocessing Instructions

R. Darin Ellis[a], *Serge Yee*[b], *Nancy Lightner*[c], *Kai Yang*[a], *and Will Jordan*[c]

[a] Department of Industrial & Systems Engineering, Wayne State University, Detroit, MI 48202, USA
RDEllis@wayne.edu
Kai.Yang@wayne.edu

[b] Department of Veterans Affairs, Veterans Engineering Resource Center, VA-Center for Applied Systems Engineering, Detroit, MI, 48201, USA
Serge.Yee@va.gov

[c] Department of Veterans Affairs, Veterans Engineering Resource Center, VA-Center for Applied Systems Engineering, Indianapolis, IN, 46222, USA
Will.Jordan@va.gov
Nancy.Lightner@va.gov

ABSTRACT

Effective and reliable reprocessing of reusable medical equipment such as endoscopes is critical for patient safety. While manufacturers provide highly detailed and technical instructions for cleaning and maintenance, these documents are not readily usable by reprocessing technicians as real-time job aids. This paper describes some of the significant human factors considerations involved in creating job-aids from manufacturer's instructions and their inclusion in the Veterans Affairs Interactive Visual Navigator[1] (IVN™) system.

Keywords: reusable medical equipment, sterile processing, work instructions

---

[1] IVN™ is the property of the Department of Veterans Affairs and Wayne State University. Any use of the content presented in this paper without the express written consent of the IVN™ Program Manager is strictly prohibited.

1 INTRODUCTION

The Spaulding (1968) classification system categorizes medical devices that come into contact with mucous membranes and do not penetrate sterile tissue as "semicritical" and those that enter sterile body sites or penetrate the vasculature during surgery as "critical" medical devices. Equipment in both of these categories presents potential risk for infection transmission, as they are reused. Cleaning and decontamination is required between patient use and the procedures followed are designed to minimize the risk of infection of reusable medical equipment (RME). Yet, the implementation of cleaning procedures is itself subject to operational risks. Clearly presented, followed, and monitored standard operating procedures (SOPs) for the reprocessing of reusable medical equipment are essential to minimize patient risk to infection. The Veterans Health Administration (VHA) Directive 2009-004 stipulates that "It is VHA policy that systematic and local standard processes are developed in compliance with manufacturer's instruction, infection prevention and control principles, and effectively communicated and deployed to staff wherever procedures using RME are performed" (Kussman, 2009). To that end, the Veterans Health Administration, in cooperation with Wayne State University researchers, have developed an automated touch screen system for presenting work instructions for reprocessing flexible endoscopes, referred to as the Interactive Visual Navigator (IVN™). This system replaces paper-and-ink work instructions that, in current practice, are developed, authored and formatted locally on a site by site basis. One major benefit of the IVN™ is the presentation of complete instructions for the selected model of equipment. However, before RME reprocessing instructions are uploaded to the IVN™, the instructions undergo a thorough assessment and guidelines-based evaluation. The result of this evaluation directs the transformation of the instructions into a version more usable by the Sterile Processing Service (SPS) technicians that are tasked with reprocessing RME. The transformed instructions are then displayed on IVN™. While flexible endoscopes are complex pieces of equipment that require cleaning, decontamination and sterilization, SPS technicians require a high school diploma and the completion of a Sterile Processing Course for employment eligibility. Several Sterile Processing Certifications are possible and some institutions require certification to maintain employment (Meridian Health, 2012). This paper describes the instruction transformation process and the results in terms of usability.

2 PROBLEM

Reprocessing instructions are contained in a variety of manuals developed by endoscope manufacturers. Some are titled Safety Manuals, others are Users' Guides and some are just called Instructions. They rarely contain just reprocessing instructions, but rather include other model-related information concerning care. An Instruction manual for one Ultrasonic Bronchofibervideoscope contains 178 pages, beginning with Chapter 1: Checking the Package Contents. The

Reprocessing instructions begin on page 70 and end with Storage and Disposal on page 136. They include a section on the Importance of cleaning, disinfection, and sterilization and precautions about wearing appropriate personal protection equipment (PPE). The first step in Reprocessing, "Inspection of reusable equipment" begins on page 86. The instructions are mingled with Caution and Warning text containing consequences of improper step completion. This particular instruction manual contains a flowchart (including document section numbers) of the endoscope reprocessing steps on page 96. It is difficult to isolate step-by-step instructions to efficiently and effectively reprocess the RME. Because adherence to these instructions is paramount, VHA facilities have resorted to laminating the applicable pages and displaying them for use. While making the instructions accessible, this process results in technicians having to visually search for the next instruction to follow.

Once the instructions are isolated, the verbiage in the OEM instructions contains language and sentence structure that may exceed the expected reading ability of SPS Technicians. Although they are high school graduates and have typically passed a Sterile Processing Course, over 150 endoscope models exist in the VHA system and each one has separate instructions to follow. The technician most likely will refer to the instructions for models they may not regularly encounter. Making sure the technician comprehends the instructions is important to the success of reprocessing. Additional factors motivating simple work instructions include:

- Variability in technician performance (reprocessing reliability)
- Use by non-native English speakers
- Use concomitant with endoscope reprocessing (IVN™ use should not compete for the technician's attention)

Reliable comprehension of reprocessing instructions reduces the risk of task confusion. To this end, instructions are written at a grade level below the technician's expected education level.

Lastly, the immediate work environment imposes its own challenges. Endoscope Technicians wear gloves, which are usually wet and contaminated with bio-burden debri. Ensuring that liquid does not compromise the quality of the instructions is a necessary consideration for the display and use of these instructions.

3 APPROACH

IVN™ was developed as a standardized method of displaying endoscope model-specific instructions for use by SPS technicians. A touch screen capable, 17-inch monitor screen with a dual core 2 processor combined into a single casing was selected. Also, to prevent bio-burden contamination internally to the computer monitor, a sealed, fan-less casing was required. The standard IVN™ screen configuration includes breadcrumbs at the top of the screen presentation to show user progress in the overall reprocessing steps. A border surrounds the textual work instructions and contains forward and backward screen arrows and an 'X' in the upper right corner to exit the reprocessing screens. See Figure 1 for example IVN™ instruction displays.

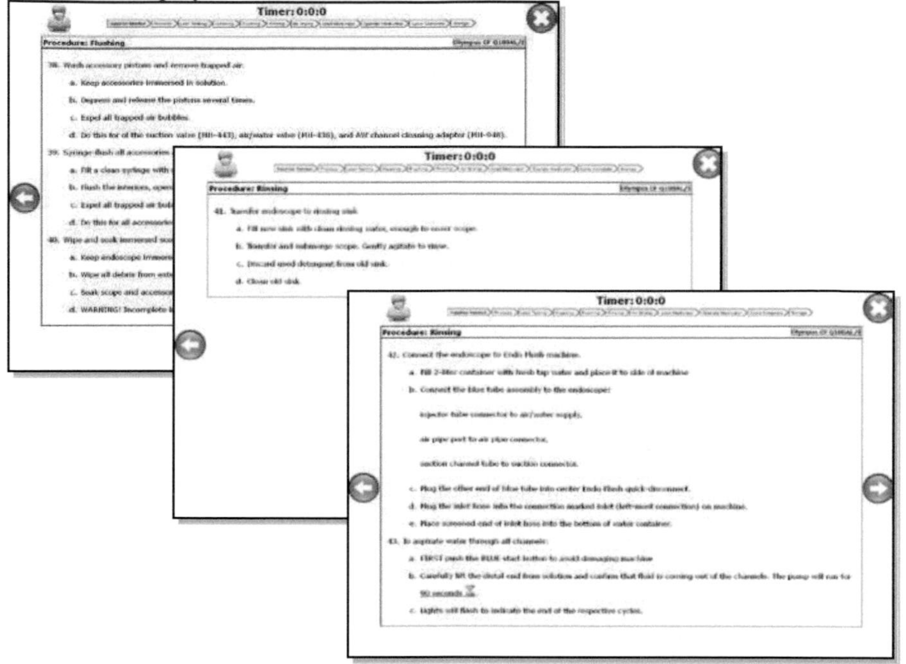

Figure 1. Work instruction screenshots from the IVN™

When evaluating the OEM instructions for suitability for use in IVN™, the following principles described in Wagner, et al. (1996) were used:

- Clear, simple language (free of vague and ambiguous words such as 'many'and 'often')
- Appropriate writing level (less than or equal to 12th grade)
- Average sentence length (less than or equal to 20 words)
- Sentence simplicity/complexity (simple sentences are easier to understand)
- Word order (subject-verb-object-predicate object-indirect object)
- Active voice (subject acts on predicate)
- Second person imperative (provide direction, such as "Soak the scope…")
- Third person indicative (used in warning and cautions, such as "Alcohol is not a high-level disinfectant…")
- Proper use of positive and negative wording (Use positive unless stating prohibitions and to correct existing or potential misconceptions)
- Consistent phrases (use the same phrase to express the same meaning throughout)
- Same wording of task steps (use the same words in all occurrences)

The Flesch-Kincaid Grade Level (Kincaid et al., 1975), an accepted measure (Crossley et al., 2007, Crossley et al., 2008, Paasche-Orlow et al., 2003) was used to evaluate the writing level of the OEM and the improved instructions. To comply with the OEM instructions, each endoscope reprocessing instructions begin with the purpose, equipment, location, a description of the overall procedure and a list of needed supplies, accompanied by an image of each supply. Step 1 is to wash hands and don a list of PPE, since that is a standard requirement for all RME reprocessing. Next, the OEM instructions are evaluated according to the principles listed above and improved by ensuring compliance to them. Rephrasing occurs to improve the observance of them. When images are included in the OEM instructions, they are made available to SPS technicians by touching an icon. Since the technicians refer to the screen periodically, white space was used between the steps to reduce text density and decrease visual search time to find the next instruction (Tullis, 1997, Wheeler, 2004).

4 EXAMPLES

Below are pairs of example work instructions. The "baseline" instructions transgress the principles discussed above, and represent the writing features found in current OEM reprocessing manuals. The "improved" instructions incorporate the human factors principles listed above. Changes between the two examples in each pair are underlined to emphasize the improvement:

Clear, simple language:

Baseline instructions: "Be sure to lower the scope in detergent solution in such a way necessary to completely submerge it and allow the scope to soak for 6 minutes in detergent".

Improved instructions: "Completely submerge the scope in detergent. Soak the scope for 6 minutes."

Replace inherently vague words with more clear ones:

Baseline instructions: "Fit endoscope properly into the washing basin, next to the disinfector".

Improved instructions: "Fit endoscope securely into the washing basin, to the left of the disinfector".

Include needed information in the immediate text:

Baseline instructions: "Brush the scope with the model recommended by the manufacturer's manual".

Improved Instructions: "Brush the scope with the ENZ-1 model brush".

Reduce unnecessary words:

Baseline instructions: "Fit adapter inside of the auxiliary port of the endoscope".

Improved instructions: "Fit adapter in the endoscope's auxiliary port".

Active voice:

Baseline instructions: "Connection hoses should be kept clear of the spray tower".

Improved instructions: "Keep connection hoses clear of the spray tower".

Positive Wording:

Baseline instructions: "Do not leave scope unattended during the leakage test. Lack of a continued stream of bubbles does not indicate a leak".

Improved instructions: "Examine scope closely during leakage test for bubbles. A continued stream of bubbles indicates a leak".

5 RESULTS

OEM reprocessing instructions are isolated and evaluated according to human factors principles. They are then manually improved upon by adhering more closely to these principles. They are then reviewed by the appropriate facility management for compliance to VHA Directive 2009-004 and approved after necessary revisions. The resulting instructions are clearer and more succinct and are presented at a reading level more reliably processed by SPS technicians. See Table 1 for an example of the change in Flesch-Kincaid Grade Level scores as measured in MS Word™. The resulting instructions are sectioned to fit within the 16 lines of text available on the screen once the standard IVN™ border is applied to the 17 inch screen.

Table 1: Comparison of OEM readability with Improved Instructions.

Manual #	Olympus Scopes Covered	Original F-K Grade	Final F-K Grade
GR7237 10	40, 140, and 240 series	12.4	7.8
GE1016 08	160 series colonsocopes and gastroscopes	12.3	7.3
GE8391 05	180 series colonoscopes and gastroscopes	12.0	7.7
GE8415 03	TJF Q180V	12.1	7.8

Figure 2 contains a completed IVN™ screen that includes links to images and a timer icon that tracks time used on the displayed page. A minimum time per page is also built into the process (5 seconds) to prevent rapid movement through the procedure and allows minimal cognizant time per page of the information displayed before allowing the technician to proceed to the next step. This minimum time should not hinder the more experienced technicians in completing the steps on the displayed page, hence five seconds was found to constitute a reasonable time to prevent frustration in experienced technicians.

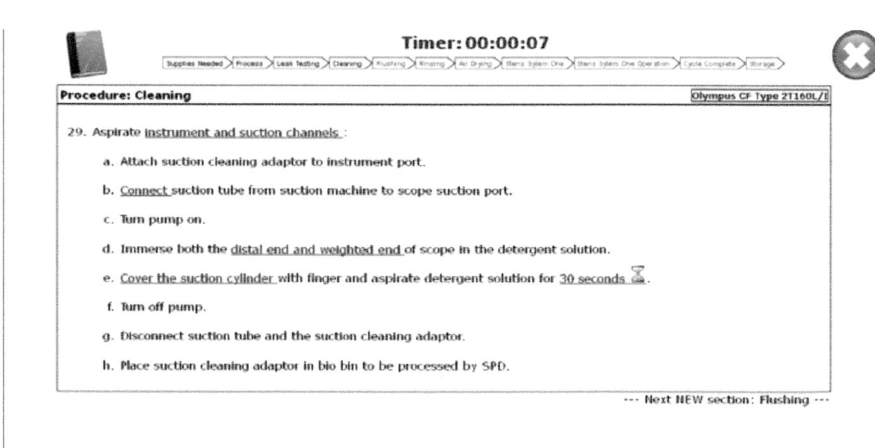

Figure 2. Completed work instruction screen from the IVN™

6 CONCLUSIONS

Improperly reprocessed RME represents a considerable risk to patient safety. Usability of text-based work instructions are a critical factor in the overall task of managing RME cleaning and maintenance. This paper has shown two major contributions to current practice. First, no guidance exists for authors of job aids or work instructions for RME cleaning technicians. The observations and examples listed above help fill that gap. Furthermore, recommendations presented are couched in the context of readily available tools and readily applicable heuristics. While this process of work instruction authoring was developed with the automated touch screen IVN™ in mind, the techniques described above are equally applicable to paper-and-ink work instructions, and the principles applied are easily transferred to related domains.

Future directions include some interesting opportunities. Further research is required to determine the extent to which RME reprocessing could benefit from its own domain-specific version of "simplified technical English." If pursued as a technical standard in the context of an industry-wide trade group, a much higher degree of standardization is attainable.

A related area of work is formal analysis of work instructions. Previous research has led to a proposal of using Extensible Markup Language (XML) to impose structure over clinical guidelines in areas such as laboratory and pathology workflows (Dart et al, 2001; Dubey & Chueh, 1998; 2000; 2001; Georg, 2005; Mea, Pittaro, & Roberto, 2004; Sedlmayr et al., 2007; Sedlmayr et al., 2006; Shiffman et al., 2000) . Authoring work instruction documents in a rigorous framework using XML will enable next-generation automated checking of consistency, completeness and correctness, and will also present opportunities for more flexible presentation and use across display formats.

ACKNOWLEDGEMENTS

The authors gratefully acknowledge the support of the Veterans Health Administration.

REFERENCES

Crossley, S.A., Dufty, D.F., McCarthy, P.M., & McNamara, D.S. (2007). Toward a new readability: A mixed model approach. In D. S. McNamara and G. Trafton (Eds.), *Proceedings of the 29th Annual Conference of the Cognitive Science Society*. Nashville, TN:Cognitive Science Society.

Crossley, S.A., Greenfield, J., McNamara, D.S. (2008). Assessing Text Readability

Using Cognitively Based Indices, *Tesol Quarterly*, 42(3), 475-493.

Dart, T., Xu, Y., Chatellier, G., & Degoulet, P. (2001). Computerization of guidelines: towards a "guideline markup language". *Studies in health technology and informatics*, *84*(Pt 1), 186–190.Drury, C. G. (1998). Case study: error rates and paperwork design, *Applied Ergonomics, 29*(3), 213-216.

Dubey, A. K., & Chueh, H. (1998). Using the extensible markup language (XML) in automated clinical practice guidelines. *Proceedings of the Annual AMIA Symposium*, 735–739.

Dubey, A. K., & Chueh, H. C. (2000). An XML-based format for guideline interchange and execution. *Proceedings of the Annual AMIA Symposium*, 205-209.

Dubey, A., & Chueh, H. (2001). Using XML Metadata to Enable the Automatic Generation and Processing of HTML FORMS from XML Documents. *Proceedings of the Annual AMIA Symposium*, 894.

Georg, G. (2005). Computerization of Clinical Guidelines: an Application of Medical Document Processing. In B. Silverman, A. Jain, A. Ichalkaranje, & L. Jain (Eds.), *Studies in Fuzziness and Soft Computing* (Vol. 184, pp. 1–30). Berlin/Heidelberg: Springer-Verlag. doi:10.1007/11311966_1

Kincaid, J. P., Fishburne, R. P. J., Rogers, R. L., Chissom, B. S., & NAVAL TECHNICAL TRAINING COMMAND MILLINGTON TENN RESEARCH BRANCH. (1975). *Derivation of New Readability Formulas (Automated Readability Index, Fog Count and Flesch Reading Ease Formula) for Navy Enlisted Personnel*. Ft. Belvoir: Defense Technical Information Center.

Kussman, M. J. (2009). Use and Reprocessing of Reusable Medical Equipment (RME) in Veterans Health Administration Facilities, VHA Directive 2009-004. Retrieved February 23, 2012 from http://www.va.gov/vhapublications/ViewPublication.asp?pub_ID–1824

Mea, Della, V., Pittaro, M., & Roberto, V. (2004). Knowledge management and modelling in health care organizations: The standard operating procedures. *Knowledge Management in Electronic Government*, 136–146.

Meridian Health. (2012). Career Opportunity: Technician SPD Certified [SPD] FT Day. Retrieved February 22, 2012 from http://jobs.meridianhealth.com/new-jersey/technician-and-technologist/technician-spd-certified-%5Bspd%5Dft-day-jobs?apstr=%26emid%3D3639.

Paasche-Orlow, M.K., Taylor, H.A., Brancati, F.L. (2003). Readability Standards for Informed-Consent Forms as Compared with Actual Readability. *The New England Journal of Medicine*, 348, 721-726.

Sedlmayr, M., Rose, T., Greiser, T., Röhrig, R., Meister, M., & Michel-Backofen, A. (2007). Automating standard operating procedures in intensive care. *Advanced Information Systems Engineering*, 516–530.

Sedlmayr, M., Rose, T., Röhrig, R., & Meister, M. (2006). A workflow approach towards GLIF execution. *Proceedings of the European Conference on Artificial Intelligence (ECAI). Riva del Garda, Italy.*

Shiffman, R. N., Karras, B. T., Agrawal, A., Chen, R., Marenco, L., & Nath, S. (2000). GEM: A proposal for a more comprehensive guideline document model using XML. *Journal of the American Medical Informatics Association*, 7(5), 488–498. doi:10.1136/jamia.2000.0070488

Spaulding, E. H. (1968). Chemical disinfection of medical and surgical materials. In Lawrence CA, Block SS, eds. Disinfection, Sterilization and Preservation, Philadelphia, PA: Lea & Febiger, 517-531.

Tullis, T.S. (1997). Screen design. In M. Helander, T.K. Landauer and P. Prabhu (Eds.), *Handbook of Human Interaction* (2nd ed., pp. 503-531).

Wagner, D., Birt, J. A., Snyder, M., Duncanson, J. P. (1996) Human Factors Design Guide For Acquisition of Commercial-Off-The-Shelf Subsystems, Non-Developmental Items, and Developmental Systems, Federal Aviation Administration, Report No. DOT/FAA/CT-96/1 retrieved February 23, 2012 from http://www.deepsloweasy.com/HFE%20resources/HFE%20Design%20Guide%204COTS.pdf.

Weller, D. (2004). The Effects of Contrast and Density on Visual Web Search. Usability News, 6(2). Retrieved February 24, 2012 from http://www.surl.org/usabilitynews/62/density.asp.

CHAPTER 65

A Proposal of the Database Schema of a Drug Composition Database to Generate Drug Package Inserts

Ryo Okuya, Hirotsugu Ishida, Keita Nabeta

Masaomi Kimura and Michiko Ohkura

Shibaura Institute of Technology
Tokyo, JAPAN
{ma11044, m110013, m709102, masaomi, ohkura }@sic.shibaura-it.ac.jp

Fumito Tsuchiya

International University of Health and Welfare
Tochigi, JAPAN
ftsuchiya@iuhw.ac.jp

ABSTRACT

We expect computerized systems in medical settings, such as prescription check systems, to prevent medical accidents caused by drugs. To create such systems, we need drug information databases. However, existing databases lack suitable data schema and inherit the problem of original data, ethical drug package inserts. Though their standpoint involves utilizing package insert information to build databases, we adopt another standpoint in this study by initially implementing a drug information database, and then generating package inserts based on the information therein. In this paper, we particularly focus on a composition database and propose its schema.

Keywords: medical safety, package inserts, data schema, cluster analysis

1 INTRODUCTION

We expect computerized systems in medical settings to prevent medical accidents caused by drugs. For example, if a computerized prescription order entry system, the task of which is to mediate prescriptions in a hospital, verifies the information of ordered drugs, it is possible to prevent prescription errors.

To create such systems, drug information databases are needed. The data and structures in most existing drug information databases are based on the descriptions in package inserts. This is because the latter are officially published documents by pharmaceutical companies and should be used as a primary source of drug data. However, since there is no strict rule on expressions used therein, the variety makes it difficult to correctly identify and extract drug information. Unfortunately, the existing databases built in this approach inherit this difficulty.

With this in mind, we adopt a new approach by initially implementing a drug information database, and then generating package inserts by inserting data in the database into a suitable template. The implemented database is unaffected by the expressions in package inserts and maintains proper drug information. Moreover, it is not necessary to create package inserts from scratch.

In this study, in order to propose data schema suitable for the drug information database, we analyzed drug information descriptions in package inserts, based on a natural language processing technique and cluster analyses on description patterns.

2 TARGET DATA

Target data are the descriptions in ethical drug package inserts published by the PMDA (Pharmaceuticals and Medical Device Agency). We used SGML (Standard General Markup Language) formatted versions of package inserts, the data of which were structured by tagging. In this study, we extracted descriptions surrounded by *composition* tags and analyzed them to find information. Figure 1 shows a sample of *composition* descriptions in SGML package inserts. As is clear, the *composition* tags surround them, and the detailed information and name are respectively enclosed in *detail* and *item* tags.

We retrieved and downloaded 11547 SGML package insert files from the PMDA website. We should note that the package inserts may include information on multiple drugs. In this study, we used 14639 sets of package insert data for individual drugs, obtained by the division of the original SGML files.

```
<composition>
   <item>Active ingredient</item>    <detail>Cabergoline</detail>
   <item>Amount</item>               <detail>0.25 mg</detail>
   <item>Inactive ingredient</item>
   <detail>Anhydrous lactose, l-leucine and magnesium stearate</detail>
</composition>
```

Figure 1 A sample composition description in SGML package inserts

3 METHODS

In this paper, we focus on the design of a composition database. In order to identify the information included in composition descriptions in package inserts, we absorb the variety of expressions and find patterns in the descriptions. These are generated by replacing certain information (*e.g.* amount, names, and so on) in the descriptions with labels (Nabeta, 2009). We apply cluster analysis to the obtained patterns and identify information commonly appearing there.

3.1 Generation of Description Patterns

Firstly, we extracted text parts from the target data. We appended text data in *item* and *detail* tags in "<BR>" delimited format, and appended the resultants delimited with the label "<
>".

Table 1 The types of words to be replaced and their corresponding labels

Drug information	Label
Brand name	[BrandName]
Active ingredient name	[ActIngreName]
Other name of active ingredient	[ActIngreOtherName]
Abbreviated name of active ingredient	[ActIngreShortName]
Number	[Number]
Unit	[Unit]
Molecular formula	[Formula]
Peculiar part of drug (e.g. support medium)	[DrugPart]
Color	[Color]
Dosage form (e.g. injection)	[Form]
Smell	[Smell]
Taste	[Taste]

Secondly, we replaced some words appearing in the text, such as active ingredient names and brand names, with labels. We replaced the active ingredient names and the amounts, expressed by combinations of numbers and units with labels, because the Ministry of Health, Labour and Welfare (MHLW) requires them to be described in the composition (MHLW, 1997). Additionally, we replaced the molecular formula of the active ingredient and the property information such as dosage forms/colors in the descriptions. Table 1 shows a list of the types of words to be replaced and their replacement labels. Moreover, we deleted unnecessary characters and delimiters such as spaces, punctuation and carriage returns in these descriptions. We call the resultants description patterns (Figure 2).

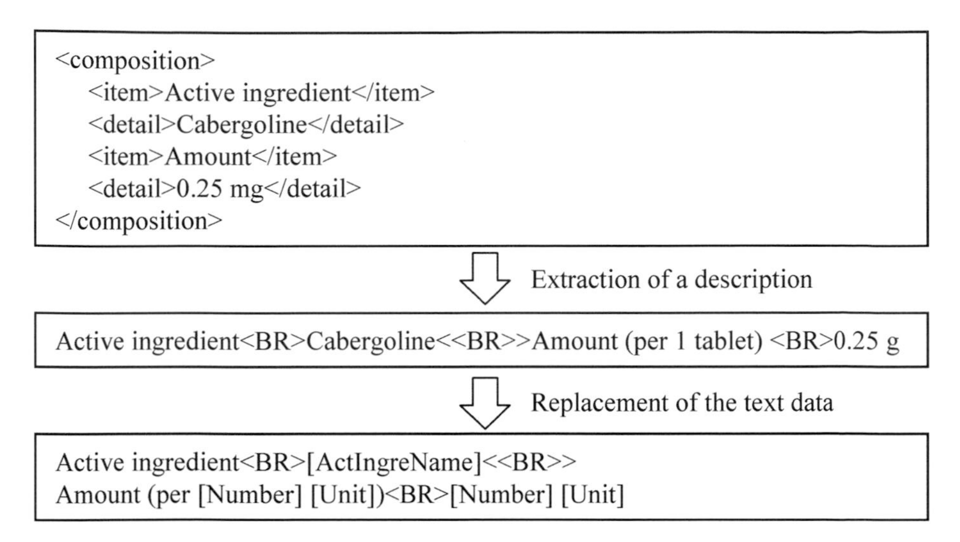

Figure 2 An example of the generation of description patterns

3.2 Cluster Analysis

In order to group generated patterns, we applied cluster analysis to them.

Cluster analysis methods classify similar data into the same groups called clusters. Their algorithms are categorized into two types, partitioning-optimization and hierarchical. Partitioning-optimization cluster analysis classifies similar data into exclusive clusters, while hierarchical cluster analysis aggregates similar data in sequence of decreasing distance between them, and provides us with the aggregation process as a tree diagram called a dendrogram. The height of a dendrogram corresponds to the distance within which data are clustered. Namely, if we cut the dendrogram at an arbitrary height, we obtain clusters within which the distances of any data are less than the height. In this study, we applied the clustering algorithm in two steps. We first applied the K-means algorithm, one of the partitioning-optimization methods, to identify the same patterns, followed by hierarchical cluster analysis to clusters in the first step, in order to determine the clusters of patterns based on a dendrogram.

The distance between patterns must be defined when employing hierarchical cluster analysis. We focus on nouns as the most numerous component of the patterns, which should be used to classify the patterns, because the same words tend to appear in similar patterns. Therefore, we create vectors representing the patterns, the elements of which denote the existence of nouns or labels. If a noun or one of the labels exists in the pattern, the element is one and otherwise, it is zero.

In order to define the vector distances, we need to tackle a problem known as the "curse of dimensionality", which is due to the high dimensionality of the vectors. In this study, we reduce the vector dimensions by applying singular value decomposition to the matrix, each of whose rows is the vector defined above.

We calculate the distance based on the cosine value of the angle between a pair of vectors. Though the cosine value is suitable for use in measuring the similarity of the vectors, it lacks the property to be a distance, since the larger cosine value means the vectors are more similar. Thus, we convert the cosine value into a distance by subtracting it from 1 and apply hierarchical cluster analysis based on this distance.

4 RESULTS

4.1 Generation of Description Patterns

We succeeded in extracting descriptions from composition tags in 14277 SGML package inserts, but failed to extract descriptions from 362 SGML data due to the lack of composition tags in the data.

The replacement of keywords in the descriptions to the labels generated 3121 patterns. During confirmation of the patterns, we found a tendency whereby the words in the descriptions of blood preparations and radiopharmaceuticals were not replaced with labels (Figure 3). The descriptions are variously expressed to describe more detailed information and are potentially grouped into numerous clusters. Therefore, we excluded the patterns of these drugs. Similarly, we also excluded the patterns of Chinese herbal drugs and crude drugs usually containing ingredients whose biochemical function is unknown. We also excluded biological drugs, because their descriptions were formularized by MHLW. Excluding the patterns of these drugs, we obtained 2685 patterns.

Active ingredient (per [Number] [Unit])
[ActIngreName] alpha (activated) (recombinant) : [Number] [Unit]
[Number] [Unit] of active ingredients correspond to 50 KIU

Figure 3 A sample of blood preparation description patterns

4.2 Cluster Analysis

Figure 4 shows a dendrogram merging the clusters obtained in the first step. Confirming the aggregated patterns, we divided the patterns into 10 clusters. The red line indicates the height to cut the dendrogram. In the following explanation, we assign numbers to clusters in order from left to right, such as Cluster 1, Cluster 2 and so on.

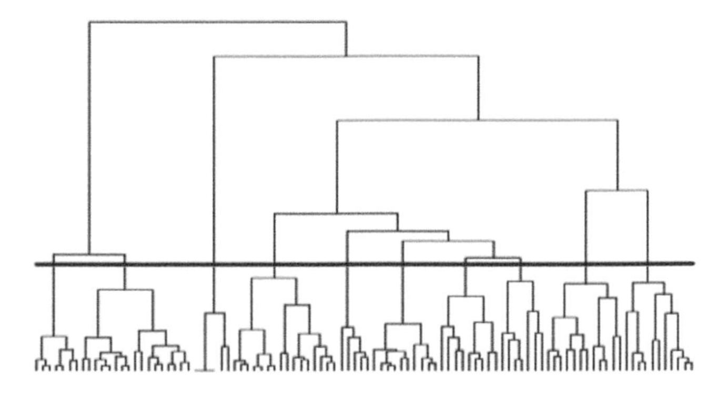

Figure 4 A dendrogram created by the hierarchical cluster analysis

Table 2 Samples of description patterns in Cluster 2

Description patterns
Active ingredient (per [Number][Unit]) [Number] [Unit] of [ActIngreName]
Ingredient amount (per [Number] [Unit]) [Number] [Unit] of [ActIngreName]
Active ingredient (per [Number][Unit]) [Number] [Unit] of [ActIngreName] [Number] [Unit] of [ActIngreName] [Number] [Unit] of [ActIngreName]
Ingredient amount (per [Number] [Unit]) [Number] [Unit] of [ActIngreName] ([Number] [Unit] of [ActIngreName]
Amount [Number] [Unit]< > Ingredient amount (per [Number] [Unit]) [Number] [Unit] of [ActIngreName]

As for Cluster 2, it includes about half the patterns. As shown in Table 3, in this cluster, most patterns mainly consist of an active ingredient name, its amount and the dosage unit of a drug expressed as "per [Number] [Unit]". Moreover, we also found patterns in which the expressions "[Number] [Unit] of [ActIngreName]" appeared repeatedly, and those whose expressions "[Number] [Unit] of [ActIngreName]" were included in parentheses. Investigating their differences, we found that the former described an active ingredient and the latter described a compound including the active ingredient.

Table 3 Typical description patterns in clusters other than Cluster 2

Number	Description patterns
1	Brand name [BrandName] < >Active ingredient and its amount (per [Number] [Unit]) [Number] [Unit] of [ActIngreName]
3	This drug which contains a specific allergen obtained by dialysis from the extraction ingredient from material (flour) is aseptic physiological saline

	This drug regulated by aseptic manipulation is [Number] [Unit] [Form] for the material weight ([Number] : [Number])
4	The following ingredients are contained per [Number] [Unit] < > Active ingredient [Number] [Unit] of [ActIngreName]
5	Active ingredient (per [Number] [Unit]) [Number] [Unit] of [ActIngreName] (recombinant)
6	This product contains the following ingredients per [Number] [Unit] < > Active ingredient and its amount [Number] [Unit] of [ActIngreName]
7	[BrandName] which contains [ActIngreName] [Number] [Unit] per [Number] [Unit] is administrated to dissolve before use < > Active ingredient [Number] [Unit] of [ActIngreName]
8	[BrandName] which contains [ActIngreName] [Number] [Unit] per [Number] [Unit] is [Form] of [Color] < > <Electrolyte concentration> (theoretical value) [Number] [Unit] of [Formula]+
9	[ActIngreName] [Number] [Unit] < > Calorie [Number] [Unit]
10	This product contains not less than [Number] [Unit] of [ActIngreName] ([Formula])

Table 4 shows typical description patterns in the clusters except for Cluster 2.

The pattern in Cluster 3 includes information about production methods such as "dialysis" and "aseptic manipulation" and about raw materials of active ingredients such as "flour".

We can see that a pattern in Cluster 5 includes supplementary information of the active ingredient, such as "recombinant" after an active ingredient name.

The pattern in Cluster 7 includes the usage of a drug.

We found that the patterns include the terms "Electrolyte concentration" in Cluster 8 and the term "Calorie" in Cluster 9.

The pattern in Cluster 10 includes the expression "not less than" described with the amount of an active ingredient. This describes the lower bound of the amount of active ingredients in products.

5 DISCUSSIONS

5.1 Discussion of the Results

Since the name, amount and dosage unit of the active ingredient tend to be described in most patterns, it is essential to define them as data attributes in a composition database. Moreover, since plural pairs of an active ingredient and its amount can appear, it is also essential to define data schema whose instances can store such pairs. As for the amount information of active ingredients, the result suggests that it can be described not as a value but a range and should thus be stored as the upper/lower limit values of the amount in the database.

Typical patterns in each cluster include calorie data, electrolyte concentration and so on. Although these are optional, they must be defined as attributes in the schema.

Conversely, the attribute of usage information found in the patterns in Cluster 7 need not be defined, because this should be described in the other section of package inserts, "dosage and administration".

5.2 Detailed analysis on rarer patterns

In Section 5.1, we focused on attributes based on frequently appearing expressions in the patterns. However, our database should also cover other kinds of information described in compositions, even if they are rarer. Therefore we additionally analyzed such patterns. In this section, we show the results and propose additional attributes necessary for a composition database.

Additional amount / concentration information of specific ingredients

Figure 5 shows the pattern including the expressions "concentration" and "amount" after the description of an active ingredient. In the original description, "iodine" was replaced with "[ActIngreName]" in this pattern. There were also other descriptions belonging to similar patterns, in which "amino acid" and "nitrogen" were replaced with "[ActIngreName]". From these findings, we found the need to define additional attributes of the amount / concentration of specific ingredients.

Amount ([Unit])
50<
>Ingredient amount
Inclusion of [Number] [Unit] of [ActIngreName] per [Number] [Unit]<
>
Concentration of [ActIngreName] ([Unit])
300<
>
Amount of [ActIngreName] per [Number] [Unit] ([Unit])
15

Figure 5 A sample of a pattern which includes the amount and concentration of particular ingredients

Detailed units

Figure 6 shows a pattern including a detailed unit at the end of the expression. Since the ways to describe this information depend on the unit and may vary, it the attribute storing this as text data must be defined.

Ingredient amount (per [Number] [Unit])
 [ActIngreName] of [Number] [Unit]   [ActIngreName] activity is described by [ActIngreName] unit, measuring by biologic assay relative to standard.

Figure 6 A sample of the pattern including information of a detailed unit

In addition, we found information on *elements, details of calories, the amount of active ingredients per use* and *the size of a drug*. We included attributes corresponding to this information.

5.3 Proposal of Database Schema

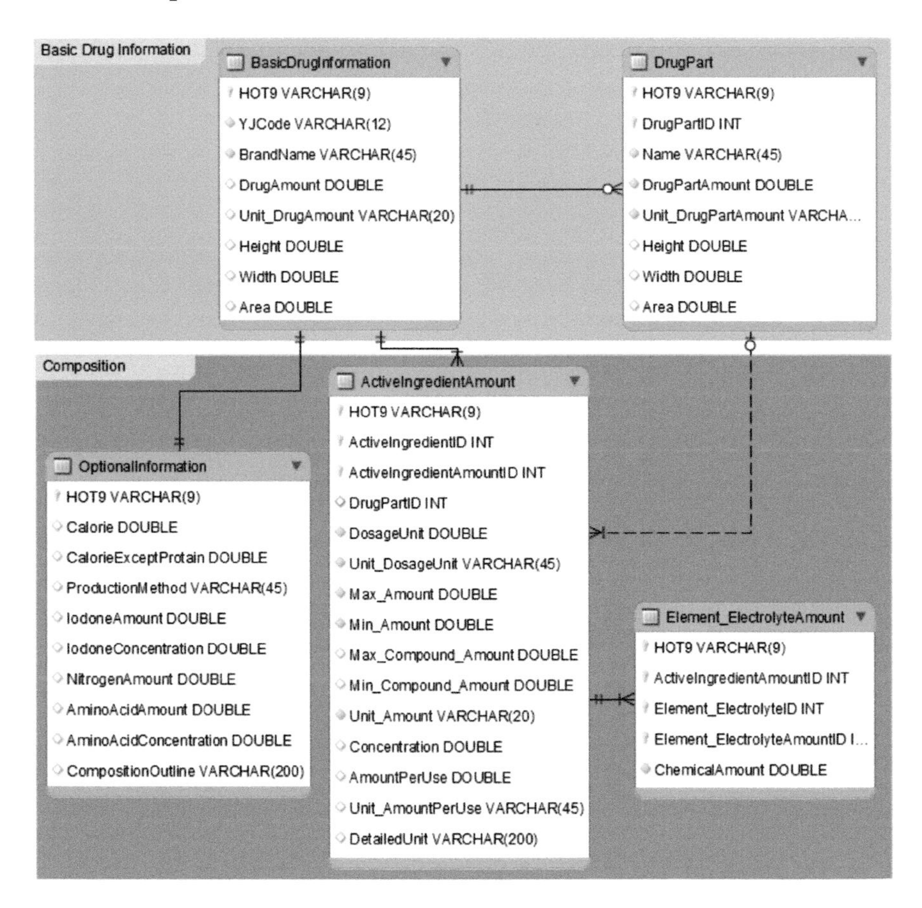

Figure 7 Entity relationship diagram based on the proposed data schema (main part)

We propose the data schema of a composition database based on the drug information attributes presented in sections 5.1 and 5.2. Essential attributes, such as *drug amount, dosage unit, active ingredient name* and *amount*, should be stored in a table as a set of drug fundamental data (Drug fundamental information table). The attributes for optional information such as *calorie data* and *the name of* a *compound,* including the active ingredient, are stored in another table known as a Composition table. Although electrolyte information is connected to the drug information in package inserts, this should be essentially related to the active ingredient information. Accordingly, we defined a table that stores the data on elements and electrolytes linked with corresponding active ingredients. Figure 7 shows the main part of the entity relation diagram based on the proposed data schema.

6 CONCLUSION

Medical drug package inserts are exclusive official drug information sources. Though there have been many efforts to utilize these as a database in computerized systems, this is hindered by variation of the expressions used in package inserts. In this study, we adopted another strategy by proposing a drug information database to generate a prototype of a package insert.

We obtained package insert data in SGML format from the PMDA website, extracted composition text data and generated description patterns from the same. In order to investigate the tendency of generated patterns, we applied combined clustering techniques to the patterns and extracted some clusters that suggested the types of information included in the composition part of package inserts.

As the results of this analysis, we found that most patterns included essential information on composition such as the dosage unit of the drug, the active ingredient name and its amount. The pairs of active ingredient name and amount thereof repeatedly appeared in the patterns. In addition, we also found that the patterns could include optional information such as calorie data, electrolyte concentration and so on. Based on these, we defined the data schema which should be used to implement the database.

In future, we will discuss a template for package inserts to be filled with the data in the proposed database. We will also propose an extended version of data schema to store the information included in the descriptions of biological drugs and that related to inactive ingredients, which was excluded in this study.

REFERENCES

Ministry of Health, Labour and Welfare 1997. A *guideline for descriptions in ethical drug package inserts*. Tokyo: MHLW.

Nabeta, K. et al: 2009, A Proposal of a Method to Extract Active Ingredient Names from Package Inserts. *Human Computer Interaction* 9: 576-585

"Pharmaceuticals and Medical Device Agency," Accessed February 2, 2012, http://www.info.pmda.go.jp/.

CHAPTER 66

Sensors as an Evaluative Tool for Independent Living

Katie Woo, Veselin Ganev, Eleni Stroulia, Ioanis Nikolaidis, Lili Liu, Robert Lederer

University of Alberta
Edmonton, AB, Canada
Email address

ABSTRACT

The anticipated increase in the world population, people's desire to live at home for as long as possible and the shortage of health-care professionals make the need to develop technological supports for people with chronic yet manageable conditions to monitor and take care of their own health. In our work in the Smart Condo™, we are developing a sensor-based infrastructure for non-intrusively monitoring and analyzing the activities of people at home. In this paper we review our technology and research methodology and we report on the findings of a case study with two participants.

Keywords: smart homes, independent living, activity monitoring

1 INTRODUCTION AND BACKGROUND

The world population over the age of 60 is increasing rapidly and is expected to triple by 2050[1] due to rising life expectancy and fertility decline[2]. According to reports from the United Nations, most seniors worldwide prefer to age in their own

---

1 United Nations Press Release - http://www.un.org/esa/population/publications/wpp2008/

2 United Nations Department of Economic and Social Affairs. http://www.un.org/esa/population/unpop.htm

homes despite possible risk to their health. In a survey undertaken by a healthcare financing administration in the U.S., 30% of those over 65 surveyed stated they would "rather die" than enter a nursing home[2]. At the same time, there is a worldwide shortage of nearly 4.3 million doctors, nurses, midwives, community-health workers and pharmacists; this need is expected to increase by a further 20% within the next two decades[3]. As a result, governments, academia and industry are undertaking research on cost-effective approaches to delivering high-quality healthcare at home, and in this research agenda, technology has a major role to play.

The Smart Condo™ project was conceived with the objective of designing and evaluating technology-based products and services in support of home-based health-care delivery. Recognizing that this challenge requires expertise from different fields of knowledge, the project involves a multi-disciplinary team consisting of faculty members and students in Occupational Therapy, Nursing, Medicine, Pharmacy, Industrial Design, and Computing Science.

Figure 1 Pictures of the various areas (bathroom, wheel-chair accessible dining area, kitchen with height-adjustable cabinets) in the Smart Condo™.

The project's current home is a dedicated space in the University of Alberta's Edmonton Clinic Health Academy (ECHA), overseen by the Health Sciences Education Research Commons (HSERC). HSERC is an umbrella organization, spanning across the health faculties of the University of Alberta, whose objective is to develop innovative curricula for the students enrolled in the programs of these faculties. To that end, HSERC relies primarily on simulation, as a pedagogical methodology for training students for competency. Given this focus, HSERC manages six simulation spaces, one among them being the Smart-Condo™ space, a one bedroom, fully functional apartment. The Smart Condo™, as shown in the pictures of Figure 1 above, has been designed to meet some basic accessibility constraints. The cabinetry can be controlled to go up (or down) to the user's preferred height; all areas are easily accessible to an occupant who uses mobility aids; the floor is covered with a non-slip surface; and the electrical outlets are in multiple heights to make them easily accessible by individuals standing or using a wheel chair.

[3] The World Health Report 2006 - Working Together For Health http://www.openmedicine.ca/article/view/248/145

Infused into the apartment and its furnishings are sensors that record a variety of environmental variables (i.e., levels of light and sound, temperature and humidity) as well as the activities of the occupant (such as motion and use of cabinetry and electrical appliances). The data gleaned by these sensors are transmitted in real time into a central repository. On-line analyses of the data stream can be performed to recognize interesting patterns, indicative of problems that should generate alerts to formal and informal caregivers. Furthermore, the data collected in the repository over time can be analyzed to recognize patterns in the daily-living activities of the occupant, such as how many times he/she usually goes to the bathroom, how frequently and how long he/she watches TV during a day, and approximately when he/she goes to bed every night. Hereafter, we will refer to the combination of the sensors embedded in the environment and the software system collecting and analyzing their data as the "Smart-Condo™ middleware".

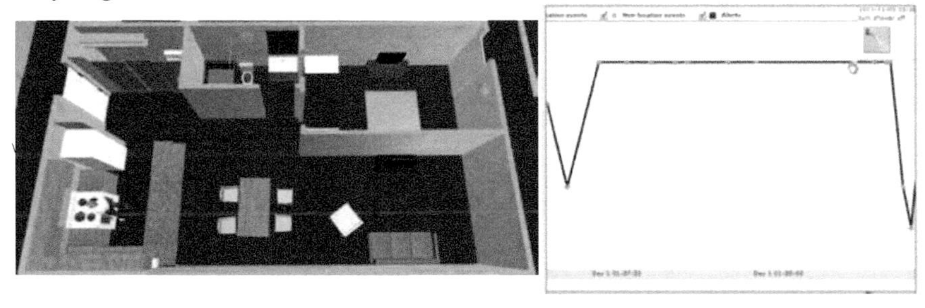

Figure 2 The 3D virtual-world model of the Smart Condo™; the avatar activities "replay" the occupant's activities, as inferred from the sensor data. The 2D visualization shows interesting events ("turn shower off") on a timeline; the different timeline levels are associated with different areas in the condo.

The raw data collected by the middleware and the results of the analyses are then visualized to better communicate the information they convey to the research team and the caregivers of the occupant. To that end, we are employing typical 2D visualizations (through web-accessible interactive front ends). In addition, and more interestingly, we are exploring the potential of virtual worlds as an immersive platform where the occupant's activities, as inferred by the middleware, can be "replayed" while being annotated with the extracted information. Intuitively, the virtual world enables a video-like simulation of the occupant's activities. Figure 2 shows a picture of the virtual-world model of the Smart Condo™.

Together, the multi-disciplinary expertise of the team members, the accessible home, and the hardware-software infrastructure embedded in the home to enable the recording and analysis of its occupant's activities constitute the three "pillars" of our simulation-based research methodology, which we discuss in this paper.

The rest of the paper is organized as follows. Section 2 provides an overview of this methodology. Section 3 discusses a specific case study, on the role of a particular technology in medication adherence that exemplifies this methodology. Section 4 summarizes the lessons we have learned through our experience thus far and lays out some future plans.

2 THE SMART-CONDO™ RESEARCH AGENDA AND METHODOLOGY

The Smart Condo™ research agenda focuses on developing and evaluating technologies, products and practices to enable older adults to live independently at home longer. At the core of this agenda, the Smart-Condo™ middleware is designed to investigate the question of *"how low-cost sensing technologies (RFIDs, QR tags, and sensors) can be used to monitor patients' at home, in order to extract clinically relevant information about their overall state of health, level of function, independence and safety"*.

Around this core question, our team is interested in studying two broad focal areas: 1) design and 2) information technology. In the *design agenda*, we are designing and manufacturing prototypes of products based on universal-design principles to support the everyday life of clients with mobility, perceptual and cognitive challenges. In our *information-technology agenda*, we are developing applications to (a) collect more and better information about the clients' activities in order to better inform their care plan; (b) support clients with impaired cognition and memory; and (c) improve the communication between clients and their family members and health-care providers. Finally, aiming to bridge our two areas of interest, we are designing and prototyping "smart products", i.e., appliances useful for daily-living activities embedded with sensors so that they can provide relevant data to our middleware about the person's health and activities.

In parallel with our research agenda, we are also developing our research methodology, according to the following three principles. First, we are committed to working in close collaboration with potential receptors of our research results. We already have an established network of such receptors, including (a) long-term elderly care facilities (such as the Shepherds' Care organization - http://www.shepherdscare.org); (b) rehabilitation hospitals (i.e., the Glenrose Rehabilitation Hospital); and (c) industrial partners who may be in a position to take on the commercialization of our products. These partners play several roles in our research program. They identify problems in their practices and services for our team to address, they make their environments available to our team to deploy and evaluate our technologies, and they guide the evaluation processes thus ensuring that our research is relevant to the community at large.

Second, we are adopting *open and extendible standards* for the development of our hardware and software systems. To that end, our middleware is developed in the service-oriented style so that new "smart devices" and analyses can be easily integrated with it. A variety of modern off-the-shelf appliances include digital sensors that emit different types of information. For example, weight scales emit data about the weight of the person using it (and since the scale is usually in a static place in the home, the event can also be used to infer the location of the occupant). The information produced by these appliances can be easily ingested in the Smart-Condo™ middleware, by developing special-purpose information services.

Third, we are committed to using mixed, i.e., quantitative and qualitative, methods for the evaluation of our work. Quantitative methods are the more appropriate approach for evaluating the effectiveness and performance of technology. For example, we have conducted a series of experiments to evaluate the precision with which our middleware tracks the actual location of the occupant. On the other hand, the ultimate evaluation of technology in this context is in its usefulness to the users it is intended to support and the clinicians whose practice it is meant to inform. To that end, we are designing descriptive and qualitative studies, using questionnaires, interviews and focus groups as data-collection instruments, to collect information from with client educators, i.e., patients who are willing to use and comment on our technologies, and clinicians associated with our partner organizations.

In the next Section, we report on a case study that we conducted in collaboration with the Glenrose Rehabilitation Hospital (GRH), which clearly exemplifies our methodology. For that study we deployed the Smart-Condo™ middleware integrated with a commercial off-the-shelf medication-adherence product (MAT), in the Independent Living Suite (ILS) of the GRH to determine the perceptions of older adults regarding this product and our technology. The secondary objective was to gain some experience on how easy the deployment of our middleware is in a new space and to obtain the feedback of clinicians on the understandability and usefulness of its reports.

3 THE ILS CASE STUDY

Smart-Condo™ Middleware Deployment in the ILS: In order to make this study possible, we had to deploy the Smart-Condo™ middleware in the GRH ILS, since we wanted to recruit GRH patients as our subjects. This task took two persons about two weeks; subsequent testing over the next two months ensured the correctness of the setup. We deployed 10 motion sensors (the ILS is about ~80 sq. m.) to make inferences about the subjects' movements. Through a special bed sensor and pressure sensors placed under the feet of the chairs and armchairs, we were able to make inferences about the use of the furniture. Through electric current sensors, we were able to collect data on the use of several small appliances, including a toaster, a coffee maker, the microwave, a television and several lamps. A special heat sensor placed right on top of the stove was used as a proxy of stovetop food preparation. Doors (within the apartment, of the fridge and of the cupboards) were instrumented with reed switches to monitor their status.

Of special concern was the instrumentation of the bathroom, since personal hygiene is an important prerequisite for independence. To that end a reed switch was used to detect flushing; a light and a motion sensor were placed within the bathroom to recognize general activity within that space. An additional motion sensor placed near the sink, serving as a proxy of sink usage. The frequency and approximate duration of the shower usage was inferred from a humidity sensor in the bathroom.

The subjects were informed about the existence of the sensors and their function and reported that they did not have any concerns about their existence. One of them said they would not have agreed to participate if there were cameras. This negative sentiment towards the use of cameras for monitoring has also been the experience of the GRH, who had already considered employing cameras in the ILS but were prevented from doing so from the very negative reaction of patients towards them. This anecdotal evidence provides some validation of our methodological choice of sensors as a less invasive mode of activity monitoring.

Research Design: We adopted a case study approach, where all participants in the trial received the exact same treatment. Purposive sampling from in-patients units at the GRH was used to identify potential participants who met the inclusion criteria. These patients had to be 65 years and over; medically frail but in stable medical condition and nearing discharge to their community dwelling; with mild to moderate cognitive impairment (Molloy, 1999). Preference was given to patients without psychosocial support or those resistant to recommendations for Community Care services. We excluded patients who were physically incapable of managing their medications, or did not speak English, or who required modified diet restrictions and injectable medications.

Procedure: Once identified as meeting the inclusion criteria, determined as medically stable and nearing discharge home, a potential patient volunteer was invited by the investigator to participate in the study. The investigator[4] provided the patient with the study information letter and returned in 1-2 days. This allowed the patient time to read the letter, formulate questions and participate without feeling pressured to participate as part of their medical care.

Once the investigator obtained written consent, participants underwent a baseline screen by the research nurse which included two validated instruments to measure levels of function: the Katz Activities of Daily Living Index (Katz et. al 1970) and the Lawton Instrumental Activities of Daily Living (Lawton & Brody, 1969) to gather information regarding the participant's current level of function. Additional information was collected from the medical chart and previously completed assessments by the rehabilitation team. The combination of validated instruments and data collection provided baseline participant characteristics to detect potential trends and enable comparisons with participants in future studies.

For safety measures, the ILS was equipped with three call bells, a Lifeline pendant/base unit and a Nursing Unit was located 15 feet from the front door of the suite. The on-duty nurses completed a visual inspection at shift change and responded to any emergencies. The investigator also completed periodic checks one to two times per day to complete the pill counts and respond to any participant concerns.

Once a participant was identified and his/her written consent was obtained, his/her pharmacist loaded three days' worth of the current medication regimen into the MAT units. The investigator and the pharmacist then programmed the unit on its online website. The pharmacist determined the therapeutic window for when each

---

[4] The term "investigator" is used to refer to one of the authors (and members of the team) who was on-site during the ILS study, and was the point person responsible for interacting with the subjects and the ILS clinicians and administrators.

drug should be taken (i.e., plus or minus 30 minutes) and any special instructions for each medication (i.e., "Take with food"). The medication regimen was then uploaded to the unit. Settings were double-checked by both the investigator and pharmacist for accuracy by pressing each of the bin buttons.

On Day One, the investigator trained the participant to use the MAT. The participant was then left to independently use the product at bedside (i.e., their hospital room) to provide a 24-hour practice period and opportunity for questions or additional training prior to entering the ILS. At the same time, basic demographic data was collected from the participant's medical records. This included age, gender, marital status, primary diagnosis, reason for admission and location of residence.

Participants were asked to spend two days and one night in the ILS in an attempt to create natural medication taking behaviors in a home-like environment. The ILS created typical extraneous distractions (i.e., noise from the television, distractions with cooking, being in a different room from the medications etc.) that created the typical competing auditory, visual and cognitive challenges to following a daily medication regimen. With the exception of the investigator pill-count checks and nursing safety checks, the participant was left alone and observed remotely by the investigator using the sensor and MAT data. This unmonitored method of usability testing served the purpose to reduce the potential influence of the investigator's presence on the participant's behavior in using the MAT product.

In the morning of Day Two, the investigator would porter the participant to the ILS suite and give him/her an "orientation tour" to increase their familiarity and comfort. Data collection started when the investigator left the suite. The participant was then left alone in the ILS to simulate living independently in a home-like environment, which included sleeping in the ILS. A participant check for a visual pill count was completed twice per day by the investigator to verify that proper doses of medications were taken and to confirm product-reporting accuracy.

Before the participant exited the ILS at the end of Day Three, the investigator conducted a semi-structured exit interview. This reduced subject desirability bias since the participant's medical team would not be present.

A report was generated within a twenty-four hour period after the end of the participant's stay in the ILS, which listed his/her medication regimen, calculated adherence rate, baseline test scores and comments. The 2D visualizations of the data collected by the various sensors and a manually produced summary (including estimates for hand-washing frequency and average time spent in each area) were also included as part of the participant activity report. The report was then distributed electronically to the medical team and a hard copy placed in the patient's chart. Finally, the members of the medical team were then asked to complete a Likert-style questionnaire to assess if the information from the MAT product assisted with discharge planning.

Findings: During the Summer of 2011, we recruited two participants for our study. Based on the data uploaded from the medication adherence product, Participant #1 was able to achieve 100 percent medication adherence with her eight medications and Participant #2 was able to achieve 90 percent medication adherence with his nine medications. Our predefined cut-off was 80 percent, so both were considered as being adherent. The investigator counted the pills in the MAT at the end of each day to verify the unit's accuracy. The participants' average response time to taking the pill (i.e., opening the bin compartment) after the electronic alert activated was one minute.

During the scheduled patient checks, the investigator made the following observations. Both participants were observed to have difficulty removing pills from each bin: they were observed opening the bin and being unable to use a tip-to-tip pinch to grasp pills. Alternatively, using one hand they tipped over the unit to empty that bin and its contents into her opposite palm to grasp one pill and re-loaded the pills back into the bin. One participant dropped some pills and later self-corrected by re-loading the dropped pill back into the bin. She was observed dropping a pill during re-loading on two occasions. The second participant had difficulties properly closing the MAT bins.

For participant #1, seven of the eight medications had a 09:00 AM dosage time, when seven red electronic lights flashed. The participant appeared anxious to take all the pills quickly due to the perceived pressure from the multiple flashing lights. She became familiar with the positional orientation of the units[5] on the counter (i.e., Unit A was in the left quadrant and Unit B in the right quadrant). When the units were re-arranged (i.e., Unit B in the left quadrant and Unit A in the right quadrant), she became momentarily confused.

Participant #2 could not hear the audible alert when the television was turned on. Inability to hear the audible alerts may account for the recorded 11-minute average response time to opening the pill bins. One of the bin buttons was difficult to push due to a sticky food substance, which may have been spilled on the unit.

It is interesting to note that, after the first participant's stay, the nurses inquired about instrumenting a walker; most patients staying in the ILS use walkers. This was done for the second patient and although the accuracy was limited by the fact that we could only work with the patient's existing walker and it was not possible to do much testing, we were still able to attach an accelerometer and get rough measurements of when the walker was in use. We found that, the patient almost never used it inside the suite but he did use it when he left the suite for his therapy.

During the exit interview with the participants, which was focused primarily on the MAT product, Participant #1 stated she had difficulty hearing the audible voice alert from the units and reiterated that she found it difficult to open. She commented that the voice alert was not clear and it was difficult to interpret the instructions, which advised the participant how many pills to take with special pill taking instructions. In terms of possibly using this specific product at home, she stated that she *"didn't really like the product...it was annoying...you couldn't make me use it at home...I prefer to use my current system but I could see how this would be helpful*

---

[5] The MAT unit consisted of multiple components, each one accommodating 4 different types of tablets.

for someone who has memory problems". She also commented that she would not pay to use this product or for monthly monitoring and it would not make her feel more independent or safer at home. Participant #2 stated that the main difficulty with the product was reading the LCD screen and hearing the audible instructions. When asked about the missed dose, Participant #2 was unaware of it. He commented that *"overall it is a good machine and I would use it at home...you would need people who would want to use this machine...the technology is helpful to save the nurses work"*. He also commented that he would pay to use this product to a maximum of $10 per month but it *"would be good"* if the government paid for it. He agreed that using a MAT product could help him feel safer at home.

Two completed team evaluations were received from the each participant's care coordinator and physical therapist. On a 5-point Likert scale, both respondents agreed that the information received from the MAT was useful. A particularly interesting finding was that, in the case of Participant #1, the care team noticed in our activity analysis report that she went too many times to the bathroom, and, especially during the night. This observation led them to pursue some potential hypothesis of pathologies that might have caused this behavior. After discussing with the patient they discovered that it was likely due to her drinking too much coffee during the day. Also, from the lights and lamp sensors we could tell whether the patients remembered to turn the lights on at night when they were going to the bathroom, which was also a concern to the nurses. The first participant actually left the bathroom light on and the door open all night, possibly to ensure she could always see on her way to the bathroom. This experience led the care team to observe that the Smart-Condo™ middleware can indeed provide information directly relevant to their task of assessing the degree to which their patients could live independently, assuming that the activity analysis can be concisely summarized.

Follow up: Our research team is co-teaching an interdisciplinary course at the graduate level open to OT, Industrial Design, Pharmacy and CS students. We presented the ILS case study, its design, procedures and findings to the students taking the course in the Fall of 2011 and one of the teams, involving students across all the above faculties, took on as their project the design of a new device. Together they developed a design consisting of a unit very similar to a typical dosette (like the ones available in drug stores), instrumented with actuators and sensors controlled through an Arduino board and a mobile application to serve as an additional reminder to the patient and their care givers. A prototype of the design has been manufactured with our 3D printer and we are looking forward to have it subjected to a future study, likely with client educators at first, since the prototype is still quite fragile. Following up on the suggestion of the ILS nurses, we are also developing a new walker design, with motion sensors embedded in it.

4 CONCLUSIONS

Around the world, there are many ongoing research and development projects aiming at developing technologies in support of elderly patients (in principle) manageable with chronic conditions. A thorough review of the literature is beyond the scope of this paper, however, the interested reader can find a list of such projects at http://gero-tech.net/smart-homes.html; in addition, a recent survey (Chan et. al 2009) reviews and analyzes much of the work in this area. In this paper we report on our own work with the development of the sensor-based Smart Condo™ middleware for activity monitoring and analysis. Our first case study with real patients as participants has indicated that indeed this technology is useful in assessing the ability of individuals to stay independently at home. Our plans for the future include the development of products whose specifications came up through our study (a new MAT and a smart walker) and the development of automated methods to summarize the analysis results to make them more accessible to care providers.

ACKNOWLEDGMENTS

The authors would like to acknowledge NSERC (HSITE), AITF and IBM for their support of the Smart Condo™ project and our colleagues at the Glenrose Rehabilitation Hospital.

REFERENCES

Chan, M., E. Campo, D. Estève, J. Y. Fourniols. Smart homes — Current features and future perspectives, Maturitas, Volume 64, Issue 2, 20 October 2009, Pages 90-97.

Boers, N. M., D. Chodos, P. Gburzynski, L. Guirguis, J. Huang, R. Lederer, L. Liu, I. Nikolaidis, C. Sadowski, and E. Stroulia. The Smart Condo™ Project: Services for Independent Living, in C. Röcker and M. Ziefle (Eds.), *E-Health, Assistive Technologies and Applications for Assisted Living: Challenges and Solutions*. Hershey, PA: IGI Global, 2010.

Ganev, V., D. Chodos, I. Nikolaidis, and E. Stroulia. The Smart Condo™: integrating sensor networks and virtual worlds. In Proceedings of the 2nd Workshop on Software Engineering for Sensor Network Applications (SESENA '11). ACM, New York, NY, USA, 49-54.

Katz, S., T. D. Downs, H. R. Cash, R. C. Grotz. Progress in the development of the Index of ADL. *Gerontologist, 10, 22.* 1970.

Lawton M. P., E. M. Brody Assessment Of Older People: Self-Maintaining And Instrumental Activities Of Daily Living. Gerontologist. 1969 Autumn;9(3):179-86.

Molloy, D. W. Standardized Mini-Mental State Examination. Troy, NY: New Grange Press; 1999.

Stroulia, E., D. Chodos, N. M. Boers, J. Huang, P. Gburzynski, and I. Nikolaidis. Software engineering for health education and care delivery systems: The Smart Condo™ project. In *Proceedings of the 2009 ICSE Workshop on Software Engineering in Health Care* (SEHC '09). IEEE Computer Society, Washington, DC, USA, 20-28.

Section X

Collaboration and Learning in Healthcare Systems

CHAPTER 67

Capturing Team Mental Models in Complex Healthcare Environments

Chelsea Kramer[1, 3], *Geneviève Dubé*[1], *Avi Parush*[2], *Esther Breton*[1], *Gilles Chiniara*[1], *Matthew Weiss*[1] *& Sébastien Tremblay*[1]

[1]Laval University, [2]Carleton University, [3]CAE Inc.
Québec, Québec; Ottawa, Ontario, Canada
Chelsea.kramer@cae.com

1 ABSTRACT

Cross-professional healthcare teams generate diverse perspectives and expectations that can affect individuals' ability to work together. Shared mental models can orchestrate a common expectation, allowing for more efficient team processes to improve team performance. To measure 'sharedness', mental models must be elicited, but the appropriateness of a selected technique will depend on the characteristics of the domain and the type of knowledge to be measured. This chapter examines two methods of mental model capture in the Operating Room (OR) and the Intensive Care Unit (ICU). In depth interviews with OR experts and observed team members captured team mental models (TMM) specific to OR tasks and associated teamwork expectations. Six recurring themes of potential TMM incongruence emerged. In the ICU, interviews served to construct a task model of emergency ICU processes that underwent a card sorting task, validated by subject matter experts. Mental model similarity was calculated using a similarity index between each pair of participants. Advantages and lessons learned from both methods are discussed.

Keywords: Team Mental Models, Healthcare, Operating Room, ICU, Teamwork, Patient Simulation, Interviews

2 INTRODUCTION

The successful delivery of patient care depends on the harmonization of people, equipment, tools and facilities, largely coordinated through training and teamwork (Weaver et al., 2010). In complex healthcare environments such as the operating room (OR) and the intensive care unit (ICU), cognitive work and expertise are distributed amongst the team (Reader, Flin, Mearns, & Cuthbertson, 2009). Cross-professional work teams provide immense coordination challenges when domains of expertise overlap (Ellingson, 2002). While the focus may be on a single patient, each team member has individual activities, motivations, and concerns (Reddy, Shabot, & Bradner, 2008). As a result, team members may have conflicting mental models of the task at hand that can affect teamwork processes. To remedy such conflict, there needs to be a common understanding of the factors that makeup good teamwork within the specific healthcare work environment.

2.1 The Importance of Mental Models in Teamwork

Healthcare teams work under evolving conditions where safety critical, non-routine events are likely to occur at any time (Burtscher, Kolbe, Wacker, & Manser, 2011). Defining important teamwork skills in these circumstances requires an understanding of the team's perception of their tasks and associated cognitive demands; their mental model. Mental models allow people to predict and explain the behavior of the world around them, to recognize and remember relationships among components of the environment, and to construct expectations for what is likely to occur next (Rasmussen, 1979; Rouse & Morris, 1985). People can have unique mental models about information that is specific to them, or shared mental models of common situations that include others. Shared team mental models (TMM) facilitate the implicit coordination of team behaviors (Mathieu, Heffner, Goodwin, Salas, & Cannon-Bowers, 2000); team members are better able to recognize the individual responsibilities and information needs of teammates, monitor their activities, diagnose deficiencies, and provide support, guidance, and information as needed (Paris, Salas & Cannon-Bowers, 2000). Implicit coordination allows for more efficient team collaboration, efficient communication strategies (Cooke et al., 2003; Macmillan, Entin, & Serfaty, 2004), and better team performance (Westli, Johnsen, Eid, Rasten, & Brattebø, 2010).

A TMM is naturally constructed through team cognitive processes, and can be understood as individual mental models merged through team interactions (Cooke & Gorman, 2009). However, a team's ability to naturally form a shared mental model is affected by their frequency of working together, and their ability and willingness to interact (Weller, Janssen, Merry, & Robinson, 2008). Healthcare teams consist of specialists from different professions (physicians and nurses) who tend to have revolving memberships. In other words, both the team doing the work, and the work of the team, changes. To examine a TMM under these dynamic healthcare conditions, it is crucial to consider the method of elicitation. Different TMM techniques require different degrees of researcher involvement, which can affect reliability, and may be more or less suitable for eliciting an individual versus a team mental model (Langan-Fox, Code, & Langfield-Smith, 2000a). This chapter presents two ways of using interviews to capture mental models in complex healthcare contexts, using the OR and the ICU as case examples.

3 OPERATING ROOM MENTAL MODELS

3.1 OR Setting

Thoracic surgery treats diseases (largely cancer-related) of organs within the thorax (i.e., chest cavity) such as the lungs, chest wall and diaphragm. A defining feature of thoracic procedures is the use of video-assisted thoracoscopic surgery (VATS), which uses a small video camera that is introduced into the patient's chest through the chest wall attached to a long flexible scope. The scope enables the surgeon(s) to view their surgical maneuvers on a large display, usually located at one end of the patient, while conducting minimally invasive operations. VATS creates a unique sociotechnical environment where the team's negotiated teamwork centers on technology. For instance, even a simple tissue biopsy involves the collaboration of two surgeons, the surgical fellow and the resident, each with a tool (e.g., clamp, suturing stapler) in both hands working side by side. As such, it is important that they share an understanding of who does what, and how to coordinate specific maneuvers using a common language. The TMM produced for the OR revolves around teamwork expectations during specific OR tasks.

3.2 Building TMM in the OR

TMM were derived from in depth interviews of a convenience sample of 12 OR team members from a large Eastern Ontario Hospital (four surgical, five anesthesia, and three nursing team members). To avoid bias from any single event, the interviews targeted task and team aspects of general thoracic procedures, avoiding direct questions related to any particular surgery. Task knowledge included the main components of thoracic surgery; while team knowledge included the interactions of individual healthcare workers. Individual mental models were then systematically organized using a thoracic task template validated by the Chiefs of Surgery and Anesthesia. The template decomposes thoracic procedures into a common surgery structure, consisting of five generic chronological phases typical of thoracic surgeries; 1) the insertion of intravenous lines and patient intubation; 2) the insertion of the scope and patient positioning; 3) the pre-surgical preparation; 4) the surgical procedure; and 5) the surgical close. TMMs were then assembled by combining individual mental models from a minimum of two or more members who worked together during a series of observed surgeries. These observations were used for a separate analysis not covered in this paper (for more detail see Kramer et al., 2010).

3.3 Emergent OR Themes

TMM were reviewed as a function of cross-professional, interdependent teamwork across the five surgical phases. Six mutually exclusive thoracic OR themes emerged: Management of Patient Ventilation, Surgical Equipment Handling, Patient Positioning, Schedule Awareness, Preoperative Pause, and Surgical Procedure. Table 1 presents the definition, summary and an example TMM cross-section of each of the themes, all depicting a different team composition.

Table 1. Summary table of analyzed TMM themes

Identified Theme	**Team Member Expectation**		
Patient Ventilation	S11	A1	A6
Airways, vents, pressure, suction, lung up/down, inflation, deflation	Patient is well-managed from both sides	A may ask S if they wants lungs down, or S may request from A	S will confirm lungs down with A, if A on the "ball" this will be done already
Equipment Handling	S11	N14	
Surgical tools, vital monitors, VATS screens, infusion pumps	If N's not usually with team, need to be more vocal, less anticipation	S's don't know equipment as well as N's, don't always ask for what they want	
Patient Positioning	S12	A9	N3
Physical layout of patient on the table prior to surgery	S and A negotiate the patient's position	A in charge, (but whoever is at head, may be AR), counts	Move when done scope, and A done with lines
Schedule Awareness	S11	A6	N3
Awareness of current's day's schedule, upcoming case schedule	Know day before if order changes, unusual not to know change in same day	AR meets pat in AM pre-surgery (delayed if N bring pat in b4 AR has seen them)	N checks with [forms containing] S's requests
Preoperative Pause	S2	A1	N3
Team briefing prior to surgical incision	After scope and move;	Done before incision, but varies.	Usually occurs when scoping,
Surgical Procedure	S2	N10	
Surgical approach, specific needs of procedure, incision manoeuvres	Towards end, S leaves tubes in chest cavity, tells N, initiates close	Good S usually tell the time that they will be closing	

Note. *S = Surgeon, N = Nurse, A = Anesthetist, R = Resident*

Planned analyses for these themes will involve systematic codification of team level expectations to infer TMM similarity/differences. Initial inspections already suggest variations of TMM within each theme. For example, a major preoperative concern among the VATS team is maintaining adequate anesthesia team support during the surgical procedure. Surgeons work intimately with the patients' vital organs and need to communicate frequently with the anesthesia team about their intentions. Given the shared thoracic workspace, it is vital that the anesthesia team be experienced in the principles and applications of selective single-lung ventilation during the operation (Hoyos, Santos, Patel, & Landreneau, 2005). A lack of shared TMM regarding the management of patient ventilation could result in threats to patient safety. Following the classification of TMM similarity for the remaining themes, the next step of this research is to infer relationships among TMM and observed teamwork processes, such as communication. Results of this analysis will be available in future work.

4 ICU MENTAL MODELS

4.1 Intensive Care Unit Setting

The Intensive care unit (ICU) is a high-risk, acute medical environment that requires cross-professional teams to provide life-saving care for critically ill patients (Reader, Flin, & Cuthbertson, 2008). The majority of patients suffer from multiple organ dysfunctions that require immediate treatment, making the outcomes of treatment interventions difficult to predict or plan accordingly (Reader, Flin, Mearns, & Cuthbertson, 2009). Like the OR, teamwork in the ICU is distributed among a cross-professional team. The typical ICU is a hectic, dynamic context where multi-tasked, frequent interruptions challenge communication and the ability for efficient and complete information transfer (Grundgeiger, Sanderson, Macdougall, & Venkatesh, 2010). Compared with domains such as surgery, the communication and teamwork skills important for safety are less well defined (Reader, Flin, & Cuthbertson, 2007). Resuscitation is a frequent ICU emergency that requires medical knowledge, clinical and non-technical skills. The TMM produced for the ICU stem from task models of emergency resuscitation.

4.2 Building Task Models of Emergency ICU

To define important teamwork skills for ICU, we must understand the team's perception of their tasks and associated cognitive demands; their mental model. Such mental models can be harvested with a task analysis, the study the demands of the system on the operator with the capabilities of the operator (Kirwan & Ainsworth, 1992). While a few studies have addressed task analysis of ICU physicians (Fackler et al., 2009); work domain analysis for ICU patients (Miller, 2004); and workflow modeling of the general ICU environment (Malhotra, Jordan, Shortliffe, & Patel, 2007), little published research has focused on emergency situations in the ICU. We conducted a task analysis during interviews with ICU subject matter experts (SME) to develop a task model of ICU crisis resuscitation. Similar to the OR study, an initial task model of ICU resuscitation was developed by a respiratory therapist. The model validation consisted of three nurses, five respiratory therapists, one physician and an orderly currently working in intensive care units of Quebec and Gatineau hospitals. Each participant had at least three years of experience in ICU. The SME ICU team modified the task hierarchy based on their own experience in a resuscitation situation.

The validated task model was transformed into a card sorting task to assess mental model similarity of teams training in an ICU simulation center. This technique is based on principles derived from cognitive categorization theory (Langan-Fox, Code & Langfield-Smith 2000). Tasks from the model were arranged as either higher order tasks (e.g. patient condition assessment, diagnostic, etc), or lower order subtasks (e.g. give medication, assess vital signs, etc.). Thus far, ten ICU teams from two Quebec hospitals have completed four resuscitation simulations at the APPRENTISS Centre in Laval University (Figure 1). The high-fidelity patient simulator (Medical Education Technologies Inc., Florida) is a computer-controlled anatomical representation of a human that responds in a coherent pathophysiological and pharmacological way to the medical team's actions, which are audio and video recorded.

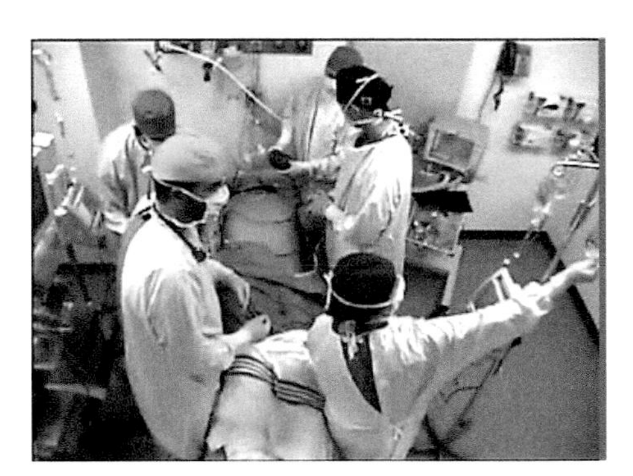

Figure 1. An ICU resuscitation simulation scenario at the APPRENTISS Centre located in Quebec City, Quebec, Canada.

4.3 Emergent ICU Themes

The analysis of ICU Crisis management and modeling of the involved tasks revealed a hierarchical structure of three high level tasks: 1) Patient evaluation, 2) Patient Diagnosis, and 3) Medical Intervention. These high level tasks corresponded to 88 mutually exclusive sub-tasks. Following the simulations at the Apprentiss Center, the ICU teams categorized the 88 subtasks among the three high-level tasks defined by SMEs. Participants also ranked these subtasks in order of importance according to their last scenario. TMM similarity will be calculated using a kappa index between each pair of participants, and a mean will be calculated for the team.

Afterwards, each participant's result will be compared to the expert model, and a similarity index obtained. The mean will be calculated to obtain a team index of accuracy of the mental model.

Given the close relationship between team performance and the extent to which a similar representation of task work is shared by all team members, it is important to ensure that that ICU teams show a high similarity index of mental (task) models – both for the intra-team similarity and team's accuracy. That said, we anticipate a higher similarity index for mental models of the higher order tasks (e.g. patient needs to be intubated), whereas the lower order tasks may vary by specialty. Ideally, a high level of similarity among the members of the ICU teams is indicative of a common understanding of the overall task, and not only their own, personal responsibilities inside the team. Results of this analysis will be available in future work

5 DISCUSSION

This chapter presented two distinct methods of capturing and assessing TMM in complex healthcare environments. In the OR, TMM were constructed from in-depth interviews with SMEs and observed surgical team members based on a typical procedure walkthrough. This ethnographic approach suited the OR because thoracic procedures tend to follow a similar timeline. As a result, the team's expectations captured during interviews could parallel that general chronology while common themes on teamwork were extracted. This will enable the comparison of TMM similarity with real time observation data such as team communication (e.g., Parush et al., 2011). In contrast, the very nature of ICU crisis implies that events are nonlinear and often unexpected. Although well-defined phases may apply for the analysis of general ICU activity, this would not make sense for a crisis situation. Accordingly, the task model of ICU resuscitation needed to consist of higher level concepts that did not rely on chronological order of events. ICU teams were then able to arrange concepts in a ranked order of importance.

The TMM technique chosen reflected the specific characteristics of interest within each domain. In the thoracic OR, the objective was to elicit TMM knowledge about interdependent teamwork. Thoracic surgery is an ideal candidate due to the constant requirement for team interaction. For example, the surgeons rely on the nurses to handle their surgical tools, and the anesthetists rely on the surgeons to manage the patient's airways and ventilation. Knowledge of interdependent teamwork is thereby readily accessible for interview within these teams in most procedures. On the other hand, high-fidelity simulations were well suited for ICU crises due to the infrequency of actual real life emergencies. Asking ICU teams to complete a card-sorting task after an actual patient crisis event would not be feasible, nor ethical to impose. The amount of data required to capture numerous team interactions during an actual patient crisis is simply not practical from a research perspective.

6 CONCLUSION

The ethnographic TMM method used for the OR provided rich, context specific information. However, a major disadvantage of this technique was the length of time required for interviews and the limited availability of healthcare workers. This process could have been reduced using similar card-sorting technique to alleviate individual interview time. On the other hand, the TMM method used for the ICU, although efficient, may be lacking interesting team knowledge not captured in the card-sorting technique. The ICU simulation study could benefit from the integration of some level of interview analysis following the simulations to help validate the mental models. Overall, examination of context specific approaches to mental model elicitation provides better means for training shared mental models and improving team performance, so long as the methodology is reasonable to execute.

7 ACKNOWLEDGMENTS

This research was supported by the Natural Sciences and Engineering Research Council of Canada Collaborative Research and Development Grant (NSERC–CR&D), the Conseil Médical du Canada, the Ontario Graduate Scholarship for Science and Technology (OGSST), and the Ottawa Hospital. The authors would also like to Kathryn Momtahan and Seneca Brandigampola for their help in the project.

8 REFERENCES

Burtscher, M. J., Kolbe, M., Wacker, J., & Manser, T. (2011). Interactions of team mental models and monitoring behaviors predict team performance in simulated anesthesia inductions. *Journal of Experimental Psychology: Applied*, *17*(3), 257-69. doi:10.1037/a0025148

Cooke, N. J., & Gorman, J. C. (2009). Interaction-Based Measures of Cognitive Systems. *Human Factors*, *3*(1), 27-46. doi:10.1518/155534309X433302.

Cooke, N. J., Kiekel, P. A., Salas, E., Stout, R. J., Bowers, C. A., & Cannon-Bowers, J. A. (2003). Measuring Team Knowledge: A Window to the Cognitive Underpinnings of Team Performance. *Group Dynamics*, *7*(3), 179-199.

Ellingson, L. L. (2002). Communication, Collaboration, and Teamwork among Health Care Professionals. *Communication Research*, *21*(3), 1-15.

Fackler, J. C., Watts, C., Grome, A., Miller, T., Crandall, B., & Pronovost, P. J. (2009). Critical care physician cognitive task analysis: an exploratory study. *Critical care (London, England)*, *13*(2), R33. doi:10.1186/cc7740

Grundgeiger, T., Sanderson, P. M., Macdougall, H. G., & Venkatesh, B. (2010). Interruption management in the intensive care unit: Predicting resumption times and assessing distributed support. *Journal of experimental psychology. Applied*, *16*(4), 317-34. doi:10.1037/a0021912

Hoyos, A. D., Santos, R. S., Patel, A., & Landreneau, R. J. (2005). Instruments and Techniques of Video-Assisted Thoracic Surgery. In T. W. Shields, J. LoCicero, R. B. Ponn, & V. W. Rusch (Eds.), *General Thoracic Surgery, 6th Edition* (6th ed., p. 524). Philiadelphia, PA: Lippincott Williams & Wilkins.

Kirwan, B., & Ainsworth, L. K. (1992). *A Guide to Task Analysis*. (B. Kirwan & L. K. Ainsworth, Eds.). Taylor & Francis, Ltd.

Kramer, C., Parush, A., Momtahan, K., & Brandigampola, S. (2010). Analysis of Cross- Professional Communication in Thoracic Operating Rooms. In V. G. Duffy (Ed.), *Advances in Human Factors and Ergonomics in Healthcare* (pp. 307-316).

Langan-Fox, J., Code, S., & Langfield-Smith, K. (2000). Team mental models: techniques, methods, and analytic approaches. *Human factors*, *42*(2), 242-71. Retrieved from http://www.ncbi.nlm.nih.gov/pubmed/11022883

Macmillan, J., Entin, E. E., & Serfaty, D. (2004). Communication Overhead: The Hidden Cost of Team Cognition. In Eduardo Salas & S. M. Fiore (Eds.), *Team cognition: Process and performance at the inter- and intra-individual level* (pp. 61-82). Washington, DC: American Psychological Association.

Malhotra, S., Jordan, D., Shortliffe, E., & Patel, V. L. (2007). Workflow modeling in critical care: piecing together your own puzzle. *Journal of biomedical informatics*, *40*(2), 81-92. doi:10.1016/j.jbi.2006.06.002

Mathieu, J. E., Heffner, T. S., Goodwin, G. F., Salas, E., & Cannon-Bowers, J. A. (2000). The influence of shared mental models on team process and performance. *The Journal of applied psychology*, *85*(2), 273-83. Retrieved from http://www.ncbi.nlm.nih.gov/pubmed/10783543

Miller, A. (2004). A work domain analysis framework for modelling intensive care unit patients. *Cognitive Technology Work*, *6*, 207-222. doi:10.1007/s10111-004-0151-5

Paris, C. R., Salas, E., & Cannon-Bowers, J. A. (2000). Teamwork in multi-person systems: a review and analysis. *Ergonomics*, *43*(8), 1052-75. Retrieved from http://www.ncbi.nlm.nih.gov/pubmed/10975173

Parush, A., Kramer, C., Foster-Hunt, T., Momtahan, K., Hunter, A., & Sohmer, B. (2011). Communication and team situation awareness in the OR: Implications for augmentative information display. *Journal of biomedical informatics*, *44*, 477-485. Elsevier Inc. doi:10.1016/j.jbi.2010.04.002

Rasmussen, J. (1979). *On the structure of knowledge: A morphology of mental models in a man-machine system context (Report No. RISWM-2192)*. Denmark: North Holland.

Reader, T. W., Flin, R. H., Mearns, K., & Cuthbertson, B. H. (2009). Developing a team performance framework for the intensive care unit. *Critical care medicine*, *37*(5), 1787-93. doi:10.1097/CCM.0b013e31819f0451

Reader, T., Flin, R., & Cuthbertson, B. (2007). Communication skills and error in the intensive care unit. *Current opinion in critical care*, *13*(6), 732-6. doi:10.1097/MCC.0b013e3282f1bb0e

Reader, T., Flin, R., & Cuthbertson, B. (2008). Factors Affecting Team Communication in the Intensive Care Unit (ICU). In C. P. Nemeth (Ed.), *Improving Healthcare Team Communication: Building on Lessons from Aviation and Aerospace* (pp. 117-133). University of Chicago, USA: Ashgate.

Reddy, M. C., Shabot, M. M., & Bradner, E. (2008). Evaluating collaborative features of critical care systems: a methodological study of information technology in surgical intensive care units. *Journal of biomedical informatics*, *41*(3), 479-87. doi:10.1016/j.jbi.2008.01.004

Rouse, W. B., & Morris, M. M. (1985). *On looking into the black box : prospects and limits in the search for mental models Dl : lFILE Copy. Psychological Bulletinource* (Vol. 100, p. 61). Atlanta, Georgia.

Weaver, S. J., Rosen, M. A., DiazGranados, D., Lazzara, E. H., Lyons, R., Salas, E., Knych, S. A., et al. (2010). Does teamwork improve performance in the operating room? A multilevel evaluation. *Joint Commission journal on quality and patient safety*, *36*(3), 133-42. Retrieved from http://www.ncbi.nlm.nih.gov/pubmed/20235415

Weller, J. M., Janssen, A. L., Merry, A. F., & Robinson, B. (2008). Interdisciplinary team interactions: a qualitative study of perceptions of team function in simulated anaesthesia crises. *Medical education*, *42*(4), 382-8. doi:10.1111/j.1365-2923.2007.02971.x

Westli, H. K., Johnsen, B. H., Eid, J., Rasten, I., & Brattebø, G. (2010). Teamwork skills, shared mental models, and performance in simulated trauma teams: an independent group design. *Scandinavian journal of trauma, resuscitation and emergency medicine*, *18*, 47. doi:10.1186/1757-7241-18-47

CHAPTER 68

Computer Supported Collaboration in Health Care

Joyram Chakraborty, Danny Ho, Anthony F. Norcio

Towson University
Towson, MD 21252, USA
jchakraborty@towson.edu

ABSTRACT

In this information age, computerization in the healthcare workplace aims to improve quality of care, reduce administrative and clinical costs, and enhance healthcare work processes. However, integration in this environment proves challenging. Computerizing healthcare processes is not merely a technological pursuit but relies on human-driven coordination of highly fluid, uncertain, and exception-filled activities and processes spanning social, technological, and political boundaries. This paper examines some of the human factor challenges of computerization in healthcare.

Keywords: Health care costs, CSCW, Human Factors

1 INTRODUCTION

In contrast to mechanized domains such as industrial manufacturing where activity processes are relatively more routine, healthcare comprises of cooperative tasks under variable time pressures, multidisciplinary roles and objectives, and a high requirement for effective communication (Xiao, Y., Hunter, W. A., Mackenzie, C. F., & Jefferies, N. J., 1996). In other words, healthcare activities are very much team and collaboration driven. Therefore, systems designers must carefully consider socio-technical requirements in order to produce effective and adoptable systems.

2 BACKGROUND

Collaborative support systems research is comprehensively captured by several major research fields. The most notable are groupware, primarily investigating technology driven design and evaluation, computer supported cooperative work (CSCW), which generally takes a social and organizational perspective to reveal systems behaviors and implications for design, and the much broader classification of computer-mediated communication (CMC).

Computerized collaboration support is a longstanding domain of research in the information systems and human factors areas, particularly in technological applications where teamwork communication and performance are crucial for system operations (Alvarez, G. & Coiera, E., 2005; Benson, Ian, Ciborra, Claudio, & Proffitt, Steve, 1990; Landgren, Jonas & Nulden, Urban, 2007). Two exemplary domains are aviation and healthcare (Ho, Danny et al., 2007; Benson, Ian, Ciborra, Claudio, & Proffitt, Steve, 1990). In both cases, networks of people, systems, and artifacts must communicate rapidly in a setting of high time pressure and demand for reliability. If copilots cannot communicate clearly with air traffic control or each other, aircraft may veer off course, or worse, collide. If doctors, nurses, and other practitioners do not achieve effective consensus on treatment plans during meetings and handoff situations, patient progress is less efficient, or worse, unintended medical errors can occur. In both cases, the complexities of a particular work process necessitate some degree of technological support (Benson, Ian, Ciborra, Claudio, & Proffitt, Steve, 1990; Pinsonneault, A. & Kraemer, K. L., 1989).

A significant challenge for computer supported cooperative work (CSCW) is sharing of knowledge and information between individuals. Multidisciplinary roles have different requirements for information and perceive them differently (Nygren, E. & Henriksson, P., 1992). Information sharing systems must strike a balance between universally transparent data and strict individual control over dissemination, designed appropriately for the information sharing activity in question.

3 UNDERSTANDING THE INFORMATION SPACE

In collaborative medical work, critical information exchanges often happen in and through shared information spaces. The notion of "shared information spaces" primarily characterizes the visual information space, where personal and public information sources (artifacts) are used to negotiate and support shared knowledge. According to Bossen, common information spaces focus on "interrelationships between information, actors, and artifacts (written documents, schedules, and computers) and cooperative work" and how artifacts support coordination of collaborative work (Bossen, C., 2002). Interactions with common information spaces can be categorized in terms of people, time, and space (Bossen, C., 2002). Persistent shared information spaces can accommodate both synchronous and asynchronous information sharing with workers co-located to the information space. The "tangible" components of medical information spaces, such as clinical notes, physical co-location, and public display boards facilitate collaborative work processes and play a role to "tailor" the work environment (Xiao, Y., 2005). Common information spaces can therefore support "divergent forms of work" or heterogeneous work, i.e. different clinical staff with diverse goals of care delivery (for example, physicians order medications and nurses administer them).

4 THE ROLE OF SENSEMAKING

Understanding and evaluating information transfer challenges must also account for the complex human cognitive activities involved in receiving, interpreting, and forming knowledge and decisions from information received (Klein, Gary, Moon, Brian, & Hoffman, Robert R., 2006). This activity of "sensemaking" has definitions in several nuances and contexts, including the search for and understanding representations to perform tasks (Russell, Daniel M., Stefik, Mark J., Pirolli, Peter, & Card, Stuart K., 1993). Weick summarized sensemaking to be about "contextual rationality", being built out of "vague questions, muddy answers, and negotiated agreements that attempt to reduce confusion" (Weick, Karl E., 1993). Sensemaking is a process, consisting of "placement of stimuli into frameworks, comprehending, redressing surprise, constructing meaning, interacting in pursuit of mutual understanding, and patterning." (Weick, Karl E., 1995). Sensemaking is highly social and helps members to achieving "common ground" (Weick, Karl E., 1995), and information technology can play a significant role to facilitate collaborative sensemaking (Bansler, Jørgen P. & Havn, Erling, 2006).

5 FRAMING THESE CHALLENGES TO MEDICAL ROUNDS

In the medical domain, care activities are highly collaborative and interdependent. Adverse events in such settings have frequently been attributed to errors related to communication among care providers (Rothschild, J. M. et al., 2005). The medical round is a forum rooted deeply in medical tradition where professionals involved in a patient's care share information and sensemake, and plays a critical role to ensure care coordination across specialties and providers. Increased availability of clinical information through computers has made it possible to provide additional information support during medical rounds.

Multidisciplinary medical rounds are widely used to improve care coordination, to establish a common shared mental model of patient status, and to make decisions and plans collaboratively (Gurses, A. P. & Xiao, Y., 2006). A number of obstacles exist that impede communication and sharing of information among care providers, such as lack of information tools supporting multiple users in noisy and interruption-prone work environments (Gurses, A. P. & Xiao, Y., 2006). Medical rounds are further complicated by undesirable long duration (Chung, R. S. & Ahmed, N., 2007), frequent interruptions (Alvarez, G. & Coiera, E., 2005), perceived "waste of time" and socio-political impacts on participation (Birtwistle, L., Houghton, J. M., & Rostill, H., 2000; Manias, E. & Street, A., 2001). Support tools for rounds can improve patient safety by supporting a shared mental model of patient status and goals, thus allowing care providers to adopt an "anticipative" mode of care management as opposed to a reactive mode due to lack of communication and corresponding lack of knowledge and decision-making support (Beuscart-Zephir, M. C., Pelayo, S., Anceaux, F., Maxwell, D., & Guerlinger, S., 2006). Computerized and paper-based artifacts are widely used for communication and care coordination (Nemeth, C., Nunnally, M., O'Connor, M., & Cook, R., 2006; Xiao, Y., 2005). Computerized handoff tools help reduce omissions and time spent in preparation for communication among care providers (Van Eaton, E. G., Horvath, K. D., Lober, W. B., Rossini, A. J., & Pellegrini, C. A., 2005). Previous reports of computer use during rounds include medical knowledge reference via the "evidence cart" (Sackett, D. L. & Straus, S. E., 1998) and access to clinical information (Fiser, D., 2004).

Despite the fact that healthcare data are increasingly stored in electronic medium, how computers may be used to synthesize and present this data to improve communication during rounds remains poorly understood. Furthermore, information needs and transfer patterns during rounds are fluid and may be difficult if not impossible to pre-specify. Multi-disciplined participants such as attending physicians, medical trainees, surgeons, and primary care nurses may be under time pressure to exchange information and plan treatment goals at a rapid pace. One reason that computers have not been extensively used during rounds may be due to these unique requirements. With the rapid growth of mobile computing and increasingly availability of clinical information and decision support on computers, computing technology holds great potential to enhance joint social and cognitive activities during group sessions such as rounds (Gurses, A. P. & Xiao, Y., 2006).

6 CONCLUSIONS

Computer supported collaboration holds great potential for improvement of collaboration information in medical settings where multiple stakeholders meet to share information under performance pressure, whether they are situated remotely or co-located. The implications for enhancing information exchange extend beyond inter-provider interactions to potentially enrich patient-provider interactions as well. As telehealthcare and other patient-centered health applications grow in terms of system types (teleconference, phone, web-based, etc.) and care settings (from the home, satellite and remote care facilities, retail and pedestrian kiosks, etc.), it will be increasing important to also empower patients to effectively communicate their medical issues to providers through synchronous, information-rich exchanges.

REFERENCES

Alvarez, G. & Coiera, E. (2006). Interdisciplinary communication: an uncharted source of medical error? J.Crit Care, 21, 236-242.

Alvarez, G. & Coiera, E. (2005). Interruptive communication patterns in the intensive care unit ward round. Int.J.Med.Inform., 74, 791-796.

Bansler, J. P. & Havn, E. (2006). Sensemaking in Technology-Use Mediation: Adapting Groupware Technology in Organizations. Comput.Supported Coop.Work, 15, 55-91.

Bardram, J. E. (1997). I Love the System - I just don't use it! GROUP, 251-260.

Benson, I., Ciborra, C., & Proffitt, S. (1990). Some social and economic consequences of groupware for flight crew. In Proceedings of the 1990 ACM conference on Computer-supported cooperative work (pp. 119-129). Los Angeles, California, United States: ACM.

Beuscart-Zephir, M. C., Pelayo, S., Anceaux, F., Maxwell, D., & Guerlinger, S. (2006). Cognitive analysis of physicians and nurses cooperation in the medication ordering and administration process. Int J Med Inform.

Birtwistle, L., Houghton, J. M., & Rostill, H. (2000). A review of a surgical ward round in a large paediatric hospital: does it achieve its aims? Med Educ., 34, 398-403.

Bossen, C. (2002). The parameters of common information spaces:: the heterogeneity of cooperative work at a hospital ward. In (pp. 176-185). New Orleans, Louisiana, USA: ACM Press.

Chung, R. S. & Ahmed, N. (2007). How surgical residents spend their training time: the effect of a goal-oriented work style on efficiency and work satisfaction. Arch Surg., 142, 249-252.

Fiser, D. (2004). Wireless technology empowers physicians. Health Manag.Technol., 25, 42-44.

Gurses, A. P. & Xiao, Y. (2006). A systematic review of the literature on multidisciplinary rounds to design information technology. J Am Med Inform Assoc, 13, 267-276.

Ho, D., Xiao, Y., Vaidya, V., & Hu, P. (2007). Communication and sense-making in intensive care: an observation study of multi-disciplinary rounds to design computerized supporting tools. AMIA.Annu.Symp.Proc., 329-333.

Klein, G., Moon, B., & Hoffman, R. R. (2006). Making Sense of Sensemaking 1: Alternative Perspectives. IEEE Intelligent Systems, 21, 70-73.

Landgren, J. & Nulden, U. (2007). A study of emergency response work: patterns of mobile phone interaction. In Proceedings of the SIGCHI conference on Human factors in computing systems (pp. 1323-1332). San Jose, California, USA: ACM.

Manias, E. & Street, A. (2001). Nurse-doctor interactions during critical care ward rounds. J Clin Nurs, 10, 442-450.

Nemeth, C., Nunnally, M., O'Connor, M., & Cook, R. (2006). Creating resilient IT: how the sign-out sheet shows clinicians make healthcare work. AMIA.Annu.Symp.Proc., 584-588.

Nygren, E. & Henriksson, P. (1992). Reading the medical record. I. Analysis of physicians' ways of reading the medical record. Comput Methods Programs Biomed, 39, 1-12.

Pinsonneault, A. & Kraemer, K. L. (1989). The impact of technological support on groups: an assessment of the empirical research. Decision Support Systems, 5, 197-216.

Rothschild, J. M., Landrigan, C. P., Cronin, J. W., Kaushal, R., Lockley, S. W., Burdick, E., Stone, P. H., Lilly, C. M., Katz, J. T., Czeisler, C. A., & Bates, D. W. (2005). The Critical Care Safety Study: The incidence and nature of adverse events and serious medical errors in intensive care. Crit Care Med, 33, 1694-1700.

Russell, D. M., Stefik, M. J., Pirolli, P., & Card, S. K. (1993). The cost structure of sensemaking. In Proceedings of the INTERACT '93 and CHI '93 conference on Human factors in computing systems (pp. 269-276). Amsterdam, The Netherlands: ACM.

Sackett, D. L. & Straus, S. E. (1998). Finding and applying evidence during clinical rounds: the "evidence cart". JAMA, 280, 1336-1338.

Van Eaton, E. G., Horvath, K. D., Lober, W. B., Rossini, A. J., & Pellegrini, C. A. (2005). A randomized, controlled trial evaluating the impact of a computerized rounding and sign-out system on continuity of care and resident work hours. J Am Coll.Surg., 200, 538-545.

Weick, K. E. (1995). Sensemaking in Organizations. Thousand Oaks, CA: Sage Publications.

Weick, K. E. (1993). The collapse of sensemaking in organizations: the Mann Gulch disaster. Adm Sci Q., 628-652.

Xiao, Y. (2005). Artifacts and collaborative work in healthcare: methodological, theoretical, and technological implications of the tangible. J Biomed Inform, 38, 26-33.

CHAPTER 69

Dispedia.de – a Knowledge Model of the Service System on the Rare Disease

Romy Elze, Michael Martin, Klaus-Peter Fähnrich

University of Leipzig
Leipzig, Germany
elze@informatik.uni-leipzig.de

ABSTRACT

In medical information sciences, the conversion of human knowledge into a structured machine-readable representation is a central topic. Especially in the field of rare diseases, converting medical information is a challenging task because of multifaceted disease progressions, the existence of many stakeholders from different sectors (i.e. social care, healthcare, therapists, and aid suppliers) and the high costs incurred by treatment and care.

This paper shows how knowledge modeling can tackle this problem. Based on the analysis of the domain of the rare disease ALS (Amyotrophic Lateral Sclerosis), the Dispedia Knowledge Model was developed as formal knowledge representation. Furthermore we showcase how the information can be transferred on the basis of the Knowledge Model.

Keywords: Knowledge Model, Health Services, Rare Disease, Information Logistics, Knowledge Engineering

1 INTRODUCTION

The personal services provided in the health sector are very knowledge-intensive. Both the service recipient and the service provider need knowledge about the service delivery. The service provider has the medical expert knowledge and the recipient, in this case the patient, needs expert explanation to assess the service and to decide for or against the medical service to be received. This means that medical services can be provided only in conjunction with medical consultation services. The time and fee structure in healthcare is usually not enough for a full consultation. This two-way exchange of knowledge creates a need for information systems, which recognize both human factors and the mechanical limits.

The patient's need for information and thus the exchange of information between doctors and patients has been increasing steadily. Medical decisions today are shared decisions between physicians and patients and even show a slight trend towards informed patients making the decision on their own (Benkenstein and Uhrich, 2010). The autonomous decision-making on the part of the patient and therefore a targeted knowledge exchange plays an important role in dealings with rare diseases. This is justified in the fact that:

- The knowledge of orphan diseases is primarily limited to a few specialized institutions (Mitchell and Borasio, 2007). This restriction often leads to excessive demands on the side of the medicating actors, who lack such specific knowledge.
- Activities are rarely coordinated centrally. Therefore, multidisciplinary care is still insufficient (Ng and Mathers, 2009).
- Due to the unpredictability of the course of a rare disease, it is impossible to preemptively provide information or define processes across the whole course.
- Due to a lack of quantitative and qualitative information, the confusion is also high on the side of the patients (Kuschel, 2006). This lack is diametrically opposed to the aim of supporting important treatment decisions in an early state of the disease by an active information policy.

To store expert knowledge in machine-readable notation and to provide this for non-experts is a conflict in which knowledge modeling can tackle the task and identify problem solving strategies (Kienreich and Strohmaier, 2006).

The knowledge model presented in this paper uses the example of the rare disease Amyotrophic Lateral Sclerosis (ALS) to show how this service system can be analyzed and modeled as a basis for further development in order to remedy the lacking factors identified above.

2 DISPEDIA KNOWLEDGE MODEL - DESCRIPTION

The objective of our work was the precise description of the health system concerning ALS, thus allowing for the subsequent specification of improvements. ALS is a degenerative disease of the nervous system (Mitchell and Borasio, 2007). Its treatment is extremely complex because of multifaceted disease progressions, the existence of many stakeholders from different sectors (i.e. social care, health care, therapists, and aid suppliers) and the high costs incurred by treatment and care. Furthermore, there is no cure for rare diseases. The care of these patients focuses on improving living conditions and maintaining self-determination.

As a first step, we conducted a comprehensive analysis comprising three elements: interviews, literature analysis, and expert reviews. The result of the analysis was a set of concepts (i.e. terms) prevalent in the analyzed domain. In a second step, these concepts and their relations were described more formally by using so-called "concept maps" (Cañas, et al., 2005). Such concept maps display the structure of concepts, allowing for a quick overview of an analyzed domain (Fig. 1).

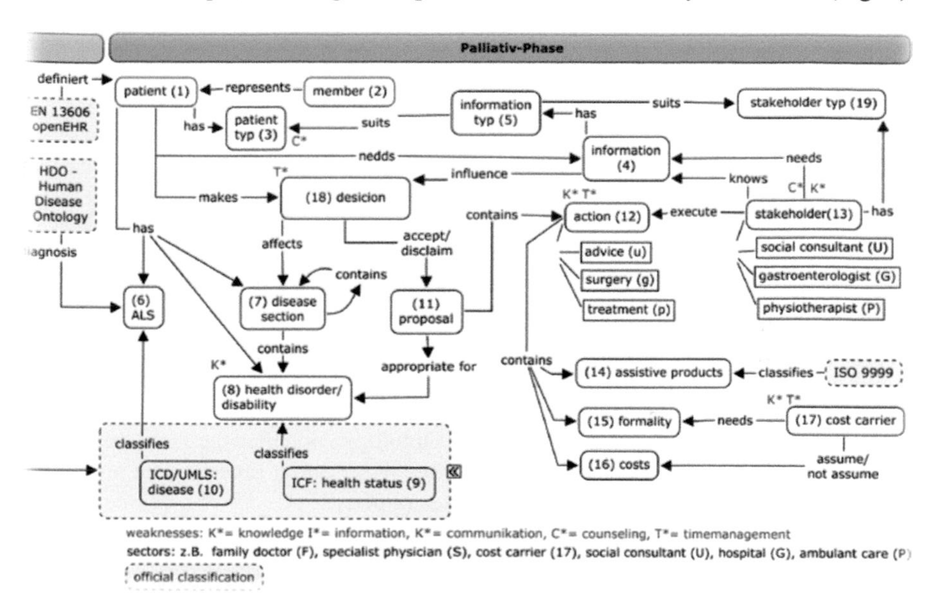

Figure 1 Concept Model of ALS Disease

The constructed Concept Model of ALS Disease demonstrates the relation between the concept patient (1) and the term of information and shows that the patient (1) has a certain disturbance of health (8). This health disorder can be approached in certain ways (11). The model also reveals that the knowledge of different treatment activities (12) lies with the executive specialists (13).

In a second step, the model was formalized into a knowledge model. In that sense, the term knowledge model is understood on different levels ranging from the possibilities of abstract knowledge presentation to specific approaches, which describe realistic problems and help specify the application of semantic technologies (Kienreich and Strohmaier, 2006). The model introduced here provides a concrete knowledge representation in the formal description language Web Ontology Language (OWL), which the semantic technologies used are based on and which is referred to below in the Dispedia Knowledge Model. The visualizations of the conceptualities within the conceptual model illustrate two separate parts of the formalized Knowledge Model (Fig. 2).

On the one hand, there is the demand for a detailed definition of the information about the patient in connection with his individual requirements (R "Requirements"). On the other hand, the representation of this information (I "Information supply") is essential as it pertains to the requirements/ standards of the patient and which can be adapted and made accessible for patients or for types of recipients.

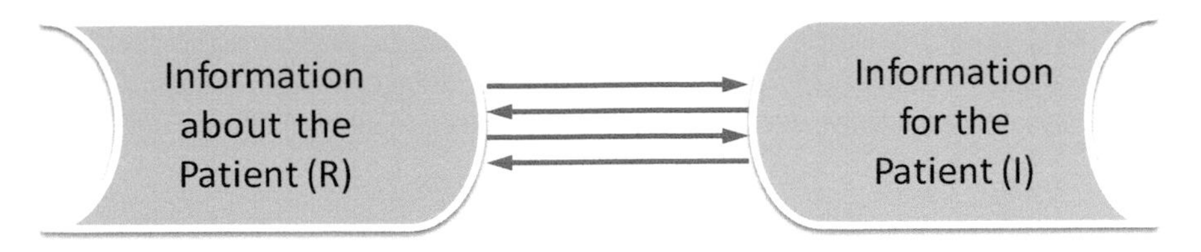

Figure 2 Information Parts of the Knowledge Model

These structural problems were also taken into consideration in the architecture of the Dispedia Knowledge Model (Fig. 3). Accordingly the Dispedia knowledge model first of all consists of a Core Ontology (www. dispedia.de), which allocates all terms and relations in the form of schematic knowledge or vocabulary. Expert knowledge and technical information (I) are stored in the Proposal Ontology (als.dispedia.de), i.e. the proposal ontology imports vocabulary from the core ontology and the disease-specific categories are instantiated. In contrast, the patient's pieces of information (R) are filed in the Case Ontology (patient.dispedia.de). Likewise, the Case Ontology imports the schemes and vocabulary of the Core Ontology as well as the classification schemes that are used for a standardized description of the patient's parameter.

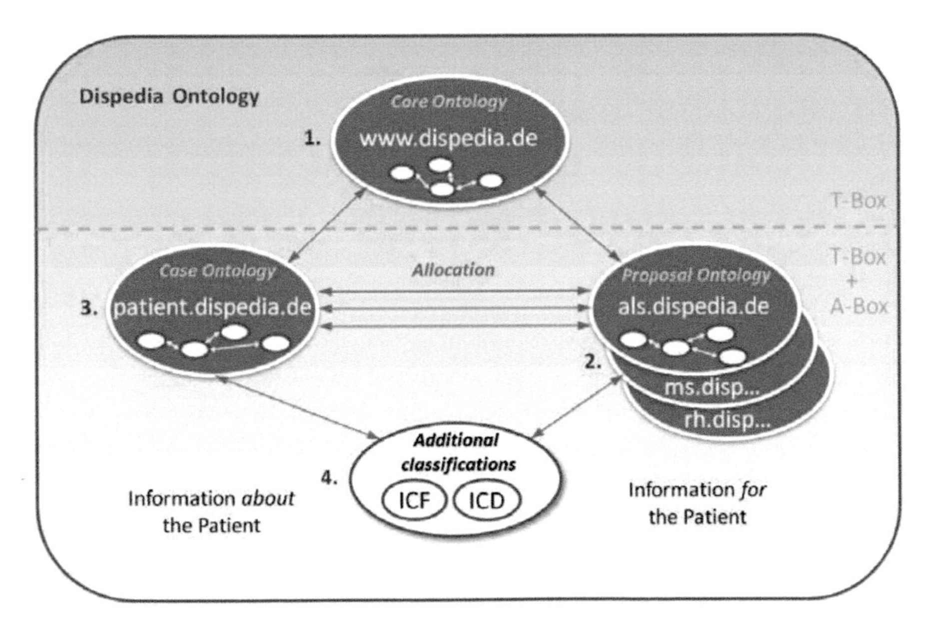

Figure 3 Architectural Overview about the Dispedia Knowledge Model

The employment of RDF/S (Manola, 2004) und OWL (W3C, 2004) for a representation of the depicted information architecture has various advantages.

- The representation of information in a coherent structure without a direct connection to specific applications. This facilitates the development of applications, which do not focus on the information logistics advised by this approach.
- The definite identification of the conceptual and concrete resources due to the use of dereferenceable URIs.
- The reutilization of the concepts in applied domains to generate an exchange of resources without a loss of information.
- Interlinking and reification of concrete resources.
- The application of established resources of the Linked Data Web, e.g. Dbpedia (Bizer, et al., 2009) and PubMed (Névéol, et al., 2011)

Furthermore, the presented information architecture facilitates the use in allotted web-environments by applying RDF/DS (distribution of resources into the three models). As part of the deployment process, the Core Ontology and the expert data-base are intended for operation in a public domain. Non-anonymized patient data will be secured from unauthorized access in private spaces by access control.

In the case that the actual amount of resources in the deployed expert knowledge base has to be increased in order to enable the availability of proposals for physicians and patients, we focus on knowledge acquisition tasks. In order to offer an authoring interface for our ontologies, we deployed OntoWiki (Heino, et al., 2009) and developed a set of plugins to improve accessibility and editability of resources as described in section 3.

3 DISPEDIA KNOWLEDGE TRANSFER

Based on the Dispedia Knowledge Model, it is possible to realize the bi-directional exchange of expert knowledge for different health states of a patient (Fig. 4). Depending on the clinical picture, the information for the patient (I) has to be stored in the respective proposal ontology in step 1. On the other hand, the health state of the patient (R) must be described. Therefore, a standardized classification can be integrated. An example of an ALS patient-specific classification is the ALS Functional Rating Scale (ALSFRS) (Cedarbaum, et al., 1999).

After that, the relevant proposals from the Proposal Ontology (e.g. percutaneous endoscopic gastrostomy) can be allocated to the specific health states (e.g. Swallowing 0: Needs supplemental tube feeding) in Step 2. This manual assignment must initially be carried out by domain experts and is stored in the Ontology Proposal.

In Step 3, the attending physician or other expert can allocate the concrete solution proposal to the concrete patient (Fig. 5). The underlying model supports the pre-selection of solutions based on the assignments made above.

Now it is on the patient to make the decision to accept the a certain proposed solution or not (Fig. 6). The doctor will then be able to see what the patient chose.

With the help of the system technology, the physician must then assign the resulting new proposals. For example, the refusal of a feeding tube then requires more detailed information on palliative care to be provided.

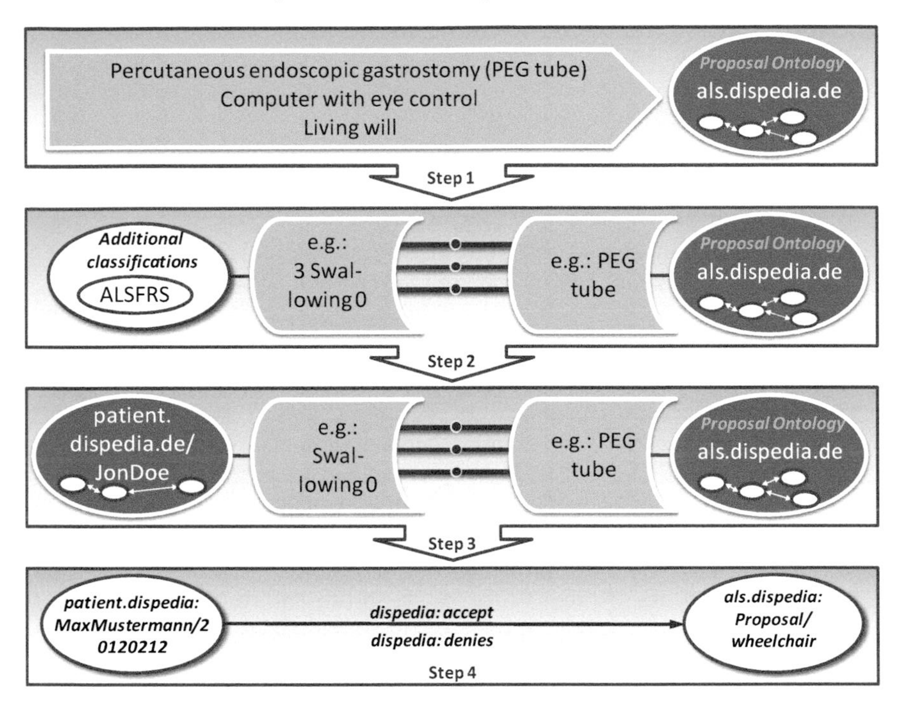

Figure 4 Domain specific expert information will allocated to the special classification and to the patient

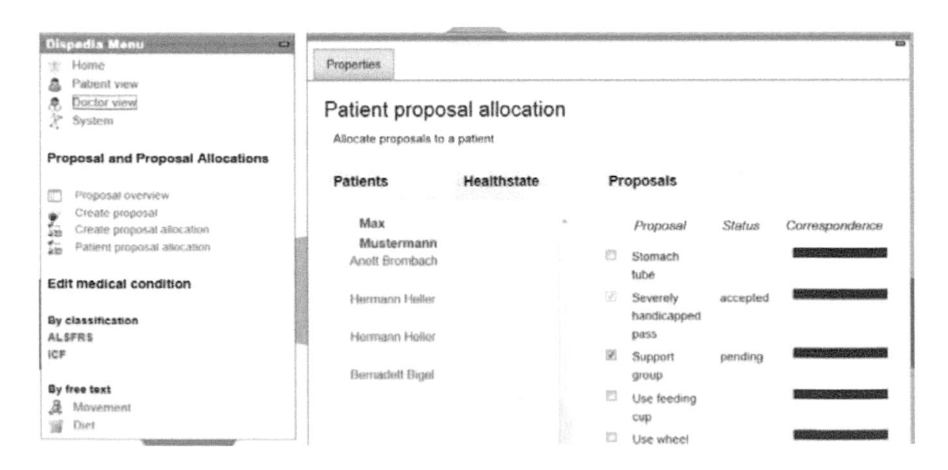

Figure 5 The doctor allocates concrete information modules to the concrete patient

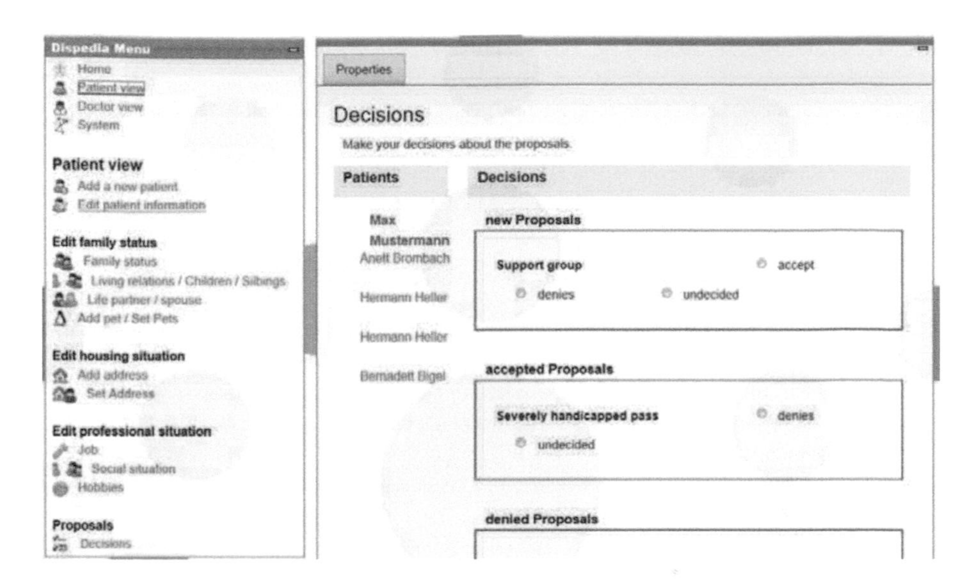

Figure 6 Illustration of how the patient becomes informed to make self-determined decisions.

2 CONCLUSIONS

One of the major goals in improving information transfer for patients with rare diseases was to develop an environment for standardized knowledge acquisition, representation, and usage. For that purpose, we analyzed the ALS domain first, engineered a set of requirements, and developed the architecture and its concepts. In the second part of our approach, we designed a process to acquire patient and expert knowledge as well as a workflow to transfer the relevant information between stakeholders. To facilitate the knowledge acquisition and deployment of designed processes, we selected and adapted OntoWiki (Heino, et al., 2009).

The defined prototypical setup of the knowledge concepts and architecture in combination with the process incorporated into OntoWiki showcases how such an information transfer can be established.

As the next step for further development, we will interlink further proposal ontologies and additional classification systems. To evaluate our approach in realistic environments, we will transform and integrate representative data from different research institutes (Charité Campus Virchow-Klinikum Neurological Clinic, German Society for Muscle Diseases).

REFERENCES

Benkenstein, M. and S. Uhrich 2010. Dienstleistungsbeziehungen im Gesundheitswesen – Ein Überblick zum Konzept „Shared Decision Making“ in der Arzt-Patienten-

Interaktion. *In:* D. Georgi & K. Hadwich (eds.) *Management von Kundenbeziehungen.* Gabler.

Bizer, C., J. Lehmann, G. Kobilarov, S. Auer, C. Becker, R. Cyganiak and S. Hellmann 2009. DBpedia - A crystallization point for the Web of Data. *Web Semantics: Science, Services and Agents on the World Wide Web,* 7, 154-165.835

Cañas, A. J., R. Carff, G. Hill, M. Carvalho, M. Arguedas, T. C. Eskridge, J. Lott and R. Carvajal 2005. Concept Maps: Integrating Knowledge and Information Visualization. *In:* S.-O. Tergan & T. Keller (eds.) *Knowledge and Information Visualization.* Springer Berlin / Heidelberg.

Cedarbaum, J., N. Stambler, E. Malta, C. Fuller, D. Hilt, B. Thurmond and A. Nakanishi 1999. The ALSFRS-R: a revised ALS functional rating scale that incorporates assessments of respiratory function. *Journal of the neurological sciences,* 169.646

Heino, N., S. Dietzold, M. Martin and S. Auer 2009. Developing Semantic Web Applications with the OntoWiki Framework

Networked Knowledge - Networked Media. *In:* T. Pellegrini, S. Auer, K. Tochtermann & S. Schaffert (eds.). Springer Berlin / Heidelberg.

Kienreich, W. and M. Strohmaier 2006. Wissensmodellierung — Basis für die Anwendung semantischer Technologien *In:* T. Pellegrini & A. Blumauer (eds.) *Semantic Web.* Springer Berlin Heidelberg.

Kuschel, F. 2006. *Die medizinische Versorgung erwachsener Patienten mit Muskelerkrankungen.* Doktor, Universitätsmedizin.

Manola, F. E., Miller 2004. Rdf primer. Available: http://www.w3.org/TR/rdfprimer/.

Mitchell, J. D. and G. D. Borasio 2007. Amyotrophic lateral sclerosis. *The Lancet,* 369, 2031-2041.757

Névéol, A., R. Islamaj Doğan and Z. Lu 2011. Semi-automatic semantic annotation of PubMed queries: A study on quality, efficiency, satisfaction. *Journal of Biomedical Informatics,* 44, 310-318.836

Ng, L. K., Fary and S. Mathers 2009. Multidisciplinary care for adults with amyotrophic lateral sclerosis or motor neuron disease. *Cochrane Database of Systematic Reviews.*

W3C 2004. OWL Web Ontology Language Overview. W3C Recommendation. Available: http://www.w3.org/TR/owl-features/.

CHAPTER 70

A Review of Sensory Feedback and Skill Learning for Rehabilitation

Biwen Zhu, David B. Kaber

Edwards P. Fitts Department of Industrial and Systems Engineering
North Carolina State University
Raleigh, NC, USA

ABSTRACT

As part of on-going advances in virtual reality (VR) and robotics technologies for rehabilitation applications, augmented sensory feedback (e.g., knowledge of result [KR] and knowledge of performance [KP]) systems have been developed to motivate patients through visual, auditory or haptic modalities. The objective of the present research was to provide a systematic review on types of augmented feedback and modalities that could produce optimum outcomes in terms of learning rate and skill retention. Based on the review, we proposed links between the constructs of a human information processing (HIP) model and motor learning processes. Our findings indicate human perceptual, cognitive and motor performance in training tasks mainly depend on the flow of information processing; whereas, skill learning (expertise) is primarily related to the formation of long-term memory (LTM) and how frequently communications occur among the perceptual process, working memory and LTM. On this basis, KP feedback was expected to support greater learning than KR since it facilitates more frequent memory communications. Similarly, auditory and haptic cues were expected to support greater learning than visual cues due to their retention strength in short-term memory stores. Future directions of empirical research were identified, including assessing the proposed model of human learning under various feedback conditions as well as the effectiveness of recommended modes of feedback in actual training systems.

Keywords: rehabilitation, motor learning, virtual reality systems, sensory feedback, human information processing

1 INTRODUCTION

Two major challenges are commonly associated with motor skill rehabilitation. First, patients are typically required to perform repetitive limb movements over time and this can be boring. Second, most physical therapies require at least one therapist to work with a patient during the training process. High training costs can occur if rehabilitation learning is slow.

To address these issues, various forms of sensory feedback have been presented to patients through VR-based task simulations or robotic-assisted exercises. The feedback typically emphasizes delivery of information related to the movement process or task outcomes. Although researchers generally agree that both types of feedback can support skill learning to some extent, no consistent feedback design guidelines exist for combinations of feedback type and modality to produce optimum outcomes in terms of learning rate and skill retention.

The objective of the present research was to provide a review of sensory feedback and motor learning for rehabilitation. We began with a focus on prior research on the use of VR or robotic technology in upper-extremity motor-control training or rehabilitation (Section 2). Following this, a rough taxonomy of augmented feedback was developed to further categorize methods as either knowledge of results (KR) or knowledge of performance (KP). The learning benefits of the two general feedback types were compared (Section 3). In Section 4, the learning utility of feedback modality (i.e., visual, auditory or kinesthetic) is reviewed. In covering both feedback type and modality, we used a theoretic model to explain how the motor learning process might be affected from a human information processing (HIP) perspective. Finally, a set of feedback design guidelines was developed based on the model of HIP and connections to motor learning. Future directions of empirical work validating the model are also described in Section 5.

2 EXISTING TRAINGING SYSTEMS

Several robotic training systems for upper limb motor rehabilitation have been developed and tested in clinical settings. For example, MIME is a bimanual robotic device for reaching and tracing tasks (Burgar, Lum, Shor & Van der Loos, 2000). During training with this device, both hands of a patient are attached to two separate robot effectors. The patient's goal is to complete the task with the less impaired limb as much as possible. In the meantime, the robot continuously moves the other more affected limb to mirror the motion of the patient-controlled limb. Both kinesthetic and visual feedback can be provided when a patient actively produces a force in the direction of movement. For instance, patients cannot only feel the robot's velocity-sensitive resistance during movement, but also object collision with the hand at the completion of movement. Feedback on segments of the movement completed and the remaining time needed to complete the task can also be presented to patients via a visual display to sustain motivation.

MIT-MANUS (Krebs et al., 1998) is another device that can guide the movement of a patient's upper limb using impedance-control. The robotic training system integrates gaming components (e.g., drawing circles, stars, squares, diamonds, and navigating through windows). Feedback on performance errors can be presented to a patient directly through a computer screen. The game aspects and feedback are regarded as key design strategies that may enhance patient motivation when exercising with MIT-MANUS, and such strategies are not easily replicated in standard physical training.

Besides robot technology, VR has also been widely used in motor skill rehabilitation to achieve a richer and a more immersive training environment. For example, Gentle/s (Loureiro et al., 2003) is a robot device for training reaching tasks for stroke patients, which integrates a haptic device and VR simulation technology for presenting various types of sensory feedback. Similar to the MIT-MANUS system, Gentle/s can provide performance feedback on motion errors through a computer display. The system can also provide collision detection feedback and target contact feedback through the haptic controller. Beyond this, the system is capable of creating different virtual scenes to represent real training scenarios (e.g., an empty room, real room or decorated room). The haptic controller and VR simulation flexibility make the Gentle/s an easy and effective tool for addressing essential human motor learning elements (e.g., rich feedback enabling enhanced training motivation and repetitive practice). Evidence has also been found that humans can learn motor skills in virtual environments and transfer them to the real world (Lintern et al., 1990). A contemporary review of VR rehabilitation technologies can be found in Holden (2005).

In summary, the above review suggests that augmented feedback is a novel feature for VR or robotic-based training systems that may support enhanced learning and intrinsic motivation to a greater extent than manual physical therapy. However, there are currently few feedback design guidelines for such systems and designers tend to overuse VR tools without consideration of limitations in human information processing. A theoretic framework for classifying the types of feedback integrated in such systems is needed along with empirical studies to assess feedback methods in terms of learning rate and skill retention.

3 FEEDBACK TYPES AND MOTOR LEARNING

In general, augmented feedback for motor training systems can be separated into KP (knowledge of performance) or KR (knowledge of results) (Magill, 2004), both of which can be presented in visual, auditory or kinesthetic forms. KP is process-related feedback and it directs trainee attention to the pattern of the movement (e.g., gymnist limb position during training). KR is result-oriented feedback and it directs trainee attention to the outcome of task performance (e.g., "you were off the target by 3 inches"). The content of KR may vary based on the task goal. The delivery mechanism may also vary with some feedback delivered based on discrete measurement (i.e., success or failure) and feedback based on continuous measurement (i.e., path deviation, etc.). Sometimes, the learning utility of KR may not be obvious because the chosen tasks already include inherent information about movement outcomes (Platz and Winter, 2001). Therefore, KP is more commonly provided to patients or athletes in real training settings, since such feedback is harder to obtain through intrinsic channels (e.g., it is difficult for a trainee to see their limb posture when shooting a basket ball).

Similar to KR, the content of KP can also vary depending on aspects of a movement that are considered to be more critical to the learning process. For example, KP can be focused on kinematic information, which involves the presentation of a pure movement process, including position information, time, velocity and patterns of coordination. It can also be focused on kinetic information which relates to the forces produced during movement. In addition, biofeedback is considered a special type of KP. During training with biofeedback, therapists make covert physiological processes (e.g., muscle activity, joint angles, body surface pressures, brain activity, etc.) explicit for patients, and train them to learn self-control of responses based on the feedback (Huang, Wolf & He, 2006). Regardless of the content, the easiest approach to presenting KP is through graphics or video replays. However, such feedback may contain irrelevant information for the learner. Therefore, additional attention directing cues may be necessary and have been suggested by some studies (Kernodle & Carlton, 1992).

3.1 Comparison of KP vs. KR in terms of motor learning

Both KR and KP have utility for motor learning and skill rehabilitation. In one study conducted by Cirstea, Ptito and Levin (2006), 37 chronic stroke patients were recruited to complete a series of reaching tasks for motor recovery. During the test, two different types of augmented feedback were delivered including: (1) KP on joint motion, and (2) KR on overall movement precision. A control group with no feedback was also included. It was found that kinematic gains in either the KR or KP groups exceeded those for the control group.

There is also some evidence that KP leads to superior learning outcomes than KR. In a follow-up study, Cirstea and Levin (2007) randomly assigned 28 chronic stroke patients to either a KR or KP group for performance of a similar reaching and pointing task. The training included 75 trials per session with a total of 10 sessions over a 2-week period. KP was delivered in terms of joint motion and KR was delivered in terms of overall movement precision. ANOVA results showed that the KP group experienced both immediate and long-term increases in joint range as compared with the KR group. The KP group also achieved better inter-joint coordination and such motor improvements were retained in follow-up evaluations.

3.2 Limitations

Several weaknesses exist in the prior research comparing KP and KR. First, synchronization of feedback delivery with task performance has not been controlled. In Cirstea and Levin's study, for example, KP was provided concurrently during the training task, whereas KR was delivered at the end of the task. This is a drawback of the experimental design as prior research has revealed that synchronization of feedback and performance can have significantly different effects on skill acquisition and long term retention. For example, Armstrong (1970b) found that concurrent kinematic feedback led to lower errors in a skill acquisition period; whereas, terminal kinematic feedback had a significantly greater learning effect during the retention period.

Second, the sensory channel through which KP or KR has been delivered has not been controlled. Again, in the Cirstea and Levin study, KP was presented in a spatial visual format; whereas, KR was presented in a verbal format. The authors did not provide a basis for their choice of feedback modality. Their assumption was that human motor adaption and learning is multisensory, highly-flexible and not dependent on the effectiveness of a particular sensory channel for conveying feedback information (Sarlegna, Gauthier & Blouin, 2007). Even if this assumption holds true, various feedback modalities could still have differences in motivating patients during the training process. Such motivational effects might also influence motor learning and skill acquisition over the long-term. As a consequence, there remains a need for greater control over this feedback parameter in future studies involving direct comparisons of KP and KR.

Third and most important, various models of learning have been developed to describe the underlying mechanisms of motor control in rehabilitation (e.g., a feedback [reactive] and feedforward [predictive] model from Wing and Whitall (2003)). However, prior research has generally failed to construct hypotheses regarding the learning utility of KR and KP based on cognitive theories. Consequently, our understanding of how sensory feedback might affect human learning process is still unclear.

3.3 A HIP perspective

One of the most well known HIP models in the human factors literature is the "pipeline" model from Wicken and Hollands (1999). This model describes several critical stages of information processing involved in human performance or skill acquisition. The first stage is called the "*short term sensory store*" (STSS), where feedback information is preserved temporarily with no attentional resources allocated. Information in this stage, however, decays quickly and the rate of decay is dependent on the feedback modality. In general, prior research has found that iconic cues decay much faster than echoic or kinesthetic cues in the STSS (Cowan, 1988).

Once attentional resources are made available and allocated, the information in the STSS is perceptually encoded in a second stage referred to as "*perception*". During this stage, stimulus information is held in working memory (WM) and a set of declarative knowledge emerges that is related to the task goal. This knowledge set is compared with operational rules stored in the long-term memory (e.g., if an object's shape is round with no corners, it probably is a circle or globe). New procedural knowledge is formulated in the event that no operational rules exist. These internal information transactions are regarded as key aspects of HIP affecting skill learning and retention. Development of LTM knowledge or motor programs for a new task cannot be achieved without them.

After stimuli are perceptually encoded in a motor control task, a decision to action is made and a movement response is selected and executed. Outcomes of the current action are then fed back to the start of the HIP loop for another cycle. Wicken assumes that each stage within the HIP loop takes some time for operation. This suggests that human performance is primarily dependent on the flow of HIP as a whole, or how quickly information can be transformed among the stages of processing. Therefore, performance might be facilitated if information can be processed rapidly through the HIP loop. However, humans may then spend less time processing and communicating information among the perceptual process, WM and LTM, and this could compromise skill learning and motor schema formation over the long run.

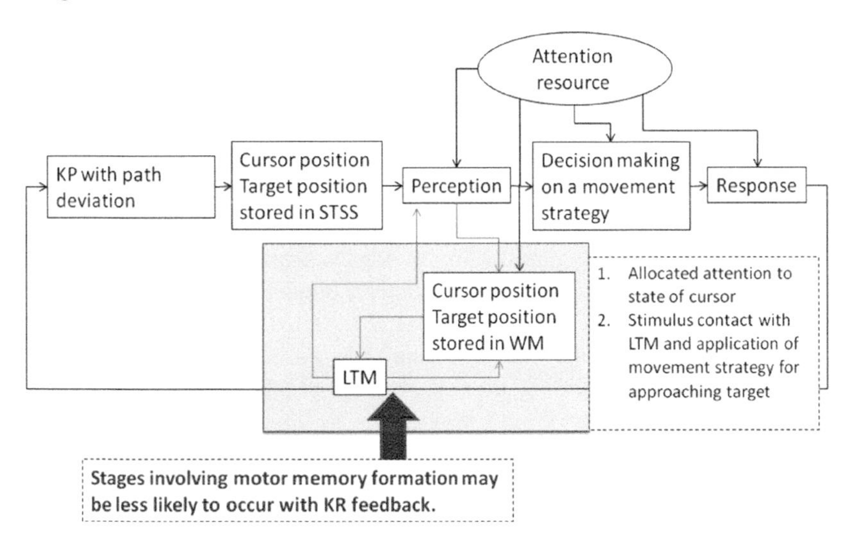

Figure 1. Illustration of human information processing in a Fitts' task with KP feedback; model based on Wicken and Hollands (1999).

This model can be used to explain the differences between KR and KP in terms of learning utility. Trainees receiving KP must process information with extensive internal communications as compared to those trained with KR. Such communications are expected to result in a stronger learning experience and outcomes over time. A simple computer-based version of a Fitts' (discrete movement) task can be used to illustrate this point (see Figure 1). Consider two groups learning to perform the Fitts' task in repeated trials with one receiving KR notifications of target hits and the other group receiving KP notifications of trajectory/path deviations. Trainees in the KP group would need to store current cursor and target positions in WM, perform mental calculations of the degree of deviation between them in a two-dimensional (2D) space, and access LTM for an appropriate movement strategy for approaching the target. These are all time-consuming processes and could slow the flow of HIP flow, in general. Trainees in the KR group, on the other hand, receive packaged or pre-coded information on movement outcomes. Such information enters the perceptual store, passes directly to WM and is ready to be used for decision making. Internal interactions with LTM are less likely to occur in this group. Therefore, KR might be good for temporary performance acquisition; however, KP was expected to be better than KR in terms of skill retention and learning due to extended information processing.

4 FEEDBACK MODALITIES AND MOTOR LEARNING

4.1 Sensory characteristics and motor learning

The visual modality is commonly used for presenting spatial information (e.g., road signs), or in situations when the auditory system is taxed due to environment noise. A disadvantage of the visual modality is that there may be too much information presented in a single display and this can cause visual overload and tunneling. In such situations, feedback through auditory and tactile modalities can provide additional information without loading the visual system by using semantic coding. The disadvantage of the auditory or haptic modality is that the content must be limited due to the transient nature of displays. Complex information, such as limb position over time, may not be effectively coded using these modalities.

Another key difference among sensory modalities is the strength of stimulus retention in the STSS. Cowan (1988) found that humans can recall unattended near-threshold auditory or kinesthetic stimuli up to 10s after presentation. However, storage of partially processed iconic stimuli (i.e., codes being used for pattern recognition) begins to fade after 0.5s and continues up to 2s. This implies that stimuli processed visual will have less of chance to be elaborately processed for perception and retention, and as a basis for action and subsequent learning.

4.2 Feedback modality and motor rehabilitation

Visual feedback has been mostly used in robotic-assisted or VR-based motor rehabilitation. For example, in Wei et al. (2005) study, 16 subjects were recruited to complete a series of reaching tasks and their arms were obscured by a projection plane. Several augmented visual cues, including subject hand position, target position, as well as a trajectory connecting the two objects indicating instantaneous error, were projected on the plane. Such feedback was regarded as helpful since it provided real time information on the state of the movement process (i.e., KP) as well as the result of movement outcomes (i.e., KR). Results showed that both cues (KP and KR) could be effectively delivered through the visual modality and they both led to an improved level of learning outcome.

Auditory feedback can also be used to represent spatial information during motor rehabilitation. For example, Johanna et al. (2009) used audio cues to reinforce the awareness of the kinematics of reaching movements (i.e., KP feedback) for hemi-paretic stroke patients. In their feedback design, information regarding the direction of the target relative to the hand was presented using semantic encoding. If a participant's hand approached the right side of the target, the intensity and frequency of a sound in the left ear would increase. Usability test results showed that such a design was accepted and valid for use across trainees.

Morris et al. (2007) investigated the use of haptic feedback in a computer-based force reproduction task for training surgery or tattooing-related operations. During training, participants were asked to follow along a virtual trajectory and to produce a sequence of one-dimensional forces with a haptic device. To support and accelerate the learning process, the haptic device was programmed to apply the opposite of the requested force pattern to a subject's hand (i.e., KP feedback), and the subject could simply focus on applying forces to keep the device in the normal plane. Furthermore, a blue bar was presented visually to indicate the instantaneous target force pattern (again, KP feedback). A green bar representing the current force applied by the subject was also available (an additional form of visual KP). Results showed that a combination of the visual and haptic assistive cues led to significantly less force recall error than the visual cue only group. Therefore, multi-modal KP may be more effective than modal feedback or the haptic cueing may provide some additional benefit over the visual KP, alone.

In summary, the above reviews reveal that augmented feedback has been delivered through all sensory modalities and, in general, the feedback was successful in supporting the human learning process. However, few studies have compared the learning utility of the various modalities. With the advance of haptic technology, kinesthetic feedback cues are gaining popularity in rehabilitation training system design. However, there is still a lack of a theoretical basis for explaining how and why haptic cues might surpass visual cues in terms of skill learning.

4.3 A HIP perspective for modality effect

To better understand the effect of feedback modality on skill learning, it is necessary to look further into the key characteristics of the STSS within the model of HIP. Cowan (1988) theorized two distinct subphases of the STSS. The first phase (STSS[1]) lasts approximately 200ms in duration and retains unanalyzed information for further processing. The second phase (STSS[2]) is regarded as part of short-term memory (STM) or WM and the information may be partially "processed" for approximately several seconds, without direct allocation of attentional resources. Cowan said that retention capability is comparable across modalities for STSS[1]. However, the substantial differences in STSS retention duration occur in Phase 2. Kinesthetic and auditory information can be retained in STSS[2] for much longer time than visual.

Such differences might mediate the degree of information exchange between the STSS and other memory stores. With greater retention time, haptic or auditory stimuli are more likely to be elaborately encoded in WM through access of LTM.

Such a process is likely to occur in STSS[2] and would lead to differences in motor memory generation. In specific, the pattern of desired muscular innervations may be trained and stored more efficiently in LTM with haptic or auditory feedback. As a consequence, haptic or auditory cues are expected to lead to greater learning outcomes than visual cues. Such phenomena are also expected to be more prominent in rehabilitation learning since seniors, or neurologically impaired patients, simply need more time to process information.

5 CONCLUSION AND FUTURE RESEARCH

In conclusion, the present research provided a review on how augmented sensory feedback could be used to support further learning in VR or robotic-based motor rehabilitation system design. Focus was placed on the type of information feedback (i.e., KR or KP) that should be delivered and how it should be presented (i.e., visual, auditory or haptic). A HIP (human information processing) perspective of manual motor skill learning was adopted and used as a basis for formulating postulates on the effectiveness of feedback type and modality. Findings of the review indicate that movement process related information (i.e., KP) may support motor learning to a greater extent than outcome related cues. Haptic and auditory modalities are also expected to support greater learning than visual cues due to their retention properties in the human memory system.

Future empirical studies should be conducted to test these expectations. For example, we can start with training of a computer-based psychomotor task with a healthy population over an extended period to determine whether there is a statistically significant influence of sensory feedback characteristics. Follow-on testing with real patients in a rehabilitation setting could then be conducted with the same objective. However, samples of patients are typically limited and data can be more challenging to analyze. For both tests, retention sessions should be included to verify learning effects.

ACKNOWLEDGMENTS

This research was supported by a grant from the National Science Foundation (NSF) (No. IIS-0905505) to North Carolina State University. The technical monitor was Ephraim Glinert. The views and opinions expressed on all pages and in all documents are those of the authors and do not necessarily reflect the views of the NSF.

REFERENCES

Armstrong, T. (1970). Training for the production of memorized movement patterns. Technical Report 26, University of Michigan (Ann Arbor), Human Performance Center.

Burgar, C., Lum, P., Shor, P., & Van der Loos, H. (2000). Development of robots for rehabilitation therapy: The Palo Alto VA/Stanford experience. *Development, 37*(6), 663-673.

Cirstea, C., Ptito, A., & Levin, M. (2006). Feedback and cognition in arm motor skill reacquisition after stroke. *Stroke, 37*(5), 1237.

Cirstea, M., & Levin, M. (2007). Improvement of arm movement patterns and endpoint control depends on type of feedback during practice in stroke survivors. *Neurorehabilitation and Neural Repair, 21*(5), 398.

Cowan, N. (1988). Evolving conceptions of memory storage, selective attention, and their mutual constraints within the human information processing system. *Psychological Bulletin*, 104 (2), 163-191.

Holden, M. (2005). Virtual environments for motor rehabilitation: review. *Cyberpsychology & Behavior, 8*(3), 187-211.

Huang, H., Wolf, S., & He, J. (2006). Recent developments in biofeedback for neuromotor rehabilitation. *Journal of NeuroEngineering and Rehabilitation, 3*(1), 11.

Johanna, R., Thomas, H., Pael, L., Djamel, B., Sylvain, H., & Agnes, R. (2009). Effect of auditory feedback differs according to side of hemiparesis: a comparative pilot study. *Journal of NeuroEngineering and Rehabilitation, 6*(45).

Kernodle, M., & Carlton, L. (1992). Information feedback and the learning multiple-degree-of-freedom activities. *Journal of Motor Behavior, 24*(2), 187.

Krebs, H., Hogan, N., Aisen, M., & Volpe, B. (1998). Robot-aided neurorehabilitation. *IEEE Transactions on Rehabilitation Engineering: a publication of the IEEE Engineering in Medicine and Biology Society, 6*(1), 75.

Lintern, G., Roscoe, S., Koonce, J., & Segal, L. (1990). Transfer of landing skills in beginning flight training. *Human Factors: The Journal of the Human Factors and Ergonomics Society, 32*(3), 319-327.

Loureiro, R., Amirabdollahian, F., Topping, M., Driessen, B., & Harwin, W. (2003). Upper limb robot mediated stroke therapy GENTLE/s approach. *Autonomous Robots, 15*(1), 35-51

Magill, R. (2004). *Motor learning and control: Concepts and applications*: McGraw-Hill.

Morris, D., Tan, H., Barbagli, F., Chang, T., & Salisbury, K. (2007). *Haptic feedback enhances force skill learning.* Paper presented at the IEEE World Haptics.

Platz, T., & Winter, T. (2001). Arm ability training for stroke and traumatic brain injury patients with mild arm paresis: a single-blind, randomized, controlled trial. *Archives of Physical Medicine and Rehabilitation, 82*(7), 961-968.

Sarlegna, F., Gauthier, G., & Blouin, J. (2007). Influence of feedback modality on sensorimotor adaptation: Contribution of visual, kinesthetic, and verbal cues. *Journal of Motor Behavior, 39*(4), 247-258.

Wei, Y., Bajaj, P., Scheidt, R., & Patton, J. (2005). *Visual error augmentation for enhancing motor learning and rehabilitative relearning.* Paper presented at the 9th International Conference on Rehabilitation Robotics, Chicago, IL.

Wickens, C., & Hollands, J. (1999). Engineering psychology and human performance: Prentice Hall.

Winstein, C., Wing, A. M., & Whitall, J. (2003). Motor control and learning principles for rehabilitation of upper limb movements after brain injury *Handbook of Neuropsychology: Plasticity and Rehabilitation* (pp. 79).

CHAPTER 71

Supporting Occupational Health and Safety at the Workplace of People with Cognitive Impairments with a Speech Driven Interaction System

Benjamin Tannert, Saeed Zare, Michael Lund, Sabrina Wilske

University of Bremen
Bremen, Germany
btannert@uni-bremen.de

ABSTRACT

People with cognitive impairments need special care to make their life safer when they perform self-determined actions. For a self-dependent conduct of life work and to give contribution to society by work is an important aim. But workplaces with machines include a considerable risk that they hurt themselves. For people with cognitive impairments it is hard to keep every work step in mind. This limitation can induce hazardous moves which can put them in danger. Therefore, in order to prevent injuries, it would be helpful to guide those persons through their working procedure. But nevertheless this assistance should be low-key and always support self-confidence of the person. We try this by providing them with a mobile system that gives hints about their working steps at the place they need this information. This mobile system also has to adapt itself to the abilities and disabilities of the individual users, because they can lose attention if they are overwhelmed or underchallenged.

Keywords: Inclusive Design, Mobile Assisted Work Training, Technology for People with Cognitive Disabilities, Sheltered Workshop, Mobile Guided Workplace

1 INTRODUCTION

Developing applications for the learning context of people with cognitive impairments is a big challenge for digital media in education. We try to observe how these people can be assisted by a mobile learning application to learn and keep in mind the steps they have to do at the workplace. Through empirical studies we will work out in detail in what way mobile learning is more helpful for this target group then traditional forms of learning.

Our intention is to protect the health of people with cognitive impairments at their workplace by developing a system that helps them to do their work autonomously and more safely. For this we work closely together with a sheltered workshop. In this participative design we embed both people with disabilities and their supervisors.

The system guides the employees step by step through their construction cycle. The application that provides this opportunity runs on mobile devices so that the user can handle it directly in the work context. This is important for our target group because their memory and attention span is more limited. Because of the variety of their cognitive abilities there will be individual guidelines for every workstation, which will be provided in different difficulty levels so that the users are neither overwhelmed nor underchallenged.

The guidelines will be prepared by the supervisors of the sheltered workshop. They have the experience to estimate in how many steps a construction cycle has to be divided for the individual employee. Therefore a storyboard-based application will be provided that allows the supervisors to make a guideline consisting of many small steps.

The contribution of this paper is to explore and discuss how mobile learning can be used for people with cognitive disabilities to foster their autonomy and enable them to lead a more independent life. The application should be able to identify the mood of the user by using speech recognition in order to adapt the level of the learning material to the current mood. In this study we give a short overview about our research process. Afterwards we describe the implementation model and the application that will be developed. Besides this, specifics in the realization and limitations are emphasized which should have a high focus during the whole research process.

2 MOLEDIWO – MOBILE LEARNING SYSTEM FOR DISABLED PEOPLE AT WORKPLACE

2.1 Related Work

In a preliminary work for our current approach, we carried out a literature review on current systems based on digital media for our target group (Suta, 2007, Traxler, 2008, Newell, 2006). In parallel, we developed "Intelligent Mobile Learning Interaction System (IMLIS)", a personalized mobile learning system for people with mental disabilities at school (Zare, 2010). We researched and analyzed the abilities to personalize the provided learning material for each user. Different workshops have been held for observing and estimating their abilities for using mobile devices. In this way we could get a better insight of their activities with mobile technology.

In order to model voice recognition we analyzed the current literature (Dai, 2008, Cichosz, 2007, Ververidis, 2006). The literature points out diverse approaches to tackle this challenge. The outcome of those approaches is still not finalized. Nevertheless we evaluated the different approaches to create a model to be implemented. Through the evaluation and in cooperation with experts of cognitive science (psychology) we listed the strength and weaknesses of each method of the different approaches. . Beside this we described in a scenario the requirements for voice recognition we need for our application. Finally we decided for each method how close this approach met our expectations. Each approach focuses at different measurable aspects of human voice and speaking, for example one approach on sound level and another on retarded oral fluency. As a result of our analysis of the existing approaches, we noticed that our target group has specific cognitive and often also physical restrictions that influence the previously mentioned measurable aspects as well. With our experts we decided that the easiest solution would be a personalized profile of a small number of these measurable aspects. This voice-emotion profile needs to be created during the registration.

In the project Vila-b (Vila-b, 2012) we analyzed the use of mobile devices for further training during work. The concept was acquired from a blended learning situation. In this context three different kinds of learning contexts (learning in work process, learning in courses, learning at a pc-workplace had been established, each with its specific learning method. The most challenging context was learning in work process. Here, in a first step the real work actions had been observed and analyzed in order to design the learning steps as part of regular work steps. A single work step could be structured into three sequences. The first sequence is planning, the second is the real action and the third is the evaluation of the result. Our solution for a mobile system that provides the possibility to learn during the work-process was designed for the target group of building workers. The results of this project showed that it is very beneficial to have access to required information directly in the context of work. Such information could be, for example, the direction for use or field reports that could solve the current problem.

2.2 The System

Mentally disabled people have to learn new things in their own individual way and due to limited memory capacities it is harder for them to keep the necessary steps of a procedure in mind. At their workplace this problem can be very dangerous because it is important to recall their working steps in mind to fulfill their work and to avoid inappropriate actions that could lead to injuries. People with cognitive impairments need to learn directly at their workstation because they are not able to abstract their work-steps and for this a low-level explanation is needed.

A characteristic that is just as important is their difficulty to stay focused. A majority of this target group can be easily distracted from their working process. The ability to maintain focus, as well as other abilities depend to their emotional status and their health constitution. Our application will consider this indication. The learning material that will be provided to the target group must be presented in a specific way, because of their cognitive impairments. Otherwise they often become overwhelmed, underchallenged, can lose their motivation or get distracted (see details 2.2.3).

An important aspect for disabled people to sustain their self-confidence is their ability to do their work autonomously. Thus people need help in many situations of their life and for this reason it would be a big enhancement if they could do their work-steps without calling for somebody's help. Unlike face-to-face guidance, the technological assistance will give them the opportunity to fulfill their tasks more self-dependent and prevent them from feeling imposed upon by their supervisors.

2.2.1 Implementation of the approach

For this context, we developed a conceptual design based on several empirical studies run during the last three years. In this time we established a close cooperation with a sheltered workshop. As a base for our research we performed different workshops in which we observed how people with cognitive impairments could cope with mobile technology and audiovisual instructions. Our research interest was to analyze with which work environment the target group can be supported to perform their work tasks in a safe and effective way. The achieved results, the given characteristics by experts from inclusive didactics and the references from the supervisors led us to the conclusion that complex actions have to be reduced and restructured. The information that is presented has to be split up into small actions that will build story units. These units were planned and organized by the supervisors through interactive storyboards. For the target group it is also important to get the instructions embedded in their work task they do at present. In dependence on the personal profile the information has to be provided and presented in a fitting way. The system prepares this information in small steps as so-called learning nuggets.

The applications control is realized through simple speech commands because of the difficulties of these users with reading and writing. In order to respect the specific characteristics of this target group we are designing an alternative way of controlling the system. We are planning to let them control the application by using their voice (either simple sound combination or short well defined words). To swap between explanations they can use their voice and say something like "next" or "previous" to get the corresponding illustrations. These explanations will be made available through a visual as well as an audio channel. In other multimedia learning environments, such as museums, technology should direct the visitor's attention to the real context situation, therefore audio guides are preferred. This kind of guide facilitates learning but also reduces the visitor's cognitive load. Respecting those related results we discussed modalities and the best way of combining sensorial channels. Finally we decided to focus on audio as a framework to bind visual information towards the real context (Mayer, 1998).

2.2.2 The system architecture and interaction pattern

The system is built of two main components. The first is the learning management system that shows all the stored learning material and gives the ability to create new ones. The second is a client application that is running on mobile devices. The planned technology for the mobile devices is handhelds. The core of the learning management system is a database. The database functionalities are completed by several interconnected modular applications. The structure is redefined for extension by additional modules.

During the registration of a new user a profile will be created which contains individual characteristics of the user with descriptions on behavioral patterns. In these patterns the individual variation of interactions and feedbacks are described. These profiles have to be created by a supervisor of the sheltered workshop and get constantly updated according to observed or traced reactions and behaviors of the user. In order to analyze this information, criteria have been implemented as rules in a decision engine.

Another module provides a complex application for speech recognition with focus on sound variation of human voice. This analyzes the differences in voice to identify the current emotional status. For example dynamic, sound level and tone pitch are important indicators for the current emotional status. The system has to manage a specific problem of human voice. Every human has a specific, individual expression varied to individual ability of articulation (blurred between words or different parts of words). Some users for example may have a loud voice that the system has to handle. Therefore, during the registration, the user has to verbalize predefined sentences in different dynamic and sound level.

The most challenging part is the main module that carries the audio-visual tutorials. For the creation of these tutorials a specific sub-application will be designed. With this application the supervisor can create storyboards of small audio-visual work units that are based on given storyboard-templates. These work units consist of three steps with well-defined clarifications. The first step is titled “context”, the second “action” and the third “test of success”. The first step describes the context and the needed initial situation where an orientation is provided. The next step shows the work-action in detail and in a tempo according to the personal profile and emotional status. In the last step a test on the outcome will be presented that shows whether the work-action had been successfully completed or not. For every instruction an audio comment can be created and connected with it. The implementation of our system isn’t finished yet and by further tests the model might be redefined.

2.2.3 Catch the mood

The integration of identification of the current mood of the user is a new approach in developing digital media for learning contexts. We try to include this aspect because the cognitive abilities of our target group are depending on their present mood. With this diagnose our software will be able to individualize and personalize the displayed learning material and the representation. Because of the fluctuation of the abilities in different situations a specific personalization is needed for people with cognitive impairments. In traditional face-to-face learning, the personalization is always considered by the teachers during the learning process. Thus the lack of personalization in digital media should be in some way compensated.

The recognition will be made by two types of feedback given by the user. Firstly, their interactions are traced and analyzed according to a stored personal profile. Secondly, the system is analyzing the spoken words of the user. By the pronunciation of their instruction, the application will be able to cognize how they feel during their usage, and how should be interacted with the user in the next steps. For this, we will implement sound patterns for different emotional status. With the first use of the system the supervisor has to run a specialized test to evaluate the individual patterns. With respect to the emotional status and the personal profile of each user, the application will be able to prepare the learning material for the user according to their current cognitive status.

2.2.4 Limitations

Although the speech recognition is in many areas really good and helpful, it is a new approach to extract the feeling and emotional conditions of somebody out of the spoken words and let this result flow into further action. The reliability and the correctness of this have to be evaluated during our research process. Furthermore for our research we try to get indications out of our evaluation with experts on a reliable combination of different approaches that are on one hand not too complex to implement and on other hand applicable for most of the situations. But the combination also enlarges the shortcomings of each method. In order to handle this we will respect these shortcomings.

In previous work, there are some results for recognizing the emotional status of speech (Dai, 2008, Cichosz, 2007, Ververidis, 2006) that show serious disadvantages. For example Keshi Dai et. al. explored how to recognize different emotional status by a combination of basic acoustic features, prosodic features with their landmark and syllable features. They achieved a result of over 90% by differ between "hot anger" and "neutral" emotional status. But the more different emotional status they use the less exact their results have been. As a first result our expert from psychology suggested to reduce the number of emotional criteria's to 3 to 5 basic emotions that influences strongly cognitive processes.

From the other side, we clearly state that the mobile learning solutions or generally learning based on digital media cannot replace face-to-face learning. They can merely be an alternative for face-to-face learning in specific situations. Such a situation could be a context where the live instructor cannot participate in person during the learning or working process. Sometimes this kind of mediatization fosters also self-reliant activities that increase personal autonomy. Those activities have the potential to empower our target group. The availability and further development of mobile technologies provides a growing potential for empowerment that should be explored. Specific technologies provide already functionalities that are preferred by our target group in comparison to real life situations. For example someone will feel less embarrassed by mobile devices than to interact with their supervisor. We learned that the most important aspect of software development for this target group is to respect the individual mental life from an emphatic point of view. Technology becomes more applicable by such an emphatic design. The use of mobile devices will lead to an enhancement of quality in the life of the disabled people but it is also an opportunity for a sheltered-workshop to provide more safety workplaces for their employees with disabilities.

3 CONCLUSION

In this paper we presented a new approach to support people with cognitive impairments in learning at the workplace based on mobile technology. We try to raise their autonomy and protect them from injuries by providing them with a tool to check their next work-steps directly at their workplace. In workshops, we observed the target group in different views and cleared that they have not the skill to abstract things, as they need, as well as special learning material that guides them step by step through their work.

During the forerunner project “IMLIS” we established cooperation with an expert for simple language. This movement propagates to express important information in a language that enables participation with people with less developed cognitive abilities. Although we designed every dialog in simple German, we got indications that for our target group the amount of scripted instructions and interactions should be avoided. The writing and reading problems that a lot of the disabled people have will be eased by a voice based computer-human interaction. In addition to this, the spoken words will be analyzed and evaluated. The extracted emotional status will be integrated into the decision algorithm that determines how the learning material will be presented. So the material with the most appropriate attributes that matches to the actual mood will be shown.

Mobile devices can enhance the motivation of learning and can be used as a catalyzer for improving learning and work performance. The communicative functionalities and social self-determined mobile activities will be a challenge for our future work. As this approach is an ongoing research process, we cannot give some conclusive results yet. At the current state of our research, this will contribute to an ongoing discussion on computer-aided support for people with disabilities and on the discussion of personalized learning applications.

4 SUMMARY

We expect that the aid provided by our system will make the workplace safer and will increase the abilities of the target group to run a self-determent work live.

The assistance of our system is based on guiding the user by showing what they have to do during their real time performance in the actual working context. As a result of this a lot of accidents can be avoided that otherwise could happen due to lack in knowing the correct use of machines.

5 ACKNOWLEDGMENTS

Our special thanks to Professor Heidi Schelhowe for motivating and guiding our research. Her expertise on didactical aspects of software development helped us to build a stable cooperation with experts of didactics and psychology. We thank also the sheltered workshop Martinshof for their support especially the supervisor Hans-Dieter Viebrock. Further thanks go to Professor Pixa-Kettner an expert for special needs education (inclusive didactics).

REFERENCES

Cichosz, J., K. Slot. 2007. *Emotion recognition in speech signal using emotion-extracting binary decision trees.* http://www.di.uniba.it/intint/DC-ACII07/Chicosz.pdf

Dai, K., H.J. Fell, J. MacAuslan. 2008. *Recognizing emotion in speech using neural networks.* IASTED International Conference on Telehealth and Assistive Technologies. Baltimore, USA.

Krannich, D., S. Zare. 2009. *Concept and Design of a Mobile Learning Support System for Mentally Disabled People at Workplace.* The International Conference on E-Learning in the Workplace - ICELW. ISBN: 978-0-615-29514-5. New York, USA.

Mayer, R.E. 1998. *A spit-attention effect in multimedia learning: Evidence for dual processing systems in working memory.* Journal of Educational Psychology, Vol 90I(2),312-320.

Nevel, A.F. 2006. A user centred approach to supporting people with cognitive dysfunction – CHI 2006.
http://people.cs.ubc.ca/~joanna/CHI2006Workshop_CognitiveTechnologies/positionPapers/22_Newell_CHIworkshop_keynote.pdf

Suta, V., L. Suta, M. Vasile. 2007. Study on the ICT Application in the Didactic Activity of Children with Mental Deficiency. *ICT in Education: Reflections and Perspectives.* Bucharest, June 14-16, 2007.

Traxler, J. 2008. Learning in a Mobile Age. In. *International Journal of Mobile and Blended Learning*, Vol. 1.

Ververidis, D., C. Kotropoulos. 2006. Emotional speech recognition: Resources, features, and methods. In. *Elsevier Speech communication*, vol. 48, no. 9.

Vila-b. 2011. Virtuelles Lernen auf der Baustelle. http://www.vila-b.de/

Zare, S. 2011. *Personalization in Mobile Learning for People with Special Needs.* HCII2011. Universal Access in Human-Computer Interaction. Orlando, Florida, USA.

Zare, S. 2010. *Intelligent Mobile Interaction: A Personalized Learning System for People with Mental Disabilities. "IMLIS".* Doctoral Thesis. Digital Media in Education Research Group. University of Bremen, Germany.

Section XI

Organization Aspects in Healthcare

CHAPTER 72

Interdisciplinary Interactions in Cross-Functional Collaborative Prototyping at a Pediatric Hospital

Catherine Campbell[1,2], *Avi Parush*[2], *Thomas Garvey*[2]

[1]Children's Hospital of Eastern Ontario
[2]Carleton University
Ottawa, Canada
ccampbell@cheo.on.ca

ABSTRACT

Insights from early research using Cross-functional Collaborative Prototyping (CFCPing) suggest that prototypes provide a fuller integration of disciplinary knowledge and expertise, helping to conduct efficient collaboration more easily and thus creating concrete product concepts and definitions more effectively. A pediatric hospital in need of innovative solutions for identified medication safety challenges provided an opportunity to apply CFCPing in industry. A series of collaborative prototyping sessions were conducted with front line pharmacy and nursing staff and industrial designers to identify new ways of storing and preparing medication on a surgical unit. Systematic observation was used to identify characteristics of the interactions between participants and the artifacts they created. Findings from this study illustrate, and lead to a better understanding of, how prototyping can support interdisciplinary interactions and design development in this context.

Keywords: interdisciplinary, concept development, prototyping, healthcare, pediatric hospital

1 INTRODUCTION

It is widely acknowledged that healthcare is a complex system, facing multi-faceted challenges that demand collaboration between experts from multiple disciplines. One such challenge is medication delivery. A pediatric hospital had identified 68 factors that could potentially contribute to medication error on a surgical unit (Parush et al., 2010). Factors related to medication preparation and tools and equipment in the medication room were highlighted as top risk contributors (Campbell et al., 2010). The need to identify new ways of storing and preparing medication was recognized as an interdisciplinary problem, requiring the knowledge of pharmacy and nursing staff as well as product designers for the development of concrete solution concepts. Thus a participatory approach was selected such that front-line staff would be directly involved in concept development.

In the last three decades, research in participatory design and interdisciplinary collaboration has been moving towards the use of generative tools to support non-designers in the communication of knowledge, ideas and experience (Sanders, 2006). Prototypes have been recognized as boundary objects (Carlile, 2002) that facilitate socio-cultural relationships within collaborative teams (Chung, 2009) and support communication and ideation. As such, they can play a key role supporting communication between disciplines during initial concept development.

Concept development occurs early in the design process and represents an uncertain time in the product development cycle. In product design, this phase includes the period after customer needs are identified, when product concepts are developed, but before technical and financial limitations are introduced and a final concept selected (Dahan & Hauser, 2002; Ulrich & Epinger, 2000; Chung, 2009).

Cross-functional-collaborative-prototyping (CFCPing) is a method developed to facilitate interdisciplinary team communication and idea generation (Chung, 2008) during the early stages of design development. It involves a team of experts with diverse disciplinary backgrounds working together to understand a design problem using physical prototypes as a medium for discussion and discovery. The method is based on three modes of action: ideation, embodiment and critique and is guided by seven principles defined in Chung's doctoral dissertation, published in 2009.

Insights from early research using CFCPing suggest prototypes provide a fuller integration of disciplinary knowledge and expertise, helping to conduct efficient collaboration more easily and thus creating concrete product concepts and definitions more effectively. Therefore CFCPing appeared to be an appropriate method for the pediatric hospital design challenge above. Having experienced limited implementation with professionals in industry, there was also a need for more empirical research exploring the use if CFCPing in a wider variety of design projects and with more diverse disciplines (Chung, 2009). The pediatric hospital case provided an opportunity to explore interdisciplinary interaction during early concept development and contribute to better understanding how CFCPing can be used to support collaborative concept development in industry. The results of this exploration are presented herein.

2 STUDY METHOD

2.1 Approach

Two collaborative concept development sessions were conducted at the hospital during which participants were presented with a specific design challenge: to address identified to risk contributors through redesign of the medication preparation and storage area on a pediatric surgical unit.

With the objective of exploring the interactive behaviour of interdisciplinary participants with each other and with prototypes, systematic observation was selected as the primary method of inquiry (Bakeman & Gottman, 1997). Observational data was collected by video recording the collaborative sessions. This was used to identify verbal and physical interactions between participants, with reference material and with prototyping materials.

2.2 Participants

Participants included hospital staff and industry professionals with expert knowledge related to the identified design problem, including: surgical unit nursing staff and management (n=9), pharmacists, pharmacy technicians (n = 7) and professional product designers from industry (n=7). Participants employed by the hospital were required to be either currently working on the surgical unit, have a professional interest in pain management or patient safety and knowledgeable about IV bolus morphine delivery, the pharmaceutical characteristics of the drug, and/or some aspects of the existing preparation and delivery process in the context of pediatric care. Designers recruited to participate in the study were professional designers with a minimum of five years of combined work experience.

2.3 Materials

Sessions were conducted in meeting rooms outside the normal work environments for all participants, placing participants on equal ground in terms of familiarity with the physical environment within which they were interacting. The available seating and work surfaces allowed participants to assemble as one large group and also break out into smaller groups for focused brainstorming, discussion and prototyping.

Each small group was provided with reference images of the current surgical unit medication preparation equipment, tools and room layout, pictures depicting the current tools and context as well as images of existing products, not currently in use at the hospital, as inspiration, and to increase awareness of available technology.

Low-tech prototyping materials, including paper, markers, post-it notes, Styrofoam, modeling clay, tapes, etc. were provided. These materials were selected because they are easy to manipulate, shortening the time required for construction (Chung, 2009, p. 111), and less intimidating, so as to facilitate the direct participation of non-designers (Muller, 1992, p. 455) who may not be familiar with prototyping materials and practices. Participants were encouraged to generate visual representations of their ideas as their group progressed from brainstorming to concept development and finished mock-ups.

2.4 Collaborative concept development sessions

The general structure of the collaborative sessions was modeled after Chung's (2009) Cross-Functional Collaborative Prototyping (CFCPing). Time and resource constraints inherent to the hospital context resulted in modifications to the original method. The most significant variation was shortening the overall duration of the exercise by opting for two cycles of small group work followed by large group discussion, rather than the recommended three cycles.

The structure of the collaborative sessions conducted in this study was as follows: (1) Introduction by lead facilitator, (2) Breakout 1: divide into small groups & brainstorm solution concepts, (3) Sharing of ideas with the larger group, (4) Breakout 2: use collaborative prototyping to develop solution concepts, (5) Presentation of solution concepts to the larger group and discussion. This process is presented in Figure 1, wherein each small group is represented by the three disciplines that participated in this study: industrial design, nursing, and pharmacy.

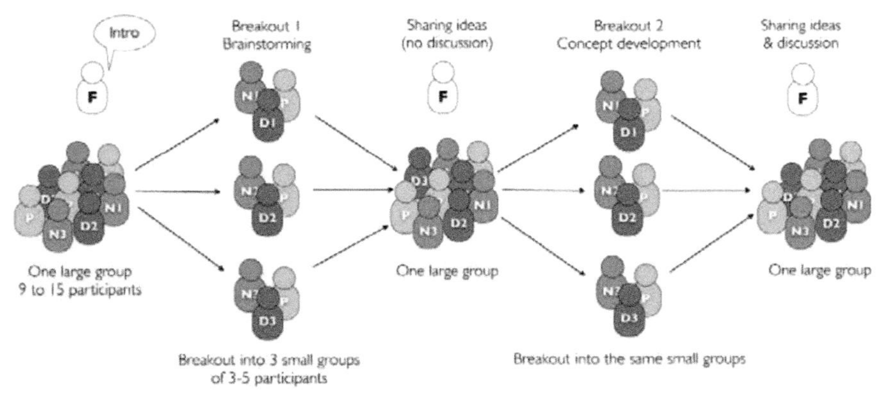

Figure 1. Visual representation of the collaborative sessions structure used in this study (From C. Campbell. 2010. Interactions in Collaborative Prototyping for Early Design Development in Pediatric Medication Safety. Master of Design thesis, Carleton University, 2010.)

Within each collaborative session, participants were divided into three interdisciplinary teams for a total of six small groups, each with two to five participants (M = 3.6). Each session was scheduled for a total of 2.5h. Based on the captured video data, the mean duration of Breakout 1 was 23 minutes (SD 3.7) and Breakout 2 was 37 minutes and 42 seconds (SD 0.52).

2.5 Systematic observation

A coding scheme for individual behaviors was initially developed based on characteristics of roles observed in literature related to group problem solving and facilitation (Schwartz et al., 2005) and collaborative design activities (Robillard et al., 1998; Lee et al., 2009; Lewis et al., 1996). Preliminary analysis of recorded sessions led to the refinement of a mutually exclusive set of codes that would allow for the full scope of individual behaviors to be captured and examined in context. The final coding scheme, summarized in Table 1, included 38 codes in three broad categories: actors (participants), actions (physical and verbal behaviours) and artifacts (tangible objects including prototyping and reference materials and intangible concepts that participants act upon).

Table 1 Coding scheme used to identify individual bahaviours

Category	Codes
Actors	*N#* (nurse), *D#* (designer), *P#* (pharmacist), *PT#* (pharmacy technician), *F* (lead facilitator), *Team*
Verbal behaviours	Information-seeking: *question, request confirmation*
	Information providing: *explain(artifact), clarify(artifact), relate to(artifact), agree, state*
	Generating ideas: *suggest(artifact)*
	Evaluating: *reject, doubt, evaluate, justify*
	Organizing: *coach, order, summarize*
	irrelevant
Physical behaviours	*gesture(artifact)*
	sketch, make/modify(artifact), write on(artifact)
Physical artifacts	*handouts, other/notes, sketch, model, materials*
Conceptual artifacts	*problem, ideas, personal experience, current context, familiar concept, plan*

Note: A detailed table with definitions is reported in C. Campbell. 2010. Interactions in Collaborative Prototyping for Early Design Development in Pediatric Medication Safety. Master of Design thesis, Carleton University, 2010.

A coding strategy, involving repeated review and verification of coding for each video clip, was implemented in an effort to minimize loss of data and strengthen reliability of a single observer.

3 FINDINGS

3.1 Collaborative prototyping for concept development

Counting and comparison of verbal and physical behaviours across groups and between disciplines was used to identify patterns in small group behavior. This analysis showed that group activity was initially dominated by verbal communication. This is consistent with observations made in previous studies by Chung (2009, p. 87-88). In Breakout 1 gesture was used to support explanation and information sharing and sketching was used to embody initial ideas generated by small interdisciplinary groups. Prototyping was used in Breakout 2 to make models of the initial concepts generated in Breakout 1 and to further develop these concepts through testing and generation of alternatives. It was observed that prototyping supported concept development by providing a tangible representation of an idea or concept that others could build on and also by allowing participants to test alternatives individually or as a group.

The average frequency of physical behaviours (gesture and all types of visual representation: prototyping, sketching and writing) exhibited by discipline is presented in Figure 2.

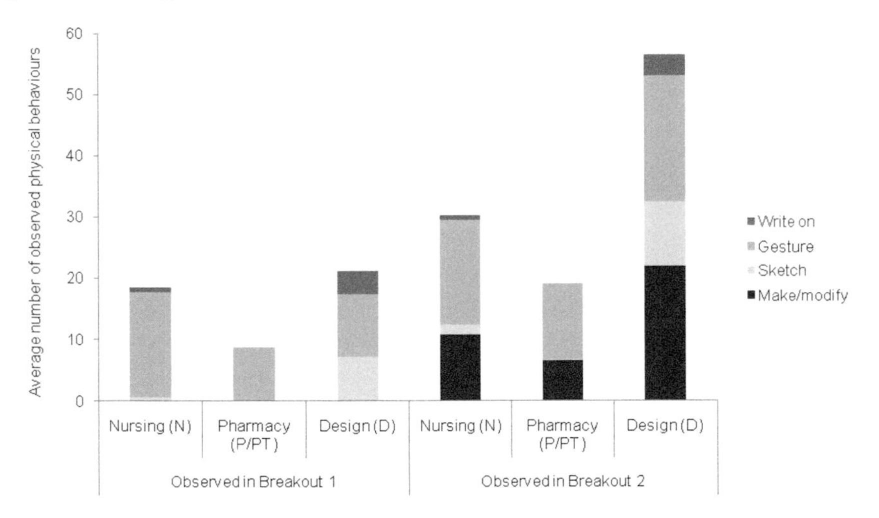

Figure 2. Average frequency of physical behaviours exhibited by discipline (From C. Campbell. 2010. Interactions in Collaborative Prototyping for Early Design Development in Pediatric Medication Safety. Master of Design thesis, Carleton University, 2010.)

While designers initiated model making in all of the small groups and were the first to modify the prototyping materials provided, healthcare workers (non-designers) quickly followed their lead, cutting and shaping cardboard and foam to build representations of ideas that had been conceptualized by the group. For example, one designer initiated prototyping by playing with a piece of paper to test different forms for a 'narcotic booth' that had been conceptualized by the group. This act motivated nursing and pharmacy staff to engage in model making and resulted in two healthcare workers becoming the self-appointed primary model makers for the group. In other groups, designers did most of the material modifications, and healthcare workers seemed to be more comfortable interacting with built prototypes by rearranging pieces and suggesting modifications or alternatives. Figure 3 presents a number of examples where quick prototypes or pieces of an existing model were used by participants to support explanation, testing and generation of alternatives. The examples and images reported here illustrate how prototyping was used to support both knowledge sharing between disciplines and concept development.

Figure 3. Observed interactions with prototypes: (Left) Nurse makes model of existing MAR to explain concept to designer; (Top right) Designer makes new elements of the model while hospital staff interact with existing model; (Bottom right) Pharmacy technician uses a piece of the built model to generate new ideas.

3.2 Identification of disciplinary roles

Counting and comparison of verbal and physical behaviours exhibited between disciplines also suggested that designers and healthcare workers assumed different roles throughout the collaborative sessions.

Designers exhibited more *coaching* behaviour and were more likely to engage in organizing group activity. They encouraged healthcare workers to participate in sketching and prototyping either explicitly or leading by example. Thus observations from this study suggest that designers are more likely to play a facilitative role.

Conversely, healthcare workers provided more explanation and examples of the current context and design problems, and were more likely to critique or reject ideas presented by others. Though the rules of creative brainstorming (Olson, 1963) were highlighted in the introduction to the collaborative session, hospital staff more frequently exhibited rejection, doubt and hasty evaluation of ideas, as illustrated in Figure 4. Observations suggested that healthcare workers lack experience with creative brainstorming and this may be one aspect of collaborative concept development process that needs to be reinforced during CFCPing sessions.

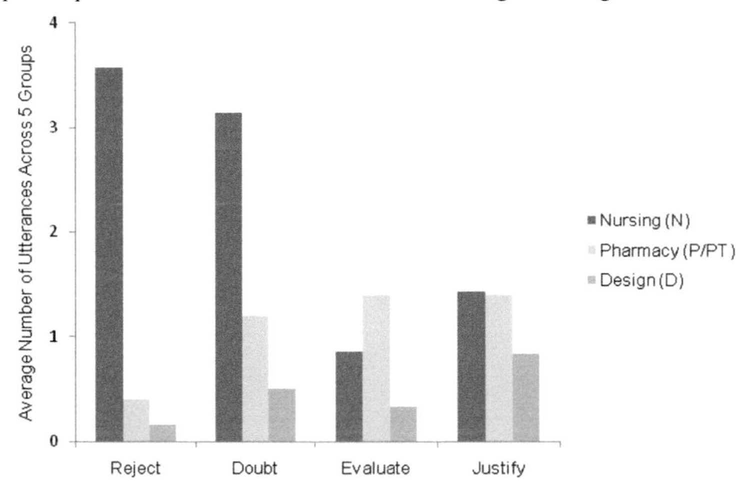

Figure 4. Average number of evaluative behaviors exhibited in Breakout 1 by discipline (From C. Campbell. 2010. Interactions in Collaborative Prototyping for Early Design Development in Pediatric Medication Safety. Master of Design thesis, Carleton University, 2010.)

4 DISCUSSION

This study of collaborative prototyping sessions conducted at a paediatric hospital brought together nursing, pharmacy and industrial designers to develop tangible solutions addressing identified risks related to morphine storage and preparation in existing medication rooms.

Systematic observation of small group behaviour allowed for a detailed look at individual contributions during collaborative sessions. Analysis of this data resulted in the identification of roles exhibited by participants and suggests that designers could potentially play a facilitation role during small group interaction: both to encourage prototyping and remind non-designers of the rules of creative brainstorming.

Reported examples where quick prototypes or pieces of an existing model were used by participants to support explanation, testing and generation of alternatives illustrate how prototyping can be used as a means to support inter-professional knowledge sharing and concept development in a healthcare setting. Whereas previous studies (Muller, 1992; Chung, 2009) have noted a hesitation by non-designers to make (prototypes) and suggest that they would feel more comfortable modifying, this study reports that with some encouragement nurses, pharmacists and pharmacy technicians engaged in both making and modifying prototypes representing medication room design concepts.

While sample size and the narrow focus of the problem investigated during these sessions means that results cannot be generalized, this is an important demonstration of how CFCPing can contribute to interdisciplinary collaboration and concept development. The collaborative sessions reported here represent empirical data supporting the use of CFCPing in a healthcare context.

The method applied in this study was shortened from the full-length CFCPing structure recommended by Chung (2009) to meet hospital resource constraints. Interestingly, at the end of the sessions participants from all disciplines commented that they would have liked more time to further develop solutions. This suggests that the full-length CFCPing schedule, including three iterations of ideation and concept development may actually be more feasible in this context than originally anticipated. More time would allow for added group discussion, critique and further development of design concepts.

The interdisciplinary nature and complexity of problems currently faced by healthcare institutions points to a need for collaborative problem solving and development of innovative solutions. This study illustrates that CFCPing is one method that can help meet this need.

ACKNOWLEDGMENTS

The authors would like to acknowledge WonJoon Chung for many conversations at the outset of this research. The contributions of the IV Morphine Study team and the healthcare workers and professional designers who participated in the collaborative sessions were invaluable. This study was made possible through funding from MITACS and the Children's Hospital of Eastern Ontario.

REFERENCES

Bakeman, R. & Gottman, J. (1997). *Observing interaction: An introduction to sequential analysis* (2nd ed.). Cambridge, UK: Cambridge University Press.

Campbell, C. (2010). *Interactions in Collaborative Prototyping for Early Design Development in Pediatric Medication Safety* (Master of Design thesis, Carleton University, 2010).

Campbell, C., Parush, A., Ellis, J., Vaillancourt, R., Lockett, J., Lebreux, D., … Pascuet, E. (2010). *Prioritizing Risk Factors Related to Morphine Administration in a Pediatric Surgical Unit* [Poster]. Presented at CAPHC'10: The Canadian Association of Paediatric Health Centers Annual Conference. Winnipeg, MB. 2010.

Carlile, P. (2002). A pragmatic view of knowledge and boundaries: Boundary objects in new product development. *Organization Science, 13*(4), 442-455.

Chung, W. (2008). *Research interest: Interdisciplinary collaboration in product development* [PowerPoint slides]. (unpublished).

Chung, W. (2009). *Cross-functional collaborative prototyping in the front end of the design process* (Doctoral dissertation, Illinois Institute of Technology, 2009).

Dahan, E., & Hauser, J. (2002). Managing a dispersed product development process. In B. Weitz & R. Wensley (Eds.), *Handbook of Marketing* (pp. 179-222). Sage Publications Ltd.

Lee, J., Popovic, V., Blackler, T., & Lee, K. (2009). User-designer collaboration during the early stage of the design process. In Proceedings from *IASDR 2009,* (pp. 1-12). Accessed November 27, 2009 from http://www.iasdr2009.org/m12.asp.

Lewis, S., Mateas, M., Palmiter, S., & Lynch, G. (1996). Ethnographic data collection and analysis for product development: Our experiences with a collaborative process. *Interactions*. doi: 10.1145/242485.242505.

Muller, M. (1992). Retrospective on a year of participatory design using the PICTIVE technique. *Computer Human Interaction,* 455-463. ACM.

Parush, A., Campbell, C., Ellis, J., Vaillancourtt, R., Lockett, J., Lebreux, D., … Pascuet, E. (2010). A human factors approach to evaluating morphine administration in a pediatric surgical unit. In V. Duffy (Ed.), *Advances in human factors and ergonomics in healthcare* (pp. 468-477). CRC Press.

Robillard, P., D'Astous, P., Detienne, F., & Visser, W. (1998). Measuring cognitive activities in software engineering. In Proceedings from *The 20th International Conference on Software Engineering* (pp. 292-300). Kyoto, Japan: IEEE Comput. Soc. doi: 10.1109/ICSE.1998.671342.

Robillard, P., Astous, P., Detienne, F., & Visser, W. (1998). An empirical method based on protocol analysis to analyze technical review meetings. In Proceedings from *The 1998 conference of the Centre for Advanced Studies on Collaborative Research* (pp. 1-12). Toronto, Canada: IBM Press.

Sanders, E. (2006). Design research in 2006. *Design Research Quarterly, 1*(September), 1-2. Accessed September 15, 2009 from www.designresearchsociety.org.

Schwartz, R., Davidson, A., Carlson, P., & McKinney, S. (2005). The skilled facilitator fieldbook: Tips, tools, and tested methods for consultants, facilitators, managers, trainers and coaches. San Francisco, CA: Jossey-Bass.

Ulrich, K. & Eppinger, S. (2000). *Product design and development*, (2nd ed). McGraw-Hill.

CHAPTER 73

Incorporating Human Systems Integration into the Evacuation of Individuals with Disabilities

Michael W. Boyce and Janan Al-Awar Smither

University of Central Florida
Applied/Experimental Human Factors Psychology Program
Orlando, FL
mboyce@knights.ucf.edu; janan.smither@ucf.edu

ABSTRACT

There is an interest within the Department of Homeland Security to evaluate evacuation planning for persons with disabilities (Sutherland, 2006). This chapter provides both a theoretical model and applied recommendations to improve emergency evacuation. It examines the relationship between Human System Integration (HSI) principles and their relevance to the field of Assistive Technology (AT). HSI, which has its foundation in systems engineering and management, aims to improve the efficiency and effectiveness of systems to maximize human performance (DHS, 2011; Booher, 2003). The International Classification of Functioning, Disability, and Health (ICF) provides a widely accepted definition of assistive technology: "any product, instrument, equipment, or technology adapted or specially designed for improving the functioning of a person with a disability" (WHO, 2001).

The research model proposed in this chapter uses the seven domains of HSI as its underpinning. These domains include manpower, personnel, training, human factors engineering, habitability, personnel survivability, and safety and health (Defense Acquisition Guidebook, 2011; DHS, 2011; Booher, 2003). The research model consists of three levels: 1) investigation of the nature of the interaction between the person with the disability and assistive technology, 2) determination of the various required roles of the human (i.e., the person with the disability, as well as individuals who are assisting persons with disabilities in the evacuation such as co-workers or healthcare practitioners), and 3) identification of requirements for the coordination between teams of individuals. By leveraging the domains of HSI and applying this discipline within the context of healthcare, this effort endeavors to produce recommendations that are understandable, usable, and have feasible best practices for practitioners to follow.

Keywords: Assistive Technology, Evacuation, Human Systems Integration, Systems Engineering

1 INTRODUCTION / BACKGROUND

Approximately, 20% of all individuals over the age of five have some form of disability. Individuals with mobility impairments are especially challenging due to their reliance on assistive technology to successfully ambulate. As of 2008, over 250,000 individuals in the United States sustained Spinal Cord Injuries. Seventy percent of these individuals use wheelchairs or similar technology (McClure, et al., 2011). Medical economics of healthcare continue to play an increasingly important role in evacuation planning (Barbera, Yeatts, and Macintyre, 2009).Assistive technology therefore can be a valuable resource by reducing workload and increasing evacuation efficiency, thereby reducing cost.

Emergency evacuation, due to its nature, is a challenging problem to tackle in the field of Human Systems Integration. This is especially true with vulnerable populations such as the elderly and the disabled who typically require increased manpower and workload during evacuation tasks (i.e. physical lifting into evacuation equipment). They also need specialized evacuation training and procedures, and are more susceptible to environmental health and safety issues such as susceptibility to smoke inhalation. The amount of research which focuses on evacuating people from hospitals, nursing homes, and assisted living buildings is limited. This is because the focus has traditionally been around isolating evacuation issues (i.e. a fire within a floor / hall) and evacuating only the patients / residents affected. However, this does not work for all emergency situations, such as when an entire building needs to be evacuated (Barbera, Yeatts, and Macintyre, 2009).

This review proposes the collaboration of professionals in assistive technology (AT) with those in human systems integration (HSI) in order to provide a model for the evacuation of hospitals, assisted living residences, and nursing homes using assistive technologies. The Human Activity Assistive Technology (HAAT) model is a research model that practitioners use to understand the relationship between individuals with disabilities and the technologies that they interface with.

This research model is broken down into the human, the activity being performed, the assistive technology, and the context which serves as an integration mechanism for the other three factors. The field of HSI originated in the 1980's as a response to reducing costs and inefficiencies in the Armed Forces. HSI is a process which takes into account human capabilities and limitations and determines how to develop a system in a cost-effective manner (U.S. Air Force, 2008; Booher 2003, DHS, 2011).

HSI is developed around seven primary domains:

- Manpower—quantity / quality of personnel required to get a job done,
- Personnel—requirements for recruitment and skillsets necessary for the selection for certain duties,
- Training—learning techniques needed for an individual to have the proficiency to perform a job,
- Human Factors Engineering—concepts and methodologies used to increase the overall usability of interfaces while reducing the potential for error. Fundamental to this domain area is the consideration given to human capabilities and limitations.
- Habitability—requirements for an individual to maintain quality of life,
- Personnel Survivability—requirements to safeguard human life, and
- Safety and Health--requirements designed to reduce environmental hazards to safety and health of users (DHS, 2011).

Both HSI and AT have a basic human-centered approach / philosophy. When compared, relationships within the HAAT model appear to be transferable to those in HSI. The human component can be compared to the human factors engineering, habitability, and personnel domains. The activity component can be compared to training and manpower, as well as requirements analysis which is an essential activity to HSI. The assistive technology component can also be represented through human factors engineering, safety and health, as well as from a more general systems engineering perspective. Finally the context can be looked at in two different ways: the first being related to habitability, survivability, and safety and health, while at the same time address the integration aspects of HSI across domains.

In an effort to provide a structured research model to emergency evacuation, this analysis takes the domains of HSI and applies them to three levels of interaction with emergency evacuation assistive technology. Being that HSI has already been proven to positively impact human performance in certain domains such as military systems (Wilson, Malone, Lockett-Reynolds, and Wilson, 2009), an expansion into healthcare could provide more support for assessing human performance in emergencies. This could therefore impact risk factors such as potential for accidents or injuries, as well as hazard avoidance and preparation.

The proposed research model identifies three levels pertaining to the human-technology relationship in emergency evacuations (See Figure 1). The first level involves the relationship between the person with the disability and the assistive technology. The second level incorporates the assistive technology relationships from level 1 and involves the various roles of the human (i.e. the person with the disability, as well as those who are assisting in the evacuation). Often times, support teams (also known as "buddy systems") are put in place beforehand where abled-bodied individuals are trained to evacuate someone with a disability should an emergency situation arise.

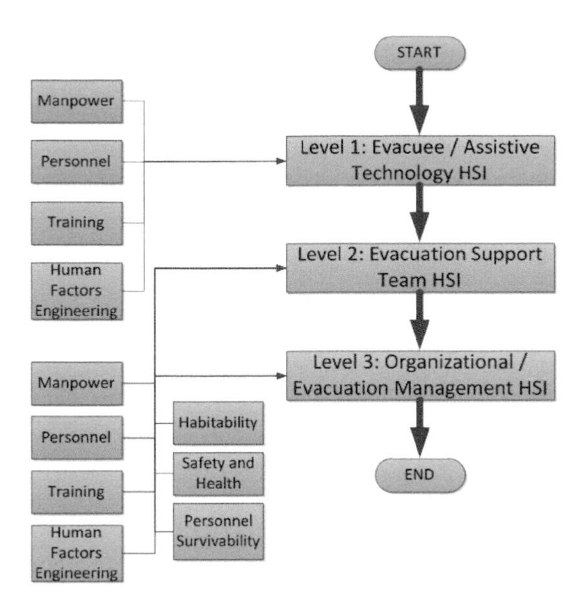

Figure 1 The proposed HSI research model for Emergency Evacuation for Persons with Disabilities

Finally, the third level takes all the outputs from levels 1 and 2 and involves the macro-level organizational relationship, under whose guidance multiple support teams would be put in place to evacuate larger groups of people such as those in a healthcare facility. The administration and execution of emergency planning can use the principles of HSI to provide opportunities for human performance improvement on a large scale.

2 THE EMERGENCY EVACUATION RESEARCH MODEL

The formulation of an HSI process within a hospital, assisted living residence, or nursing home begins with the development of an integrated product team (IPT). Integrated product teams consist of domain experts who work together to provide human performance assessments within a given environment. In addition to an HSI professional, the IPT could have individuals from safety, clinical practitioners (such as physical therapists and nurses), as well as management personnel, social workers, and emergency response coordinators (Defense Acquisition Guidebook, 2011).

Together these individuals develop key performance parameters (KPPs). The key performance parameters are thresholds related to the performance goals of the environment (Defense Acquisition Guidebook, 2011).

1.1 Human Peformance Requirements Analysis

Prior to beginning an HSI effort with any assistive technology system, it is important to recognize that the primary objective of HSI, which comes from the field of systems engineering, is to provide insight into design through the use of human performance requirements. According to the DHS Human Systems Integration, best practice on requirements analysis, can include: "…requirements for human capability, capacity, proficiency, competence, availability, utilization, accommodation, survivability, and safety and health" (2011). Requirements serve as the building blocks to any HSI process and assist in determining the necessary HSI activities associated with that product. The process of defining requirements is called Human Performance Requirements Analysis (HPRA).

A full discussion of HPRA is beyond the scope of this chapter. However it is important to involve HSI Subject Matter Experts (SMEs) when attempting to perform HPRA. The HSI SME can assist in developing a task model for the assistive technology which can identify the appropriate requirements for a given system. The reader is referred to the DHS HSI Best Practices (2011) along with Malone and Carson (2003) for more information.

1.2 Level 1: Evacuee / Assistive Technology HSI

As stated previously, level 1 relates to the nature of the interaction between the person with the disability and the technology. There are four HSI domains that are particularly relevant to this level: Manpower, Personnel, Training, and Human Factors Engineering. The reason these domains are chosen is because they place emphasis on the human-technology relationship without examining external factors such as the environment. This information can then serve as input to level 2 which encompasses factors beyond the individual.

When HSI SMEs asess an evacuee's **manpower** requirements, they are evaluating the number of resources or individuals required to sucessfully complete an evacuation using assisitive devices based on functional needs. In gathering this information the HSI SMEs may use information from the disabled persons themselves, as well as input from clincians. Manpower needs can greatly increase the cost of an effort. Therefore appropriate time should be taken with each user (or class of users) to determine how much support is really needed. This manpower assessment would also include an assessment of workload which involves breaking down the tasks against the tolerance capabilities of both the people with disabilities and the assistive technology.

Personnel as defined by Booher (2003), is the specific skills, attributes, experiences, or other aspects of an individual to maximize performance capability. In the context of a person with a disability, this can be identifying the specific skills necessary for an individual to accomplish an evacuation (such as the ability to transfer themselves into an assistive device). Through identifying skills, it is possible to also identify performance gaps or potential risks between the operational device requirements and abilities of an individual with a disability. This information can then provide input into level 2 in constructing the evacuation support teams.

Training for the individual with the disability on how the device works can serve as a last line of defense with inexperienced support personnel. Even if the individual with the disability cannot communicate due to functional limitations, having a solution to allow the person with the disability to instruct quickly and effectively can contribute to a positive outcome. The individual with the disability is a valuable participant in evacuation device testing and evaluation. This can take the form of drills or simulations where the user provides qualitative input on the human performance of the system and also provides recommendations for improvement. Additionally, due to the infrequent nature of emergency evacuation, refresher training on proper device operation is necessary to minimize skill degredation.

With regard to **Human Factors Enginneering**, there needs to be a clear understanding of what the goals of the product are related to the user. In many cases the goal is transporting down flights of stairs in an emergency situation. Furthermore, there needs to be an understanding of what the individual's functional capabilities and needs are. It is important when assessing the needs of the user to look beyond medical diagnosis and get the perspective from the persons with a disability themselves. Assessment tools include MPT – Matching Person and Technology Framework, PIADS – The Psychosocial Impact of Assistive Devices Scale, and the OT FACT – a functional assessment measure (Hersh, 2011). Furthermore, an assessment of the user interface of the device is needed. An HSI SME could examine several different types of user interfaces such as physical interfaces. These refer to handles, displays, labels, and controls that an individual with a disability uses to interact with the device (DHS, 2011).

1.3 Level 2: Evacuation Support Team HSI

Level 1, as mentioned above, focuses on the relationship between the individual with a disability and the assistive technology. Level 2 takes this a step further in that in addition to this relationship it also includes the able-bodied individuals who will be assisting in the evacuation. For the purpose of this chapter these able-bodied individuals are called the evacuation support team. Evacuation support teams, also known as buddy systems, are used to assist in accomodating functional needs in evacuations. The outputs from level 1 (i.e. user needs, workload estimates, and skillsets) serve to provide information for level 2.

Manpower, in the context of the support team, refers to ensuring availability of resources by designating a group of individuals to assist in the evacuation of an individual with a disability. Depending on how much support an individual needs, there may be redundancy to accommodate both availability and fatigue. One strategy that can be implemented is using a primary and secondary (or backup) support team, in the event that one or more of the resources are not available at the time of the evacuation. Furthermore this assessment should also consider, based on the length of the evacuation, whether there is enough manpower to effectively rotate shifts to reduce fatigue.

When making **personnel** selections, the skill mix and capabilities of the support teams needs' are very important. This goes beyond capacity for physical workload, highlighting any medically-specific requirements that a person with a disability might have (i.e. if they need an oxygen tank or other type of life support technology). If a particular skillset is needed (i.e. specialized experience for a patient who is under psychiatric care but also happens to have a disability), it must be incorporated or else the human performance of the system may fail. In addition to looking at the medical requirements of the individual with the disability, any existing medical conditions of the support teams should be highlighted as well.

In developing **training**, each member of the support team should have a clear understanding of the various roles they could possibly assume in the event of an emergency. The team should also have a clear understanding of operational instructions to accomplish evacuation of an individual with a disability. The training should address both the human performance capabilities of the person with the disability (i.e. assisting with a transfer to the assistive device) and the capabilities of others within the support team (i.e. capability to handle a certain amount of physical load). Futhermore, training and retraining needs to occur on a regular basis to account for the fact that individuals frequently change in and out of positions in an organization.

Human Factors Enginneering at level 2 relates to the design considerations which assist the evacuation support teams. There are many different types of interfaces that can affect performance. Of specific emphasis in this stage are those which can require input from multiple support personnel. Each team will undergo operational testing and evaluation with regard to these various types of interfaces. This testing and evaluation from an HSI standpoint will consist of both human centered design and function based approaches. The human centered design approach focuses on making the product efficient and effective for all parties involved. The functional based approach, on the other hand, focuses on successful outcome of the task in a given time period (DHS, 2011).

Habitability are factors that contribute to the quality of the work environments and the lives of the employees which it supports. In the case of an emergency evacuation, proper hygeine standards must be planned for and maintained. Proper hygeine measures include consideration of the usage of examination gloves for lifting / assisting the person with the disability. If the habitability conditions for the support team are poor, they could be less likely to provide assistance to other employees. Habitability is also important in terms of team cohesion. The more satisfied an individual is, the less likely he or she will to go to another position, resulting in an increased retention of team members as well as increased improvements through practice and drills.

Survivability factors are those that may be faced by the team which could have life threatening consequences to them. In the event that the assistive technology becomes unusable due to the conditions of the surrounding environment, alternative evacuation plans can be developed to ensure the team's safety. There is also a need to minimize physical and mental fatigue to facilitate optimal team and human performance. In emergency evacuation scenarios, injuries can occur due to complex tasks, taxing operations, sleep deprivation or the nature of the high stress environment (U.S. Air Force, 2008).

Safety and Health are factors, which in the context of a support team, minimize risk of injury, illness or disability. Issues that fall under this category include topics such as noise, atmospheric hazards (such as oxygen deficiency) and human factors issues which can cause injury (i.e. repetitive stress injury). Due to the fact that every emergency evacuation situation is unique, often there aren't ample data about one type of scenario. However it is possible, to have an idea of the potential hazards that could be present in a given type of situation.

2.4 Level 3: Organizational / Evacuation Management HSI

Level 2, as mentioned above, focuses on the relationship between the individual with a disability, assistive technology, and the evacuation support team. Level 3 looks at facilitating organizational evacuation procedures with multiple individuals with disabilities and multiple evacuation teams. The outputs from levels 1 and 2 (i.e. hygiene, survivability planning, and safety and health procedures) serve to provide information for level 3.

When considering the **manpower** for level 3, it is important determine how the operational context of evacuation will work with regard to manning. The manpower strategy that is determined should be incorporated and validated into the long term emergency evacuation planning (DHS, 2011). Looking at existing evacuation plans / data can help determine where improvements can be implemented. In addition, an analysis of alternatives can be developed to support the manpower determination (Defense Acquisition Guidebook, 2011, DHS, 2011).

Target audience descriptions (TAD) can assist in making organizational **personnel** decisions in an evacuation. The TAD identifies the cognitive, physical, and sensory abilities of users. This includes the capabilities and limitations of both the different groups of patients and the different groups of support personnel. By defining specifically the individuals who are a part of the system, an organization can make better educated decisions on skill mix (U.S Air Force, 2008).

When developing **training** opportunities, including actual patients can assist in increasing readiness and performance. Developing benchmarks can assist in monitoring training program progress. Realism can be increased by including cooperative training such as incorporating neighboring facilities, law enforcement personnel, and community organizations when feasible (DHS, 2011).

When performing **Human Factors Engineering**, the organization can develop a clear understanding of the human performance requirements as well as how designing to those requirements can improve user effectiveness and efficiency. Additionally discussing how errors will be identified, assessed and handled to mitigate the impact of human error on the evacuation process can assist in more efficient decision making. Finally simulation can be leveraged to evaluate human-in-the-loop decision making and workload under representative conditions (DHS, 2011).

One key piece to the **habitability** domain is fostering a positive environment and encouraging employees to state their concerns and suggestions. Also it is important to reach out to individuals with disabilities and ensure that the person being evacuated is in agreement with the evacuation procedures that have been developed for them.

One of the weaknesses in developing organizational policies is that such policies often lack adaptability. It is essential to understand that many different types of situations cause evacuation (e.g. bomb threats, fires, chemical / biological hazardous outbreaks), and that the differences between these types may have different effects on **survivability**. Each one of these situations requires different evacuation plans and presents different speeds at which an evacuation must occur. It may also be the case that the major avenue for evacuation is not an option due to damage sustained to the facility.

The development of the process to eliminate and control **safety and health** risks can assist greatly in an evacuation. Additionally organizations can develop a tracking plan toward safety and health risks. This along with a risk reduction plan can identify potential areas of improvement and appropriate safety and health actions that need to be implemented. The focus consistently rests with the health and safety of the persons involved in the human-technology system (DHS, 2011).

3 CONCLUSIONS

This review, although theoretical in nature, strives to develop concrete recommendations for each of the HSI domains across the three levels. Ideally this work should involve an HSI SME, as he or she can assist in clearly defining human performance requirements through each aspect of the emergency evacuation. Here are some actions that practitioners can take to assist in the process of evacuation:

- Learn as much as possible about the functional capabilities of the individual with the disability and include that knowledge when assessing workload and the capabilities of an evacuation system.
- Identify performance gaps or potential risks between the operational device requirements and abilities of an individual with a disability. Additionally, look to see if there are potential risks for evacuation support personnel.
- Perform drills or simulations on a regular basis where individuals with disabilities provide input and recommendations to improve performance. Provide refresher training to minimize skill decrement.
- Assess the capabilities of the individual with the disability and the support team personnel, as well as the usability of the device using questionnaires and other assessment measures.
- Understand different types of situations that cause evacuation, such as bomb threats, fires, chemical / biological hazardous outbreaks as they may have different effects on evacuation procedures.
- Create a risk reduction plan which can assess potential areas of improvement and the appropriate safety and health actions that need to be implemented.

A common question that is asked with regard to evacuation is a justification of the additional costs of preparation. However through efficient and effective training it is possible to increase human performance, increase safety, and create a higher morale for an organization at large. Further research can focus on identifying the strengths and weaknesses of the model and updating it accordingly. Through organizational leadership and implementation, the benefits of HSI can be recognized. This, in turn, could save the lives of potential victims next time an emergency strikes.

REFERENCES

Assistive Technology Act of 2004, Pub. L. 108-364, 118 Stat. 1707 (2004)

Barbera, J. A., Yeatts, D. J., and Macintyre, A. G. 2009. Challenge of hospital emergency preparedness: analysis and recommendations. *Disaster Med Public Health Prep, 3*(2 Suppl), S74-82.

Booher, H. R. 2003. *Handbook of Human Systems Integration*, 1 Ed: Wiley.

Cook, A. M., Polgar, J. M., and Livingston, N. J. 2010. Need and Task-Based Design and Evaluation. *Design and Use of Assistive Technology* 1:41-47 Springer New York.

Defense Acquisition Guidebook, Available at: https://acc.dau.mil/adl/enUS/350719/file/49150/DEFENSE%20ACQUISITION%20GUIDEBOOK%2007-29-2011.pdf [Accessed November 20, 2011].

DHS. 2011. *Human Systems Integration: Best Practices.*

Fox, M. H., White, G. W., Rooney, C., and Rowland, J. L. 2007. Disaster Preparedness and Response for Persons With Mobility Impairments. *Journal of Disability Policy Studies* 17(4): 196-205.

Hersh M. A., 2011. The Design and Evaluation of Assistive Technology Products and Devices Part 3: Outcomes of Assistive Product Use. *International Encyclopedia of Rehabilitation.* Available at: http://cirrie.buffalo.edu/encyclopedia/en/article/312/ [Accessed November 20, 2012].

Malone, T.B. and Carson, F.P., 2003. HSI Top-Down Requirements Analysis. *Naval Engineers Journal,* 1152(2).

McClure, L. A., Boninger, M. L., Oyster, M. L., Roach, M. J., Nagy, J., and Nemunaitis, G. 2011. Emergency Evacuation Readiness of Full-Time Wheelchair Users With Spinal Cord Injury. *Archives of Physical Medicine and Rehabilitation*, 92(3), pp. 491-498.

Sutherland, D. W. 2006. "Remarks at the National Hurricane Conference, April 14, 2006". Available at: http://www.dhs.gov/xabout/structure/editorial_0842.shtm [Accessed November 20, 2011].

Wilson, D. P., Malone, T. B., Lockett-Reynolds, J., and Wilson, E. L., 2009. A Vision for Human Systems Integration in the U.S. Department of Homeland Security. *Proceedings of the Human Factors and Ergonomics Society Annual Meeting.* 53(1839). p. 1840-1841.

World Health Organization. 2001. *International Classification of Functioning, Disability and Health.* Available at: <http://www.who.int/classifications/icfbrowser> [Accessed November 20, 2011].

U.S Air Force, 2008. Air Force Human Systems Integration Handbook. Air Force 711 Human Performance Wing, Directorate of Human Performance Integration, Human Performance Optimization Division, Washington, DC: Government Printing Office.

CHAPTER 74

Management of Medical Equipment Reprocessing Procedures: A Human Factors/System Reliability Perspective

R. Darin Ellis[a], *Mahtab Jahanbani Fard*[a], *Kai Yang*[a], *Will Jordan*[b], *Nancy Lightner*[b] *and Serge Yee*[c]

[a] Department of Industrial & Systems Engineering, Wayne State University, Detroit, MI 48202, USA
RDEllis@wayne.edu
Mahtab.Jahanbanifard@wayne.edu
Kai.Yang@wayne.edu
[b] Department of Veterans Affairs, Veterans Engineering Resource Center, VA-Center for Applied Systems Engineering, Indianapolis, IN, 46222, USA
Will.Jordan@va.gov
Nancy.Lightner@va.gov
[c] Department of Veterans Affairs, Veterans Engineering Resource Center, VA-Center for Applied Systems Engineering, Detroit, MI, 48201, USA
Serge.Yee@va.gov

ABSTRACT

Effective reprocessing of reusable medical equipment is essential for patient safety; however current practices rely primarily on complex manufacturers' instructions. This paper describes an information system, the Interactive Visual

Navigator[1](IVN™), designed to manage and present user-friendly reprocessing instructions to the cleaning technician, and to provide support and feedback on cleaning task times. The system provides time tracking throughout the cleaning process. Deviations from normal processing times could indicate deviations from normal processes, thus the system provides technicians with useful feedback. With a more user-centered systems approach to process design and monitoring, overall system reliability will result.

Keywords: reusable medical equipment, sterile processing, time and motion study

1 INTRODUCTION

New minimally invasive medical techniques have improved outcomes in a wide variety of diagnostic and surgical procedures (Galloro, 2012; Najarian, Fallahnezhad, & Afshari, 2011; Teoh, Chiu, & Ng, 2010). Advances in medical equipment have enabled and supported these techniques. This complex equipment requires increased precision to develop and manufacture, making it inherently expensive. To reduce the overall cost of using minimally invasive equipment, such as flexible endoscopes, they are designed with reuse in mind.

The goal of reprocessing reusable medical equipment (RME) is, following a standard usage of the equipment, to restore it to a sterile or high-level disinfected state (Malchesky, Chamberlain, Scott-Conner, Salis, & Wallace, 1995). Manufacturers provide instructions for the maintenance of their equipment, and facilities follow these instructions while conducting reprocessing operations. Manufacturers' instructions, however, are developed to ensure thoroughness and technical precision from regulatory and product reliability perspectives. This results in highly complex and sometimes difficult-to-follow instructions that have the potential to cause the unintended consequence of decreasing compliance.

The Veterans Health Administration (VHA) Directive 2009-004 stipulates that the Network Director is responsible for "Ensuring identification of accountability, responsibility and documented performance at each step in the process" and "Ensuring assigned staff are collecting data, conducting an analyses, and taking required actions to ensure safe and effective use of RME" (Kussman, 2009). To follow this directive, the Veterans Administration, working with Wayne State University researchers, has developed an automated touch screen system for presenting work instructions and collecting data related to reprocessing flexible endoscopes. This system, referred to as the Interactive Visual Navigator

---

[1] IVN™ is the property of the Department of Veterans Affairs and Wayne State University. Any use of the content presented in this paper without the express written consent of the IVN™ Program Manager is strictly prohibited.

(IVN™™™), replaces paper-and-ink work instructions that, in current practice, are displayed and followed on a site by site basis. One major benefit of the IVN™ is that it collects data on the time spent on each instruction display screen.

IVN™ displays reprocessing instructions on a touch screen for use in Sterile Processing Service (SPS) decontamination and/or preparation and sterilization areas. Reprocessing instructions developed by endoscope original equipment manufacturers (OEM) are typically embedded in safety or other types of manuals. Before they are presented on the IVN™ screen, analysts (including representatives of the SPS) isolate the step-by-step reprocessing instructions in OEM instructions and evaluate them for adherence to human factors principles described by Wagner, et al. (1996). They improve the usability of the instructions by changing the voice from passive to active, for example, while retaining the accuracy and completeness of the OEM instructions. The improved instructions are displayed on a 17 inch touch screen as shown in Figure 1. In order to use IVN™, technicians must first login using their unique id and password. They enter the unique endoscope identifier to display the correct reprocessing instructions. They touch the arrows to advance or revisit instructions as they are presented on the screen.

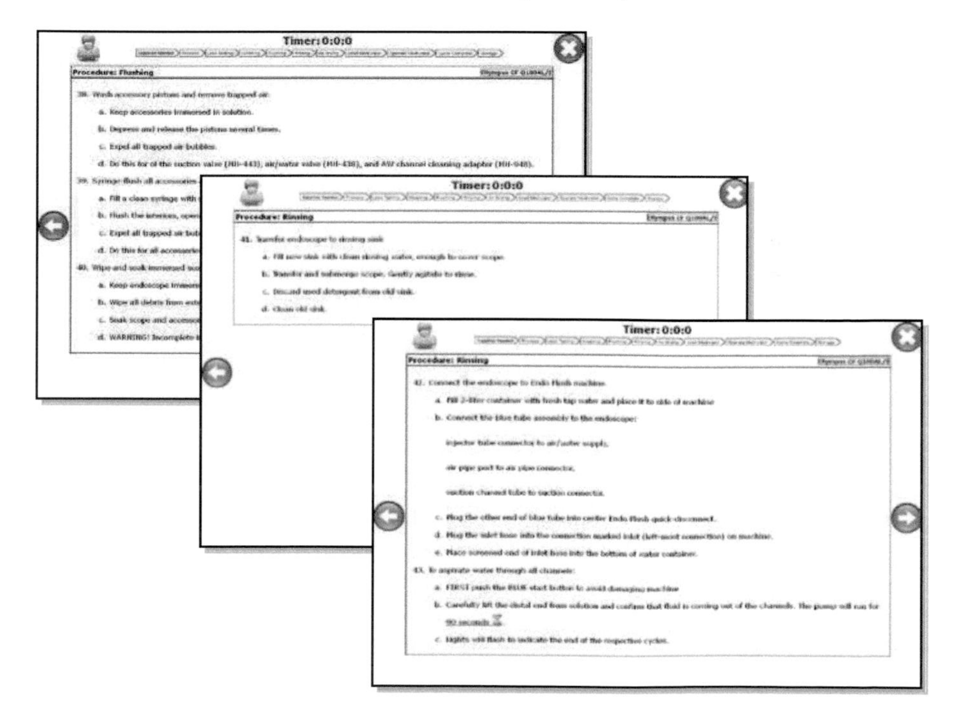

Figure 1. Work instruction screenshots from the IVN™.

This paper describes the data collection and processing opportunities included in the IVN™ system. It focuses on the system reliability benefits of standardized work and the monitoring of cleaning task time as a surveillance and audit method to ensure cleaning procedures are followed. While numerous error and adverse event surveillance methods are described in the literature (Murff, Patel, Hripcsak, & Bates, 2003), the literature contains no reports of using task performance times as a quality and reliability monitoring metric.

2 STANDARD WORK & SYSTEM RELIABILITY IN HEALTH CARE

Quantitative analysis of task performance in the field of industrial engineering goes back over 100 years to the days of 'scientific management' (Taylor, 1911). While some analysis of health care systems have utilized Taylor's time and motion study principles, the techniques often rely on statistical sampling of activity patterns (Patton, 2011; Abbey, Chaboyer, & Mitchell, 2012), or exhaustive minute by minute activity logging (e.g. a work diary) (Finkler, Knickman, Hendrickson, Lipkin Jr., & Thompson, 1993), rather than rigorous development of task processing time probability distributions. This is possibly because standardized work instructions do not usually exist for processes. Standardization is a key element to traditional time measurement (Konz, 2001; Meyer & Stewart, 2002; Freivalds & Niebel, 2009) and if this element does not exist, sampling-based time and motion data only give a limited perspective of the work environment. This results in data with questionable reliability, since special occurrences are not incorporated. A key component of the IVN™ system is the standardization of work instructions.

Experts in the healthcare literature have pointed out the benefits of standardized work. Johnson (2005) and Lefton (2008) link standardization solely to issues such as operational efficiency and cost control. At the care outcomes level, evidence-based medicine and use of clinical guidelines are used (Wenneberg, 2002; Davis, Weinstein, & Galantowicz, 2003). Investigations of procedural compliance for RME reprocessing procedures have noted exceptions to accepted practice (Jackson & Ball, 1997; Schaefer, Jhung, & Dahl, 2010). However, these investigations looked at issues such as employee training, rather than solutions such as task-element level standardization and performance feedback.

In terms of the effects of individualized feedback on performance in a health care setting, Cuschieri (2006) points out that feedback on task errors can play a positive role in performance, in that the actor is given a chance to understand the "cause of the error, how to correct and, more importantly, how to avoid it." One specific method of providing feedback to the operator is to give them a rating of their task times relative to "normal." Based on the expected time for each step in reprocessing, based on a corpus of time records, tasks with outlying times are easily identified. Analysts may use multivariate data analysis to study operator-to-operator and scope model-to-model variation in processing times. As standard reprocessing times for models are determined, these times will provide improvement in the scheduling of equipment reprocessing, optimizing resource availability to the point of use.

Carroll and Rudolph (2006) point out several challenges to system reliability and patient safety in health care systems. One such challenge is organizational goals for increased 'production' such as improved service times, wait times and resource turnaround times. The effect of such time pressure on tasks is well documented and results in behavior such as adopting sub-optimal heuristics (Goodie & Crooks, 2004). Additionally, they note that safety is sometimes difficult to define and measure, compared to bottom-line issues like service times, and is often viewed as imposed by external entities (e.g. regulators or accrediting agencies). Also relevant to the present topic, they argue that health care systems are a complex environment, with challenges such as frequent novel problems and a high level of interruptions and distractions. The designers of IVN™ considered these issues and included measures to accommodate them. Although IVN™ provides reports on the task time performance, recommendations for presentation methods to the SPS Technician are outside the scope of the current research. Reason (2000) critiqued traditional human-in-the-loop system reliability analysis for its focus on individuals and human error at the "sharp end" of performance. Human reliability, in this framework, is characterized by human choices and behavior, particularly by avoidance of unsafe acts and procedural violations. He argued instead for a system approach that identifies potential causes of error and sets up defenses against them. One way to develop defenses is to use tools such as Healthcare Failure Modes and Effects Analysis (HFMEA) (DeRosier, Stalhandske, Bagian, & Nudell, 2002). Deviations of task time from expected task times could point out areas for discussion and inclusion in prospective reliability analysis techniques like HFMEA.

3 APPROACH

As shown on Figure 1, during endoscope reprocessing each IVN™ screen presents guidance for the specific endoscope model. Screens were designed such that, wherever possible, sub-task elements are presented together in a 'chunk' (Miller, 1956) appropriate to the task goal. While there is some variation in instructions from scope to scope, there is considerable consistency in these instruction chunks that the process leverages to support development of highly skilled 'habit'-like behavior in the cleaning process. See Fisk and Rogers (1992) for a discussion of the role of consistency in skilled performance. This 'chunking' requirement was traded off against display layout and perceptual considerations including a design goal of less than 16 lines per screen, and physical job constraints, such as the requirement for hand operation of screen navigation. The variation in the number, scope and requirements of task elements gives rise to the situation that not all screens have the same amount of work, nor do they take the same amount of time to complete.

An important aspect to IVN™ is the timer feature, as shown in Figure 2 and Figure 3. Figure 2 contains a completed IVN™ screen on which step 29 e indicates that the technician must aspirate the detergent solution for 30 seconds. The text "30 seconds" is linked and an hourglass icon presents to the right of that instruction. When the hourglass is selected, the screen changes to the display in Figure 3, with a timer that counts down to zero.

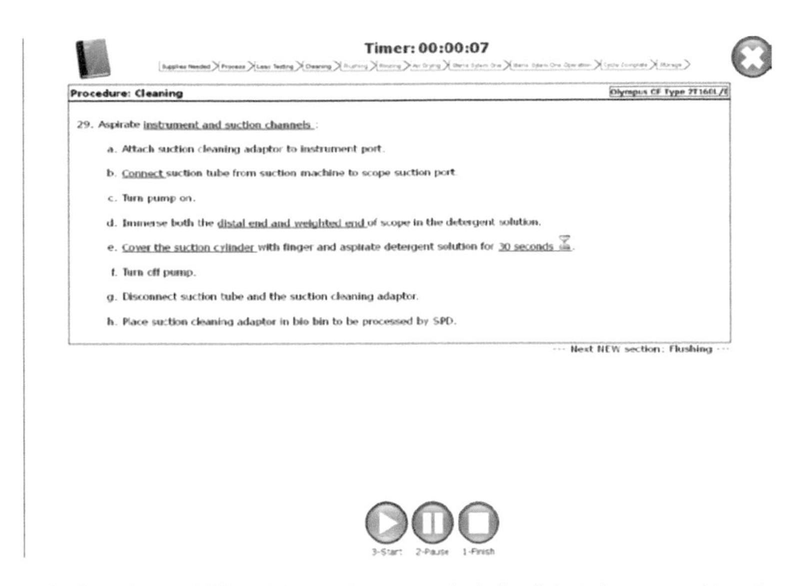

Figure 2. Completed IVN™ work instruction screen including links to images and to a timer.

If a technician attempts to progress to the next slide without first engaging the timer, a warning message will appear to remind the technician to perform the timed task. This message will display instead of allowing the technician to proceed to the next slide of instructions. IVN™ locks forward progress until the timer counts to zero. The screen color also shifts from white to grey so that the technician readily identifies that forward progress is halted. This allows the technician to perform other tasking or reviews while the timing function completes. Once the timer finishes counting to zero, the screen will revert to the white background to indicate that forward progress with the procedure can now resume.

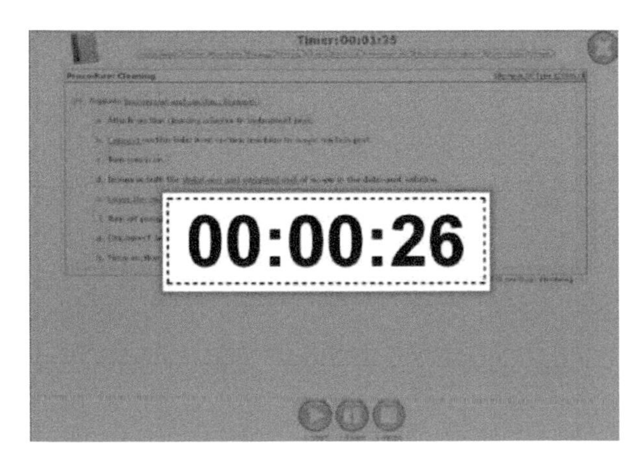

Figure 3. IVN™ screen displayed when hourglass icon selected.

Tasks, of course, do not always proceed according to the standardized work plan. In the case of traditional time industrial engineering time study (Freivalds & Niebel, 2009; Groover, 2007; Meyer & Stewart, 2002), the recording observer would track exceptions to the standard method in the form of 'wild readings' (i.e. statistical outliers) or 'foreign elements' (e.g. interruptions, etc.) IVN™ provides for technician feedback to explain variations to the 'normal' procedure in the comments and exceptions fields. Comments are notations entered into a slide by the technician, unaccompanied by procedure cancellation or any evidence of equipment damage. The technician can continue the reprocessing session beyond entering the comment. The technician can use the field to note interruptions, etc. that might have an influence on the normal task time, but would not have an influence on the overall quality and reliability of the reprocessing procedure. Exceptions, on the other hand, are automated messages generated by the IVN™ system as a result of a process cancellation, as well as comments entered in the last slide registered into the database before the procedure. In this situation, the technician must stop reprocessing, begin a different method, or wait for the repair of equipment before continuing. An example exception is when an Automated Equipment Reprocessor (AER) fails and the technician must switch to manual cleaning. Another example is when an endoscope fails a leak test. This requires the technician to complete reprocessing, but the endoscope is then sent to the manufacturer for diagnosis and repair.

4 RESULTS

IVN™ collects data throughout each reprocessing session. IVN™ stores technician sign-on, the endoscope model selected, and processing time by step in an off-site server. In IVN™ version one, two reports are generated: a Time Tracking report, and a Report by Endoscope Model. Figure 4 contains an example time tracking module report titled "Scope Operation Detail". The five columns in the main table body of the report page contain information on the following: technician identification, slide title for each reprocessing task step, exception, comments, and time used. Although the operators' names are hidden, the identifiers of two technicians are displayed, indicating who was involved in what steps. Time used is a sum of "working" time and does not take into account when a technician indicates they are halting reprocessing for a period of time. Both pause time and work time are included in the overall session duration noted in the top right hand corner of the time report screen.

A report that displays historical reprocessing data for a selected endoscope model is also available. This report is intended to facilitate analysis of performance of the displayed work instructions. In addition, several reports are planned for future IVN™ versions.

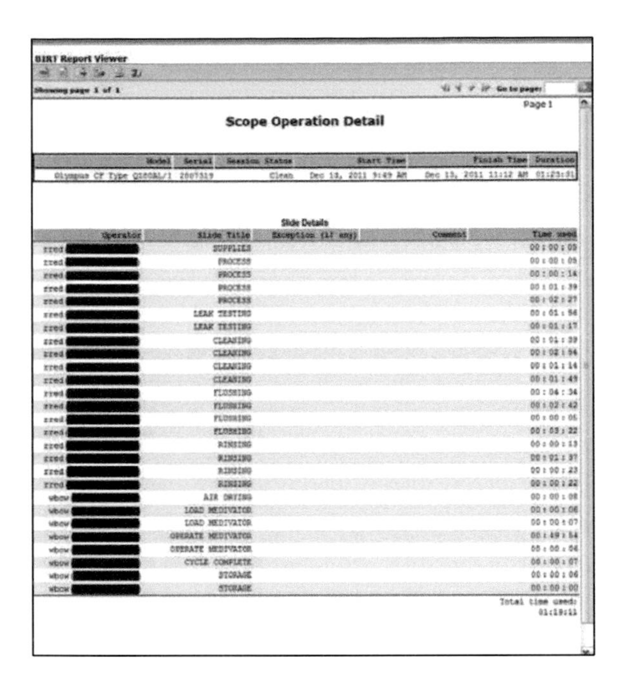

Figure 4. Screen shot of the IVN™ time tracking module.

5 CONCLUSIONS

System reliability is a complex issue that is made even more so by the myriad considerations of human factors engineering. The approach presented herein does not substitute for the broad array of techniques in current practice. However, it does offer a complementary and novel technique to the management of adherence to best practices in reusable medical equipment reprocessing. When a time for a processing step is too short or too long, this event signals that follow-up action is required. The time data, in and of itself, does not present conclusive evidence that a task element in the cleaning process was mishandled. Task time feedback is one element available to monitor procedures and seek improvements to the process.

As this is the first version of IVN™, significant work remains for future improvements in its functionality. Several new console reports are planned for future implementation. One report will focus on the entire scope cleaning history. This will make it possible to track equipment malfunction to stressful cleaning practices (for example, excessive soak times). The system will also enable users to build comprehensive and in-depth models of technician competencies, and tracking the frequency and most recent cleaning session on a model-by-model basis.

Significant opportunities exist to integrate the IVN™ with other advanced hospital information technology systems. The Veterans Health Administration is currently deploying Radio-Frequency IDentification (RFID) and other technologies to monitor equipment using Real-Time Locating Systems (RTLS) to boost utilization and improve lifecycle management for RME. When these systems are used to complement the IVN™, RTLS can add a layer of time tracking to validate the sum of the individual task times for the overall reprocessing task.

ACKNOWLEDGEMENTS

The authors gratefully acknowledge the support of the Veterans Health Administration.

IVN™ is the property of the Department of Veterans Affairs and Wayne State University. Any use of the content presented in this paper without the express written consent of the IVN™ Program Manager is strictly prohibited.

REFERENCES

Abbey, M., Chaboyer, W., & Mitchell, M. (2012). Understanding the work of intensive care nurses: A time and motion study. *Australian Critical Care , 25* (1), 13-22.

Carroll, J. S., & Rudolph, J. W. (2006). Design of high reliability organizations in health care. *Quality and safety in health care , 15* (suppl 1), i4-i9.

Cuschieri, A. (2006). Nature of human error: Implications for surgical practice. *Annals of surgery , 244* (5), 642-648.

Davis, J. T., Weinstein, S., & Galantowicz, M. E. (2003). Reducing variability in patient care: renewed focus for the pediatric cardiac surgeon in the twenty first century. *Progress in pediatric cardiology , 18* (2), 159-162.

DeRosier, J., Stalhandske, E., Bagian, J. P., & Nudell, T. (2002). Using Health Care Failure Mode and Effect Analysis™: The VA National Center for Patient Safety's prospective risk analysis system. *Joint commission journal on quality and patient safety , 28* (5), 248-267.

Finkler, S. A., Knickman, J. R., Hendrickson, G., Lipkin Jr., M., & Thompson, W. G. (1993). A Comparison of work-sampling and time-and-motion techniques for studies in health services research. *Health services research , 28* (5), 577-597.

Fisk, A. D., & Rogers, W. A. (1992). Cognitive approaches to automated instruction. In J. W. Regian, & V. J. Shute, *Cognitive approaches to automated instruction* (pp. 171-193). Hillsdale, NJ, USA: Lawrence Erlbaum Associates.

Freivalds, A., & Niebel, B. W. (2009). *Niebel's methods, standards, and work design* (12th Edition ed.). Boston, MA, USA: McGraw-Hill Higher Education.

Galloro, G. (2012). High technoogy imaging in digestive endoscopy. *World journal of gastrointestinal endscopy , 4* (2), 22-26.

Goodie, A. S., & Crooks, C. L. (2004). Time-pressue effects on performance in a base-rate task. *The Journal of General Psychology , 131*, 18-28.

Groover, M. P. (2007). *Work systems and the methods, measurement, and management of work.* Upper Saddle River, NJ, USA: Pearson Prentice Hall.

Jackson, F. W., & Ball, M. D. (1997). Correction of deficiencies in flexible fiberoptic sigmoidoscope cleaning and disinfection technique in family practice and internal medicine offices. *Archives of family medicine*, *6* (6), 578-582.

Johnson, D. (2005, November). A roadmap to optimal sterile processing workflow. *Healthcare purchasing news*, pp. 30-35.

Kussman, M. J. (2009). Use and Reprocessing of Reusable Medical Equipment (RME) in Veterans Health Administration Facilities, VHA Directive 2009-004. Retrieved February 23, 2012 from http://www.va.gov/vhapublications/ViewPublication.asp?pub_ID=1824

Lefton, R. (2008, July). Reducing variation in healthcare delivery: the lack of clinical and administrative standardization across the provider and payer spectrum creates variation and inefficiency.

Malchesky, P. S., Chamberlain, V. C., Scott-Conner, C., Salis, B., & Wallace, C. (1995). Reprocessing of reusable medical devices. *ASAIO Journal*, *41* (2), 146-151.

Meyer, F. E., & Stewart, J. R. (2002). *Motion and time study for lean manufacturing* (3rd Edition ed.). Columbus, OH, USA: Prentice Hall.

Miller, G.A. (1956). The Magical Number Seven, Plus or Minus Two: Some Limits on Our Capacity for Processing Information. *The Psychological Review,* 63, 81-97.

Murff, H. J., Patel, V. L., Hripcsak, G., & Bates, D. W. (2003). Detecting adverse events for patient safety research: a review of current methodologies . *Journal of Biomedical Informatics*, *36* (3), 131-143.

Najarian, S., Fallahnezhad, M., & Afshari, E. (2011). Advances in medical robot systems with specific applications in surgery - a review. *Journal of medical engineering and technology*, *35* (1), 19-33.

Patton, M. W. (2011). *Developing a time and motion study for a lean healthcare environment.* Retrieved Feb 27, 2012, from Masters Thesis, University of Kentucky UKnowledge: http://uknowledge.uky.edu/gradschool_theses/163

Reason, J. (2000). Human error: models and management. *BMJ*, *320*, 768-770.

Schaefer, M. K., Jhung, M., & Dahl, M. (2010). Infection control assessment of ambulatory surgical centers. *JAMA*, *303*, 2273-2279.

Taylor, F. W. (1911). *The Principles of Scientific Management.* New York, NY, USA: Harper & Brothers.

Teoh, A. Y., Chiu, P. W., & Ng, E. K. (2010). Current developments in natural orifices transluminal endoscopic surgery: An evidence based review. *World journal of gastroenterology*, *16* (38), 4792-4799.

Wagner, D., Birt, J. A., Snyder, M., Duncanson, J. P. (1996) Human Factors Design Guide For Acquisition of Commercial-Off-The-Shelf Subsystems, Non-Developmental Items, and Developmental Systems, Federal Aviation Administration, Report No. DOT/FAA/CT-96/1 retrieved February 23, 2012 from http://www.deepsloweasy.com/HFE%20resources/HFE%20Design%20Guide%204COTS.pdf.

Wenneberg, J. (2002). Unwarranted variations in healthcare delivery: implications for medical centers. *BMJ*, *325* (7370), 961-964.

CHAPTER 75

Measuring Employee Safety Culture in Rural Nursing Home Settings: A Literature Review

Yulin Wang, Barbara Millet

Texas Tech University
Lubbock, TX
barbara.millet@ttu.edu

ABSTRACT

A strong safety culture is critical for healthcare settings, where the impact of potential errors is high. Research efforts have explored assessment to measure culture, to monitor workplace health, to measure progress on implementing improvement efforts, and to proactively identify new initiatives for improvement. This research, however, is targeted mostly for urban healthcare settings. Adoption of these assessment instruments may not be feasible given the unique challenges faced in rural healthcare settings due to factors, such as (1) a large proportion of aging patients living in rural areas, (2) rural nursing homes serving as major care units, and (3) issues with staff shortage and low technological access. The main objective of this review was to identify existing healthcare employee safety culture instruments and to explore the applicability of using such tools in rural nursing homes. A systematic literature review was conducted to identify tools that allow measurement of employee safety culture in rural nursing home settings. We reviewed studies published between 1987 to February 2012 and searched CINAHL, Medline and PubMed databases. Search keywords included, but were not limited to, "employee safety", "safety culture", "healthcare", "rural healthcare", and/or "nursing homes". Thirty-five articles met the selection criteria with 8 instruments identified in the review. These instruments have been predominately developed for measuring safety culture, safety climate, or employee perception of work safety in urban hospital settings. However, measuring safety culture may be significantly influenced by factors such as the type of healthcare center, the associated job titles for the healthcare employees, and rurality. This review found no existing tool that can be directly used to measure safety culture in rural healthcare settings and even less for rural nursing homes. Future research is needed in developing an instrument that measures employee safety culture with aims at improving safety for employees of rural nursing home.

Keywords: rural healthcare, nursing homes, safety culture, safety climate, employee safety

1 INTRODUCTION

Healthcare safety is defined as "the avoidance or reduction to acceptable limits of actual or potential harm from healthcare management or the environment in which healthcare is delivered" (ANHPC, 2001). It is a critical element of quality healthcare (Steinwachs & Hughes, 2008). Safety has been a topic of research priority in recent years, due to the high risks of morbidity and mortality in healthcare settings (Colla, Bracken, Kinney, & Week, 2005).

In the United States, healthcare is mainly provided by facilities such as hospitals, medical centers, and nursing homes (U.S. Bureau of Labor Statistics, 2011). While hospitals and medical centers often provide specialized treatments and diagnoses to patients, nursing homes mainly provide nursing care or rehabilitation services for injured, disabled, or sick elderly persons (U.S. Social Security Administration, 2003). In contrast to hospitals and medical centers that recruit differing levels of medical providers such as doctors, physicians, specialists and surgeons, the majority of employees in nursing homes are nursing staff, including nursing assistants (NAs), certified medical aids (CMAs), as well as licensed practical nurses (LPNs) and registered nurses (RNs). These providers mainly communicate with physicians via telephone (Zimmerman, & Antonova, 2007).

The only patients in nursing homes are elderly people. Due to (1) an increase of population, (2) the aging of the population, and (3) elderly patients accounting for a larger proportion of the total U.S. population living in non-metropolitan or rural areas (Colello, 2007), nursing homes have become major healthcare providers in rural America. It is expected that the demand for nursing care will rise dramatically in the next two decades given that advances in medical technology increase life expectancy and in turn increase needed care required for dealing with the medical impairments of those who live longer (Zimmerman, & Antonova, 2007). However, there is already a shortage of nurses (Goodin, 2003; Ewart et al., 2004). As the shortage continues and the number of patients increases, U.S. healthcare will face a serious challenge in the years ahead to avoid a decline in quality nursing home care, specifically, for those in rural areas (Zimmerman, & Antonova, 2007).

The nurse shortage is attributed to overloaded work, low job satisfaction, perceived low salaries, poor social image, and other factors (Goodin, 2003; Zimmerman, & Antonova, 2007). Among these factors, employee safety has become a topic of interest in healthcare research as a safe work environment will minimize nursing work-related injuries and illnesses.

Safety culture is defined as "the product of individual and group values, attitudes, perceptions, competencies and patterns of behavior that determine the commitment to, and the style and proficiency of, an organization's health and safety management" (Sexton et al., 2006, p. 2). A strong safety culture is critical to improve employee safety. A strong culture of safety in nursing homes emphasizes employee involvement and manager commitment to safety.

In order to achieve a strong safety culture and in turn provide quality healthcare, effective tools are needed to identify gaps and safety hazards in existing settings and to monitor current operation and performance as the basis for adjustments and future developments. Overall, valid methods and instruments to evaluate safety culture in healthcare settings are lacking (Mearns & Flin, 1999; Wiegmann, et al., 2002; Nieva & Sorra, 2003; Pronovost & Sexton, 2005). While hospitals are making strides to assess their safety culture, nursing homes still lag behind in the development and execution of safety culture assessment. Instruments developed for hospital settings may not be appropriate for direct adoption in nursing home assessments as nursing homes differ from hospitals in many ways including the patients served, employee structure, care provided, and the safety principles driven by the organizations (Castle & Sonon, 2006; Bonner, Perera, Castle, & Handler, 2008) that are vastly different for nursing homes and hospitals. Furthermore, existing methods that measure healthcare safety culture or climate are targeted mostly for urban healthcare settings. Few studies have investigated the unique challenges faced by rural healthcare employees. This review identifies existing instruments in measuring healthcare safety culture or climate and explores applicability of these tools in assessing safety culture in rural nursing homes.

2 METHODS

We conducted a systematic literature review to address the question regarding safety culture assessment in rural nursing homes. The Cochrane Handbook was used as a guide for this review as it provides instruction for conducting systematic reviews in healthcare (Leeflang, Deeks, & Gatsonis, 2008). The online databases of CINAHL, MEDLINE, and PubMed were used for reviewing the academic literature. Search keywords included, but were not limited to, "employee safety", "safety culture", "healthcare", "rural healthcare", and/or "nursing homes". "Safety climate" was also included as a keyword to expand search results. The types of publication were limited to academic journals published from 1987 (when the National Advisory Committee on Rural Health and Human Services (NACRHHS) was chartered to provide expert recommendations to the Department of Health and Human Services) to February 2012.

After the initial search, title and abstracts of the articles were read and the following criteria were applied in including articles for further review: 1. written in English, 2. empirical studies focused on healthcare settings in western countries, 3.

targeted to employee safety, 4. relevant to safety culture or climate assessment, and/or 5. conducted at an organizational level (e.g. safety culture, safety climate, job stress, or job satisfaction) rather than focused on any one specific safety factor. On further review, a we read the full articles and developed a summary table of study specific information including the geographic locations (rural vs. urban), facility characteristics of the healthcare settings (e.g. amount of beds and number of nurses), study sample size, job titles of health workers, data collection methods (e.g. chart reviews, focus groups, and mail surveys), instrument development (i.e. self-developed or based on previous instruments), data analysis methods (e.g. descriptive statistics and factorial analysis) and study outcomes. In order to expand the search results, we also examined the references of the articles and selected those meeting the inclusion criteria for further review.

3 RESULTS

The initial search returned 214 articles. All articles titles and abstracts were reviewed. Based on the inclusion criteria previously defined, 21 articles were selected for full-article review. References from these articles were also selected for review based on the same inclusion criteria. At the final review, 35 articles met the selection criteria with 8 instruments identified that measured safety culture, safety climate, or employee perception of work safety in healthcare settings.

3.1 Instruments

The tools identified are the Job Content Questionnaire (JCQ: Karasek, 1985; Karasek & Theorell, 1990; Karasek et al., 1998), the Maslach Burnout Inventory (MBI: Maslach & Jackson, 1986), the Exposure to Disruptive Behaviors Instruments (EDBI: Middleton, Stewart, & Richardson, 1999), the Nursing Work Index-Revised (NWI-R: Aiken & Patrician, 2000), the Individual Workload Perception Scale (IWPS: Cox, 2003; Cox et al., 2006), and the Nursing Stress Scale (NSS: Gray-Toft & Anderson, 1981), the Nurse Stress Index (NSI: Harris, 1989). In addition, the Hospital Safety Climate Scale (HSCS) was included as it was developed to evaluate the safety climate in a hospital conducting a bloodborne pathogen risk management program (Gershon et al., 2000).

The JCQ is a highly reliable survey with 49 items and 5 subscales. The JCQ measures social and psychological characteristics of jobs by assessing the jobs' psychological demands, decision latitude, social support, physical demands, and job insecurity. These then allow identification of job-related illness and safety hazards (Karasek, 1985). The JCQ is one of the most widely used instruments among all the instruments identified in this review. It has been used across industries and has been translated for use in different countries (Karasek et al., 1998).

The MBI is another widely used tool across fields. This tool has 22 survey items with 3 subscales (emotional exhaustion, depersonalization, and personal accomplishment). Research findings suggested that employees who suffer from burnout firstly may experience exhaustion, then depersonalization and reduced personal accomplishment (Maslach, 1993). Cross-sectional results have deemed the MBI to be of satisfactory reliability (Pinikahana & Happell, 2004).

The EDBI is a 59 item survey with 3 subscales and measures long-term health-care providers' exposures to disruptive behaviors linked with job stress (Middleton et al., 1999). Assessments included staff perceptions of care giving relationships (23 items), disruptive behaviors (20 items), and job characteristics (16 items).

The NWI-R was developed by Aiken and Patrician (2000) and then later adapted by Krebs, Madigan, and Tullai-McGuinness (2008) to assess the organizational culture and nursing safety practice in hospitals. The NWI-R measurements involved manager support, nursing decision freedom, teamwork quality, staffing level, and opportunity for nurses to work in highly specialized units.

The IWPS also focuses on the nursing staff but is specific to pediatric facilities (Cox, 2003; Cox et al., 2006). The IWPS is a 32 item tool with 5 subscales. This tool has been used to assess nurse perceptions of management support, peer support, unit support, nurses' workloads, and their intent to stay.

The NSS is a 34 item survey developed by Gray-Toft and Anderson (1981). The NSS was designed to identify major sources of stress experienced by nurses. Based on the survey results, 7 stressors were identified including death and dying of patients, conflict with physicians, inadequate preparation, lack of support, conflict with other nurses, heavy workloads, and uncertainty concerning treatment.

Harris' (1989) NSI measures hospital nurses' perceptions of stress from six aspects. These stressors include managing workload, organizational support and involvement, dealing with patients and relatives, physical working conditions, home/work conflict, and confidence and competence in the role (1989). The weighted influence of each stressor was also assessed (Flanagan & Flanagan, 2002). It was found that time pressure and organizational support and involvement were the two major stressors. In addition, an inverse relationship between stress and job satisfaction was also identified, indicating that reducing job stress is critical in increasing nurses' job satisfaction.

Lastly HSCS is a survey with 20 items and 6 subscales (Gershon et al., 2000). The HSCS was developed to measure safety climate in a medical center. The safety climate within the working unit was found to be associated with employees' safe work practice and exposure incidents. The assessment was validated by its correlations between employees' safe work practices and their exposure to blood and other body fluids.

Seven additional instruments were identified in this review. These instruments include the Teamwork and Safety Climate Survey (Hutchinson et al., 2006), the Strategies for Leadership Survey (Pronovost et al., 2003), the Safety Climate Scale (Pronovost et al., 2003), the Hospital Unit Safety Climate (Blegen, Pepper, & Rosse, 2005), the Safety Climate Survey (Kho, Carbone, Lucas, & Cook, 2005), the Hospital Survey on Patient Safety Culture (AHRQ, 2003), and the Nursing Home Survey on Patient Safety Culture (AHRQ, 2011). As the focus of this review was on employee safety culture, these additional instruments were minimally included in this review.

3.2 Content Validity and Factor Structures

All 8 instruments conducted exploratory factor analysis to identify latent factors (i.e. subscales) before using these instruments in healthcare settings. Some of their factor structures were validated by subsequent studies that adapted the original instruments and conducted confirmatory factory analysis. Comparison across instruments identified overlapping factors. This comparison provides researchers' consensus on the factors that may influence employee safety culture or climate in healthcare. For instance, management support is measured in 5 of 8 instruments (JCQ, NWI-R, IWPS, NSI, and HSCS) to assess perception of safety showing evidence that a supportive management is a key condition to promote employee perception of safety (Karasek, 1985; Aiken & Patrician, 2000; Cox, 2003; Harris, 1989; Gershon et al., 2000).

Structural differences, however, are found between the hospital safety climate instruments (as in HSCS) and instruments used to measure nurses' perception of work safety (as in JCQ, MBI, EDBI, NWI-R, IWPS, NSS, and NSI). For example, nursing workload (assessed in JCQ, IWPS, EDBI, & NSI), nurse decision latitude (assessed in JCQ, EDBI, & NWI-R), and balance between family and work (assessed in NSI) overlapped in most instruments measuring nurse perception of work safety but were not included in instruments assessing safety climate in hospitals. The instruments measuring safety climate in hospitals were mainly focused on safe working practices and communication openness of medical errors as they relate to improving patient safety. This supports claims of a lack of balance in measuring safety culture or climate for patients in comparison to employees. In contrast, instruments assessing work safety, such as those that measure nursing workload and job stress, are often set apart from assessments of patient safety. In fact, no instruments measuring nursing perception of work safety (JCQ, EDBI, IWPS, NSS, and NSI) described herein have any items measuring patient safety.

3.3 Internal Reliability

Instrument internal reliability is often measured by Cronbach's alpha, which is an indicator of test reliability. All the studies describing the nursing safety instruments, as well as the Hospital Safety Climate Scale (HSCS) reported high internal reliability with a Cronbach's alpha values greater than 0.7.

3.4 Outcome Validity

Possible safety outcomes in healthcare include safe working practices, employee and patient injuries, medical errors, and other organizational outcomes such as litigation costs (Flin, 2007). Studies should examine the correlation between safety outcomes and the results from safety culture or climate assessments to evaluate whether the instruments used are reflective of the actual healthcare settings. Among instruments measuring hospital safety climate, we found only one instrument, the Hospital Safety Climate Scale (HSCS) that compared the safety climate scores with actual employee incidents reported in the targeted working areas. The comparison indicated that the higher the score measured, the safer the work practices and the less likely that blood or body fluid exposure were reported by the employees. For instruments measuring nurse perception of work safety, we found that only the Exposure to Disruptive Behaviors Instrument (EDBI) and the Nursing Stress Scale (NSS) had validated their measuring outcomes.

4 DISCUSSIONS

Overall, this review found only a few instruments that focus on employee perceptions of work safety in healthcare settings and no tools that measure employee safety culture for nursing homes. Additionally, this review identifies two major issues with the current tools. These include an interchangeable use of the safety culture and safety climate concepts, as well as a lack of instrument cross-sectional validity.

4.1 Concepts Differentiations: Safety Culture or Climate?

As concluded by Gershon et al. (2000) in their study of the Hospital Safety Climate Scale (HSCS), we also found that the terms and instruments used across the literature for measuring safety culture and safety climate are used interchangeably. This is especially true of the instruments that assess patient safety climate in hospitals. For instance, Pronovost et al. (2003) used the Safety Climate Scale (SCS) to evaluate the culture of safety.

Previous studies suggest that safety culture or climate measures in healthcare could take lessons from other industries, specifically in learning from aviation on how to measure safety culture or climate and manage risks (Lewis, Vaithianathan, Guyhirst, & Bagian, 2011; Pronovost et al., 2009). Aviation and healthcare settings are similar given that both provide services to human beings rather than producing materials (as in heavy industries) and that in both employees do not only interact with equipments but also cooperate with their teammates to serve people (Lewis et al., 2011; Pronovost et al., 2009). From this observation, we rely herein on definitions for healthcare safety culture and safety climate as supplied from an aviation study of aviation safety culture and safety climate.

Wiegmann, Zhang, Thaden, Sharma, and Mitchell (2002) define safety climate as the "temporal state of safety culture. It is therefore situationally based, refers to the perceived state of safety at a particular place at a particular time, is relatively unstable, and subject to change depending on the features of the current environment or prevailing conditions" (p. 10). Furthermore, Wiegmann et al. (2002) define safety culture as follows:

> Safety culture is the enduring value and priority placed on worker and public safety by everyone in every group at every level of an organization. It refers to the extent to which individuals and groups will commit to personal responsibility for safety, act to preserve, enhance and communicate safety concerns, strive to actively learn, adapt and modify (both individual and organizational) behavior based on lessons learned from mistakes, and be rewarded in a manner consistent with these values (p. 8).

As can be noted in the definitions provided, measures of safety culture and safety climate are vastly different. Therefore, using these terms interchangeably is misleading.

4.2 Cross-Sectional Validity

Organizational differences must be considered when adopting existing instruments across healthcare settings. According to Yule (2003), safety culture is embedded within an organization and represents the organization's behaviors. Most industrial organizations have highly hierarchical management structures with well-developed reporting systems as is apparent in aviation (Caves, 1980; Barach, 2000; U.S. Government Accountability Office, 2011). However, leadership clarification and report systems in healthcare settings are still problematic. These introduce ambiguity into the existing safety culture measurements (Scott-Cawiezell et al., 2004). Specifically, nursing homes rarely have the staff or the expertise in monitoring environmental workplace safety and quality of care, as nursing homes use traditional quality assurance models to monitor quality of care and address problems on an individual level instead of on a systems basis (Hughes & Lapane, 2006). Moreover, safety culture studies in industry are focused on the workers rather than the products, while healthcare safety culture is closely related to the outcomes, that is, the patients. Therefore, direct adoption of instruments developed by other industries may not be suitable for use in healthcare settings.

Currently, no instrument has been developed to specifically measure safety culture in rural nursing homes. Considerable care should be taken when directly adopting safety culture instruments from previous studies given the organizational differences between urban hospitals and medical centers and rural nursing homes. Most safety climate instruments identified in our review were initially developed for urban hospitals, in which the cross-level communications between nurses and doctors or physicians were measured as a key indicator of safety. However, as described previously, the majority of employees in nursing homes are nurses and nurses' aides. Physicians and doctors are rarely on site (Zimmerman & Antonova, 2007). Therefore, cross-level communications between nurses and doctors or physicians in nursing homes is minimal.

Furthermore, instruments measuring nurses' perception of work safety in urban hospitals consider low decision latitude of nurses as a key factor leading to low job satisfaction (Gardulf et al., 2005; Morgan, Semchuk, Stewart, & D'Arcy, 2002).

However, nurses in rural nursing homes play different roles in comparison to their peers in urban hospitals as they often act not only as nursing specialists but also as healthcare generalists (Lauder, Reynolds, Reilly, Angus, 2001; Ricketts, 2000). Nurses in rural nursing homes may need to make many decisions, daily and on their own, to care for their patients. From this standpoint, decision latitude as an indicator could reflect differing assessments of nurses' perception of work safety in rural nursing homes compared to urban hospitals. Moreover, as illustrated in section 3.2, nursing workload (assessed in JCQ, IWPS, EDBI, & NSI), nurse decision latitude (assessed in JCQ, EDBI, & NWI-R), and family-work confliction (assessed in NSI) were identified as safety stressors in nursing safety studies but were not measured in hospital safety climate instruments. For instance, the NSI (Harris, 1989) identifies work-family confliction as a job stressor for nurses. Specifically, for nurses working in small rural areas the closeness between the nurses, their patients, and their families, due to their knowing each other within their community could often lead to stress (Davis & Droes, 1993). However, such potential stressors are not investigated in the existing hospital instruments.

As evident, instruments measuring safety culture or climate in urban healthcare settings may misrepresent the organizational structures of rural nursing homes. These instruments may amplify factors that are not necessarily associated with safety culture or climate in nursing homes while ignoring factors that may truly influence nurses' perception of work safety in rural nursing homes. Therefore, direct adoption of instruments designed for urban hospitals is not recommend for use in measuring safety culture in rural nursing homes. Instruments aimed at assessing safety culture in rural nursing homes should be focused on the organization characteristics of rural nursing homes and identify casual factors associated with rural nurses' perception of work safety.

5 CONCLUSION

In this review, we found no existing tool that could be directly used for quantitative measurement of employee safety culture or safety climate in rural healthcare settings. Existing instruments have been developed for measuring safety culture or climate for urban hospitals, as well as some specific to specialized hospital units or medical centers. However, measuring safety culture may be significantly influenced by factors such as the type of healthcare center, the associated job titles for the healthcare employees, and rurality. Some factors have been identified that relate to nurses' perception of work safety in rural nursing, but there is still a lack of comprehensive measurement to identify casual factors as well as their influences on safety culture for rural nursing homes. Therefore, effective instruments need to be developed, based on which data could be collected to investigate factors and their influences, and evaluated over time.

REFERENCES

Agency for Healthcare Research and Quality. (2003). Hospital Survey on Patient Safety. Retrieved from: http://www.ahrq.gov/qual/patientsafetyculture/hospscanform.pdf

Agency for Healthcare Research and Quality. (2011). Nursing Home Survey on Patient Safety. Retrieved from: http://www.ahrq.gov/qual/nhsurvey08/nhsurvey.pdf

Aiken, H., Clarke, P., & Sloane, M. (2000). Hospital staffing, organization, and quality of care: cross-national findings. *Nursing Outlook*, 50(5): 187-194.

Australian National Health Performance Committee. (2001). Retrieved from: http://www.aihw.gov.au/sqhc-definitions/.

Barach, P. & Small, D. (2000). Reporting and preventing medical mishaps: lessons from non-medical near miss reporting systems, *BMJ,* 320(7237): 759–763.

Blegen, A., Pepper, A. & Rosse, J. (2005). Safety climate on hospital units: A new measure. *Advances in Patient Safety*, 4: 429 – 433.

Bonner, F., Castle, G., Perera, S., & Handler, M. (2008).Patient safety culture: A review of the nursing home literature and recommendations for practice. *Annals of Long Term Care: Clinical Care and Aging*, 16: 18–22.

Castle, N. & Sonon, K. (2006). A culture of patient safety in nursing home, *Qal Saf Health Care*, 15: 405-408.

Caves, E. (1980). Industrial Organization, Corporate Strategy and Structure, *Journal of Economic Literature*, 18 (1): 64-92.

Colello, J. (2007). Where Do Older Americans Live? Geographic Distribution of the Older Population, *CRS report for Congress*. Retrieved November 5, 2011, from http://aging.senate.gov/crs/aging5.pdf.

Colla, B., Bracken, C, Kinney, M., & Week, B. (2005). Measuring patient safety climate: a review of surveys, *Qual Saf Health Care*, 14: 364–366.

Cox, S. (2003). *Individual work environment perception scale user's manual.* Kansas City, MO: Children's Mercy Hospitals and Clinics.

Cox, S., Teasley, L., Zeller, A., Lacey, R., Parsons, L., Carroll, A., & Ward-Smith, S. (2006). Know staff's "intent to say." *Nursing Management*, 37: 13-15.

Davis, J. & Droes, S. (1993). Community health nursing in rural and frontier counties. *Nursing Clinics of North America*, 28 (1), 159–169.

Ewart, W., Marcus, L., Gaba, M., Bradner, H., Medina, L., & Chandler, B. (2004). The critical care medicine crisis: a call for federal action: a white paper from the critical care professional societies, *Chest*, 125: 1518–1521.

Flanagan, A. & Flanagan, J. (2002). An analysis of the relationship between job satisfaction and job stress in correctional nurses, *Research in Nursing & Health*, 25: 282–294.

Gardulf, A., Söderström, A.-L., Orton, M.-L., Eriksson E., Arnetz, B. & Nordström, G. (2005). Why do nurses at a university hospital want to quit their jobs? *Journal of Nursing Management*, 13, 329–337.

Gershon, R., R, Karkashian, D., Grosch, W., Murphy, R., Escamilla-Cejudo, A., Flanagan, A., Bernacki, E., Kasting, C., & Martin, L. (2000). Hospital safety climate and its relationship with safe work practices and workplace exposure incidents, *American Journal of Infection Control*, 28(3):211-221.

Gray-Toft, P., & Anderson, G. (1981). The nursing stress scale: Development of an instrument, *Journal of Behavioral Assessment*, 3(1): 11-23.

Goodin, J., (2003). The nursing shortage in the United States of American: an integrative review of the literature, *Journal of advanced nursing,* 43(4): 335-343.

Harris, E. (1989). The nurse stress index, *Work & Stress,* 3(4): 335-346.

Hughes, M. & Lapane, L. (2006). Nurses' and nursing assistants' perceptions of patient safety culture in nursing homes, *International Journal for Quality in Health Care*, 18(4): 281–286.

Hutchinson, A., Cooper, L., Dean, E., McIntosh, A., Patterson, M., Stride, B., Laurence, E., & Smith, M. (2006). Use of a safety climate questionnaire in UK health care: factor structure, reliability and usability. Qual Saf Health Care. 15(5):347-353.

Karasek, A. (1985). *Job Content Questionnaire and user's guide.* Lowell: University of Massachusetts Lowell, Department of Work Environment.

Karasek, A. & Theorell, T. (1990). Healthy work: Stress, productivity, and the reconstruction of working life. New York: Basic Books.

Karasek, A., Theorell, T., Schwartz, J., Schnall, L., Pieper, F., & Michela, L. (1998). Job characteristics in relation to the prevalence of myocardial infarction in the US Health Examination Survey (HES) and Nutrition Examination Survey (HANES). *American Journal of Public Health*, 78: 910-918.

Kho, E., Carbone, M., Lucas, J., & Cook, J. (2005). Safety Climate Survey: reliability of results from a multicenter ICU survey, *Qual Saf Health Care*, 14(4): 273–278.

Krebs, P., Madigan, A., & Tullai-McGuinness, S. (2008). The rural nurse work environment and structural empowerment, Policy Politics Nursing Practice, 9: 28-39.

Lauder, W., Reynolds, W., Reilly, V., & Angus, N. (2001). The role of district nurses in caring for people with mental health problems who live in rural settings, *Journal of Clinical Nursing*, 10(3): 337-344.

Leeflang, G., Deeks,J., & Gatsonis, C. (2008). Systematic Reviews of Diagnostic Test Accuracy, *Annals of Internal Medicine*, 149(12): 889-897.

Lewis, H., Vaithianathan, R., Hockey, M., Hirst, G., Bagian, P. (2011). Counterheroism, common knowledge, and ergonomics: concepts from aviation that could improve patient safety, *The Milbank Quarterly*, 89 (1): 4–38.

Maslach, C., & Jackson, E. (1986). *Maslach Burnout Inventory manual* (2nd ed.). Palo Alto, CA: Consulting Psychologists Press.

Maslach, C. (1993). Burnout: A multidimensional perspective. In W. Schaufeli, C. Maslach, & T.Marek (Eds.), Professional Burnout: Recent Developments in Theory and Research (pp. 19–32).Washington, DC: Taylor & Francis.

Mearns, J. & Flin, R. (1999). Assessing the state of organizational safety-culture or climate? *Current Psychology: Developmental, Learning, Personality*, 18: 5–17.

Middleton. J., Stewart, N., & Richardson, S. (1999). Caregiver distress related to disruptive behaviors on special care units vs. traditional long-term care units. *Journal of Gerontological Nursing,* 25: 11–19.

Nieva, F. & Sorra, J. (2003). Safety culture assessment: a tool for improving patient safety in healthcare organizations, *Qual Saf Health Care*, 12: 17–23.

Pinikahana, J. & Happell, B. (2004). Stress, burnout and job satisfaction in rural psychiatric nurses: A Victorian study, Aust. J. Rural Health, 12: 120–125.

Pronovost, J., Weast, B., Holzmueller, G., Rosenstein, J., Kidwell, P., Haller, B., Feroli, R., Sexton, B., & Rubin, R. (2003). Evaluation of the culture of safety: survey of clinicians and managers in an academic medical center, *Qual Saf Health Care*, 12(6):405-410.

Pronovost, P. & Sexton, B. (2005). Assessing safety culture: guidelines and recommendations, *Qual Saf Health Care*, 14: 231-233.

Pronovost, J., Goeschel, A., Olsen, L., Pham, C., Miller, R., Berenholtz, M., Sexton, B., Marsteller, A., Morlock, L., Wu, W.,Loeb, M., & Clancy, M. (2009). Reducing health care hazards: lessons from the commercial aviation safety team, *Health Aff,* 28(3): 479–489.

Ricketts, C. (2000). The changing nature of rural health care, *Annu. Rev. Public Health*, 21:639–657.

Sexton, B., Helmreich, L., Neilands, B., Rowan, K., Vella, K., Boyden, J., Roberts, R., & Thomas, J. (2006). The Safety Attitudes Questionnaire: psychometric properties, benchmarking data, and emerging research, *BMC Health Services Research*, 6: 44.

Scott-Cawiezell, J., Schenkman, M., Moore, L., Vojir, C., Connolly, P., Pratt, M., & Palmer, L. (2004). Exploring nursing home staff's perceptions of communication and leadership to facilitate quality improvement, *J Nurs Care Qual,* 19(3): 242-252.

Steinwachs, M., & Hughes, G. (2008). Health Services Research: Scope and Significance. In: R. G. Hughes (Ed.). *Patient Safety and Quality: An Evidence-Based Handbook for Nurses.* Retrieved Feb 07, 2012, from http://www.ncbi.nlm.nih.gov/books/NBK2660/.

U.S. Government Accountability Office. (2010). Report to Congressional Requesters: Design and Implementation Considerations for Safety Reporting Systems. Retrieved from: http://www.gao.gov/assets/310/309358.pdf.

U.S. Bureau of Labor Statistics. (2011). *Occupational injury and illness classification manual Career Guide to Industries: Healthcare*. Retrieved from http://www.bls.gov/oco/cg/cgs035.htm.

U.S. Social Security Administration. (2003). Requirements for, and assuring quality of care in, skilled nursing facilities. In The Social Security Act as amended with related enhancements (Vol. 1, Sec. 1819) [42 U.S.C. 1395i-3].

CHAPTER 76

Ergonomics and Occupational Health and Safety: a Cost-Benefit Analysis Model

Ramos, D. G., Arezes, P., Afonso, P.

University of Minho, Department of Production and Systems
Guimaraes, Portugal
gramos@det.uminho.pt

ABSTRACT

The economic costs of occupational safety and health problems place a considerable burden on the competitiveness of both the public entities and the enterprises. Ergonomic problems at the workplace and bad work organization are part of the contributing risk factors to the abovementioned occupational safety and health problems. There are also considerable social costs of workplace accidents related to ergonomic and work organization risk factors.

This paper aims to propose a model for the application of Cost-Benefit Analysis in Occupational Health and Safety, taking into account namely ergonomic aspects involved. A previous qualitative exploratory study was undertaken using the Delphi methodology to get some input from an expert panel in order to determine the most important factors which should be included in the model of cost-benefit analysis. This model permits to perform economic evaluations of risks and prevention initiatives from both the company and the society perspectives. It is an important tool to support managers and experts to make an economic analysis before the beginning of any ergonomic intervention project.

Keywords: Ergonomics, Occupational Health and Safety, Cost-Benefit Analysis Model

1 INTRODUCTION

According to ISO Guide 73:2009 (Risk Management – Vocabulary), the process of risk management consists on the systematic application of management policies, procedures and practices to the activities of communicating, consulting, establishing the context, and identifying, analyzing, evaluating, treating, monitoring and reviewing risk. ISO 31000:2009 indicates the phases of the process of risk management: a) communication and consultation, b) establishing the context, c) risk assessment (includes risk identification, risk analysis and risk evaluation), d) risk treatment, e) monitoring and review.

Communication and consultation with external and internal stakeholders should take place during all stages of the risk management process. When establishing the context, the organization articulates its objectives and defines the external and internal parameters to be taken into account in risk management. Risk assessment is an overall process of risk identification, risk analysis and risk evaluation. Risk treatment involves selecting one or more options for modifying risks, and implementing those options. Both monitoring and review should be a planned part of the risk management process and involve regular checking or surveillance. It can be periodic or ad hoc. Responsibilities for monitoring and review should be clearly defined.

ISO/IEC 31010 (2009) indicates the applicability of techniques and tools used in risk management and assessment, and includes namely cost/benefit analysis (CBA). This technique is very well fitted to occupational risk management, as the decisions concerning the measures to adopt following a risk assessment need to be carefully based on a cost/benefit analysis not only in terms of the particular company but also in terms of the impact all the stakeholders and especially on the society. CBA is also very useful in the evaluation of an ergonomic program (Karwowski and Warras, 2003).

This study aims developing a model for cost/benefit analysis in occupational health and safety.

2 COST/BENEFIT ANALYSIS IN OCCUPATIONAL HEALTH AND SAFETY

Carrying out an occupational risk assessment it is necessary to take into account the associated costs and benefits. However, only a cost-benefit analysis (CBA) can capture all impacts resulting from work accidents and/or from prevention measures regarding Occupational Health and Safety (OHS). The CBA is used to determine whether a project is feasible, from the standpoint of social welfare, by the sum of the costs and benefits, discounted over time (Cullis and Jones, 2009). Firstly, from the company's perspective, it is necessary to consider costs and revenues which result from the implementation of each particular measure identified in an OHS exercise. Nevertheless, this financial evaluation of the measures is not enough. It is also important to understand the real contribution of each solution to society. Such economic evaluation includes the analysis of the related externalities (some positive and other negative to society).

Cost/benefit analysis can be used for risk evaluation where total expected costs are weighed against the total expected benefits in order to choose the best or most profitable option. It is an implicit part of many risk evaluation systems and of an ergonomic program. It can be qualitative or quantitative or involve a combination of quantitative and qualitative elements. Quantitative CBA aggregates the monetary value of all costs and all benefits to all stakeholders that are included in the scope and adjusts for different time periods in which costs and benefits accrue. The Net Present Value (NPV) which is produced becomes an input to decisions about risk. A positive NPV associated with an action would normally mean the action should occur. However, for some negative risks, particularly those involving risks to human life or damage to the environment the ALARP (*as low as reasonably practicable*) principle may be applied. This system divides risks into three regions: a level above which negative risks are intolerable and should not be taken except in extraordinary circumstances; a level below which risks are negligible and need only to be monitored to ensure they remain low; and a central band where risks are made as low as reasonably practicable. Towards the lower risk end of this region, a strict cost/benefit analysis may apply but where risks are close to intolerable, the expectation of the ALARP principle is that treatment will occur unless the costs of treatment are grossly disproportionate to the benefit gained.

Benefit/Cost ratio is also a simple method that calculates the ratio between the benefits of a project and the cost to implement it. It can be applied to projects related to occupational health and safety and ergonomics. The idea is that an ergonomics program that pays for itself is a good investment. According to Karwowski and Warras (2003), the economic benefit of an ergonomic solution can be calculated by looking at the cost of injuries associated with ergonomic problems, based on the assumption that implementing an ergonomic program will generate solutions, which by its turn will prevent future injuries.

3 DELPHI STUDY

3.1 Principles of Delphi method

The Delphi method was developed by Dalkey and Helmer (1963) of RAND Corporation in 1950 for a project sponsored by the U.S. military. It was created as part of a post-war movement in order to predict the possible effects of technological development in economic and social regeneration. The word Delphi refers to the sacred site of the most revered oracle in ancient Greece, where forecasts and advice was sought from the gods through intermediaries in this oracle.

The Delphi methodology is an exploratory study that allows gathering the views of the considered participants, typically a panel of experts on the study domain, which is called the Delphi panel. This process is carried out by conducting a series of questionnaires, or rounds, on a specific subject (Ramos et al., 2011a, 2011b).

Thus, a selected group of experts is asked in successive rounds anonymous and with maximum autonomy of the participants, with the goal of reaching consensus

(Linstone and Turrof, 1975; Landeta, 1999; Vergara, 2005). In this research method the results depend strongly on the quality of the questionnaire and selection of experts. Some of the characteristics of the Delphi methodology are the anonymity of the participants, the statistical representation of the distribution of results and the use of the feedback from the group to review the answers in a later round.

According to Skulmoski et al. (2007), the process can be considered as concluded when the answers are near the consensus, according to appropriate statistical methods.

3.2 Development of Delphi study

In this study, the subject to be examined is the application of "Cost-Benefit in OHS".

The questionnaire included a total of 51 questions divided into 5 sections: 1) Occupational Risk Assessment; 2) Costs and Benefits Analysis of Occupational Risks; 3) Financial Assessment; 4) Economic Evaluation and 5) Externalities.

For each question, the expert could choose the answer in a scale 1-5 (1=very low, 2=low, 3=medium, 4=high, 5=very high); the possibility of answering "no opinion" was also available. The variables studied are discrete, categorical and qualitative of ordinal type. In each section there was also a question about the degree of familiarity of the expert with each section (with three options. low, intermediate, high). The detailed structure of the questionnaire has been presented in a previous publication (Ramos et al., 2011a).

The following statistical parameters have been obtained: median, mode and interquartile range. The interquartile range (IQR) is a measure of the dispersion and the median is the obtained value for 50% of observations. Thus, an IQR of 1 or less means that more than 50% of all opinions fall on a given point on the scale (von der Gracht and Darkow, 2010).

An initial expert panel has been established, which included a total of 29 experts, including 13 Academics, 8 OHS Professionals/Technicians and 8 OHS Consultants/Auditors. All of them were initially contacted. From the initial 29 experts, 23 of these experts have confirmed their interest to participate in the study. The first Delphi round started in September 2011. The successive questionnaires were used to reduce the "interquartile interval", a measure of the deviation of the opinion of an expert from the opinion of the whole panel (median).

3.3 Results of the Delphi study

In the first round we have had the participation of 20 experts. The results showed a good agreement (IQR≤1) in 34 of the 51 questions. In the second round we have had the participation of 19 experts. The results showed a good agreement (IQR≤1) in 39 of the 51 questions. In the third round we have had the participation of 14 experts. The results showed a good agreement (IQR≤1) in 43 of the 51 questions. The evolution of the opinion of the panel along the three rounds is presented on Table 1.

Table 1 – Consensus Criterion

Section	IQR ≤ 1 in Round 1	IQR ≤ 1 in Round 2	IQR ≤ 1 in Round 3
1. Occupational Risk Assessment	50.0%	58.3%	66.7%
2. Costs and Benefits Analysis of Occupational Risks	60.0%	80.0%	100.0%
3. Financial Assessment	83.3%	83.3%	100.0%
4. Economic Evaluation	75.0%	83.3%	75.0%
5. Externalities	72.7%	81.8%	90.9%

In order to find the most relevant topics to be included in the Cost/Benefit analysis model, the following methodology has been adopted. We considered the questions that the expert panel found extremely important (median equals to 4.5 or 5) with IQR=1 (9 questions have been selected) and also the questions that the expert panel found important (rating 4) but within a very high consensus (IQR=0) (further 6 questions have been selected).

4 A MODEL FOR COST/BENEFIT ANALYSIS IN OCCUPATIONAL HEALTH AND SAFETY

The model CBAOHS (Cost/Benefit Analysis in Occupational Health and Safety) that has been developed has considered the results of the Delphi study, taking into account the importance of each item in terms of the degree of relevance and consensus obtained.

Firstly, in a CBAOHS model, accident costs should be separated in terms of direct and indirect costs. Particularly, following experts' opinion, it should be made a "rigorous and comprehensive assessment of the direct costs of workplace accidents".

Beyond the analysis of accident costs, such a model should permit to evaluate and compare preventive measures, which often involve ergonomic actions. The costs of costs of preventive measures which could be undertaken, measured in monetary units, are balanced with the tangible benefits of preventive measures (possible productivity improvements and costs avoided both for the company and the society).

The experts' panel recognized the relevance of a CBA for the evaluation of occupational hazards from the company's perspective as well as to assess the impact of its implementation from the standpoint of social welfare. Thus, also intangible costs should be estimated. Particularly, the experts' panel emphasized the "implications in the family stability", "impact in terms of morbidity and mortality", and the "the use of measures expressed in terms of quality adjusted life years (QALY)".

Finally, responses highlighted that "the CBA can be used to promote, support, subsidize or to legislate in a reasonable manner the measures of prevention of occupational hazards". Indeed, translating risks and prevention measures in terms of firms', employees' and society costs may be used to support the reduction of the negative externalities through public measures such as taxes, fines, legislation, among others. Responses showed a high consensus on the direct calculation of the externalities based on the caused damages. For prevention measures with benefits in several years they should be considered all "future cash flows" for a correct evaluation of these measures.

The first step in a CBA model for OHS is to keep the reports on occupational accidents. There are several types of accidents. Simonds and Grimaldi (1956), for example, separated them into four groups: accidents with lost time, accidents without lost time, accidents requiring first aid and accidents requiring a doctor's intervention. From these records, a list of the most common or important hazards and causes of accidents is identified.

The reduction of the impact of occupational accidents on humans is the most important reason to support the investment on prevention actions but economic reasons are better explanations for such investment (Ale et al., 2008). Jallon et al. (2011b) state that to convince a firm to invest in measures for accident prevention, the impact of injuries, must be expressed in figures. Managers understand the language of costs because they reduce their profits. On the other hand, the accident-cost calculation must be accurate enough to allow him to justify a health and safety prevention related investment.

For example, in the Netherlands, cost effectiveness is used to reduce risk after a quantitative analysis of risk (Jongejan et al., 2006). We propose an eight-step methodology for the economic analysis of occupational risks and related prevention measures, as shown in Table 2. Firstly, to compute the expected value of accidents costs (steps 1 to 4); secondly, to perform a cost-benefit analysis of prevention measures (steps 5 to 8).

Table 2 – Eight step methodology

Cost of accidents	CBA of preventive measures
1. Identification of hazards and causes of accidents 2. Identification of relevant costs 3. Measurement of costs 4. Computation of costs (expected value)	5. Identification of prevention measures 6. Computation of investment costs 7. Estimation of cost reductions (expected value) 8. Cost-benefit analysis

In most developed countries, the direct costs of occupational injuries and illnesses are covered by private or government insurance, which compensates the victim and covers medical and hospitalization expenses. Employers pay an annual premium for this insurance, which, as a rule, reflects the risks posed by the company's activities. This annual premium is the only clearly identifiable expense linked to poor working conditions in the company.

Beyond the traditional classification of occupational accident costs in "visible" or direct costs versus "invisible" or hidden components (Gosselin, 2004), external costs to the company which affect both employees and the society, should also be considered. Direct and indirect costs to the company (e.g. absence from work), as well as external costs to the worker and the society (e.g. medical costs) justify prevention measures (Arbeidsinspectie, 2002).

The identification and measurement of relevant costs is important because global accidents' cost is much higher than direct costs to the companies (e.g. compensations, medical care and new equipment). For Gavious et al. (2009), the real challenge is to develop a reliable evaluation of indirect costs, which are for example, the cost of time spent in relation to medical care, the reduced production of the injured worker after returning to work, the cost of supervision and investigation, the overall reduced production, the cost of replacement, etc. (LaBelle, 2000).

Indirect costs of accidents are also defined as "hidden" costs, which are not recognized by employers because they are not captured by the company's accounting system (Jallon et al., 2011a). The inclusion of these costs in a cost-benefit analysis of prevention initiatives is important to justify such interventions. However, these costs are hard to identify and measure and little information on indirect costs is available (see Sun et al., 2006).

According to Jallon et al. (2011a), three main approaches can be followed to collect and compute indirect costs: top-down, bottom-up and local. The "top-down" approach uses aggregated statistics and previous studies to calculate indirect average costs by accident or injury (e.g. Biddle, 2004). The "bottom-up" approach is based on surveys or interviews in organizations to estimate the average cost by accident type or industry sector (e.g. Rikhardsson and Impgaard, 2004). Finally, the "local" approach is based on local (in company) data collection to obtain an indirect cost that is company specific (LaBelle, 2000). The decision-maker needs to strike an appropriate balance between time spent on data collection and the level of precision required for the indirect cost calculation.

Furthermore, there are also external costs to the company that should be also addressed, namely costs to the injured worker and family (e.g. income reduction) and other costs to society (e.g. public medical care).

Total costs are calculated as the sum of the costs that are internal to the firm (including direct and indirect costs) and the costs that are external to the firm (including costs to the worker and family and costs to society).

Following Jallon et al. (2011b), a cost-benefit analysis should permit to answer to the following questions: What OHS investments should be made? How much should be spent on preventive measures? When should we make a given investment? What business value can we expect from a prevention investment? Only a cost-benefit analysis supports the complete assessment of a prevention investment.

The second part of the presented eight-step methodology is the CBA of the selected or alternative preventive measures. Assuming that costs and benefits will be spread in time, initial investment should be compared with future benefits (costs avoided) and eventually with future additional costs (e.g. maintenance and replacement costs). Future benefits and costs need to be updated using an appropriate discount rate, so that a valid comparison can be obtained. Thus, all costs and benefits must be expressed in present value. The present value of all costs and benefits can be combined to produce a Net Present Value (NPV). A positive NPV means that the prevention measure is beneficial, which means that the income generated outweigh the costs incurred. Eventually, a negative NPV limited by an allowable maximum could be acceptable for the company.

From the standpoint of view of the society, it is expected an economic net present value. In this case, the (social) discount rate that should be considered is lower than the (financial) discount rate for the firm. Following the opinion of the experts' panel, it will be not relevant to compute the Internal Rate of Return (IRR) and the payback period.

Thus, the economic analysis of the preventive measures will be expressed in terms of NPV or as a ratio between the present value of benefits and the present value of costs (Benefit-Cost Ratio: BC). These indicators for both the firm and the society standpoint of view are presented in Table 3.

Table 3 – NPV and BC Ratio from the firm and the society perspective

Indicators	Description
Net Present Value (financial)	$-Investment+\sum_{t=1}^{n}\frac{DirectCostsavoided}{(1+i)^t}+\sum_{t=1}^{n}\frac{IndirectCostsavoided}{(1+i)^t}+\sum_{t=1}^{n}\frac{Otherbenefits}{(1+i)^t}-\sum_{t=1}^{n}\frac{O\&MCosts}{(1+i)^t}$
Net Present Value (economic)	$-Investment+\sum_{t=1}^{n}\frac{NetInternalCostsavoided}{(1+i)^t}+\sum_{t=1}^{n}\frac{WorkerCostsavoided}{(1+i)^t}+\sum_{t=1}^{n}\frac{SocietyCostsavoided}{(1+i)^t}$
Benefit-Cost Ratio (financial)	$\frac{\sum_{t=1}^{n}\frac{Costsavoided}{(1+i)^t}+\sum_{t=1}^{n}\frac{Otherbenefits}{(1+i)^t}}{Investment+\sum_{t=1}^{n}\frac{O\&MCosts}{(1+i)^t}}$
Benefit-Cost Ratio (economic)	$\frac{\sum_{t=1}^{n}\frac{NetInternalCostsavoided}{(1+i)^t}+\sum_{t=1}^{n}\frac{ExternalCostsavoided}{(1+i)^t}}{Investment+\sum_{t=1}^{n}\frac{O\&MCosts}{(1+i)^t}}$

* with i: discount rate, t: year, n: lifetime of the impact of the preventive measure.

This model is currently being tested with data from a case study involving a hospital. Results obtained within this case study will soon be presented. This will allow addressing eventual limitations of the model.

5 CONCLUSIONS

This paper proposed an approach and a model for the application of Cost Benefit Analysis (CBA) in Occupational Health and Safety (OHS). This model permits to perform economic evaluations of risks and prevention initiatives (including namely an ergonomic program) from both the company and society perspectives. Nevertheless, like all the techniques and tools, CBA has strengths and limitations (ISO/IEC 31010:2009).

The most important and relevant strengths of CBA are: a) it allows costs and benefits to be compared using a single metric (money); b) it provides transparency of decision making; c) it requires detailed information to be collected on all possible aspects of the decision.

The relevant limitations of this approach are: a) quantitative CBA can yield dramatically different numbers, depending on the methods used to assign economic values to non-economic benefits; b) in some applications it is difficult to define a valid discount rate for future costs and benefits; c) benefits which accrue to a large population are difficult to estimate, particularly those relating to public goods which are not exchanged in markets; d) the practice of discounting means that benefits gained in the long term have negligible influence on the decision depending on the discount rate chosen. In fact, the method becomes inappropriate for consideration of risks affecting future generations unless very low or zero discount rates are set.

ACKNOWLEDGMENTS

The authors would like to acknowledge all the experts that have participated in the Delphi panel.

REFERENCES

Ale, B.J.M., Baksteen, H., Bellamy, L.J., Bloemhof, A., Goossens, L., Hale, A., Mud, M.L., Oh, J.I.H., Papazoglou, I.A., Post, J., Whiston, J.Y. (2008). Quantifying occupational risk: The development of an occupational risk model. *Safety Science*, 46, 176-185.

Arbeidsinspectie (Labour Inspectorate). 2002. *Naar een risicomodel arbeidsongevallen* (Towards a risk model for occupational accidents). Ministry of Social Affairs & Employment, Labour Inspectorate, The Hague.

Biddle, E. (2004). The economic cost of fatal occupational injuries in the United States. *Contemporary Economic Policy*, 22, 370 –381.

Cullis, J., Jones, P. (2009). *Public Finance & Public Choice: Analytical Perspectives*. Third edition. Oxford University Press. Oxford. UK.

Dalkey, N. C. & Helmer, O. (1963). An experimental application of the Delphi Method to the use of experts. *Management Science*, 9(3), 458 - 468.

Gavious, A., Mizrahi, S., Shani, Y., & Minchuk, Y. (2009). The costs of industrial accidents for the organization: Developing methods and tools for evaluation and cost-benefit analysis of investment in safety. *Journal of Loss Prevention in the Process Industries*, 22(4), 434– 438.

Gosselin, M. (2004). Analyse des avantages et des coûts de la santé et de la sécurité au travail en entreprise: Développement de l'outil d'analyse. Rapport de recherche R-375. Montréal, QC: IRSST.

ISO Guide 73 (2009). Risk management – Vocabulary.

ISO 31000 (2009). Risk management – Principles and guidelines.

ISO/IEC 31010 (2009). Risk management – Risk assessment techniques.

Jallon, R., Imbeau, D., Marcellis-Warin, N. (2011a). Development of an indirect-cost calculation model suitable for workplace use. *Journal of Safety Research*, 42, 149–164.

Jallon, R., Imbeau, D., Marcellis-Warin, N. (2011b). A process mapping model for calculating indirect costs of workplace accidents. *Journal of Safety Research*, 42, 333–344.

Jongejan, R.B., Ale, B.J.M., Vrijling, J.K. (2006). FN-criteria for risk regulation and probabilistic design. In: Proceedings of Probabilistic Safety Analysis and Management (PSAM8). New Orleans, 14–19 May. ASME. New York.

Karwowski, W. Marras W. 2003. *Occupational ergonomics: design and management of work systems*. ISBN 0-8493-1801-7. CRC Press. New York.

LaBelle, J. E. (2000). What do accidents truly cost? Determining Total Incident Costs. *Professional Safety*, 45 (4), 38 – 42.

Linstone, H., Turrof, M. (1975). *The Delphi method, techniques and applications*. Addison Wesley Publishing.

Ramos, D. Arezes, P. Afonso, P. (2011a). Externalidades em segurança ocupacional: a importância da análise custo/benefício. SHO 2011 – International Symposium on Occupational Safety and Hygiene. University of Minho – School of Engineering. Guimarães, 10 and 11 February 2011. Conference proceedings ISBN: 978-972-99504-7-6, pg 534-538.

Ramos, D. Arezes, P. Afonso, P. (2011b). Cost-benefit analysis in occupational health and safety. Oral presentation at ICOPEV 2011 – International Conference on Project Economic Evaluation. University of Minho – School of Engineering. Guimarães, 28 and 29 April 2011. Conference proceedings ISBN: 978-989-97050-1-2, pg 251-258.

Rikhardsson, P. M., & Impgaard, M. (2004). Corporate cost of occupational accidents: an activity-based analysis. *Accident Analysis & Prevention*, 36(2), 173– 182.

Simonds, R. H., & Grimaldi, J. V. (1956). *Safety management*. Homewood, IL: Irwin.

Skulmoski, J. G., Hartman, T. F., Krahn, J. (2007). The Delphi Method for Graduate Research. *Journal of Information Technology Education*, volume 6.

Sun, L., Paez, O., Lee, D., Salem, S., & Daraiseh, N. (2006). Estimating the uninsured costs of work-related accidents, part I: a systematic review. *Theoretical Issues in Ergonomics Science*, 7(3), 227 –245.

Vergara, S. (2005). Métodos de Pesquisa em Administração. São Paulo. Editora Atlas.

von der Gracht, H. A., Darkow, I.-L. (2010). Scenarios for the logistics services industry: A Delphi-based analysis for 2025. *Int. J. Production Economics*, 127, 46 –59.

CHAPTER 77

Developing Interactive and Emergency Response Devices for People with Disabilities and Their Canine Assistants

Chakris Kussalanant, John Takamura, Dosun Shin, Winslow Burleson

Arizona State University
Tempe, USA
Chakris.Kussalanant@asu.edu

ABSTRACT

This study explores the extent to which we can trust and rely on dogs to use technology in order to perform critical behaviors that enhance health, safety, and well-being. Through the use of such technologies, humans and dogs foster a more robust and supportive environment than either assistive technologies or canine assistants can provide alone. We implement a user-centered approach focused on natural observation, respondent field survey, scenario-based design prototyping, and ethnographic case studies, including interviews and photo journals as key methods. This approach presents opportunities for further advances in human-animal interaction, new strategies for advancing assistive technologies, and richer human computer interaction (HCI) experiences. These scenarios are increasingly inclusive of dogs as users of and contributors to more fulfilling interactions with technology.

Keywords: Animal Computer Interaction, Assistive Technology, Pets, Mobile Systems, Smart Homes.

INTRODUCTION

In the few instances in which researchers analyzed or modeled technologies after dogs, it has largely been in the context of developing a robotic analogue or substitute (Hamblong, 2006; Melson, 2006; Nguyen & Kemp, 2008). For example, research has focused on replacing the mechanical functions of an assistive dog with a robotic arm programmed with door opening abilities, (Nguyen & Kemp, 2008) assessing to what extent Sony's AIBO robot can substitute as a pet (Kerepesi, et. al. 2006; Melson, 2009). Such investigations have yet to articulate the notion of service canines as a delivery agent for technological services that improve their owner's capabilities in everyday activities. Similar efforts that attempt to mimic human social interactions via investigations of human robot and human-virtual agent interactions face ongoing challenges, most notably the awkward interactions of "near-human" encounters, known as the uncanny valley effect (Mori, 1970; 2005).

This paper explores opportunities for HCI that involve dogs taking on a pivotal role in technologically enhanced services to people with a wide range of disabilities. It helps us understand to what extent we can trust and rely on dogs to perform critical behaviors that enhance health, safety, and well-being. It also presents and informs the design of technologies that can enhance human dog interactions, e.g., through systems that empower humans and dogs to use remote and digitally generated signals and reactions, particularly when it is difficult or impossible for them to initiate these. In the process, it identifies multiple real world scenarios that benefit from the affordances provided by canine assistants and mobile technology.

Investigations of animals interacting with technology and human-animal interaction mediated by technology is a nascent research area (Mancini, 2011; Noz & An, 2011; Paldanius et. al. 2011; Weilenmann & Juhlin, 2011). Paldanius et. al. 2011) explored the use of GPS technologies by hunters. They observed the behavior of dogs and hunters as well as the use of several different devices intended to help hunters keep track of their dogs over long distances. These observations led to ideation and sketches of scenarios for human-dog interactions, including the tracking of family exercise using a mobile device and remote feeding of the dog via internet and webcams (Paldanius et. al. 2011). Noz and An developed Cat Cat Revolution, a game of cat and mouse for the iPad that explores how cats use this multimodal ubiquitous computing interface. Their study raises the issue of applying HCI principles to pets and presents animals as participants in digital games (Noz & An, 2011). Mancini's *Animal-Computer Interaction: a Manifesto* makes a compelling case of how humans have applied technology to capitalize on the behaviors of animals, whether to experiment with rats, milk cows, or track endangered species over long ranges (Mancini, 2011).

Given the variety of shared behaviors and benefits dogs afford humans, it stands to reason that humans would consider ways to model assistive technologies based on the innate behaviors of canines and allow these ancient companions to become users of our technology, not just as an exercise to further test dog cognition, but in an effort to expand the reciprocal nature of the human-dog bond.

Our findings on the relationship between people and service dogs provide opportunities for the field of Animal-Computer Interaction (ACI) (Mancini, 2011).

While communication technology to connect people has become ubiquitous, there are fewer products that allow people to communicate with their dogs, much less allow dogs to adopt the role of a technology user. This paper explores the perceptions of service dog recipients and analyzes four case studies of people with different types of disabilities and their integration of technologies to communicate or enhance the services provided by their dogs. The aim of this study is to examine how these end-user innovations and experiments have provided affordances for human users. It serves to uncover unmet needs and describes scenarios that can inform dog inclusive product development that is supportive of human pet interaction. This study takes a user-centered approach that utilizes natural observation, surveys, and ethnographic case studies, including interviews and photo journals as key methods of this study (Robson, 2011).

METHODS

The investigation of human canine interactions within the context of AAT, the role of technology, and opportunities for novel and useful interaction design was qualitative. The research strategy was hybrid, including a pilot study, respondent field survey and ethnographic case studies.

The pilot study was conducted with the intent to evaluate survey and interview questions, as well as assess the receptivity of the population. Five people, ages 26-65, all of them owners of canine assistants living with various degrees of deafness and blindness, participated in the study. Two of these participants volunteered to participate in unstructured interviews. Two additional expert interviews were conducted and provided invaluable information on training methods for canine assistants and insights into the nature of important technologies for people living with disabilities. The study involved a basic 10-question survey that included several open-ended questions. These questions also guided the semi-structured interviews. The responses and experiences of the study led to the development of a 19-questions respondent field survey and useful contacts within the service dog community, including dog trainer Robyn Abels, executive director of Power Paws; Dr. Melissa Loree, director of education at Canine Assistants; and Ralynn McGuire, a technology consultant for people with disabilities.

Based on the findings of the pilot study and advice from these experts, ethnographic case studies were selected as the most appropriate research strategy, due to the extreme and unique nature of the population. The unit of analysis was holistic (Yin, 2003). The purpose of these studies was to advance a grounded theory approach, through the use of observation and ethnographic data gathering (Yin, 2003), to better understand interactions involving AAT and assistive technologies.

In contrast to most research, we did not maintain anonymity of our participants; in fact, with IRB approval, every effort was made to eliminate anonymity in the recruitment of case studies participants. The rationale for full disclosure is that it may serve to produce two productive outcomes: help the investigators and the research audience to recall and match previous information and experiences to better understand the case report; and, adds a level of transparency not achieved by a series of anonymous or abstracted explanations (Yin, 2003).

Semi-structured interviews were used as a method because they allowed the researcher to have freedom in the sequencing of questions and provided an opportunity to pursue spontaneous commentary from participants (Robson, 2011). This method of data gathering was pivotal in every phase of this study and helped identify key benefits that service dogs provide to humans and gaps that exist in traditional dog training. Due to lack of resources and distances, most interviews were conducted over the phone or via email. The focus of these interviews depended greatly on the type of participant (service dog recipient, expert or parent of a recipient). The interviews were recorded for analysis. Interview and open-ended question responses were transcribed and compiled in order to discern any similarities in language use, discover recurrent themes and reflect on their content, (Miles & Huberman, 1984; Robson, 2011).

Photography and video are important because they can capture details that can be taken for granted about a participant's community or life that may prompt discussion (Clark-IbáÑez, 2004; Harper, 2002). Each focused interview provided not just background records for all participants, but also details on the ways they integrated their service dogs and technology into their routines.

RESPONDENT FIELD SURVEY RESULTS

A total of 42 participants completed the online survey portion of this study. Of the participants, 82% were women. The survey was geared for an adult population and respondents ranged from 18-66 years of age. In contrast to most research in AAT, which focuses on elderly people, 37% of participants were under the age of 40, with 8 participants in age range 18-25. Regarding the conditions listed, 27 participants (64%) claimed to have a condition not included in the list of N disabilities, but 9 participants (21%) checked they suffer from paraplegia or mobility impairment. All participants had service dog at the time of the survey.

From a list of 10 essential services provided by a canine assistant the number one choice was companionship (19%), closely followed by emotional support (18%) and assistance with everyday tasks (17.%). Only 4% of respondents selected emergency management or crisis management as a service provided by their dog. The most common interaction participants shared with their dogs was talking to them (26.3%), closely followed by play activities (25.7%). When asked to select the single most important benefit of owning a canine assistant, 9 participants (21%) claimed overall confidence, 7 participants (17%) felt safer, 5 participants (12%) experienced greater efficiency in daily tasks and another 5 participants (12%) perceived greater socialization with others.

Answers to open questions within the survey and alternative answers under the category of "other" offered a further glimpse into the background, circumstances, and relationship participants have with their canine assistants. The majority of respondents described their relationship with their dog as "loving" or related personable ways. For example, one young female participant described her relationship with her dog, saying:

"Bizmark is more than just my service animal, he is my best friend. Bizmark has saved my life about 12 times in the last year. He helps me walk, he is my feet, he has a harness, and I can hold on to it so I have not had any injuries... Bizmark alerts me and others of my diabetes status and it's so, so cool.... He knows when my sensor goes off. He turns his head away if my sugars are too high or low and that indicates he doesn't like my blood sugar."

Independence and mobility are the two most important benefits participants gained from being paired with their dogs. Indirectly many connected these benefits with the idea of their loved ones feeling more comfortable that they were not entirely alone. Therefore, companionship is an indirect benefit that comes simply from owning the dog. For as many as six hearing impaired participants, having a service dog has meant not just having an alternative alert to an impending seizure, but also to their surroundings and events; "notification of water running, microwave timer, emergency sirens, picking-up dropped keys, credit card and money." Elaborating beyond the services contained in the survey, 5 participants shared that their dogs were trained to provide seizure alert:

"Joey wears a 'Medic Alert' tag on his collar that matches mine and 'Medic Alert' patches on his vest. Should I have a seizure it instantly alerts people I have a known medical condition and I am not just "passing out". Second, I have a medical implant, a Vagus Nerve Stimulator (VNS). It is on auto, but during a seizure can be manually activated with an extra amount of energy by placing a magnet in front of it. Joey will wear a magnet on his collar and be trained to lay across my chest during a seizure to activate [the] implant."

An important function of the survey was to gather the attitude of participants towards technology and what kinds of technologies they would use in everyday activities. Understanding the type of technologies used within the community is an important precursor that can inform design of the next generation of support. Twenty-two respondents said they rely heavily on computer software, 13 used electronic wheelchairs and four used hearing aids or audio players. However, 17 people indicated the use of "other" devices, including cochlear implants, mouth sticks, videophones, and digital talkers. The majority of users expressed using technology almost exclusively to communicate. A majority 69% (29 people) agreed that they would invest in technology that would allow them to monitor their dog's health. More than half of respondents (22 people) said they would trust their dog to assist them in case of an emergency and another five believed their dogs could perform basic services in the case of an emergency. Only one respondent said they would not trust their dog in such a situation.

Based on the survey results, independence and mobility are the two most important benefits participants gained from being paired with their dogs. Indirectly many connected these benefits with the idea of their loved ones feeling more comfortable that they were not entirely alone. The use of customized mobile devices and software solutions are common, but not overwhelmingly prevalent, suggesting this community of users relies more on their dogs than on technology.

CASE STUDIES

Two of the interviews and observations were conducted in the homes of the participants, while the other two interviews were conducted and recorded over the phone and relied on photo journals. All participants had a service dog or were the handler of a dog on behalf of a recipient. All interviews started with charting background information about the recipient of the service dog and the reasons for choosing supportive technologies.

In the first case, overwhelmed parents Wes and Sherrine Hayward from Gilbert, AZ acquired a Labrador retriever service dog named Omega for their 14-year old son Dayton, who suffers from cerebral palsy and unidentified violent seizures. To control his seizures and help provide continuous comfort to Dayton, doctors implanted a vagus nerve stimulation (VNS) device in his chest six years ago.

However, the parents still needed to take care of another three children. In order to give some respite to Sherrine from monitoring Dayton, especially at nighttime, the Haywards outfitted Omega's collar with the magnetic key that activates Dayton's VNS should he suffer a seizure between the device's oscillations.

Omega does not stimulate socialization, mobility or developmental abilities. Although Dayton is the recipient of the dog, Omega is a service dog for the entire family. While Omega has access to four other children in the household (a fourth child was born during the course of this study), Omega spends a considerable amount of time by Dayton's feet or a short distance away from the living room and the dog will only move away from Dayton on command.

By serving as a close companion to Dayton and bringing her body close to Dayton, Omega can now activate and adjust the intensity of the seizure mitigating VNS device with the magnet in her collar.

In case study two, parents Jerry and Lisa McMillan acquired a German shepherd service dog named Caleb to aid and protect their seven-year old twins James and Eric, who were diagnosed with autism from birth. As in the first case, the handler of the dog was not the recipient. Caleb's handler is Jennifer Fair, the twins' aunt and caretaker for the past three years.

Although the children are always together, they interact quite differently with the dog. James has developed an engaged and sensitive relationship with the dog, while Eric has unfortunately withdrawn from Caleb and sometimes displays aggressive behavior towards the dog. Eric continues to be tethered to Caleb around the home or when in public, James no longer needs to be tied to the dog. For both children, the dog is a source of socialization opportunities in public and at school.

Caleb's primary functions was to serve as a physical anchor and literally help herd the children when at home and in public. To achieve this objective, the McMillan family trained Caleb through the use of a remote stimulation Dogtra collar. The device continues to help Fair reinforce training and obedience, particularly making sure Caleb stays on paths and stays close to the children.

During the course of this study, James' diagnosis was upgraded to Asperger Syndrome, while Eric became legally blind and developed added health problems.

Key benefits in this case were greater independence and added safety.

Case study three analyzed the relationship between Iraq veteran Jeffrey Adams and his Labrador retriever service dog Sharif. Adams lost 97% of his left leg to a roadside bomb while on patrol in Baghdad, Iraq, on Nov. 7, 2004. He was 25 years old at the time.

Adams took a medical retirement as a lieutenant first class from the U.S. Army and came back stateside to complete his education, graduating from Louisiana State University, and marrying his college sweetheart Katie. After his college graduation, Adams received a prosthetic titanium C-leg issued by Walter Reed Army Medical Center (WRAMC) in December 2005. During the course of his physical therapy at the Military Advanced Training Center (MATC) in Washington D.C., Adams developed an affinity for adaptive sports and requested he be given a service dog. MATC physical therapist Harvey Naranjo helped pair Adams with Sharif.

Adams feels that any arguments to diminish the value of service dogs as an ignorant position: "I would say they are pretty close minded. If they are intelligent enough to understand how a device is able to aid or help a person that they should just see the canine companion as a device. Another assistance like the prosthetic assists me in walking. Well, Sharif assists me in doing activities around the house…so just look at it as an object. The leg is a device that I use, the dog is a device that also gives me other capabilities."

Adams has a high range of mobility and independence. He also has a social network. The key benefits reported from this case were supports for everyday tasks, especially picking-up objects, companionship and greater confidence.

Case study four captures the relationship Anne and Wayne Wicklund of Fountain Hills, Arizona have with their white German shepherd service dog named Snow. Wayne is 63 years old, a veteran of Vietnam and an entrepreneur. At the age of 52 he discovered he was deaf.

Having bought Snow as a puppy, purely for their enjoyment, the dog began alerting Wayne of events around him. The couple conducted serious research and underwent training to train Snow as a full service dog for Wayne.

Over the past 11 years, Snow has not only fulfilled his obligations as a companion and alert dog for Wicklund, but has garnered national attention for his skills. Although the Wicklunds have access to a wide range of modern technologies, they are adamant about the value of service dogs and do not believe they could substitute Snow's services with gadgets.

Wicklund has become frail and cannot walk at a brisk pace or exercise Snow anymore, so the couple bought a TRX Scooter so that Wicklund can 'walk' his dog for an extended amount of time every day. This technology and his interactions with Snow serve to keep Wayne more active and alert.

Although Wayne has a wife and social network, Snow has at times protected him from countless mishaps while outside their home. Snow nudges, bumps and even may raise a paw to catch Wayne's attention or to alert him to something requiring his attention. Snow serves as an alert and guide dog for Wayne if he ever becomes disoriented or weak. Wicklund uses technology not to substitute his dog, but to enhance the reciprocal supportive relationship he shares with Snow.

DISCUSSION

All of the participants in the case studies learned about service dogs through word of mouth and worked with a non-profit organization to receive or paid a nominal fee for a trained dog. Each of them had some form of physical and/or cognitive disability. They enjoyed a family and maintained some form of extended social network. Each has access to modern technological tools such as mobile phones, Internet and in several instances, highly specialized medical devices.

Participants frequently reported using dogs for specific services because it was easier than using a technology or performing an action themselves. For example, Jeffrey Adams explained that once he returns home from work he takes off his prosthetic leg and uses crutches to move around the house and on occasion relies on Sharif to do certain things: "If I drop something, it is easier to ask Sharif to bring it to me waist high than to try and retrieve it myself."

In addition to providing the primary services that the assistive canine was trained to support, each of the case studies indicates that the dogs also provide additional affordances, e.g., beneficial actions, experiences, or opportunities for their humans.

While there is not yet definitive evidence the presence of a canine assistant can help revert neurodevelopmental disorders, the McMillan case is consistent with research findings that suggest the use of a service dog in certain cases can help improve the psychomotor development and socialization of children suffering from severe forms of autism spectrum disorder (Burrows, Adams, & Spiers, 2008; Ng, James, & McDonald, 2000).

The case of Wicklund and his dog Snow is significant because it exemplifies how most respondent field survey participants expressed concerned for the welfare of their service dogs. Wicklund reciprocates Snow's service through his supportive attention to Snow's health needs. He repurposed a TRX Scooter as a way to walk his dog. The nudging, bumping and anchoring behaviors exhibited by Snow match the research reports made in other studies on the ways service dogs help constrain bolting behaviors and wandering, as well as alerting recipients of their surroundings (Burrows, Adams, & Spiers, 2008, Naderi, Miklósi, Dóka, & Csányi, 2001).

There are clear differences between the case studies participants with respect to needs, behaviors and the ways they use technology to enhance their relationship and mediate communication with their dogs. In all of the four case studies, some technologies played the dual role of supporting interaction between the service dog and the recipient, while adding to the role of the dog as a personal assistant and enhancing the experience of interacting with the dog. The McMillan family's use of a Dogtra collar to control Caleb's ongoing training and mitigate his unwanted behaviors is similar to actions reported regarding the use of GPS collars to enhance communication between humans and dogs in hunting (Paldanius et. al., 2011; Weilenmann & Juhlin, 2011). As Dayton Hayward's parents took it upon themselves to modify Omega's training and enabled her to intervene when Dayton experiences a violent seizure, they demonstrated how new roles and tasks associated with the use of novel technology can empower dogs and their families.

It is clear from the results of the survey that most owners of service dogs see a tremendous value in the human-dog bond and would never relinquish it, if given a choice. While each of the case study participants advocated and defended their choice of a service dog to support an important set of needs, Lt. Adams was without a doubt the most adamant defender of the concept. He argues that people, "should just see the canine companion as a device… The leg is a device that I use, the dog is a device that also gives me other capabilities." Adams' explanation sums-up the sentiments most study participants experience, but at times struggle to describe — assistive canines can be equated to a device or assistive technology and in many circumstances represent a more effective approach than many high tech solutions.

The Hayward family specifically trained Omega to reliably activate the VNS device on a highly consistent basis. The family is representative of the 64% of survey respondents that trusts their service dog to act on their behalf in emergency situations. In the McMillan case, the Dogtra collar was used to modify Caleb's behavior and ensure the dog would stay on paths and not move far from Eric. In the near future, there will likely be opportunities for field proximity (e.g., signal strength of Bluetooth, IR, etc.) to allow for a virtual tethering, vs. physical tethering, and thus improve social affordances for children like Eric and James, as they grow older. In the Wicklund case, as the owner can no longer walk without the scooter and since the dog was well trained to heal at the owners left side, Wayne was readily able to increase his activity and companionship while experiencing no major issues in using the scooter. The case of Adams serves to further illustrate that even when a user is highly familiar with technology, the use of a canine assistant has benefits beyond what assistive technologies can provide.

IMPROVING AFFORDANCES

Guided by findings from the survey and case studies, we designed and implemented a pair of wearable devices that can be used by both dogs and their owners (Fig. 1, 2 and 3).

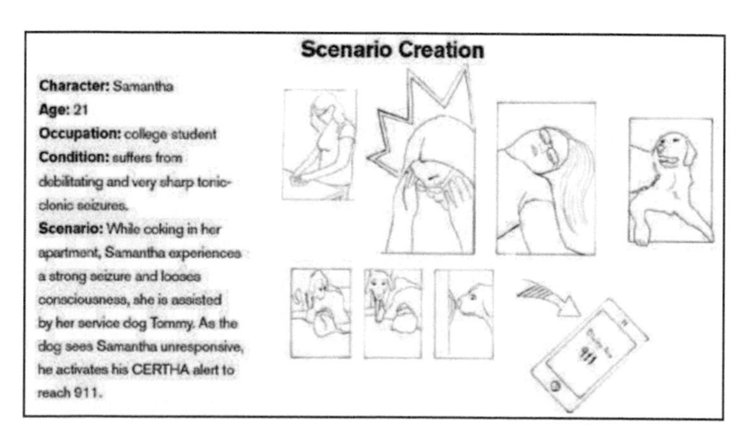

Figure 1: Scenario sketch in which user experiences a sharp tonic-clonic seizure

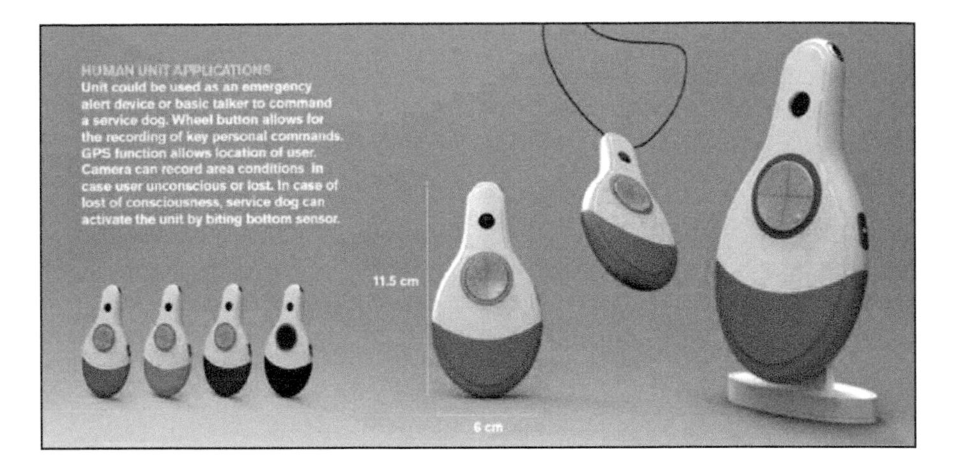

Figure 2: Human unit has basic kinetic and tactile feedback, panic and command buttons.

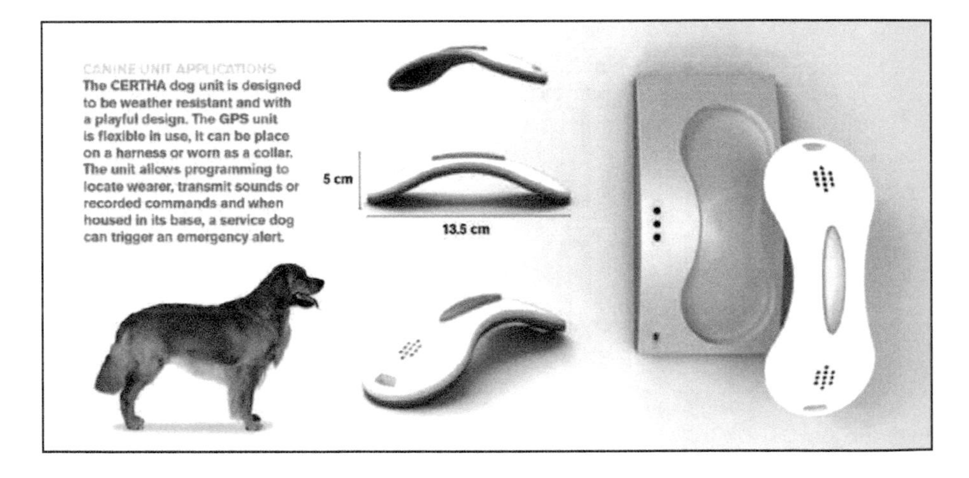

Figure 3: Wearable prototype concept that would allow dog to trigger an alarm in case human companion is unconscious.

The first devise is a customized emergency alert device and basic 2-way radio to remotely command a service dog. The owner wears the small, (6cm x 11.5cm), flat, pear shaped device as a pendant. The top is designed with a wireless camera that can record and transmit surroundings and conditions incase the user is unconscious or lost. In the center, a circular button allows for the recording of key personal commands. A GPS function can locate the user. The bottom of the device has a soft rubber cover and is equipped with a sensor that a service dog can bite to activate the alert system in case of emergencies. The color palate of each device is tailored to canine dichromatic vision (dogs mainly see yellows, blues, white, black and gray).

The second device docked on a wall or worn by the dog on a harness or collar is shaped like a flattened hourglass. This small (5cm x 13.5 cm) flexible unit is weather resistant and equipped with GPS. A long oval button is centered on the device allowing a service dog to activate an emergency alert. These devices, and related future research, propose improvements to affordances technologies within ACI and Animal Assisted Therapy contexts (Nimer & Lundahl, 2007).

CONCLUSIONS

While there is a great deal that could be derived by studying a broader range of participants, disabilities, canine training activities, and the diverse and innovate uses of technologies, there have been several key contributions of this investigation. Collectively, the findings of the pilot study, expert interviews, respondent field survey, and case studies, reiterate the conclusions of multiple studies describing the value of animal assisted therapies and in particular the impacts of service dogs.

The unique contribution of this investigation is evidence that people are repurposing and hacking current technologies, and training there dogs in innovative ways, to mediate human-canine communication, improve the affordances provided by assistive canines, and demonstrate reciprocity to enhance their relationship with their canine companions. This user centered approach and its findings serve to foster a greater appreciation of the opportunities to create supportive systems that embrace the richness and benefits of human-canine interactions with technology.

ACKNOWLEDGMENTS

The authors would like to acknowledge the invaluable help of the following experts and non-profits during the course of this study: Robyn Abels, executive director of Power Paws; Dr. Melissa Loree, director of education at Canine Assistants; Joanne Mauger, CEO of Handi Dogs; and Ralynn McGuire, technology consultant for people with disabilities.

A special thanks to all the people who participated in this study.

REFERENCES

Burrows, K.E., Adams, C.L., & Spiers, J. Qualitative health research. Sentinels of Safety: Service Dogs Ensure Safety and Enhance Freedom and Well-Being for Families With Autistic Children. Qualitative Health Research, 18(12), 2008, 1642-1649.

Clark-IbáÑez, M. (2004). Framing the social world with photo-elicitation interviews. American Behavioral Scientist, 47(12), 1507-1527.

Diamond, J. Guns, germs and steel: the fates of human societies. New York: W.W. Norton (2005).

Dogtra. (n.d.). Dogtra.com. Retrieved from Dogtra Key Features: http://www.dogtra.com/

Hambling, D. (2006). Robotic 'Pack Mule' Displays Stunning Reflexes.

Harper, D. (2002). Talking about pictures: a case for photo elicitation. Visual Studies, 17(1), 13-26.

Lawson, S.H., Wells, D., & Strong, V. (2004). Using support dogs to inform assistive technology: Towards an artificial seizure alert system. Home and Electronic Assistive Technology.

Mancini, C. (2011). Animal-computer interaction (aci): a manifesto. Interactions, 18(4), 69-73.

Melson, G. F. (2009). Robotic Pets in Human Lives: Implications for the Human - Animal Bond and for Human Relationships with Personified Technologies. Journal of Social Issues , 65(3) 545-567.

Miles, M.B., & Huberman, A.M. (1984). Drawing valid meaning from qualitative data: toward a shared craft. Educational Researcher, 13(5), 20-30.

Naderi, Sz., Miklósi, A., Dóka, A., & Csányi, V. (2001). Co-operative interactions between blind persons and their dogs. Applied Animal Behaviour Science, 74(1), 59-80.

Nguyen, H., & Kemp, C.C. (2008). Bioinspired assistive robotics: service dogs as a model for human-robot interaction and mobile manipulation. IEEE RAS/EMBS International Conference on Biomedical Robotics and Biomechatronics (BIOROB).

Nimer, J., & Lundahl, B. (2007). Animal-assisted therapy: a meta-analysis. Anthrozoos, 20(3), 225-238.

Noz, F. , & An, J. (2011). Cat cat revolution: an interspecies gaming experience. In Tan, D., Fitzpatrick, G., Gutwin, C., Begole, B., & Kellogg, W.A. (Eds.), Proceedings of the 2011 annual conference on Human factors in computing systems (pp. 2661-2664). New York, NY: Association for Computing Machinery.

Paldanius, M. , Kärkkäinen, T. , Väänänen-Vainio-Mattila, K. , Juhlin, O. , & Häkkilä, J. (2011). Communication technology for human-dog interaction: exploration of dog owners' experiences and expectations. In Tan, D., Fitzpatrick, G., Gutwin, C., Begole, B., & Kellogg, W.A. (Eds.), Proceedings of the 2011 annual conference on Human factors in computing systems (pp. 2641-2650). New York, NY: Association for Computing Machinery.

Pongrácz, P., Miklósi, Á., Timár-Geng, K., Csányi, V. 2003. Preference for copying unambiguous demonstrations in dogs. Journal of Comparative Psychology, 117: 337-343.

Robson, C. (2011). Real world research. Great Britain: Wiley-Blackwell.

Strong, V., Brown, S.W., & Walker, R. (1999). Seizure-alert dogs — fact or fiction? Seizure, 8(1), 62-65.

Strong, V., Brown, S, Huyton, M., & Coyle, H. (2001). Effect of trained seizure alert dogs on frequency of tonic–clonic seizures. Seizure, 11(6), 402-405.

Weilenmann, A. , & Juhlin, O. (2011). Understanding people and animals: the use of a positioning system in ordinary human-canine interaction. In Tan, D., Fitzpatrick, G., Gutwin, C., Begole, B., & Kellogg, W.A. (Eds.), Proceedings of the 2011 annual conference on Human factors in computing systems (pp. 2631-2640). New York, NY: Association for Computing Machinery.

Wobber, B., Hare, B., Koler-Matznick, J., Wrangham, R., & Tomasello, M., (2009). Breed differences in domestic dogs' (canis familiaris) comprehension of human communicative signals. Interaction Studies, 10(2), 206–224.

Yin, R.K. Case study research: design and methods. Thousand Oaks, California: Sage Publications Inc. (2003).

CHAPTER 78

Knowledge Management of Healthcare Team Using Clinical-Pathway – Case Study in Hospital of Miyazaki University

Tomoyoshi Yamazaki[1), Muneou Suzuki[1) Kenji Araki[1) Yukiko Kai[2)

1) Hospital of Miyazaki University, Medical Informatics
2) Miyazaki University, Medical School
Miyazaki, Japan
yama-cp@med.miyazaki-u.ac.jp
suzukim@med.miyazaki-u.ac.jp
taichan@med.miyazaki-u.ac.jp
kaiyukik@fc.miyazaki-u.ac.jp

ABSTRACT

Healthcare using clinical-pathway(CP) is a knowledge-intensive service by collaboration work of inter-professionals. Clinical-pathways(CPs) are using by many healthcare team-work as a tool for performing the healthcare process, sharing and utilizing knowledge by each professional. Aim of this paper is construction of theoretical model to knowledge management(KM) of healthcare team-work with CPs. The model was constructed by case study which uses CPs actively in the healthcare process by collaboration team-work of inter-professionals. Theoretical model implications are as follows. First, this model provides patients with individual care by CP. Second, patients are comprised in the healthcare team-work members with this model. Third, this model consists of four phases, "Integrating," "Implementing," "Expressing," "Evaluating." At each phase, useful knowledge to CP is created by the interaction between different knowledge. Fourth, this dynamic model produces CP as unified knowledge from diverse knowledge. Therefore, this model indicates evolution of integrated knowledge in healthcare team-work (include patients) with CP. In this theoretical model, CPs are suggested to be an effective tool for KM(KM) in healthcare team-work.

Keywords: clinical-pathway, healthcare team, integrated knowledge

1 INTRODUCTION

The current healthcare is asked to lower costs, and improvement in the quality of continuous care is also required simultaneously. Furthermore, healthcare is a knowledge intensive service provided by collaboration of diverse professionals, such as medical doctors, nurses, and pharmacists. Therefore, in many healthcare organizations (HCOs), management based on KM used in the industrial world is being carried out (Bose, 2003).

From the latter half of the 90s, CPs began to be used as a tool for performing optimization of healthcare resources and enhancement of care quality by HCOs (Every, 2000). Now, CPs are used as a tool for carrying out KM in many HCOs.

Team-works are an increasing focus point on the importance of clinical competence and collaboration to meet the challenges of modern healthcare. Integral to that progress by healthcare team-works is recognition that the workforce must meet new expectations and demands (D'Amour, 2005). However, theoretical model explaining detailed KM about healthcare team-work with CP is few. New knowledge needs to be created for continuous enhancement of quality of healthcare treatment by team-work with CP, and a theoretical model for this is required. (Vanhaecht, 2006).

The aim of this research is filling in current gaps in this knowledge, through construction of a theoretical model of systematic knowledge creation in the healthcare team-work process according to professionals' collaboration with CPs.

2 CLINICAL-PATHWAYS

CP applies critical-path idea methods (used in process control in industry) to the healthcare process as a management tool, and was developed in the United States in 1985 (Zander, 1988). Nowadays, the purpose of a CP is to improve the quality of care, decrease risks, increase patient's satisfaction and increase the efficiency in the use of healthcare resources. (Bleser, 2006).

Such CPs are structured instruments which lead to optimal inter-professionals patient care. Practice of CPs in clinical site involves all healthcare professionals, physicians, nurse staff, physiotherapists, social workers, etc. It can be thought of as a visualization of the patient healthcare process. The development and implementation of CPs are multi-faceted and resource-intensive processes involving all concerned parties. CPs are used in healthcare in many countries (Zander, 2002).

Sharing and integration of the knowledge of diverse professionals are important for implementation of a successful healthcare process with CPs. CPs establish optimal resource utilization and improve communication among doctors, nurses, and other staff (Coffey, 2005). However, in the healthcare process using CPs, it is difficult to respond to patients' individuality (Shi, 2008).

CPs are structured tools which formalize various pre-determined steps in the healthcare processes and make them predictable. CPs are static documents. Therefore, they do not have flexibility or adaptability to accommodate dynamic changes in a patient's conditions in clinical practices (Chu, 2001).

3 KNOWLEDGE MANAGEMENT(KM)

KM is a business concept. New knowledge is created by the interaction of tacit knowledge and explicit knowledge. Nonaka & Takeuchi described a knowledge-spiral that creates new knowledge. In particular, it models how tacit knowledge can be externalized and discussed, thereby making it explicit. It stimulates questioning and creative thinking and values the externalization of tacit knowledge in order to be able to implement change(Nonaka & Takeutci, 1995).

KM is defined as the process by which an organization creates, capture, acquires, and used knowledge to support and improve the performance of the organization. Therefore, KM can be said to involve a conscious effort to incorporate strategies and practices that ensure maximum use knowledge in organizations with purpose of advancing goals (Hansen, 1999).

4 TEAM-WORK IN HEALTHCARE

Team work is most important component in healthcare KM. Team work is required for discussing patients' condition, clinical decision-making, surgical operations. And, team work supposes a particular customized from of knowledge, since it allows transmitting tacit knowledge and experiences, beyond the content of the document of healthcare records. Many of knowledge which team work needs are based on the experience and the intuition of professionals. (Casas-Valdes, 2008).

Compartmentalization of healthcare services derives from medical specialization associated within traditional acute care systems. Therefore, fragmentation of healthcare service is resulting in the prioritization of service provider needs over patient needs. Consequent service fragmentation is associated with authoritative and hierarchical structures, resulting in the prioritization of provider needs over client needs, unnecessary time expenditure in coordinating care, diminished professional autonomy, and poor team communication. (Thylerfors, 2005).

Team-work is characterized by practitioners working together "in harmony", the intended outcome being to benefit patients and the team-work members as a whole. Team-work is an increasing focus point on the importance of clinical competence and collaboration to meet the challenges of modern healthcare. Integral to that progress by healthcare team-work is recognition that the workforce must meet new expectations and demands (D'Amour, 2005).

5 TYPE OF HEALTHCARE TEAM-WORK

Healthcare team-works by professionals are roughly classifying into three types, multidisciplinary and interdisciplinary.

In a multidisciplinary team-work, healthcare provider tend to treat patients independently and to share information with each other, while the patient may be a mere recipient of healthcare. Multidisciplinary team-work in the healthcare process is providing a juxtaposition of disciplines that is additive, not integrative. Generally, this team director is usually a medical doctor. In addition, in this team type, the patients cannot participate in a team (Klein, 1990).

Interdisciplinary healthcare team-work is a synthesis of two or more disciplines, establishing a new level of discourse and integration knowledge. Interdisciplinary team work is carried out by collaboration among different professionals. Interdisciplinary team-work, defined as partnerships between two or more healthcare professionals who achieve shared decision making according to patient centered goals and values, optimization of the composite team's knowledge, skills, and perspectives, as well as mutual respect and trust among all team members. Leader of this team-work is changed responding to the situation of patient. In this team-work type, the patients and their families are members of the team (Forbes, 1993).

6 KM IN HEALTHCARE

KM is used in many HCOs, because, healthcare is a knowledge intensive service provided by professionals. And, when HCO introduces KM into their process management, it is important to take into consideration the culture inherent to each expertise (Russ, 2005).

In KM, sharing and utilizing scientific evidence of explicit knowledge is required for implementation of evidence based medicine (EBM). However, there is no combination with a scientific basis about explicit knowledge acquired by clinical experience which each professional has accumulated, so, carrying out effective clinical practices is difficult. Therefore, KM which can share and utilizing, both explicit and tacit knowledge is required of HCO (Sandars, 2006).

In modern health care systems, healthcare providers face ever new challenges with regard to quality and cost of care, as well as to satisfaction and training of professionals. In order to solve these challenges, introduction of KM in healthcare process management is effective (Kitchiner, 1996).

The core of a team-work in modern healthcare is changing from the doctor to the patient. Accordingly, within a team, KM which can create the optimal healthcare process for the patient by various professionals will be required (Metaxiotis, 2006).

The one of major reason of teamwork barriers in healthcare practices is poor communications among diverse healthcare professionals. The poor communication arises by differences in inherent professional's occupational culture background. As professional cultures differ among specialties, a significant number of miscommunication can arise, particularly in emergency situations where events become unpredictable and a high level of coordination between healthcare professionals becomes mandatory (Haller, 2008).

7 KM WITH CP IN HEALTHCARE

CPs are developed through collaborative efforts of doctors, nurses, pharmacists, and others to improve the quality and value of patient care. CPs are prepared using clinical guidelines' based on EBM of the visualized knowledge. But, healthcare professionals' context knowledge is essential in using a clinical guideline. The production process of a clinical guideline is based on agreement formed by this discussion, so, the care team can provide optimal healthcare treatment. CPs are a tool for utilizing diverse knowledge from healthcare professionals (Mitton, 2007).

CPs express clearly that tacit knowledge is a part of the healthcare processes. Therefore, knowledge processes are included in CPs. CPs are carrying out the interaction of different healthcare professionals' knowledge. In addition, promotion of the knowledge communication between healthcare professionals by CPs activities is important for excellent healthcare KM (Yamazaki, 2010).

8 RESEARCH APPROACH

We carried out the case study in Hospital of Miyazaki University(HoMU), which use CPs activity for healthcare process management.

Case study by the team constituted from diverse researcher was performed from November-08 to January-09. The team consisted of researchers of social science, knowledge engineering, education technology, and business. Methods of research included analysis of documents relevant to CPs activity, and interviews with CPs directors, CPs producers, and CPs users. Documents of CPs activities in healthcare team-work were classified into categories of the cause and moment of introduction, the feature of activity, management of activity, and analyzed the contents of each category. The contents of document analysis were verified by participant observation by each researcher of team to healthcare team-work with CPs activities.

The purpose of these interviews was to obtain information about the intentions and interpretations of team-work with CPs which were not obtained from document analysis. By analysis of the data obtained from our investigation, it is possible to observe the healthcare process with CPs. The extraction of the knowledge process from this healthcare team-work with CPs is team's main goal.

9 CASE: THE CP ACTIVITY OF HoMU

9.1 The background and the characteristics of CP activity

In 2004, CPs were introduced at this hospital. CPs were introduced for cost management of the healthcare processes using comprehensive medicine system.

The purpose of CPs was confined to increasing the efficiency of the healthcare processes cost management at first. Therefore, production of CP was considering cost management and production efficiency as important. For these purpose, production and implementation of CP were limited to specific healthcare division and professionals. In 2006, enhancement of the healthcare quality by standardization of the healthcare process was added to the production purpose of CP by the wish of the healthcare professionals of this hospital. As a result, production and implementation of CP were popularizing step by step in this hospital. However, CP activity rule of this hospital is not that all the staff members participate.

9.2 Detail of the CP activity

CP activity was a healthcare process constituted production, implementation, improvement.

(i) Production process;

Production of CPs was mainly nurses. Medical doctors did not participate in creation directly but cooperation was carried out. In production process, the healthcare target "outcome" was set up as a milestone of the patients' condition. In addition, guidelines and the medical records were referred to in CP production.

(ii) Implementation process;

Many clinical business routines with CPs were assessed by healthcare professionals for "outcome." In this process, the pharmacist and the nutritionist also used CP. Diverse healthcare professionals explained the concrete healthcare processes by CP to patients. The objective assessment of "outcome" was important in the implementation process of CPs. The condition of a patient whose "outcome" assessment was different from the usual case was called "variance." In using CPs, response of "variance" by healthcare professionals adapted for a patient's individuality. It is important to necessarily write down the medical records of a patient's condition which could not be assessed by "outcome." The intuition obtained by observation by professionals was also recorded.

(iii) Improvement process;

Improving of CP by "variance" analysis was not carried out yet. As this reason, the rule which analyzed "variance" has not completed. However, improving by many medical records, and good practices from the other hospital CP was carried out. Setup of the subjective hypotheses by professionals was effective for the creation of new "outcome." Professionals' new knowledge was required for creation of new "outcome."

In this study, from CP documents, the participant observation team extracted the useful many improving points of CP, and provided to CP producers. The CP director of this hospital asked the team for divers view points from CP documents. The team provided the CP producers with many improving points of CP documents. The CP producers in this study are improving CPs using the provided improving points.

9.3 Healthcare team-work in CP activity

The research result is shown as follows.

(i) CP activity ;

a) Production of CPs was mainly nurses and other professionals have not participated.

b) CP producer was mainly referring to medical records by both setup of "outcome" and improving process.

c) CP users performed response of patient's individuality by "variance" in clinical business routines.

d) The pharmacists and the nutritionists were also included in CP users besides the medical doctors and the nurses.

e) CP users certainly recorded a patient's individuality on medical records

f) By CP contents, the patients can understand the complex healthcare processes.

(ii) Team-work with CP;

a) CP producers assumed that CP users consist of diverse professionals.

b) CP user used its experience for response of patient's condition.

c) Because response of patient's condition was individual, the CP user acquires a new experience.

d) In this research, the new observation team which consists of different professionals provided the important matter of improving to the CP producers. The CP producers are improving CPs using the provided improving points by the team.

10 RESEARCH FINDING

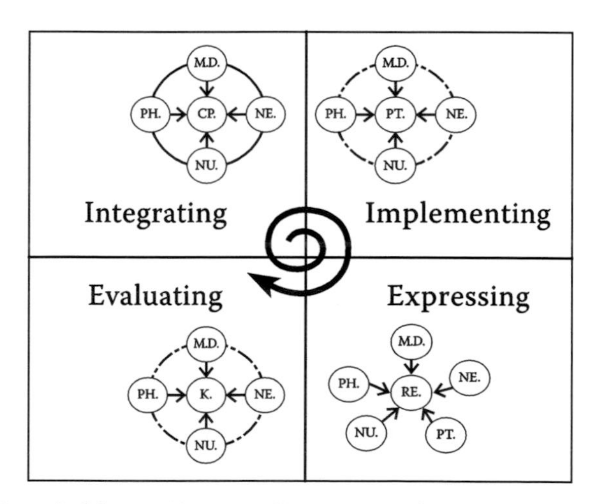

MD: medical doctor, PT.: patient, NU.: Nurse, PH: pharmacist, NE: nutritionist, CP.: clinical-pathway, RE.: medical records

Figure1 The knowledge process of healthcare team-work with CP

From the research result and literature reviews, a theoretical model which shows the knowledge process of healthcare team-work with CP was constructed (Fig. 1).

The theoretical model shows the interaction in the healthcare team-work process of clinical-pathways, integration knowledge, and professional knowledge. The professional knowledge used in the healthcare process in team-work is contextual knowledge which each professional has, such as know-how, skill, and empirical knowledge. The integration knowledge used in the healthcare process in team-work is objective knowledge, such as the clinical-pathways and the contents of medical records.

The theoretical model shows the interaction in the healthcare team-work process of CPs, integration knowledge, and professional knowledge. The professional knowledge used in the healthcare process in team-work is contextual knowledge which each professional has, such as know-how, skill, and empirical knowledge. The integration knowledge used in the healthcare process in team-work is objective knowledge, such as the CPs and the contents of medical records.

The characteristics of the theoretical model, including the interaction of professional knowledge and integration knowledge, include:

(i) Healthcare process in team-work by divers professionals is shown four phases of knowledge process which carries out the interaction of different knowledge.

(ii) Each "phases are "Integrating," "Implementing," "Expressing," and "Evaluating" and connected spirally.

(iii) "Integrating" interacts between present integration knowledge such as clinical-pathway, and CP producers' professional knowledge, and this phase creates new CP as new integration knowledge.

(iv) "Implementing" interacts between new integration knowledge such as created CP, and CP users' professional knowledge, and this phase acquires new empirical knowledge as new professional knowledge for CP users and patients.

(v) "Expressing" interacts between new integration knowledge such as new empirical knowledge recorded on medical records, and CP users' professional knowledge, and this phase visualizes new empirical knowledge as new professional knowledge from CP users and patients.

(vi) "Evaluating" interacts between new integration knowledge such as visualized new empirical knowledge, and observers' professional knowledge, this phase provides new integration knowledge by observers of diverse viewpoints to CP producers for CP improving.

(vii) Spiral risen "Integrating" interacts between provided integration knowledge by observers, and CP producers' professional knowledge, and this phase improves CP with new integration knowledge by CP producers.

In this model, new integration knowledge is created because observers evaluate individual professional knowledge such as empirical knowledge. Therefore, this model is extracting the knowledge which can be used universally from the individual knowledge such as diverse professionals' contextual knowledge responding to patients' state. This model is useful to healthcare team-work which provides individual healthcare services for diverse patients.

12 CONCLUSION

The theoretical model which can create new knowledge continuously was constructed from case study of HoMU. Clinical-pathways clearly expressed that integration knowledge is used by team-work in the healthcare processes. Therefore, CPs are an effective tool for healthcare team-work which consists of diverse professionals for standard healthcare. However, Healthcare is asked for individual service.

CPs activity which added observers of diverse viewpoints to the member provides the knowledge process which can respond to individual patients' condition.

This knowledge process consists of four phases in the theoretical model, "Integrating," "Implementing," "Expressing," and "Evaluating" and connected spirally. At each phase, useful knowledge to healthcare team-work with CPs for responding individual condition of patients is created by interaction between integration knowledge and professional knowledge. Therefore, this knowledge process by constructed theoretical model in this study indicates evolution of integration knowledge in healthcare team (including patients) by CPs.

In the future, more studies about other hospitals should be conducted to refine the theoretical model.

ACKNOWLEDGMENTS

The authors with to acknowledge the healthcare professionals of HoMU.

REFERENCES

Bleser, DE L., R. Depereitere, and De K. Waele. 2006. Defining Pathways. *Journal of Nursing Management* 13: 553-563.

Bose, R. 2003. Knowledge Management-Enabled Health Care Management Systems: Capabilities, Infrastructure, and Decision-Support. *Expert systems with Applications* 24: 59-71.

Casas-Valdes, A., D. Oramas. 2008. Theoretical Aspects of Knowledge Management in Evidence-Based Medicine. ACIMED 17.

Chu, S. 2001. Reconceptualising Clinical Pathway System Design. *Collegian* 3: 33-36.

Coffey, R. J. Richards, and C. Remmert. 2005. An Introduction to Critical Paths. *Quality Management in Health Care* 14: 46-55.

D'Amour, D., M. Ferrada-Videla. 2005. The Conceptual Basis for Interprofessional Collaboration: Core Concepts and Theoretical Frameworks. *Journal of Interprofessional Care* 19: 116-131

Every, N., R. Becker, and S. Kopecky. 2000. Critical Pathways a Review. *Circulation* 101: 461-465.

Forbes, E., V. Fitzsimons. 1993. Education: The Key for Holistic Interdisciplinary Collaboration. *Holistic Nursing Practice* 7: 16-29.

Haller, G., M. Morales, and R. Pfister. 2008. Improving Interprofessional Teamwork in Obstetrics: A Crew Resource Management Based Training Programme. *Journal of Interprofessional Care* 22: 545-548.

Hansen, M., N. Nohira, and T. Tierney. 1999. What's Your Strategy for Managing Knowledge? *Harvard Business Review* 77: 106-116.

Kitchiner, D., C. Davidson, and P. Bundred. 1996. Integrated Care Pathways: Effective Tools for Continuous Evaluation of Clinical Practice. *Journal of Evaluation in Clinical Practice* 2: 65-69.

Klein, J. 1990. *Interdisciplinarity: History, Theory, and Practice*. Detroit: Wayne State University.

Metaxiotis, Kostas. 2006. Healthcare Knowledge Management. In. *Encyclopedia of Knowledge Management*, ed. D. Schwartz. London, IDEA GROUP REFERENCE, 204-210.

Mitton, C., C.Adair, and E. Mckenzie. 2007. Knowledge Transfer and Exchange: Review and Synthesis of the Literature. *Milbank Quarterly* 85: 729-768.

Nonaka, I., H. Takeuchi. 1995. *The Knowledge-Creating Company: How Japanese Companies Create the Dynamics of Innovation*. New York: Oxford University Press.

Russ, M., J. Jones. 2005. A Typology of Knowledge Management Strategies for Hospital Preparedness: What Lesson Can be Learned? *International Journal of Emergency Management* 2: 319-342.

Sandars, J., R. Heller. 2006. Improving the Implementation of Evidence-Based Practice: A knowledge management perspective. *Journal of Evaluation in Clinical Practice* 12: 342-346.

Shi, J., Su, Q., Zhao, Z., (2008) "Critical Factors for the Effectiveness of Clinical Pathway in Improving Care Outcomes", Service Systems and Service Management, 2008 International Conference on: 1-6.

Thylerfors, I., O. Persson, and D. Hellstrom. 2005. Team Types, Perceived Efficiency and Team Climate in Swedish Crossprofessional Teamwork. *Journal of Interprofessional Practice* 14: 14-19.

Yamazaki, T., K. Umemoto. 2010. Knowledge Management of Healthcare by Clinical-Pathways. *Journal of Information & Knowledge Management* 9: 119-125.

Vanhaecht, K., K. Witte, and R. Depreitere. 2006. Clinical Pathway Audit Tool: Systematic review. *Journal of Nursing Management* 14: 529-537.

Zander, K. 1988. Nursing Case Management: Strategic Management of Cost and Quality Outcomes. *Journal of Nursing Administration* 18: 23-30.

Zander, K. 2002. Integrated care pathways: eleven international trends. *Journal of INTEGRATED CARE PATHWAYS* 6: 101-107.

Index